I A	II A	III B	IV B	V B	VI B	VII B	VII	VII	VII	I B	II B	III A	IV A	V A	VI A	VII A	0
1 **H** Hydrogen																	2 **He** Helium
3 **Li** Lithium	4 **Be** Beryllium											5 **B** Boron	6 **C** Carbon	7 **N** Nitrogen	8 **O** Oxygen	9 **F** Fluorine	10 **Ne** Neon
11 **Na** Sodium	12 **Mg** Magnesium											13 **Al** Aluminum	14 **Si** Silicon	15 **P** Phosphorus	16 **S** Sulfur	17 **Cl** Chlorine	18 **Ar** Argon
19 **K** Potassium	20 **Ca** Calcium	21 **Sc** Scandium	22 **Ti** Titanium	23 **V** Vanadium	24 **Cr** Chromium	25 **Mn** Manganese	26 **Fe** Iron	27 **Co** Cobalt	28 **Ni** Nickel	29 **Cu** Copper	30 **Zn** Zinc	31 **Ga** Gallium	32 **Ge** Germanium	33 **As** Arsenic	34 **Se** Selanium	35 **Br** Bromine	36 **Kr** Krypton
37 **Rb** Rubidium	38 **Sr** Strontium	39 **Y** Yttrium	40 **Zr** Zirconium	41 **Nb** Niobium	42 **Mo** Molybdenum	43 **Tc** Technetium	44 **Ru** Ruthenium	45 **Rh** Rhodium	46 **Pd** Palladium	47 **Ag** Silver	48 **Cd** Cadmium	49 **In** Indium	50 **Sn** Tin	51 **Sb** Antimony	52 **Te** Tellurium	53 **I** Iodine	54 **Xe** Xenon
55 **Cs** Cesium	56 **Ba** Barium	57 **La** Lanthanum	72 **Hf** Hafnium	73 **Ta** Tantalum	74 **W** Tungsten	75 **Re** Rhenium	76 **Os** Osmium	77 **Ir** Iridium	78 **Pt** Platinum	79 **Au** Gold	80 **Hg** Mercury	81 **Tl** Thallium	82 **Pb** Lead	83 **Bi** Bismuth	84 **Po** Polonium	85 **At** Astatine	86 **Rn** Radon
87 **Fr** Francium	88 **Ra** Radium	89 **Ac** Actinium															

Lanthanides	58 **Ce** Cerium	59 **Pr** Praseodymium	60 **Nd** Neodymium	61 **Pm** Promethium	62 **Sm** Samarium	63 **Eu** Europium	64 **Gd** Gadolinium	65 **Tb** Terbium	66 **Dy** Dysprosium	67 **Ho** Holmium	68 **Er** Erbium	69 **Tm** Thulium	70 **Yb** Ytterbium	71 **Lu** Lutetium
Actinides	90 **Th** Thorium	91 **Pa** Protactinium	92 **U** Uranium	93 **Np** Neptunium	94 **Pu** Plutonium	95 **Am** Americium	96 **Cm** Curium	97 **Bk** Berkelium	98 **Cf** Califomium	99 **Es** Einsteinium	100 **Fm** Fermium	101 **Md** Mendelevium	102 **No** Nobelium	103 **Lr** Lawrencium

The ART of
ANALOG LAYOUT

The ART of ANALOG LAYOUT

Alan Hastings

Prentice Hall

PRENTICE HALL
Upper Saddle River, NJ 07458

Library of Congress Cataloging-in-Publication Data
Hastings, Alan (Ray Alan)
The art of analog layout / Alan Hastings.
p. cm.
Includes bibliographical references and index.
ISBN 0-13-087061-7
1. Integrated circuits—Design and construciton. 2. Layout (Printing) I. Title.

TK7874.H3926 2001
621.3815—dc21 00-045307

Vice president and editorial director, ECS: **Marcia Horton**
Publisher: **Tom Robbins**
Associate editor: **Alice Dworkin**
Editorial assistant: **Jessica Power**
Production editor: **Carlisle Communications, Ltd.**
Executive managing editor: **Vince O'Brien**
Managing editor: **David A. George**
Art director: **Jayne Conte**
Cover design: **Joseph Sengotta**
Art editor: **Adam Velthaus**
Manufacturing manager: **Trudy Pisciotti**
Manufacturing buyer: **Dawn Murrin**
Assistant vice president of production and manufacturing, ESM: **David W. Riccardi**

Printed in the United States of America

10 9 8 7 6 5 4 3 2

ISBN 0-13-087061-7

Prentice-Hall International (UK) Limited, *London*
Prentice-Hall of Australia Pty. Limited, *Sydney*
Prentice-Hall Canada Inc., *Toronto*
Prentice-Hall Hispanoamericana, S.A., *Mexico*
Prentice-Hall of India Private Limited, *New Delhi*
Prentice-Hall of Japan, Inc., *Tokyo*
Pearson Education Asia Pte. Ltd., *Singapore*
Editora Prentice-Hall do Brasil, Ltda., *Rio de Janeiro*

For My Father

Contents

Preface

An integrated circuit reveals its true appearance only under high magnification. The intricate tangle of microscopic wires covering its surface, and the equally intricate patterns of doped silicon beneath it, all follow a set of blueprints called a *layout.* The process of constructing layouts for analog and mixed-signal integrated circuits has stubbornly defied all attempts at automation. The shape and placement of every polygon require a thorough understanding of the principles of device physics, semiconductor fabrication, and circuit theory. Despite thirty years of research, much remains uncertain. What information there is lies buried in obscure journal articles and unpublished manuscripts. This textbook assembles this information between a single set of covers. While primarily intended for use by practicing layout designers, it should also prove valuable to circuit designers who desire a better understanding of the relationship between circuits and layouts.

The text has been written for a broad audience, some of whom have had only limited exposure to higher mathematics and solid-state physics. The amount of mathematics has been kept to an absolute minimum, and care has been taken to identify all variables and to use the most accessible units. The reader need only have a familiarity with basic algebra and elementary electronics. Many of the exercises assume that the reader also has access to layout editing software, but those who lack such resources can complete many of the exercises using pencil and paper.

The text consists of fourteen chapters and five appendices. The first two chapters provide an overview of device physics and semiconductor processing. These chapters avoid mathematical derivations and instead emphasize simple verbal explanations and visual models. The third chapter presents three archetypal processes: standard bipolar, silicon-gate CMOS, and analog BiCMOS. The presentation focuses upon development of cross sections and the correlation of these cross sections to conventional layout views of sample devices. The fourth chapter covers common failure mechanisms and emphasizes the role of layout in determining reliability. Chapters Five and Six cover the layout of resistors and capacitors. Chapter Seven presents the principles of matching, using resistors and capacitors as examples. Chapters Eight through Ten cover the layout of bipolar devices, while chapters Eleven and Twelve cover the layout and matching of field-effect transistors. Chapters Thirteen and Fourteen cover a variety of advanced topics, including device mergers, guard rings, ESD protection structures, and floorplanning. The appendices include a list of acronyms, a discussion of Miller indices, sample layout rules for use in working the exercises, and the derivation of formulas used in the text.

Alan Hastings

Acknowledgments

The information contained in this text has been gathered through the hard work of many scientists, engineers, and technicians, the vast majority of whom must remain unacknowledged because their work has not been published. I have included references to as many fundamental discoveries and principles as I could, but in many cases I have been unable to determine original sources.

I thank my colleagues at Texas Instruments for numerous suggestions. I am especially grateful to Ken Bell, Walter Bucksch, Lou Hutter, Clif Jones, Jeff Smith, Fred Trafton, and Joe Trogolo, all of whom have provided important information for this text. I am also grateful for the encouragement of Bob Borden, Nicolas Salamina, and Ming Chiang, without which this text would never have been written.

1 Device Physics

Before 1960, most electronic circuits depended upon vacuum tubes to perform the critical tasks of amplification and rectification. An ordinary mass-produced AM radio required five tubes, while a color television needed no fewer than twenty. Vacuum tubes were large, fragile, and expensive. They dissipated a lot of heat and were not very reliable. So long as electronics depended upon them, it was nearly impossible to construct systems requiring thousands or millions of active devices.

The appearance of the bipolar junction transistor in 1947 marked the beginning of the solid-state revolution. These new devices were small, cheap, rugged, and reliable. Solid-state circuitry made possible the development of pocket transistor radios and hearing aids, quartz watches and touch-tone phones, compact disc players and personal computers.

A *solid-state device* consists of a crystal with regions of impurities incorporated into its surface. These impurities modify the electrical properties of the crystal, allowing it to amplify or modulate electrical signals. A working knowledge of device physics is necessary to understand how this occurs. This chapter covers not only elementary device physics but also the operation of three of the most important solid-state devices: the junction diode, the bipolar transistor, and the field-effect transistor. Chapter 2 explains the manufacturing processes used to construct these and other solid-state devices.

1.1 SEMICONDUCTORS

The inside front cover of the book depicts a long-form periodic table. The elements are arranged so those with similar properties group together to form rows and columns. The elements on the left-hand side of the periodic table are called *metals*, while those on the right-hand side are called *nonmetals*. Metals are usually good conductors of heat and electricity. They are also malleable and display a characteristic metallic luster. Nonmetals are poor conductors of heat and electricity, and those that are solid are brittle and lack the shiny luster of metals. A few elements in the middle of the periodic table, such as silicon and germanium, have electrical properties that lie midway between those of metals and nonmetals. These elements are called *semiconductors.* The differences between metals, semiconductors, and nonmetals result from the electronic structure of their respective atoms.

Every atom consists of a positively charged nucleus surrounded by a cloud of electrons. The number of electrons in this cloud equals the number of protons in the nucleus, which also equals the atomic number of the element. Therefore a carbon atom has six electrons because carbon has an atomic number of six. These electrons occupy a series of *shells* that are somewhat analogous to the layers of an onion. As electrons are added, the shells fill in order from innermost outward. The outermost or *valence* shell may remain unfilled. The electrons occupying this outermost shell are called *valence electrons.* The number of valence electrons possessed by an element determines most of its chemical and electronic properties.

Each row of the periodic table corresponds to the filling of one shell. The leftmost element in the row has one valence electron, while the rightmost element has a full valence shell. Atoms with filled valence shells possess a particularly favored configuration. Those with unfilled valence shells will trade or share electrons so that each can claim a full shell. Electrostatic attraction forms a chemical bond between atoms that trade or share electrons. Depending upon the strategy adopted to fill the valence shell, one of three types of bonding will occur.

Metallic bonding occurs between atoms of metallic elements, such as sodium. Consider a group of sodium atoms in close proximity. Each atom has one valence electron orbiting around a filled inner shell. Imagine that the sodium atoms all discard their valence electrons. The discarded electrons are still attracted to the positively charged sodium atoms, but, since each atom now has a full valence shell, none accepts them. Figure 1.1A shows a simplified representation of a sodium crystal. Electrostatic forces hold the sodium atoms in a regular lattice. The discarded valence electrons wander freely through the resulting crystal. Sodium metal is an excellent electrical conductor due to the presence of numerous free electrons.[1] These same electrons are also responsible for the metallic luster of the element and its high thermal conductivity. Other metals form similar crystal structures, all of which are held together by metallic bonding between a sea of free valence electrons and a rigid lattice of charged atomic cores.

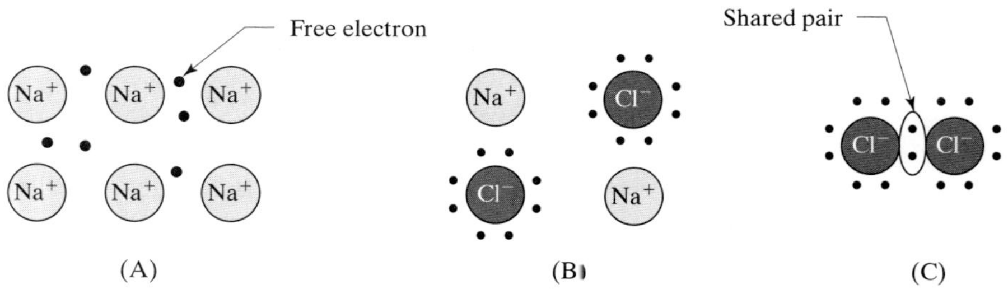

FIGURE 1.1 Simplified illustrations of various types of chemical bonding: metallically bonded sodium crystal (A), ionically bonded sodium chloride crystal (B), and covalently bonded chlorine molecule (C).

Ionic bonding occurs between atoms of metals and nonmetals. Consider a sodium atom in close proximity to a chlorine atom. The sodium atom has one valence electron, while the chlorine atom is one electron short of a full valence shell. The sodium atom can donate an electron to the chlorine atom and by this means both can achieve filled outer shells. After the exchange, the sodium atom has a net posi-

[1] Some metals conduct by means of holes rather than electrons, but the general observations made in the text still apply.

tive charge and the chlorine atom a net negative charge. The two charged atoms (or *ions*) attract one another. Solid sodium chloride thus consists of sodium and chlorine ions arranged in a regular lattice, forming a crystal (Figure 1.1B). Crystalline sodium chloride is a poor conductor of electricity, since all of its electrons are held in the shells of the various atoms.

Covalent bonding occurs between atoms of nonmetals. Consider two chlorine atoms in close proximity. Each atom has only seven valence electrons, while each needs eight to fill its valence shell. Suppose that each of the two atoms contributes one valence electron to a common pair shared by both. Now each chlorine atom can claim eight valence electrons: six of its own, plus the two shared electrons. The two chlorine atoms link to form a molecule that is held together by the electron pair shared between them (Figure 1.1C). The shared pair of electrons forms a *covalent bond.* The lack of free valence electrons explains why nonmetallic elements do not conduct electricity and why they lack metallic luster. Many nonmetals are gases at room temperature because the electrically neutral molecules exhibit no strong attraction to one another and thus do not condense to form a liquid or a solid.

The atoms of a semiconductor also form covalent bonds. Consider atoms of silicon, a representative semiconductor. Each atom has four valence electrons and needs four more to complete its valence shell. Two silicon atoms could theoretically attempt to pool their valence electrons to achieve filled shells. In practice this does not occur because eight electrons packed tightly together strongly repel one another. Instead, each silicon atom shares one electron pair with each of four surrounding atoms. In this way, the valence electrons are spread around to four separate locations and their mutual repulsion is minimized.

Figure 1.2 shows a simplified representation of a silicon crystal. Each of the small circles represents a silicon atom. Each of the lines between the circles represents a covalent bond consisting of a shared pair of valence electrons. Each silicon atom can claim eight electrons (four shared electron pairs), so all of the atoms have full valence shells. These atoms are linked together in a molecular network by the covalent bonds formed between them. This infinite lattice represents the structure of the silicon crystal. The entire crystal is literally a single molecule, so crystalline silicon is strong and hard, and it melts at a very high temperature. Silicon is normally a poor conductor of electricity because all of its valence electrons are used to form the crystal lattice.

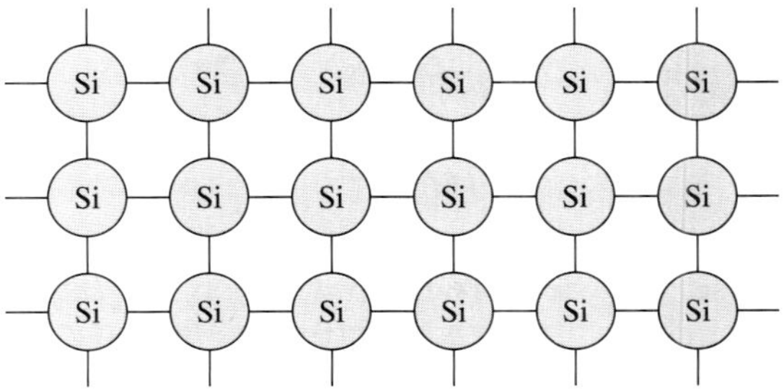

FIGURE 1.2 Simplified two-dimensional representation of a silicon crystal lattice.

A similar macromolecular crystal can theoretically be formed by any group-IV element,[2] including carbon, silicon, germanium, tin, and lead. Carbon, in the form of diamond, has the strongest bonds of any group-IV element. Diamond crystals

---

[2] The group-III, IV, V, and VI elements reside in columns III-B, IV-B, V-B, and VI-B of the long-form periodic table. The group-II elements may fall into either columns II-A or II-B. The A/B numbering system is a historical curiosity and the International Union of Pure and Applied Chemists (IUPAC) has recommended its abandonment; see J. Hudson, *The History of Chemistry* (New York: Chapman and Hall, 1992), pp. 122–137.

are justly famed for their strength and hardness. Silicon and germanium have somewhat weaker bonds due to the presence of filled inner shells that partially shield the valence electrons from the nucleus. Tin and lead have weak bonds because of numerous inner shells; they typically form metallically bonded crystals instead of covalently bonded macromolecules. Of the group-IV elements, only silicon and germanium have bonds of an intermediate degree of strength. These two act as true semiconductors, while carbon is a nonmetal, and tin and lead are both metals.

1.1.1. Generation and Recombination

The electrical conductivity of group-IV elements increases with atomic number. Carbon, in the form of diamond, is a true insulator. Silicon and germanium have much higher conductivities, but these are still far less than those of metals such as tin and lead. Because of their intermediate conductivities, silicon and germanium are termed *semiconductors.*

Conduction implies the presence of free electrons. At least a few of the valence electrons of a semiconductor must somehow escape the lattice to support conduction. Experiments do indeed detect small but measurable concentrations of free electrons in pure silicon and germanium. The presence of these free electrons implies that some mechanism provides the energy needed to break the covalent bonds. The statistical theory of thermodynamics suggests that the source of this energy lies in the random thermal vibrations that agitate the crystal lattice. Even though the average thermal energy of an electron is relatively small (less than 0.1 electron volt), these energies are randomly distributed, and a few electrons possess much larger energies. The energy required to free a valence electron from the crystal lattice is called the *bandgap energy.* A material with a large bandgap energy possesses strong covalent bonds and therefore contains few free electrons. Materials with lower bandgap energies contain more free electrons and possess correspondingly greater conductivities (Table 1.1).

TABLE 1.1 Selected properties of group-IV elements.[3]

Element	Atomic Number	Melting Point, °C	Electrical Conductivity $(\Omega cm)^{-1}$	Bandgap Energy, eV
Carbon (diamond)	6	3550	$\sim 10^{-16}$	5.2
Silicon	14	1410	$4 \cdot 10^{-6}$	1.1
Germanium	32	937	0.02	0.7
White Tin	50	232	$9 \cdot 10^{4}$	0.1

A vacancy occurs whenever an electron leaves the lattice. One of the atoms that formerly possessed a full outer shell now lacks a valence electron and therefore has a net positive charge. This situation is depicted in a simplified fashion in Figure 1.3. The ionized atom can regain a full valence shell if it appropriates an electron from a neighboring atom. This is easily accomplished since it still shares electrons with three adjacent atoms. The electron vacancy is not eliminated; it merely shifts to the

[3] Bandgap energies for Si, Ge: B. G. Streetman, *Solid State Electronic Devices*, 2nd ed. (Englewood Cliffs, NJ: Prentice-Hall, 1980), p. 443. Bandgap for C: N. B. Hanny, ed., *Semiconductors* (New York: Reinhold Publishing, 1959), p. 52. Conductivity for Sn: R. C. Weast, ed., *CRC Handbook of Chemistry and Physics,* 62nd ed. (Boca Raton, FL: CRC Press, 1981), pp. F135–F136. Other values computed. Melting points: Weast, pp. B4–B48.

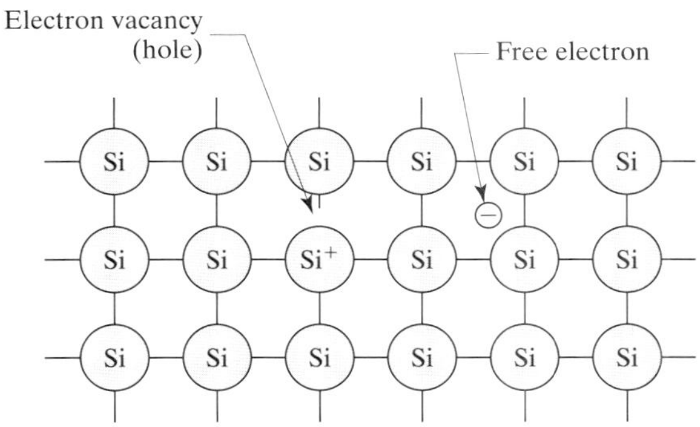

FIGURE 1.3 Simplified diagram of thermal generation in intrinsic silicon.

adjacent atom. As the vacancy is handed from atom to atom, it moves through the lattice. This moving electron vacancy is called a *hole.*

Suppose an electric field is placed across the crystal. The negatively charged free electrons move toward the positive end of the crystal. The holes behave as if they were positively charged particles and move toward the negative end of the crystal. The motion of the holes can be compared to bubbles in a liquid. Just as a bubble is a location devoid of fluid, a hole is a location devoid of valence electrons. Bubbles move upward because the fluid around them sinks downward. Holes shift toward the negative end of the crystal because the surrounding electrons shift toward the positive end.

Holes are usually treated as if they were actual subatomic particles. The movement of a hole toward the negative end of the crystal is explained by assuming that holes are positively charged. Similarly, their rate of movement through the crystal is measured by a quantity called *mobility.* Holes have lower mobilities than electrons; typical values in bulk silicon are $480 cm^2/V{\cdot}sec$ for holes and $1350 cm^2/V{\cdot}sec$ for electrons.[4] The lower mobility of holes makes them less efficient charge carriers. The behavior of a device therefore relies upon whether its operation involves holes or electrons.

A free electron and a hole are formed whenever a valence electron is removed from the lattice. Both particles are electrically charged and move under the influence of electric fields. Electrons move toward positive potentials, producing an electron current. Holes move toward negative potentials, producing a hole current. The total current equals the sum of the electron and the hole currents. Holes and electrons are both called *carriers* because of their role in transporting electric charge.

Carriers are always generated in pairs since the removal of a valence electron from the lattice simultaneously forms a hole. The generation of electron-hole pairs can occur whenever energy is absorbed by the lattice. Thermal vibration produces carriers, as do light, nuclear radiation, electron bombardment, rapid heating, mechanical friction, and any number of other processes. To consider only one example, light of a sufficiently short wavelength can generate electron-hole pairs. When a lattice atom absorbs a photon, the resulting energy transfer can break a covalent bond to produce a free electron and a free hole. Optical generation will occur only if the photons have enough energy to break bonds, and this in turn requires light of a sufficiently short wavelength. Visible light has enough energy to produce electron-hole pairs in most semiconductors. Solar cells make use of this phenomenon to convert sunlight into electrical current. Photocells and solid-state camera detectors also employ optical generation.

---

[4] Streetman, p. 443.

Just as carriers are generated in pairs, they also recombine in pairs. The exact mechanism of carrier recombination depends on the nature of the semiconductor. Recombination is particularly simple in the case of a *direct-bandgap semiconductor.* When an electron and a hole collide, the electron falls into the hole and repairs the broken covalent bond. The energy gained by the electron is radiated away as a photon (Figure 1.4A). Direct-bandgap semiconductors can, when properly stimulated, emit light. A *light-emitting diode* (LED) produces light by electron-hole recombination. The color of light emitted by the LED depends on the bandgap energy of the semiconductor used to manufacture it. Similarly, the so-called *phosphors* used in manufacturing glow-in-the-dark paints and plastics also contain direct-bandgap semiconductors. Electron-hole pairs form whenever the phosphor is exposed to light. A large number of electrons and holes gradually accumulate in the phosphor. The slow recombination of these carriers causes the emission of light.

Silicon and germanium are *indirect-bandgap semiconductors.* In these semiconductors, the collision of a hole and an electron will not cause the two carriers to recombine. The electron may momentarily fall into the hole, but quantum mechanical considerations prevent the generation of a photon. Since the electron cannot shed excess energy, it is quickly ejected from the lattice and the electron-hole pair reforms. In the case of an indirect-bandgap semiconductor, recombination can only occur at specific sites in the lattice, called *traps,* where flaws or foreign atoms distort the lattice (Figure 1.4B). A trap can momentarily capture a passing carrier. The trapped carrier becomes vulnerable to recombination because the trap can absorb the liberated energy.

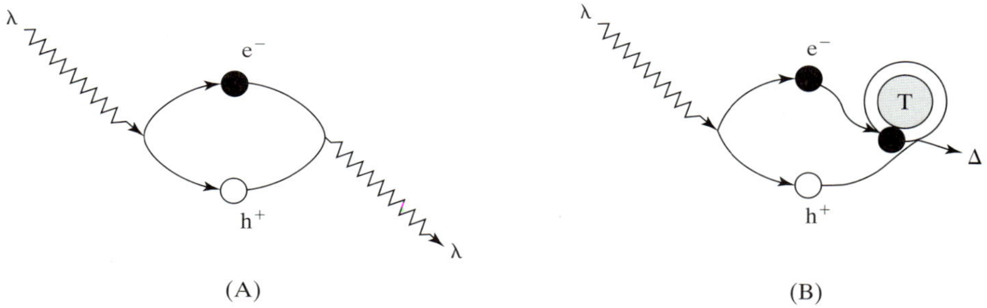

FIGURE 1.4 Schematic representations of recombination processes: (A) direct recombination, in which a photon, λ, generates a hole, h^+, and an electron, e^-, that collide and re-emit a photon; and (B) indirect recombination, in which one of the carriers is caught by a trap, T, and recombination takes place at the trap site with the liberation of heat, Δ.

Traps that aid the recombination of carriers are called *recombination centers.* The more recombination centers a semiconductor contains, the shorter the average time between the generation of a carrier and its recombination. This quantity, called the *carrier lifetime,* limits how rapidly a semiconductor device can switch on and off. Recombination centers are sometimes deliberately added to semiconductors to increase switching speeds. Gold atoms form highly efficient recombination centers in silicon, so high-speed diodes and transistors are sometimes made from silicon containing a small amount of gold. Gold is not the only substance that can form recombination centers. Many transition metals such as iron and nickel have a similar (if less potent) effect. Some types of crystal defects can also serve as recombination centers. Solid-state devices must be fabricated from extremely pure single-crystal materials in order to ensure consistent electrical performance.

1.1.2. Extrinsic Semiconductors

The conductivity of semiconductors depends upon their purity. Absolutely pure, or *intrinsic,* semiconductors have low conductivities because they contain only a few thermally generated carriers. The addition of certain impurities greatly increases the number of available carriers. These *doped*, or *extrinsic*, semiconductors can approach the conductivity of a metal. A lightly doped semiconductor may contain only a few parts per billion of dopant. Even a heavily doped semiconductor contains only a few hundred parts per million due to the limited solid solubility of dopants in silicon. The extreme sensitivity of semiconductors to the presence of dopants makes it nearly impossible to manufacture truly intrinsic material. Practical semiconductor devices are, therefore, fabricated almost exclusively from extrinsic material.

Phosphorus-doped silicon is an example of an extrinsic semiconductor. Suppose a small quantity of phosphorus is added to a silicon crystal. The phosphorus atoms are incorporated into the crystal lattice in positions that would otherwise have been occupied by silicon atoms (Figure 1.5). Phosphorus, a group-V element, has five valence electrons. The phosphorus atom shares four of these with its four neighboring atoms. Four bonding electron pairs give the phosphorus atom a total of eight shared electrons. These, combined with the one remaining unshared electron, result in a total of nine valence electrons. Since eight electrons entirely fill the valence shell, no room remains for the ninth electron. This electron is expelled from the phosphorus atom and wanders freely through the crystal lattice. Each phosphorus atom added to the silicon lattice thus generates one free electron.

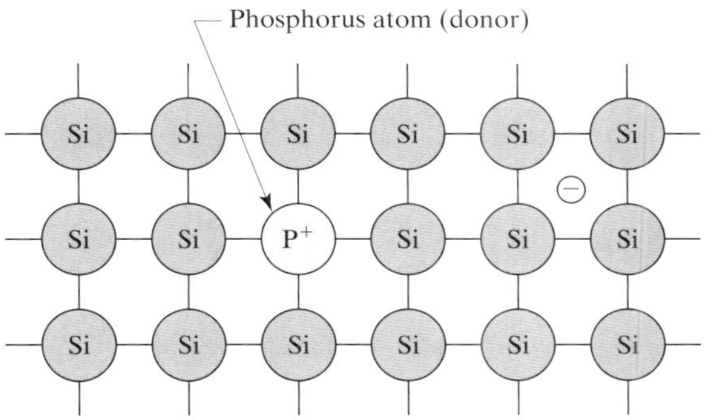

FIGURE 1.5 Simplified crystal structure of phosphorus-doped silicon.

The loss of the ninth electron leaves the phosphorus atom with a net positive charge. Although this atom is ionized, it does not constitute a hole. Holes are electron vacancies created by the removal of electrons from a filled valence shell. The phosphorus atom has a full valence shell despite its positive charge. The charge associated with the ionized phosphorus atom is therefore immobile.

Other group-V elements will have the same effect as phosphorus. Each atom of a group-V element that is added to the lattice will produce one additional free electron. Elements that donate electrons to a semiconductor in this manner are called *donors*. Arsenic, antimony, and phosphorus are all used in semiconductor processing as donors for silicon.

A semiconductor doped with a large number of donors has a preponderance of electrons as carriers. A few thermally generated holes still exist, but their numbers actually diminish in the presence of extra electrons. This occurs because the extra electrons increase the probability that the hole will find an electron and recombine. The large number of free electrons in N-type silicon greatly increases its conductivity (and greatly reduces its resistance).

A semiconductor doped with donors is said to be *N-type*. Heavily doped N-type silicon is sometimes marked N+, and lightly doped N-type silicon N–. The plus and minus symbols denote the relative numbers of donors, not electrical charges. Electrons are considered the *majority carriers* in N-type silicon due to their large numbers. Similarly, holes are considered the *minority carriers* in N-type silicon. Strictly speaking, intrinsic silicon has neither majority nor minority carriers because both types are present in equal numbers.

Boron-doped silicon forms another type of extrinsic semiconductor. Suppose a small number of boron atoms are added to the silicon lattice (Figure 1.6). Boron, a group-III element, has three valence electrons. The boron atom attempts to share its valence electrons with its four neighboring atoms, but, because it has only three, it cannot complete the fourth bond. As a result, there are only seven valence electrons around the boron atom. The electron vacancy thus formed constitutes a hole. This hole is mobile and soon moves away from the boron atom. Once the hole departs, the boron atom is left with a negative charge caused by the presence of an extra electron in its valence shell. As in the case of phosphorus, this charge is immobile and does not contribute to conduction. Each atom of boron added to the silicon contributes one mobile hole.

FIGURE 1.6 Simplified crystal structure of boron-doped silicon.

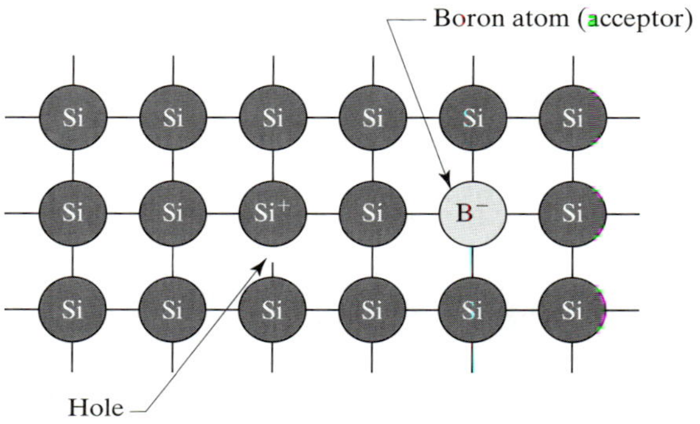

Other group-III elements can also accept electrons and generate holes. Technical difficulties prevent the use of any other group-III elements in silicon fabrication, but indium is sometimes used to dope germanium. Any group-III element used as a dopant will *accept* electrons from adjoining atoms, so these elements are called *acceptors.* A semiconductor doped with acceptors is said to be *P-type.* Heavily doped P-type silicon is sometimes marked P+, and lightly doped P-type silicon P–. Holes are the majority carriers and electrons are the minority carriers in P-type silicon. Table 1.2 summarizes some of the terminology used to describe extrinsic semiconductors.

TABLE 1.2 Extrinsic semiconductor terminology.

Semiconductor Type	Dopant Type	Typical Dopants for Silicon	Majority Carriers	Minority Carriers
N-type	Donors	Phosphorus, arsenic, and antimony	Electrons	Holes
P-type	Acceptors	Boron	Holes	Electrons

A semiconductor can be doped with both acceptors and donors. The dopant present in excess determines the type of the silicon and the concentration of the carriers. It is thus possible to invert P-type silicon to N-type by adding an excess of donors. Similarly, it is possible to invert N-type silicon to P-type by adding an excess of acceptors. The deliberate addition of an opposite-polarity dopant to invert the type of a semiconductor is called *counterdoping.* Most modern semiconductors are made by selectively counterdoping silicon to form a series of P- and N-type regions. Much more will be said about this practice in the next chapter.

If counterdoping were taken to extremes, the entire crystal lattice would consist of an equal ratio of acceptor and donor atoms. The two types of atoms would be present in exactly equal numbers. The resulting crystal would have very few free carriers and would appear to be an intrinsic semiconductor. Such *compound semiconductors* actually exist. The most familiar example is *gallium arsenide,* a compound of gallium (a group-III element) and arsenic (a group-V element). Materials of this sort are called III-V compound semiconductors. They include not only gallium arsenide but also gallium phosphide, indium antimonide, and many others. Many III-V compounds are direct-bandgap semiconductors, and some are used in constructing light-emitting diodes and semiconductor lasers. Gallium arsenide is also employed to a limited extent for manufacturing very high-speed solid-state devices, including integrated circuits. II-VI compound semiconductors are composed of equal mixtures of group-II and group-VI elements. Cadmium sulfide is a typical II-VI compound used to construct photosensors. Other II-VI compounds are used as phosphors in cathode ray tubes. A final class of semiconductors includes IV-IV compounds such as silicon carbide, recently used on a small scale to fabricate blue LEDs.

Of all the semiconductors, only silicon possesses the physical properties required for high-volume, low-cost manufacture of integrated circuits. The vast majority of solid-state devices are fabricated in silicon, and all other semiconductors are relegated to niche markets. The remainder of this text, therefore, focuses upon silicon integrated circuits.

1.1.3. Diffusion and Drift

The motion of carriers through a silicon crystal results from two separate processes: *diffusion* and *drift.* Diffusion is a random motion of carriers that occurs at all times and places, while drift is a unidirectional movement of carriers under the influence of an electric field. Both of these processes contribute to conduction in semiconductors.

Diffusion closely resembles Brownian motion. That is, individual carriers move through the semiconductor until they collide with lattice atoms. The collision process scatters the carriers through unpredictable angles. After a very few collisions, the motion of the carriers becomes completely randomized. The carriers wander aimlessly about, tracing a sort of drunkard's walk (Figure 1.7A).

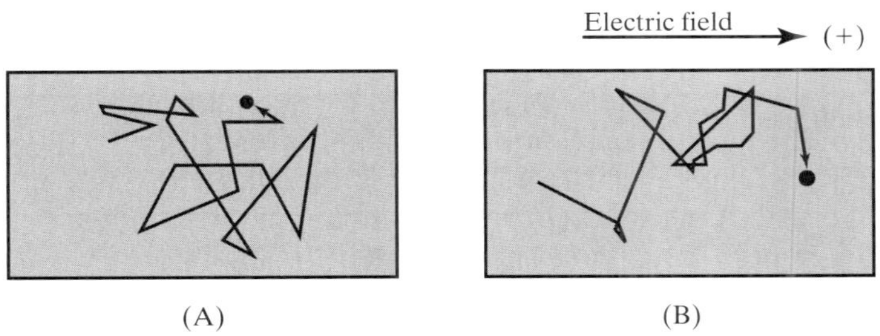

FIGURE 1.7 Comparison of conduction mechanisms for an electron: diffusion (A) and drift superimposed on diffusion (B). Notice the gradual motion of the electron toward the positive potential.

The diffusion of carriers through a semiconductor is analogous to the diffusion of dye molecules through still water. When a drop of concentrated dye falls into water, the dye molecules all initially occupy a small volume of liquid. The molecules gradually diffuse from regions of higher concentration to regions of lower concentration. Eventually the dye becomes distributed uniformly throughout the solution. Similarly, the diffusion of carriers across concentration gradients produces a *diffusion current.*

Unless some mechanism constantly adds more carriers, diffusion eventually redistributes them uniformly throughout the silicon and the diffusion current subsides.

The motion of a carrier under the influence of an electric field is called *drift.* Although the carrier still collides with the lattice and thus moves in a random drunkard's walk, it gradually drifts in a specific direction (Figure 1.7B). This subtle bias is caused by the electric field. No matter what direction the carrier moves, the field always and relentlessly acts upon it. If the carrier is moving opposite the field, its motion is retarded; if it is moving with the field, its motion is accelerated. Frequent collisions prevent the carrier from building up any appreciable velocity, but a subtle overall motion appears. Electrons move toward positive potentials, even if slowly and erratically. Holes likewise move toward negative potentials. A simple analogy to drift consists of the motion of the steel ball in a pinball machine. Although bumpers and pegs may divert the ball in any direction, the tilt of the board causes it to eventually move downward. Similarly, an electric field biases carriers toward motion in a specific direction, producing a *drift current.*

1.2 PN JUNCTIONS

Uniformly doped semiconductors have few applications. Almost all solid-state devices contain a combination of multiple P- and N-type regions. The interface between a P-type region and an N-type region is called a *PN junction,* or simply a *junction.*

Figure 1.8A shows two pieces of silicon. On the left is a bar of P-type silicon, and on the right is a bar of N-type silicon. No junction is present as long as the two are not in contact with one another. Each piece of silicon contains a uniform distribution of carriers. The P-type silicon has a large majority of holes and a few electrons; the N-type silicon has a large majority of electrons and a few holes.

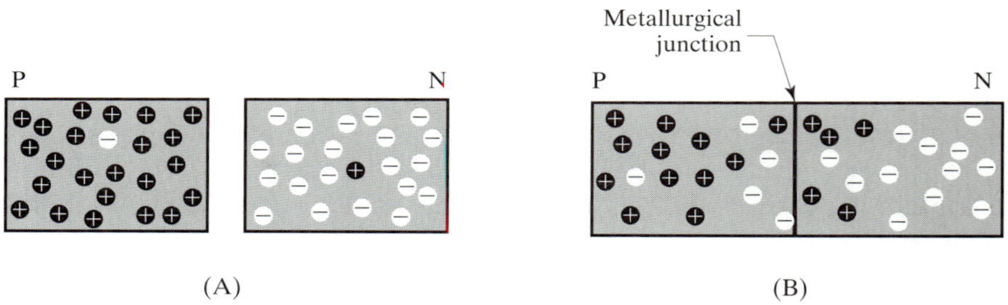

FIGURE 1.8 Carrier populations in silicon before the junction is assembled (A), and afterward (B).

Now, imagine the two pieces of silicon are brought into contact with one another to form a junction. No physical barrier to the motion of the carriers remains. There is a great excess of holes in the P-type silicon, and a great excess of electrons in the N-type silicon. Some of the holes diffuse from the P-type silicon to the N-type. Likewise, some of the electrons diffuse from the N-type silicon to the P-type. Figure 1.8B shows the result. A number of carriers have diffused across the junction in both directions. The concentration of minority carriers on either side of the junction has risen above that which would be produced by doping alone. The excess of minority carriers produced by diffusion across a junction is called the *excess minority carrier concentration.*

1.2.1. Depletion Regions

The presence of excess minority carriers on either side of a junction has two effects. First, the carriers produce an electric field. The extra holes in the N-type silicon represent a positive charge, while the excess electrons in the P-type silicon represent a negative charge. Thus an electric potential develops across the PN junction that biases the N-side of the junction positive with respect to the P-side.

When carriers diffuse across the junction, they leave equal numbers of ionized dopant atoms behind. These atoms are rigidly fixed in the crystal lattice and cannot move. On the P-side of the junction lie ionized acceptors that produce a negative charge. On the N-side of the junction lie ionized donors that produce a positive charge. An electric potential again develops that biases the N-side of the junction positive with respect to the P-side. This potential adds to the one produced by the separation of the charged carriers.

Carriers tend to drift in the presence of an electric field. Holes are attracted to the negative potential on the P-side of the junction. Similarly, electrons are attracted to the positive potential on the N-side of the junction. The drift of carriers thus tends to oppose their diffusion. Holes diffuse from the P-side of the junction to the N-side and drift back. Electrons diffuse from the N-side of the junction to the P-side and drift back. Equilibrium occurs when the drift and diffusion currents are equal and opposite. The excess minority carrier concentrations on either side of the junction also reach equilibrium values, as does the voltage potential across the junction.

The voltage difference across a PN junction in equilibrium is called its *built-in potential,* or its *contact potential.* In a typical silicon PN junction, the built-in potential can range from a few tenths of a volt to as much as a volt. Heavily doped junctions have larger built-in potentials than lightly doped ones. Because of the higher doping levels, more carriers diffuse across the heavily doped junction and thus a larger diffusion current flows. In order to restore equilibrium, a larger drift current is also needed and thus a stronger electric field develops. Heavily doped junctions therefore have larger built-in potentials than lightly doped ones.

Although the built-in potential is quite real, it cannot be measured with a voltmeter. This apparent paradox can be explained by a closer examination of a circuit containing a PN junction and a voltmeter (Figure 1.9). The two probes of the meter are made of metal, not silicon. The points of contact between the metal probes and the silicon also form junctions, each of which has a contact potential of its own. Because the silicon beneath the two probes has different doping levels, the two contact potentials of the probe points are unequal. The difference between these two contact potentials exactly cancels the built-in potential of the PN junction, and no current flows in the external circuit. This situation must occur because any current flow would constitute a free energy source, or a sort of perpetual motion machine. The cancellation of the built-in potentials ensures that energy cannot be extracted from a PN junction in equilibrium and thus prevents a violation of the laws of thermodynamics.

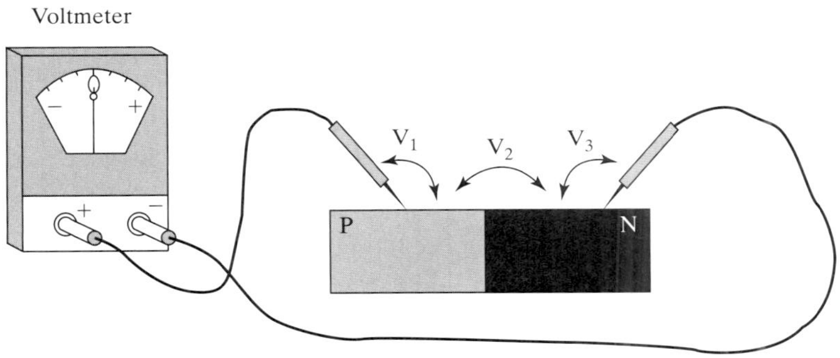

FIGURE 1.9 Demonstration of the impossibility of directly measuring a built-in potential. Contact potentials V_1 and V_3 exactly cancel built-in potential V_2.

The built-in potential has two causes: the separation of ionized dopant atoms and the separation of charged carriers. The carriers are free to move, but the dopant atoms are rigidly fixed in the crystal lattice. If the dopant atoms could move, they would be drawn together by their opposite charges. They remain separated because they are anchored to the lattice. The region occupied by these charged atoms is subject to a strong electric field. Any carrier that enters this region must move quickly or it will be swept out again by the field. As a result, this region contains very few carriers at any given instant in time. This region is sometimes called a *space charge layer* because of the presence of the charged dopant atoms. More commonly, it is called a *depletion region* because of the relatively low concentration of carriers found there.

If the depletion region contains few carriers, then the excess minority carriers must pile up on either side of it. Figure 1.10 graphically shows the resulting distributions of excess minority carriers. The concentration gradients cause these carriers to diffuse into the electrically neutral regions beyond the junction. The electric field produced by the separation of these charged carriers pulls them back toward the junction. An equilibrium is soon established, resulting in steady-state distributions of minority carriers resembling those shown in Figure 1.10.

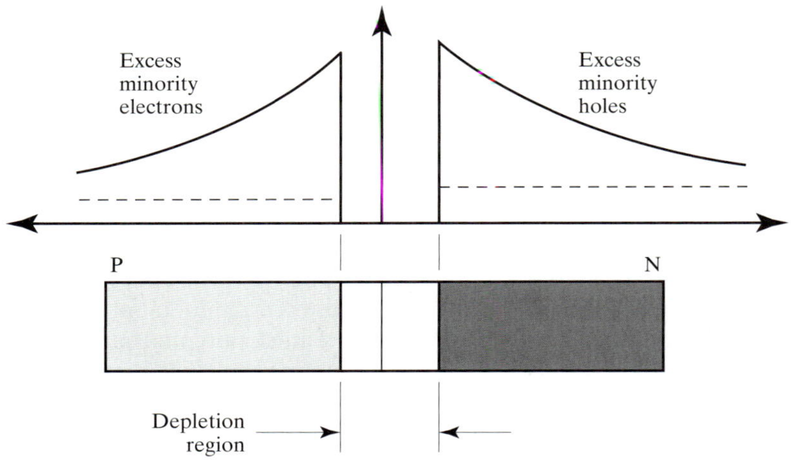

FIGURE 1.10 Diagram of excess minority carrier concentrations on either side of a PN junction in equilibrium.

The behavior of a PN junction can be summarized as follows: the diffusion of carriers across the junction produces excess minority carrier concentrations on either side of a depletion region. The separation of ionized dopant atoms causes an electric field to form across the depletion region. This field prevents most of the majority carriers from crossing the depletion region, and the few that do are eventually swept back to the other side by the electric field.

The thickness of a depletion region depends on the doping of the two sides of the junction. If both sides are lightly doped, then a substantial thickness of silicon must deplete in order to uncover enough dopant atoms to support the built-in potential. If both sides are heavily doped, then only a very thin depletion region need be uncovered to produce the necessary charges. Therefore heavily doped junctions have thin depletion regions and lightly doped junctions have thick ones. If one side of the junction is more heavily doped than the other, then the depletion region will extend further into the lightly doped side. In this case, a substantial thickness of lightly doped silicon must be uncovered to yield enough ionized dopants. Only a thin layer of heavily doped silicon need be uncovered to yield a counterbalancing charge. Figure 1.10 illustrates this case since the N-side of the junction is more lightly doped than the P-side.

1.2.2. PN Diodes

A PN junction forms a very useful solid-state device called a *diode.* Figure 1.11 shows a simplified diagram of the structure of a PN diode. The diode has, as its name suggests, two terminals. One terminal, called the *anode,* connects to the P-side of the junction. The other terminal, called the *cathode,* connects to the N-side. These two terminals are used to connect the diode to an electrical circuit. The schematic symbol for a diode consists of an arrowhead representing the anode and a perpendicular line representing the cathode. Diodes conduct current preferentially in one direction—that indicated by the arrowhead.

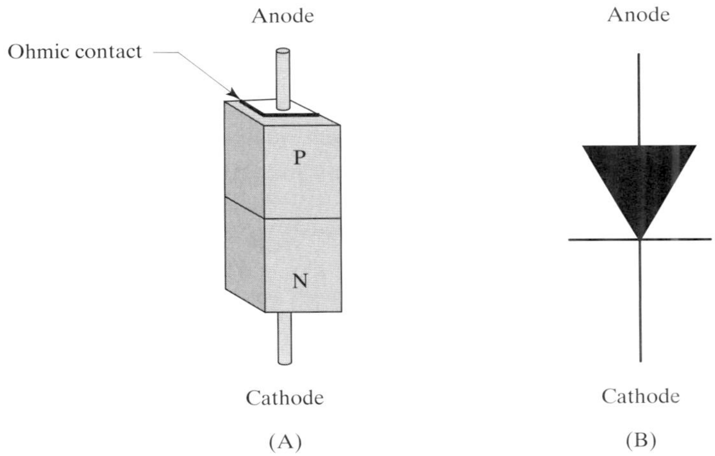

FIGURE 1.11 PN junction diode: simplified structure (A) and standard schematic symbol (B).

To illustrate how the diode operates, imagine that an adjustable voltage source has been connected across it. If the voltage source is set to zero volts, then the diode is under *zero bias.* No current will flow through a zero-biased diode. If the voltage source is set to bias the anode negative with respect to the cathode, then the diode is *reverse biased.* Very little current flows through a reverse-biased diode. If the voltage source is set to bias the anode positive with respect to the cathode, then the diode is *forward biased* and a large current flows. This accords with the simple mnemonic: *current flows with the arrow, not against it.* Devices that conduct in only one direction are called *rectifiers*. They find frequent application in power supplies, radio receivers, and signal processing circuits.

Diode rectification depends upon the presence of a junction. Each of the three bias conditions can be explained by an appropriate analysis of carrier flows across this junction. The case of the zero-bias diode is particularly simple since it is identical to the case of the equilibrium junction already discussed. The only potential present across the junction is the built-in potential. When the diode is connected in a circuit, the contact potentials of the leads touching the silicon balance the built-in potential of the junction. Thus no current flows in the circuit.

The behavior of a reverse-biased diode is also simple to explain. The reverse bias makes the N-side of the junction even more positive with respect to the P-side. The voltage seen across the junction increases, so excess minority carriers continue to be swept back across it and majority carriers continue to be held on their respective sides of the junction. The increased voltage across the junction causes the ionization of additional dopant atoms on either side, so the depletion region widens as the reverse bias increases.

The behavior of a forward-biased junction is somewhat more complex. The voltage applied to the terminals opposes the built-in potential. The voltage across the junction therefore lessens and the depletion region thins. The drift currents caused by the electric field are simultaneously reduced. More and more majority carriers make the transit across the depletion region without being swept back by the electric field. Figure 1.12 shows graphically the overall flow of carriers: holes are injected across the junction from anode to cathode (left to right), while electrons are injected across the junction from cathode to anode (right to left). In the illustrated diode, the hole current across the junction outweighs the electron current because the anode is more heavily doped than the cathode and there are more majority holes available in the anode than there are majority electrons in the cathode. Once these carriers have been injected across the junction, they become minority carriers and recombine with majority carriers present on the other side. Currents are drawn in from the terminals in order to replenish the supply of majority carriers in the neutral silicon. This illustration is somewhat simplified since it only shows the general flow of carriers through the diode. Some of the carriers injected across the junction are swept back by the electric field before they can recombine. Such carriers do not contribute to the net current flow through the diode, so they are not illustrated. Likewise, the tiny numbers of thermally generated minority carriers that cross the junction are not shown since they form an insignificant portion of the overall current flow through a forward-biased diode.

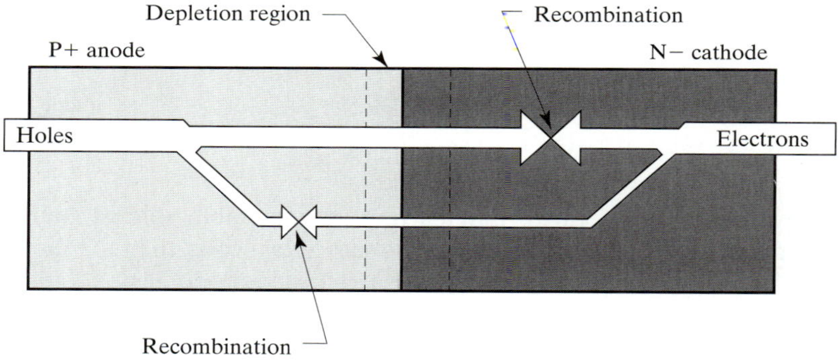

FIGURE 1.12 Carrier flow in a forward-biased PN junction.

The current through a forward-biased diode depends exponentially upon the applied voltage (Figure 1.13). About 0.6V suffices to produce substantial forward conduction in a silicon PN junction at room temperature.[5] Because diffusion is caused by the thermal motions of carriers, higher temperatures cause an exponential increase in diffusion currents. The forward current through a PN junction thus increases exponentially as temperature increases. Expressed another way, the forward bias required to sustain a constant current in a silicon PN junction decreases by approximately 2mV/°C.

Figure 1.13 also shows a low level of current flow when the diode is reverse-biased. This current flow is called *reverse conduction* or *leakage*. Leakage currents are produced by the few minority carriers thermally generated in the silicon. The electric field opposes the flow of majority carriers across a reverse-biased junction,

[5] The most widely quoted value is 0.7V, but in practice a typical integrated circuit base-emitter junction under microamp-level bias at 25°C exhibits a value nearer 0.6V than 0.7V.

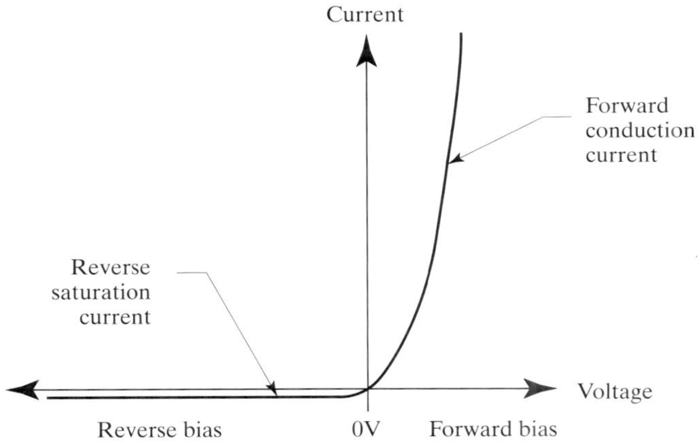

FIGURE 1.13 Diode conduction characteristics. The current scale is greatly magnified to show the reverse saturation current, which typically equals no more than a few picoamps at 25°C.

but it aids the flow of minority carriers. The application of a reverse bias sweeps these minority carriers across the junction. Because the rate of generation of minority carriers in the bulk silicon is essentially independent of electric fields, the leakage current does not vary much with reverse bias. Thermal generation does increase with temperature, and leakage currents are therefore temperature-dependent. In silicon, leakage currents double approximately every eight degrees Celsius. At high temperatures, the leakage currents begin to approach the operating currents of the circuit. The maximum operating temperature of a semiconductor device is therefore limited by leakage current. A maximum junction temperature of 150°C is widely accepted for silicon-integrated circuits.[6]

1.2.3. Schottky Diodes

Rectifying junctions can also form between a semiconductor and a metal. Such junctions are called *Schottky barriers.* The behavior of a Schottky barrier is somewhat analogous to that of a PN junction. Schottky barriers can, for example, be used to construct *Schottky diodes,* which behave much like PN diodes. Schottky barriers can also form in the contact regions of an integrated circuit's interconnection system.

The *work function* of a material equals the amount of energy required to remove an electron from it. Each material has its own characteristic work function that depends upon the properties of its crystal lattice as well as its composition. When two materials with different work functions are brought into contact, the electrons in each material have different initial energies. A voltage difference called the *contact potential* therefore exists between the two materials. Consider the case of a PN junction. The two semiconductors on either side of the junction have the same lattice structure. The contact potential of a PN junction, or its *built-in potential*, depends only upon doping. In the case of a Schottky barrier, the different lattice structures of the metal and the semiconductor also contribute to the contact potential.

A typical rectifying Schottky barrier results when aluminum metal touches lightly doped N-type silicon (Figure 1.14B). The carriers must redistribute in order to counterbalance the contact potential. Electrons diffuse from the semiconductor into the metal, where they pile up to form a thin film of negative charge. This exodus of electrons from the silicon leaves behind a zone of ionized dopant atoms that form a

---

6 Integrated circuits can be built that work at 200°C, but many standard design practices do not apply. See R. J. Widlar and M. Yamatake, "Dynamic Safe-Area Protection for Power Transistors Employs Peak-Temperature Limiting," *IEEE J. Solid-State Circuits*, SC-22, #1, 1987, p. 77–84.

depletion region (Figure 1.14A). The electric field generated by the depletion region draws electrons from the metal back into the semiconductor. Equilibrium occurs when the drift and diffusion currents are equal. The potential difference across the Schottky barrier now equals the contact potential. Few minority carriers exist on the semiconductor side of the Schottky barrier, so the Schottky diode is called a *majority-carrier device.*

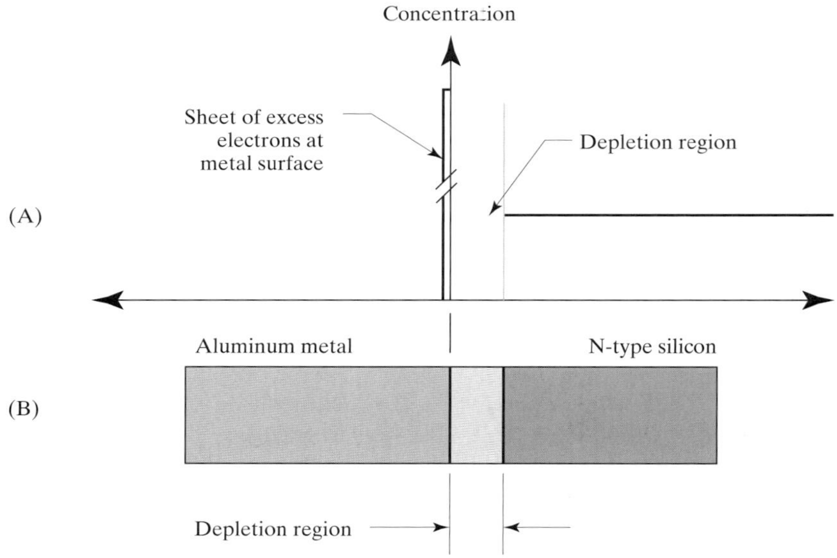

FIGURE 1.14 Diagram of excess carrier concentrations on either side of the Schottky barrier (A) and cross-section of the corresponding Schottky structure (B).

The behavior of a Schottky diode under bias can be similarly analyzed. The N-type silicon forms the *cathode* of the diode, while the metal plate forms the *anode.* The case of a zero-biased Schottky diode is identical to the case of the equilibrium Schottky barrier analyzed above. A reverse-biased Schottky has an external voltage connected in order to bias the semiconductor positively with respect to the metal. The resulting voltage difference adds to the contact potential. The depletion region widens to counterbalance the increased voltage difference, equilibrium is restored, and very little current flows through the diode.

A forward-biased Schottky diode has an external voltage connected in order to bias the metal positively with respect to the semiconductor. The resulting voltage difference across the junction opposes the contact potential, and the width of the depletion region shrinks. Eventually the contact potential is entirely offset, and a depletion region attempts to form on the metal side of the junction. The metal, being a conductor, cannot support an electric field, and no depletion region can form to oppose the externally applied potential. This potential begins to sweep electrons across the junction from the semiconductor into the metal, and a current flows through the diode.

Schottky diodes exhibit current-voltage characteristics similar to those of a PN diode (Figure 1.13). Schottky diodes also exhibit leakage currents caused by low levels of minority carrier injection from the metal into the semiconductor. These conduction mechanisms are accelerated by high temperatures, producing temperature dependencies similar to those of a PN diode.

Despite many apparent similarities, there are a few fundamental differences between Schottky diodes and PN junction diodes. Schottky diodes are majority-carrier devices since they rely primarily upon majority-carrier conduction. At high current densities a few holes do flow from the metal to the semiconductor, but these contribute only a small fraction of the total current. Schottky diodes do not support large excess minority-carrier populations. Since the switching speed of a diode is a function of the time required for excess minority carriers to recombine, Schottky diodes can switch very rapidly. Some types of Schottky diodes also exhibit lower forward bias voltages than PN diodes. This combination of low forward-voltage drop and highly efficient switching make Schottky diodes very useful devices.

Schottky diodes can also be formed to P-type silicon, but the forward biases required for conduction are usually quite low. This renders P-type Schottky diodes rather leaky and they are therefore rarely used.[7] Most practical Schottky diodes result from the union between lightly doped N-type silicon and a class of materials called *silicides*. These substances are definite compounds of silicon and certain metals, for example platinum and palladium. Silicides exhibit very stable work functions and therefore form Schottky diodes that have consistent and repeatable characteristics.

1.2.4. Zener Diodes

Under normal conditions, only a small current flows through a reverse-biased PN junction. This leakage current remains approximately constant until the reverse bias exceeds a certain critical voltage, beyond which the PN junction suddenly begins to conduct large amounts of current (Figure 1.15). The sudden onset of significant reverse conduction is called *reverse breakdown,* and it can lead to device destruction if the current flow is not limited by some external means. Reverse breakdown often sets the maximum operating voltage of a solid-state device. However, if appropriate precautions are taken to limit the current flow, a junction in reverse breakdown can provide a fairly stable voltage reference.

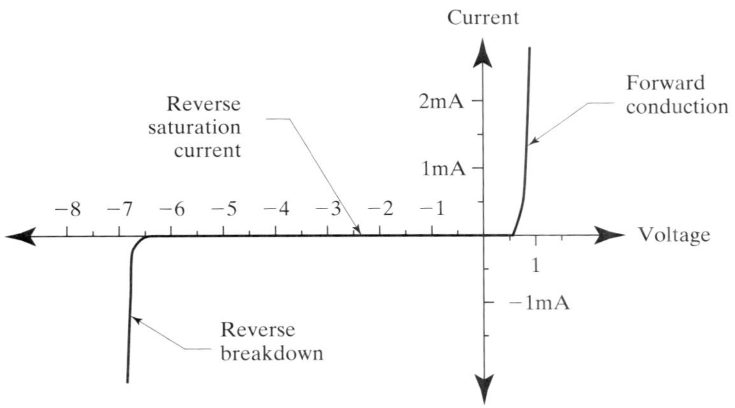

FIGURE 1.15 Reverse breakdown in a PN junction diode.

One of the mechanisms responsible for reverse breakdown is called *avalanche multiplication.* Consider a PN junction under reverse bias. The width of the depletion region increases with bias, but not fast enough to prevent the electric field from intensifying. The intense electric field accelerates the few carriers crossing the depletion region to extremely high velocities. When these carriers collide with lattice atoms, they knock loose valence electrons and generate additional carriers. This

---

[7] For example, compare the differences in work functions for platinum with respect to N-type silicon (0.85V) and P-type silicon (0.25V): R. S. Muller and T. I. Kamins, *Device Electronics for Integrated Circuits,* 2nd ed. (New York: John Wiley and Sons, 1986), p. 157.

process is aptly named because a single carrier can spawn literally thousands of additional carriers through collisions, just as a single snowball can start an avalanche.

The other mechanism behind reverse breakdown is called *tunneling*. Tunneling is a quantum-mechanical process that allows particles to move short distances regardless of any apparent obstacles. If the depletion region is thin enough, then carriers can leap across it by tunneling. The tunneling current depends strongly on both the depletion region width and the voltage difference across the junction. Reverse breakdown caused by tunneling is called *Zener breakdown.*

The reverse breakdown voltage of a junction depends on the width of its depletion region. Wider depletion regions produce higher breakdown voltages. As previously explained, the more lightly doped side of a junction sets its depletion region width and therefore its breakdown voltage. When the breakdown voltage is less than five volts, the depletion region is so thin that Zener breakdown predominates. When the breakdown voltage exceeds five volts, avalanche breakdown predominates. A PN diode designed to operate in reverse conduction is called either a *Zener diode* or an *avalanche diode,* depending on which of these two mechanisms predominates. Zener diodes have breakdown voltages of less than five volts, while avalanche diodes have breakdown voltages of more than five volts. Engineers traditionally call all breakdown diodes Zeners regardless of what mechanism underlies their operation. This can lead to confusion because a 7V Zener conducts primarily by avalanche breakdown.

In practice, the breakdown voltage of a junction depends on its geometry as well as its doping profile. The above discussion analyzed a *planar junction* consisting of two uniformly doped semiconductor regions intersecting in a planar surface. Although some real junctions approximate this ideal, most have curved sidewalls. The curvature intensifies the electric field and reduces the breakdown voltage. The smaller the radius of curvature, the lower the breakdown voltage. This effect can have a dramatic impact on the breakdown voltages of shallow junctions. Most Schottky diodes have sharp discontinuities at the edge of the metal-silicon interface. Electric field intensification can drastically reduce the measured breakdown voltage of a Schottky diode unless special precautions are taken to relieve the electric field at the edges of the Schottky barrier.

Figure 1.16 shows schematic symbols for all of the diodes discussed above. The PN junction diode uses a straight line to denote the cathode, while the Schottky diode and Zener diode are indicated by modifications to the cathode bar. In all cases, the arrow indicates the direction of conventional current flow through the forward-biased diode. In the case of the Zener diode, this arrow can be somewhat misleading because Zeners are normally operated in reverse bias. To the casual observer, the symbol may thus appear to be inserted "the wrong way around."

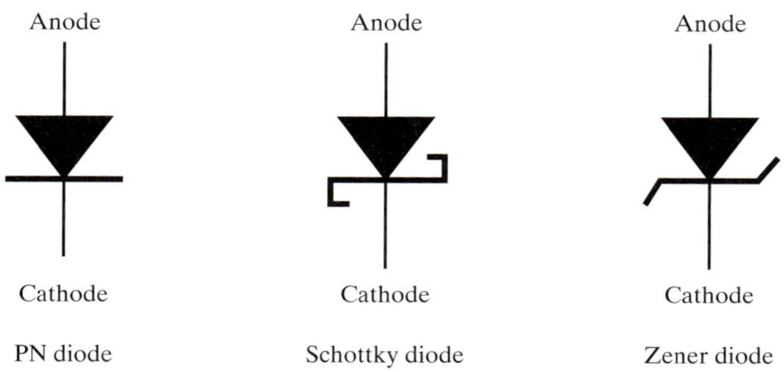

FIGURE 1.16 Schematic symbols for PN junction, Schottky, and Zener diodes. Some schematics show the arrowheads unfilled or show only half the arrowheads.

1.2.5. Ohmic Contacts

Contacts must be made between metals and semiconductors in order to connect solid-state devices into a circuit. These contacts would ideally be perfect conductors, but in practice they are *Ohmic contacts* that exhibit a small amount of resistance. Unlike rectifying contacts, these Ohmic contacts will conduct current equally well in either direction.

Schottky barriers can exhibit Ohmic conduction if the semiconductor material is doped heavily enough. The high concentration of dopant atoms thins the depletion region to the point where carriers can easily tunnel across it. Unlike normal Zener diodes, Ohmic contacts can support tunneling at very low voltages. Rectification does not occur since the carriers can effectively bypass the Schottky barrier by tunneling through it.

An Ohmic contact can also form if a Schottky barrier's contact potential causes surface accumulation rather than surface depletion. In accumulation, a thin layer of majority carriers forms at the semiconductor surface. In the case of an N-type semiconductor, this layer consists of excess electrons. The metal is a conductor and therefore cannot support a depletion region. A thin film of charge thus appears at the surface of the metal to counterbalance the accumulated carriers in the silicon. The lack of a depletion region on either side of the barrier prevents the contact from supporting a voltage differential, and any externally applied voltage will sweep carriers across the junction. Carriers can flow in either direction, so this type of Schottky barrier forms an Ohmic contact rather than a rectifying one.

In practice, rectifying contacts form to lightly doped silicon, and Ohmic contacts form to heavily doped silicon. The exact mechanism behind Ohmic conduction is unimportant since all Ohmic contacts behave in essentially the same manner. A lightly doped silicon region can be Ohmically contacted only if a thin layer of more heavily doped silicon is placed beneath the contact. Contact resistances of less than $50\Omega/\mu m^2$ can be obtained if a heavily doped silicon layer is used in combination with a suitable metal system. This resistance is small enough that it can be neglected for most applications.

Any junction between dissimilar materials exhibits a contact potential equal to the difference between the work functions of the materials. This rule applies to Ohmic contacts as well as to PN junctions and rectifying Schottky barriers. If all the contacts and junctions are held at the same temperature, then the sum of the contact potentials around any closed loop will equal zero. Contact potentials are, however, strong functions of temperature. If one of the junctions is held at a different temperature than the others, then its contact potential will shift and the sum of the contact potentials will no longer equal zero. This *thermoelectric effect* has significant implications for integrated circuit design.

Figure 1.17 shows a block of N-type silicon contacted on either side by aluminum. If one end of the block is heated, then a measurable voltage develops across the

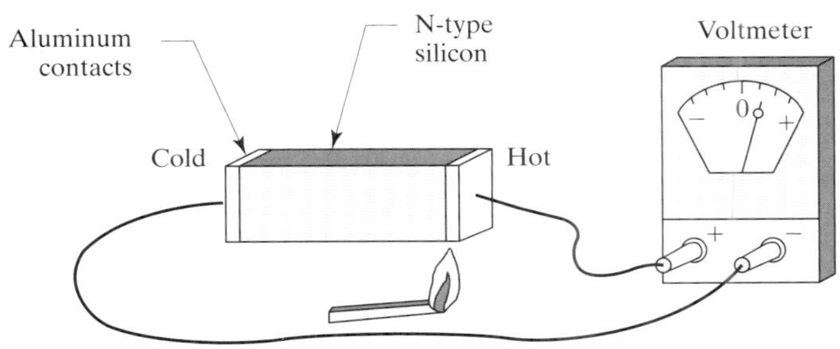

FIGURE 1.17 The thermoelectric effect produces a net measurable voltage if the two contacts are held at different temperatures.

block due to the mismatch between the two contact potentials. This voltage drop is typically 0.1–1.0mV/°C.[8] Many integrated circuits rely upon voltages matching within a millivolt or two, so even small temperature differences are enough to cause such circuits to malfunction.

1.3 BIPOLAR JUNCTION TRANSISTORS

While diodes are useful devices, they cannot amplify signals, and almost all electronic circuits require amplification in one form or another. One device that can amplify signals is called a *bipolar junction transistor* (BJT).

The structures of the two types of bipolar junction transistors are shown in Figure 1.18. Each transistor consists of three semiconductor regions called the *emitter, base,* and *collector.* The base is always sandwiched between the emitter and the collector. An NPN transistor consists of an N-type emitter, a P-type base, and an N-type collector. Similarly, a PNP transistor consists of a P-type emitter, an N-type base, and a P-type collector. In these simplified cross-sections, each region of the transistor consists of a uniformly doped section of a rectangular bar of silicon. Modern bipolar transistors have somewhat different cross-sections, but the principles of operation remain the same.

Figure 1.18 also shows the symbols for the two types of transistors. The arrowhead placed on the emitter lead indicates the direction of conventional current flow through the forward-biased emitter-base junction. No arrow appears on the collector lead even though a junction also exists between the collector and the base. In the simplified transistors of Figure 1.18, the emitter-base and collector-base junctions appear to be identical. One could apparently swap the collector and emitter leads without affecting the behavior of the device. In practice, the two junctions have different doping profiles and geometries and are not interchangeable. The emitter lead is distinguished from the collector lead by the presence of the arrowhead.

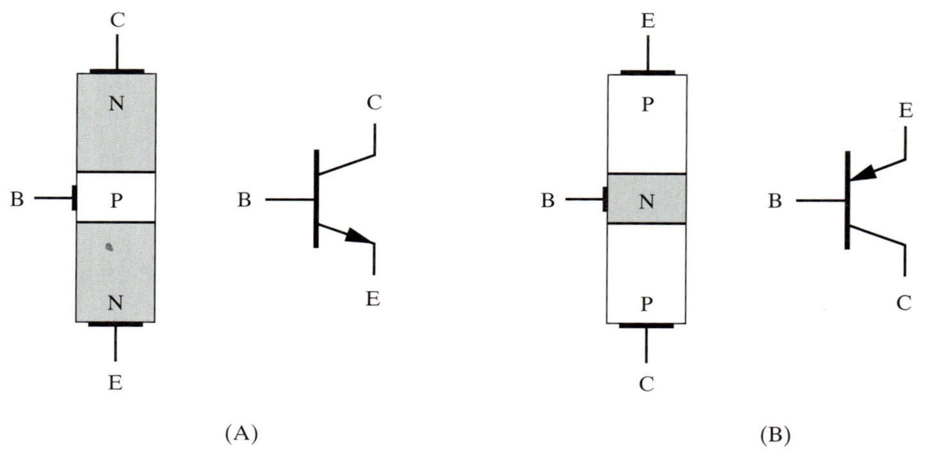

FIGURE 1.18 Structures and schematic symbols for the NPN transistor (A) and the PNP transistor (B).

A bipolar junction transistor can be viewed as two PN junctions connected back-to-back. The base region of the transistor is very thin (about 1–2μm wide). When the two junctions are placed in such close proximity, carriers can diffuse from one junction to the other before they recombine. Conduction across one junction therefore affects the behavior of the other junction.

---

[8] Lightly-doped silicon exhibits a higher Seebeck voltage; these values are taken from Widlar, *et al.*, p. 79.

Figure 1.19A shows an NPN transistor with zero volts applied across the base-emitter junction and five volts applied across the base-collector junction. Neither junction is forward biased, so very little current flows through any of the three terminals of the transistor. A transistor with both junctions reverse biased is said to be in *cutoff.* Figure 1.19B shows the same transistor with ten microamps of current injected into its base. This current forward biases the base-emitter junction to a potential of about 0.65V. A collector current a hundred times larger than the base current flows across the base-collector junction even though this junction remains reverse biased. This current is a consequence of the interaction between the forward-biased base-emitter junction and the reverse-biased base-collector junction. Whenever a transistor is biased in this manner, it is said to operate in the *forward active* region. If the emitter and collector terminals are interchanged so that the base-emitter junction becomes reverse-biased and the base-collector junction becomes forward-biased, the transistor is said to operate in the *reverse active* region. In practice, transistors are seldom operated in this manner.

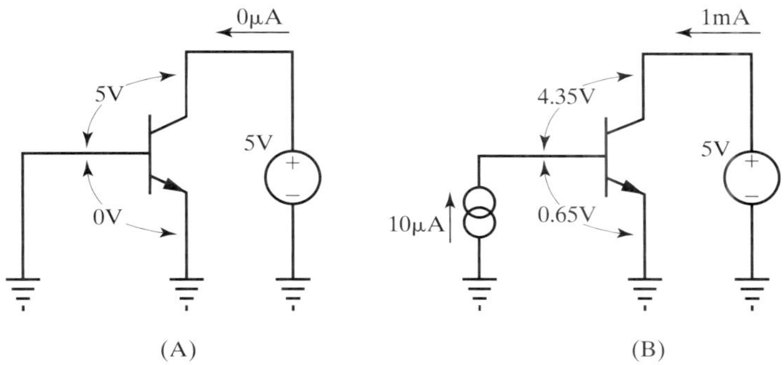

FIGURE 1.19 An NPN transistor operating in cutoff (A) and in the forward active region (B).

Figure 1.20 helps explain why collector current flows across a reverse-biased junction. Carriers flow across the base-emitter junction as soon as it becomes forward biased. Most of the current flowing across this junction consists of electrons injected from the heavily doped emitter into the lightly doped base. Most of these electrons diffuse across the narrow base before they recombine. The base-collector junction is reverse biased, so very few majority carriers can flow from the base into the collector. The same electric field that opposes the flow of majority carriers actually aids the flow of minority carriers. The electrons are minority carriers in the base, so they are swept across the reverse-biased base-collector junction into the collector. Here they again become majority carriers flowing toward the collector terminal. The collector current consists of the electrons that successfully complete the journey from emitter to collector without recombining in the base.

Some of the electrons injected into the base do not reach the collector. Those that do not reach the collector recombine in the base region. Base recombination consumes holes that are replenished by a current flowing in from the base terminal. Some holes are also injected from the base into the emitter, where they rapidly recombine. These holes represent a second source of base terminal current. These recombination processes typically consume no more than 1% of the emitter current, so only a small base current is required to maintain the forward bias across the base-emitter junction.

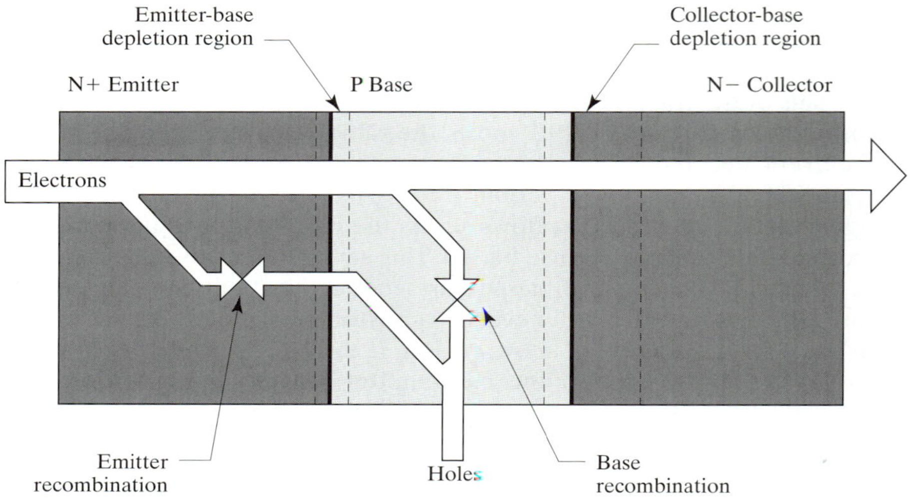

FIGURE 1.20 Current flow in an NPN transistor in the forward-active region.

1.3.1. Beta

The current amplification achieved by a transistor equals the ratio of its collector current to its base current. This ratio has been given various names, including *current gain* and *beta*. Likewise, different authors have used different symbols for it, including β and h_{FE}. A typical integrated NPN transistor exhibits a beta of about 150. Certain specialized devices may have betas exceeding 10,000. The beta of a transistor depends upon the two recombination processes illustrated in Figure 1.20.

Base recombination occurs primarily within the portion of the base between the two depletion regions, which is called the *neutral base* region. Three factors influence the base recombination rate: neutral base width, base doping, and the concentration of recombination centers. A thinner neutral base reduces the distance that the minority carriers must traverse and thus lessens the probability of recombination. Similarly, a more lightly doped base region minimizes the probability of recombination by reducing the majority carrier concentration. The *Gummel number* Q_B measures both of these effects. It is calculated by integrating the dopant concentration along a line traversing the neutral base region. In the case of uniform doping, the Gummel number equals the product of the base dopant concentration and the width of the neutral base. Beta is inversely proportional to the Gummel number.

The switching speed of transistors depends primarily on how quickly the excess minority carriers can be removed from the base, either through the base terminal or through recombination. Gold-doping is sometimes used to deliberately increase the number of recombination centers in bipolar junction transistors. The elevated recombination rate helps speed transistor switching, but it also reduces transistor beta. Few analog integrated circuits are built on gold-doped processes because of their low betas.

Bipolar transistors typically use a lightly doped base and a heavily doped emitter. This combination helps ensure that almost all of the current injected across the base-emitter junction consists of carriers flowing from emitter to base and not *vice-versa.* Heavy doping enhances the recombination rate in the emitter, but this has little impact since so few carriers are injected into the emitter in the first place. The ratio of current injected into the emitter to that injected into the base is called the *emitter injection efficiency.*

Most NPN transistors use a wide, lightly doped collector in combination with a heavily doped emitter and a thin, moderately doped base. The light collector doping allows a wide depletion region to form in the neutral collector. This permits a high collector operating voltage without avalanching the collector-base junction. The asymmetric doping of emitter and collector helps explain why bipolar transistors do not operate well when these terminals are swapped.[9] A typical integrated NPN transistor with a forward beta of 150 has a reverse beta of less than 5. This difference is primarily due to the drastic reduction in emitter injection efficiency caused by the substitution of a lightly doped collector for a heavily doped emitter.

Beta also depends upon collector current. Beta is reduced at low currents by leakage and by low levels of recombination in the depletion regions. At modest current levels, these effects become insignificant and the beta of the transistor climbs to a peak value determined by the mechanisms discussed above. High collector currents cause beta to roll off due to an effect called *high-level injection.* When the minority carrier concentration approaches the majority carrier concentration in the base, extra majority carriers accumulate to maintain the balance of charges. The additional base majority carriers cause the emitter injection efficiency to decrease, which in turn causes beta to decrease. Most transistors are operated at moderate current levels to avoid beta roll-off, but power transistors must often operate in high-level injection because of size constraints.

The behavior of the PNP transistor is very similar to that of the NPN transistor. The beta of a PNP transistor is lower than that of an NPN of comparable dimensions and doping profiles, because the mobility of holes is lower than that of electrons. In many cases, the performance of the PNP is further degraded because of a conscious choice to optimize the NPN transistor at the expense of the PNP. For example, the material used to construct the base region of an NPN is often used to fabricate the emitter of a PNP. Since the resulting emitter is rather lightly doped, emitter injection efficiency is low, and the onset of high-level injection occurs at moderate current levels. Despite their limitations, PNP transistors are very useful devices, and most bipolar processes support their construction.

1.3.2. I-V Characteristics

The performance of a bipolar transistor can be graphically depicted by drawing a family of curves that relate base current, collector current, and collector-emitter voltage. Figure 1.21 shows a typical set of curves for an integrated NPN transistor. The vertical axis measures collector current I_C, while the horizontal axis measures collector-to-emitter voltage V_{CE}. A number of curves are superimposed upon the same graph, each representing a different base current I_B. This family of curves shows a number of interesting features of the bipolar junction transistor.

In the *saturation region,* the collector-emitter voltage remains so small that the collector-base junction is slightly forward-biased. The electric field that sweeps minority carriers across the collector-base junction still exists, so the transistor continues to conduct current. The collector-emitter voltage remains so low that Ohmic resistances in the transistor (particularly those in the lightly doped collector) become significant. The current supported in saturation is therefore less than that supported in the forward active region. The saturation region is of particular interest to integrated circuit

---

9 This is only part of the explanation. The effective base width of the transistor also increases when it is operated in the reverse active mode.

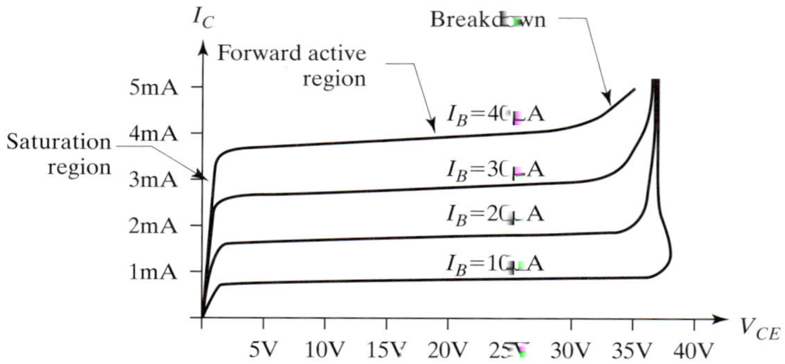

FIGURE 1.21 Typical I-V plot of an NPN transistor.

designers because the forward biasing of the collector-base junction injects minority carriers into the neutral collector. Section 8.1.4 discusses the effects of saturation upon integrated bipolar transistors in greater detail.

The collector-emitter voltage in the forward active region is large enough to reverse bias the collector-base junction. Ohmic drops in the collector no longer significantly reduce the electric field across the collector-base junction, so the current flow through the transistor now depends solely upon beta. The slight upward tilt to the current curves results from the *Early effect.* As the reverse bias on the collector-base junction increases, the depletion region at this junction widens and consequently the neutral base narrows. Since beta depends on base width, it increases slightly as the collector-emitter voltage rises. The Early effect can be minimized by using a very lightly doped collector, so the depletion region extends primarily into the collector rather than into the base.

Beyond a certain collector-emitter voltage, the collector current increases rapidly. This effect limits the maximum operating voltage of the transistor. In the case of a typical integrated NPN transistor, this voltage equals some 30V–40V. The increased current flow results from either one of two effects, the first of which is avalanche breakdown. The collector-base junction will avalanche if it is sufficiently reverse-biased. A wide lightly doped collector region can greatly increase the avalanche voltage rating, and discrete power transistors can achieve operating voltages of more than a thousand volts.

The second limiting mechanism is *base punchthrough.* Punchthrough occurs when the collector-base depletion region reaches all the way through the base and merges with the base-emitter depletion region. Once this occurs, carriers can flow directly from emitter to collector, and current is limited only by the resistance of the neutral collector and emitter. The resulting rapid increase in collector current mimics the effects of avalanche breakdown.

Base punchthrough is often observed in high-gain transistors. For example, *super-beta* transistors use an extremely thin base region to obtain betas of a thousand or more. Base punchthrough limits the operating voltage of these devices to a couple of volts. Super-beta transistors also display a pronounced Early effect because of the encroachment of the collector-base depletion region into the extremely thin neutral base. General-purpose transistors use wider base regions to reduce the Early effect, and their operating voltages are usually limited by avalanche instead of base punchthrough (Section 8.1.2).

1.4 MOS TRANSISTORS

The bipolar junction transistor amplifies a small change in input current to provide a large change in output current. The gain of a bipolar transistor is thus defined as the ratio of output to input current (beta). Another type of transistor, called a *field-*

effect transistor (FET), transforms a change in input voltage into a change in output current. The gain of an FET is measured by its *transconductance,* defined as the ratio of change in output current to change in input voltage.

The field-effect transistor is so named because its input terminal (called its *gate*) influences the flow of current through the transistor by projecting an electric field across an insulating layer. Virtually no current flows through this insulator, so the gate current of a FET transistor is vanishingly small. The most common type of FET uses a thin silicon dioxide layer as an insulator beneath the gate electrode. This type of transistor is called a *metal-oxide-semiconductor* (MOS) transistor, or alternatively, a *metal-oxide-semiconductor field-effect transistor* (MOSFET). MOS transistors have replaced bipolars in many applications because they are smaller and can often operate using less power.

The MOS transistor can be better understood by first considering a simpler device called a *MOS capacitor.* This device consists of two electrodes, one of metal and one of extrinsic silicon, separated by a thin layer of silicon dioxide (Figure 1.22A). The metal electrode forms the *gate,* while the semiconductor slab forms the *backgate* or *body.* The insulating oxide layer between the two is called the *gate dielectric.* The illustrated device has a backgate consisting of lightly doped P-type silicon. The electrical behavior of this MOS capacitor can be demonstrated by grounding the backgate and biasing the gate to various voltages. The MOS capacitor of Figure 1.22A has a gate potential of 0V. The difference in work functions between the metal gate and the semiconductor backgate causes a small electric field to appear across the dielectric. In the illustrated device, this field biases the metal plate slightly positive with respect to the P-type silicon. This electric field attracts electrons from deep within the silicon up toward the surface, while it repels holes away from the surface. The field is weak, so the change in carrier concentrations is small and the overall effect upon the device characteristics is minimal.

Figure 1.22B shows what occurs when the gate of the MOS capacitor is biased positively with respect to the backgate. The electric field across the gate dielectric strengthens and more electrons are drawn up from the bulk. Simultaneously, holes are repelled away from the surface. As the gate voltage rises, a point is reached where more electrons than holes are present at the surface. Due to the excess electrons, the surface layers of the silicon behave as if they were N-type. The apparent reversal of doping polarity is called *inversion* and the layer of silicon that inverts is called a *channel.* As the gate voltage increases still further, more electrons accumulate at the surface and the channel becomes more strongly inverted. The voltage at which the channel just begins to form is called the *threshold voltage* V_t. When the voltage difference between gate and backgate is less than the threshold voltage, no channel forms. When the voltage difference exceeds the threshold voltage, a channel appears.

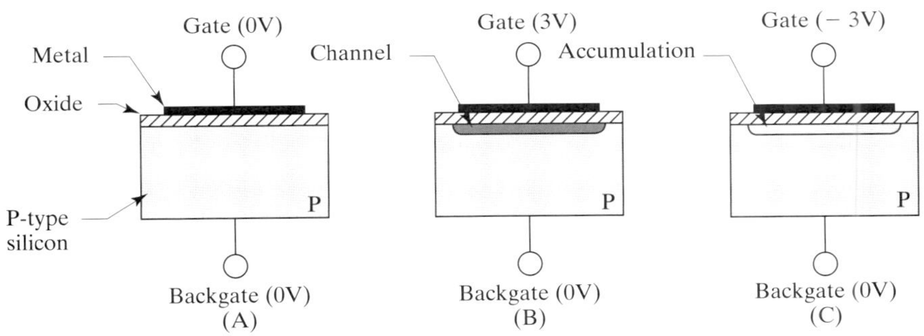

FIGURE 1.22 MOS capacitor: (A) unbiased (V_{BG} = 0V), (B) inversion (V_{BG} = 3V), (C) accumulation (V_{BG} = –3V).

Figure 1.22C shows what happens if the gate of the MOS capacitor is biased negatively with respect to the backgate. The electric field now reverses, drawing holes toward the surface and repelling electrons away from it. The surface layers of silicon appear to be more heavily doped, and the device is said to be in *accumulation*.

The behavior of the MOS capacitor can be utilized to form a true MOS transistor. Figure 1.23A shows the cross section of the resulting device. The gate, dielectric, and backgate remain as before. Two additional regions are formed by selectively doping the silicon on either side of the gate. One of these regions is called the *source* and the other is called the *drain*. Imagine that the source and backgate are both grounded and that a positive voltage is applied to the drain. As long as the gate-to-backgate voltage remains less than the threshold voltage, no channel forms. The PN junction formed between drain and backgate is reverse-biased, so very little current flows from drain to backgate. If the gate voltage exceeds the threshold voltage, a channel forms beneath the gate dielectric. This channel acts like a thin film of N-type silicon shorting the source to the drain. A current consisting of electrons flows from the source across the channel to the drain. In summary, drain current will only flow if the gate-to-source voltage V_{GS} exceeds the threshold voltage V_t.

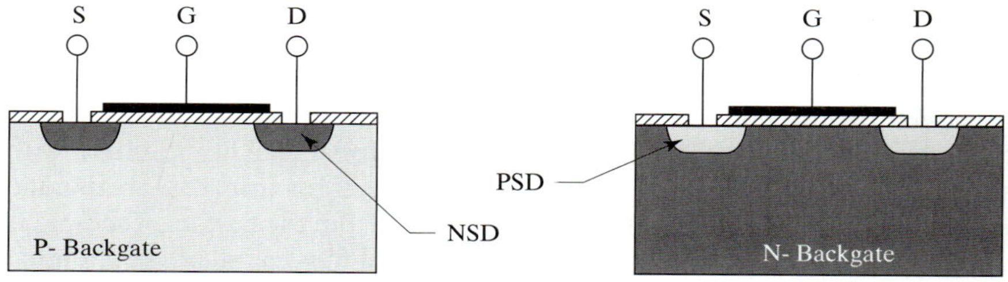

(A) NMOS (B) PMOS

FIGURE 1.23 Cross sections of MOSFET transistors: NMOS (A) and PMOS (B). In these diagrams, S = Source, G = Gate, and D = Drain. The backgate connections, though present, are not illustrated.

The source and drain of a MOS transistor are interchangeable, as both are simply N-type regions formed in the P-type backgate. In many cases, these two regions are identical and the terminals can be reversed without changing the behavior of the device. Such a device is said to be *symmetric*. In a symmetric MOS transistor the labeling of source and drain becomes somewhat arbitrary. By definition, carriers flow out of the source and into the drain. The identity of the source and the drain therefore depends on the biasing of the device. Sometimes the bias applied across the transistor fluctuates and the two terminals swap roles. In such cases, the circuit designer must arbitrarily designate one terminal the drain and the other the source.

Asymmetric MOS transistors are designed with different source and drain dopings and geometries. There are several reasons why transistors may be made asymmetric, but the result is the same in every case. One terminal is optimized to function as the drain and the other as the source. If source and drain are swapped, then the performance of the device will suffer.

The transistor depicted in Figure 1.23A has an N-type channel and is therefore called an *N-channel MOS transistor*, or NMOS. *P-channel MOS* (PMOS) transistors also exist. Figure 1.23B shows a sample PMOS transistor consisting of a lightly doped N-type backgate with P-type source and drain regions. If the gate of this transistor is biased positive with respect to the backgate, then electrons are drawn to the surface and holes are repelled away from it. The surface of the silicon accumulates, and no channel forms. If the gate is biased negative with respect to the backgate, then holes are drawn to the surface, and a channel forms. The PMOS transistor thus has a negative threshold voltage. Engineers often ignore the sign of the threshold voltage since it is normally positive for NMOS transistors and negative for PMOS transistors. An engineer might say, "The PMOS V_t has increased from 0.6V to 0.7V" when in actuality the PMOS V_t has shifted from –0.6V to –0.7V.

1.4.1. Threshold Voltage

The *threshold voltage* of a MOS transistor equals the gate-to-source bias required to just form a channel with the backgate of the transistor connected to the source. If the gate-to-source bias is less than the threshold voltage, then no channel forms. The threshold voltage exhibited by a given transistor depends on a number of factors, including backgate doping, dielectric thickness, gate material, and excess charge in the dielectric. Each of these effects will be briefly examined.

Backgate doping has a major effect on the threshold voltage. If the backgate is doped more heavily, then it becomes more difficult to invert. A stronger electric field is required to achieve inversion, and the threshold voltage increases. The backgate doping of an MOS transistor can be adjusted by performing a shallow implant beneath the surface of the gate dielectric to dope the channel region. This type of implant is called a *threshold adjust implant* (or *V_t adjust implant*).

Consider the effects of a V_t adjust implant upon an NMOS transistor. If the implant consists of acceptors, then the silicon surface becomes more difficult to invert and the threshold voltage increases. If the implant consists of donors, then the surface becomes easier to invert and the threshold decreases. If enough donors are implanted, the surface of the silicon can actually become counterdoped. In this case, a thin layer of N-type silicon forms a permanent channel at zero gate bias. The channel becomes more strongly inverted as the gate bias increases. As the gate bias is decreased, the channel becomes less strongly inverted and at some point it vanishes. The threshold voltage of this NMOS transistor is actually negative. Such a transistor is called a *depletion-mode NMOS,* or simply a *depletion NMOS.* In contrast, an NMOS with a positive threshold voltage is called an *enhancement-mode NMOS,* or *enhancement NMOS.* The vast majority of commercially fabricated MOS transistors are enhancement-mode devices, but there are a few applications that require depletion-mode devices. A depletion-mode PMOS can also be constructed. Such a device will have a positive threshold voltage.

Depletion-mode devices should always be explicitly identified as such. One cannot rely on the sign of the threshold voltage to convey this information, because many engineers customarily ignore threshold polarities. Therefore, one should say "a depletion-mode PMOS with a threshold of 0.7V," rather than a PMOS with a threshold of 0.7V." Many engineers would interpret the latter statement as indicating an enhancement PMOS with a threshold of –0.7V rather than a depletion PMOS with a threshold of +0.7V. Explicitly referring to depletion-mode devices as such eliminates any possibility of confusion.

Special symbols are often used to distinguish between different types of MOS transistors. Figure 1.24 shows a representative collection of these symbols.[10] Symbols A and B are the standard symbols for NMOS and PMOS transistors, respectively. These symbols are not commonly used in the industry; instead symbols

---

[10] Symbols A, B, E, F, G, and H are used by various authors; see A. B. Grebene, *Bipolar and MOS Analog Integrated Circuit Design* (New York: John Wiley and Sons, 1984), pp. 112–113; also P. R. Gray and R. G. Meyer, *Analysis and Design of Analog Integrated Circuits,* 3rd ed. (New York: John Wiley and Sons, 1993), p. 60. The *J. Solid State Circuits* also uses three-terminal MOS symbols but differentiates PMOS devices by placing a bubble on their gate leads.

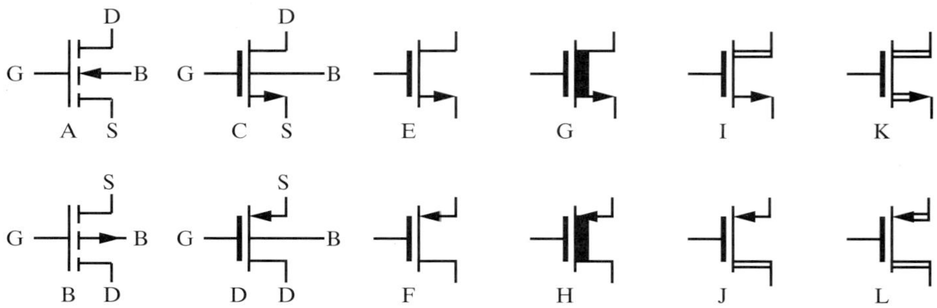

FIGURE 1.24 MOSFET symbols: A, B: standard symbols; C, D: industry symbols (four-terminal); E, F: industry symbols (three-terminal); G, H: depletion-mode devices; I, J: asymmetric high-voltage MOS symbols; K, L: symmetric high-voltage MOS symbols.

C and D are preferred for NMOS and PMOS transistors, respectively. These symbols intentionally resemble NPN and PNP transistors. This convention helps highlight the essential similarities between MOS and bipolar circuits. Symbols E and F are sometimes employed when the backgates of the transistors connect to known potentials. Every MOS transistor has a backgate, so this terminal must always connect to something. Symbols E and F are potentially confusing, because the reader must infer the backgate connections. These symbols are nonetheless very popular because they make schematics much more legible. Symbols G and H are often used for depletion-mode devices, where the solid bar from drain to source represents the channel present at zero bias. Symbols I and J are sometimes employed for asymmetric transistors with high-voltage drains, and symbols K and L are used for symmetric transistors with high-voltage terminations for both source and drain. There are many other schematic symbols for MOS transistors; the ones shown in Figure 1.24 form only a representative sample.

Returning to the discussion of threshold voltage, the dielectric also plays an important role in determining the threshold voltage. A thicker dielectric weakens the electric field by separating the charges by a greater distance. Thus, thicker dielectrics increase the threshold voltage while thinner ones reduce it. In theory, the material of the dielectric also affects the electric field strength. In practice, almost all MOS transistors use pure silicon dioxide as the gate dielectric. This material can be grown in extremely thin films of exceptional purity and uniformity; no other material has comparable properties. Alternate dielectric materials therefore have very limited application.[11]

The gate electrode material also affects the threshold voltage of the transistor. As mentioned above, an electric field appears across the gate oxide when the gate and backgate are shorted together. This field is produced by the difference in work functions between the gate and backgate materials. Most practical transistors use heavily doped polysilicon for the gate electrode. The work function of polysilicon can be varied to a limited degree by changing its doping.

A potentially troublesome source of threshold voltage variation comes from the presence of excess charges in the gate oxide or along the interfaces between the oxide and the silicon surface. These charges may consist of ionized impurity atoms, trapped carriers, or structural defects. The presence of trapped electric charge in the dielectric or along its interfaces alters the electric field and therefore the threshold voltage. If the amount of trapped charge varies with time, temperature, or applied bias, then the threshold voltage will also vary. This subject is discussed in greater detail in Section 4.2.2.

---

[11] A few devices have been fabricated using high-permittivity materials such as silicon nitride for the gate dielectric. Some authors use the term *insulated-gate field effect transistor* (IGFET) to refer to all MOS-like transistors, including those with non-oxide dielectrics.

1.4.2. I-V Characteristics

The performance of an MOS transistor can be graphically illustrated by drawing a family of I-V curves similar to those used for bipolar transistors. Figure 1.25 shows a typical set of curves for an enhancement NMOS. The source and backgate were connected together to obtain these particular curves. The vertical axis measures drain current I_D, while the horizontal axis measures drain-to-source voltage V_{DS}. Each curve represents a specific gate-to-source voltage V_{GS}. The general character of the curves resembles that of the bipolar transistor shown in Figure 1.21, but the family of curves for an MOS transistor are obtained by stepping gate voltage, while those for a bipolar transistor are obtained by stepping base current.

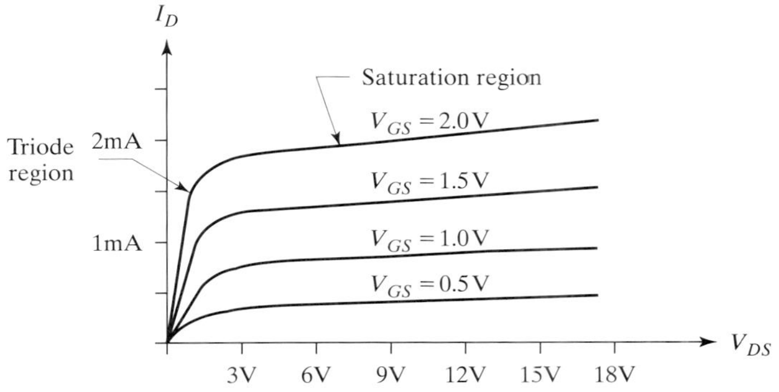

FIGURE 1.25 Typical I-V plot of an NMOS transistor.

At low drain-to-source voltages the MOS channel behaves resistively, and the drain current increases linearly with voltage. This region of operation is called the *linear region* or *triode region.* This roughly corresponds to the saturation region of a bipolar transistor. The drain current levels off to an approximately constant value when the drain-to-source voltage exceeds the difference between the gate-to-source voltage and the threshold voltage. This region is called the *saturation region,* and it roughly corresponds to the forward active region of a bipolar transistor. The term *saturation* thus has very different meanings for MOS and bipolar transistors.

The behavior of the MOS transistor in the linear region is easily explained. The channel acts as a film of doped silicon with a characteristic resistance that depends upon the carrier concentration. The current increases linearly with voltage, exactly as one would expect of a resistor. Higher gate voltages produce larger carrier concentrations and therefore lessen the resistance of the channel. PMOS transistors behave similarly to NMOS transistors, but since holes have lower mobilities than electrons, the apparent resistance of the channel is considerably greater. The effective resistance of an MOS transistor operating in the triode region is symbolized $R_{DS(on)}$.

MOS transistors saturate because of a phenomenon called *pinch-off.* While the drain-to-source voltage remains small, a depletion region of uniform thickness surrounds the channel (Figure 1.26A). As the drain becomes more positive with respect to the source, the depletion region begins to thicken at the drain end. This depletion region intrudes into the channel and narrows it. Eventually the channel depletes all the way through and it is said to have *pinched off* (Figure 1.26B). Carriers move down the channel propelled by the relatively weak electric field along it. When they reach the edge of the pinched-off region, they are sucked across the depletion region by the strong electric field. The voltage drop across the channel does not increase as the drain voltage is increased; instead the pinched-off region widens. Thus, the drain current reaches a limit and ceases to increase.

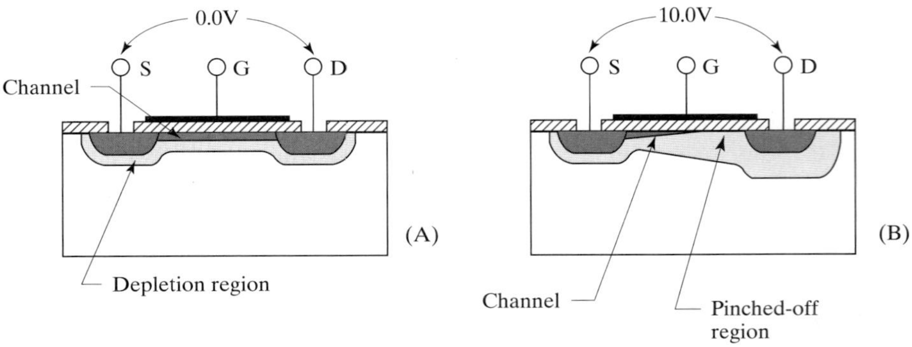

FIGURE 1.26 Behavior of a MOS transistor under bias: (A) V_{DS} = 0V (triode region); (B) V_{DS} = 10V (saturation region).

The drain current curves actually tilt slightly upward in the saturation region. This tilt is caused by *channel length modulation*, which is the MOS equivalent of the Early effect. Increases in drain voltage cause the pinched-off region to widen and the channel length to shorten. The shorter channel still has the same potential drop across it, so the electric field intensifies and the carriers move more rapidly. The drain current thus increases slightly with increasing drain-to-source voltage.

The I-V curves of Figure 1.25 were obtained with the backgate of the transistor connected to the source. If the backgate is biased independently of the source, then the apparent threshold voltage of the transistor will vary. If the source of an NMOS transistor is biased above its backgate, then its apparent threshold voltage increases. If the source of a PMOS transistor is biased below its backgate, then its threshold voltage decreases (it becomes more negative). This *backgate effect*, or *body effect,* arises because the backgate-to-source voltage modulates the depletion region beneath the channel. This depletion region widens as the backgate-to-source differential increases, and it intrudes into the channel as well as into the backgate. A high backgate-to-source differential will thin the channel, which in turn raises the apparent threshold voltage. The intrusion of the depletion region into the channel becomes more significant as the backgate doping rises, and this in turn increases the magnitude of the body effect.

MOS transistors are normally considered majority carrier devices, which conduct only after a channel forms. This simplistic view does not explain the low levels of conduction that occur at gate-to-source voltages just less than the threshold voltage. The formation of a channel is a gradual process. As the gate-to-source voltage increases, the gate first attracts small numbers of minority carriers to the surface. The concentration of minority carriers rises as the voltage increases. When the gate-to-source voltage exceeds the threshold, the number of minority carriers becomes so large that the surface of the silicon inverts and a channel forms. Before this occurs, minority carriers can still move from the source to the drain by diffusion. This *subthreshold conduction* produces currents that are much smaller than those that would flow if a channel were present. However, they are still many orders of magnitude greater than junction leakages. Subthreshold conduction is typically significant only when the gate-to-source voltage is within about 0.3V of the threshold voltage. This is sufficient to cause serious "leakage" problems in low-V_t devices. Some electrical circuits actually take advantage of the exponential voltage-to-current relationship of subthreshold conduction, but these circuits cannot operate at temperatures much in excess of 100°C because the junction leakages become so large that they overwhelm the tiny subthreshold currents.

As with bipolar transistors, MOS transistors can break down by either avalanche or punchthrough. If the voltage across the depletion region at the drain becomes so large that avalanche multiplication occurs, the drain current increases rapidly. Similarly, if the entire channel pinches off, then the source and drain will be shorted by the resulting depletion region and the transistor will punch through.

The operating voltage of an MOS transistor is often limited to a value considerably below the onset of avalanche or punchthrough by a long-term degradation mechanism called *hot carrier injection.* Carriers that traverse the pinched-off portion of the drain are accelerated by the strong electric field present here. The carriers can achieve velocities far beyond those normally associated with room-temperature thermal diffusion, so they are called *hot carriers.* When these carriers collide with atoms near the silicon surface, some of them are deflected up into the gate oxide, and a few of these become trapped. Slowly, over a long period of operation, the concentration of these trapped carriers increases and the threshold voltage shifts. Hot hole injection occurs less readily than hot electron injection because the lower mobility of holes limits their velocity and therefore their ability to surmount the oxide interface. For this reason, NMOS transistors are frequently limited to lower operating voltages than PMOS transistors of similar construction. Various techniques have been devised to limit hot carrier injection (Section 12.1).

1.5 JFET TRANSISTORS

The MOS transistor represents only one type of field-effect transistor. Another is the *junction field-effect transistor* or JFET. This device uses the depletion regions surrounding reverse-biased junctions as a gate dielectric. Figure 1.27A shows a cross-section of an N-channel JFET. This device consists of a bar of lightly doped N-type silicon called the *body* into which two P-type diffusions have been driven from opposite sides. The thin region of N-type silicon remaining between the junctions forms the *channel* of the JFET. The two diffusions act as the *gate* and the *backgate* and the opposite ends of the body form the *source* and the *drain.*

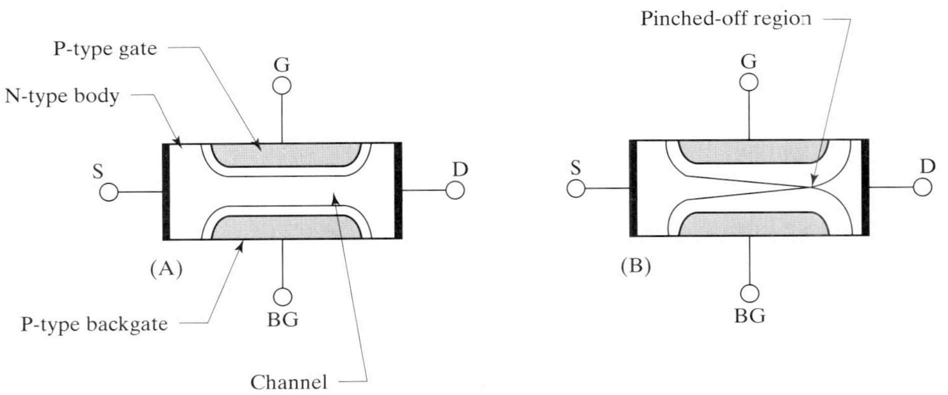

FIGURE 1.27 Cross sections of an N-channel JFET transistor operating in the linear region (A) and in saturation (B). In both diagrams, S = Source, D = Drain, G = Gate, and BG = Backgate.

Suppose that all four terminals of the N-channel JFET are grounded. Depletion regions form around the gate-body and backgate-body junctions. These depletion regions extend into the lightly doped channel, but they do not actually touch one another. A channel therefore exists from the drain to the source. If the drain voltage rises above the source voltage, then a current flows through the channel from drain to source. The magnitude of this current depends on the resistance of the channel, which in turn depends on its dimensions and doping. As long as the drain-to-source voltage remains small, it does not significantly alter the depletion regions bounding the channel. The resistance of the channel therefore remains constant and the drain-to-source voltage varies linearly with drain current. Under these conditions, the JFET is said to operate in its *linear region.* This region of operation corresponds to the linear (or triode) region of an MOS transistor. Since a channel forms at $V_{GS} = 0$, the JFET resembles a depletion-mode MOSFET rather than an enhancement-mode one.

The depletion regions at the drain end of the JFET widen as the drain voltage increases. The channel becomes increasingly constricted by the encroachment of the opposing depletion regions. Eventually the depletion regions meet and pinch off the channel (Figure 1.27B). Drain current still flows through the transistor even though the channel has pinched off. This current originates at the source terminal and consists of majority carriers (electrons). These carriers move down the channel until they reach the pinched-off region. The large lateral electric field across this region draws the carriers across into the neutral drain.

Further increases in drain voltage have little effect once the channel has pinched off. The pinched-off region widens slightly, but the dimensions of the channel remain about the same. The resistance of the channel determines the magnitude of the drain current, so this also remains approximately constant. Under these conditions, the JFET is said to operate in *saturation.*

The gate and backgate electrodes also influence the current that flows through the channel. As magnitudes of the gate-body and backgate-body voltages increase, the reverse biases across the gate-body and backgate-body junctions slowly increase. The depletion regions that surround these junctions widen and the channel constricts. Less current can flow through the constricted channel, and the drain-to-source voltage required to pinch the channel off decreases. As the magnitudes of the gate and backgate voltages continue to increase, eventually the channel will pinch off even at $V_{DS} = 0$. Once this occurs, no current can flow through the transistor regardless of drain-to-source voltage, and the transistor is said to operate in *cutoff.*

Figure 1.28 shows the I-V characteristics of an N-channel JFET whose gate and backgate electrodes have been connected to one another. Each curve represents a different value of the gate-to-source voltage V_{GS}. The drain currents are at their greatest when $V_{GS} = 0$, and they decrease as the magnitude of the gate voltage

FIGURE 1.28 Typical I-V plot of an N-JFET transistor with $V_T = -8V$.

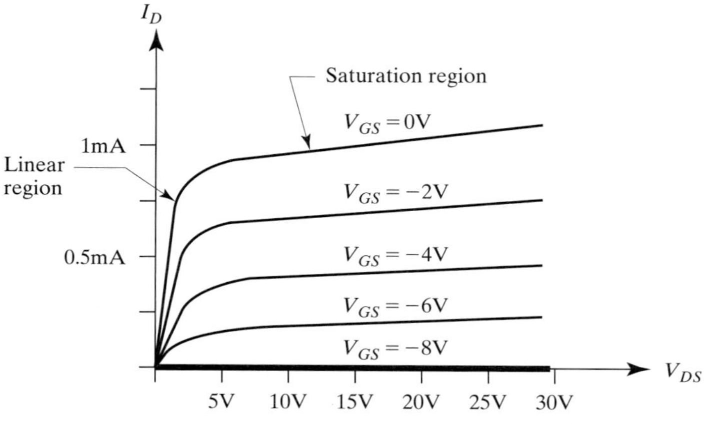

increases. Conduction ceases entirely when the gate voltage equals the *turnoff voltage* V_T. The turnoff voltage qualitatively corresponds to the threshold voltage of an MOS transistor. The comparison must not be taken too far, however, as the conduction equations of the two devices differ considerably.

The drain current curves of the N-JFET tilt slightly upward in saturation due to *channel length modulation.* This effect is analogous to that which occurs in MOS transistors. The pinched-off region of the JFET lengthens as the drain-to-source voltage increases. Any increase in the length of the pinched-off region produces a corresponding decrease in the length of the channel. The effect of channel length modulation is usually quite small because the channel length greatly exceeds the length of the pinched-off region.

The source and drain terminals of a JFET can often be interchanged without affecting the performance of the device. The JFET structure of Figure 1.27A is an example of such a *symmetric* device. More complex JFET structures sometimes exhibit differences in source and drain geometries that render them *asymmetric.*

Almost all JFET structures short the gate and backgate terminals. Consider the device of Figure 1.27A. The channel is bounded on the left by the source, on the right by the drain, on the top by the gate, and on the bottom by the backgate. The drawing does not show what bounds the channel on the front or the rear. In most cases, these sides of the channel are also bounded by reverse-biased junctions that are extensions of the gate-body and backgate-body junctions. This arrangement necessitates shorting the gate and backgate.

Figure 1.29 shows the conventional schematic symbols for N-channel and P-channel JFET transistors. The arrowhead on the gate lead shows the orientation of the PN junction between the gate and the body of the device. The symbol does not explicitly identify the source and drain terminals, but most circuit designers orient the devices so that the drain of an N-JFET and the source of a P-JFET lie on top.

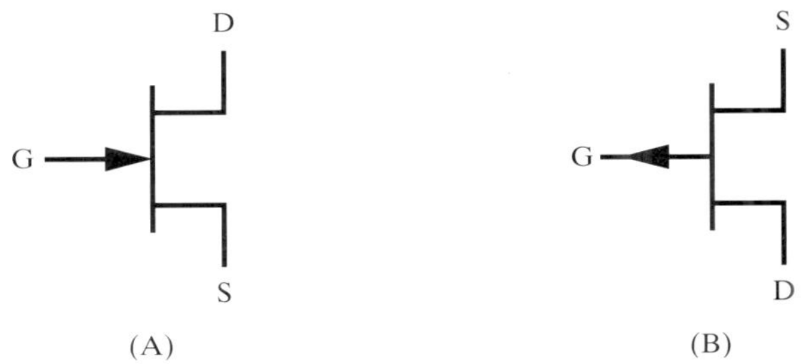

FIGURE 1.29 Symbols for an N-channel JFET (A) and a P-channel JFET (B).

1.6 SUMMARY

Device physics is a complex and ever-evolving science. Researchers constantly develop new devices and refine existing ones. Much of this ongoing research is highly theoretical and therefore lies beyond the scope of this introductory text. The functionality of most semiconductor devices can be satisfactorily explained using relatively simple and intuitive concepts.

This chapter emphasizes the role of majority and minority carrier conduction across PN junctions. If a junction is reverse-biased, then the majority carriers on either side of it are repelled and a depletion region forms. If the junction is forward-biased, then majority carriers diffuse across and recombine to create a net current flow across the PN junction. The PN junction diode employs this phenomenon to rectify signals.

When two junctions are placed in close proximity, carriers emitted by one junction can be collected by the other before they can recombine. The bipolar junction transistor (BJT) consists of just such a pair of closely spaced junctions. The voltage across the base-emitter junction of the BJT controls the current flowing from collector to emitter. If the transistor is properly designed, then a small base current can control a much larger collector current. The BJT therefore serves as an amplifier capable of transforming weak signals into much stronger ones. Thus, for example, a BJT can amplify a weak signal picked up by a radio receiver into a signal strong enough to drive a loudspeaker.

The metal-oxide-semiconductor (MOS) transistor relies upon electrical fields projected across a dielectric to modulate the conductivity of a semiconductor material. A suitable voltage placed upon the gate of a MOS transistor produces an electric field that attracts carriers up from the bulk silicon to form a conductive channel. The gate is insulated from the remainder of the transistor, so no gate current is required to maintain conduction. MOS circuitry can thus potentially operate at very low power levels.

The junction diode, the bipolar junction transistor, and the MOS transistor are the three most important semiconductor devices. Together with resistors and capacitors, they form the vast majority of the elements used in modern integrated circuits. The next chapter will examine how these devices are fabricated in a production environment.

1.7 EXERCISES

1.1. What are the relative proportions of aluminum, gallium, and arsenic atoms in intrinsic aluminum gallium arsenide?

1.2. A sample of pure silicon is doped with exactly 10^{16} atoms/cm^3 of boron and exactly 10^{16} atoms/cm^3 of phosphorus. Is the doped sample P-type or N-type?

1.3. The *instantaneous* velocity of carriers in silicon is almost unaffected by weak electric fields, yet the *average* velocity changes dramatically. Explain this observation in terms of drift and diffusion.

1.4. A layer of intrinsic silicon 1μ thick is sandwiched between layers of P-type and N-type silicon, both heavily doped. Draw a diagram illustrating the depletion regions that form in the resulting structure.

1.5. A certain process incorporates two different N+ diffusions that can be combined with a P– diffusion to produce Zener diodes. One of the resulting diodes has a breakdown voltage of 7V, while the other has a breakdown voltage of 10V. What causes the difference in breakdown voltages?

1.6. When the collector and emitter leads of an integrated NPN transistor are swapped, the transistor continues to function but exhibits a greatly reduced beta. There are several possible reasons for this behavior; explain at least one.

1.7. If a certain transistor has a beta of 60, and another transistor has a base twice as wide and half as heavily doped, then what is the approximate beta of the second transistor? What other electrical characteristics of the devices will vary, and how?

1.8. A certain MOS transistor has a threshold voltage of –1.5V. If a small amount of boron is added to the channel region, the threshold voltage shifts to –0.6V. Is the transistor PMOS or NMOS, and is it an enhancement or a depletion device?

1.9. If a depletion PMOS transistor has a threshold voltage of 0.5V when constructed using a 200Å oxide, will this threshold voltage increase or decrease if the oxide is thickened to 400Å?

1.10. A certain NMOS transistor has a threshold voltage of 0.5V; the gate-to-source voltage V_{GS} of the transistor is set to 2V, and the drain-to-source voltage V_{DS} is set to 4V.

What is the relative effect of doubling the gate-to-source voltage versus doubling the drain-to-source voltage, and why?

1.11. A certain silicon PN junction diode exhibits a forward voltage drop of 620mV when operated at a forward current of 25μA at a temperature of 25°C. What is the approximate forward drop of this diode at –40°C? At 125°C?

1.12. Two JFET transistors differ only in the separation between their gate and backgate; in one transistor these two regions are twice as far apart as in the other transistor. In what ways do the electrical properties of the two transistors differ?

2 Semiconductor Fabrication

Semiconductor devices have long been used in electronics. The first solid-state rectifiers were developed in the late nineteenth century. The galena crystal detector, invented in 1907, was widely used to construct crystal radio sets. By 1947, the physics of semiconductors was sufficiently understood to allow Bardeen and Brattain to construct the first bipolar junction transistor. In 1959, Kilby constructed the first integrated circuit, ushering in the era of modern semiconductor manufacture.

The impediments to manufacturing large quantities of reliable semiconductor devices were essentially technological, not scientific. The need for extraordinarily pure materials and precise dimensional control prevented early transistors and integrated circuits from reaching their full potential. The first devices were little more than laboratory curiosities. An entire new technology was required to mass produce them, and this technology is still rapidly evolving.

This chapter provides a brief overview of the process technologies currently used to manufacture integrated circuits. Chapter 3 then examines three representative process flows used for manufacturing specific types of analog integrated circuits.

2.1 SILICON MANUFACTURE

Integrated circuits are usually fabricated from *silicon,* a very common and widely distributed element. The mineral *quartz* consists entirely of silicon dioxide, also known as *silica.* Ordinary sand is chiefly composed of tiny grains of quartz and is therefore also mostly silica.

Despite the abundance of its compounds, elemental silicon does not occur naturally. The element can be artificially produced by heating silica and carbon in an electric furnace. The carbon unites with the oxygen contained in the silica, leaving more-or-less pure molten silicon. As this cools, numerous minute crystals form and grow together into a fine-grained gray solid. This form of silicon is said to be *polycrystalline* because it contains a multitude of crystals. Impurities and a disordered crystal structure make this *metallurgical-grade polysilicon* unsuited for semiconductor manufacture.

Metallurgical-grade silicon can be further refined to produce an extremely pure semiconductor-grade material. Purification begins with the conversion of the crude silicon into a volatile compound, usually trichlorosilane. After repeated distillation, the extremely pure trichlorosilane is reduced to elemental silicon using hydrogen gas. The final product is exceptionally pure, but still polycrystalline. Practical integrated circuits can only be fabricated from single-crystal material, so the next step consists of growing a suitable crystal.

2.1.1. Crystal Growth

The principles of crystal growing are both simple and familiar. Suppose a few crystals of sugar are added to a saturated solution that subsequently evaporates. The sugar crystals serve as seeds for the deposition of additional sugar molecules. Eventually the crystals grow to be very large. Crystal growth would occur even in the absence of a seed, but the product would consist of a welter of small intergrown crystals. The use of a seed allows the growth of larger, more perfect crystals by suppressing undesired nucleation sites.

In principle, silicon crystals can be grown in much the same manner as sugar crystals. In practice, no suitable solvent exists for silicon, and the crystals must be grown from the molten element at temperatures in excess of 1400°C. The resulting crystals are at least a meter in length and ten centimeters in diameter, and they must have a nearly perfect crystal structure if they are to be useful to the semiconductor industry. These requirements make the process technically challenging.

The usual method for growing semiconductor-grade silicon crystals is called the *Czochralski process.* This process, illustrated in Figure 2.1, uses a silica crucible charged with pieces of semi-grade polycrystalline silicon. An electric furnace raises the temperature of the crucible until all of the silicon melts. The temperature is then reduced slightly and a small seed crystal is lowered into the crucible. Controlled cooling of the melt causes layers of silicon atoms to deposit upon the seed crystal. The rod holding the seed slowly rises so that only the lower portion of the growing crystal remains in contact with the molten silicon. In this manner, a large silicon crystal can be pulled centimeter-by-centimeter from the melt. The shaft holding the crystal rotates slowly to ensure uniform growth. The high surface tension of molten silicon distorts the crystal into a cylindrical rod rather than the expected faceted prism.

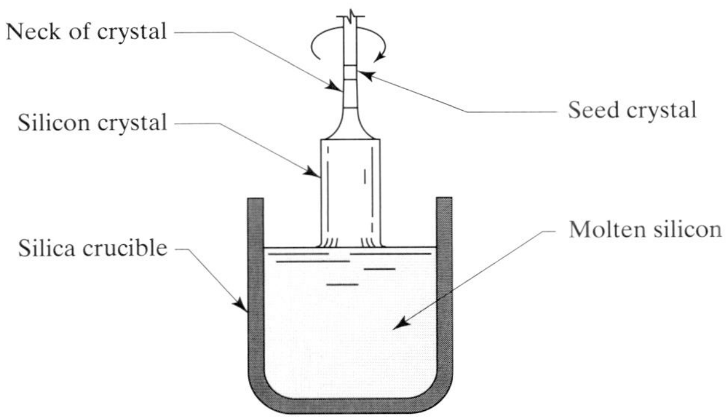

FIGURE 2.1 Czochralski process for growing silicon crystals.

The Czochralski process requires careful control to provide crystals of the desired purity and dimensions. Automated systems regulate the temperature of the melt and the rate of crystal growth. A small amount of doped polysilicon added to the melt sets the doping concentration in the crystal. In addition to the deliberately introduced impurities, oxygen from the silica crucible and carbon from the heating elements dissolve in the molten silicon and become incorporated into the growing crystal. These impurities subtly influence the electrical properties of the resulting silicon. Once the crystal has reached its final dimensions, it is lifted from the melt and is allowed to slowly cool to room temperature. The resulting cylinder of monocrystalline silicon is called an *ingot.*

Since integrated circuits are formed upon the surface of a silicon crystal and penetrate this surface to no great depth, the ingot is customarily sliced into numerous thin circular sections called *wafers.* Each wafer yields hundreds or even thousands of integrated circuits. The larger the wafer, the more integrated circuits it holds and the greater the resulting economies of scale. Most modern processes employ either 150mm (6″) or 200mm (8″) wafers. A typical ingot measures between one and two meters in length and can provide hundreds of wafers.

2.1.2. Wafer Manufacturing

The manufacture of wafers consists of a series of mechanical processes. The two tapered ends of the ingot are sliced off and discarded. The remainder is then ground into a cylinder, the diameter of which determines the size of the resulting wafers. No visible indication of crystal orientation remains after grinding. The crystal orientation is experimentally determined and a flat stripe is ground along one side of the ingot. Each wafer cut from it will retain a facet, or *flat,* which unambiguously identifies its crystal orientation.

After grinding the flat, the manufacturer cuts the ingot into individual wafers using a diamond-tipped saw. In the process, about one-third of the precious silicon crystal is reduced to worthless dust. The surfaces of the resulting wafers bear scratches and pockmarks caused by the sawing process. Since the tiny dimensions of integrated circuits require extremely smooth surfaces, one side of each wafer must be polished. This process begins with mechanical abrasives and finishes with chemical milling. The resulting mirror-bright surface displays the dark gray color and characteristic near-metallic luster of silicon.

2.1.3. The Crystal Structure of Silicon

Each wafer constitutes a slice from a single silicon crystal. The underlying crystalline structure determines how the wafer splits when broken. Most crystals tend to part along *cleavage planes* where the interatomic bonding is weakest. For example, a diamond crystal can be cleaved by sharply striking it with a metal wedge. A properly oriented blow will split the diamond into two pieces, each of which displays a perfectly flat cleavage surface. If the blow is not properly oriented, then the diamond shatters. Silicon wafers also show characteristic cleavage patterns that can be demonstrated using a scrap wafer, a pad of note paper, and a wooden pencil. Place the wafer on the notepad, and place the pad in your lap. Take a wooden pencil and press down in the center of the wafer using the eraser. The wafer should split into either four or six regular wedge-shaped fragments, much like sections of a pie (Figure 2.2). The regularity of the cleavage pattern demonstrates that the wafer consists of monocrystalline silicon.

Figure 2.3 shows a small section of a silicon crystal drawn in three dimensions. Eighteen silicon atoms lie wholly or partially within the boundaries of an imaginary cube called a *unit cell.* Six of these occupy the centers of each of the six faces of the cube. Eight more atoms occupy the eight vertices of the cube. Two unit cells placed side-by-side share four vertex atoms and a single face-centered

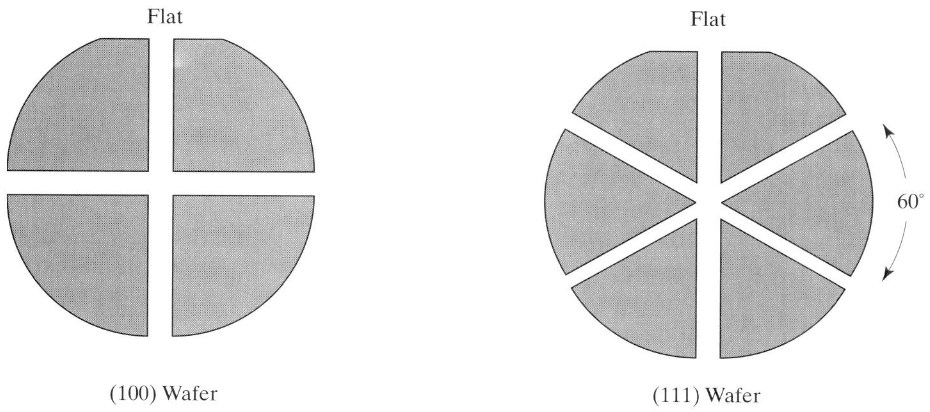

FIGURE 2.2 Typical fracture patterns for (100) and (111) silicon wafers. Some wafers possess a second, smaller flat that denotes crystal orientation and doping. These *minor flats* have not been illustrated.

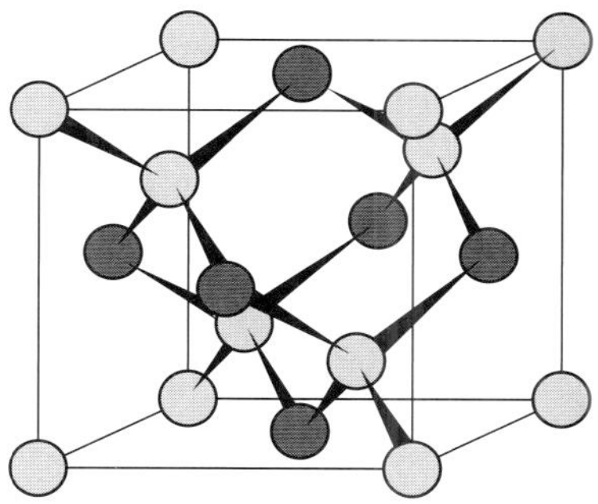

FIGURE 2.3 The diamond lattice unit cell displays a modified face-centered cubic structure. The face-centered atoms are shown in dark gray for emphasis.

atom. Additional unit cells can be placed on all sides to extend the crystal in all directions.

When the sawblade slices through a silicon ingot to form a wafer, the orientation of the resulting surface with respect to the unit cell determines many of the wafer's properties. A cut could, for example, slice across a face of the unit cell or diagonally through it. The pattern of atoms exposed by these two cuts differ, as do the electrical properties of devices formed into the respective surfaces. However, not all cuts made through a silicon crystal necessarily differ. Because the faces of a cube are indistinguishable from one another, a cut made across any face of the unit cell looks the same as cuts made across other faces. In other words, planes cut parallel to any face of a unit cube expose similar surfaces.

Because of the awkwardness of trying to describe various planes verbally, a trio of numbers called *Miller indices* are assigned to each possible plane passing through the crystal lattice (Appendix B). Figure 2.4 shows the two most important planar orientations. A plane parallel to a face of the cube is called a *(100) plane,* and a plane slicing diagonally through the unit cube to intersect three of its vertices is called a *(111) plane.* Silicon wafers are generally cut along either a (100) plane or a (111) plane. Although many other cuts exist, none of these have much commercial significance.

A trio of Miller indices enclosed in brackets denotes a direction perpendicular to the indicated crystal plane. For instance, a (100) plane has a [100] direction perpendicular to it and a (111) plane has a [111] direction perpendicular to it. Appendix B discusses how Miller indices are computed and explains the meaning of the different symbologies used to represent them.

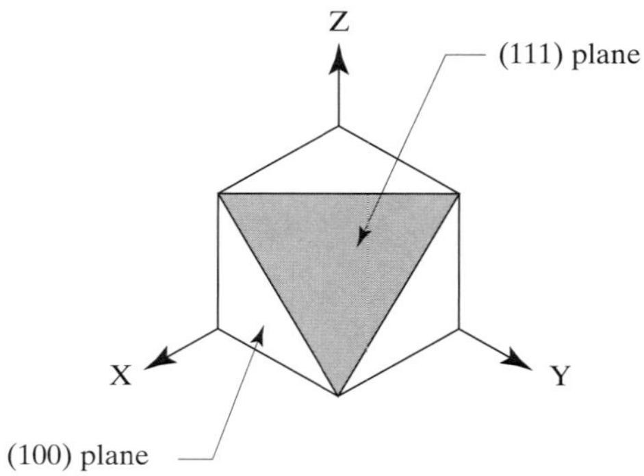

FIGURE 2.4 Identification of (100) and (111) planes of a cubic crystal.

2.2 PHOTOLITHOGRAPHY

The production of silicon wafers constitutes only the first step in the fabrication of integrated circuits. Many of the remaining steps deposit materials on the wafer or etch them away again. A variety of sophisticated deposition and etching techniques exist, but most of these are not *selective.* A nonselective, or *blanket,* process affects the entire surface of the wafer rather than just portions of it. The few processes that are selective are so slow or so expensive that they are useless for high-volume manufacturing. A technique called *photolithography* allows photographic reproduction of intricate patterns that can be used to selectively block depositions or etches. Integrated circuit fabrication makes extensive use of photolithography.

2.2.1. Photoresists

Photolithography begins with the application of a photosensitive emulsion called a *photoresist.* An image can be photographically transferred to the photoresist and a developer used to produce the desired masking pattern. The photoresist solution is usually *spun* onto the wafer. As shown in Figure 2.5, the wafer is mounted on a turntable spinning at several thousand revolutions per minute. A few drops of photoresist solution are allowed to fall onto the center of the spinning wafer, and centrifugal force spreads the liquid out across the surface. The photoresist solution adheres to the wafer and forms a uniform thin film. The excess solution flies off the edges of the spinning wafer. The film thins to its final thickness in a few seconds, the solvent rapidly evaporates, and a thin coating of photoresist remains on the wafer. This coating is baked to remove the last traces of solvent and to harden the photoresist to allow handling. Coated wafers are sensitive to certain wavelengths

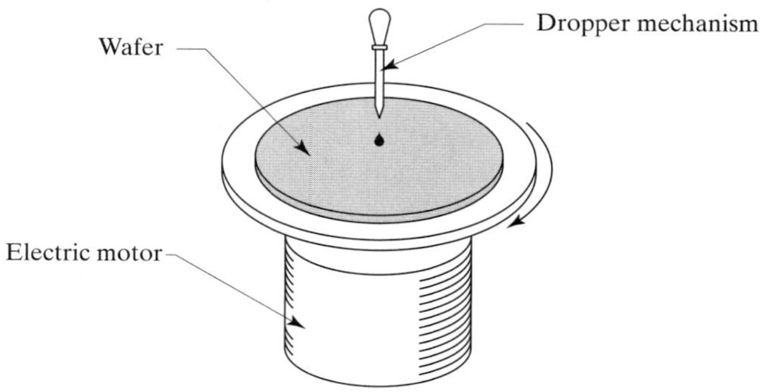

FIGURE 2.5 Application of photoresist solution to a wafer by spinning.

of light, particularly ultraviolet (UV) light. They remain relatively insensitive to other wavelengths, including those of red, orange, and yellow light. Most photolithography rooms therefore have special yellow lighting systems.

The two basic types of photoresists are distinguished by what chemical reactions occur during exposure. A *negative resist* polymerizes under UV light. The unexposed negative resist remains soluble in certain solvents, while the polymerized photoresist becomes insoluble. When the wafer is flooded with solvent, unexposed areas dissolve and exposed areas remain coated. A *positive resist,* on the other hand, chemically decomposes under UV light. These resists are normally insoluble in the developing solvent, but the exposed portions of the resist are chemically altered in order to become soluble. When the wafer is flooded with solvent, the exposed areas wash away while the unexposed areas remain coated. Negative resists tend to swell during development, so process engineers generally prefer to use positive resists.

2.2.2. Photomasks and Reticles

Modern photolithography depends upon a type of projection printing conceptually similar to that used to enlarge photographic negatives. Figure 2.6 shows a simplified illustration of the exposure process. A system of lenses collimates a powerful UV light source, and a plate called a *photomask* blocks the path of the resulting light beam. The UV light passes through the transparent portions of the photomask and through additional lenses that focus an image on the wafer. The apparatus in Figure 2.6 is called an *aligner* since it must ensure that the image of the mask aligns precisely with existing patterns on the wafer.

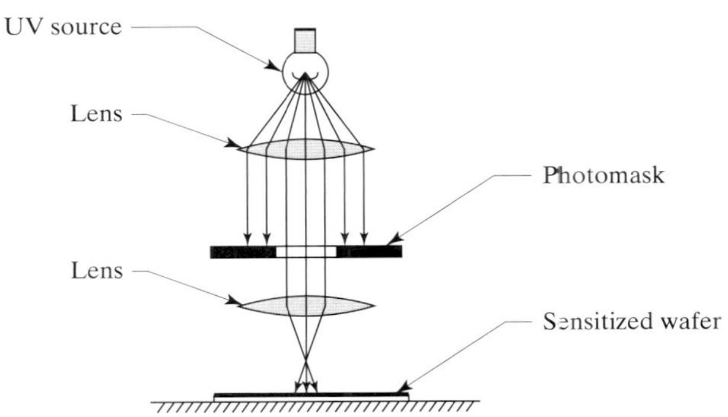

FIGURE 2.6 Simplified illustration of photomask exposure using an aligner.

The transparent plate used as the substrate of a photomask must be dimensionally stable or the pattern it projects will not align with those projected by previous masks. These plates most often consist of fused silica (often erroneously called quartz). After a thin layer of metal is applied to one surface of the plate, any one of various highly precise—but extremely slow and costly—methods are used to pattern the photomask. The image on the photomask is usually five or ten times the size of the image projected onto the wafer. Photographic reduction shrinks the size of any defects or irregularities in the photomask and therefore improves the quality of the final image. This type of enlarged photomask is called either a 5X or a 10X *reticle* depending on the degree of magnification employed.

A reticle can be used to directly pattern a wafer, but there are mechanical difficulties in doing so. The size of the photomask that an aligner can accept is limited by mechanical considerations, including the difficulty of constructing large lenses of the required accuracy. As a result, most commercial aligners accept a photomask about the same size as the wafer. A 5X reticle that could pattern an entire wafer in one shot would be five times the size of the wafer, and would therefore not fit in the aligner. Practical 5X or 10X reticles are constructed to expose only a small rectangular portion of the final wafer pattern. The reticle must be stepped across the wafer and exposures made at many different positions in order to replicate the pattern across the entire wafer. This process is called *stepping,* and an aligner designed to step a reticle is called a *stepper.* Steppers are slower than ordinary aligners and are therefore more costly.

There is a faster method of exposing wafers that can be used for integrated circuits that do not require extremely fine feature sizes. The reticle can be stepped, not onto a sensitized wafer, but instead onto another photomask. This photomask now bears a 1X image of the desired pattern. The resulting photomask, called a *stepped working plate,* can expose an entire wafer in one shot. Stepped working plates make photolithography faster and cheaper, but the results are not as precise as directly stepping the reticle onto the wafer.

Even the tiniest dust speck is so large that it will block the transfer of a portion of the image and ruin at least one integrated circuit. Special air filtration techniques and protective garments are routinely used in wafer fabs, but some dust gets past all of these precautions. Photomasks are often equipped with *pellicles* on one or both sides to prevent dust from interfering with the exposure. Pellicles consist of thin transparent plastic films mounted on ring-shaped spacers that hold them slightly above the surface of the mask. Light passing through the plane of a pellicle is not in focus, so particles on the pellicle do not appear in the projected image. The pellicle also hermetically seals the surface of the mask and thereby protects it from dust.

2.2.3. Patterning

The exposed wafers are sprayed with a suitable developer, typically consisting of a mixture of organic solvents. The developer dissolves portions of the resist to uncover the surface of the wafer. A deposition or etch affects only these uncovered areas. Once the selective processing has been completed, the photoresist can be stripped away using solvents. Alternatively, the photoresist can be chemically destroyed by reactive ion etching in an oxygen ambient (Section 2.3.2). This procedure is called *ashing.*

Many important fabrication processes require masking layers that can withstand high temperatures. Since most practical photoresists are organic compounds, they are clearly unsuited to this task. Two common high-temperature masking materials are silicon dioxide and silicon nitride. These materials can be formed by the reaction of appropriate gases with the silicon surface. A photoresist can then be applied and patterned and an etching process used to open holes in the oxide or nitride film. Modern processing techniques make extensive use of oxide and nitride films for masking high-temperature depositions and diffusions.

2.3 OXIDE GROWTH AND REMOVAL

Silicon forms several oxides, the most important of which is *silicon dioxide* (SiO_2). This oxide possesses a number of desirable properties that together are so valuable that silicon has become the dominant semiconductor. Other semiconductors have better electrical properties, but only silicon forms a well-behaved oxide. Silicon dioxide can be grown on a silicon wafer by simply heating it in an oxidizing atmosphere. The resulting film is mechanically rugged and resists most common solvents, yet it readily dissolves in hydrofluoric acid. Oxide films are superb electrical insulators and are useful not only for insulating metal conductor patterns but also for forming the dielectrics of capacitors and MOS transistors. Silicon dioxide is so important to silicon processing that it is universally known as *oxide*.

2.3.1. Oxide Growth and Deposition

The simplest method of producing an oxide layer consists of heating a silicon wafer in an oxidizing atmosphere. If pure dry oxygen is employed, then the resulting oxide film is called a *dry oxide*. Figure 2.7 shows a typical oxidation apparatus. The wafers are placed in a fused silica rack called a *wafer boat*. The wafer boat is slowly inserted into a fused silica tube wrapped in an electrical heating mantle. The temperature of the wafers gradually rises as the wafer boat moves into the middle of the heating zone. Oxygen gas blowing through the tube passes over the surface of each wafer. At elevated temperatures, oxygen molecules can actually diffuse through the oxide layer to reach the underlying silicon. There oxygen and silicon react, and the layer of oxide gradually grows thicker. The rate of oxygen diffusion slows as the oxide film thickens, so the growth rate decreases with time. As Table 2.1 indicates, high temperatures greatly accelerate oxide growth. Crystal orientation also affects oxidation rates, with (111) silicon oxidizing significantly faster than (100) silicon.[1] Once the oxide layer has reached the desired thickness (as gauged by time and temperature), the wafers are slowly withdrawn from the furnace.

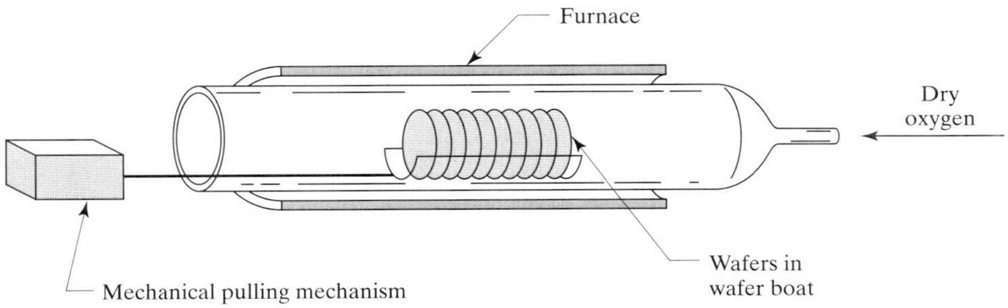

FIGURE 2.7 Simplified diagram of an oxidation furnace.

TABLE 2.1 Times required to grow 0.1μm of oxide on (111) silicon.[2]

Ambient	800°C	900°C	1000°C	1100°C	1200°C
Dry O_2	30 hr	6 hr	1.7 hr	40 min	15 min
Wet O_2	1.7 hr	20 min	6 min		

Dry oxide grows very slowly, but it is of particularly high quality because relatively few defects exist at the oxide-silicon interface. These defects, or *surface states*, interfere with the proper operation of semiconductor devices, particularly MOS transistors. The density of surface states is measured by a parameter called the *surface state charge*, or Q_{ss}. Dry oxide films that are thermally grown on (100) silicon have especially low surface state charges and thus make ideal dielectrics for MOS transistors.

---

[1] W. R. Runyan and K. E. Bean, *Semiconductor Integrated Circuit Processing Technology* (Reading, MA: Addison-Wesley, 1994), p. 84ff.

[2] Calculated from R. P. Donovan, "Oxidation," in R. M. Burger and R. P. Donovan, eds., *Fundamentals of Silicon Integrated Device Technology* (Englewood Cliffs, NJ: Prentice-Hall, 1967), pp. 41, 49.

Wet oxides are formed in the same way as dry oxides, but steam is injected into the furnace tube to accelerate the oxidation. Water vapor moves rapidly through oxide films, but hydrogen atoms liberated by the decomposition of the water molecules produce imperfections that may degrade the oxide quality.[3] Wet oxidation is commonly used to grow a thick layer of *field oxide* where no active devices will be built. Dry oxidations conducted at higher-than-ambient pressures can also accelerate oxide growth rates.

Sometimes an oxide layer must be formed on a material other than silicon. For instance, oxide is frequently employed as an insulator between two layers of metallization. In such cases, some form of deposited oxide must be used rather than the grown oxides previously discussed. Deposited oxides can be produced by various reactions between gaseous silicon compounds and gaseous oxidizers. For example, silane gas and nitrous oxide react to form nitrogen gas, water vapor, and silicon dioxide. Deposited oxides tend to possess low densities and large numbers of defect sites, so they are not suitable for use as gate dielectrics for MOS transistors. Deposited oxides are still acceptable for use as insulating layers between multiple conductor layers, or as protective overcoats.

Oxide films are brightly colored due to *thin-film interference.* When light passes through a transparent film, destructive interference between transmitted and reflected wavefronts causes certain wavelengths of light to be selectively absorbed. Different thicknesses of films absorb different colors of light. Thin-film interference causes the iridescent colors seen in soap bubbles and films of oil on water. The same effect produces the vivid colors visible in microphotographs of integrated circuits. These colors are helpful in distinguishing various regions of an integrated circuit under a microscope or in a microphotograph. The approximate thickness of an oxide film can often be determined using a table of oxide colors.[4]

2.3.2. Oxide Removal

Figure 2.8 illustrates the procedure used to form a patterned oxide layer. The first step consists of growing a thin layer of oxide across the wafer. Next, photoresist is applied to the wafer by spinning. A subsequent oven bake drives off the final traces

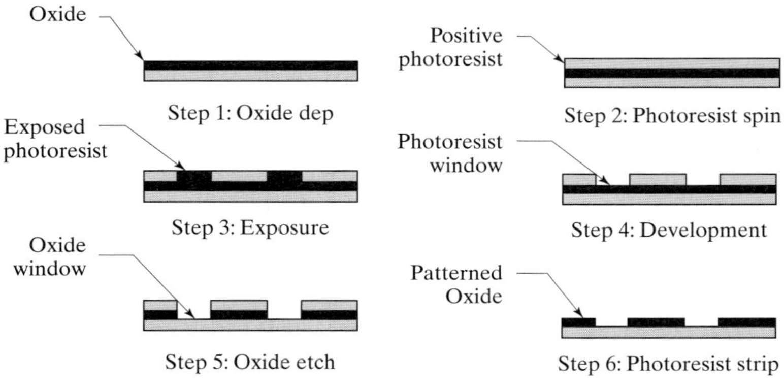

FIGURE 2.8 Steps in oxide growth and removal.

[3] Hydrogen incorporation due to wet oxidation conditions reduces the concentration of dangling bonds, but it increases the fixed oxide charge. The differences between wet and dry oxidation are therefore not as simplistic as the text may suggest.

[4] For a table, see W. A. Pliskin and E. E. Conrad, "Nondestructive Determination of Thickness and Refractive Index of Transparent Films," *IBM J. Research and Development,* Vol. 8, 1964, pp. 43–51.

of solvent and hardens the photoresist for handling. After photolithographic exposure, the wafer is developed by spraying it with a solvent that dissolves the exposed areas of photoresist to reveal the underlying oxide. The patterned photoresist serves as a masking material for an oxide etch. Having served its function, the photoresist is finally stripped away to leave the patterned oxide layer.

Oxide can be etched by either of two methods. *Wet etching* employs a liquid solution that dissolves the oxide, but not the photoresist or the underlying silicon. *Dry etching* uses a reactive plasma to perform the same function. Wet etches are simpler, but dry etches provide better linewidth control.

Most wet etches employ solutions of buffered hydrofluoric acid (HF). This highly corrosive substance readily dissolves silicon dioxide, but it does not attack either elemental silicon or organic photoresists. The etch process consists of immersing the wafers in a plastic tank containing the hydrofluoric acid solution for a specified length of time, followed by a thorough rinsing to remove all traces of the acid. Wet etches are *isotropic* because they proceed at the same rate laterally as well as vertically. The acid works its way under the edges of the photoresist to produce sloping sidewalls similar to those shown in Figure 2.9A. Since the etching must continue long enough to ensure that all openings have completely cleared, some degree of overetching inevitably occurs. The acid continues to erode the sidewalls as long as the wafer remains immersed. The extent of sidewall erosion varies depending upon etching conditions, oxide thickness, and other factors. Because of these variations, wet etching cannot provide the tight linewidth control required by modern semiconductor processes.

There are several types of dry etching processes.[5] One called *reactive ion etching* (RIE) employs plasma bombardment to erode the surface of the wafer. A silent electrical discharge passed through a low-pressure gas mixture forms highly energetic molecular fragments called *reactive ions*. The etching apparatus projects these ions downward onto the wafer at high velocities. Because the ions impact the wafer at a relatively steep angle, etching proceeds vertically at a much greater rate than laterally. The *anisotropic* nature of reactive ion etching allows the formation of nearly vertical sidewalls such as those shown in Figure 2.9B. Figure 2.10 shows a simplified diagram of a reactive ion etching apparatus.

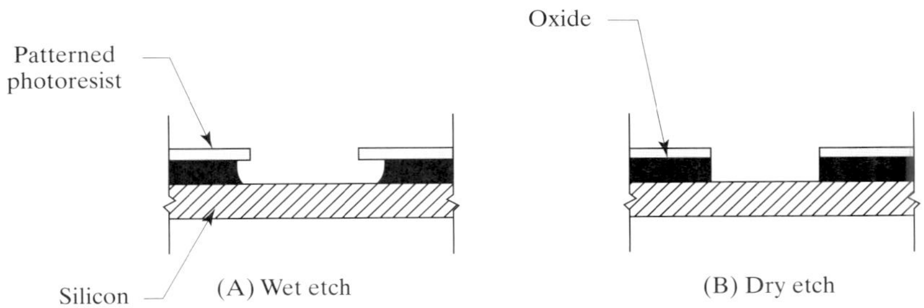

FIGURE 2.9 Comparison of isotropic wet etching (A) and anisotropic dry etching (B). Note the undercutting of the oxide caused by wet etching.

The etch gas employed in the RIE system generally consists of an organohalogen compound such as trichloroethane, perhaps mixed with an inert gas such as argon. The reactive ions formed from this mixture selectively attack silicon dioxide in preference to either photoresist or elemental silicon. Different mixtures of etch gases

[5] Reactive ion etching is actually only one of three forms of dry etching, the other two being plasma etching and chemical vapor etching. RIE is among the most useful because it produces highly anisotropic etching characteristics. See Runyan, *et al.*, pp. 269–272.

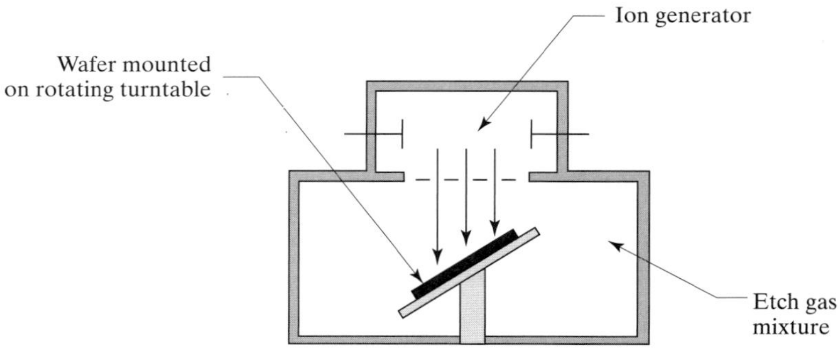

FIGURE 2.10 Simplified diagram of reactive ion etching apparatus.

have been developed that allow anisotropic etching of silicon nitride, elemental silicon, and other materials.

Modern processes rely on dry etching to obtain tight control of submicron geometries that cannot be fabricated in any other way. The increased packing density and higher performance of these structures more than compensate for the complexity and cost of dry etching.

2.3.3. Other Effects of Oxide Growth and Removal

During a typical processing sequence, the wafer is repeatedly oxidized and etched to form successive masking layers. These multiple masked oxidations cause the silicon surface to become highly nonplanar. The resulting surface irregularities are of great concern because modern fine-line photolithography has a very narrow depth of field. If the surface irregularities are too large, then it becomes impossible to focus the image of the photomask onto the resist.

Consider the wafer in Figure 2.11. A planar silicon surface has been oxidized, patterned, and etched to form a series of oxide openings (Figure 2.11A). Subsequent thermal oxidation of the patterned wafer results in the cross-section shown in Figure 2.11B. The opening that is left from the previous oxide removal initially oxidizes very rapidly, while the surfaces already coated with an oxide layer oxidize more slowly. The silicon surface erodes by about 45% of the oxide thickness grown.[6] The silicon under the previous oxide opening therefore recedes to a greater depth than the surrounding silicon surfaces. The thickness of oxide in the old opening will always be less than that of the surrounding surfaces since these already have some oxide on them when growth begins. The differences in oxide thickness and in the depths of the silicon surfaces combine to produce a characteristic surface discontinuity called an *oxide step.*

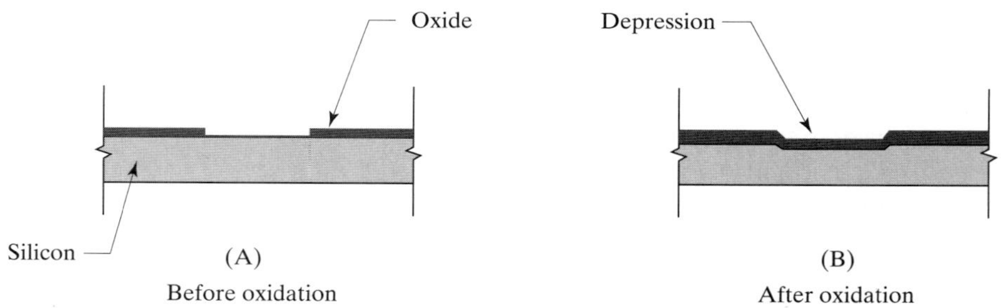

FIGURE 2.11 Effects of patterned oxidation on wafer topography.

[6] This value is the inverse of the *Pilling-Bedworth ratio,* which equals 2.2: G. E. Anner, *Planar Processing Primer* (New York: Van Nostrand Reinhold, 1990), p. 169.

The growth of a thermal oxide also affects the doping levels in the underlying silicon. If the dopant is more soluble in oxide than in silicon, during the course of the oxidation it will tend to migrate from the silicon into the oxide. The surface of the silicon thus becomes depleted of dopant. Boron is more soluble in oxide than in silicon, so it tends to segregate into the oxide. This effect is sometimes called *boron suckup*. Conversely, if the dopant dissolves more readily in silicon than in oxide, then the advancing oxide-silicon interface pushes the dopant ahead of it and causes a localized increase in doping levels near the surface. Phosphorus (like arsenic and antimony) segregates into the silicon, so it tends to accumulate at the surface as oxidation continues. This effect is sometimes called *phosphorus pileup* or *phosphorus plow*. The doping profiles of Figures 2.12A and 2.12B illustrate boron suckup and phosphorus plow, respectively. In both cases, the pre-oxidization doping profiles were constant and the varying dopant concentrations near the surface are solely due to segregation. The existence of these segregation mechanisms complicates the task of designing dopant profiles for integrated devices.

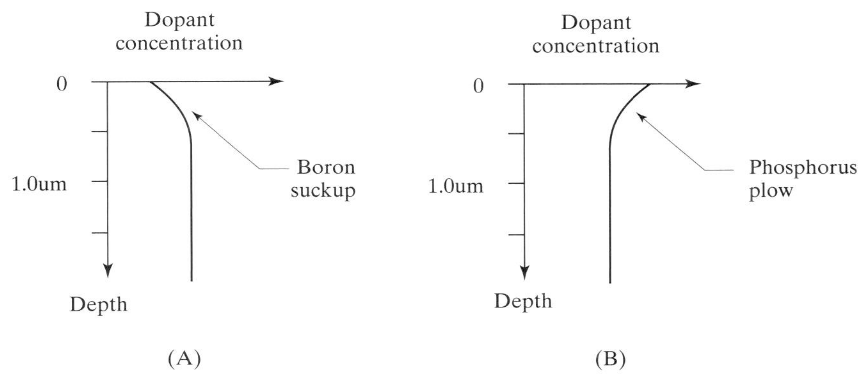

FIGURE 2.12 Oxide segregation mechanisms: (A) boron suckup and (B) phosphorus plow.[7]

The doping of silicon also affects the rate of oxide growth. A concentrated N+ diffusion tends to accelerate the growth of oxide near it by a process called *dopant-enhanced oxidation*. This occurs because the donors interfere with the bonding of atoms at the oxide interface, causing dislocations and other lattice defects. These defects catalyze oxidation and thus accelerate the growth of the overlying oxide. This effect can become quite significant when a heavily doped N+ deposition occurs early in the process, before the long thermal drives and oxidations. Figure 2.13 shows a wafer in which a long thermal oxidation has been

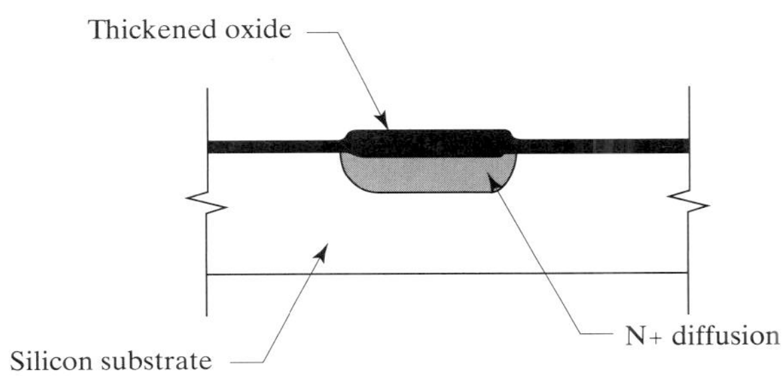

FIGURE 2.13 Effects of dopant-enhanced oxidation.

[7] A. S. Grove, O. Leistiko, and C. T. Sah, "Redistribution of Acceptor and Donor Impurities During Thermal Oxidation of Silicon," *J. Appl. Phys.*, Vol. 35, #9, 1964, pp. 2695–2701.

conducted after the deposition of an N+ region. The oxide over the N+ diffusion is actually thicker than the oxide over adjacent regions. Dopant-enhanced oxidation can be used to thicken the field oxide in order to reduce its capacitance per unit area. Thus, a capacitor formed over a deep-N+ diffusion will exhibit less parasitic capacitance between its bottom plate and the substrate than will a capacitor formed over lightly doped regions.

2.3.4. Local Oxidation of Silicon (LOCOS)

A technique called *local oxidation of silicon* (LOCOS) allows the selective growth of thick oxide layers.[8] The process begins with the growth of a thin *pad oxide* that protects the silicon surface from the mechanical stresses induced by subsequent processing (Figure 2.14). Chemical vapor deposition produces a nitride film on top of the pad oxide. This nitride is patterned to expose the regions to be selectively oxidized. The nitride blocks the diffusion of oxygen and water molecules, so oxidation only occurs under the nitride windows. Some oxidants diffuse a short distance under the edges of the nitride, producing a characteristic curved transition region called a *bird's beak.*[9] Once oxidation is complete, the nitride layer is stripped away to reveal the patterned oxide.

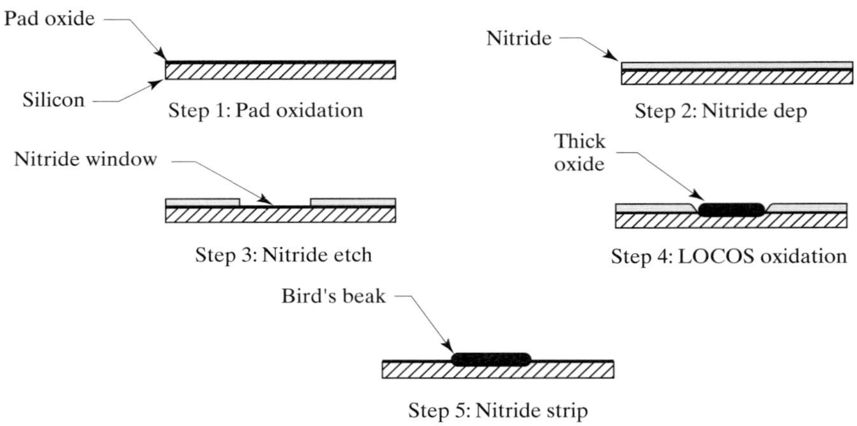

FIGURE 2.14 Local oxidation of silicon (LOCOS) process.

CMOS and BiCMOS processes employ LOCOS to grow a thick *field oxide* over electrically inactive regions of the wafer. The areas not covered by field oxide are called *moat* regions because they form shallow trenches in the topography of the wafer. A very thin, high-quality gate oxide subsequently grown in the moat regions forms the gate dielectric of the MOS transistors.

A mechanism called the *Kooi effect* complicates the growth of gate oxide.[10] The water vapor typically used to accelerate LOCOS oxidation also attacks the surface of the nitride film to produce ammonia, some of which migrates beneath the pad oxide near the edges of the nitride window. There it reacts with the underlying silicon to form silicon nitride again (Figure 2.15). Since these nitride deposits lie

---

8 "LOCOS: A New I.C. Technology," *Microelectronics and Reliability,* Vol. 10, 1971, pp. 471–472.

9 E. Bassous, H. N. Yu, and V. Maniscalco, "Topology of Silicon Structures with Recessed SiO_2," *J. Electrochem. Soc.,* Vol. 123, #11, 1976, pp. 1729–1737.

10 E. Kooi, J. G. van Lierop, and J. A. Appels, "Formation of Silicon Nitride at a Si-SiO_2 Interface during Local Oxidation of Silicon and during Heat-Treatment of Oxidized Silicon in NH_3 Gas," *J. Electrochem. Soc.,* Vol. 123, #7, 1976, pp. 1117–1120.

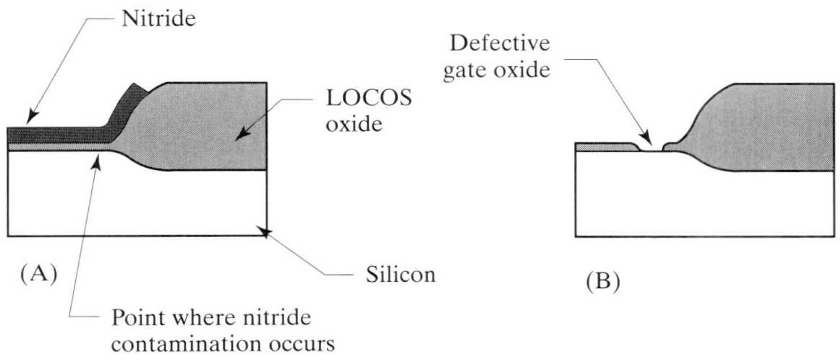

FIGURE 2.15 The Kooi effect is caused by nitride that grows under the bird's beak (A), preventing formation of gate oxide during subsequent oxidation (B).

beneath the pad oxide, they remain even after the LOCOS nitride is stripped. Removing the pad oxide prior to growing the gate oxide does not eliminate these deposits because this etch is selective to oxide, not to nitride. During gate oxidation, the nitride residues act as an unintentional LOCOS mask that retards oxide growth around the edges of the moat region. The gate oxide at these points may not be sufficiently thick to withstand the full operating voltage. The Kooi effect can be circumvented by first growing a thin oxide layer and then stripping it away. Because silicon nitride slowly oxidizes, this *dummy gate oxidation* removes the nitride residues and improves the integrity of the true gate oxide grown immediately afterward.

2.4 DIFFUSION AND ION IMPLANTATION

Discrete diodes and transistors can be fabricated by forming junctions into a silicon ingot during crystal growth. Suppose that the silicon ingot begins as a P-type crystal. After a short period of growth, the melt is counterdoped by the addition of a controlled amount of phosphorus. Continued crystal growth will now produce a PN junction embedded in the ingot. Successive counterdopings can produce multiple junctions in the crystal, allowing the fabrication of *grown-junction* transistors. Integrated circuits cannot be grown because there is no way to produce differently doped regions in different portions of the wafer. Even the manufacture of simple grown-junction transistors presents a challenge, because the thickness and planarity of grown junctions are difficult to control. Each counterdoping also raises the total dopant concentration. Some properties of silicon (such as minority carrier lifetime) depend upon the total concentration of doping atoms, not just upon the excess of one dopant species over the other. The repeated counterdopings therefore progressively degrade the electrical properties of the silicon.

Historically, the grown junction process was soon abandoned in favor of the much more versatile *planar process*. This process is used to fabricate virtually all modern integrated circuits as well as the vast majority of modern discrete devices. Figure 2.16 shows how a wafer of discrete diodes can be fabricated using planar processing. A uniformly doped silicon crystal is first sliced to form individual wafers. An oxide film grown on these wafers is photolithographically patterned and etched. A dopant source spun onto the patterned wafers touches the silicon only where the oxide has been previously removed. The wafers are then heated in a furnace to drive the dopant into the silicon, which forms shallow counterdoped regions. The finished wafer can be diced to form hundreds or thousands of individual diodes. The planar process does not require multiple counterdopings of the silicon ingot, thereby allowing more precise control of junction depths and dopant distributions.

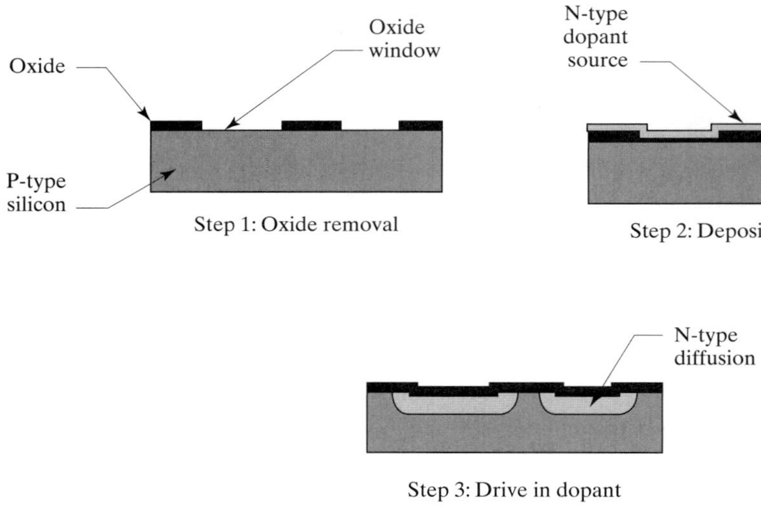

FIGURE 2.16 Formation of diffused PN-junction diodes using the planar process.

2.4.1. Diffusion

Dopant atoms can move through the silicon lattice by thermal diffusion in much the same way as carriers move by diffusion (Section 1.1.3). The heavier dopant atoms are more tightly bound to the crystal lattice, so temperatures of 800°C to 1250°C are required to obtain reasonable diffusion rates. Once the dopants have been driven to the desired junction depth, the wafer is cooled and the dopant atoms become immobilized within the lattice. A doped region formed in this manner is called a *diffusion*.

The usual process for creating a diffusion consists of two steps: an initial *deposition* (or *predeposition*) and a subsequent *drive* (or *drive-in*). Deposition consists of heating the wafer in contact with an external source of dopant atoms. Some of these diffuse from the source into the surface of the silicon wafer to form a shallow heavily doped region. The external dopant source is then removed and the wafer is heated to a higher temperature for a prolonged period of time. The dopants introduced during deposition are now driven down to form a much deeper and less concentrated diffusion. If a very heavily doped junction is required, then it is usually unnecessary to strip the dopant source from the wafer, and the deposition and subsequent drive can be conducted as a single operation.

Four dopants find widespread use in silicon processing: *boron, phosphorus, arsenic,* and *antimony.*[11] Only boron is an acceptor; the other three are all donors. Boron and phosphorus diffuse relatively rapidly, while arsenic and antimony diffuse much more slowly (Table 2.2). Arsenic and antimony are used where slow rates of

TABLE 2.2 Representative junction depths, in microns (10^{20} atoms/cm^3 source, 10^{16} atoms/cm^3 background, 15 min deposition, 1 hr drive).[12]

Dopant	950°C	1000°C	1100°C	1200°C
Boron	0.9	1.5	3.6	7.3
Phosphorus		0.5	1.6	4.6
Antimony			0.8	2.1
Arsenic			0.7	2.0

[11] These dopants were chosen because they readily ionize and because they are sufficiently soluble in silicon to form heavily doped diffusions. See F. A. Trumbore, "Solid Solubilities of Impurity Elements in Germanium and Silicon," *Bell Syst. Tech. J.,* Vol. 39, #1, 1960, pp. 205–233.

[12] Calculated using diffusivities from R. S. Muller and T. I. Kamins, *Device Electronics for Integrated Circuits,* 2nd ed. (New York: John Wiley and Sons, 1986), p. 85.

diffusion are advantageous—for example, when very shallow junctions are desired. Even boron and phosphorus do not diffuse appreciably at temperatures below 800°C, necessitating the use of special high-temperature diffusion furnaces.

Figure 2.17 shows a simplified diagram of a typical apparatus for conducting a phosphorus diffusion. A long fused silica tube passes through an electric furnace that is constructed to produce a very stable heating zone in the middle of the tube. After the wafers are loaded into a wafer boat, they are slowly pushed into the furnace by means of a mechanical arrangement that controls the insertion rate. Dry oxygen is blown through a flask containing liquid phosphorus oxychloride ($POCl_3$, often called "pockle"). A small amount of $POCl_3$ evaporates and is carried by the gas stream over the wafers. Phosphorus atoms released by the decomposition of the $POCl_3$ diffuse into the oxide film, forming a doped oxide that acts as a deposition source. When enough time has passed to deposit sufficient dopant in the silicon, the wafers are removed from the furnace and the doped oxide is stripped away (a process called *deglazing*). The wafers are then reloaded into another furnace, where they are heated to drive the phosphorus down to form the desired diffusion. If a very concentrated phosphorus diffusion is desired, then the wafers need not be removed for deglazing prior to the drive. With suitable modifications to the dopant source, this apparatus can diffuse any of the four common dopants.

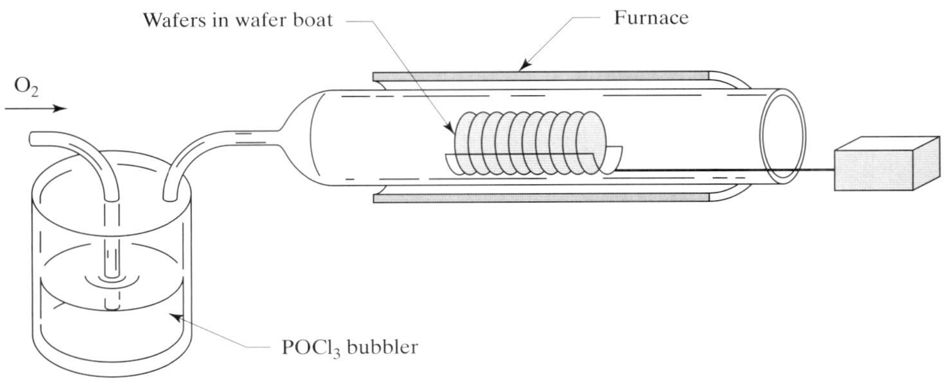

FIGURE 2.17 Simplified diagram of a phosphorus diffusion furnace using a $POCl_3$ source.

Many alternate deposition sources have been developed. A gaseous dopant such as diborane (for boron) or phosphine (for phosphorus) can be injected directly into the carrier gas stream. Thin disks of boron nitride placed between silicon wafers can serve as a solid deposition source for boron. In a high-temperature oxidizing atmosphere, a little boron trioxide outgases from these disks to the adjacent wafers. Various proprietary *spin-on glasses* are also sold as dopant sources. These consist of doped oxide dispersed in a volatile solvent. After the solution is spun onto a wafer, a brief bake drives out the solvent and leaves a doped oxide layer on the wafer. This so-called *glass* then serves as a dopant source for the subsequent diffusion.

None of these deposition schemes are particularly well controlled. Even with gaseous sources (which can be precisely metered) nonuniform gas flow around the wafer inevitably produces doping variations. For less-demanding processes such as standard bipolar, any of these schemes can give adequate results. Modern CMOS and BiCMOS processes require more accurate control of doping levels and junction depths than conventional deposition techniques can achieve. Ion implantation can provide the necessary accuracy at the expense of much more complex and costly apparatus.

2.4.2. Other Effects of Diffusion

The diffusion process suffers from a number of limitations. Diffusions can only be performed from the surface of the wafer, limiting the geometries that can be fabricated. Dopants diffuse unevenly, so the resulting diffusions do not have constant doping profiles. Subsequent high-temperature process steps continue the drive of previously deposited dopants, so junctions formed early in the process are driven substantially deeper during later processing. Dopants out-diffuse under the edges of the oxide windows, spreading the diffusion pattern. Diffusions interact with oxidizations due to segregation mechanisms, resulting in depletion or enhancement of surface doping levels. Diffusions even interact with one another since the presence of one doping species alters the diffusion rates of others. These and other complications make the diffusion process far more complex than it might at first appear.

Diffusion can produce only relatively shallow junctions. Practical drive times and temperatures limit junction depths to about fifteen microns. Most diffusions will be much shallower. Since diffusions are typically patterned using an oxide mask, the cross section of a diffusion generally resembles that shown in Figure 2.18. The dopant diffuses out in all directions at roughly the same rate. The junction moves laterally under the edges of the oxide window a distance equal to about 80% of the junction depth.[13] This lateral movement, known as *outdiffusion,* causes the final size of the diffused region to exceed the drawn dimensions of the oxide window. Outdiffusion is not visible under the microscope since the changes in oxide color caused by thin film interference correspond to the locations of oxide removals and not to the positions of the final junctions.

FIGURE 2.18 Cross section and doping profile of a typical planar diffusion.

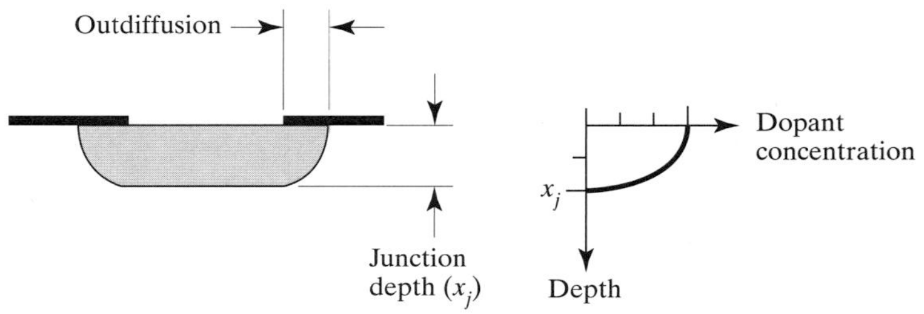

The doping level of a diffusion varies as a function of depth. Neglecting segregation mechanisms, dopant concentrations are highest at the surface and gradually lessen with depth. The resulting *doping profile* can be theoretically predicted and experimentally measured. Figure 2.18 shows the theoretical doping profile for a point in the center of the oxide window. This profile assumes that oxide segregation remains negligible, which is not always the case. Boron suckup may substantially reduce the surface doping of a P-type diffusion and can cause a lightly doped diffusion to invert to become N-type. Phosphorus pileup will not cause surface inversion, but it still affects surface doping levels.

As mentioned above, the rate of diffusion can be altered by the presence of other doping species. Consider an NPN transistor with a heavily doped phosphorus emitter diffused into a lightly doped boron base. The presence of high concentrations of donors within the emitter causes lattice strains that spawn defects. Some of these

---

[13] See D. P. Kennedy and R. R. O'Brien, "Analysis of the Impurity Atom Distribution Near the Diffusion Mask for a Planar p-n Junction," *IBM J. of Research and Development,* Vol. 9, 1965, pp. 179–186.

defects migrate to the surface, where they cause dopant-enhanced oxidation. Other defects migrate downward, where they accelerate the diffusion of boron in the underlying base region. This mechanism, called *emitter push,* results in a deeper base diffusion under the emitter than in surrounding regions (Figure 2.19A).[14] The presence of NBL beneath a diffusion may reduce the junction depth due to the intersection of the tail of the updiffusing NBL with the base diffusion. This effect is sometimes called *NBL push* in analogy with the better-known emitter push, even though the underlying mechanisms are quite different (Figure 2.19B). NBL push can interfere with the layout of accurate diffused resistors.

A similar mechanism accelerates the diffusion of dopants under an oxidization zone. The oxidation process spawns defects, some of which migrate downward to enhance the rate of diffusion of dopants beneath the growing oxide. This mechanism is called *oxidization-enhanced diffusion.*[15] It affects all dopants, and it can produce significantly deeper diffusions under a LOCOS field oxide than under adjacent moat regions (Figure 2.19C).

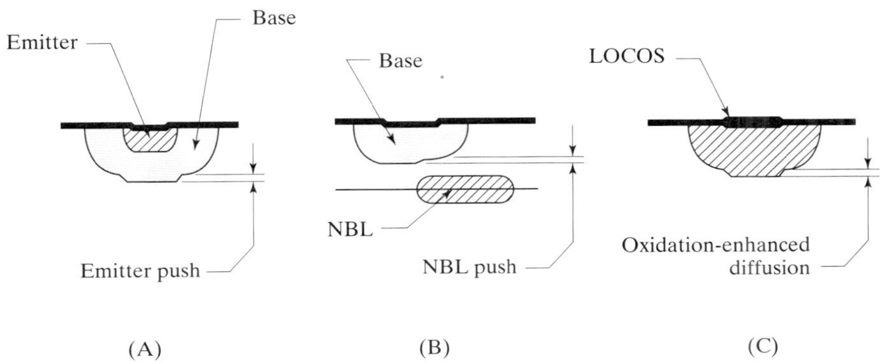

FIGURE 2.19 Mechanisms that can alter diffusion rates include emitter push (A), NBL push (B), and oxidation-enhanced diffusion (C).

Even the most sophisticated computer programs cannot always predict actual doping profiles and junction depths because of the many interactions that occur. Process engineers must experiment carefully to find the proper recipe for manufacturing a given combination of devices on a wafer. The more complicated the process, the more complex these interactions become and the more difficult it is to find a suitable recipe. Since process design takes so much time and effort, most companies use only a few processes to manufacture all of their products. The difficulty of incorporating new process steps into an existing recipe also explains the reluctance of process engineers to modify their processes.

2.4.3. Ion Implantation

Due to the limitations of conventional diffusion techniques, modern processes make extensive use of *ion implantation.* An ion implanter is essentially a specialized particle accelerator used to accelerate dopant atoms so that they can penetrate the silicon crystal to a depth of several microns.[16] Ion implantation does not require high

---

[14] A. F. W. Willoughby, "Interactions between Sequential Dopant Diffusions in Silicon—A Review," *J. Phys. D: Appl. Phys.,* Vol. 10, 1977, pp. 455–480.

[15] K. Taniguchi, K. Kurosawa, and M. Kashiwagi, "Oxidation Enhanced Diffusion of Boron and Phosphorus in (100) Silicon," *J. Electrochem. Soc.,* Vol. 127, #10, 1980, p. 2243–2248.

[16] The depth of the implant depends on implant energy. The implants discussed in this section involve energies of no more than several hundred keV. Some modern CMOS processes now employ multi-MeV implants to achieve significantly deeper profiles (5–10μm).

temperatures, so a layer of patterned photoresist can serve as a mask against the implanted dopants. Implantation also allows better control of dopant concentrations and profiles than conventional deposition and diffusion. However, large implant doses require correspondingly long implant times. Ion implanters are also complex and costly devices. Many processes use a combination of diffusions and implantations to reduce overall costs.

Figure 2.20 shows a simplified diagram of an ion implanter. An ion source provides a stream of ionized dopant atoms that are accelerated by the electric field of a miniature linear accelerator. A magnetic analyzer selects the desired species of ion, and a pair of deflection plates scans the resulting ion beam across the wafer. A high vacuum must be maintained throughout the system, so the entire apparatus is enclosed in a steel housing.

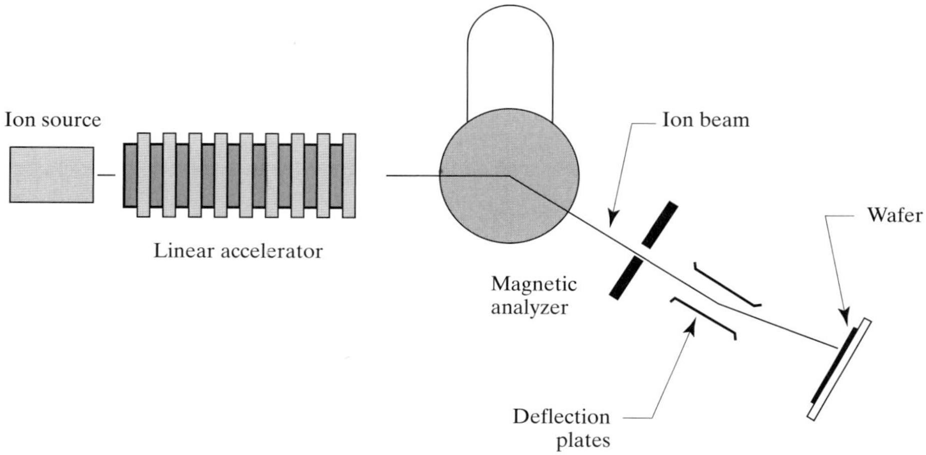

FIGURE 2.20 Simplified diagram of an ion implanter.[17]

Once the ions enter the silicon lattice, they immediately begin to decelerate due to collisions with surrounding atoms. Each collision transfers momentum from a moving ion to a stationary atom. The ion beam rapidly spreads as it sheds energy, causing the implant to spread out (*straggle*) in a manner reminiscent of outdiffusion. Atoms are also knocked out of the lattice by the collisions, causing extensive lattice damage that must be repaired by *annealing* the wafer at moderate temperatures (800°C to 900°C) for a few minutes. The silicon atoms become mobile and the intact silicon crystal structure around the edges of the implant zone serves as a seed for crystal growth. Damage progressively anneals out from the sides of the implant zone toward the center. Dopants added by ion implantation will redistribute by thermal diffusion if the wafer is subsequently heated to a sufficiently high temperature. Therefore, a deep lightly doped diffusion can be created by first implanting the required dopants and subsequently driving them down to the desired junction depth.

The dopant concentration provided by ion implantation is directly proportional to the *implant dose,* which equals the product of ion beam current and time. The dose can be precisely monitored and controlled, which allows for much better repeatability than conventional deposition techniques do. The doping profile is determined by the energy imparted to individual ions, a quantity called the *implant energy.* Low-energy implants are very shallow, while high-energy implants actually place most of the dopant atoms beneath the surface of the silicon. Ion implantation

[17] The scheme shown is but one of several; see Anner, p. 313ff.

can be used to counterdope a subsurface region to form a *buried layer.* Because of practical limitations on implant energy, these buried layers are usually quite shallow.

Ion implantation is somewhat anisotropic. The edges of an implant, especially a shallow low-energy one, do not spread as much as those produced by thermal diffusion. This aids in the manufacture of *self-aligned* structures that greatly improve the performance of MOS transistors. Figure 2.21 illustrates the creation of self-aligned MOS source/drain regions by ion implantation. A layer of polysilicon has been deposited and patterned on top of a thin gate oxide. The polysilicon not only forms the gate electrodes for MOS transistors but also simultaneously serves as an implant mask. The polysilicon blocks the implant from the region beneath the gate electrode, forming precisely aligned source and drain regions. The alignment of the source and drain with the gate is limited only by the small amount of straggle caused by the spreading of the ion beam. If self-aligned implants were not used, then photolithographic misalignments would occur between the gate and the source/drain diffusions, and the resulting overlap capacitances would substantially reduce the switching speed of the MOS transistors.

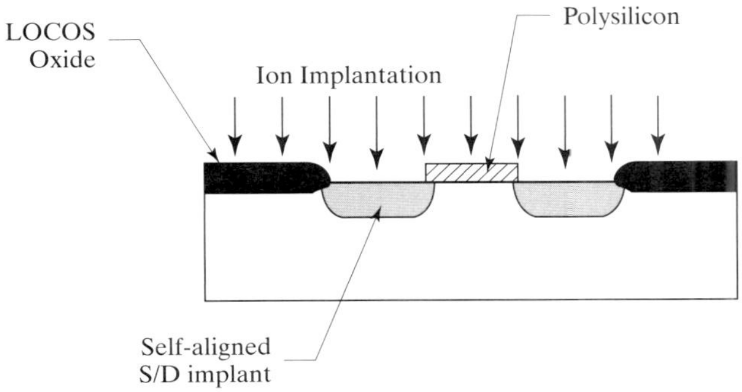

FIGURE 2.21 Self-aligned source and drain regions formed by ion implantation.

When the silicon lattice is viewed from certain angles, interstices between columns of silicon atoms, called *channels,* become visible. These disappear from view when the crystal is turned slightly. Channels are visible in both the (100) and the (111) silicon surfaces when these are viewed perpendicularly. If the ion beam were to impinge perpendicularly upon a (100) or a (111) surface, then ions could move deep into the crystal before scattering would commence. The final dopant distribution would depend critically upon the angle of incidence of the ion beam. To avoid this difficulty, most implanters project the ion beam onto the wafer at an angle of about 7°.

2.5 SILICON DEPOSITION

Films of pure or doped silicon can be chemically grown on the surface of a wafer. The nature of the underlying surface determines whether the resulting film will be monocrystalline or polycrystalline. If the surface consists of exposed monocrystalline silicon, then this serves as a seed for crystal growth and the deposited film will also be monocrystalline. If the deposition is conducted on top of an oxide or nitride film, then no underlying crystalline lattice will exist to serve as a seed for crystal nucleation, and the deposited silicon will form a fine-grained aggregate of polycrystalline silicon (*poly*). Modern integrated circuits make extensive use of both monocrystalline and polycrystalline deposited silicon films.

2.5.1. Epitaxy

The growth of a single-crystal semiconductor film upon a suitable crystalline substrate is known as *epitaxy*. The substrate normally consists of a crystal of the same material as the semiconductor that is to be deposited, but this need not always be the case. High-quality monocrystalline silicon films have been grown on wafers of synthetic sapphire or spinel, as these materials possess a crystal structure that is enough like silicon to allow crystal nucleation. The cost of synthetic sapphire or spinel wafers so greatly exceeds the cost of similar-sized silicon wafers that the vast majority of epitaxial depositions consist of silicon films grown on silicon substrates.

There are several different methods of growing epitaxial (*epi*) layers. One relatively crude method consists of pouring molten semiconductor material on top of the substrate, allowing it to crystallize for a short period of time, and then wiping the excess liquid off. The wafer surface can then be reground and polished to form an epitaxial layer. Obvious drawbacks to this *liquid-phase epitaxy* include the high cost of regrinding the wafer and the difficulty of producing a precisely controlled epi thickness.

Most modern epitaxial depositions use *low pressure chemical vapor deposited* (LPCVD) epitaxy. Figure 2.22 shows a simplified diagram of an early type of LPCVD epi reactor. The wafers are mounted on an inductively heated carrier block, and a mixture of dichlorosilane and hydrogen passes over them. These gases react at the surface of the wafers to form a slow-growing layer of monocrystalline silicon. The rate of growth can be controlled by adjusting the temperature, pressure, and gas mixture used in the reactor. No polishing is required to render the epitaxial surface suitable for further processing, as vapor-phase epitaxy faithfully reproduces the topography of the underlying surface. The epitaxial film can also be doped *in situ* by adding small amounts of gaseous dopants such as phosphine or diborane to the gas stream.

There are several benefits of growing an epitaxial layer on the starting wafer. For one, the epi layer need not have the same doping polarity as the underlying wafer. For example, an N– epitaxial layer can be grown on a P– substrate—an arrangement commonly employed for standard bipolar processes. Multiple epitaxial layers can be grown in succession and the resulting stack can be used to form transistors or other devices. The potential of epitaxy is limited chiefly by the slow rate of epi growth and by the expense and complexity of the required equipment, which are much greater than Figure 2.22 suggests.

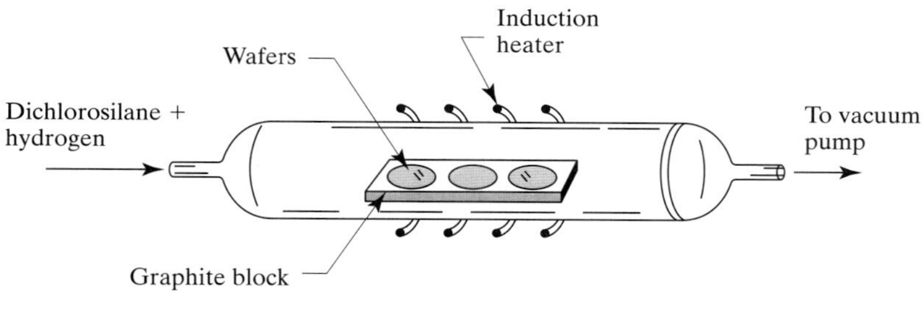

FIGURE 2.22 Simplified diagram of an epi reactor.[18]

[18] The horizontal tube reactor shown here has long been obsolete; see C. W. Pearce, "Epitaxy," in S. M. Sze, ed., *VLSI Technology* (New York: McGraw-Hill, 1983), pp. 61–65.

Epitaxy also allows the formation of *buried layers*. An N+ buried layer constitutes a key step in most bipolar processes since it allows the construction of vertical NPN transistors with low collector resistances. Figure 2.23 depicts the growth of such an N-buried layer (NBL). Arsenic and antimony are the preferred dopants for forming an NBL because their slow diffusion rates minimize the outdiffusion of the buried layer during subsequent high-temperature processing. Antimony is often chosen instead of arsenic because it exhibits less tendency to spread laterally during epitaxy (an effect called *lateral autodoping*).[19] Buried layer fabrication begins with a lightly doped P-type wafer. This wafer is oxidized, and windows are patterned in the resulting oxide layer. Either arsenic or antimony is implanted through the windows, and the wafer is briefly annealed to eliminate the resulting implant damage. Thermal oxidation occurs during this anneal, and discontinuities form around the edges of the oxide windows. Next, all oxide is stripped from the wafer, and an N– epitaxial layer is deposited. The resulting structure consists of patterned N+ regions buried under an epitaxial layer.

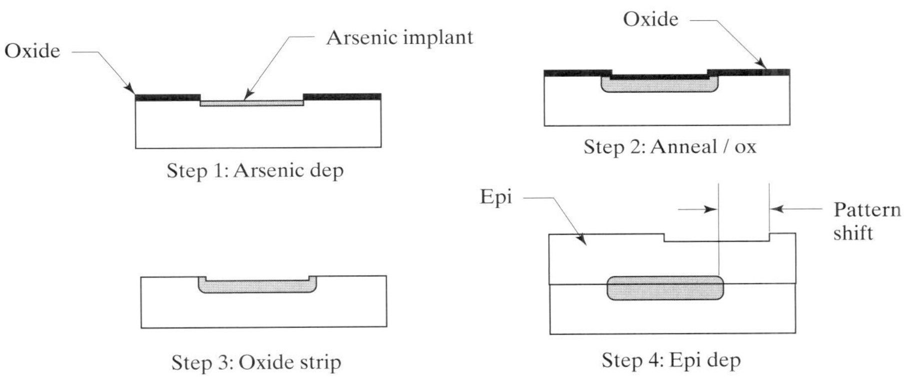

FIGURE 2.23 Formation of an N-buried layer (NBL), showing pattern shift.

As mentioned previously, oxidization during the anneal of the NBL causes slight surface discontinuities to form around the edges of the oxide window. The epitaxial layer faithfully reproduces these discontinuities in the final surface of the wafer. Under a microscope, the resulting step forms a faintly visible outline called the *NBL shadow*. Subsequent photomasks are aligned to this discontinuity. An alternative alignment method uses infrared light to image the NBL doping through the overlying silicon, but this requires more complicated equipment.

The accretion of silicon atoms at the edge of the NBL shadow during epitaxy displaces it laterally, an effect called *pattern shift* (Figure 2.23).[20] The magnitude of shift depends on many factors, including temperature, pressure, gas composition, substrate orientation, and tilt (See Section 7.2.3). When other layers are aligned to the NBL shadow, these must be offset to compensate for pattern shift.

---

[19] M. W. M. Graef, B. J. H. Leunissen, and H. H. C. de Moor, "Antimony, Arsenic, Phosphorus, and Boron Autodoping in Silicon Epitaxy," *J. Electrochem. Soc.,* Vol. 132, #8, 1985, pp. 1942–1954.

[20] M. R. Boydston, G. A. Gruber, and D. C. Gupta, "Effects of Processing Parameters on Shallow Surface Depressions During Silicon Epitaxial Deposition," in *Silicon Processing,* American Society for Testing and Materials STP 804, 1983, pp. 174–189.

2.5.2. Polysilicon Deposition

If silicon is deposited on an amorphous material, then no underlying lattice exists to align crystal growth. The resulting silicon film consists of an aggregate of small intergrown crystals. This *poly* film has a granular structure with a grain size dependent upon deposition conditions and subsequent heat treatment. The grain boundaries of the poly represent lattice defects, which can provide sneak paths for leakage currents. Therefore, PN junctions are not normally fabricated from poly. Polysilicon is often used to construct the gate electrodes of self-aligned MOS transistors because, unlike aluminum, it can withstand the high temperatures required to anneal the source/drain implants. In addition, the use of poly has led to better control of MOS threshold voltages due to the ability of phosphorus-doped polysilicon to immobilize ionic contaminants (Section 4.2.2). Suitably doped poly can be used to fabricate very narrow resistors that exhibit fewer parasitics than diffused devices. Heavily doped polysilicon can also be used as an additional metallization layer for signals that can tolerate the insertion of considerable resistance in the signal path.

A patterned poly layer is produced by first depositing polysilicon across the wafer using an apparatus similar to that employed for epitaxy (see Figure 2.22). The wafer is then coated with photoresist, patterned, and etched to selectively remove the polysilicon. Modern processes usually employ dry etching rather than wet etching because of the importance of precisely controlled gate dimensions.

2.6 METALLIZATION

The active components of an integrated circuit consist of diffusions, ion implantations, and epitaxial layers grown in or on a silicon substrate. When this processing is complete, the resulting components are connected to form the integrated circuit, using one or more layers of patterned wiring. This wiring consists of layers of metal and polysilicon separated by insulating material, usually deposited oxide. These same materials can also be used to construct passive components such as resistors and capacitors.

Figure 2.24 illustrates the formation of a typical *single-level-metal* (SLM) interconnection system. After the final implantations and diffusions, a layer of oxide is grown or deposited over the entire wafer, and selected areas are patterned and etched to create oxide windows exposing the silicon. These windows will form *contacts* between the metallization and the underlying silicon. Once these contacts have been opened, a thin metallic film can be deposited and etched to form the interconnection pattern.

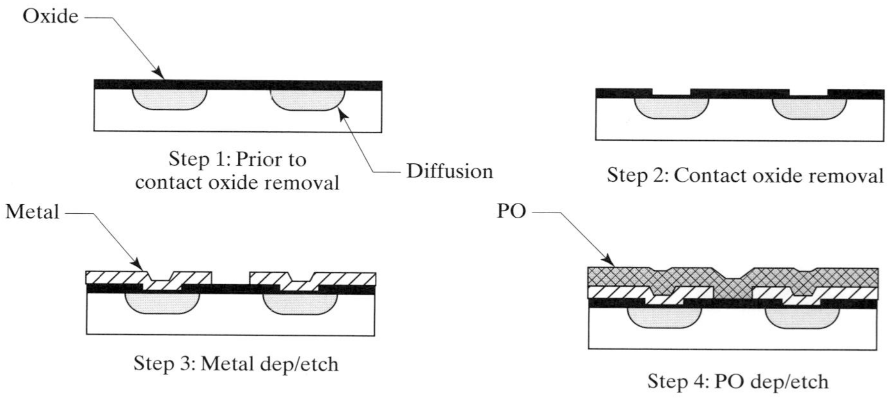

FIGURE 2.24 Formation of a single-level metal system.

Exposed aluminum wiring is vulnerable to mechanical damage and chemical corrosion. An oxide or nitride film deposited over the completed wafer serves as a *protective overcoat* (PO). This layer acts as a conformal seal similar in principle to the plastic conformal coatings sometimes applied to printed circuit boards. Windows etched through the overcoat expose selected areas of the aluminum metallization so that bondwires can be attached to the integrated circuit.

The process illustrated in Figure 2.24 fabricates only a single aluminum layer. Additional layers of metallization can be sequentially deposited and patterned to form a multilevel metal system. Multiple metal layers increase the cost of the integrated circuit, but they allow denser packing of components and therefore reduce the overall die size. The savings in die area often compensate for the cost of the extra processing steps. Multiple metal layers also simplify interconnection and reduce layout time.

CMOS processes frequently employ low-resistivity polysilicon to form the gate electrodes of self-aligned MOS transistors. This material can serve as a free additional layer of interconnect. Even the lowest-sheet poly still has many times the resistance of aluminum, so the designer must take care to avoid routing high-current or high-speed signals in poly. Advanced processes may add a second and even a third layer of polysilicon. These additional layers are used to fabricate different types of MOS transistors, to form the plates of capacitors, and to construct polysilicon resistors. Each of these additional poly layers can be pressed into service as another layer of interconnect.

2.6.1. Deposition and Removal of Aluminum

Most metallization systems employ aluminum or aluminum alloys to form the primary interconnection layers. Aluminum conducts electricity almost as well as copper or silver, and it can be readily deposited in thin films that adhere to all of the materials used in semiconductor fabrication. A brief period of heating will cause the aluminum to alloy into the silicon to form low-resistance contacts.

Aluminum is usually deposited by *evaporation* using an apparatus similar to the one shown in Figure 2.25. The wafers are mounted in a frame that holds their exposed surfaces toward a crucible containing a small amount of aluminum. When the crucible is heated, some of the aluminum evaporates and deposits on the wafer surfaces. A high vacuum must be maintained throughout the evaporation system to prevent oxidation of the aluminum vapor prior to its deposition upon the wafers. The illustrated evaporation system can only handle pure aluminum, but somewhat more complex systems can also evaporate selected aluminum alloys.

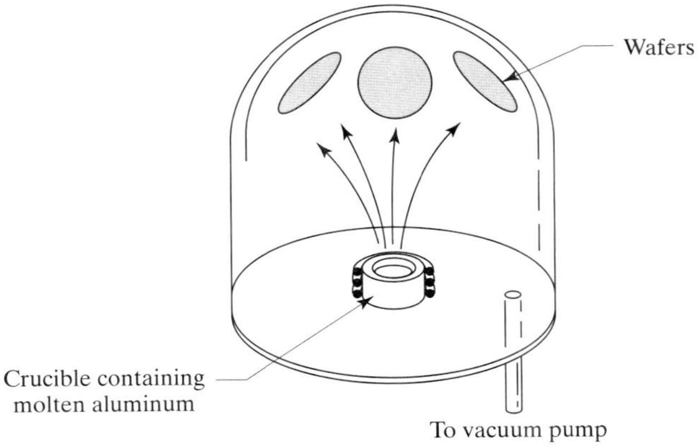

FIGURE 2.25 Simplified diagram of an aluminum evaporation apparatus.

Aluminum and silicon alloy at moderate temperatures. A brief period of heating will form an extremely thin layer of aluminum-doped silicon beneath the contact openings. This process, called *sintering,* can achieve Ohmic contact to P-type silicon because aluminum acts as an acceptor. The aluminum-silicon alloy forms a shallow, heavily doped P-type diffusion that bridges between the metal and the P-type silicon. Less obviously, Ohmic contact also occurs when aluminum touches heavily doped N-type silicon. Junctions form beneath these contacts, but their depletion regions are so thin that carriers can surmount them by quantum tunneling. Rectification will occur if the donor concentration falls too low, so Ohmic contact cannot be established directly between aluminum and lightly doped N-type silicon. The addition of a shallow N+ diffusion will enable Ohmic contact to these regions.

Sintering causes a small amount of aluminum to dissolve in the underlying silicon. Some silicon simultaneously dissolves in the aluminum metal, eroding the silicon surface. Some diffusions are so thin that erosion can punch entirely through them, causing a failure mechanism called *contact spiking.* Historically this was first observed in conjunction with the emitter diffusion of NPN transistors, so it is also called *emitter punchthrough.*[21] Contact spiking can be minimized by replacing pure aluminum metallization with an aluminum-silicon alloy. If the deposited aluminum is already saturated with silicon, then—at least in theory—it cannot dissolve any more. In practice, the silicon content of the alloy tends to separate during sintering to leave an unsaturated aluminum matrix. Careful control of sinter time and temperature will minimize this effect.

Another failure mechanism was encountered in high-density digital logic. As the dimensions of the integrated circuits were progressively reduced, the current density flowing through the metallization increased. Some devices eventually exhibited open-circuit metallization failures after many thousands of hours of operation at elevated temperatures. When the faulty units were examined, some of their leads contained unexpected breaks. These were eventually found to result from a failure mechanism called *electromigration.*[22] Carriers flowing through the metal collide with the lattice atoms. At current densities in excess of several million amps per square centimeter, these impacts become so frequent that the metal atoms begin to move. The displacement of the atoms causes voids to form between individual grains of the polycrystalline metal aggregate. Eventually these voids grow together to form a gap across the entire lead, causing an open-circuit failure (Section 4.1.2). The addition of a fraction of a percent of copper to the aluminum alloy improves electromigration resistance by an order of magnitude. Most modern metal systems therefore employ either aluminum-copper-silicon or aluminum-copper alloys.

2.6.2. Refractory Barrier Metal

The feature sizes of integrated circuits have steadily shrunk as ever-increasing numbers of components have been packed into approximately the same amount of silicon real estate. In order to obtain the necessary packing density, the sidewalls of contact and via openings have become increasingly steep. Evaporated aluminum does not deposit isotropically; the metal thins where it crosses oxide steps (Figure 2.26A). Any reduction in the cross-sectional area of a lead raises the current density and accelerates electromigration. A variety of techniques have been developed to improve step coverage on very steep sidewalls like those formed by reactive ion etching of thick oxide films.

---

[21] M. D. Giles, "Ion Implantation," in S. M. Sze, ed., *VLSI Technology* (New York: McGraw-Hill, 1983), pp. 367–369.

[22] J. R. Black, "Physics of Electromigration," *Proc. 12th Reliability Phys. Symp.,* 1974, p. 142.

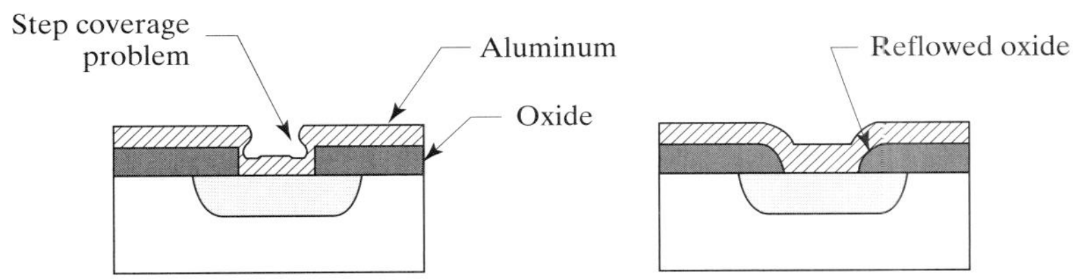

FIGURE 2.26 Step coverage of evaporated aluminum without reflow (A) and with reflow (B).

The step coverage of evaporated aluminum can be greatly increased by moderating the angle of the sidewalls. This can be achieved by heating the wafer until the oxide melts and slumps to form a sloped surface. This process is called *reflow* (Figure 2.26B). Pure oxide melts at too high a temperature to allow reflow, so phosphorus and boron are added to the oxide to reduce its melting point. The resulting doped oxide film is called either a *phosphosilicate glass* (PSG) or a *borophosphosilicate glass* (BPSG), depending on the choice of additives.

Reflow cannot be performed after aluminum has been deposited, because it cannot tolerate the temperatures required to soften PSG or BPSG. Therefore, while reflow can help improve the step coverage of first-level metal, it must be supplemented by other techniques in order to successfully fabricate a multilevel metal system. One option consists of using metals that deposit isotropically upon steeply sloped sidewalls, such as molybdenum, tungsten, or titanium. These *refractory barrier metals* possess extremely high melting points and are thus unsuited for evaporative deposition. A low-temperature process called *sputtering* can successfully deposit them. Figure 2.27 shows a simplified diagram of a sputtering apparatus. The wafers rest on a platform inside a chamber filled with low-pressure argon gas. Facing them is a plate of refractory barrier metal forming one of a pair of high-voltage electrodes. Argon atoms bombard the refractory metal plate. This knocks atoms loose that then deposit on the wafers to form a thin metallic film.

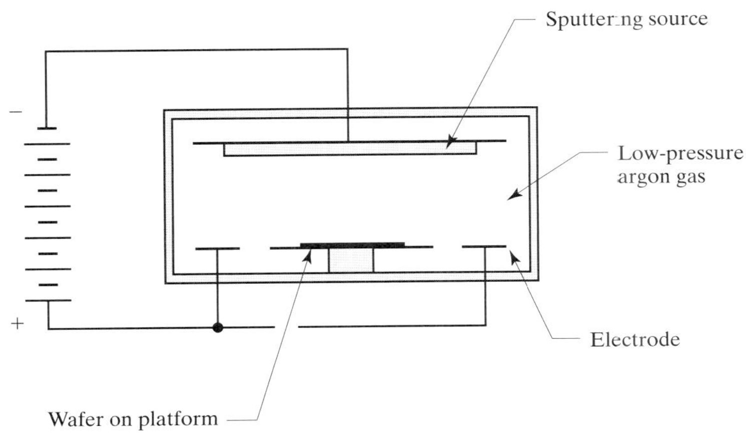

FIGURE 2.27 Simplified diagram of a sputtering apparatus.

The sputtered refractory barrier metal film not only provides superior step coverage, but also virtually eliminates emitter punchthrough.[23] If step coverage were the only criterion for choosing a metal system, then aluminum could be entirely replaced by refractory barrier metal. Unfortunately, refractory metals have relatively

[23] T. Hara, N. Ohtsuka, K. Sakiyama, and S. Saito, "Barrier Effect of W-Ti Interlayers in Al Ohmic Contact Systems," *IEEE Trans. Electron Devices,* Vol. ED-34, #3, 1987, pp. 593–597.

high resistivities and cannot be deposited in thick films as easily as aluminum can. Most metal systems therefore employ a sandwich of both materials. A thin layer of refractory metal beneath the aluminum ensures adequate step coverage in the contacts where the aluminum metal drastically thins. Elsewhere the aluminum reduces the electrical resistance of the metal leads. The relatively short sections of refractory barrier metal in the contacts do not contribute much resistance to the overall interconnection system.

Refractory barrier metals are extremely resistant to electromigration, so the thinning of aluminum on the sidewalls of contacts and vias does not represent an electromigration risk. Refractory barrier metal also tends to suppress classical electromigration failures by bridging any open circuits that develop in the aluminum metallization. Aluminum displaced by electromigration can still short adjacent leads, so refractory barrier metal cannot be relied on to supplement the current-carrying capacity of aluminum wiring except on the sidewalls of contact and via openings.

As mentioned, refractory barrier metal virtually eliminates emitter punchthrough. The degree of alloying between silicon and refractory metals is negligible and the aluminum cannot penetrate the barrier metal to contact the silicon. Most refractory barrier metal systems therefore employ aluminum-copper alloys rather than aluminum-copper-silicon, because aluminum-silicon alloying cannot occur.

2.6.3. Silicidation

Another modification of the standard metallization flow involves the addition of a silicide. Elemental silicon reacts with many metals, including platinum, palladium, titanium, and nickel, to form compounds of definite composition. These *silicides* can form both low-resistance Ohmic contacts and, in the case of certain silicides, stable rectifying Schottky barriers. Thus silicidation not only improves contact resistance—which can be a problem with barrier metal systems—but also allows the formation of Schottky diodes at no extra cost. Silicides have much lower resistivities than even the most heavily doped silicon, so they can also be used to reduce the resistance of selected silicon regions. Many MOS processes employ silicided poly (also called *clad poly*) to form the gates of high-speed MOS transistors. Some of these processes also clad the source/drain regions of the transistors to reduce their resistance. Since most silicides are relatively refractory, their deposition does not preclude subsequent high-temperature processing. Silicided gates can thus be used to form self-aligned source/drain regions.

Figure 2.28 shows the steps required to deposit a platinum silicide layer in selected regions of the wafer. Immediately after the contacts are opened, a thin film of

FIGURE 2.28 Silicidation process, showing both silicided contacts and silicided poly.

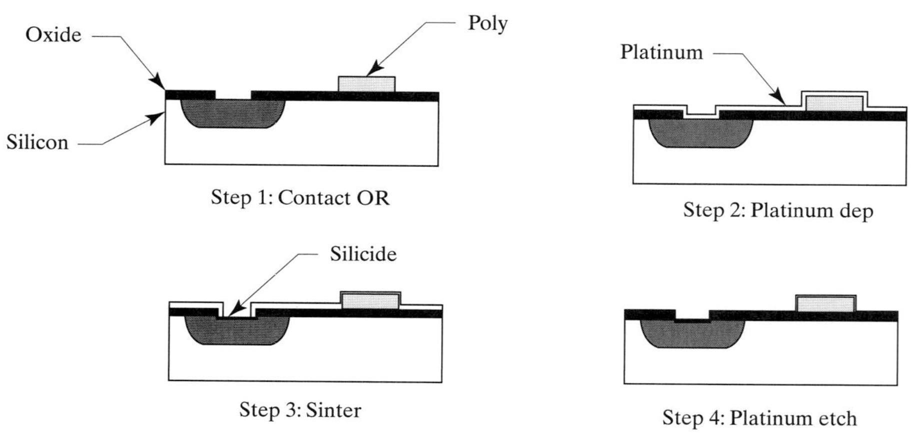

platinum metal is deposited across the entire wafer. The wafer is then heated to cause the portions of the platinum film in contact with silicon to react to form platinum silicide. The unreacted platinum can be removed using a mixture of acids called aqua regia. This procedure silicides both contact openings and any exposed polysilicon. If desired, an additional masking step can select which regions should receive silicide. Processes employing clad poly must incorporate a silicide block mask to fabricate polysilicon resistors. If this were not done, silicidation would turn all of the poly into a low-resistance material.

A typical silicided metal system consists of a lowermost layer of platinum silicide, an intermediate layer of refractory barrier metal,[24] and a topmost layer of copper-doped aluminum. The resulting sandwich exhibits low electrical resistance, high electromigration immunity, stable contact resistance, and precisely controlled alloying depths. The three layers required to obtain all of these benefits are more costly than a simple aluminum alloy metallization, but the performance benefits are substantial.

2.6.4. Interlevel Oxide, Interlevel Nitride, and Protective Overcoat

Figure 2.29 shows a cross section of a typical modern metallization system. The first layer of material above the silicon consists of thermally grown oxide. Upon this oxide lies a patterned polysilicon layer that will eventually form the gates of MOS transistors. On top of this poly lies a thin deposited oxide layer called a *multilevel oxide* (MLO) that serves to insulate the poly and to thicken the thermal oxide layer. Contact openings are etched through the MLO and thermal oxide to contact the silicon, and through the MLO to contact the poly. Following reflow, the contact openings are silicided to reduce contact resistance. Above the MLO lies the first layer of metal, consisting of a thin film of refractory barrier metal and a much thicker layer of copper-doped aluminum. Above the first metal layer lies another deposited oxide layer called an *interlevel oxide* (ILO), which insulates the first metal from the overlying second metal. Vias are etched through the ILO. On top of this lies the second layer of metal, again consisting of refractory barrier metal and copper-doped aluminum. The topmost and final layer consists of a compressive nitride film, which serves as a *protective overcoat* (PO). This metallization system has a total of six layers (one poly, two metals, MLO, ILO, and PO) and requires five masking steps (poly, contact, metal-1, via, metal-2, and PO). Some advanced processes may employ as many as three layers of polysilicon and five layers of metal.

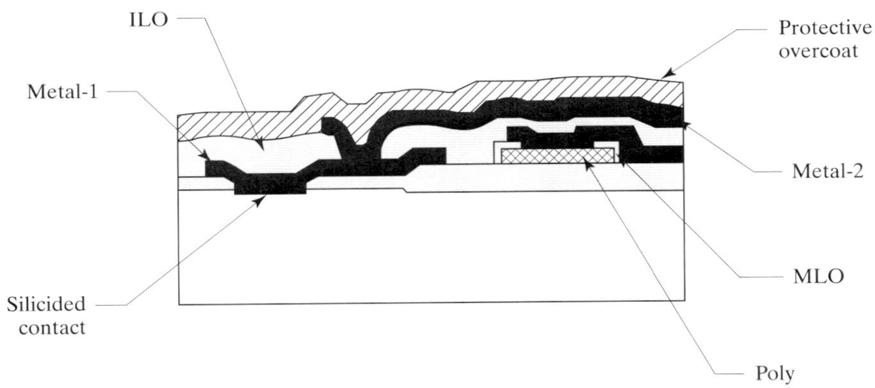

FIGURE 2.29 Cross section of a double-metal, single-poly metallization system.

[24] The addition of a refractory barrier metal prevents the platinum silicide from reacting with the aluminum. This is not required for most refractory silicides; see Sze, pg. 409.

Interlevel oxide layers are normally produced by low-temperature deposition—for example, by the reaction of silane and nitrous oxide or by the decomposition of tetraethoxysilane (TEOS). A relatively thick ILO helps minimize parasitic capacitances between layers of the conductor sandwich, but it can cause step coverage problems in via openings. As previously discussed, reflow is not possible once aluminum has been deposited, so a refractory barrier metal is often used to improve the step coverage of the second metal layer.

An excellent capacitor can be formed between two layers of metal or polysilicon. A thin insulating dielectric deposited between the plates completes the capacitor. The thinner this dielectric, the greater the resulting capacitance per unit area. One technique for forming a capacitor consists of depositing one polysilicon layer, oxidizing this to form a thin dielectric, and depositing a second polysilicon layer to complete the capacitor. Any region where the two poly layers overlap will form a capacitor consisting of two poly plates separated by the thin oxide dielectric. Oxide forms an ideal capacitor dielectric because it is a nearly perfect insulator, and very thin oxide films can be grown with little risk of pinholes or other defects. The capacitance achievable with oxide dielectrics is limited by the rupture voltage of the oxide; the thicker oxide layers required to withstand higher voltages have proportionately smaller capacitances per unit area.

One way to boost the capacitance per unit area for a given operating voltage consists of using a material with a higher dielectric constant. Silicon nitride, with a dielectric constant that is 2.3 times that of oxide, is a common choice for fabricating high capacitance-per-unit-area films. Unfortunately, nitride films are more prone to pinhole formation than are oxide films of equivalent thickness. Therefore oxide and nitride films are sometimes combined to form a stacked dielectric with a dielectric constant between that of oxide and nitride. A typical oxide-nitride-oxide stacked dielectric can achieve a dielectric constant about twice that of oxide.

The protective overcoat consists of a thick deposited oxide or nitride film coating the entire integrated circuit. It insulates the uppermost metal layer from the outside world, so that (for example) a particle of conductive dust will not short two adjacent leads. The overcoat also helps to ruggedize the integrated circuit, a necessary precaution since the aluminum metallization is soft and deforms under pressure. The protective overcoat also helps to block the entrance of contaminants. Both the aluminum metallization and the underlying silicon are vulnerable to certain types of contaminants that can penetrate the plastic encapsulation. A properly formulated protective overcoat can form a barrier to these contaminants. Heavily doped phosphosilicate glasses are sometimes used as protective overcoats, but most modern processes have switched to compressive nitride films, which offer superior mechanical hardness and chemical resistance.

2.7 ASSEMBLY

Wafer fabrication ends with the deposition of a protective overcoat, but there remain a number of manufacturing steps required to complete the integrated circuits. Since most of these steps require less-stringent cleanliness than wafer fabrication, they are usually performed in a separate facility called an *assembly/test site.*

Figure 2.30 shows a diagram of a typical finished wafer. Each of the small squares on the wafer represents a complete integrated circuit. This wafer contains approximately 300 integrated circuit dice arrayed in a rectangular pattern by the step-and-repeat process that created the stepped working plates. A few locations in the array are occupied by process control structures and test dice rather than actual integrated circuits.

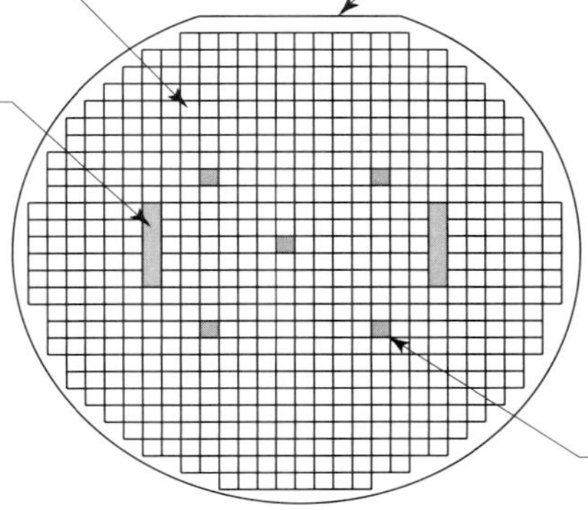

FIGURE 2.30 Pattern for a typical wafer created from a stepped working plate.

Process control structures consist of extensive arrays of transistors, resistors, capacitors, and diodes, as well as more specialized structures such as strings of contacts and vias. The wafer fab uses these structures to evaluate the success or failure of the manufacturing process. Automated testing equipment gathers data on every wafer, and any that fail to meet specifications are discarded. The data are also analyzed for statistical trends so that corrections can be implemented before the variances become large enough to cause yield losses. The process control structures are standardized, and the same ones are used for a wide range of products.

Test dice are used by design engineers to evaluate prototypes of an integrated circuit. Unlike process control structures, test dice are specific to a given product and in most cases are actually variations on the layout of the integrated circuit. A dedicated test metal mask allows probing of specific components and subcircuits that would be difficult to access on the finished die. Sometimes a test contact or protective overcoat mask is also used, but in almost every case the test die shares the same diffusion masks as the integrated circuit. Test dice are normally created by adding a few more layers (e.g., test metal, test nitride) to the database containing the layout of the integrated circuit. These layers create a separate set of reticles that are used to expose a few selected spots on the stepped working plate. The wafer in Figure 2.30 contains only five test die locations. These locations become unnecessary when testing has been completed. Sometimes a new set of masks is created that replaces the test dice with product dice to gain an extra percent or two of yield. In other cases, the tiny increase in die yield cannot justify fabrication of new masks, so the test dice remain on production material.

Figure 2.30 depicts a wafer produced from a set of stepped working plates. Wafers created by *direct-step-on-wafer* (DSW)[25] processing rarely include any test dice because at least one test die must be included in every exposure. This would result in twenty or more test dice per wafer, which would consume a corresponding amount of area. If test dice are included in a DSW design, then the production mask set will almost certainly be modified to replace them with product dice to improve the die yield.

As mentioned previously, all completed wafers are tested to determine whether the processing was performed correctly. If the wafers pass this test, then each die is

[25] The acronym *DSW* has also been used to stand for *direct slice write*, a process by which a scanned electron beam directly exposes the photoresist on a wafer. This process, more commonly called *direct-write-on-wafer* (DWW), is strictly of academic interest because it is too slow to have any practical application in silicon processing. However, it is frequently used to fashion photomasks.

individually tested to determine its functionality. The high-speed automated test equipment typically requires less than three seconds to test each die. The percentage of good dice depends on many factors, most notably the size of the die and the complexity of the process used to create it. Most products yield better than 80% and some yield in excess of 90%. High yields are obviously desirable because every discarded die represents lost profit. The equipment that tests the wafers also marks those that fail the test. Marking is usually done by placing a drop of ink on each defective die, but some modern systems eliminate the need for inking by remembering the location of the bad dice electronically.

Wafer-level testing, or *wafer probing,* requires contact to specific locations on the interconnection pattern of the integrated circuit. These locations are exposed through holes in the protective overcoat, allowing contact to be made with the help of an array of sharp metal needles, or *probes.* These probes are mounted on a board called a *probe card.* The automated test machine lowers the probe card until electrical continuity is established. The integrated circuit is tested, the card is lifted, and positioning servos move the wafer to align the next die underneath the probes.

Once the wafer has been completely tested, the individual dice are sawn apart using a diamond-tipped sawblade. Another automated system then selects the good dice from the scribed wafer for mounting and bonding. The rejected dice (including the remains of the process control structures and the test dice) are discarded.

2.7.1. Mount and Bond

Many manufacturers now offer unmounted integrated circuit dice, but the sales of such *bare dice* are seldom large. Most customers simply do not have the equipment or expertise needed to handle bare dice, let alone to package them. Packaging therefore falls in the province of integrated circuit manufacturing.

The first step in packaging an integrated circuit is mounting it on a *leadframe.* Figure 2.31 shows a somewhat simplified diagram of a leadframe for an eight-pin *dual-in-line package* (DIP), complete with a chip mounted on it. The leadframe itself consists of a rectangular *mount pad* that holds the die and a series of lead fingers that will eventually be trimmed to form the eight leads of the DIP. Leadframes usually come in strips, so several dice can be handled as a single assembly. They are either stamped out of thin sheets of metal, or they are etched using photographic techniques similar to those used to pattern printed circuit boards. The lead frame usually consists of copper or a copper alloy, often plated with tin or a tin-lead alloy. Copper is not an ideal material for leadframes because it has a different coefficient

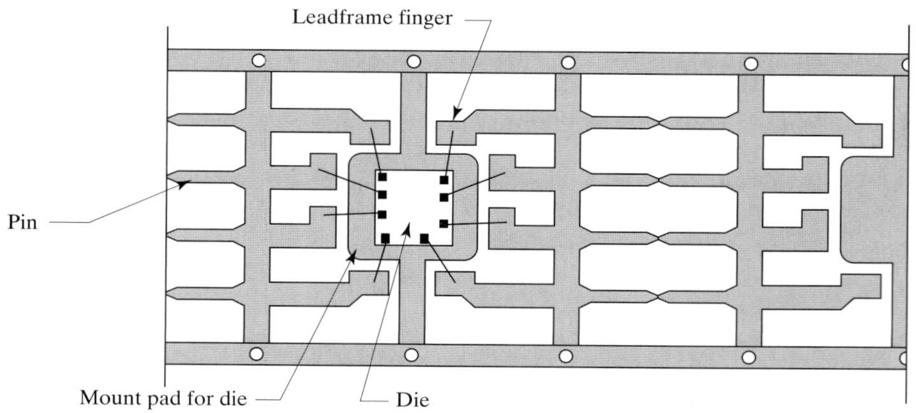

FIGURE 2.31 Simplified diagram of a leadframe for an 8-pin DIP.

of thermal expansion than silicon. As the packaged part is heated and cooled, differential expansion of the die and the leadframe sets up mechanical stresses injurious to the performance of the die. Unfortunately, most of the materials that possess coefficients of expansion similar to silicon have inferior mechanical and electrical properties. Some of these materials are occasionally used for low-stress packaging of specialty parts; a nickel-iron alloy called *Alloy-42* is probably the most commonly encountered (Section 7.2.6).

The die is usually mounted to the leadframe using an epoxy resin. In some cases, the resin may be filled with silver powder to improve thermal conductivity. Epoxy is not entirely rigid, and this helps reduce the stresses produced by thermal expansion of the leadframe and die. Alternate methods exist that provide superior thermal union between the silicon and the leadframe, but at the cost of greater mechanical stress. For example, the backside of the die can be plated with a metal or metal alloy and soldered to the leadframe. Alternatively, a rectangle of gold foil called a *gold preform* can be attached to the leadframe; heating the die causes it to alloy with the gold preform to create a solid mechanical joint. Solder connections and gold preforms both allow excellent thermal contact between the die and the leadframe. Both also produce an electrical connection that can be used to connect the substrate of the die to a pin. Conductive epoxies improve thermal conductivity, but they cannot always be trusted to provide electrical connectivity.[26]

After the dice are mounted on leadframes, the next step is attaching bondwires to them. Bonding can only be performed in areas of the die where the metallization is exposed through openings in the protective overcoat; these locations are called *bondpads.* The probe card used for wafer probing makes contact to the bondpads for purposes of testing, but the probes may also make contact to a few pads that will not receive bondwires. Those pads reserved for testing purposes are usually called *testpads* to distinguish them from actual bondpads. Testpads are often made smaller than bondpads since probe needles can usually land in a smaller zone than can bondwires.

Bonding is performed by high-speed automated machines that use optical recognition to determine the locations of the bondpads. These machines typically employ one-mil (25μm) gold wire for bonding, although gold wires as small as 0.8 mil or as large as 2.0 mil are in common use. Aluminum wires up to ten mils in diameter can also be employed, although these require rather different bonding machinery. Only one diameter and type of wire can be bonded at a time, so few dice use more. The most common arrangement consists of one-mil gold bondwires on all pads. Multiple one-mil wires bonded in parallel can carry higher currents or provide lower resistances without requiring a second bonding pass for larger-diameter wire.

The most common technique for bonding gold wire is called *ball bonding.*[27] Since aluminum wire cannot be ball-bonded, an alternate technique called *wedge bonding* has been developed for it. Figure 2.32 shows the essential steps of the ball-bonding process.

The bonding machine feeds the gold wire through a slender tube called a *capillary.* A hydrogen flame melts the end of the wire to form a small gold sphere, or *ball* (Figure 2.32, Step 1). Once a ball has been formed, the capillary presses down against the bondpad. The gold ball deforms under pressure, and the gold and aluminum fuse together to form a weld (Step 2). Next, the capillary lifts and moves to the vicinity of

---

[26] R. L. Opila and J. D. Sinclair, "Electrical Reliability of Silver Filled Epoxies for Die Attach," *23rd International Reliability Physics Symp.*, 1985, pp. 164–172.

[27] B. G. Streetman, *Solid State Electronic Devices,* 2nd ed. (Englewood Cliffs, NJ: Prentice-Hall, 1980), pp. 368–370.

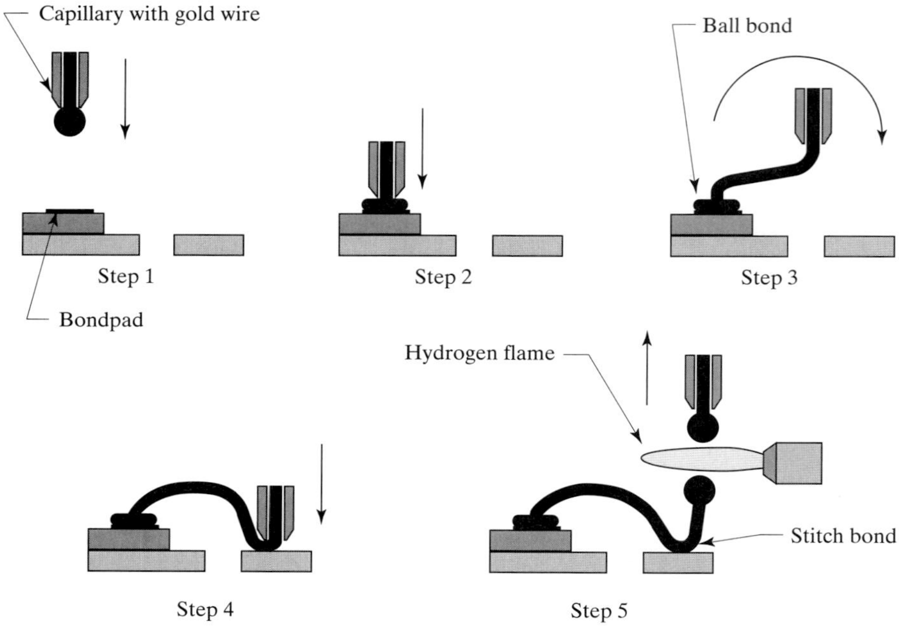

FIGURE 2.32 Steps in the ball-bonding process.

the lead finger (Step 3). The capillary again descends, smashing the gold wire against the lead finger. This causes the gold to alloy to the underlying metal to produce a weld (Step 4). Since no ball is present at this point, the resulting bond is called a *stitch bond* rather than a ball bond. Finally, the capillary lifts up from the lead finger and the hydrogen flame passes through the wire, causing it to fuse into two (Step 5). The bond is now complete and another ball has been formed on the wire protruding from the capillary, allowing the process to be repeated. An automated bonding machine can perform these steps ten times a second with great precision. The extreme speed and unerring accuracy of these machines produce huge economies of scale, and the entire bonding process costs no more than a penny or two.

Aluminum wire cannot be ball-bonded because the hydrogen flame would ignite the fine aluminum wire. Instead, a small, wedge-shaped tool is used to supplement the capillary. When the capillary brings the wire into proximity with the bondpad, the tool smashes it against the pad to create a stitch bond. The process is repeated at the lead finger, and the tool is then held down against the lead finger while the capillary moves up. The tension created in the aluminum wire snaps it at its weakest point, immediately adjacent to the weld. The process can then be repeated as many times as necessary.

The ball-bonding process requires a square bondpad approximately three times as wide as the diameter of the wire. Thus a one-mil gold wire can be attached to a square bondpad about three mils across. Wedge bonding is more selective, and usually requires bondpads that are rectangular in shape. These bondpads must lie in the same direction as the wedge tool. They are typically twice as wide and four times as long as the wire is thick. The exact rules for wedge bondpads can become quite complex, particularly for thicker aluminum wires.

Figure 2.31 shows a die mounted on a leadframe after the bonding process is complete. The bondwires connect various bondpads to their respective leads. Although the wires are quite small compared to the pins, each is still capable of carrying an amp of current.

2.7.2. Packaging

The next step in the assembly process is injection molding. A mold is clamped around the leadframe and heated plastic resin is forced into the mold from below. The plastic wells up around the die, lifting the wires away from it in gentle loops. Injection from the side or from the top usually smashes the wires against the integrated circuit and is therefore impractical. The plastic resin employed for integrated circuits cures rapidly at the temperatures used in molding and, once cured, it forms a rigid block of plastic.

When the molding process is complete, the leads are trimmed and formed to their final shapes. This is done in a mechanical press using a pair of specially shaped dies that simultaneously trim away the links between the individual leads and bend them to the required shape. Depending on the material of the leadframe, solder dipping or plating may be required to prevent oxidation and contamination of the pin surfaces. The completed integrated circuits are now labeled with part numbers and other designation codes (these usually include a code identifying the date of manufacture and the lot number). The completed integrated circuits are tested again to ensure that they have not been damaged by the packaging process. Finally, the completed devices are packaged in tubes, trays, or reels for distribution to customers.

2.8 SUMMARY

Modern semiconductor processing takes advantage of the properties of silicon to manufacture inexpensive integrated circuits in huge volumes. Photolithography allows the reproduction of intricate patterns hundreds or thousands of times across each wafer, leading to enormous economies of scale.

Junctions can be formed by one of three means: epitaxial deposition, diffusion, or ion implantation. Low-pressure chemical-vapor-deposited (LPCVD) epitaxial layers can produce extremely high-quality silicon films with precisely controlled dopant concentrations. Diffusion of dopants from a surface source allows the formation of vast numbers of junctions using only a single photolithographic masking step. Ion implantation allows similar but more costly patterning of junctions with superior control of doping levels and distributions.

Many materials can also be deposited on the surface of the wafer. These include polycrystalline silicon (poly), oxide, nitride, and any of numerous metals and metal alloys. Typical semiconductor processes combine several diffusions into the bulk silicon with several depositions of materials onto the resulting wafer. The next chapter examines how the various techniques of semiconductor fabrication are combined to manufacture three of the most successful integrated circuit processes.

2.9 EXERCISES

2.1. When pressure is applied to the center of an unknown wafer, it splits into six segments. What can be definitely concluded from this observation? What may be reasonably conjectured?

2.2. Draw a diagram similar to that in Figure 2.4 illustrating the relationship between the (100) and (110) planes of a cubic crystal (refer to Appendix B, if necessary).

2.3. Suppose a photomask consists of a single opaque rectangle on a clear background. A negative resist is used in combination with this mask to expose a sensitized wafer. Describe the pattern of photoresist left on the wafer after development.

2.4. Suppose a wafer is subjected to the following processing steps:

a. Uniform oxidation of the entire wafer surface.

b. Opening of an oxide window to the silicon surface.

c. An additional period of oxidation.
d. Opening of a smaller oxide window in the middle of the region patterned in step b.
e. An additional period of oxidation.

Draw a cross section of the resulting structure, showing the topography of both the silicon and the oxide surfaces. The drawing need not be made to scale.

2.5. Suppose a wafer is uniformly doped with equal concentrations of boron and phosphorus atoms. After a prolonged oxidation, will the surface of the silicon be N-type or P-type? Why?

2.6. Suppose that a wafer is uniformly doped with 10^{16} atoms/cm^3 of boron. This wafer is then subjected to the following processing steps:

a. Uniform oxidation of the entire wafer surface.
b. Opening of an oxide window to the silicon surface.
c. Deposition of boron and phosphorus, each at a source concentration of 10^{20} atoms/cm^3, using a fifteen-minute deposition and a one-hour drive at 1000°C.

Assuming that the two dopants do not interact with each other or with the oxide, draw a cross section of the resulting structure. Indicate the approximate depths of any junctions formed.

2.7. Phosphorus is diffused into a lightly doped wafer through an oxide window 5μm square. If the resulting junction is 2μm deep, then what is the width of the phosphorus diffusion at the surface?

2.8. Most ion implantation systems position the accelerator so that the ion beam impacts the wafer surface at a slight angle (often 7°). Explain the reason for this feature.

2.9. If the surface of the oxide layer covering a wafer is ground perfectly smooth, different regions of the wafer still exhibit different colors, but the NBL shadow vanishes. Explain these observations.

2.10. Draw a cross section of the following metallization system:

a. 1μm-wide contacts through 5kÅ oxide silicided with 2kÅ platinum silicide.
b. First-level metal consisting of 2kÅ RBM and 6kÅ copper-doped aluminum.
c. 1μm-wide vias 3kÅ deep through highly planarized ILO.
d. Second-level metal consisting of 2kÅ RBM and 10kÅ copper-doped aluminum.
e. 10kÅ protective overcoat.

Assume that the silicide surface is level with the surrounding silicon surface, and that the aluminum metal thins 50% on the sidewalls of the contacts and vias. The drawing should be made to scale.

2.11. Suppose that a die measures 60 by 80 mils, where one mil is one thousandth of an inch. Approximately how many of these dice could be fabricated on a 150mm-diameter wafer? Assuming that 70% of these potential dice actually work, and that a finished wafer costs $250, compute the cost of each functional die.

3 Representative Processes

Semiconductor processing has evolved rapidly over the past fifty years. The earliest processes produced only discrete components, mainly switching diodes and bipolar transistors. The first practical integrated circuits appeared in 1960.[1] These consisted of a few dozen bipolar transistors and diffused resistors connected to form simple logic gates. By modern standards, these early integrated circuits were terribly slow and inefficient. Refinements were soon made, and by the mid-1960s bipolar integrated logic offered clear advantages over discrete logic. The first analog integrated circuits appeared at about the same time; these consisted of matched transistor arrays, operational amplifiers, and voltage references. The standard bipolar process that was created to support these products remains in use today.

Integrated bipolar logic was fast but power-hungry. MOS integrated circuits held out the promise of a low-power alternative, but the metal-gate MOS processes of the 1960s suffered from unpredictable threshold voltage shifts. This problem was eventually conquered through the development of polysilicon-gate MOS processes in the early 1970s. MOS logic soon replaced bipolar logic and created vast new markets for microprocessors and dynamic RAM chips. Analog CMOS circuits of this era touted greatly reduced operating currents but provided only mediocre performance, so standard bipolar remained the process of choice for high-performance analog integrated circuits.

By the mid-1980s, customers were demanding the integration of both digital and analog functions onto a single mixed-signal integrated circuit. A new generation of merged bipolar-CMOS (BiCMOS) processes were soon developed specifically for mixed-signal design. Although these processes are complex and costly, they offer a level of performance unachievable by other means. The world of analog integrated circuits is dominated by these three processes: standard bipolar, polysilicon-gate CMOS, and analog BiCMOS. This chapter will analyze the implementation of a representative process of each type.

---

1 J. S. Kilby, "Invention of the Integrated Circuit," *IEEE Trans. on Electron Devices,* Vol. ED-23, #7, 1976, pp. 648–654.

3.1 STANDARD BIPOLAR

Standard bipolar was the first analog integrated circuit process. Over the years, it has produced many classic devices, including the 741 operational amplifier, the 555 timer, and the 431 voltage reference. Even though these parts represent thirty-year-old technology, they are still manufactured in vast quantities today.

Standard bipolar is seldom used for new designs. CMOS offers lower supply currents, BiCMOS provides superior analog performance, and various advanced bipolar processes yield faster switching speeds. But the knowledge gained through first developing and then refining standard bipolar will never become obsolete. The same devices reappear in every new process, along with many of the same parasitic mechanisms, design tradeoffs, and layout principles. This chapter will therefore begin with an overview of standard bipolar.

3.1.1. Essential Features

Standard bipolar was shaped by a conscious decision to optimize the NPN transistor at the expense of the PNP. This decision rested on the observation that the NPN transistor employs electron conduction while the PNP transistor relies on hole conduction. The lower mobility of holes reduces both the beta and the switching speed of PNP transistors. Given equivalent geometries and doping profiles, an NPN will outperform a PNP by more than 2:1. Several additional processing steps are required to optimize both types of transistors simultaneously, so early processes optimized NPN transistors and avoided PNP transistors altogether. This decision met the requirements of bipolar logic, consisting as it does of NPN transistors, resistors, and diodes. When analog circuits were first constructed using standard bipolar, several types of PNP transistors were cobbled together from existing process steps. Although these transistors performed relatively poorly, they sufficed to design many useful circuits.

The standard bipolar process employs *junction isolation* (JI) to prevent unwanted currents from flowing between devices that are formed on the same substrate.[2] The components reside in a lightly doped N-type epitaxial layer deposited on top of a lightly doped P-type substrate (Figure 3.1). A deep-P+ *isolation diffusion* driven down to contact the underlying substrate provides isolation between components. Regions of N-epi separated from one another by isolation are called *tanks* or *tubs*. If the isolation is biased to a potential equal to or below the lowest-voltage tank, then reverse-biased junctions surround every tank. The substrate forms the floor of these tanks, and isolation diffusions form their sidewalls.

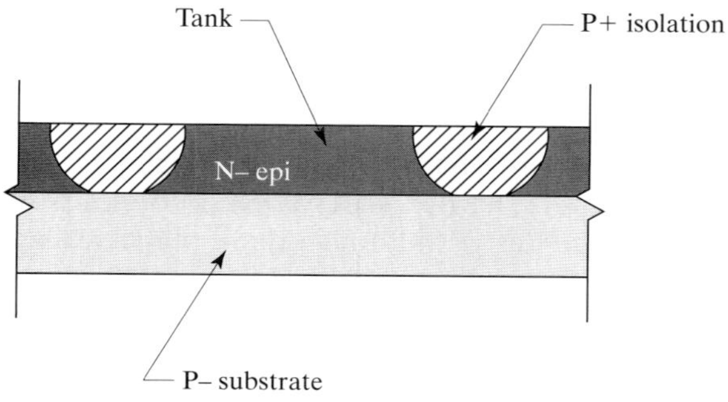

FIGURE 3.1 Cross section of the junction isolation system employed for standard bipolar.

---

[2] R. N. Noyce, U.S. Patent #2,981,877, 1961.

Junction isolation has several significant drawbacks. The reverse-biased isolation junctions exhibit enough capacitance to slow the operation of many circuits. High temperatures can cause significant leakage currents, as can exposure to light or ionizing radiation. Unusual operating conditions can also forward bias the isolation junctions and inject minority carriers into the substrate. Despite these difficulties, junction-isolated processes can successfully fabricate most circuits. Junction isolation is also considerably cheaper than any of its alternatives.

3.1.2. Fabrication Sequence

The baseline standard bipolar fabrication sequence consists of eight masking operations. The significance of each step can be illustrated best by presenting the entire flow from starting material to finished wafer. Representative cross sections will be used to illustrate each step. When examining these cross sections (and all others in this text), keep in mind that the vertical scale of the drawings has been exaggerated by a factor of two to five for clarity. The lateral dimensions of a typical integrated device are so much greater than its vertical dimensions that a true-scale diagram would be virtually illegible. The cross sections therefore exaggerate the vertical scale by a factor of two to five. The substrate is also much thicker than depicted; the additional silicon serves to strengthen the wafer against warping and breakage.

Starting Material

Standard bipolar integrated circuits are fabricated on a lightly doped (111)-oriented P-type substrate. The wafers are usually cut several degrees off-axis to minimize distortion of the NBL shadow (*pattern distortion*).[3] The use of (111) silicon helps suppress a parasitic PMOS transistor that is inherent in the standard bipolar process. The N-epi forms the backgate of this parasitic, while a lead crossing the field oxide above the tank acts as its gate electrode. A base region within the tank forms the source, and the drain consists of the P+ isolation (Figure 3.2). When the base diffusion is biased to a high voltage relative to the metal lead, a channel forms and allows current to flow from the base to the isolation. The threshold voltage of an MOS transistor formed under thick field oxide is called a *thick-field threshold*. The use of (111) silicon artificially elevates the PMOS thick-field threshold by introducing positive surface-state charges along the oxide-silicon interface.

N-Buried Layer

The first processing step consists of growing a thin layer of oxide across the wafer. Photoresist spun onto this oxide is patterned using the N-buried layer (NBL) mask.

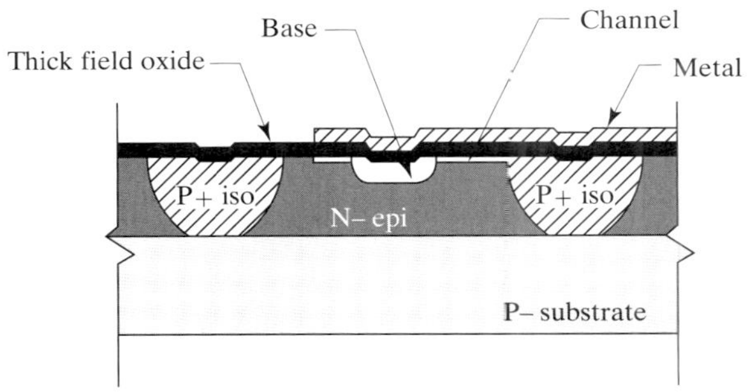

FIGURE 3.2 Parasitic PMOS formation in standard bipolar.

[3] W. R. Runyan and K. E. Bean, *Semiconductor Integrated Circuit Processing Technology* (Reading, MA: Addison-Wesley, 1994), p. 331.

After an oxide etch opens windows to the silicon surface, ion implantation or thermal deposition transfers an N-type dopant into the wafer. The N-buried layer customarily consists of either arsenic or antimony because the low diffusivities of these elements limit up-diffusion during subsequent processing. The brief drive following the deposition serves two purposes: first, it anneals out lattice damage; and second, it grows a small amount of oxide that forms a slight discontinuity at the silicon surface (Figure 3.3). This discontinuity will later produce an NBL shadow to which other masks can align.

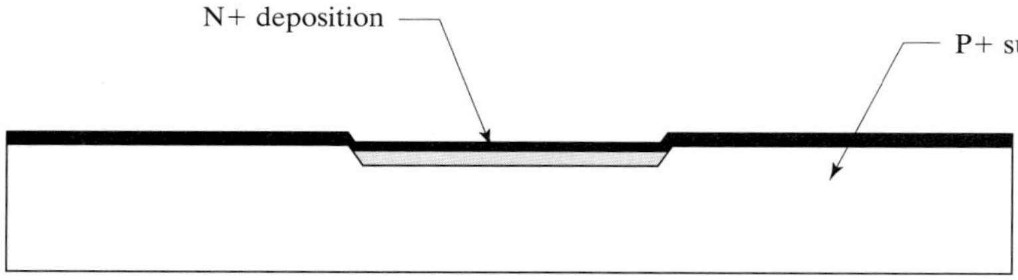

FIGURE 3.3 Wafer after anneal of NBL implant.

Epitaxial Growth

The oxide layer that remains on the wafer is stripped prior to the growth of some 10 to 25μm of lightly doped N-type epi. Surface discontinuities propagate upward during epitaxial deposition at approximately a 45° angle along the axis. Upon completion of epitaxial growth, the NBL shadow will have shifted laterally a distance approximately equal to the thickness of the epi (Figure 3.4).

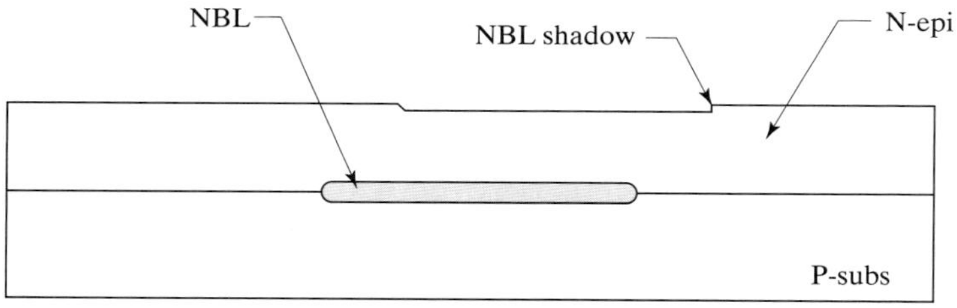

FIGURE 3.4 Wafer after epitaxial deposition. Note the pattern shift exhibited by the NBL shadow.

Isolation Diffusion

The wafer is again oxidized, coated with photoresist, and patterned using the *isolation* mask. This mask must be aligned to the NBL shadow using a deliberate offset to correct for pattern shift. A heavy boron deposition followed by a high-temperature drive forces the isolation diffusion partway down through the epi layer. Oxidation also occurs during this drive, covering the isolation windows with a thin layer of thermal oxide. The drive stops before the isolation junction reaches the substrate, since later high-temperature processing steps (primarily the deep-N+ drive) will force the diffusion the rest of the way down. Figure 3.5 shows the wafer after this partial drive.

Deep-N+

A deep-N+ diffusion (sometimes called a *sinker*) allows low-resistance connection to the NBL. First a photoresist is applied and patterned using the deep-N+ mask. A heavy phosphorus deposition followed by a high-temperature drive forms the deep-N+ sinkers. The drive not only causes the deep-N+ to diffuse down to meet the upward-diffusing NBL but also completes the isolation drive. Sufficient time is

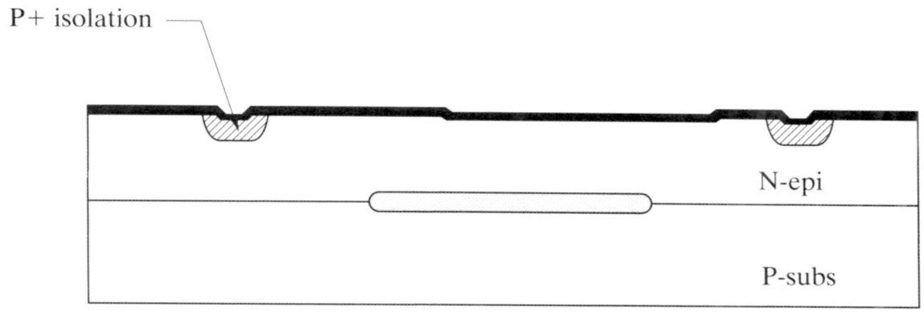

FIGURE 3.5 Wafer after isolation deposition and partial drive.

allowed to overdrive the junctions by about 25%. Without this overdrive, the bottom of the isolation and deep-N+ diffusions would be very lightly doped. The overdrive simultaneously reduces the vertical resistance through both the isolation and the deep-N+ sinkers. The deep-N+ drive also forms the thick field oxide.

Both deep-N+ and isolation diffusions approach their final junction depths during the deep-N+ drive (Figure 3.6). These junctions will diffuse slightly deeper during subsequent processing, but all of the later diffusions are fairly shallow compared to deep-N+ and isolation, and therefore the tank regions appear fully formed in Figure 3.6. The NBL regions are normally spaced some distance inside the isolation diffusion to increase the tank breakdown voltage. Otherwise the N+/P+ junction formed by the intersection of NBL and isolation would avalanche at about 30V.

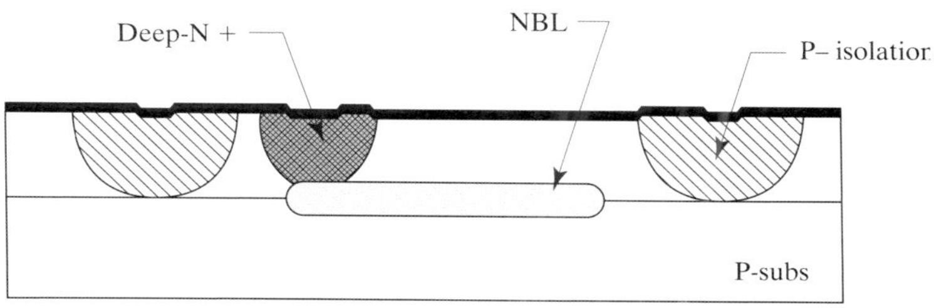

FIGURE 3.6 Wafer after isolation drive.

Base Implant

Next, photoresist spun onto the wafer is patterned using the base mask. An oxide etch clears windows through the field oxide to the silicon surface. A light boron implant conducted through these openings counterdopes the N-epi to form the base regions of the NPN transistors. Ion implantation allows precise control of base doping and thus minimizes process-derived beta variation. The subsequent drive anneals implant damage and sets the base junction depth. Oxide grown during this drive serves as a mask for the subsequent emitter deposition. Base is also implanted across the isolation regions to increase surface doping. This practice, called *base-over-isolation* (BOI), substantially increases the NMOS thick-field threshold without requiring the use of a separate channel stop. Figure 3.7 shows a cross section of the wafer following the base drive.

Emitter Diffusion

The wafer is again coated with photoresist and patterned using the emitter mask. A subsequent oxide etch exposes the silicon surface in regions where NPN emitters will form and in regions where Ohmic contact must be made to the N-epi or the deep-N+ diffusion. A very concentrated phosphorus deposition forms the emitter.

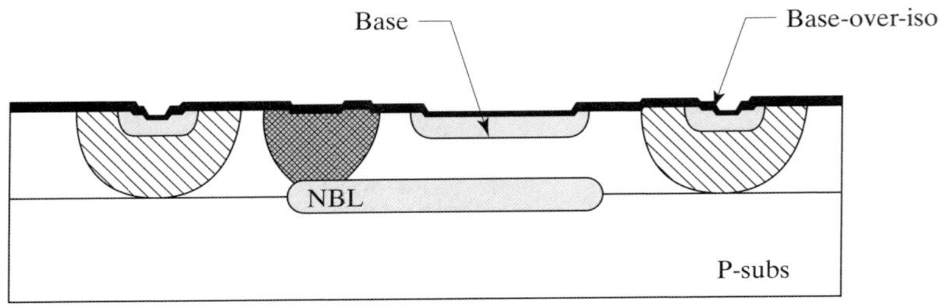

FIGURE 3.7 Wafer after base drive.

A $POCl_3$ source is often used since precise control of emitter doping is unnecessary. A brief drive sets the final emitter junction depth and thereby determines the width of the active base region of the NPN transistors.

An oxide film grown over the emitter diffusion insulates it from subsequent metallization. Some processes employ dry oxidation for this step, but the short oxidation time results in a *thin emitter oxide* vulnerable to electrostatic discharges (Section 4.1.1). Alternatively, a wet oxidation can grow a *thick emitter oxide* that possesses a higher rupture voltage. Figure 3.8 shows a cross section of the wafer after the emitter drive.

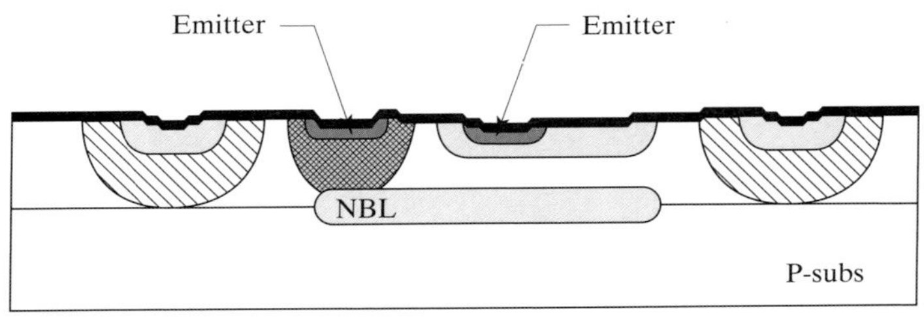

FIGURE 3.8 Wafer after emitter drive.

Many older processes also incorporate an *emitter pilot* step to provide a means of adjusting NPN beta. A dummy wafer that is inserted into the wafer lot before base implant and removed after emitter deposition is used to conduct an experimental emitter drive. By monitoring the performance of the base-emitter junction formed on the pilot wafer, the actual drive can be adjusted on a lot-by-lot basis to target the desired NPN beta.

Contact

All diffusions are now complete. The remaining steps form the metallization and apply the protective overcoat. The first step in this sequence forms contacts to selected diffusions. The wafer is again coated with photoresist, patterned using the contact mask, and etched to expose bare silicon. This process is sometimes called *contact OR,* in which OR stands for *oxide removal.*

Metallization

A layer of aluminum-copper-silicon alloy is evaporated or sputtered across the entire wafer. This metal system typically incorporates 2% silicon to suppress emitter punchthrough and 0.5% copper to improve electromigration resistance. Standard bipolar employs relatively thick metallization, typically at least 10kÅ (1.0μm) thick, to reduce interconnection resistance and decrease vulnerability to electromigration. The metallized wafer is patterned using the metal mask and etched to form the interconnection system.

Protective Overcoat

Next, a thick layer of *protective overcoat* (PO) is deposited across the entire wafer. Compressive nitride protective overcoats provide excellent mechanical and chemical protection. Some processes use a *phosphosilicate-doped glass* (PSG) layer either beneath a compressive nitride or as a replacement for it. Since the deposition of the protective overcoat occurs at moderate temperatures, it also sinters the aluminum metallization.

Finally, a layer of photoresist is applied and patterned using the PO mask. A special etch opens windows through the protective overcoat to expose areas of metallization for bonding. This composes the final fabrication step; the wafer is now complete. Figure 3.9 shows a fully processed wafer (the illustrated cross section does not include a bondpad opening).

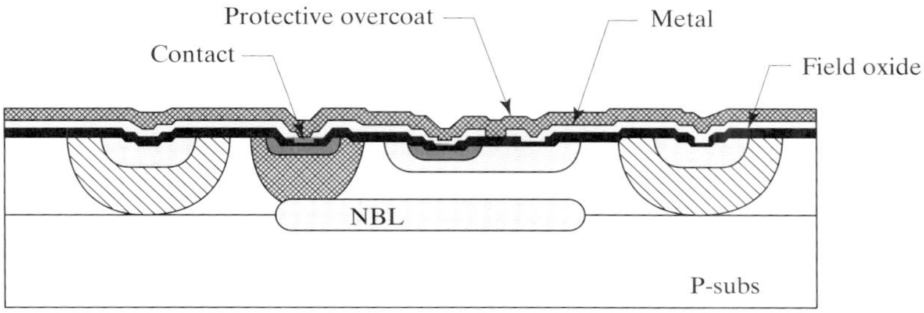

FIGURE 3.9 Completed standard bipolar wafer.

3.1.3. Available Devices

Standard bipolar was originally developed to provide bipolar NPN transistors and diffused resistors. A number of other devices can also be fabricated using the same process steps, including two types of PNP transistors, several types of resistors, and a capacitor.[4] These devices form a basic component set suitable for fabricating a wide variety of analog circuits. Section 3.1.4 will examine several additional devices that require extensions to the baseline process.

NOTE: The dimensions of standard bipolar devices are often specified in mils. A *mil* equals 0.001 inches. The relevant conversion factors are 1mil = 25.4μm, and $1\text{mil}^2 \cong 645\mu\text{m}^2$. Another obscure unit of measurement sometimes used to specify junction depths is the *sodium line*, which equals one-half of the wavelength of the sodium spectrum D-line (1 line = 0.295μm).[5]

NPN Transistors

Figure 3.10 shows a representative layout and cross section of a minimum-area NPN transistor. The collector of the NPN consists of an N-epi tank, and the base and emitter are fabricated by successive counterdopings. The carriers flow vertically from emitter to collector through the thin base region underneath the emitter diffusion. The difference between the base and emitter junction depths determines the effective base width. Since these dimensions are controlled entirely by diffusion processing, they are not subject to photolithographic misalignment, allowing a base width substantially smaller than the alignment tolerance. For example, a process with a 5μm minimum feature size can easily fabricate a 2μm base width.

---

[4] For a general overview of the standard bipolar process, see N. Doyle, "LIC Technology," *Microelectronics and Reliability*, Vol. 13, 1974, pp. 315–324.

[5] G. E. Anner, *Planar Processing Primer* (New York: Van Nostrand Reinhold, 1990), pp. 107–108.

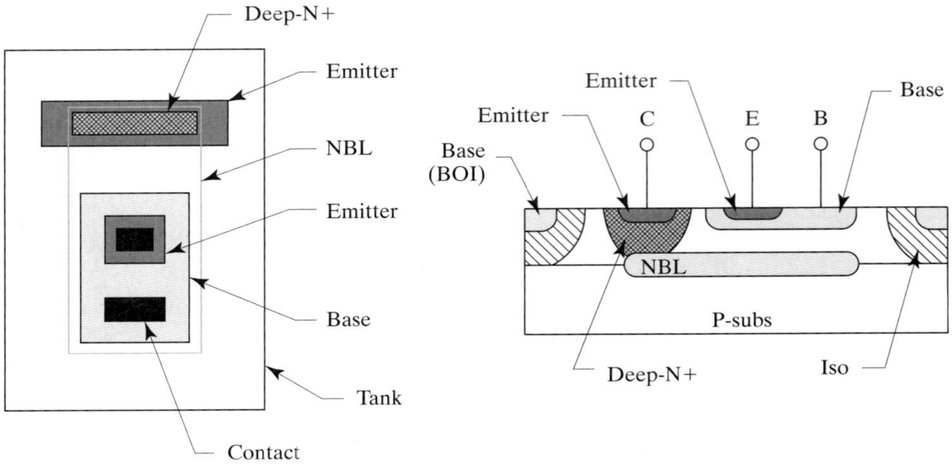

FIGURE 3.10 Layout and cross section of an NPN transistor with deep-N+ and NBL.[6]

The collector consists of lightly doped N-type epi lying on top of heavily doped NBL. The lightly doped epi allows the formation of a wide collector-base depletion region without excessive intrusion into the neutral base. This enables the transistor to support high operating voltages while simultaneously minimizing the Early effect (Section 8.2). NBL and deep-N+ create a low-resistance pathway to the portion of the epi layer beneath the transistor's active base. By this means, the collector resistance of a minimum NPN can be reduced to less than 100Ω and the collector resistance of a power NPN can be reduced to less than 1Ω.

The high concentration of donors in the NBL effectively halts the downward growth of the collector-base depletion region. The distance between the bottom of the base diffusion and the top of the NBL determines the maximum operating voltage of the NPN transistor. Thicker epi layers allow higher operating voltages, up to a limit set by the breakdown of the base diffusion sidewall (typically 50 to 80V). The maximum operating voltage of a bipolar process is usually specified in terms of the avalanche voltage of the NPN collector to emitter with base open (V_{CEO}). Depending on epi thickness and doping, this voltage can range from less than 10V to more than 100V.

The vertical NPN is the best device fabricated in the standard bipolar process. It consumes relatively little area and offers reasonably good performance. Circuit designers try to use as many of these transistors as possible. Table 3.1 lists typical device parameters for a minimum-emitter NPN transistor in a 40V standard bipolar process.

TABLE 3.1 Typical vertical NPN device parameters.

Parameter	Nominal Value
Drawn emitter area	100μm^2
Peak current gain (beta)	150
Early voltage	120V
Collector resistance, in saturation	100Ω
Collector current range for maximum beta	5μA–2mA
V_{EBO} (Emitter-base breakdown, collector open)	7V
V_{CBO} (Collector-base breakdown, emitter open)	60V
V_{CEO} (Collector-emitter breakdown, base open)	45V

[6] In many standard bipolar processes, the isolation spacings are much wider than this illustration suggests; see Appendix C.1 for a brief discussion and some typical spacing rules.

The NPN transistor can also act as a diode, the characteristics of which depend upon the terminals chosen to form the anode and cathode. The least series resistance and fastest switching speeds occur when the base and collector form the anode and the emitter forms the cathode. This configuration is sometimes called a *CB-shorted diode*, or a *diode-connected transistor*. Its only serious drawback consists of a low breakdown voltage, equal to the V_{EBO} of the transistor, or about 7V. On the other hand, the relatively low V_{EBO} allows a suitably-connected transistor to serve as a useful Zener diode. The breakdown voltage of this structure varies somewhat due to doping variations and surface effects, so a tolerance of at least $\pm$ 0.3V should be allowed.

PNP Transistors

The standard bipolar process cannot fabricate an isolated vertical PNP because it lacks a P-type tank. A non-isolated vertical PNP transistor, called a *substrate PNP*, can be constructed using the substrate as a collector. The collector of this device always connects to the substrate potential of the die, which usually consists of either ground or the negative supply rail. Figure 3.11 shows a representative layout and cross section of this device.

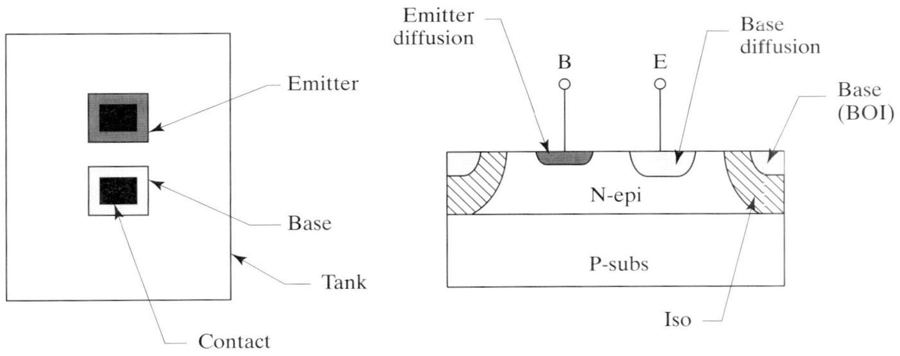

FIGURE 3.11 Layout and cross section of a substrate PNP transistor. The substrate forms the collector and is contacted through substrate contacts (not shown).

The base of the substrate PNP consists of an N-tank, and the emitter is fabricated from base diffusion. The collector current must exit through the substrate and the isolation. The collector contact does not have to reside next to the substrate PNP since all isolation regions interconnect electrically through the substrate. The resistance of the isolation and substrate are, however, substantial. Substrate contacts placed adjacent to the transistor help extract the collector current and thus minimize voltage drops in the substrate (*substrate debiasing*) that might otherwise impair circuit performance (Section 4.4.1).

The difference between the final epi thickness and the base junction depth determines the base width of a substrate PNP. As in the case of the vertical NPN, the base width is unaffected by photolithographic tolerances. NBL must be left out of the substrate transistor because its presence severely reduces beta. Deep-N+ therefore serves no useful function in a substrate PNP. An emitter diffusion placed beneath the collector contact ensures the surface doping concentration necessary to achieve Ohmic contact, while simultaneously thinning the oxide. The epi thicknesses and doping concentrations of the standard bipolar process are calculated to optimize the vertical NPN transistor, but the substrate PNP performs respectably (Table 3.2). The choice of names for the emitter and base diffusions is somewhat unfortunate, as the *emitter* of a substrate PNP consists of *base* diffusion.

TABLE 3.2 Typical PNP device parameters.

Parameter	Lateral PNP	Substrate PNP
Drawn emitter area	100μm^2	100μm^2
Drawn base width	10μm	N/A
Peak current gain (beta)	50	100
Early voltage	100V	120V
Typical operating current for maximum beta	5–100μA	5–200μA
V_{EBO} (Emitter-base breakdown, collector open)	60V	60V
V_{CBO} (Collector-base breakdown, emitter open)	60V	60V
V_{CEO} (Collector-emitter breakdown, base open)	45V	45V

The lack of an isolated collector limits the versatility of the substrate PNP. Another transistor, called a *lateral PNP,* trades off device performance for isolation. Figure 3.12 shows a representative layout and cross section of a minimum-geometry lateral PNP transistor. Both the collector and the emitter regions of the lateral PNP consist of base diffusions formed into an N-tank. As in the case of the substrate PNP, this tank serves as the base of the transistor. Transistor action in the lateral PNP occurs laterally outward from the central emitter to the surrounding collector. The separation of the two base diffusions sets the base width of the transistor. The emitter and collector of the lateral PNP are said to *self-align* because a single masking operation forms both regions. The base width of the lateral PNP can be precisely controlled because photolithographic misalignment does not occur between self-aligned diffusions. The effective base width of the transistor is considerably less than the drawn base width because of outdiffusion. This consideration limits the drawn base width to a minimum of about twice the base junction

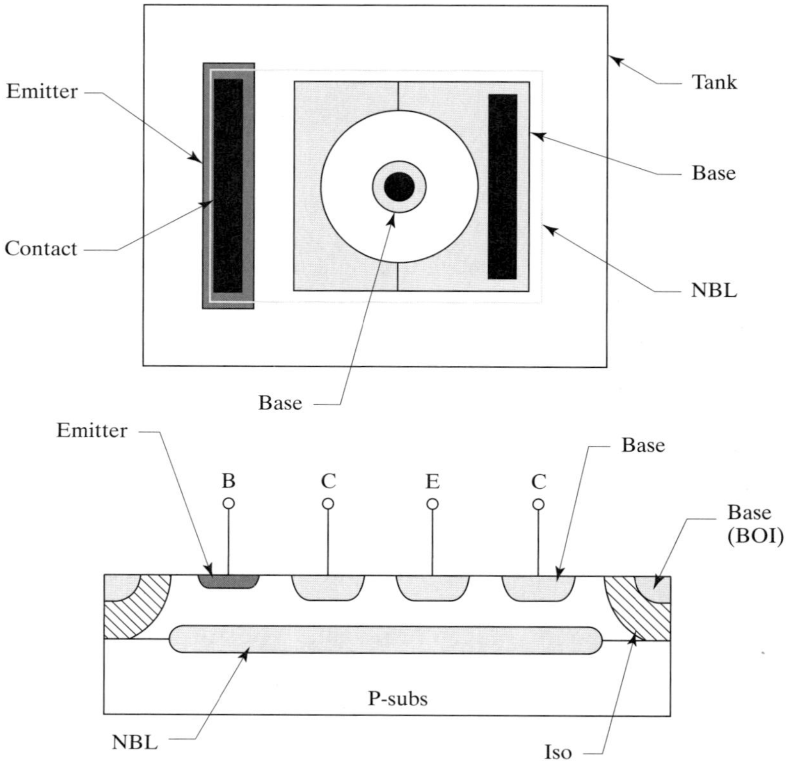

FIGURE 3.12 Layout and cross section of a lateral PNP transistor. The collector of the transistor appears twice on the cross section because it encircles the emitter.

depth. Narrow-base lateral PNP transistors exhibit low Early voltages and low-voltage punchthrough breakdown, so wider base widths are often employed.

Some percentage of the carriers injected by the lateral PNP's emitter will actually flow to the substrate rather than to the intended collector. This undesired conduction path forms a parasitic substrate PNP transistor. Unless this parasitic is somehow suppressed, most of the current injected by the emitter will find its way to the substrate, and the lateral PNP will exhibit very low apparent beta. For reasons that will be explained in Section 8.2.3, NBL largely blocks substrate injection and therefore boosts the lateral PNP beta.

Lateral PNP transistors have lower effective betas than their Gummel numbers would indicate. A large number of recombination centers reside at the oxide-silicon interface, especially in (111) silicon. The surface recombination rate thus far exceeds that in the bulk. Much of the current flow in the lateral PNP occurs near the surface and is therefore subject to these elevated recombination rates.[7] Certain layout techniques can minimize surface effects and, with care, betas of fifty or more can be obtained. Lateral PNP transistors are also quite slow, due mainly to large parasitic junction capacitances associated with the base terminal.

Neither the lateral nor the substrate PNP transistor forms a true complement to the vertical NPN. Both are useful devices, but each has its drawbacks and limitations. Circuit designers tend to avoid routing active signal paths through PNP devices (especially laterals) because of their poor frequency response, but most analog circuits still contain PNP transistors in supporting roles. Table 3.2 lists typical device parameters for PNP transistors formed on a 40V standard bipolar process.

Resistors

Standard bipolar does not include any diffusions intended specifically for fabricating resistors, but several types of resistors can be made using layers intended for other purposes. Typical examples include base, emitter, and pinch resistors, all three of which employ the relatively shallow base and emitter diffusions to achieve tighter spacings.

Each of the materials used to construct resistors possesses a characteristic *sheet resistance,* defined as the resistance measured across a square of the material contacted on opposite sides. Sheet resistance is customarily given in units of *Ohms per square* (Ω/□). It can be calculated from the thickness of the material and its doping concentration, but in the case of diffusions, nonuniform doping complicates these calculations (Section 5.2). In practice, sheet resistances are best determined by measuring sample resistors with known geometries constructed from the desired materials. Typical values for silicon diffusions range from 5 to 5000Ω/□.

A *base resistor* consists of a strip of base diffusion isolated by an N-tank and connected so that it will reverse-bias the base-epi junction (Figure 3.13). Connecting the tank to the more positive end of the resistor will ensure isolation. Alternatively, the tank can be connected to any point in the circuit biased to a higher voltage than the resistor. If a base resistor should forward-bias into its tank, a parasitic substrate PNP will inject current from the resistor into the substrate. NBL can help suppress this parasitic PNP should the base-epi junction momentarily forward-bias. Deep-N+ need not be added because the tank terminal does not draw significant current. Most standard bipolar processes produce base resistors with sheet resistances of 150 to 250Ω/□.

---

[7] R. S. Muller and T. I. Kamins, *Device Electronics for Integrated Circuits,* 2nd ed. (New York: John Wiley and Sons, 1986), pp. 366–368.

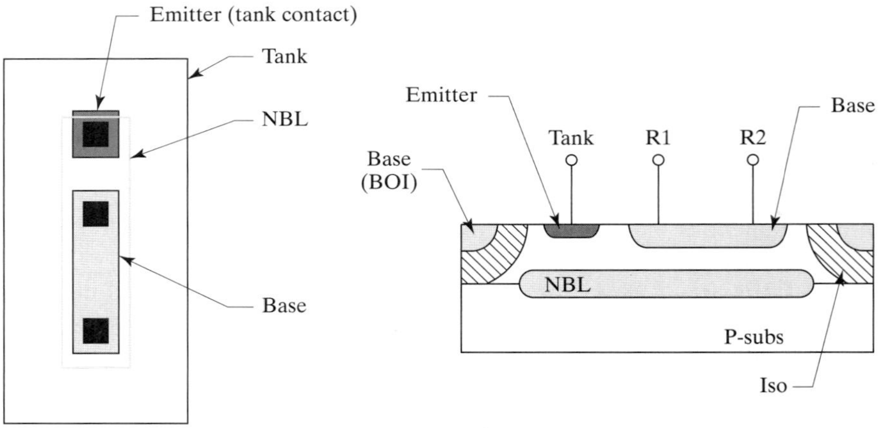

FIGURE 3.13 Layout and cross section of a base resistor.

An *emitter resistor* consists of a strip of emitter diffusion isolated by a base diffusion enclosed within an N-tank (Figure 3.14). The base region is connected so that it will reverse-bias the emitter-base junction, while the tank is biased to reverse-bias the base-epi junction. The simplest way to achieve this goal consists of tying the base to the low-voltage end of the resistor and the tank to the high-voltage end. Various other connections are also feasible, so long as neither junction forward-biases. NBL is usually added to help suppress parasitic substrate PNP action. The emitter sheet resistance is relatively low (typically less than 10Ω/□), and the breakdown of the emitter-base junction limits the differential voltage across the resistor to about 6V.

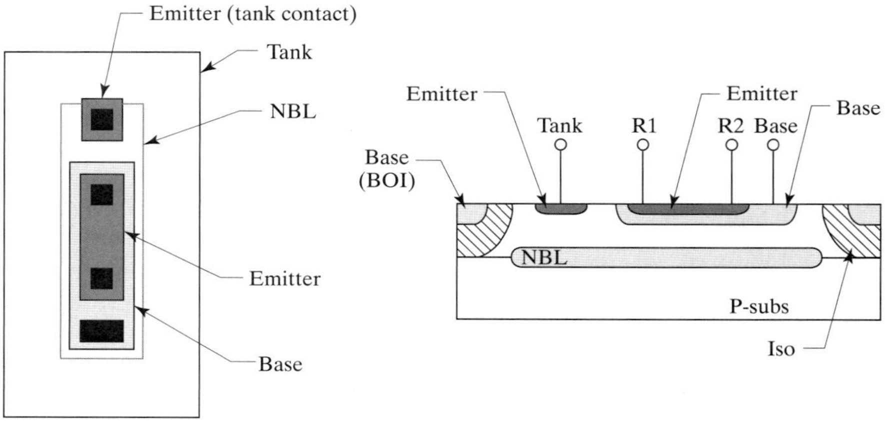

FIGURE 3.14 Layout and cross section of an emitter resistor (note the presence of bias terminals for both the tank and the base region).

A *pinch resistor* consists of a combination of base and emitter diffusions (Figure 3.15). The emitter forms a plate overlapping the middle of a thin strip of base.[8] Contacts occupy the ends of the base strip, which project from under the emitter plate. The tank and emitter plate are both N-type and are therefore electrically united. A tank contact biases both to a voltage slightly more positive than the resistor in order to ensure isolation. The body of the resistor consists of the portion of the base diffusion beneath the emitter plate. This *pinched base* is thin

[8] R. P. O'Grady, "The 'Pinch' Resistor in Integrated Circuits," *Microelectronics and Reliability,* Vol. 7, 1968, pp. 233–236.

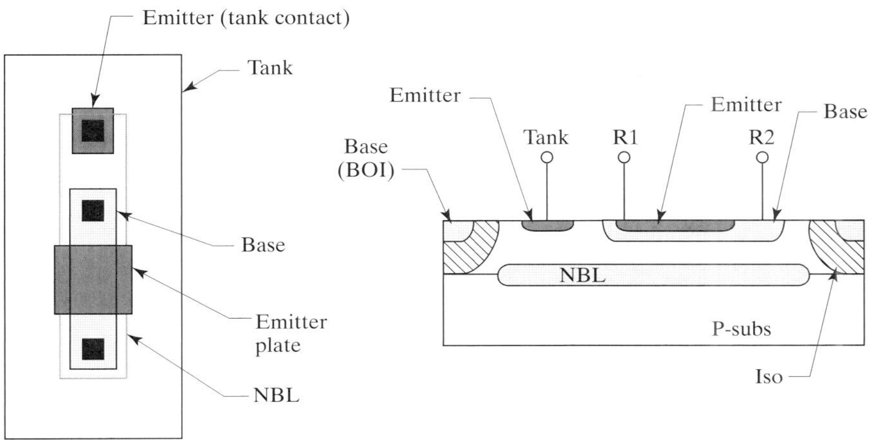

FIGURE 3.15 Layout and cross section of a base pinch resistor.

and lightly doped and therefore its resistance may exceed 5000Ω/□. Emitter-base breakdown limits the differential voltage across the resistor to about 7V. Pinch resistors are notoriously variable—much more so than either emitter or base resistors. Worst of all, these resistors exhibit severe voltage modulation. The intrusion of depletion regions into the neutral base tends to further pinch the resistor, causing it to act much like a JFET (Section 12.3). Pinch resistors find application in startup circuits and other noncritical roles, but their many drawbacks prohibit more widespread use. Table 3.3 compares the performance of emitter, base, and pinch resistors.

TABLE 3.3 Typical resistor device parameters.

Parameter	Emitter	Base	Pinch
Sheet resistance	5Ω/□	150Ω/□	3000Ω/□
Minimum drawn width	8μm	8μm	8μm
Breakdown voltage	7V	50V	7V
Variability (15μm width)	±20%	±20%	±50% or more

Capacitors

Standard bipolar was not intended to support capacitors. All of its oxide layers are so thick that they cannot be used to fabricate any but the smallest capacitors. However, the depletion region of a base-emitter junction exhibits a capacitance on the order of 0.8fF/μm^2 (0.5pF/mil^2), which can be used to construct a so-called *junction capacitor* (Figure 3.16). This capacitor consists of a base diffusion overlapping an emitter diffusion, both placed in a common tank. The emitter diffusion shorts to the tank, and the base-tank capacitance therefore adds to the base-emitter capacitance. The emitter plate must be biased positively with respect to the base plate to maintain a reverse bias across the base-emitter junction, and the differential voltage across the capacitor must not exceed the emitter-base breakdown voltage (about 7V). The resulting capacitance depends on bias and varies substantially (±50% or more). Junction capacitors are frequently used for compensating feedback loops, where their high capacitance per unit area makes up for their excessive variability.

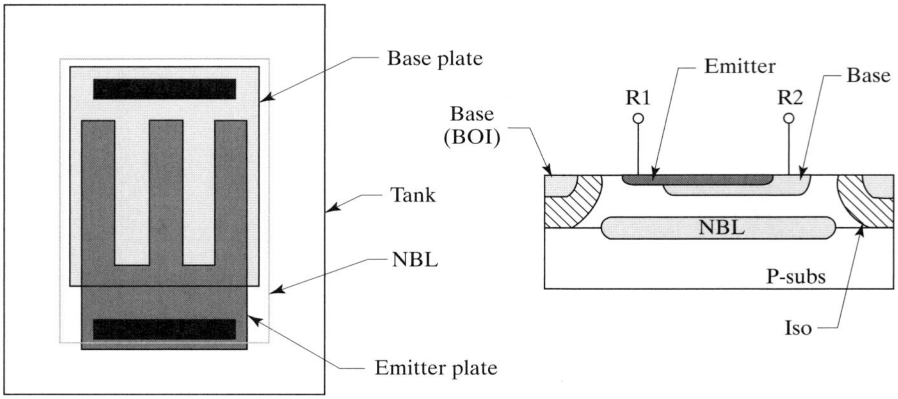

FIGURE 3.16 Layout and cross section of a junction capacitor.

3.1.4. Process Extensions

Standard bipolar has spawned a large number of process extensions, five of which are discussed in this section. They include up-down isolation, double-level metal, Schottky diodes, high-sheet resistors, and super-beta transistors. The benefits of each extension must be weighed against the increased cost and process complexity.

Up-down Isolation

Standard bipolar employs deep-P+ junction isolation driven down through the epitaxial layer to the underlying substrate. Outdiffusion increases the width of the isolation by 20μm or more, limiting how closely components can be packed together. One means of reducing outdiffusion uses a *P-buried layer* (PBL) to supplement the P+ isolation. The resulting *up-down isolation* consists of an isolation diffusion drawn coincident with the PBL. The isolation diffuses down from the surface, while the PBL diffuses up from the epi-substrate interface (Figure 3.17). Each only has to cross half the distance of a regular isolation diffusion, and outdiffusion is therefore cut approximately in half.

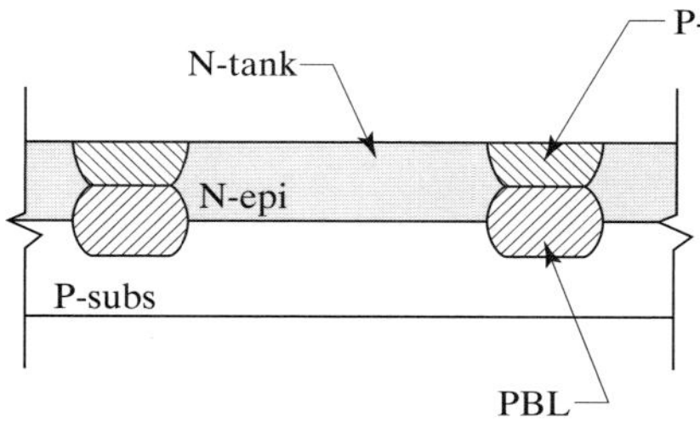

FIGURE 3.17 Cross section of a typical up-down isolation system.

Up-down isolation does have one significant drawback. Process considerations limit the PBL implant dose, so the final PBL becomes very lightly doped, and vertical resistance through the up-down isolation greatly exceeds that of conventional top-down isolation. The PBL also requires an additional masking step and diffusion. Up-down isolation saves 15 to 20% die area, so a case-by-case analysis should be performed to determine if it is worthwhile.

Double-level Metal

Standard bipolar originated as a *single-level metal* (SLM) process. The lack of a second metal layer greatly complicated lead routing. Instead of crossing wires by means of jumpers, diffusions were employed to form low-value resistors, called *crossunders* or *tunnels* (Section 13.3.2). Many devices can be custom-tailored to incorporate crossunders at the cost of compromising device performance and increasing die area. Single-level routing requires a deep understanding of device and circuit operation and an intuitive sense of topological connectivity. Most layout designers require years to master these skills.

Double-level metal (DLM) can be added to a standard bipolar process at the cost of two extra masks: vias and metal-2. The thickness of the first metal layer is often reduced to simplify planarization. Double-level metal is a useful, if somewhat costly, option. Lead routing no longer requires the use of customized devices, allowing component standardization and a considerable reduction of layout time and effort. Since metallization consumes a great deal of area, double-level metal can also reduce die area by up to 30%. These benefits are so attractive that manufacturers now routinely employ double-level metal for all new designs.

Schottky Diodes

Standard bipolar originally used silicon-doped aluminum metallization. Modern processes usually employ a combination of silicidation and refractory barrier metallization to ensure adequate step coverage while maintaining low contact resistance. Along with its more obvious benefits, silicidation also offers the opportunity to fabricate reliable Schottky diodes. Although aluminum forms a rectifying Schottky barrier to lightly doped N-type silicon, the forward voltage of the resulting diodes varies unpredictably depending on annealing conditions. Certain silicides, most notably those of platinum and palladium, produce Schottky barriers with very stable and repeatable properties. The forward voltages of these Schottky diodes lie slightly below that of a moderately doped PN-junction diode, so they can serve as antisaturation clamps (Section 8.1.4).

Schottky diodes require contacts formed through thick-field oxide. A contact etch that just penetrates the field oxide will severely overetch the base and emitter contacts. Overetching can be prevented by performing two consecutive oxide removals, the first of which thins the field oxide over the Schottky contacts and the second of which creates the actual contact openings. The fabrication of Schottky diodes thus requires an additional masking step.

Figure 3.18 shows a typical Schottky diode layout. The anode consists of a rectifying contact to an N-epi tank while the cathode Ohmically contacts the same tank with the assistance of emitter diffusion. The addition of NBL and deep-N+ to the cathode

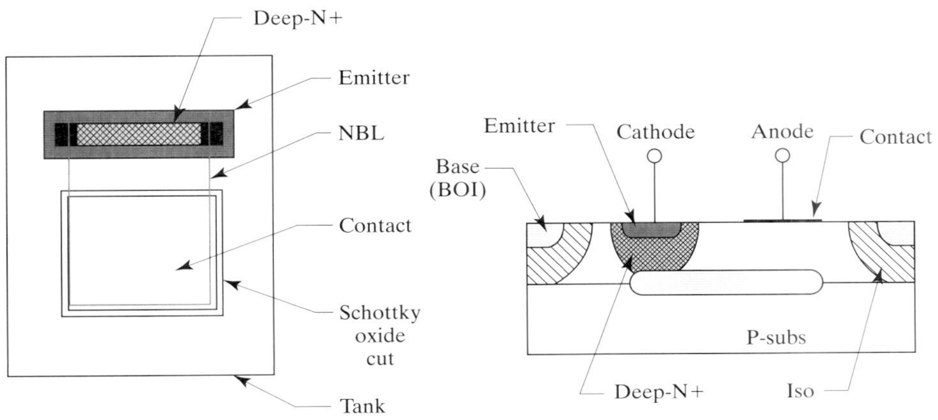

FIGURE 3.18 Layout and cross section of a Schottky diode.

greatly reduces the series resistance of the diode. The anode employs two concentric oxide removals, the larger of which thins the field oxide and the smaller of which forms the actual contact. This two-stage process not only eliminates overetching of base and emitter contacts but also improves the step coverage of metallization to the Schottky contact. This structure readily scales to provide diodes of any size. Electric field intensification at the exposed edges of the Schottky contact opening limits this simple structure's breakdown voltage and causes excessive reverse-bias leakage. Section 10.1.3 discusses methods of circumventing these problems.

High Sheet Resistors

Accurate resistors in standard bipolar are usually fabricated from base diffusion. Because the sheet resistance of this material rarely exceeds 200Ω/□, a typical die can incorporate a total of only 200 to 500kΩ of base resistance. Low-current circuits require more resistance than base diffusion can provide and more precision than pinch resistors can achieve. The only solution to this dilemma consists of adding a process extension to fabricate a precisely controlled high-sheet resistance (HSR) material.

The *high-sheet implant* consists of a shallow, lightly doped P-type implant. Depending on dose and junction depth, the sheet resistance of this implant can range from 1 to 10kΩ/□. The larger sheet values suffer from surface depletion effects that cause resistance variations, so most processes employ HSR implants of 1 to 3kΩ/□.

Figure 3.19 shows the layout and cross section of a typical HSR resistor. The body of the resistor consists of high-sheet implant, but small regions of base diffusion at either end of the resistor ensure Ohmic contact. The resistor occupies an N-tank and is isolated by the reverse-biased HSR-tank junction. The tank is often connected to the more positive end of the resistor, just as in the case of the base resistor. High-sheet resistors require one additional mask step and one dedicated implant. The cost of this process extension can usually be justified if the circuit includes more than 100 to 200kΩ of resistance.

FIGURE 3.19 Layout and cross section of a high-sheet resistor.

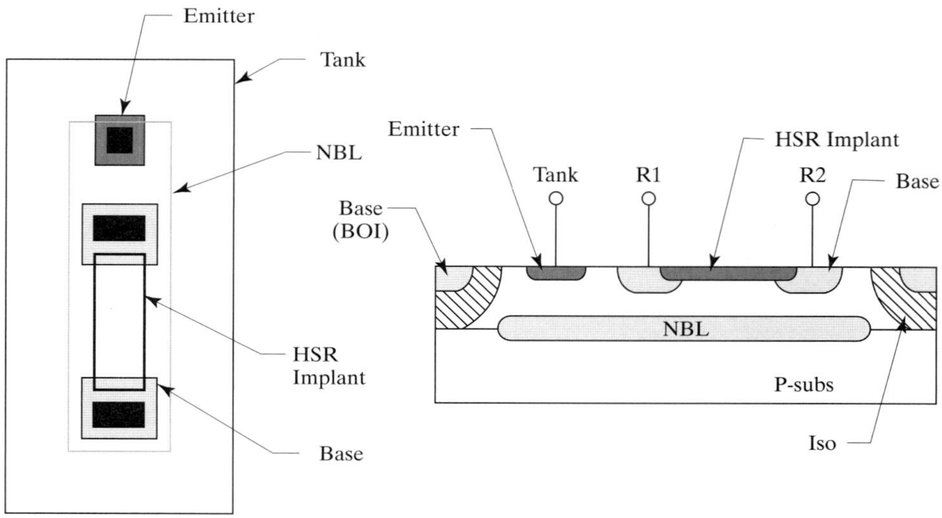

Super-beta Transistors

The standard bipolar NPN provides a reasonable compromise between high beta and adequate operating voltage. Beta can be greatly increased by narrowing the base width. If the process incorporates two separate emitter implants, then one can be optimized for beta and the other for operating voltage. The resulting *super-beta transistors* can achieve betas of 1000 to 3000 (Section 8.3.1). The use of an extremely thin and lightly doped base causes considerable depletion region intrusion into the neutral base and hence compromises not only operating voltage but also Early voltage. These highly specialized devices find application only in a limited range of circuits. For example, they have been used to fabricate bipolar operational amplifiers with extremely low input bias currents.

3.2 POLYSILICON-GATE CMOS

With the addition of two masking steps, standard bipolar can fabricate metal-gate PMOS transistors similar to those fabricated by early MOS processes (Figure 3.20). An N-tank serves as a backgate for the PMOS transistor; the backgate contact incorporates emitter diffusion to ensure Ohmic contact. None of the oxide layers in standard bipolar is sufficiently thin to serve as a gate oxide, necessitating the addition of a special masking step. Aluminum metal forms the gate electrode, while the source and drain consist of shallow P+ implants. Since standard bipolar does not include any suitable implant, another masking step patterns a special P-type source/drain (PSD) implant.

Practical MOS transistors require threshold voltages that lie within relatively narrow limits. Threshold voltages of less than 0.5V cause excessive leakage, while those of more than 1.5V unnecessarily reduce the available transconductance. The threshold voltage of an unadjusted (or *natural*) PMOS usually lies between 2 and 4V, necessitating a threshold adjust implant to shift it to the desired target of about 1V. The threshold voltage must lie within approximately ±0.5V of target. The metal gate PMOS transistor has difficulty maintaining even this minimal degree of control. The excess surface state charges introduced by using (111) silicon constitute one source of threshold variation; mobile ion contamination (Section 4.2.2) represents another.

Metal-gate MOS transistors also suffer from excessive overlap capacitance. The gate electrode is patterned by a different mask than the source and drain diffusions are. The gate must therefore overlap the source and drain sufficiently to form a continuous channel, even in the presence of photolithographic misalignments. The overlap between the gate and source causes a gate-to-source capacitance C_{gs}, while the

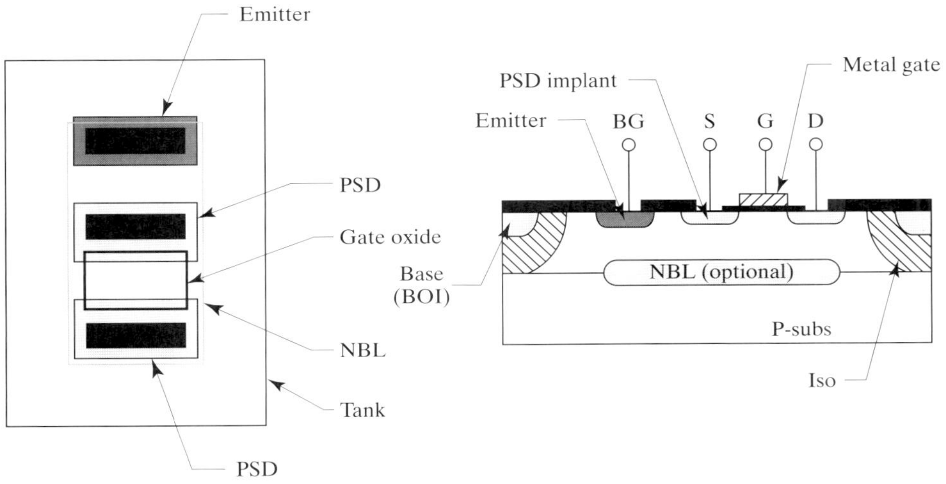

FIGURE 3.20 Layout and cross section of a metal-gate PMOS transistor constructed in standard bipolar.

overlap between gate and drain causes a gate-to-drain capacitance C_{gd}. These parasitic capacitances slow the transistor because they must be charged and discharged during switching. The gate-to-drain capacitance is particularly deleterious because the voltage gain of the transistor multiplies its apparent value (a phenomenon called the *Miller effect*). These parasitic overlap capacitances must be minimized in order to allow the construction of high-speed logic circuitry.

A complementary NMOS transistor would greatly enhance the utility of the metal-gate PMOS process. Taken together, NMOS and PMOS transistors would allow the construction of versatile *complementary MOS* (CMOS) circuits. Unfortunately, standard bipolar cannot easily fabricate NMOS transistors because they require a lightly doped P-type backgate that does not exist in this process. The NMOS threshold voltage on suitably doped (111) silicon is slightly negative, so a threshold-adjust implant is required to form an enhancement device. Yet another masking step is required to raise the thick-field threshold in the lightly doped P-type backgate around the NMOS transistors to prevent parasitic channel formation. The relatively poor performance of metal-gate CMOS cannot justify the cost of five additional masking steps, especially when a nine-mask polysilicon-gate CMOS process can fabricate vastly superior transistors. The following section examines the construction and performance of this alternate process.

3.2.1. Essential Features

The polysilicon-gate CMOS process is optimized to form complementary PMOS and NMOS transistors on a common substrate. It does not support the construction of bipolar transistors and it offers only a limited range of passive components. Originally intended solely for manufacturing CMOS logic gates, with slight modifications this process can also fabricate a limited variety of analog circuits.

A key difference between polysilicon-gate CMOS and standard bipolar lies in the choice of substrate material. Standard bipolar employs (111) silicon to enhance the thick-field threshold by increasing the surface state density, while polysilicon-gate CMOS uses (100) silicon to reduce the surface state density in order to improve threshold voltage control. A second major innovation lies in the use of polysilicon rather than aluminum as the gate material. Polysilicon can safely withstand the high temperatures required to anneal the source/drain implants, so it can act as a mask to form self-aligned sources and drains. The effects of mobile ion contamination can also be minimized by doping the polysilicon gate with phosphorus. Poly gates thus offer not only faster switching speeds but also better control of threshold voltages.

The choice of threshold voltages forms one of the few differences between analog and digital CMOS processes. Most digital CMOS processes[9] target threshold voltages between 0.8V and 0.9V with a variation of about ±0.2V. Analog CMOS designers favor maximizing headroom by targeting threshold voltages around 0.7V ± 0.2V. Since (100) silicon dictates the use of threshold-adjust implants in either case, the threshold voltages can often be retargeted by simply changing the implant dosage. Analog CMOS also rules out the use of blanket silicidation (Section 3.2.4). Neither of these requirements fundamentally modify the polysilicon-gate CMOS process.

---

[9] The information provided in the text applies to processes with operating voltages of 5V or more. Lower-voltage processes require smaller threshold voltages and tighter control. For example, a 3V process will typically target a threshold voltage of 0.6 ± 0.15V.

3.2.2. Fabrication Sequence

The baseline polysilicon-gate CMOS fabrication sequence consists of nine masking operations. The processing steps required to fabricate a finished wafer will be presented in the order in which they are performed. The cross sections used to illustrate this process employ a vertical exaggeration of between two and five, just as did the cross sections previously presented for standard bipolar.

Starting Material

CMOS integrated circuits are normally fabricated on a P-type (100) substrate doped with as much boron as possible in order to minimize substrate resistivity. This precaution helps provide a degree of immunity to *CMOS latchup* by minimizing substrate debiasing (Section 4.4). CMOS processes do not require NBL, so substrate doping is limited only by solid solubility.

Epitaxial Growth

The first step of the CMOS process consists of growing a lightly doped P-type epitaxial layer on the substrate. This epitaxial layer, typically some 5 to 10μm thick, is considerably thinner than the one used for standard bipolar. NMOS transistors are formed directly into the epi layer, which serves as their backgate. Since this process needs no buried layers, epi-coated wafers can be stockpiled to serve as starting material for all types of products. Standard bipolar does not allow this economy of scale, since each product requires a uniquely patterned NBL.

In theory, CMOS processes do not require epitaxy since the MOS transistors can be grown directly into a P– substrate. Epitaxial processing increases costs, but it also improves latchup immunity by allowing the use of a P+ substrate. In addition, the electrical properties of the epitaxial layer are more precisely controllable than those of Czochralski silicon, resulting in better control of MOS transistor parameters.

N-well Diffusion

After the wafer has been thermally oxidized, a layer of photoresist that has been spun onto it is patterned using the N-well mask. An oxide etch opens windows through which ion implantation deposits a controlled dose of phosphorus. A prolonged high-temperature drive creates a deep lightly doped N-type region called an *N-well* (Figure 3.21). The N-well for a typical 20V CMOS process has a junction depth of about 5μm. Thermal oxidation during the well drive covers the exposed silicon with a thin layer of *pad oxide.*

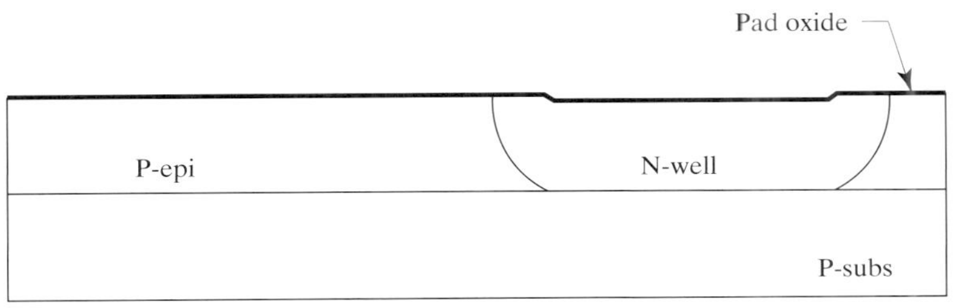

FIGURE 3.21 Wafer after N-well drive.

In an *N-well CMOS process,* such as that illustrated in Figure 3.21, NMOS transistors occupy the epi, and PMOS transistors reside in the well. The increased total dopant concentration caused by counterdoping the well slightly degrades the mobility of majority carriers within it. The N-well process therefore optimizes the performance of the NMOS transistor at the expense of the PMOS transistor. As a side effect, the N-well process also produces the grounded substrate favored by most circuit designers.

A *P-well CMOS process* uses an N+ substrate, an N– epitaxial layer, and a P-well. NMOS transistors are formed in the P-well and PMOS transistors in the epi. This process optimizes the PMOS transistor at the expense of the NMOS transistor, but the NMOS still outperforms its counterpart because electrons are more mobile than holes. A P-well process requires that the substrate connect to the highest-voltage supply instead of ground. Designs that employ multiple power supplies often have difficulty biasing an N-type substrate because of ambiguities in the sequencing of the supplies.

Both P-well and N-well CMOS processes exist. The N-well process offers a slightly better NMOS transistor, and it allows the use of a grounded substrate. N-well CMOS is also upwardly compatible with BiCMOS technology, as will become apparent later in this chapter. The N-well process has therefore been chosen to illustrate CMOS technology.

Inverse Moat

The CMOS process employs a thick-field oxide for much the same reasons as standard bipolar: it increases the thick-field threshold voltages, and it reduces parasitic capacitance between the metallization and the underlying silicon. Unlike standard bipolar, CMOS processes employ LOCOS technology to selectively grow the field oxide, leaving only a thin pad oxide over the regions where active devices will be formed. The locally oxidized regions of the die are called *field regions,* while the areas protected from oxidation are called *moat regions.*

The LOCOS process uses a patterned nitride layer formed by first depositing nitride across the entire wafer, then patterning this nitride using the inverse moat mask, and finally employing a selective etch to remove the nitride over the field regions (Figure 3.22). The photomask used for this step is called the *inverse moat* mask because it consists of a color reverse of the moat regions. In other words, the mask codes for areas where moat is absent, not where it is present.

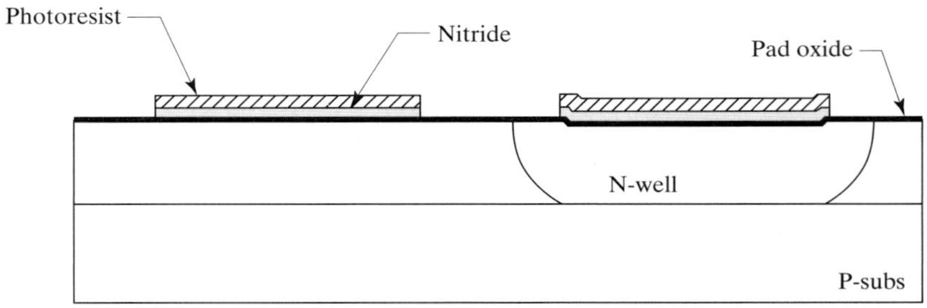

FIGURE 3.22 Wafer after nitride deposition and inverse moat pattern.

The nitride layer used for LOCOS must lie on top of a thin oxide layer (the *pad oxide*) because the conditions of nitride growth induce mechanical stresses that can cause dislocations in the silicon lattice. The pad oxide provides mechanical compliance and prevents strains produced by the nitride growth from damaging the underlying silicon.

Channel Stop Implants

The CMOS process deliberately minimizes threshold voltages in order to produce practical MOS transistors. The LOCOS field oxide will raise the thick-field thresholds, but not by enough to support supply voltages of more than a few volts. Practical CMOS processes nearly always require additional measures to ensure that the thick field thresholds exceed the operating voltages. Dopants are usually selectively implanted beneath the field regions to raise the threshold voltages of the thick-field transistors. P-epi field regions receive a P-type *channel stop implant,* while N-well field regions receive an N-type channel stop implant. The formation of channel stops thus requires two successive ion implantations.

Various techniques have been developed to produce channel stops. The method presented here involves the use of a blanket boron implant followed by a patterned phosphorus implant. The boron implant uses the photoresist left from patterning the LOCOS nitride. This mask exposes the field regions where channel stops will be deposited, so all of these regions receive the blanket boron implant (Figure 3.23A). This step sets the thick-field threshold in the epi regions.

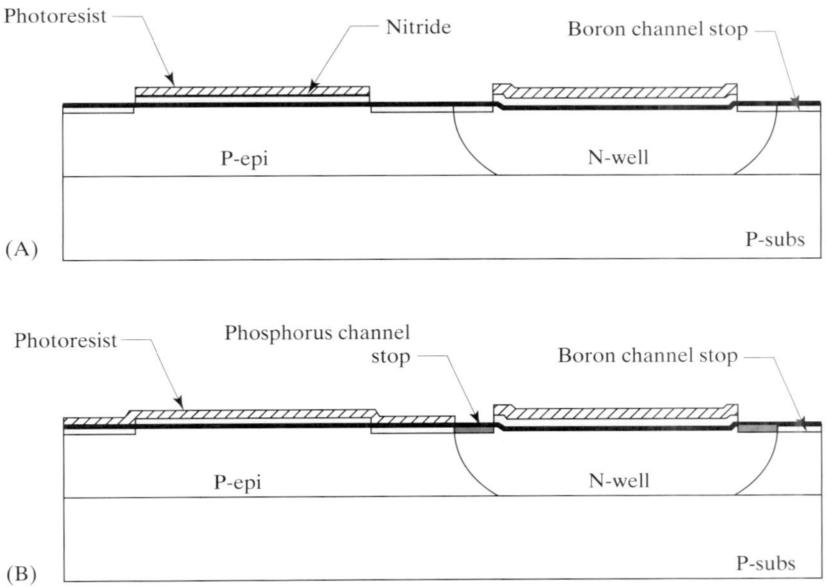

FIGURE 3.23 Wafer after blanket boron channel stop implant (A) and after selective phosphorus channel stop implant (B).

The wafer is again coated with photoresist immediately after the boron implant. The previous photoresist can remain in place since the channel stop implant will not affect the moat regions that lie beneath it. The recoated wafer is patterned using the channel stop mask, exposing only N-well field regions. The subsequent phosphorus implant counterdopes the previous blanket boron implant and raises the NMOS thick-field threshold above the maximum operating voltage (Figure 3.23B). Following the phosphorus implant, all photoresist is stripped from the wafer in preparation for LOCOS oxidation.

LOCOS Processing and Dummy Gate Oxidation

Steam is often used to increase the rate of LOCOS oxidation; alternately the furnace pressure can be raised to five or ten times atmospheric. After LOCOS oxidation, a suitable etchant strips away the remnants of the nitride block mask. Figure 3.24 shows a cross section of the resulting wafer. The curved transition region, called a *bird's-beak,* at the edges of the moat results from oxidants diffusing under the edges of the nitride film.

The *Kooi effect* (Section 2.3.4) causes nitride deposits to form underneath the pad oxide around the edges of the moat. These deposits can potentially cause gate oxide integrity failures, but they can be eliminated by a dummy gate oxidation. A

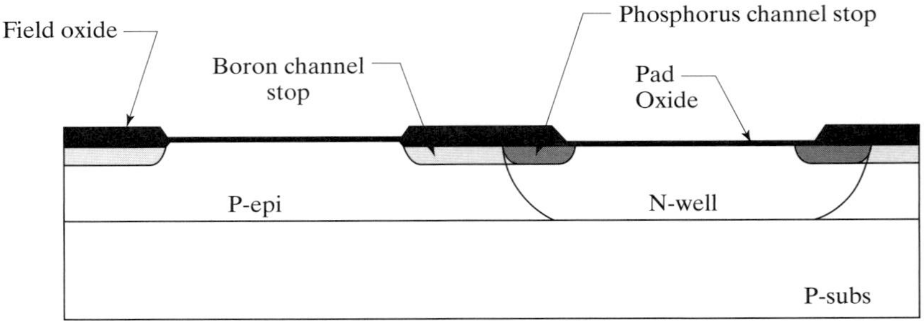

FIGURE 3.24 Wafer after LOCOS oxidation and nitride strip.

brief etch strips away the thin pad oxide without substantially eroding the thick-field oxide. Next, a brief dry oxidation grows a thin layer of oxide called a *dummy gate oxide* (or *sacrificial gate oxide*) in the moat regions. Any nitride deposits that remain will gradually oxidize. All of the nitride will be consumed if the dummy gate oxidation continues for a sufficient length of time.

Threshold Adjust

The use of (100) silicon helps stabilize the threshold voltages of the MOS transistors, but the backgate dopings and gate electrode materials preclude the achievement of usable threshold voltages without threshold adjust implants. For example, the unadjusted PMOS threshold might range from –1.5 to –1.9V, while the NMOS threshold might range from –0.2 to 0.2V. One or two threshold adjust implants (also known as *V_t adjusts*) retarget the threshold voltages to the desired targets, usually 0.7V for NMOS and –0.7V for PMOS transistors.

Two methods exist for adjusting threshold voltages. The first method employs two separate implants, one to set the PMOS V_t and the other the NMOS V_t. The use of two implants allows independent optimization of both thresholds. Many processes do not require this degree of flexibility. These processes can use a single V_t adjust to simultaneously reduce the PMOS threshold and increase the NMOS threshold. If this implant is properly performed, a nominal threshold voltage of 0.7 to 0.9V can be obtained for both types of MOS transistors. Figure 3.25 illustrates this approach.

After the wafer has been coated with photoresist, the V_t adjust mask is used to open windows over areas where MOS transistors will form. The boron V_t adjust implant penetrates the dummy gate oxide to dope the underlying silicon. After the V_t adjust implant, the dummy gate oxide is stripped away to reveal bare silicon in the moat regions.

The true gate oxidation employs dry oxygen to minimize excess charge incorporation due to surface states and fixed oxide charges. This oxidation must be very brief, because gate oxides are exceedingly thin. A 10V MOS transistor typically requires a 300Å (0.03μm) gate oxide, while a 3V transistor may employ an oxide less than 100Å (0.01μm) thick. This gate oxide will form the dielectric of the MOS

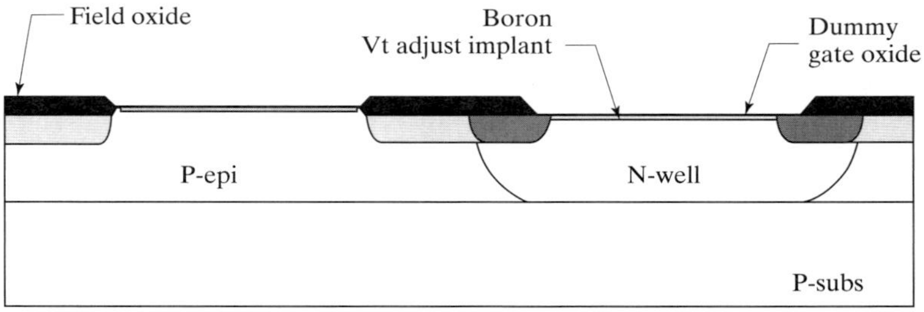

FIGURE 3.25 Wafer after V_t adjust implant.

transistors; it also covers the regions where source and drain implants will later occur.

Polysilicon Deposition and Patterning

The polysilicon layer used to form gate electrodes is heavily doped with phosphorus to reduce its resistivity to about 20 to 40Ω/□. Although gate leads do not conduct DC current, switching transients do draw substantial AC current, and low-resistance gate polysilicon greatly improves the switching speeds of MOS circuitry. Phosphorus doping simultaneously adjusts the work function of the polysilicon to produce threshold voltages compatible with a single-step V_t adjust. Phosphorus-doped gate polysilicon also minimizes threshold voltage variation due to mobile ions, allowing threshold voltage control of ±0.1 to 0.2V. While it is possible to dope polysilicon during deposition, most processes first deposit intrinsic polysilicon and subsequently dope it using conventional deposition or implantation techniques.

The deposited polysilicon layer must now be patterned using the poly mask (Figure 3.26). Modern submicron processes can fabricate polysilicon gates less than 0.5μm long, and any variation in gate length directly affects the transconductance of the resulting transistors. Thus, the patterning and etching of poly form the most critical photolithographic steps of a CMOS process. The simple process discussed here produces a minimum channel length of about 2μm and therefore does not require as high a degree of precision as submicron processes, but polysilicon patterning still remains its most challenging photolithographic step.

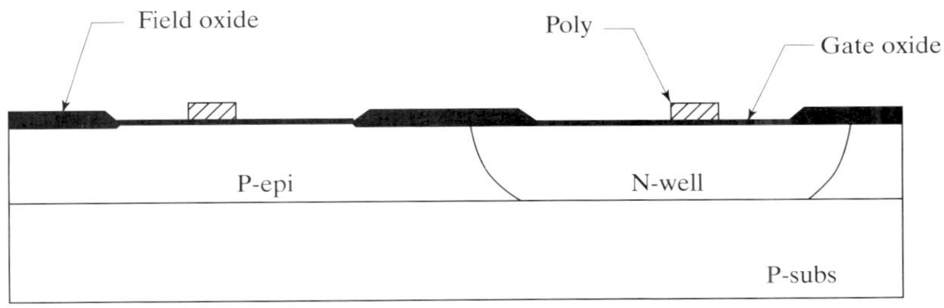

FIGURE 3.26 Wafer after polysilicon deposition and pattern. For simplicity, the channel stop and threshold adjust implants do not appear in this or subsequent cross sections.

Source/Drain Implants

The completed polysilicon gates now act as masks that self-align the source/drain implants for both the PMOS and NMOS transistors. These implants can be performed in either order. In the illustrated process, the N-type source/drain (NSD) implant occurs first, followed by the P-type source/drain (PSD) implant.

The NSD implant begins with the application of photoresist to the wafer, followed by patterning using the NSD mask. Shallow, heavily doped N-type regions are then formed by implanting arsenic through the exposed gate oxide. The polysilicon gate blocks this implant from the regions directly underneath the gate and therefore minimizes the gate/source and gate/drain overlap capacitances. Once the NSD implant has been completed, the photoresist residue is stripped from the wafer. The PSD implant begins with the application of a second photoresist layer patterned using the PSD mask. A shallow, heavily doped P-type region is formed by implanting boron through the exposed gate oxide. As with the NSD implant, the PSD implant self-aligns to the polysilicon and the PMOS transistors also exhibit minimal overlap capacitance. Following the PSD implant, photoresist is again stripped from the wafer.

A brief anneal activates the implanted dopants and slightly thickens the oxide over the source and drain regions. This anneal is the final high-temperature step of the process, corresponding to the emitter drive of standard bipolar. Figure 3.27 shows a cross section of the wafer following the source/drain anneal.

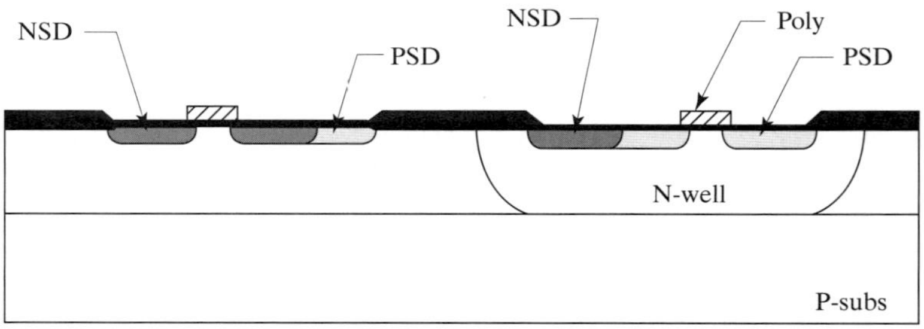

FIGURE 3.27 Cross section of the wafer after NSD and PSD implants and anneals. The backside contact implants abut the source implants to save space.

Contacts

Despite further oxidation during the source/drain anneal, the oxide covering the moat regions remains thin and therefore vulnerable to oxide rupture. Most processes deposit a *multilevel oxide* (MLO) before contact patterning. The MLO thickens the oxide over the moat regions and at the same time coats and insulates the exposed polysilicon pattern. Metal leads can now run over moat regions and polysilicon gates without risk of oxide rupture.

After the wafer is again coated with photoresist, the contact regions are patterned using the contact mask. Ohmic contacts form to the heavily doped source and drain without difficulty, but the backgate regions are far too lightly doped to allow direct Ohmic contact. The addition of NSD and PSD implants in the vicinity of the backgate contacts overcomes this difficulty. Contacts opened over polysilicon allow contact to the gate electrodes.

Metallization

The shallow NSD and PSD diffusions are vulnerable to junction spiking. Most CMOS processes employ a combination of contact silicidation and refractory barrier metallization to ensure reliable contact to the source/drain regions. After formation of silicide in the contact openings, a thin film of refractory metal sputtered over the wafer precedes a much thicker layer of copper-doped aluminum. The metallized wafer is coated with photoresist and patterned, using the metal mask. A suitable etchant then removes unwanted metal to form the interconnection pattern. Most processes also include a second layer of metallization. In such a process, another layer of oxide deposited over the first metal pattern insulates it from the second metal pattern. This second deposited oxide is usually called an *interlevel oxide* (ILO). Some form of planarization minimizes the nonplanarities caused by the first metal pattern to ensure adequate second metal step coverage. Vias etched through the ILO connect to a second metal layer deposited and patterned in much the same way as the first.

Protective Overcoat

A protective overcoat is now deposited over the final layer of metallization, both to provide mechanical protection and to prevent contamination of the die. The protective overcoat must resist penetration by mobile ions, so it normally consists of either a thick phosphosilicate glass (PSG), a compressive nitride layer, or both.

After coating with photoresist, the wafer is patterned using the *protective overcoat* (PO) mask. A suitable etchant removes the overcoat over selected areas of metallization and allows attachment of bondwires to the integrated circuit. This composes the final fabrication step; the wafer is now complete. Figure 3.28 shows a cross section of the resulting wafer, with only a single level of metal for simplicity. No bondpad openings exist in the illustrated portion of the die. This cross section includes an NMOS transistor on the left and a PMOS transistor on the right.

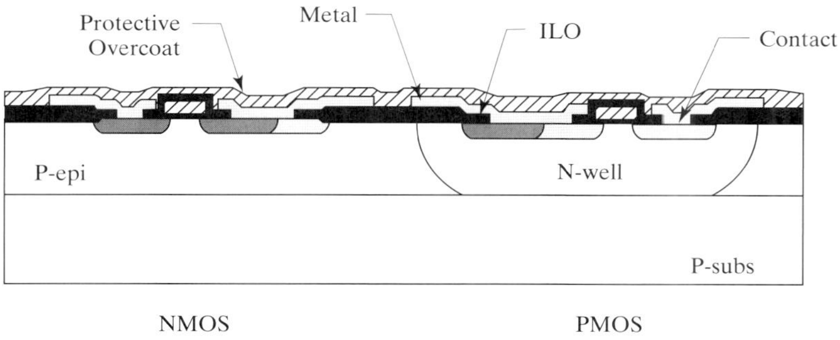

FIGURE 3.28 Cross section of the completed polysilicon-gate CMOS wafer.

3.2.3. Available Devices

Polysilicon-gate CMOS was originally developed to provide relatively low-voltage NMOS and PMOS transistors. The same process steps can fabricate several other components, including natural MOS transistors, a substrate PNP, several types of resistors, and a capacitor. Together these components allow the construction of a considerable variety of analog circuits. Section 3.2.4 examines process extensions that allow higher operating voltages and denser circuit packing.

NMOS Transistors

Figure 3.29 shows a representative layout and cross section of an NMOS transistor. The source and drain regions consist of NSD implants that self-align to the polysilicon gate. Since the backgate of the NMOS consists of the P-epi (and by extension the substrate), any substrate contact on the die will serve as a backgate terminal for the transistor. Many layouts actually include separate backgate contacts immediately adjacent to each NMOS transistor even though these are not strictly necessary. The close proximity of these backgate contacts improves CMOS latchup immunity, and this arrangement ensures that an adequate number of substrate contacts are distributed throughout the layout. In cases where the source of the NMOS transistor connects to the substrate potential, a very compact layout can be achieved by butting the PSD substrate contact against the NSD source. PSD and NSD cannot abut one another if they connect to different potentials because the resulting P+/N+ junction leaks excessively.

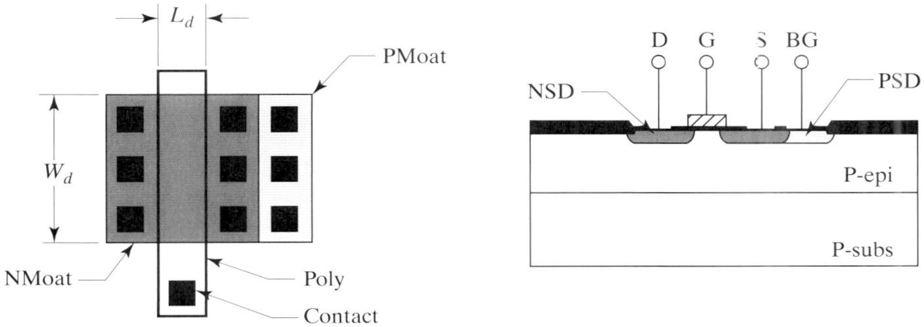

FIGURE 3.29 Layout and cross section of an NMOS transistor. The source and backgate of this transistor are shorted together using metal (not shown).

Figure 3.29 illustrates the common practice of coding CMOS transistors using drawing layers called *NMoat* and *PMoat.* These layers do not correspond to individual masks but rather to combinations of several masks. A figure drawn on the NMoat layer simultaneously produces figures on both the NSD and moat masks. Similarly, a figure drawn on PMoat simultaneously generates figures on both the PSD and moat masks. The use of PMoat and NMoat drawing layers simplifies the layout by reducing the number of figures required to draw a transistor.

The arrays of small square contacts shown in Figure 3.29 are characteristic of CMOS processes. The etch rate of oxide windows varies somewhat depending upon their size and shape, and these variations become particularly severe for very small openings. Many processes therefore allow only a single size of contact opening, usually consisting of a minimum-dimension square. Larger contacts must consist of arrays of minimum contacts rather than larger oxide openings.

In Figure 3.29, W_d and L_d denote the *drawn width* and *drawn length* of the NMOS transistor, respectively. The names given to these two dimensions may seem counterintuitive since the length of the gate is actually the width of the drawn polysilicon strip, but the channel length is customarily defined as the separation between its source and drain regions. The transconductance of an MOS transistor is approximately proportional to the ratio of the channel width divided by the channel length (W_d/L_d). Short channel lengths generate more transconductance per unit area, but analog circuits often employ longer channels to reduce channel length modulation.

Hot electron degradation limits the simple NMOS transistor of Figure 3.29 to relatively low operating voltages. The depletion region across the pinched-off portion of the channel accelerates electrons to high velocities. Some of these *hot electrons* collide with lattice atoms and ricochet out of the channel into the overlying gate oxide, where they become trapped. Hot electron injection causes a gradual shift in threshold voltage due to the slow accumulation of a fixed oxide charge. Eventually the threshold voltage shifts so far that the circuit ceases to meet parametric specifications.

Hot electron injection only occurs when the NMOS transistor operates in saturation with a relatively large drain-to-source bias. In the linear region the drain-to-source voltage is too small to produce hot electrons, and in cut-off no conduction occurs. NMOS transistors used as switches experience hot electron injection only during switching transients. If the switching frequency remains fairly low, then the total quantity of hot electrons produced over the operating lifetime of the integrated circuit remains acceptably small. Two different operating voltages are often specified for NMOS transistors. Junction breakdown and punchthrough limit the *blocking voltage rating,* which applies to transistors used as switches and as components of low-frequency digital logic. The somewhat lower *operating voltage rating* determined by the onset of hot electron degradation applies to transistors that operate for an appreciable length of time in saturation (as do the majority of analog transistors).

Increasing the length of the transistor reduces the impact of hot electron injection. Although hot electrons still appear at the drain end of the transistor, the trapped oxide charge forms only in the vicinity of the drain, and the majority of the transistor's gate oxide remains unaffected. Long-channel NMOS transistors thus exhibit somewhat less hot electron degradation than short-channel transistors.

The V_t adjust implant shifts the NMOS transistor threshold voltage from about zero volts to approximately 0.7V. The V_t adjust mask can block the implant from an NMOS transistor to form a *natural NMOS* possessing a relatively low threshold voltage. Natural NMOS transistors are used in certain analog circuits where the normal threshold is inconveniently large. Table 3.4 lists the device characteristics of NMOS and PMOS transistors constructed on a typical 2μm, 10V analog CMOS process.

Parameter	NMOS	PMOS
Minimum channel length	2μm	2μm
Gate oxide thickness	400Å	400Å
Threshold voltage (adjusted)	0.7V	–0.7V
Threshold voltage (natural)	0V	–1.4V
Transconductance (W_d/L_d = 10/10)	50μA/V²	20μA/V²
Operating voltage	7V	15V
Blocking voltage	15V	15V

TABLE 3.4 Typical polysilicon-gate CMOS device parameters.[10]

PMOS Transistors

Figure 3.30 shows a representative layout and cross section of a PMOS transistor. This device resides in an N-well that acts as its backgate. Any number of PMOS transistors can occupy the same well as long as their backgates all tie to the same potential. The relatively deep N-well outdiffuses substantially and the layout dimensions associated with it become quite large. Merging PMOS transistors in the same tank therefore saves substantial area. While the backgate of an NMOS transistor inherently connects to substrate, the backgate of a PMOS transistor can connect to any potential greater than or equal to its source. The N-well backgate thus provides an extra degree of freedom that analog designers frequently employ to enhance circuit performance.

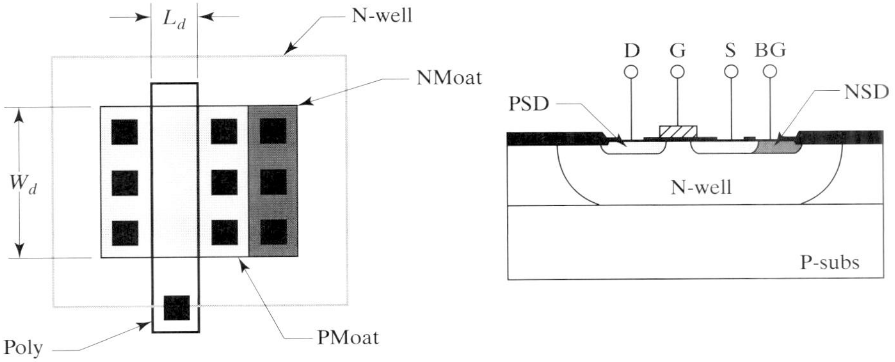

FIGURE 3.30 Layout and cross section of a PMOS transistor. The backgate and source of this transistor are shorted together using metal (not shown).

PMOS transistors are subject to *hot hole degradation,* but this causes fewer problems than hot electron degradation because holes are less mobile than electrons. A higher electric field and therefore a larger drain-to-source voltage is required to accelerate holes to velocities sufficient to inject charge into the oxide. Junction avalanche and punchthrough often limit PMOS transistors to voltages where hot hole degradation remains unimportant. Higher-voltage PMOS transistors will encounter hot-carrier problems similar to those of NMOS transistors.

A *natural PMOS* transistor can be fabricated by blocking the V_t adjust implant from the channel region of the device. Natural PMOS transistors possess inconveniently high threshold voltages, usually in excess of a volt. These transistors see only

[10] These parameters are roughly analogous to those of the 3μm Advanced LinCMOS™ process described in R. K. Hester, L. Hutter, L. Le Toumelin, J. Lin, and Y. Tung, "Linear CMOS Technology," *TI Technical Journal,* Vol. 8, #1, 1991, pp. 29–41. (The trademark belongs to Texas Instruments.) The transconductance figures given in this text are those used in the Schichman-Hodges equation $I_d = 1/2\, k\, (W/L)\, (V_{gs} - V_t)^2$.

occasional use in analog circuit design. Table 3.4 lists typical device parameters for a 2μm, 10V PMOS transistor.

Substrate PNP Transistors

The only bipolar transistor available in an N-well process is a substrate PNP. Figure 3.31 shows a typical layout and cross section for this device. The emitter consists of a PSD implant formed in an N-well that acts as the base of the transistor. A slug of NSD provides Ohmic contact to the N-well. The collector of this device consists of the P+ substrate and P-epi surrounding the well.

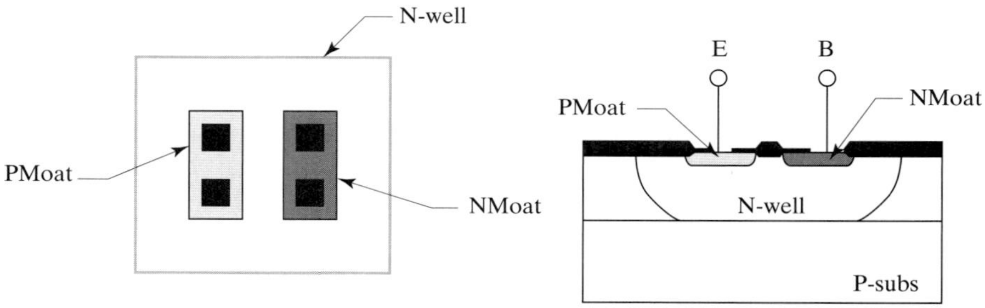

FIGURE 3.31 Layout and cross section of a substrate PNP transistor. The collector is connected by means of substrate contacts (not illustrated).

Although CMOS processes do not deliberately optimize bipolar components, the substrate PNP can still provide reasonably good performance. Its beta may approach that of a standard bipolar device (50 to 100), but it often drops to much lower values if the emitter diffusion is silicided, as is the case in a clad-moat process. Since this transistor injects current into the substrate, care must be taken to provide adequate substrate contact. The main component of collector resistance consists of the lightly doped P-epi layer interposed between the P+ substrate and the PSD diffusion beneath the substrate contacts. A large substrate contact area is necessary to prevent substrate debiasing. A typical CMOS integrated circuit incorporates enough substrate contact in the scribe street to accommodate 10 to 20mA of substrate current. If higher substrate currents will occur, then unused areas of the die should be filled with substrate contacts.

Although a lateral PNP transistor can theoretically be constructed in an N-well process, the absence of NBL encourages substrate injection, and only a small fraction of the total emitter current reaches the intended collector. These transistors exhibit extremely low apparent gains, rendering them of limited use to the analog circuit designer.

Resistors

The most useful of the four resistors available in polysilicon-gate CMOS consists of doped polysilicon (Figure 3.32). Although gate polysilicon exhibits a sheet resistance of only 20 to 30Ω/□, very narrow widths and spacings allow substantial resistance per unit area. A 2μm process can produce polysilicon resistors as area-efficient

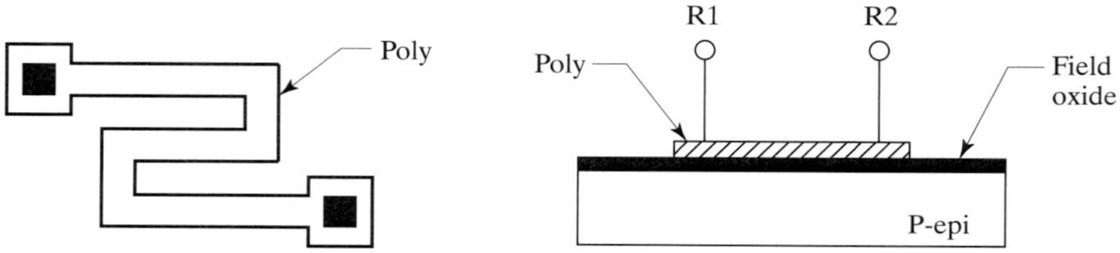

FIGURE 3.32 Layout and cross section of a poly resistor.

as standard bipolar base resistors. Submicron processes can provide remarkable amounts of resistance from narrow polysilicon traces, but the tolerances and matching of such resistors leave much to be desired. In a clad-poly process, a silicide block mask becomes necessary to obtain sufficient sheet resistance to construct practical poly resistors.

The polysilicon resistor of Figure 3.32 consists of a strip of poly deposited on top of field oxide. Contacts at either end allow it to be connected into the circuit. Oxide completely isolates this resistor, enabling it to be biased in any manner desired. Poly resistors can withstand large voltages relative to the substrate (100V or more) and can operate below substrate potential or above the positive supply voltage. The thick-field oxide also reduces parasitic capacitance between the resistor and the underlying substrate. Oxide isolation has one drawback: it isolates the resistor thermally as well as electrically. A poly resistor that dissipates sufficient power will experience permanent resistance variations due to self-induced annealing. Extreme power dissipation will melt or crack polysilicon long before diffused resistors of similar dimensions suffer damage. This behavior allows the construction of polysilicon fuses for wafer-level trimming, but it makes poly resistors undesirable for certain specialized applications, such as ESD protection.

Figure 3.33A shows the layout and cross section of an NSD resistor formed by contacting either end of a strip of NSD diffusion. NSD typically has a sheet resistance of 30 to 50Ω/□. Avalanche breakdown of the relatively shallow NSD diffusion also limits the operating voltage of this resistor, typically to no more than 10 to 15V. A similar resistor can also be constructed from PSD (Figure 3.33B). This resistor consists of a strip of PSD diffusion contained in an N-well region. The well must be biased above the resistor to maintain isolation. The well is therefore connected either to the more positive end of the resistor or to a high-voltage node (for example, the positive supply). PSD resistors also suffer from limited sheet resistance and a relatively low breakdown voltage.

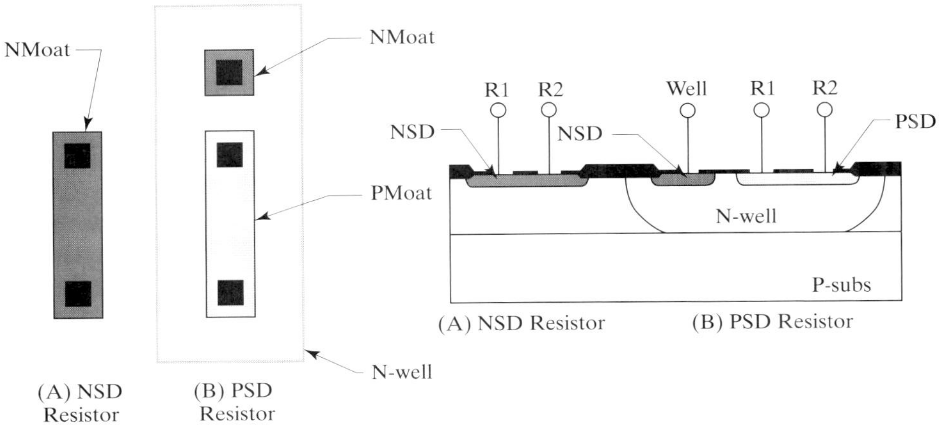

FIGURE 3.33 Layout and cross section of an NSD resistor (A) and a PSD resistor (B).

Another type of resistor consists of a strip of N-well with contacts at either end. NSD placed beneath these contacts prevents the formation of rectifying Schottky barriers. The large spacings required to allow for outdiffusion partially offset the high sheet resistance of the well (typically some 2 to 5kΩ/□). Well resistors are notoriously variable. Slight differences in doping and outdiffusion, voltage modulation of the depletion regions, and surface effects can all cause significant variations in well resistance. Proper layout precautions can help minimize these sources of variability, but most designers prefer to employ narrow polysilicon resistors. Table 3.5 summarizes the properties of the four types of resistors available in a typical CMOS process.

TABLE 3.5 Typical resistor device parameters.

Parameter	Poly	PSD	NSD	Well
Sheet resistance	20Ω/□	50Ω/□	50Ω/□	2000Ω/□
Minimum drawn width	2μm	3μm	3μm	5μm
Breakdown voltage	>100V	15V	15V	50V
Variability (5μm width)	±30%	±20%	±20%	±50% (10μm)

Capacitors

The gate oxide used to fabricate MOS transistors can also be employed to construct capacitors. One plate of the capacitor consists of doped polysilicon and the other of a diffusion, typically N-well. Figure 3.34 shows the layout and cross section of one type of MOS capacitor. The capacitance of this device typically ranges from 0.5 to 1.6fF/μm^2 (0.3 to 1.0pF/mil^2), depending on oxide thickness. The tight control of gate oxide thickness characteristic of modern MOS processes results in typical tolerances of ± 20% as long as the well electrode remains at least 1V above the poly electrode. Failure to maintain adequate bias across the capacitor will cause a dramatic drop in capacitance (Section 6.2.2). The main drawbacks of these capacitors consist of excessive bottom-plate parasitic junction capacitance and series resistance, and of nonlinearity effects at certain voltages.

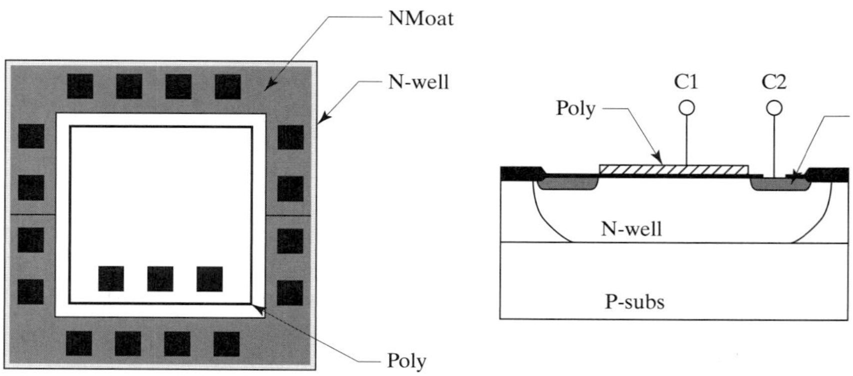

FIGURE 3.34 Layout and cross section of a PMOS capacitor.

3.2.4. Process Extensions

CMOS process extensions tend to focus on improving the PMOS and NMOS transistors rather than on constructing additional devices. One set of extensions seeks to provide higher operating voltages by suppressing hot carrier degradation. Another set of extensions focuses on reducing the size of the transistors to improve packing. Unlike standard bipolar, little emphasis is placed on providing a large number of specialized components because most CMOS fabs build primarily digital products. The large expenditure of time and money required to construct additional devices cannot be justified by the relatively small volume of analog CMOS products. Analog BiCMOS presents an entirely different picture because this process caters specifically to analog design. Many of the features and process extensions of BiCMOS can be retrofitted into a CMOS process if sufficient economic incentive exists.

Double-level Metal

The early CMOS processes used one metal and one poly layer (a combination sometimes called *1½-level metal*). Polysilicon can be used to jumper most signals, so routing is much easier than for single-level metal. Autorouting software still cannot efficiently route 1½-level metallization, so most modern CMOS processes add a second layer of metal. Analog CMOS layouts can benefit from using this second metal layer to increase packing density. Since planarization becomes more difficult as device sizes shrink, CMOS metallization tends to be substantially thinner than that used for standard bipolar. Higher-power CMOS circuits such as output drivers often benefit from combining multiple metal layers to form a thicker conductor.

Double-level metal adds two mask steps to the process: one for vias and one for metal-2. An *interlevel oxide* (ILO) deposited between the two metal layers provides insulation, and some form of planarization improves planarity for the second-level metal. Although these extra processing steps increase the cost of the wafer, most CMOS fabs routinely employ double-level metal for the majority of their products. Some processes use three, four, or even five metal layers to further reduce the area required for interconnection in high-density autorouted logic.

Silicidation

CMOS processes make extensive use of silicides. In addition to contacts, gate poly is often silicided to reduce its sheet resistance. Some processes even silicide source and drain diffusions to reduce their parasitic resistance. Processes with submicron feature sizes may use all of these forms of silicidation. Older processes with larger feature sizes cannot construct sufficiently fast transistors to justify siliciding their gates and source/drain regions, but they usually employ silicided contacts to prevent contact spiking.

A Schottky diode can be formed on an N-well CMOS process that employs platinum or palladium silicides. The Schottky anode consists of a silicided contact while the cathode consists of N-well contacted by means of an NSD diffusion. The lack of a buried layer increases the internal resistance of this diode and prevents it from being used for high-current applications. The exposed edges of the silicide limit the operating voltage, but techniques exist for suppressing edge breakdown without creating any additional process steps (Section 10.1.3). CMOS processes do not require a Schottky contact mask since a moat region can serve the same function. Note that some silicides do not form usable Schottky barriers. For example, titanium silicide produces such a low forward voltage that the resulting Schottky diodes leak excessively.

Silicidation of the gate poly reduces its sheet resistance to about 2Ω/□, which is too low for constructing most resistors. Poly resistors can still be formed in a silicided-gate process by adding a *silicide block mask*. This mask patterns the silicide layer, either by preventing metal deposition over resistors or by allowing its removal in these areas prior to sintering. The use of a silicide block mask complicates the silicidation process, but is necessary for most analog designs.

Some processes also silicide NSD and PSD diffusions to form so-called *clad moats*. These cannot be used as resistors since their sheet resistances approximate that of their silicide layer (typically about 2Ω/□). A silicide block mask can prevent the silicidation of selected PSD and NSD regions to allow their use as resistors. The silicidation of emitter regions also reduces the beta of substrate PNP transistors (Section 8.3.3). A silicide block mask can sometimes improve substrate PNP beta to the point where this device becomes a practical component.

Lightly Doped Drain (LDD) Transistors

If MOS transistors must operate at high drain-to-source voltages, then precautions become necessary to prevent hot carrier degradation. Typical NMOS transistors with 3μm channel lengths can indefinitely withstand 5 to 10V, while PMOS transistors of the same dimensions can withstand 15 to 20V. Voltages beyond these limits require alternative structures.

The strong electric field across the pinched-off channel of a saturated MOS transistor causes hot carrier degradation. The electric field intensity diminishes if the depletion region can somehow be widened. In a conventional transistor, the depletion region cannot intrude to any significant extent into the heavily doped drain. If the drain diffusion is more lightly doped, then the depletion region can extend into the drain as well as into the channel and the electric field intensity will decrease. Such *lightly doped drain* (LDD) transistors can withstand substantially higher drain-to-source voltages than can conventional *singly doped drain* (SDD) devices.

LDD transistors actually use two drain diffusions, one forming a lightly doped *drift region* near the edge of the gate, and the other forming a more heavily doped region beneath the contact. The heavily doped diffusion reduces the drain resistance of the structure and allows the transistor to retain most of the performance of a conventional SDD device. One process for forming LDD transistors uses an *oxide sidewall spacer* to self-align the two drain diffusions, enabling precise control of the width of the drift region.[11] Figure 3.35 illustrates the steps required to fabricate this structure. Immediately after patterning the polysilicon gate, a shallow implant self-aligned to the edges of the gate polysilicon deposits the lightly doped drain. The wafer is coated with a thick layer of isotropically deposited oxide. The oxide at the edges of the polysilicon gate is deeper than the oxide over adjacent regions of the wafer. An anisotropic dry etch removes most of the deposited oxide, but some remains along the edges of the gates even after the planar surfaces have entirely cleared. The etch is halted so that filaments of oxide remain along either side of the gate; these form the desired oxide sidewall spacers. These spacers are approximately as wide as the polysilicon gate is thick. A second drain implant self-aligned to the edges of the oxide sidewall spacers forms the heavily doped portions of the LDD structure. The width of the lightly doped drift region approximately equals the width of the sidewall spacer, typically about 0.5μm.

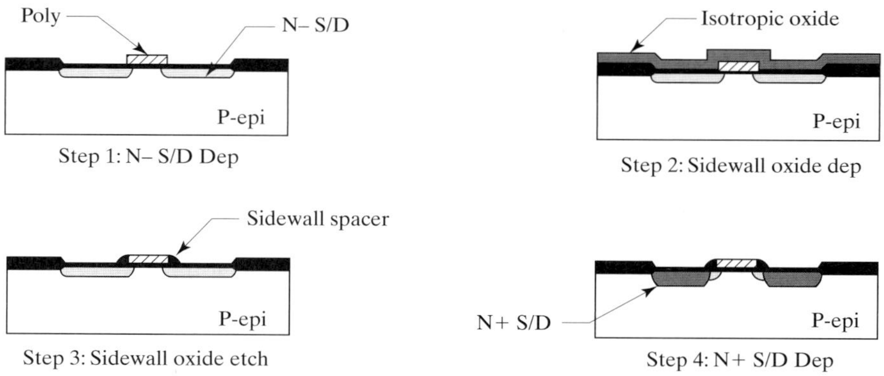

FIGURE 3.35 Steps required to fabricate an LDD NMOS transistor.

Only the drain terminal of the NMOS transistor requires an LDD structure, but no simple way exists to block the formation of the sidewall spacer, so the source terminal of the NMOS also receives an LDD structure. The resulting transistor is *symmetric* in the sense that source and drain can be interchanged without affecting device performance. The PMOS transistor also receives the oxide sidewall spacers,

[11] R. H. Eklund, R. A. Haken, R. H. Havemann, and L. N. Hutter, "BiCMOS Process Technology," in A. R. Alvarez, ed., *BiCMOS Technology and Applications,* 2nd ed. (Boston: Kluwer Academic, 1993), pp. 90–95.

but no lightly doped diffusion. For reasons discussed in section 12.1.1, a channel forms underneath these sidewall spacers. The transistor, therefore, appears to have a slightly longer channel than its drawn dimensions suggest.

The LDD process described previously forms transistors that are suitable for drain-to-source voltages of 10 to 20V. No additional masks are required if all transistors receive both source/drain implants (N– S/D and N+ S/D). Purchasing an additional mask to selectively block the N– S/D implant may provide some additional benefits. Short-channel transistors do not require LDD processing because they break down by punchthrough before hot carrier generation becomes significant. The presence of N– S/D in short-channel NMOS transistors therefore serves no useful purpose. The transistors can pack more densely if the drawn channel dimensions can be reduced without causing premature punchthrough. One way to achieve this goal consists of selectively blocking the N– S/D implant from the short-channel devices. By leaving out the N– S/D, the drawn channel length can be decreased by 0.5 to 1.0μm without affecting the effective dimensions of the device. The purchase of separate N– S/D and N+ S/D masks can make a significant impact on high-density, low-voltage logic circuitry that does not require LDD transistors.

Extended-Drain, High-Voltage Transistors

Oxide sidewall spacers allow the construction of fairly conventional MOS transistors that can withstand drain-to-source voltages of 10 to 20V. Higher voltages require a different approach to constructing the lightly doped drain region to protect against avalanche breakdown as well as hot carriers. Practical high-voltage MOS transistors can be constructed using only the existing masks of the standard N-well polysilicon-gate CMOS process. These *extended-drain* devices do not self-align, so they are inherently long-channel devices that exhibit substantial overlap capacitance. Even so, they suffice for constructing many high-voltage circuits.

Figure 3.36 shows a sample high-voltage, extended-drain NMOS. This transistor uses an N-well region as an extremely lightly doped drain. Since the well is both relatively deep and very lightly doped, it possesses a breakdown voltage in excess of 30V. A plug of NSD provides Ohmic contact to the drain. The source of the transistor consists of an NSD region without the addition of N-well. Since source and drain employ different diffusions, this device is an *asymmetric* MOS transistor.

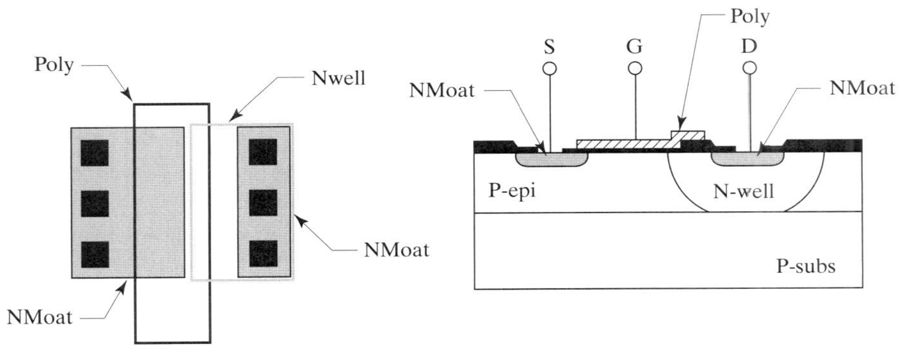

FIGURE 3.36 Layout and cross section of an extended-drain NMOS. The backgate is common to the substrate.

The gate oxide of a high-voltage MOS transistor presents something of a dilemma. The 15V CMOS gate oxide is 300 to 400Å thick and can safely withstand 20 to 25V. Higher voltages will rupture the oxide and destroy the device. A separate gate oxidation can form a thicker oxide for the high-voltage devices, but increasing the oxide thickness decreases the device transconductance. The best solution consists of thickening the gate oxide only over the lightly doped drain where the highest field stresses occur (Figure 3.36). The thin gate oxide remains in place over the channel and ensures a high device transconductance.

High-voltage, extended-drain PMOS transistors use the P-type channel-stop as a lightly doped drain. These devices are discussed in Section 12.1.2.

3.3 ANALOG BiCMOS

The incessant demand for higher levels of integration has driven the evolution of ever more complex and costly processes. Not only must more devices fit into the same die area, but the performance of these devices must steadily improve to satisfy new applications. By the early 1980s, customers demanded *mixed-signal* integrated circuits incorporating both analog and digital subsystems on a common substrate. A typical mixed-signal integrated circuit contains 90 to 95% digital circuitry and 5 to 10% analog circuitry.[12] CMOS logic overwhelmingly outperforms bipolar logic in packing density and power requirements, so the first attempts to fashion mixed-signal circuits employed unmodified CMOS processes. Analog CMOS circuitry had been designed for decades, so few manufacturers envisioned any difficulty in building the last percentages of the mixed-signal system. But these manufacturers soon discovered that difficulty does not correspond to component count. Although the analog components compose only a few percent of the total devices, they often consume the majority of the design effort. The inferior performance of analog CMOS requires even more design resources to compensate for process inadequacies.

After a few years of failures and qualified successes, most manufacturers began to realize that the analog portions of a mixed-signal system require tailor-made components. The true benefits of using such components lie not in improved performance, but in faster cycle times and higher probabilities of success. The late 1980s saw the development of a new generation of processes specifically aimed at the construction of mixed-signal integrated circuits. These *analog BiCMOS* processes are usually based on CMOS process flows, but are augmented by the addition of bipolar transistors, high-sheet poly resistors, and other special devices.

3.3.1. Essential Features

Analog BiCMOS processes are characterized by their complexity. Most require at least fifteen masks, and the more exotic variants use upward of thirty masks. The penalties of complexity include higher wafer costs, longer manufacturing times, and lower process yields. Set against these disadvantages are the benefits of higher-performance analog circuitry, reduced design effort, and faster design cycles. By the mid-1990s, the majority of new analog designs used some form of analog BiCMOS processing.

Because analog BiCMOS is still an evolving technology, the different manufacturers have not yet settled upon a common definition of the process. Most agree that practical BiCMOS must consist of a standard CMOS flow with the addition of a minimum number of steps to construct adequate bipolar transistors. Most also agree that deep P+ isolation is undesirable because it requires a prolonged high-temperature drive. One alternative form of isolation uses the N-well to form the collector of the NPN transistor. The base and emitter then consist of successive

[12] Reckoned by device count, not area. The relatively few analog devices on a mixed-mode integrated circuit tend to take up an inordinate amount of die area because analog circuitry cannot take full advantage of the reduced dimensions of modern transistors while retaining adequate performance.

counterdopings of the well. The collector-epi junction serves to isolate this style of bipolar transistor, thus the name *collector-diffused-isolation* (CDI).[13] Figure 3.37 shows an NPN transistor constructed using CDI.

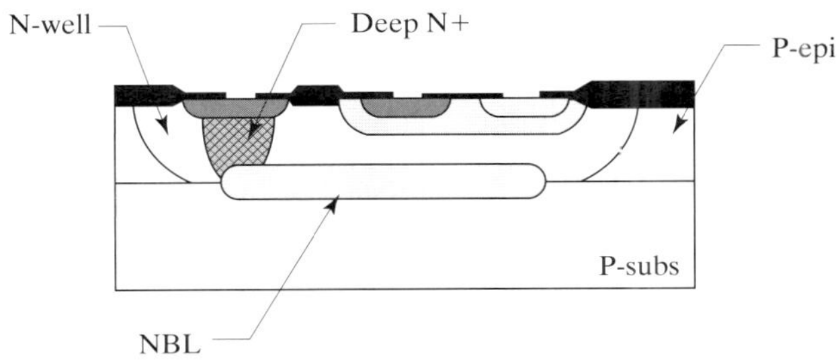

FIGURE 3.37 Details of collector-diffused-isolation (for an NPN transistor). Only the portions of the CDI system related to the collector are labeled.

The specific steps required to construct bipolar transistors on a CMOS process have been hotly debated for some years. While no consensus has yet emerged, analysis shows that three masks are probably needed: NBL, deep-N−, and base.

NBL is the most reluctantly conceded mask, but also the most vital. Some CMOS fabs do not possess epi reactors and lack the experience necessary to implement buried layers. Predictably, attempts have been made to eliminate NBL, and equally predictably, the results have been unsatisfactory. NBL greatly reduces NPN collector resistance—one of the major parasitics of this device. NBL also provides higher NPN operating voltages on the thin epi favored by modern CMOS processing because it blocks vertical punchthrough breakdown. Furthermore, NBL suppresses parasitic substrate PNP action. Without it, lateral PNP transistors become almost impossible to construct. Any practical analog BiCMOS process will almost certainly include NBL.

Standard bipolar uses a deep-N+ sinker to further reduce the collector resistance of power NPN transistors. Although a power MOS transistor can substitute for a power NPN in many applications, deep-N+ remains necessary for creating fully effective minority carrier injection guard rings. Furthermore, due to the graded nature of the well, CDI NPN transistors often exhibit excessive vertical resistance between the collector contact and the NBL. These transistors saturate prematurely, limiting low-voltage operation, complicating device modeling, and causing undesired substrate injection. Most BiCMOS processes include deep-N+, if only as a process extension.

The base diffusion sets the NPN transistor gain, breakdown voltage, and Early voltage. Some processes have attempted to construct NPN transistors using layers scavenged from other devices, with mixed results. Attempts to construct NPN transistors using the diffusions that normally form DMOS transistors have succeeded in some processes but not in others. Questions still remain as to the suitability of the DMOS NPN for applications requiring matching or conducting large currents (Section 12.2.2). Extended-base transistors that use the P-epi or a shallow P-well as a base region have also been successfully constructed, but these devices have several drawbacks. They require more area than conventional CDI devices, especially for small devices, and they often have low Early voltages because they lack a drift region (see Section 8.3.2).

---

13 B. T. Murphy, S. M. Neville, and R. A. Pedersen, "Simplified Bipolar Technology and Its Application to Systems," *IEEE J. Solid-State Circuits*, Vol. SC-5, #1, 1970, pp. 7–14.

3.3.2. Fabrication Sequence

This section discusses an analog BiCMOS process based on N-well polysilicon-gate CMOS.[14] The N-well provides collector-diffused isolation; NBL, deep-N+, and base are added to create bipolar transistors. Double-level metal has been added to simplify interconnection. This process requires fifteen masks, one of which is used twice for a total of sixteen masking operations.

Starting Material

The substrate material chosen for analog BiCMOS consists of a P+ (100) substrate cut several degrees off the crystal axis to minimize pattern distortion. The use of NBL in conjunction with a P+ substrate dictates the insertion of an additional epitaxial deposition into the process. Without this step, the NBL would directly contact the substrate to form an N+/P+ junction with a very low breakdown voltage. A lightly doped P-epi layer some 20μm thick therefore resides between substrate and NBL. Three factors determine the thickness of this first epi layer: the up-diffusion of the underlying substrate, the down-diffusion of the NBL, and the width of the depletion region required to withstand the maximum anticipated operating voltage (typically 30 to 50V). The first epi deposits on an unpatterned wafer, so epi-coated wafers may be stockpiled for use as starting material. One could alternatively sacrifice the P+ substrate to eliminate the need for a first epi deposition, but the use of a lightly doped substrate makes the process very susceptible to latchup and substrate debiasing (Section 4.4.1).

N-buried Layer

A brief thermal oxidation grows a thin layer of oxide across the wafer. This oxide is patterned using the N-buried layer (NBL) mask, and an oxide etch opens windows to the silicon surface. Ion implantation deposits an N-type dopant, either arsenic or antimony, in these windows. A brief drive conducted in an oxidizing ambient anneals out lattice damage and causes the formation of the surface discontinuity required for subsequent mask alignment.

Epitaxial Growth

After the NBL anneal, the oxide is stripped and the wafers are returned to the epitaxial reactor for deposition of a second P-epi layer. Surface discontinuities propagate upward through the epi at approximately a 45° angle along a [100] axis determined by the tilt of the wafer. After epitaxial growth, the NBL shadow will have shifted laterally a distance approximately equal to the thickness of the second epi; typically about 10μm. Figure 3.38 shows the wafer after the second epi deposition.

During the growth of the second epi, the reactant gases leach some of the NBL dopant from the surface of the wafer and redeposit it elsewhere, a process called *lateral autodoping.*[15] This mechanism can cause the formation of a thin layer of N-type silicon at the interface between the first and second epi layers, shorting adjacent wells. Autodoping can be limited by using antimony as the dopant or by conducting the epitaxy at reduced pressure. In either case, the BiCMOS NBL is liable to be somewhat more lightly doped than that used for standard bipolar.

---

[14] Eklund, *et al.,* pp. 120ff. Also see J. Erdeljac, B. Todd, L. Hutter, K. Wagensohner, and W. Bucksch, "A 2.0 micron BiCMOS Process Including DMOS Transistors for Merged Linear ASIC Analog/Digital/Power Applications," *Proceedings 1992 Applied Power Elect. Conf.,* 1992, pp. 517–522.

[15] M. W. M. Graef, B. J. H. Leunissen, and H. H. C. de Moor, "Antimony, Arsenic, Phosphorus, and Boron Autodoping in Silicon Epitaxy," *J. Electrochem. Soc.,* Vol. 132, #8, 1985, pp. 1942–1954.

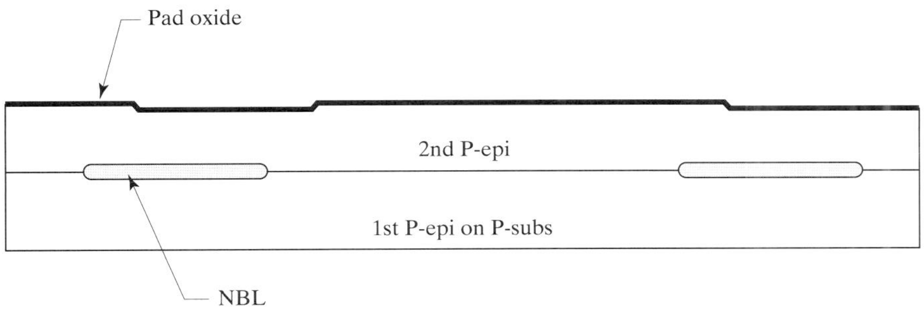

FIGURE 3.38 Wafer after second epitaxial deposition. The P+ substrate is not explicitly shown, but it lies beneath the first P-epi layer.

N-well Diffusion and Deep-N+

A thin layer of oxide is now grown and patterned using the N-well mask. Ion implantation deposits phosphorus, which is subsequently driven down to form the well diffusion. This drive stops before the well and NBL collide to permit the timely insertion of the deep-N+ deposition into the process flow. Additional oxide grown during the well drive allows patterning of the subsequent deep-N+ diffusion. After a heavy dose of phosphorus is implanted, the drive continues until the well and the deep-N+ both overlap the NBL by about 25% of their respective junction depths in order to minimize vertical resistance. Figure 3.39 shows a cross section of the wafer after the drive.

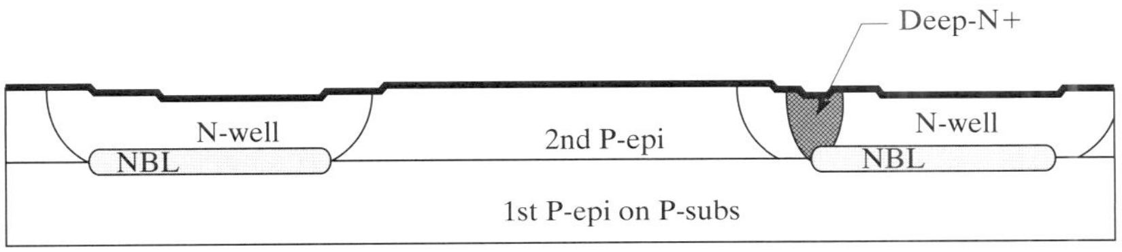

FIGURE 3.39 Wafer after N-well and deep-N+ deposition and drive.

The N-well diffusion influences a number of device parameters for both PMOS and bipolar transistors. Compromises must be made between the two types of devices to the detriment of either or both. For example, short-channel PMOS transistors require a moderately doped well to suppress punchthrough while bipolar transistors need a lightly doped well to form their collector drift region. The doping of the well therefore targets a compromise value that dictates a minimum channel length of 2 to 3μm. If one desires to fabricate shorter channel lengths, then additional wells must be added to allow independent optimization of the MOS and bipolar components of the process.

Base Implant

A uniform, thin layer of pad oxide is grown after the remnants of the previous oxide patterns have been stripped from the wafer. The wafer is patterned using the base mask, and boron is implanted through the pad oxide to form P-type regions, which are subsequently annealed under an inert ambient. Later high-temperature processing steps complete the base drive and set the final base-junction depth. Figure 3.40 shows the wafer after the base anneal. Triple counterdoping degrades the beta of the CDI NPN by raising the total base dopant concentration and hence the recombination rate in the neutral base. This can be partially offset by using a relatively lightly doped base implant. The final base-sheet resistance therefore equals several times that of standard bipolar.

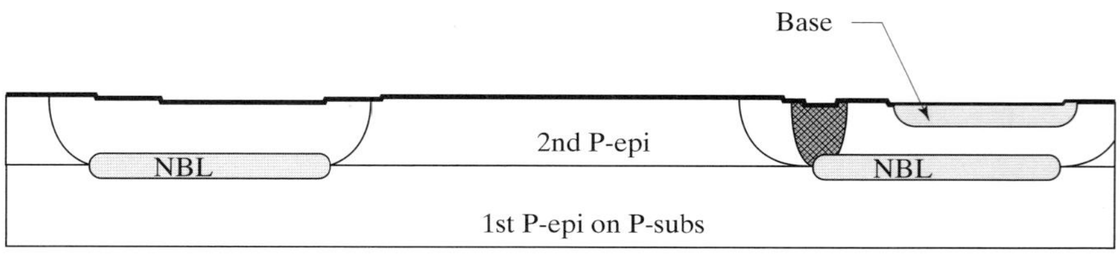

FIGURE 3.40 Wafer after base implant and anneal.

Inverse Moat

Analog BiCMOS uses the same LOCOS technology employed by polysilicon-gate CMOS. A thick layer of LPCVD nitride, patterned using the inverse moat mask, is etched to expose the regions where field oxide will eventually form (Figure 3.41). As in the case of polysilicon-gate CMOS, the MOS transistors occupy moat regions that are not covered by thick-field oxide. Moat regions also serve two additional purposes:

- Base regions are enclosed by moats to prevent oxidation-enhanced diffusion from overdriving the base.
- Schottky contacts are enclosed by moat to allow their etching to proceed simultaneously with base and emitter contacts.

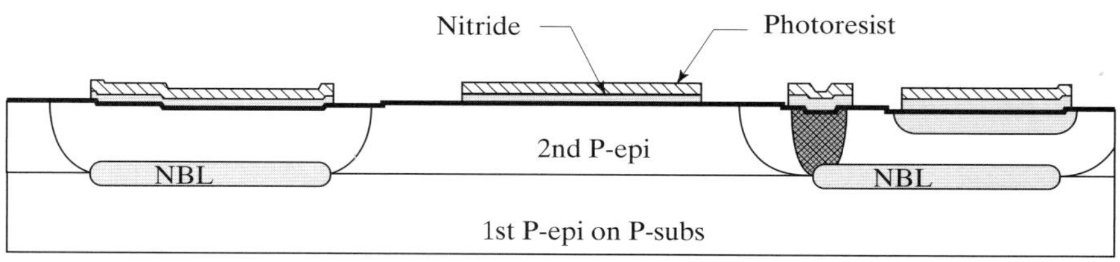

FIGURE 3.41 Wafer after nitride deposition and inverse moat etch.

The inverse moat mask consists of a color reverse of a combination of NMoat, PMoat, base, and Schottky contact drawing layers. Some processes generate the inverse moat mask automatically during pattern generation; other processes require the layout designer to code moat geometries over some or all of the aforementioned drawing layers.

Channel Stop Implants

Since analog BiCMOS uses (100) silicon, channel stop implants are required to raise the thick-field threshold above the operating voltage. The channel stop strategy for analog BiCMOS parallels that for polysilicon-gate CMOS. A blanket boron channel stop adjusts the thick-field threshold over the P-epi, while a patterned phosphorus channel stop sets the thick-field threshold over well regions. The boron channel stop is implanted using the patterned photoresist left from the inverse moat masking operation. A second coat of photoresist is applied and patterned, using the channel stop mask. This mask exposes only the regions of N-well that will ultimately lie beneath the thick-field oxide. A phosphorus implant offsets the previously deposited boron and increases the surface concentration in these well regions. Figure 3.42 illustrates the wafer cross section after the completion of both channel stop implants and the subsequent photoresist removal.

LOCOS Processing and Dummy Gate Oxidation

The LOCOS oxidation employs either steam or elevated pressures to increase the rate of oxide growth. Afterward, the nitride layer and the underlying pad oxide are stripped away. A dummy gate oxidation removes any lingering nitride residue (Figure 3.43).

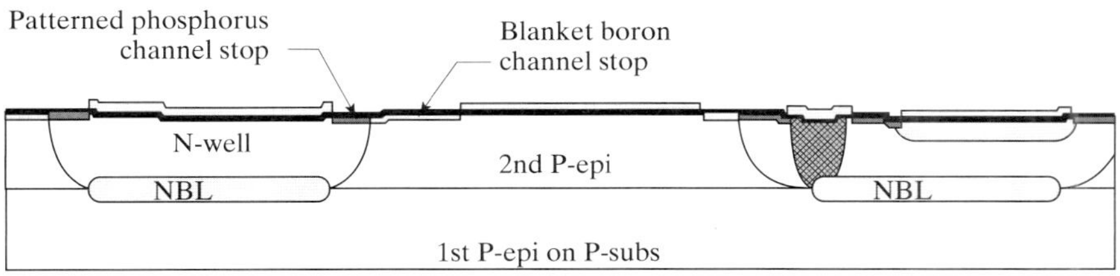

FIGURE 3.42 Wafer after channel stop implants.

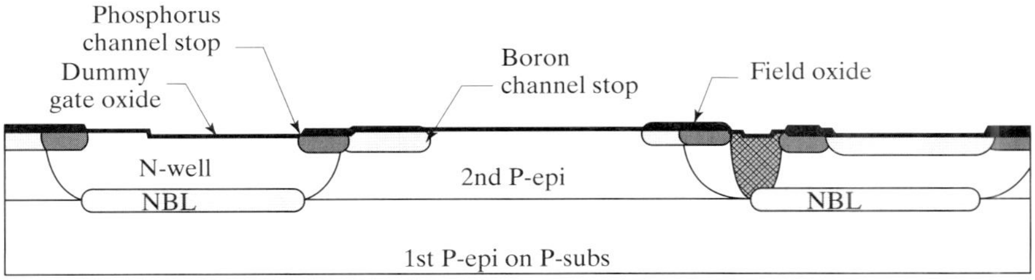

FIGURE 3.43 Wafer after LOCOS oxidation, nitride strip, and dummy gate oxidation.

Threshold Adjust

A single boron V_t adjust implant simultaneously raises the NMOS threshold and lowers the PMOS threshold. With suitable well and epi doping concentrations, both of these thresholds can be simultaneously adjusted to the desired target of 0.7 ± 0.2V. These MOS threshold voltages approximately coincide with the base-emitter voltages of bipolar transistors, although differences in the underlying physics preclude any true matching between the two types of devices.

The threshold implant process consists of coating the wafer with photoresist, patterning the resist with the V_t adjust mask, and implanting the necessary dose of boron through the dummy gate oxide. The final gate dielectric consists of some 300Å of high-quality dry oxide grown after the removal of the dummy gate oxide (Figure 3.43). Figure 3.44 shows a cross section of the resulting wafer.

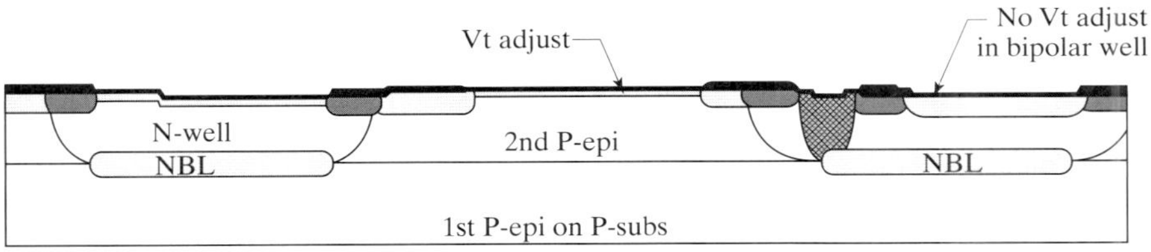

FIGURE 3.44 Wafer after V_t adjust implant.

Polysilicon Deposition and Pattern

The gates of the MOS transistors consist of heavily doped N-type polysilicon formed by depositing intrinsic polysilicon and subsequently doping it with a blanket phosphorus deposition. The patterning step uses the poly-1 mask, so named because a second poly layer may be added as a process option. Figure 3.45 shows the wafer after poly-1 pattern.

Source/Drain Implants

Analog BiCMOS typically produces bipolar transistors with 10 to 20V breakdown voltages. The MOS transistors should ideally be capable of withstanding similar voltages. The breakdown voltages of PSD and NSD can be raised to 15 to 20V without difficulty. Punchthrough can be averted by increasing the minimum channel

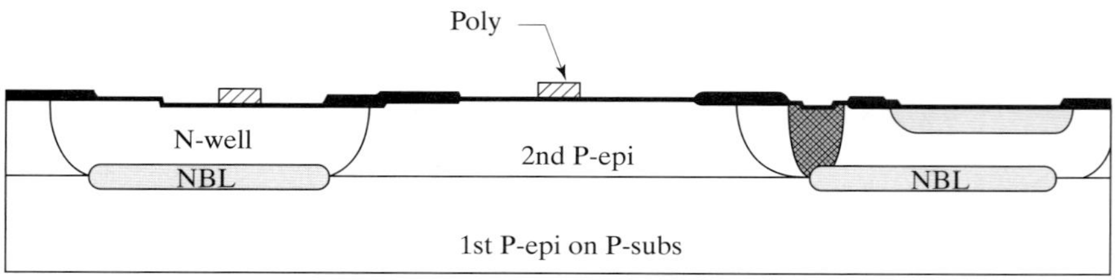

FIGURE 3.45 Wafer after polysilicon deposition and pattern. The channel stop and threshold adjust implants do not appear in this or subsequent cross sections.

length to 2 to 3μm. The addition of a lightly doped drain will suppress hot electron generation in the NMOS transistors.

Photoresist is spun onto the wafer and patterned using the N– S/D mask. A phosphorus implant forms lightly doped source and drain regions that self-align to the polysilicon gate. Isotropic deposition of oxide and subsequent anisotropic etching form sidewall spacers on either side of the gates. A second photoresist layer, patterned using the N+ S/D mask, defines a heavier and somewhat deeper N+ S/D implant that aligns to the edges of the oxide sidewall spacers. The lightly doped drain consists of that portion of the N– S/D implant that lies beneath the sidewall spacers. If all NMOS transistors receive LDD structures, then the N– S/D mask can be reused for the N+ S/D implant.

A 10 to 20V PMOS transistor does not require a lightly doped drain, eliminating the need for a P– S/D implant. The P+ S/D implant occurs after the formation of the sidewall spacers, so the PMOS transistor channel length increases by twice the width of the sidewall spacer (Figure 3.46). This increase in width can be offset by reducing the drawn length of the PMOS gate proportionately.

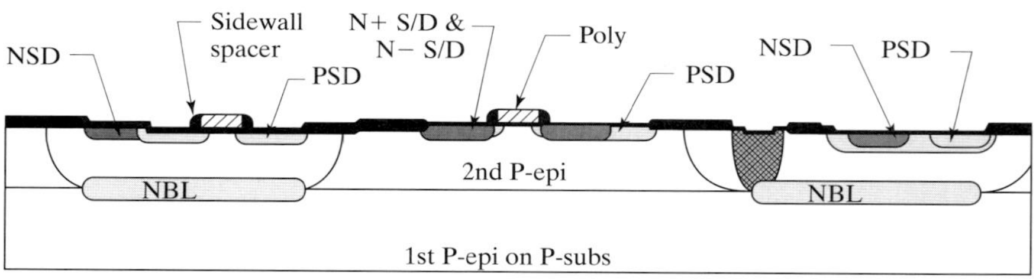

FIGURE 3.46 Wafer after N– S/D, N+ S/D, and P+ S/D implants.

Metallization and Protective Overcoat

The double-level metal flow requires five masks: contact, metal-1, via, metal-2, and protective overcoat. The contacts are silicided to control resistance and, coincidentally, to allow the formation of Schottky diodes. Refractory barrier metal ensures adequate step coverage across the nearly vertical sidewalls of contacts and vias. Copper-doped aluminum minimizes electromigration susceptibility, and a thick layer of compressive nitride protective overcoat provides a mechanical and chemical barrier between the metallization and the encapsulation. Figure 3.47 shows the cross section of a completed wafer, which includes NPN, NMOS, and PMOS transistors.

Process Comparison

Analog BiCMOS uses all of the same steps as polysilicon-gate CMOS. Three additional masks (NBL, deep-N+, and base) are inserted at opportune points in the process as shown by the shaded entries in Table 3.6. NBL deposition must occur before the growth of the second epi. Deep-N+ and base must occur early in the process because these deep diffusions require high temperatures and long drive times.

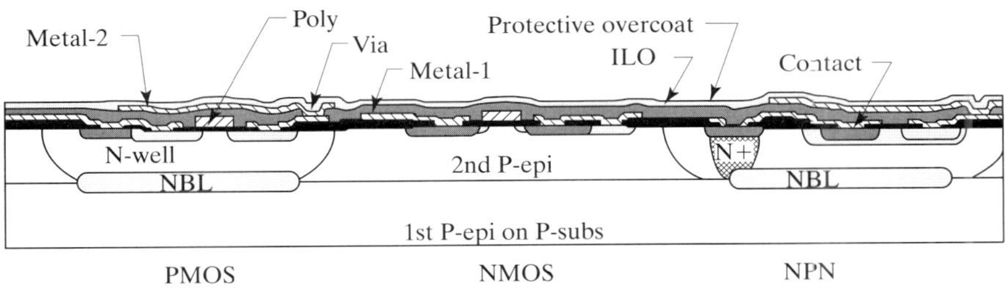

FIGURE 3.47 Completed wafer, showing metal-1 and metal-2 layers.

TABLE 3.6 Comparison of analog BiCMOS and polysilicon-gate CMOS processes.

Mask	Polysilicon-gate CMOS	Analog BiCMOS
		First epi growth
1. NBL		NBL deposition/anneal
	Epi growth	Second epi growth
2. N-well	N-well deposition/drive	N-well deposition/drive
3. Deep-N+		Deep-N+ deposition/drive
	Pad oxidation	Pad oxidation
4. Base		Base implant/anneal
5. Inverse moat	Nitride deposition/pattern Blanket boron channel stop	Nitride deposition/pattern Blanket boron channel stop
6. Channel stop	Patterned phosphorus channel stop	Patterned phosphorus channel stop
	LOCOS oxidation Nitride/pad oxide strip Dummy gate oxidation	LOCOS oxidation Nitride/pad oxide strip Dummy gate oxidation
7. V_t Adjust	Threshold adjust implant True gate oxidation	Threshold adjust implant True gate oxidation
8. Poly-1	Polysilicon deposition Poly implant/anneal	Polysilicon deposition Poly implant/anneal
9. NSD	N– S/D implant Sidewall spacer formation	N– S/D implant Sidewall spacer formation
10. NSD (again)	N+ S/D implant	N+ S/D implant
11. PSD	P+ S/D implant MLO deposition	P+ S/D implant MLO deposition
12. Contact	Contact OR Platinum sputter/sinter/etch	Contact OR Platinum sputter/sinter/etch
13. Metal-1	1st metal deposition/etch ILO deposition/planarization	1st metal deposition/etch ILO deposition/planarization
14. Via	Via etch	Via etch
15. Metal-2	2nd metal deposition/etch	2nd metal deposition/etch
16. PO	PO deposition/etch	PO deposition/etch

3.3.3. Available Devices

All of the devices available in polysilicon-gate CMOS also exist in analog BiCMOS. These devices include LDD NMOS and SDD PMOS transistors, four types of resistors (poly-1, PSD, NSD, and N-well), gate oxide capacitors, and Schottky diodes. Extended drain transistors can also be created without using any additional masks. Sections 3.2.3 and 3.2.4 discuss the preceding devices in detail. Additional analog

BiCMOS components include a CDI NPN transistor, lateral and substrate PNP transistors, and base resistors.

NPN Transistors

Figure 3.48 shows the layout and cross section of a minimum-geometry NPN transistor. The collector of the NPN consists of an N-well region into which the base and emitter (consisting of NSD) are successively diffused. The inclusion of NBL beneath the active region of the transistor and the addition of a deep-N+ sinker help minimize collector resistance. NSD implanted on top of the sinker ensures Ohmic contact. In a similar manner, PSD allows contact to the lightly doped base. The general appearance of this transistor closely resembles that of the standard bipolar transistor in Figure 3.10. There are, however, several subtle differences.

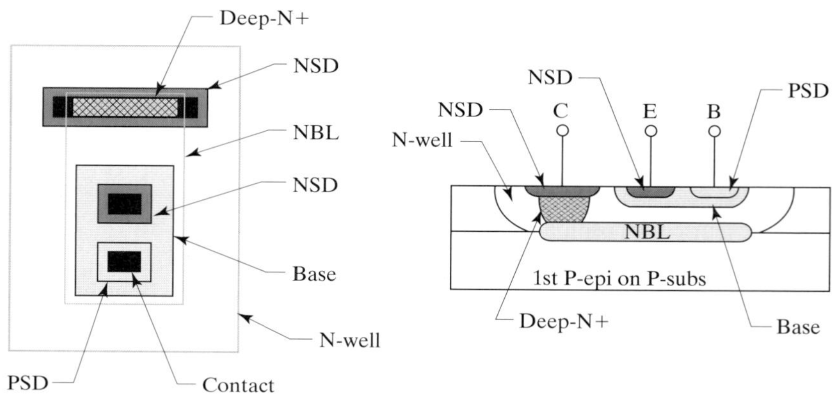

FIGURE 3.48 Layout and cross section of an NPN transistor with deep-N+ and NBL.

The use of three successive counterdopings increases recombination in the neutral base and therefore decreases the gain of the BiCMOS NPN transistor. Lighter base doping partially compensates for this effect. Conventional circuit design techniques require a minimum beta of about fifty. Allowing for a tolerance of ±50%, the nominal beta target becomes seventy-five, which is only about half that offered by a standard bipolar NPN transistor. Higher betas could be achieved, but only at the expense of degrading other device characteristics.

The graded well exhibits an extremely high collector resistance if deep-N+ is not placed beneath the collector contact. This resistance causes a soft transition from the saturation to the forward active region, which resembles the effects of quasisaturation (Section 8.3.2). The transistor may also intrinsically saturate even when the terminal voltages seem to indicate that saturation cannot occur (Section 8.1.4). These problems can be prevented by adding a deep-N+ sinker. Transistors used as Zeners do not require deep-N+, even in analog BiCMOS, because the collectors of these devices do not conduct significant current.

The analog BiCMOS NPN transistor does not perform as well as the standard bipolar NPN transistor, but it still serves for most applications (Table 3.7). Analog BiCMOS also allows much smaller emitter areas than standard bipolar. This benefit does not translate into a proportional reduction in transistor area since many other spacings contribute to the size of the final device. Still, the minimum-geometry analog BiCMOS transistor requires only about 30% of the room of its standard bipolar counterpart.

PNP Transistors

Analog BiCMOS supports both substrate and lateral PNP transistors. Figure 3.49 shows the layout and cross section of a representative substrate PNP transistor. This device is

Parameter	Standard Bipolar	Analog BiCMOS
Drawn emitter area	100μm^2	16μm^2
Peak current gain (beta)	150	80
Early voltage	120V	120V
Collector resistance, in saturation	100Ω	200Ω
Typical operating current range for maximum beta	5μA–2mA	1–200μA
V_{EBO} (Emitter-base breakdown, collector open)	7V	8V
V_{CBO} (Collector-base breakdown, emitter open)	60V	50V
V_{CEO} (Collector-emitter breakdown, base open)	45V	25V

TABLE 3.7 NPN device parameters for standard bipolar and analog BiCMOS.

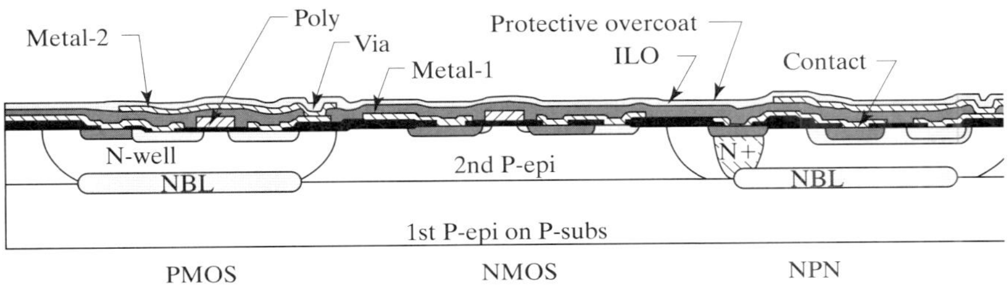

FIGURE 3.49 Layout and cross section of a typical substrate PNP transistor. The substrate acts as the collector.

constructed by implanting PSD into an N-well region. The PSD implant forms the emitter of the transistor while N-well forms the base. A small NSD plug ensures Ohmic contact to the lightly doped N-well. The emitter of the substrate PNP cannot consist of base diffusion because this reaches so deeply into the well that it compromises the breakdown voltage of the transistor. The characteristics of the analog BiCMOS substrate PNP broadly resemble those of a substrate PNP formed in standard bipolar.

Figure 3.50 shows a layout and cross section of a lateral PNP transistor constructed in analog BiCMOS. This device employs the base diffusion to form the emitter and collector, which both reside in an N-well region that forms the base. The addition of NBL to the transistor serves several purposes. First, it acts as a depletion stop, which allows the lateral PNP to withstand higher operating voltages without suffering punchthrough. Second, NBL blocks punchthrough breakdown and allows the base implant to be used in place of the shallower PSD implant. The deeper base implant enhances emitter sidewall injection and therefore raises the beta of the device. NBL also helps to minimize substrate injection. The graded nature of the well reduces the effectiveness of the NBL as a minority carrier barrier, but substantial benefits remain. Without NBL, the lateral transistor would exhibit an apparent beta of less than 10; with NBL the beta can easily exceed 100.

Additional implants must surround the contacts since both the N-well and the base diffusion are too lightly doped to allow direct Ohmic contact. A rectangle of PSD encloses both the emitter and the collector, but this implant only penetrates the moat regions that form the actual collector and emitter regions of the device. The thick LOCOS oxide blocks the PSD implant from reaching the base of the transistor and prevents the P-type implant from shorting the emitter and collector. Similarly, an NSD plug allows Ohmic contact to the N-well. Deep-N+ is not normally

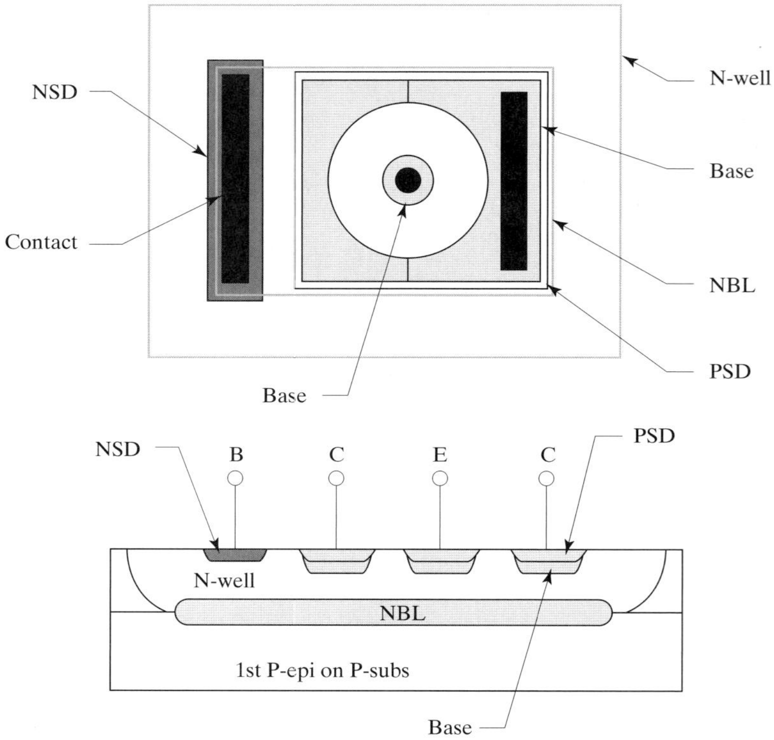

FIGURE 3.50 Layout and cross section of a lateral PNP transistor.

required, although it may become necessary in larger transistors that conduct significant base current.

The minimum-geometry lateral PNP transistor can attain a peak beta well in excess of 100. Fine-line photolithography allows a narrower base width than standard bipolar and greatly reduces the area of a minimum emitter. As the emitter area decreases, the ratio of periphery to area increases. This enhances the desired lateral injection of carriers at the expense of undesired vertical injection. The graded nature of the well and the presence of the channel stop implant generate a doping gradient that forces carriers away from the surface, diminishing surface recombination losses. The use of (100) silicon instead of (111) silicon also reduces surface recombination by minimizing the surface state charge. All of these effects together produce a transistor having a beta of up to 500.

The use of a lightly doped base implant for the emitter reduces emitter injection efficiency and causes a corresponding reduction in beta. Large PNP transistors consist of arrays of many minimum emitters to achieve a high area-to-periphery ratio. Table 3.8 lists several important parameters for both types of PNP transistors available in analog BiCMOS.

TABLE 3.8 Typical PNP device parameters for analog BiCMOS.

Parameter	Lateral PNP	Substrate PNP
Drawn emitter area	16μm^2	16μm^2
Drawn base width	5μm	—
Peak current gain (beta)	120	100
Early voltage	80V	100V
Typical operating current for maximum beta	1–20μA	1–50μA
V_{EBO} (Emitter-base breakdown, collector open)	45V	45V
V_{CBO} (Collector-base breakdown, emitter open)	45V	45V
V_{CEO} (Collector-emitter breakdown, base open)	30V	45V

Resistors

Analog BiCMOS base resistors consist of rectangles of base material occupying an N-well. Contact is made to either end of the resistor through a PSD plug. The well contact contains an NSD implant to increase the surface doping of the N-well. As with base resistors in standard bipolar, the addition of NBL serves not only to block possible minority carrier injection into the well during transients, but also to raise the operating voltage by preventing punchthrough between the resistor and the underlying substrate. Figure 3.51 shows the cross section and layout of a typical base resistor. The analog BiCMOS base diffusion is relatively lightly doped and typically exhibits a sheet resistance of 500Ω/□. While the high sheet resistance provides compactness, it also complicates resistor layout because the diffusion becomes vulnerable to surface depletion effects (Section 5.3.3). Few designers actually employ base resistors; most prefer to pay the small additional cost for a process extension for high-sheet polysilicon resistors. Pinch resistors are also available, but they offer little or no advantage over minimum-width poly resistors.

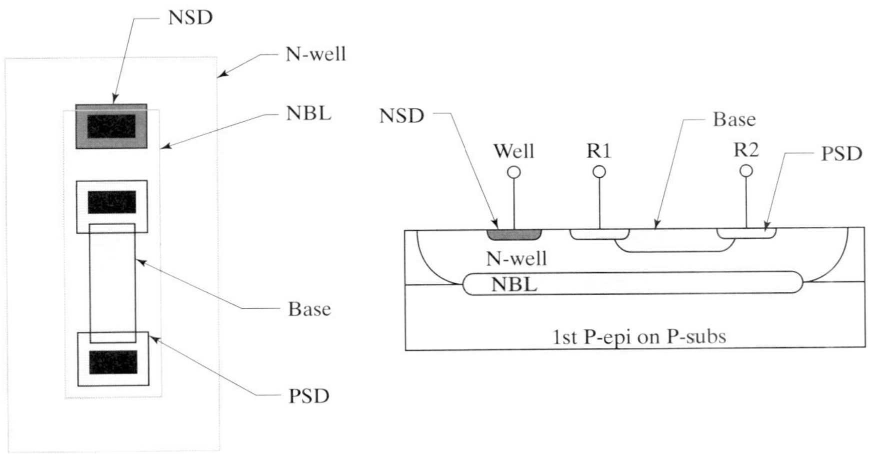

FIGURE 3.51 Layout and cross section of a base resistor.

3.4 SUMMARY

This chapter has examined three representative processes in detail: the standard bipolar process used to construct early analog integrated circuits, a polysilicon-gate CMOS process favored for digital logic, and an analog BiCMOS process, which merges the best of these two technologies onto a common substrate. Variations of these three processes fabricate the bulk of the low-cost, high-volume analog integrated circuits available today.

Standard bipolar is the oldest of the many available bipolar processes, most of which offer much faster switching speeds. The newer processes often replace junction isolation with partial or complete dielectric isolation to simultaneously reduce component area and minimize parasitic capacitance. Diffusions consist of much thinner layers, often created through innovative use of polysilicon as a doping source. Self-alignment and improved dimensional control allow the construction of far smaller geometries. These improvements have increased switching speeds by two orders of magnitude, allowing modern bipolar transistors to operate at speeds approaching fifty gigahertz.[16] The highly specialized processes used to obtain such performance are fully as complex and costly as the most sophisticated of CMOS processes.

[16] These speeds are only obtained using specialized processing techniques, such as the formation of silicon-germanium layers within the base of the transistors.

CMOS processes have undergone their share of evolutionary advances. Gate lengths have shrunk relentlessly as digital designers have sought to pack ever larger numbers of transistors onto limited amounts of silicon real estate. Once gate lengths fall below 1μm, a variety of short-channel effects dictate the use of much more elaborate structures. The backgate doping levels must increase to combat premature punchthrough and elaborate implantation strategies become necessary to confine the channel to the desired region immediately beneath the gate oxide. Cost and complexity have soared as processes push steadily deeper into the submicron regime seeking the elusive ultimate limits of device performance.

BiCMOS technologies seek to merge the best of both worlds. They are therefore heir to both the complexities of submicron CMOS and the elaborate bipolar structures required to obtain high switching speeds. Almost any bipolar or CMOS innovation can be merged into a BiCMOS process, and most have been.

3.5 EXERCISES

Refer to Appendix C for layout rules and process specifications.

3.1. Lay out the NPN transistor shown in Figure 3.10. Use a minimum-size square emitter and a minimum contact overlap of the deep-N+ sinker. Allow room for all necessary metallization.

3.2. Draw a cross section of the NPN transistor shown in Exercise 3.1 to scale, assuming an epi thickness of 10μm, NBL up-diffusion from the epi-substrate junction of 3μm, NBL down-diffusion from the epi-substrate junction of 4μm, an isolation junction depth of 12μm, a deep-N+ junction depth of 9μm, a base junction depth of 2μm, and an emitter junction depth of 1μm. Assume 80% outdiffusion where necessary. Oxide nonplanarities and deposited layers need not be shown.

3.3. Lay out the lateral PNP transistor shown in Figure 3.12. Use the minimum possible basewidth. Allow room for all necessary metallization, including a circular metal field plate connecting to the emitter contact and overhanging the inner edge of the collector region by 2μm.

3.4. Lay out a 500Ω base resistor following the example in Figure 3.13. Make the base resistor 8μm wide, and widen the contacts as much as the width of the resistor allows.

3.5. Lay out a 25kΩ base pinch resistor following the example in Figure 3.15. Assume that all of the resistance comes from the portion of the base beneath the emitter plate. Make the base strip 8μm wide and overlap the emitter plate over the base strip by at least 6μm. NBL should overlap the base strip in the pinched region by at least 2μm.

3.6. Lay out a fingered junction capacitor similar to the one shown in Figure 3.16. Make each of the three emitter fingers 50μm long. The emitter plate should overlap the base by at least 6μm; minimize all other dimensions.

3.7. Lay out the Schottky diode shown in Figure 3.18, assuming a contact opening that is 25 by 25μm. Overlap metal over the Schottky contact layer (SCONT) by no less than 4μm.

3.8. Lay out a 20kΩ HSR resistor. Make the width of the HSR resistor 8μm. The contacts should have the same width as the HSR resistor body. Assume that the base heads contribute negligible resistance and compute the value of the resistor based only on the drawn length of the HSR segment between the base heads.

3.9. Lay out an NMOS transistor with a drawn width of 10μm and a drawn length of 4μm following the example in Figure 3.29. Allow room for all necessary metallization.

3.10. Draw a cross section of the NMOS shown in Exercise 3.9 to scale, assuming a well junction depth of 6μm, PSD and NSD junction depths of 1μm, a gate oxide thickness of 350Å (0.035μm), and a polysilicon thickness of 3kÅ. Ignore the V_t adjust and the channel stop implants. Assume 80% outdiffusion where necessary. Assume the silicon surface is planar, and ignore details of the metallization system.

3.11. Lay out a PMOS transistor with a drawn width of 7μm and a drawn length of 15μm following the example in Figure 3.30. Assume NBL is not used. Include all necessary metallization.

3.12. Lay out a PSD resistor with a value of 200Ω following the example in Figure 3.33B. Make the resistor minimum width, and abut the well contact with one end of the resistor to save space.

3.13. Lay out a 3pF poly capacitor following the example in Figure 3.34. Include all necessary metallization. Assume that contacts and vias can reside on top of the poly plate.

3.14. Lay out the BiCMOS NPN transistor shown in Figure 3.48. The NBL should overlap the base region by at least 2μm. Use minimum emitter dimensions and include all necessary metallization.

3.15. Draw a cross section of the NPN from Exercise 3.14 to proper scale, assuming the dimensions given in Exercise 3.10. Assume NBL diffuses upward by 3μm and downward by 2μm. In addition, assume a first epi thickness of 7μm, a deep-N+ junction depth of 5μm, and a base junction depth of 1.5μm. Ignore channel-stop implants and the effects of LOCOS field oxidation on surface planarity. Assume that the silicon surface is planar, and ignore the details of the metallization system.

3.16. Lay out the BiCMOS lateral PNP transistor shown in Figure 3.50. Assume a minimum basewidth. Unlike the device from Exercise 3.3, this transistor does not require a metal field plate over the base region. NBL should overlap the outer edge of the collector by at least 1.0μm. Include all necessary metallization.

3.17. Lay out the resistor-transistor logic NOR gate shown in Figure 3.52A using standard bipolar layout rules. Place Q_1 and Q_2 in the same tank, and use as small a plug of deep-N+ as possible to contact the collector of this tank. Assume the emitters of Q_1 and Q_2 are both minimum-size. Place R_1 in its own tank. Provide at least one substrate contact. Surround this contact with base; this base region may touch, but not extend into, adjacent tanks. Label all inputs and outputs.

3.18. Lay out the CMOS NOR gate shown in Figure 3.52B using poly-gate CMOS layout rules. The W and L values for each transistor are shown on the schematic in the form of a fraction; 5/3 indicates a drawn width of 5μm and a drawn length of 3μm. Place all PMOS transistors in the same well, and connect this well to V_{DD}. Provide at least one substrate contact. Bring all inputs and outputs up to second-level metal, and label them appropriately.

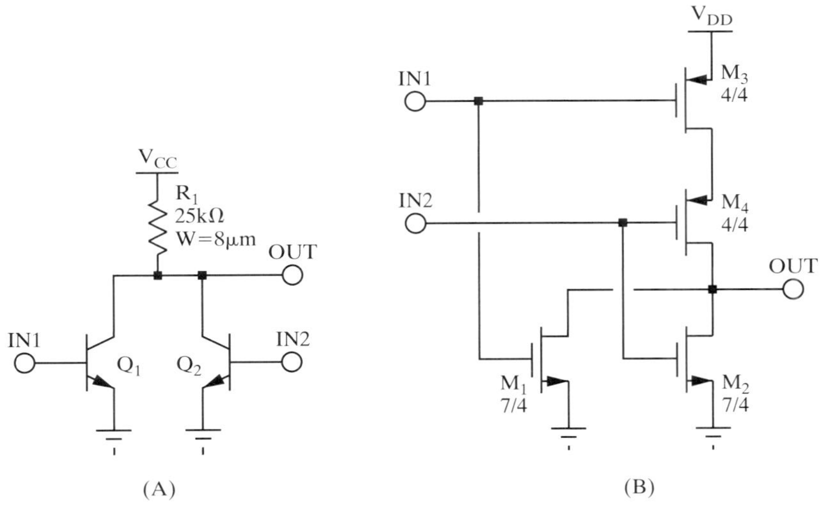

FIGURE 3.52 Circuits for Exercises 3.17 and 3.18.

4 Failure Mechanisms

Integrated circuits are incredibly complex devices, and few of them are perfect. Most contain subtle weaknesses and flaws, which predispose them toward eventual failure. Such components can fail catastrophically and without warning after operating perfectly for many years. Engineers have traditionally relied on quality assurance programs to uncover hidden design flaws. Operation under stressful conditions can accelerate many failure mechanisms, but not every design flaw can be found by testing. The designer must therefore find and eliminate as many of these flaws as possible.

The layout of an integrated circuit contributes to many types of failures. If the designer knows about potential weaknesses, then safeguards can be built into the integrated circuit to protect it against failure. This chapter discusses a number of failure mechanisms that can be partially or entirely prevented by layout precautions.

4.1 ELECTRICAL OVERSTRESS

The term *electrical overstress* (EOS) refers to failures caused by the application of excessive voltages or currents to a component. Layout precautions can minimize the probability of three common types of EOS failures. *Electrostatic discharge* (ESD) is a form of electrical overstress caused by static electricity. The addition of special protective structures to vulnerable bondpads can minimize ESD failures. *Electromigration* is a slow wearout mechanism caused by excessive current densities; it can eventually cause open circuits or shorts between adjacent leads. Electromigration failures can be prevented by making leads wide enough to handle the maximum operating currents. The *antenna effect* is an unusual failure mechanism caused by charge accumulation on gate electrodes during etching or ion implantation. The problems posed by the antenna effect can be minimized by following a few simple design guidelines.

4.1.1. Electrostatic Discharge (ESD)

Almost any form of friction can generate static electricity. For example, if you shuffle across a carpet in dry weather and then touch a metal doorknob, a visible spark will leap from finger to doorknob. The human body acts as a capacitor, and the act of shuffling across a carpet charges this capacitance to a potential of 10,000V or more. When a finger is brought near the doorknob, the sudden discharge creates a visible spark and a perceptible electrical shock. A discharge of as little as 50V will destroy the gate dielectric of a typical integrated MOS transistor. Voltages this low produce neither visible sparks nor perceptible electrical shock. Almost any human or mechanical activity can produce such low-level electrostatic discharges.

Proper handling precautions will minimize the risks of electrostatic discharge. ESD-sensitive components (including integrated circuits) should always be stored in static-shielded packaging. Grounded wrist straps and soldering irons can reduce potential opportunities for ESD discharges. Humidifiers, ionizers, and antistatic mats can minimize the buildup of static charges around workstations and machinery. These precautions reduce but do not eliminate ESD damage, so manufacturers routinely include special ESD protection structures on-board integrated circuits. These structures are designed to absorb and dissipate moderate levels of ESD energy without damage.

Special tests can measure the vulnerability of an integrated circuit to ESD. The two most common test circuits are called the human body model and the machine model.[1] The *human body model* (HBM) employs the circuit shown in Figure 4.1A. When the switch is pressed, a 150pF capacitor charged to a specified voltage discharges through the integrated circuit to ground. A 1.5kΩ series resistor limits the peak current through the part. Ideally each pair of pins would be independently tested for ESD susceptibility, but most testing regimens only specify a limited number of pin combinations to reduce test time. Each pair of pins is subjected to a series of positive and negative pulses; for example, five positive and five negative. Modern integrated circuits are routinely expected to survive 2kV HBM. Specific pins on certain parts may be required to survive up to 25kV HBM.

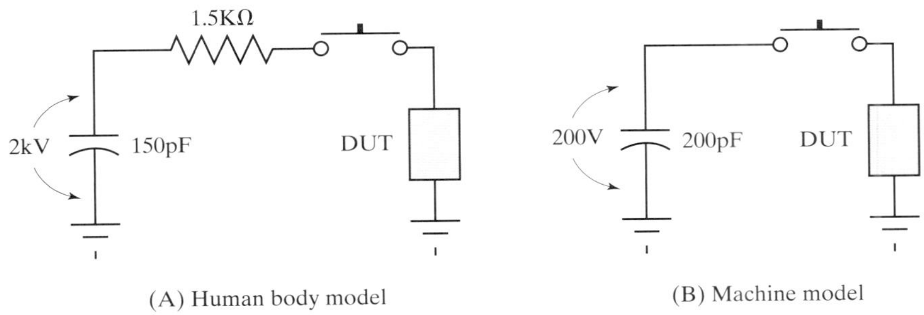

FIGURE 4.1 Representative ESD tests: 2kV human body model (A) and 200V machine model (B). In both circuits, DUT stands for *device under test.*

Figure 4.1B shows the circuit employed for the *machine model* (MM). A 200pF capacitor charged to a specified voltage discharges through the integrated circuit to ground. The test circuit contains no intentional series resistance, but practical testers incorporate some 20 to 40Ω in wiring, switches, sockets, and so forth. This, together with the series inductance of the wiring, limits the peak currents generated during testing. The machine model forms a much harsher test than the human body model; few parts can survive more than 500V under these conditions.

A third ESD test called the *charged device model* (CDM) is gradually replacing the machine model. The charged device model charges the integrated circuit package to a specified static potential and then discharges one pin to a low-impedance ground. Researchers believe that this model provides more accurate modeling of

[1] R. Lahri, J. Shibley, D. Merrill, H. Wang, and B. Bastani, "Process Reliability," in A. R. Alvarez, ed., *BiCMOS Technology and Applications,* 2nd ed. (Boston: Kluwer Academic, 1993), p. 170ff.

factory handling conditions than either the human body or the machine model. A typical testing regimen will specify 1 to 1.5kV CDM testing.

Effects

Electrostatic discharge causes several different forms of electrical damage, including gate oxide rupture, gate oxide degradation, and avalanche-induced junction leakage.[2] In extreme cases, ESD discharges can even vaporize metallization or shatter the bulk silicon.

Less than 50V can rupture the gate dielectric of a typical MOS transistor. The rupture occurs in nanoseconds, requires little or no sustained current flow, and is for all intents and purposes irreversible. The rupture typically shorts the gate and the backgate of the damaged transistor.[3] Capacitors that use oxide or nitride dielectrics are also vulnerable to this failure mechanism. An ESD discharge that strikes a pin connecting only to gates or to capacitors may cause oxide rupture in these devices. If the pin connects to any diffusions, then these usually avalanche before the gate oxide ruptures.

The integrity of a gate oxide can be compromised by an ESD event that does not actually rupture it. The weakened oxide can then fail at any time, perhaps after hundreds or thousands of hours of flawless operation. Sometimes the failure does not occur until the product has been delivered to the customer. Testing cannot screen out this type of delayed ESD failure; instead, vulnerable dielectrics must be protected against excessive voltages.

Although junctions are considerably more robust than dielectrics, they can still suffer ESD damage. An avalanching junction dumps a large amount of energy into a small volume of silicon. Extreme current densities can sweep metallization through contacts to short out underlying junctions. Excessive heating can also physically damage junctions by melting or shattering the silicon. These catastrophic forms of junction damage most often manifest themselves as short circuits. Avalanched junctions that do not fail outright usually exhibit increased leakage that may or may not cause the overstressed unit to fail parametric testing.

Preventative Measures

All vulnerable pins must have ESD protection structures connected to their bondpads. Some pins can resist ESD and therefore do not require additional protection. Examples include pins connected to substrate and to large diffusions, such as the collectors of power NPN transistors. These large junctions disperse and absorb the ESD energy before it can damage other circuitry. The power supply pins of most integrated circuits connect to a multitude of diffusions, and thus are also quite robust.

Pins connecting to relatively small diffusions, particularly those connected to the bases or emitters of small NPN transistors, are vulnerable to ESD-induced junction damage. Avalanching the base-emitter junction of an NPN transistor permanently degrades its beta. A circuit designer can sometimes eliminate the vulnerable junctions by rearranging the circuit. ESD protection should be added to any pin connecting to a base-emitter junction, or more generally, to any pin connecting to a relatively small diffusion. Because ESD vulnerabilities are difficult to predict, cautious designers add protection devices to all pins that might be even remotely vulnerable.

---

[2] Some of these are discussed in A. Amerasekera, W. van den Abeelen, L. van Roozendaal, M. Hannemann, and P. Schofield, "ESD Failure Modes: Characteristics, Mechanisms, and Process Influences," *IEEE Trans. Electron Devices,* Vol. 39, #2, 1992, pp. 430–436.

[3] C. M. Osburn and D. W. Ormond, "Dielectric Breakdown in Silicon Dioxide Films on Silicon, II. Influence of Processing and Materials," *J. Electrochem. Soc.,* Vol. 119, #5, 1972, pp. 597–603.

Pins that connect only to gates of MOS transistors or to deposited capacitor electrodes are extremely vulnerable to ESD-induced dielectric rupture. Special gate protection structures should be placed on all such pins. These structures usually include a significant amount of series resistance (typically 500Ω to 5kΩ) and therefore cannot be used on pins that must conduct more than a fraction of a milliamp.

Processes that employ thin emitter oxides are also susceptible to ESD-induced rupture. This vulnerability can be eliminated by ensuring that leads that connect to external bondpads do not cross any emitter region to which they do not connect. Alternatively, ESD structures similar to those used for protecting gates can protect the vulnerable bondpads. Most modern versions of the standard bipolar process employ thick emitter oxides, which eliminate the need for these precautions.[4]

Considerable ingenuity is often required to formulate successful ESD structures for analog integrated circuits. A dozen or more protection circuits are often required to satisfy the large range of voltages and the several types of vulnerable devices found in analog circuits. Section 13.4.3 discusses several commonly employed ESD structures and shows how these can be modified to meet a variety of special requirements.

4.1.2. Electromigration

Electromigration is a slow wearout phenomenon caused by extremely high current densities. The impact of moving carriers with stationary metal atoms causes a gradual displacement of the metal. In aluminum, electromigration only becomes a concern when current densities approach $5 \cdot 10^5 A/cm^2$. Although this may seem a tremendous current density, a minimum-width lead in a submicron process can experience electromigration at currents of only a few milliamps.[5]

Effects

Despite its homogenous appearance, aluminum metal is a polycrystalline material. The individual crystals, or *grains*, normally abut one another. Electromigration causes metal atoms to gradually move away from the grain boundaries, forming voids between adjacent grains. This causes a decrease in the lead's effective cross-sectional area and raises the current density seen by the remainder of the lead. Additional voids form and gradually coalesce until they ultimately sever the lead.

The addition of refractory barrier metal changes the observed modes of failure. Since the refractory metal is relatively resistive, most of the current initially flows through the aluminum. Once voiding finally severs the aluminum, the underlying refractory metal bridges the gap and continues to conduct current. Refractory metals strongly resist the effects of electromigration, so the lead will not completely fail. Instead, the formation of voids in the aluminum causes the lead's resistance to gradually and somewhat erratically increase. More ominously, aluminum metal displaced by voiding sometimes shorts adjacent leads together. The cross-sectional area of the aluminum portion of a lead therefore determines how much current it can safely conduct, regardless of the presence or absence of refractory barrier metal.

Refractory barrier metal is often used to prevent electromigration failures in contacts and vias because it ensures electrical continuity across steep sidewalls after the thin aluminum metallization at these points succumbs to electromigration-induced voiding. Lateral extrusion does not normally occur in contacts or vias since

---

[4] Even thick emitter oxides can rupture under certain conditions; see "Dielectric Breakdown of Emitter Oxide," *Semiconductor Reliability News,* Vol. IV, #1, 1992, p. 1.

[5] Assuming a lead width of one micron and a thickness of 5000Å, a current of 2.5mA will produce a current density of $5 \cdot 10^5 A/cm^2$.

a contiguous sheet of metal covers the entire structure. Likewise, resistance changes caused by voiding are usually small compared to the inherent resistance of the contact or via structure. Where these resistance changes cannot be tolerated, the designer can insert additional contacts or vias to help reduce the current density.

Preventative Measures

The first line of defense against electromigration consists of process improvements. Aluminum metallization is now routinely doped with 0.5 to 4% copper to enhance electromigration resistance.[6] Copper accumulates at the grain boundaries, where it inhibits voiding by increasing the activation energy required to dislodge metal atoms from the lattice. Copper-doped aluminum exhibits five to ten times the current handling capability of pure aluminum.[7] The electromigration resistance of leads can be further improved by using compressively stressed protective overcoats that confine the metal under pressure and inhibit void formation. Refractory barrier metal can also help prevent electromigration failures in contacts and vias. Most manufacturers do not rely upon refractory metal to protect other oxide steps because of the risk of lateral extrusion. Instead, the leads are designed so that the aluminum portion of the metallization can handle the full current over all portions of the metal pattern except at contact and via openings.

Processing techniques can minimize electromigration, but there remains some maximum current density that cannot be exceeded without risking eventual metallization failure. The design rules for each process thus define a maximum allowed current per unit width. Typical values are 2mA/μm (50mA/mil) for leads that do not cross oxide steps and 1mA/μm (25mA/mil) for those that do. These values depend on the thickness of the metallization and its composition, and on the anticipated operating temperature (Section 14.3.3). Consider a lead that must conduct 50mA following the electromigration limits specified above. If this lead routes across field oxide in order to avoid oxide steps, then it need be only 25μm wide; otherwise its width must increase to 50μm. The lead cannot widen abruptly at the oxide steps because the current only gradually flows out from a narrow lead into a wider one. The wider lead should extend beyond the oxide step in either direction for a distance at least twice its greatest width.

Excessive current can also cause bondwires to overheat and fail. In practice, a typical one-mil gold bondwire that is 50mil (1.25mm) in length can safely conduct about an amp, while a similar aluminum wire can conduct about 750mA. If the anticipated currents exceed these limits, then the design will require larger-diameter bondwires or multiple bondwires placed in parallel (Section 14.3.3).

4.1.3. The Antenna Effect

Dry etching uses intense electrical fields to generate an ionizing plasma. During the etching of the gate poly and the oxide sidewall spacers, electrostatic charges may accumulate on the gate poly. The resulting voltages may become so large that current flows through the gate oxide. Although the amount of energy involved is usually inadequate to rupture the gate oxide, it still degrades its dielectric strength. The amount of degradation is proportional to the total charge that passes through the gate oxide divided by the total gate oxide area (Section 11.1.2). Each poly region collects an electrostatic charge proportional to its own area. A small gate oxide

---

6 I. Ames, F. M. d'Heurle, and R. E. Horstmann, "Reduction of Electromigration in Aluminum Films by Copper Doping," *IBM J. of Research and Development,* Vol. 14, #4, 1970, pp. 461–463.

7 S. S. Iyer and C. Y. Ting, "Electromigration Study of Al-Cu/Ti/Al-Cu System," *Proc. International Reliability Physics Symp.,* 1984, pp. 273–278. See also Lahri, *et al.,* p. 166.

region connected to a large poly geometry can suffer disproportionate damage. This mechanism is sometimes called the *antenna effect* because the large poly area acts as an antenna to collect the charge that flows through the vulnerable gate oxide. Gate oxide damage due to the antenna effect has also been observed during ion implantation of the source/drain regions.[8,9]

The magnitude of the antenna effect is proportional to the ratio between exposed conductor area and gate oxide area. During the patterning of the polysilicon, the poly is the exposed conductor. Similarly, during the patterning of the first level of metal, the metal is the exposed conductor. Separate area ratios must be computed for each conductor layer. One also computes separate ratios for PMOS and NMOS gate oxides because the two may not break down at exactly the same voltage. Conductor/gate area ratios of several hundred are usually required to produce significant damage. Most layouts contain few geometries with ratios this large, so antenna-effect damage is usually limited to a few locations on the die. Figure 4.2A shows an example of a layout that can produce a conductor-gate area ratio large enough to trigger this type of failure. The gate lead of NMOS transistor M_1 has been elongated to facilitate connection to transistor M_2. The elongated lead has sufficient area to endanger the small gate oxide region of transistor M_1. This vulnerability can be eliminated by inserting a metal jumper in the poly lead next to transistor M_1 (Figure 4.2B). This jumper drastically reduces the area of the poly geometry connected to M_1's gate oxide, which in turn reduces the conductor/gate area ratio to a safe value.

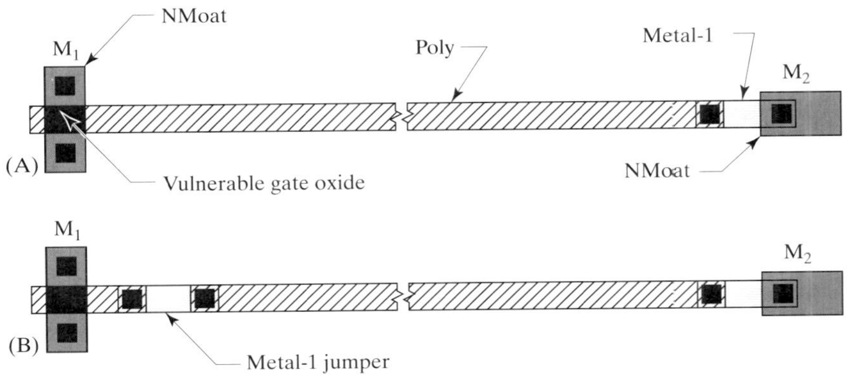

FIGURE 4.2 A layout susceptible to the antenna effect (A) can be made immune by the addition of a metal jumper (B).

Electrostatic damage can also occur during the etching of metal layers. Metal regions connecting to diffusions rarely pose any problem because these diffusions provide paths through which the electrostatic charges can leak away. The topmost layer of metal is almost immune to antenna effects because every geometry on this layer connects to a diffusion somewhere on the die, but the lower metal layers do not necessarily connect to diffusions until the top metal layer is in place. During the etching of a lower metal layer, any geometry that does not connect to a diffusion through layers already present can potentially collect a damaging electrostatic charge.

Antenna effects on lower metal layers can be eliminated by inserting short jumpers on the top metal layer to minimize the conductor area of the lower metal

---

[8] T. Watanabe and Y. Yoshida, "Dielectric Breakdown of Gate Insulation Due to Reactive Ion Etching," *Solid State Technology*, Vol. 27, #4, 1984, pp. 263–266.

[9] K. Markus, C. M. Osburn, P. Magill, and S. M. Bobbio, "Thin-oxide Degradation Along Feature Edges During Reactive Ion Etching of Polysilicon Gates," *J. Vac. Sci. Technol. A*, Vol. 12, #4, 1994, pp. 1339–1345.

layer attached to small gate oxide regions. This solution resembles that shown in Figure 4.2B. In cases where top-metal jumpers are not feasible, damage can still be avoided by ensuring that the lower-metal lead connects to a diffusion. If the layout does not include any conveniently located diffusion, consider adding a minimum-size NMoat/P-epi or PMoat/N-well diode (Section 10.2). These diodes may affect circuit operation, so they must not be added without consulting the circuit designer.

4.2 CONTAMINATION

Integrated circuits are vulnerable to certain types of contaminants. Assuming that the device has been properly manufactured, very low levels of contaminants will initially exist inside the plastic encapsulation. Plastic mold compounds have been carefully formulated to provide the highest possible degree of resistance to penetration by external contaminants, but no plastic is entirely impregnable. Contaminants seep in along the interface between the metal pins and the plastic, or they directly penetrate the plastic itself. Two major contamination issues faced by modern plastic-encapsulated dice are *dry corrosion* and *mobile ion contamination.*

4.2.1. Dry Corrosion

The aluminum metal system will corrode if exposed to ionic contaminants in the presence of moisture. Only trace amounts of water are necessary to initiate this so-called *dry corrosion.* Since moisture and ionic contaminants are both ubiquitous, integrated circuits must depend on their encapsulation to protect them. Early mold compounds had little moisture resistance. Newer compounds are more impermeable, but given enough time, moisture will eventually penetrate any plastic package.[10]

All modern integrated circuits are covered with a protective overcoat that acts as a secondary moisture barrier. Unfortunately, openings must be made through this overcoat to allow bondwires to be attached to the die. Fuse trim schemes often require additional openings in the protective overcoat. All of these openings represent potential pathways for contaminants to reach the die.

Effects

Water alone cannot corrode aluminum, but many ionic substances dissolve in water to form relatively corrosive solutions. Phosphosilicate glasses containing more than about 5% phosphorus represent a corrosion risk, as moisture can leach phosphorus from the glass to produce phosphoric acid.[11] This acid rapidly attacks and dissolves aluminum, causing open circuit failures. Many modern processes use nitride or oxynitride protective overcoats to ensure that moisture cannot reach the phosphosilicate glass that lies beneath. Alternatively, the phosphorus content of the glass can be reduced by using a combination of boron and phosphorus as dopants. Both of these elements reduce the softening point of a glass, so a *borophosphosilicate glass* (BPSG) will require less phosphorus to achieve the same softening point as a phosphosilicate glass.

Halogen ions in water solution can also corrode aluminum.[12] Common salt, or sodium chloride, provides an abundant source of chloride ions. Moisture seeping

---

[10] J. E. Gunn and S. K. Malik, "Highly Accelerated Temperature and Humidity Stress Technique (HAST)," *Reliability Physics, 19th Annual Proc.,* 1981, pp. 48–51.

[11] W. M. Paulson and R. W. Kirk, "The Effects of Phosphorus-Doped Passivation Glass on the Corrosion of Aluminum," *Proc. Reliability Physics Symposium,* 1974, pp. 172–179.

[12] M. M. Ianuzzi, "Reliability and Failure Mechanisms of Non-hermetic Aluminum SIC's in an Environment Contaminated with Cl_2," *(sic), IEEE Trans. Comp. Hyb. Man. Tech.,* 6, 1983, pp. 191–201.

into the package of an integrated circuit can transport chloride ions to the surface of the die where they can begin to corrode the aluminum metal system. Significant levels of bromides do not normally occur in the environment, but plastic encapsulants often contain organobromine flame retardants. These flame retardants begin to decompose and release bromide ions at temperatures in excess of about 250°C, which limits the storage and soldering temperatures these packages can safely withstand.[13]

Preventative Measures

Although contamination may seem completely beyond the control of the layout designer, several measures can be taken to minimize vulnerabilities in the protective overcoat. The designer should minimize the number and size of all PO openings. A production die should not include any openings that are not absolutely necessary for its manufacture. If the designer wishes to include additional testpads for evaluation, then these should occupy a special test mask. When the part is released to production, the test mask should be replaced by a production PO mask that seals the test pads under protective overcoat.

Metal should overlap bondpad openings on all sides by an amount sufficient to account for misalignment. The metal bondpads will then protect the underlying oxide from the entry of moisture and other contaminants. Openings made for polysilicon or metal fuses should be made as small as possible, and no circuitry of any sort except the fuse element itself should appear within the opening.

4.2.2. Mobile Ion Contamination

Many potential contaminants dissolve in silicon dioxide at elevated temperatures, but most lose their mobility at normal operating temperatures because they become bound into the oxide macromolecule. The alkali metals are exceptions to this rule and remain mobile in silicon dioxide even at room temperature.[14] Of these so-called *mobile ions,* sodium is by far the most common and the most troublesome.

Effects

Mobile ion contamination induces parametric shifts, most noticeably in MOS transistor threshold voltages. Figure 4.3A shows the gate oxide of an NMOS transistor contaminated by sodium during manufacture. The positively charged sodium ions

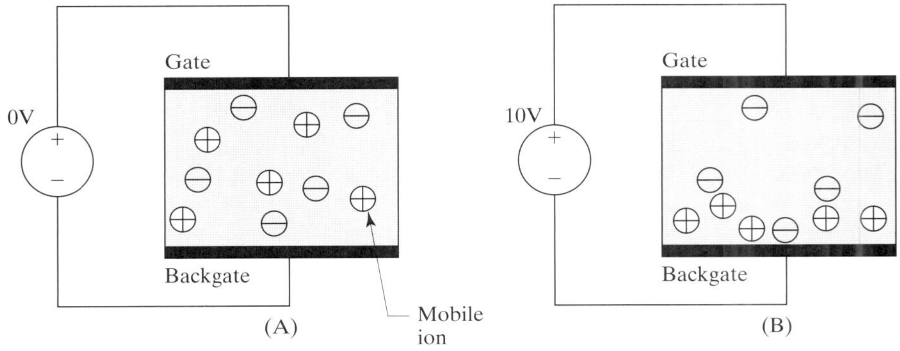

FIGURE 4.3 Behavior of mobile ions under bias: ions that were randomly distributed through the oxide (A) shift in unison in response to a positive gate bias (B).

[13] T. Raymond, "Avoiding Bond Pad Failure Mechanisms in Au-Al Systems," *Semiconductor International,* Sept. 1989, pp. 152–158.

[14] Actually, only lithium, sodium, and potassium (and perhaps hydrogen) qualify as mobile ions in silicon. The heavier alkali metals rubidium and cesium are far less mobile: B. E. Deal, "The Current Understanding of Charges in the Thermally Oxidized Silicon Structure," *J. Electrochem. Soc.,* Vol. 121, #6, 1974, pp. 198C–205C.

are initially distributed throughout the oxide. An equal number of negatively charged ions (anions) are also introduced. Unlike the sodium ions, these anions remain rigidly locked into the oxide macromolecule.[15]

Figure 4.3B shows the same gate dielectric after an extended period of operation under a positive gate bias. The positively charged gate electrode has repelled the mobile sodium ions down toward the oxide-silicon interface. Since the negative ions do not move, the redistribution of the sodium ions results in a net separation of charges within the oxide.[16] The presence of positive charges near the channel of the NMOS transistor decreases its threshold voltage. The magnitude of the threshold voltage shift depends on sodium ion concentration, gate bias, temperature, and time. Many analog circuits require that threshold voltages match within a few millivolts. Even low concentrations of mobile ions can produce shifts of this magnitude.

Mobile ion contamination can produce long-term failures when the slow drift of threshold voltages eventually causes a circuit to exceed its parametric limits. If the faulty devices are removed from operation and are baked at 200°C for a few hours, the mobile ions redistribute and the threshold shifts vanish. This treatment is only temporary; the threshold drift returns as soon as electrical bias is restored. Although analog circuits are particularly susceptible to parametric shifts caused by mobile ions, even digital circuits will eventually fail if the threshold voltages shift too far. Early metal-gate CMOS logic was plagued by threshold voltage shifts caused by severe sodium contamination.

Preventative Measures

Some mobile ions inevitably become incorporated in an integrated circuit during manufacture. This source of contamination can be minimized by using purer chemicals and improved processing techniques. MOS processes typically take extraordinary steps to ensure process cleanliness, but these alone cannot entirely eliminate threshold voltage variations.

Manufacturers of metal gate CMOS attempted to stabilize threshold voltages by adding phosphorus to the gate oxide.[17,18] Phosphorus stabilization had the desired effect of immobilizing alkali metal contaminants, but it also introduced a new problem. The electrically charged phosphate groups shift slightly under strong electrical fields even though they are bound to the oxide macromolecule. Phosphosilicate glasses therefore exhibit the same problem that they were intended to cure! All is not lost, though, because this *dielectric polarization* is not as severe a problem as mobile ion contamination. The threshold shift caused by a given voltage bias remains relatively small—a few tens of millivolts. The threshold shifts caused by dielectric polarization are also much more predictable than those produced by mobile ions, so circuit designers can predict whether a given circuit configuration will be adversely affected or not.[19] The threshold voltage shifts were finally eliminated altogether by using phosphorus-doped polysilicon gates rather than phosphorus-doped gate oxides. Phosphorus-doped polysilicon immobilizes

---

[15] Negative countercharges may not always exist, particularly if the contaminants enter the oxide after manufacture. Threshold shifts still occur because of image effects produced by the presence of electrical charges within the insulating oxide.

[16] N. E. Lycoudes and C. C. Childers, "Semiconductor Instability Failure Mechanisms Review," *IEEE Trans. of Reliability,* Vol. R-29, #3, 1980, pp. 237–249.

[17] M. Kuhn and D. J. Silversmith, "Ionic Contamination and Transport of Mobile Ions in MOS Structures," *J. Electrochem. Soc.,* Vol. 118, 1971, pp. 966–970.

[18] S. R. Hofstein, "Stabilization of MOS Devices," *Solid-State Electronics,* Vol. 10, 1967, pp. 657–670.

[19] E. H. Snow and B. E. Deal, "Polarization Phenomena and Other Properties of Phosphosilicate Glass Films on Silicon," *J. Electrochem. Soc.,* Vol. 113, #3, 1966, pp. 263–269.

alkali metals in much the same way as phosphorus stabilization without the added complication of dielectric polarization.

Moisture seeping into the integrated circuit's package can transport sodium in from the outside environment. Improved packaging materials can slow, but not stop, the ingress of sodium ions. The protective overcoat serves as a further barrier to mobile ions and can prevent them from reaching the vulnerable oxide layers in contact with the silicon. Protective overcoats typically consist of either silicon nitride, which is relatively impermeable to mobile ions, or phosphorus-doped glasses, which can immobilize them. The protective overcoat therefore serves as a final line of defense against impurities entering the die from outside.

Any opening through the protective overcoat represents a potential route for mobile ion contamination to enter the die. The metallization normally seals the bondpad openings, but scars left by probe needles can puncture the metal and expose the interlevel oxide (ILO) beneath. A minimum number of probe pads should be used, and these should be placed around the periphery of the die as far from sensitive analog circuitry as possible. Fuse openings through the protective overcoat also represent vulnerable points that should be kept far away from analog circuitry.

The scribe street surrounding the die typically consists of bare silicon because other materials either fracture or clog the saw blade. Contaminants can seep laterally into exposed oxide layers abutting the scribe street. Special structures, called *scribe seals*, placed around the periphery of the die can slow the ingress of contaminants. Figure 4.4A illustrates a typical scribe seal for a single-level-metal CMOS process. The first component of this scribe seal consists of a narrow contact strip surrounding the active area of the die. This contact must be a continuous ring uninterrupted by any gaps in order for it to block the lateral movement of mobile ions through the field oxide. A P-type diffusion placed underneath this contact allows it to double as a substrate contact. This arrangement is very convenient, as the metal plate forming part of the scribe seal also carries the substrate lead around the periphery of the die. The scribe seal also provides a guaranteed minimum area of substrate contacts.

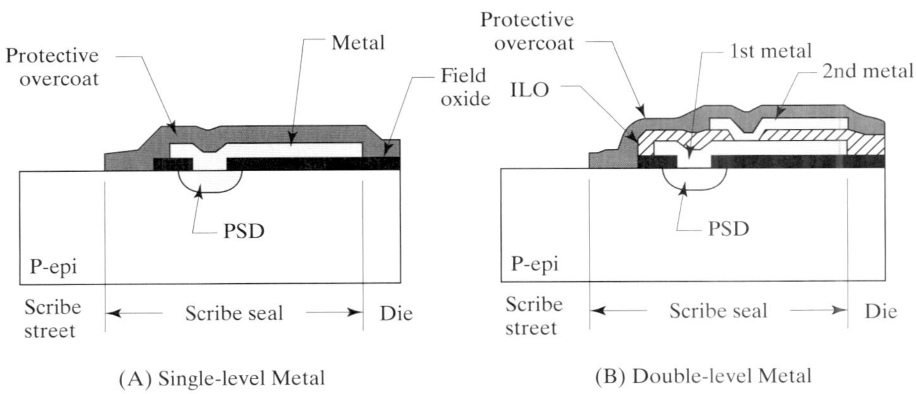

FIGURE 4.4 Scribe seals for single- and double-level-metal variants of a CMOS or BiCMOS process. Depending on the manufacturer, various diffusions may also be placed over the scribe street.

The scribe seal also contains a second contamination barrier formed by flapping the protective overcoat into the scribe street directly on top of the exposed silicon. Any mobile ions attempting to penetrate the scribe seal must first surmount this flap-down and next pass the continuous contact ring before reaching the active regions of the die. Most processes prohibit direct contact between nitride and silicon because compressive stresses in the nitride spawn defects in the silicon lattice. The flap-down of protective overcoat over the scribe street is permitted because the scribe street does not contain any active circuitry that could be damaged by defects. Nitride should still not touch exposed silicon inside the active area of the die because defects spawned by the damaged silicon can propagate for some distance and may affect adjacent components.

Figure 4.4B shows a scribe seal for a double-level-metal CMOS process. This seal includes a third barrier consisting of a continuous via ring placed just inside the contact ring. This via ring helps prevent contaminants from entering the ILO between the two metal layers. In a triple-level-metal process, a second via ring would be added to protect the second ILO layer between metal-2 and metal-3.

The scribe seals shown in Figure 4.4 can protect almost any die, but the substrate contacts may require different diffusions depending on the process flow. For example, standard bipolar would substitute a combination of P-isolation and base for the PSD rings of Figure 4.4. The P-isolation would probably extend to the edge of the active die because most designers surround the die with a ring of substrate contacts placed underneath the grounded metallization. The functionality of the scribe seals remains the same regardless of the exact diffusions used.

4.3 SURFACE EFFECTS

Surface regions of high electric field intensity can inject hot carriers into the overlying oxide. Surface electric fields can also induce the formation of parasitic channels. Both of these mechanisms are referred to as *surface effects* because they occur at the interface between the silicon and the overlying oxide.

4.3.1. Hot Carrier Injection

Carriers are always in constant motion due to the random thermal vibrations responsible for their diffusion. Electric fields may produce a slow drift of carriers in one direction, but the resulting drift velocity is usually much smaller than the instantaneous velocities produced by thermal agitation. Consequently, electric fields rarely increase the instantaneous velocities of carriers by any perceptible amount (Figure 4.5A). Only at extremely high electric field intensities does the drift of carriers become so rapid that the instantaneous velocities actually increase (Figure 4.5B). The resulting carriers are called *hot carriers* because they move at speeds normally achieved only at elevated temperatures.

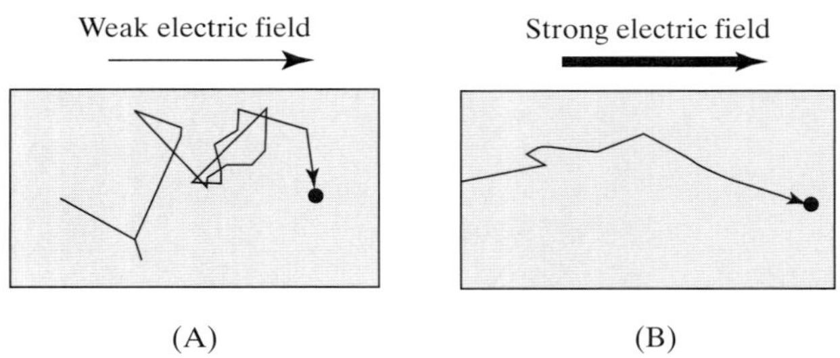

FIGURE 4.5 A weak electric field causes an overall drift of carriers but does not materially affect their instantaneous velocity (A), while a strong electric field actually increases the instantaneous velocity of the carriers (B).

Effects

MOS devices can generate hot carriers when operated in saturation at high drain-to-source voltages. As the drain-to-source voltage increases, the pinched-off portion of the channel slowly grows wider. The increase in width cannot keep pace with the applied voltage, so the electric field intensifies as the voltage increases. At high voltages, the electric field becomes large enough to generate hot carriers near the drain end of the transistor (Figure 4.6). NMOS transistors generate hot electrons, while

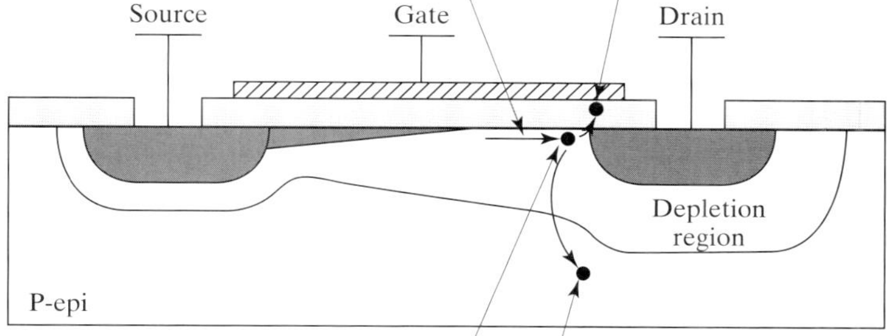

FIGURE 4.6 Simplified diagram showing the mechanism responsible for hot electron injection in an NMOS transistor.

PMOS transistors generate hot holes. Because of differences in effective mobilities, hot carrier production begins at substantially lower voltages in NMOS transistors than in PMOS transistors of similar dimensions. For example, if a 3μm NMOS experiences hot electron injection at 10V, then the equivalent PMOS would probably not experience significant hot hole injection below 20V.

Hot carriers produced at the drain end of the transistor collide with lattice atoms, and some few of the recoiling carriers travel upward into the overlying oxide. Most of these carriers pass through the oxide and return to the silicon, but a few become trapped at defect sites within the oxide. These trapped carriers represent a fixed oxide charge that gradually increases in magnitude as more carriers become trapped. This charge shifts the threshold voltage of the MOS transistor and can in turn affect the performance of the overall circuit.

Parametric shifts caused by hot carriers can be partially or completely reversed by baking the unbiased units at temperatures of 200 to 250°C for several hours. These temperatures impart sufficient thermal energy to the trapped carriers to free them and allow them to return to the silicon. The parametric shifts vanish as the fixed oxide charge dissipates. As in the case of mobile ions, the apparent cure is only temporary. As soon as bias is restored, hot carrier generation resumes and the threshold voltages begin to drift again.

Avalanching junctions also produce large numbers of hot carriers. Avalanche occurs near the surface in most diffused junctions because the dopant concentrations are highest there (Figure 4.7A). Some of the hot carriers produced by the avalanche process travel into the overlying oxide. In the case of a base-emitter junction, these carriers predominantly consist of hot holes that add to the positive fixed oxide charge.[20] As this charge increases, it induces a gradual widening of the depletion region at the surface (Figure 4.7B). The avalanche voltage slowly increases during

[20] G. Blasquez, G. Barbottin, and V. Boisson, "A Review of Passivation-Related Instabilities in Modern Silicon Devices," in G. Barbottin and A. Vapaille, eds., *Instabilities in Silicon Devices, Volume 2: Silicon Passivation and Related Instabilities* (Amsterdam: North-Holland, 1989), pp. 459–460. Blasquez, *et al.* state that Zener walkout in P+/N–junctions spontaneously reverses because some or all of the hot electron charge is not permanently trapped in the oxide macromolecule and can consequently dissipate over time even at room temperature. This effect is usually not observed in N+/P–junctions, presumably because the charge consists largely of holes.

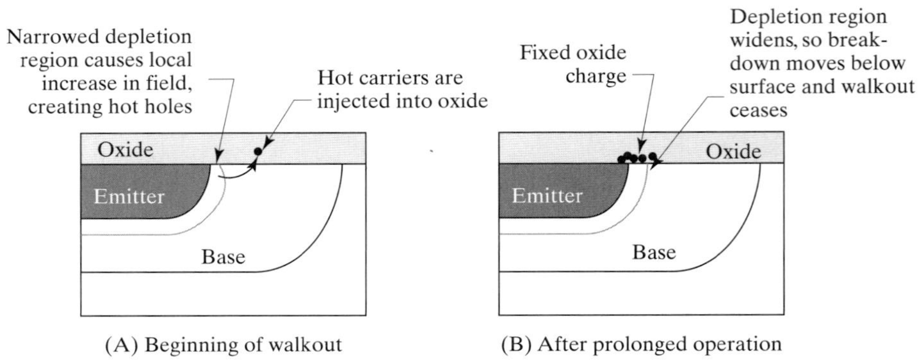

FIGURE 4.7 Simplified diagrams showing Zener walkout mechanism: (A) initial condition of junction, in which hot carrier production occurs near the surface; (B) condition of junction after extended period of operation.

operation, a phenomenon called *Zener walk-out.*[21,22] If a junction diode's reverse breakdown is observed using a curve tracer, the knee of the breakdown curve will gradually "walk out" to higher and higher voltages due to the gradual widening of the depletion region at the surface in response to the accumulation of a fixed oxide charge. Since trapped oxide charges cause walk out, an unbiased high-temperature bake will at least partially reverse it. Depending on processing conditions, emitter-base Zeners can exhibit up to 200mV of walkout.[23] Experimental evidence suggests that the magnitude of Zener walkout diminishes when the process incorporates refractory barrier metal and silicided contacts,[24] although the mechanism responsible for this improvement is not apparent.

Preventative Measures

A *lightly doped drain* (LDD) structure can reduce or even eliminate hot carrier generation in a MOS transistor. Section 3.2.4 discusses the implementation of lightly doped drain structures in a typical polysilicon-gate CMOS process. If no lightly doped drain structure exists, or if the operating voltage exceeds the capabilities of the available structure, then the circuit must be redesigned to reduce the electrical stress imposed on the MOS transistors.

Transistors used as switches generate relatively few hot carriers. Such devices operate either fully on, in which case they are in the linear region, or fully off, in which case they are in cutoff. In neither case does current flow across a large drain-to-source voltage differential. Hot carriers are only generated during brief switching transitions between the two operating states. The average rate of hot carrier generation drops to a minute fraction of the value associated with continuous conduction, and the operating lifetime of the part increases by orders of magnitude. Transistors can withstand voltages far beyond the onset of hot carrier generation as long as switching transitions remain infrequent.

Long channel devices also gain some measure of protection against hot carrier effects. Hot carriers are still produced, but only in the vicinity of the drain. The rest of the channel remains unaffected, minimizing the overall impact of hot carriers on transistor parameters. A few extra volts of operating margin can often be obtained by increasing the channel length a few microns.

---

[21] J. F. Verwey, J. H. Aalberts, and B. J. de Maagt, "Drift of the Breakdown Voltage in Highly Doped Planar Junctions," *Microelectronics and Reliability,* Vol. 12, 1973, pp. 51–56.

[22] R. W. Gurtler, "Avalanche Drift Instability in Planar Passivated p-n Junctions," *IEEE Trans. on Electron Devices,* Vol. ED-15, #12, 1968, pp. 980–986.

[23] W. Bucksch, "Quality and Reliability in Linear Bipolar Design," *TI Technical Journal,* Nov. 1987, pp. 61–69.

[24] W. Bucksch, unpublished manuscript, 1988.

Ordinary base-emitter Zener diodes are surface devices and can therefore exhibit as much as several hundred millivolts of Zener walkout. Attempts have been made to minimize walkout in surface Zeners, but none have been notably successful. The Zener voltage can only be stabilized if the avalanche breakdown is confined to a subsurface region in order to keep hot carriers away from the vulnerable oxide-silicon interface. Such structures are usually called *buried Zeners* (Section 10.1.2).

4.3.2. Parasitic Channels and Charge Spreading

Any conductor placed above the silicon surface can potentially induce a *parasitic channel.* If the conductor bridges two diffusions, then a leakage current can flow through the channel from one diffusion to the other. Most parasitic channels are relatively long and cannot conduct much current, but even small currents can cause parametric shifts in low-power analog circuitry. Channels can sometimes form even in the absence of a conductor due to a mechanism called *charge spreading.* The addition of channel stops or field plates can suppress parasitic channel formation and so protect vulnerable circuitry.

Effects

Both PMOS and NMOS parasitic channels exist. A PMOS parasitic channel can form across any lightly doped N-type region, such as an N-tank in a standard bipolar process or an N-well in a CMOS or BiCMOS process. An NMOS parasitic channel can form across any lightly doped P-type region, such as the P-epi of a CMOS or BiCMOS process, or the lightly doped P-type isolation of standard bipolar processes. Both of these types of parasitic channels can cause a great deal of trouble.

PMOS parasitic channels can form underneath leads crossing lightly doped N-type regions. Consider a metal lead that crosses an N-tank containing a base diffusion (Figure 4.8A). The lead acts as the gate of a PMOS transistor and the N-tank as its backgate. The base region forms the source of the transistor and the isolation serves as its drain. A channel will form if the voltage difference between the lead and the base region exceeds the threshold voltage of the parasitic MOS transistor.[25] Since the thick-field oxide serves as the gate dielectric of this transistor, its threshold voltage is called the *PMOS thick-field threshold.* If the process has a 40V PMOS thick-field threshold, then the base must be biased at least 40V above the lead in order for a channel to form beneath the lead. A similar condition applies to any other potential parasitic PMOS: the P-type region serving as the source must rise

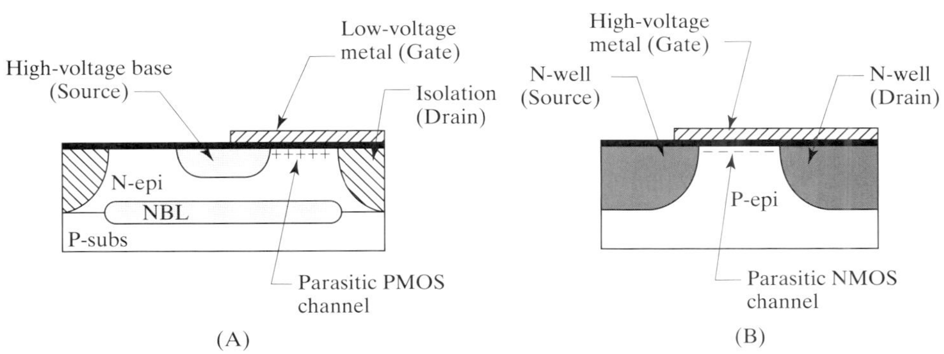

FIGURE 4.8 Parasitic PMOS in a standard bipolar process (A) and parasitic NMOS in an N-well CMOS process (B).

[25] "Bipolar Field Inversion," *Semiconductor Reliability News,* Vol. 3, #1, 1991, p. 7.

above the conductor acting as the gate by a voltage in excess of the PMOS thick-field threshold.

NMOS parasitic channels can form underneath leads crossing lightly doped P-type regions. Figure 4.8B shows a parasitic NMOS channel forming on an N-well CMOS die. This channel forms beneath the lead crossing the lightly doped P-epi. The lead acts as the gate and the P-epi as the backgate. Two adjacent wells serve as the source and the drain. A channel will form if the voltage difference between the gate and the source exceeds the NMOS thick-field threshold. In this case, the voltage on the lead must exceed the voltage on the N-well acting as the source by an amount equal to or greater than the NMOS thick-field threshold. Similar conditions apply to any other potential parasitic NMOS: the conductor serving as the gate must rise above the N-type region serving as the source by a voltage equal to or greater than the NMOS thick-field threshold.

The thick-field threshold voltages of a process depend on a number of factors, including conductor material, oxide thickness, substrate crystal orientation, doping levels, and processing conditions. Most processes quote only one value for the thick-field threshold, this being a minimum value obtained from a worst-case combination of conductors and diffusions. Other processes have undergone more extensive characterization to determine separate thick-field voltages for each combination of conductor and diffusion.

Designers sometimes invoke the body effect (Section 1.4.2) as justification for approaching or even exceeding the thick-field threshold. The body effect increases the apparent threshold voltage of the transistor when the backgate-source junction is reverse-biased. For example, the backgate of the parasitic PMOS in Figure 4.8A is probably biased to a higher voltage than the base. Unfortunately, backgate biasing cannot be relied on for any significant aid. The body effect is most significant in heavily doped backgates, whereas the backgate of a parasitic MOS is usually rather lightly doped. Furthermore, the threshold shift produced by the body effect varies as the square root of the backgate-to-source bias, so even a large backgate bias may not buy more than a few volts of margin.

Engineers once believed that channels could only form beneath conductors, but experience has shown otherwise. Channels can form whenever a suitable source and drain exist, even if no conductor exists to act as a gate. The mechanism underlying the formation of such channels is called *charge spreading*, and although some details still remain unclear, the basic principles are well understood.[26,27] The oxide and nitride films covering an integrated circuit are nearly perfect insulators. Electric current cannot flow through an insulator, but static electrical charges can accumulate on the surface of an insulator or along the interface between two dissimilar insulators. These static charges are not entirely immobile and can slowly shift or spread under the influence of electrical fields. In integrated circuits, the interface between the protective overcoat and the plastic encapsulation is susceptible to this phenomenon. If a nitride protective overcoat is used, then the oxide-nitride interface is also vulnerable. The rate of movement of such charges depends on temperature and on the presence of contaminants. Higher temperatures greatly accelerate charge spreading, as does the presence of even trace amounts of moisture.[28]

Charge spreading requires the presence of static electrical charges at the insulating interface. Experience has shown that these charges do exist and that they consist primarily of electrons, but the mechanisms that generate them are not fully understood.

---

[26] D. G. Edwards, "Testing for MOS IC Failure Modes," *IEEE Trans. Rel.,* R-31, 1982, pp. 9–17.

[27] Lycoudes, *et al.,* p. 240ff.

[28] E. S. Schlegel, G. L. Schnable, R. F. Schwarz, and J. P. Spratt, "Behavior of Surface Ions on Semiconductor Devices," *IEEE Trans. on Electron Devices,* Vol. ED-15, #12, 1968, pp. 973–980.

Hot carrier injection certainly contributes to charge spreading, but integrated circuits that do not produce hot carriers still exhibit charge spreading. Various hypothetical mechanisms have been postulated to account for these experimental observations. In practice, the source of the static charge is less important than its consequences.

Figure 4.9A shows a cross section of a standard bipolar die susceptible to charge spreading. The base region inside the tank is biased above the PMOS thick-field threshold, and therefore acts as the source of a parasitic PMOS transistor. The tank containing this base region is also, of necessity, biased above the PMOS thick-field threshold. Electrons present in the overlying insulating layers will tend to migrate toward the positively charged tank. Eventually, enough electrons may accumulate over the tank to induce a channel (Figure 4.9B). In effect, the static charge generated by charge spreading behaves as the gate electrode of an MOS transistor.

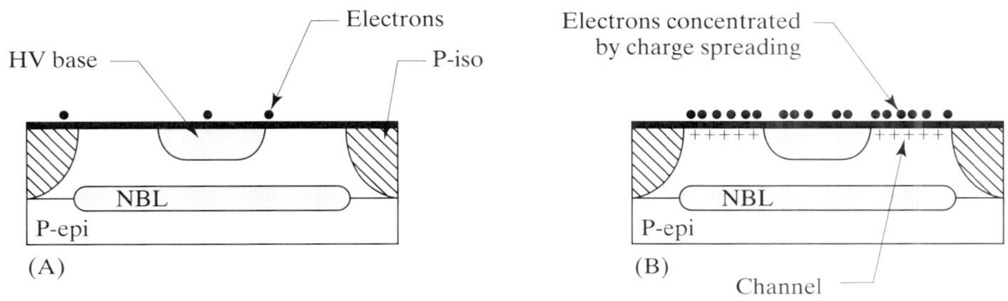

FIGURE 4.9 Cross section of a standard bipolar structure susceptible to charge spreading: (A) before and (B) after an extended period of operation under bias.

Standard bipolar appears to be more susceptible to charge spreading than does CMOS, probably because of less stringent process cleanliness. CMOS processes must minimize ionic contamination to maintain threshold voltage control; no such requirement exists for standard bipolar. The absence of excessive numbers of mobile ions gives CMOS and BiCMOS processes a certain degree of immunity to charge spreading.

Charge spreading produces parasitic PMOS transistors because it involves the accumulation of negative charges. The sources of these parasitic transistors consist of any P-regions that operate at voltages exceeding the PMOS thick-field threshold. The most vulnerable devices contain large, high-voltage P-regions operating at low currents—for example matched high-voltage HSR resistors. Failures tend to occur after long periods of high-temperature operation under bias. Moisture increases the mobility of surface charges, so environmental tests designed to detect moisture sensitivity often uncover charge spreading problems. The resulting parametric shifts resemble those produced by hot carrier injection in that they can be partially or completely reversed by baking the unbiased units at 200 to 250°C for several hours. The high temperature causes the accumulated static charges to disperse and restores an equilibrium between mobile ions and their fixed countercharges. This treatment does not constitute a permanent cure because the parametric drifts resume as soon as bias is restored.

Preventative Measures (Standard Bipolar)

NMOS channel formation can be suppressed in standard bipolar by coding base over all isolation regions. This *base-over-isolation* (BOI) requires no additional die area because the spacings required by the isolation are much larger than those required by the base. The BOI can therefore coincide with the isolation, or even slightly overlap it. Not all standard bipolar processes employ base-over-isolation; some already have a sufficiently heavily doped isolation diffusion to suppress channel formation.

Standard bipolar devices are susceptible to the formation of PMOS channels through charge spreading. Any tank that contains a P-type diffusion biased above the PMOS thick-field threshold requires protection in the form of field plates, channel stops, or a combination of both. Conservative designers usually derate the thick-field threshold of standard bipolar to account for this process's known propensity for charge spreading. For example, a designer might field plate and channel stop high-voltage P-regions operating above 30V even though the process has a rated PMOS thick-field threshold of 40V.

Figure 4.10 shows an example of a high-voltage HSR resistor vulnerable to parasitic channel formation. The tank containing the resistor connects to the positive supply to ensure isolation. A lead must route across the tank to connect to some adjacent low-voltage circuitry. A PMOS channel will form beneath this lead as soon as the voltage difference between the resistor and the lead rises above the PMOS thick-field threshold.

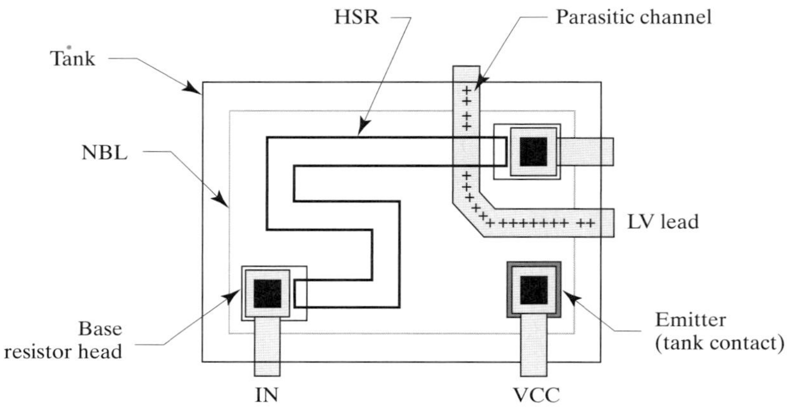

FIGURE 4.10 Example of a circuit susceptible to parasitic PMOS channel formation.

CMOS processes use channel stop implants to raise the thick-field thresholds. Standard bipolar does not include channel stop implants, but an emitter can be coded over selected regions of the N-tank to serve the same purpose. Figure 4.11A shows how emitter diffusions can disrupt the parasitic channels formed beneath a low-voltage lead. Each of the two minimum-width emitter strips disrupts a channel that would otherwise conduct current from the resistor to the isolation.

The emitter bars in Figure 4.11A extend slightly beyond the leads in either direction. These extensions will sever the channel even if the metal and the emitter misalign. The electric field also fringes out to either side of the lead. These fringing

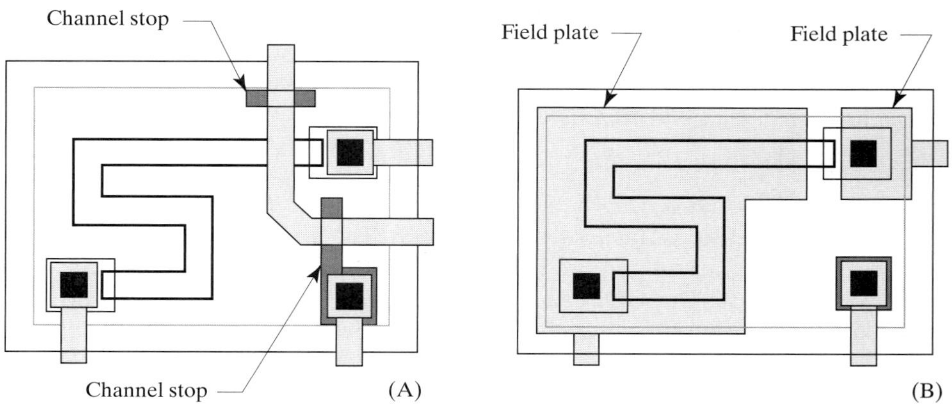

FIGURE 4.11 Two methods for preventing parasitic PMOS channels: (A) channel stops prevent channel formation beneath leads but do not stop charge spreading and (B) field plates provide relatively complete coverage, except possibly in the gap between the plates.

fields rarely extend laterally more than two or three times the oxide thickness, so the overlap of the emitter bar over the lead should equal the maximum photolithographic misalignment plus twice the oxide thickness. Assuming a two-level misalignment of 1μm and a 10kÅ thick-field oxide, the emitter bar should extend about 3 to 5μm beyond the lead on either side. These emitter bars are often called *channel stops,*[29] but they should not be confused with the blanket *channel stop implants* used in CMOS and BiCMOS processes. Channel stops are sometimes called *guard rings,* although this term is more properly applied to minority carrier guard rings (Section 4.4.2).

The channel stops in Figure 4.11A cannot, by themselves, prevent charge spreading. Even if a channel stop entirely encircles a serpentine resistor like that in Figure 4.11, parasitic channels can still form between its turns. If additional channel stops are placed between the turns, parasitic effects could still alter the effective width of the resistor by inverting the silicon along its edges. Some other technique must be used to supplement channel stops, especially for high-voltage diffused resistors.

Field plating can provide comprehensive protection against both parasitic channel formation and charge spreading. A field plate consists of a conductive electrode placed above a vulnerable diffusion and biased to inhibit channel formation.[30] Figure 4.11B shows an HSR resistor with field plates added. The low-voltage lead has been rerouted and a large plate of metal has been placed over the body of the resistor and connected to its positive terminal. The metal lead connecting to the negative end of the resistor has also been enlarged to protect the head of the resistor protruding beyond the main field plate. Both of these field plates must overlap the resistor enough to allow for outdiffusion, misalignment, and fringing fields. Assuming a two-level misalignment of 1μm, a maximum outdiffusion of 2μm, and a maximum fringing distance of 2μm, the total overlap must equal 5μm. Since the field plate consists only of metal, it can extend to fill the required area without enlarging either the resistor or its tank.

A field plate operates by providing an intentional gate for at least a portion of the MOS channel. This gate is biased to prevent the gate-to-source voltage of the parasitic transistor from exceeding the thick-field threshold. The presence of the conductive plate prevents the accumulation of static charges and thus suppresses charge spreading. Field plates also prevent modulation of carrier concentrations in the underlying silicon by acting as electrostatic shields. They therefore provide excellent protection against all types of electrostatic interactions, including conductivity modulation and noise coupling from overlying leads.

Most field plates contain gaps in which channels can still form. In the resistor of Figure 4.11B, a gap remains between the two field plates covering the resistor. Two methods exist for blocking these gaps. One method consists of flaring, or *flanging*, the ends of the field plate to elongate the channel as much as possible (Figure 4.12A). The close proximity of the parallel field plates induces a lateral electric field that sweeps static charges out of this region. The longer the potential channel, the greater the margin of safety provided by the flanges. The second method bridges the gaps between the field plates with short channel stops (Figure 4.12B). The emitter strips used for this purpose must overlap the field plates sufficiently to account for misalignment. This technique combines the strengths of a field plate with those of a channel stop to provide ironclad protection at all points.

---

[29] J. Trogolo and S. Sutton, "Surface Effects and MOS Parasitics," unpublished report, 1988, p. 13ff.

[30] Trogolo, *et al.*, p. 13ff.

The resistors in Figures 4.11 and 4.12 illustrate another important principle of field plating: the field plate biased to the highest potential should cover as much of the resistor as possible. If the low-voltage field plate were extended, it would provide less protection to the high-voltage end of the resistor. If the voltage difference between the tank and the field plate exceeds the thick-field threshold, then the field plate will actually induce channel formation. The field plate should extend from the high-voltage terminal of a vulnerable resistor to encompass as much of the resistor body as possible. The low-voltage terminal of the resistor should have just enough field plating to cover and protect the contact head. Matched resistors may require a slightly different field-plating strategy (Section 7.2.8).

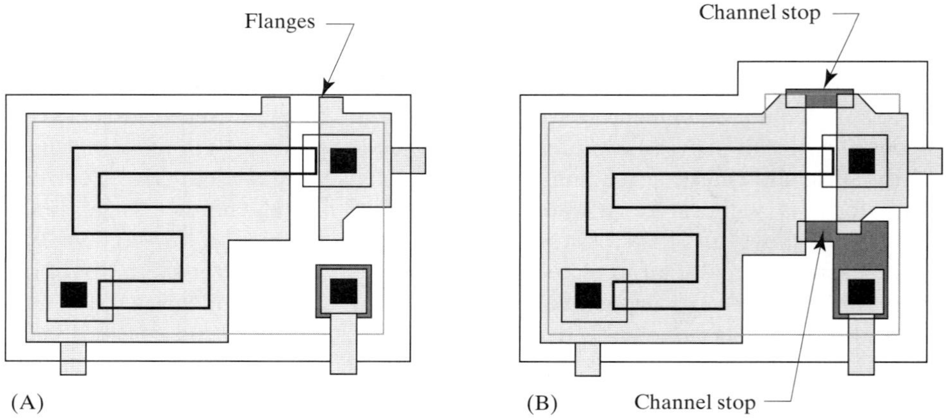

FIGURE 4.12 Improved field plating schemes: (A) flanged field plates and (B) combination of field plates and channel stops.

Figure 4.13 shows an interesting situation that sometimes occurs when laying out resistors. The two terminals of this device connect to a high potential and a low potential, respectively. The high-voltage end of the resistor needs protection against charge spreading, but the low-voltage end does not. Since the voltage drops linearly along the resistor, the field plate has been terminated partway down its length. Partial field plates should extend well beyond the point where the voltage drops below the thick-field threshold. In the case of Figure 4.13, as much of the resistor as possible has been field plated even though much of it apparently serves no useful function. The large safety margin obtained by this means costs nothing and helps ensure that the device will work even under worst-case conditions.

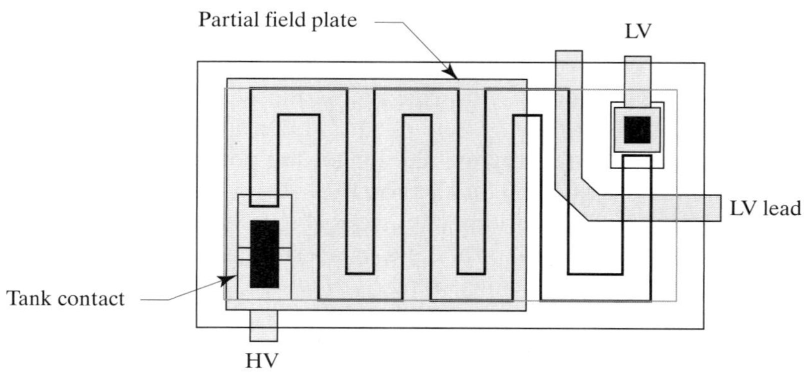

FIGURE 4.13 Example of partial field plating.

Figure 4.14 shows another example of selective field plating involving a multiple-collector lateral PNP transistor. The emitter, base, and one collector operate at voltages in excess of the thick-field threshold, while the remaining collector operates at

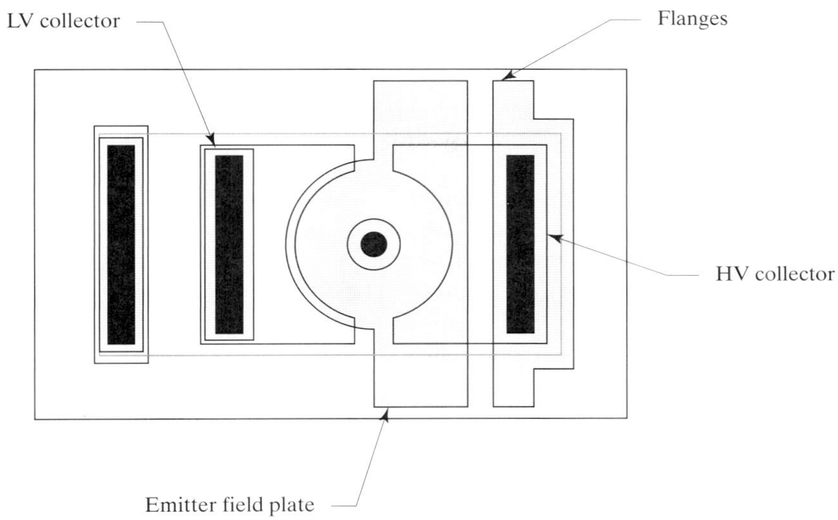

FIGURE 4.14 Field-plated, split-collector, lateral PNP with one low-voltage collector and one high-voltage collector. Flanging suppresses parasitic formation in the gaps between field plates.

a low voltage. The emitter field plate extends out from the emitter across the exposed surface of the base to a point just beyond the inner edge of the collector. The field plate need only overlap the collector by an amount equal to the maximum misalignment minus outdiffusion; certainly no more than 2 to 3μm. A second field plate extends outward from the high-voltage collector to block any parasitic channel that might form between the collectors, or from collector to isolation. No field plate surrounds the low-voltage collector since it does not require one. The field plates have been flanged to ensure that channels cannot form in the gaps. Channel stops could be added, but these would increase the size of the tank and are probably unnecessary.

To summarize, any P-type region biased in excess of the thick-field threshold acts as the source of a parasitic PMOS transistor. Field plates and channel stops ensure that no parasitic channels form from a high-voltage P-type region to any adjacent P-type diffusion. Field plating protects most of the device, while channel stops or flanges protect gaps left in the field plating. The official thick-field threshold of standard bipolar processes should be derated by 25% to provide an additional margin of safety against charge spreading. The following chapters describe additional examples of field plates and channel stops where appropriate.

Preventative Measures (CMOS and BiCMOS)

CMOS and BiCMOS processes usually incorporate channel stop implants to raise the thick-field threshold above the nominal operating voltage. The voltage rating of an N-well CMOS process is usually defined by NSD/P-epi breakdown, PSD/N-well breakdown, or gate oxide rupture, but some structures can withstand much higher voltages. The high N-well/P-substrate breakdown allows PMOS transistors to operate at elevated backgate voltages. These transistors will function normally as long as the drain-to-source voltage does not exceed the PSD/N-well breakdown voltage or the N-well punchthrough voltage. Similarly, an extended-drain NMOS using N-well as a lightly doped drain can withstand the full N-well/P-substrate breakdown voltage applied to its drain terminal.

The lightly doped N-well inverts in much the same manner as the lightly doped N-epi tanks of standard bipolar. Any N-well region containing a P-type diffusion biased above the thick-field threshold becomes vulnerable. As before, PMOS parasitic channels can be suppressed by field plates, channel stops, or a combination of both. The flanged field plate (Figure 4.12A) is especially attractive because the tighter CMOS layout rules allow a narrower gap between the flanges. The stronger lateral electric field makes close-spaced flanges particularly effective at preventing the accumulation of static charges. Flanged fieldplates are the method of choice for suppressing parasitic PMOS channels in high-voltage CMOS and BiCMOS structures.

Charge spreading is less prevalent in CMOS processes than bipolar ones, probably because of improved process cleanliness. Many CMOS designers consequently take a rather cavalier approach to field plates and channel stops. Such indiscretion is unwise considering the greatly reduced operating currents characteristic of modern CMOS designs. However, one special case does exist where charge spreading can be safely ignored. Most CMOS processes list a lower thick-field threshold for poly than for metal because the oxide layer beneath the poly consists only of thick-field oxide and MLO, while that beneath the metal contains an added layer of deposited ILO. Static charges can only accumulate at the interface between two dissimilar materials, so the lower thick-field thresholds associated with thinner oxides do not have any significance for charge spreading. Charge spreading becomes significant only at voltages beyond the highest thick-field threshold listed for the process.

Poly leads can induce parasitic channels if they run across an N-well containing a P-diffusion biased above the poly thick-field threshold. Figure 4.15A shows a typical example of a vulnerable structure consisting of the poly gate lead from a high-voltage PMOS transistor extending across the well and into the surrounding isolation. The well forms the backgate of the parasitic PMOS, the poly acts as the gate, the sources are the PSD regions of the PMOS transistor, and the drain is the P-epi isolation. As long as the backgate potential does not exceed the metal thick-field threshold, the channel can be interrupted by stopping the polysilicon lead short of the drawn edge of the well (Figure 4.15B). The channel can form only beneath the polysilicon lead, so a complete channel cannot form if the lead does not bridge the gap between source and drain. Charge spreading is unlikely to occur so long as the voltages involved do not exceed the highest thick-field threshold of the process. The minimum spacing between the poly and the drawn edge of the N-well should equal the photolithographic misalignment allowance, plus an extra 2 to 3μm to account for fringing fields. The outdiffusion of the well does not

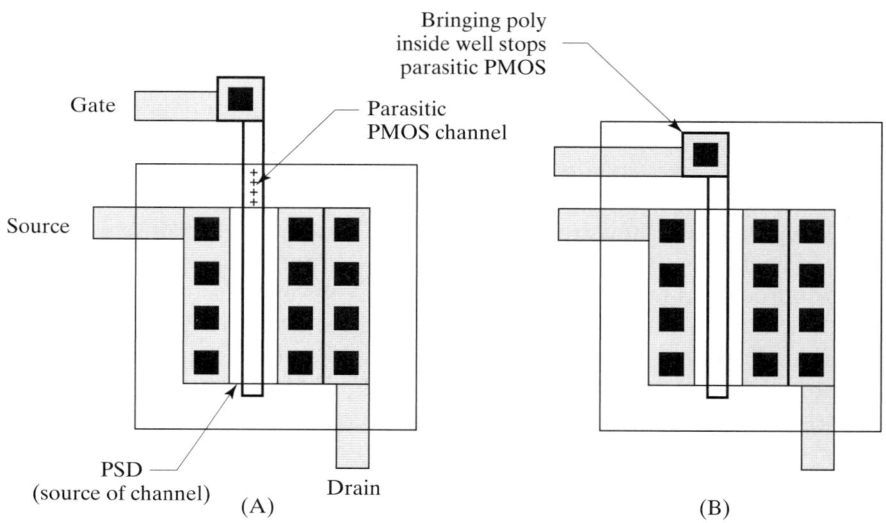

FIGURE 4.15 The parasitic PMOS channel beneath a poly lead (A) can be eliminated by pulling poly inside the well (B).

provide any margin of safety because it becomes very lightly doped beyond its drawn boundaries.

The lightly doped P-type epi can also invert if a high-voltage lead runs across it (Figure 4.8B). The source and drain of this parasitic NMOS consist of two adjacent N-wells; the high-voltage lead acts as the gate, and the P-epi acts as the backgate. A parasitic channel will form if the voltage differential between the lead and an adjacent well exceeds the NMOS thick-field threshold. A channel stop can be inserted by running a thin bar or ring of PSD material down the center of the P-epi beneath the high-voltage lead. The PSD should extend beyond either edge of the lead by an amount sufficient to account for misalignment, plus an additional 2 to 3μm to allow for fringing effects. In many cases, the N-well to N-well spacing can accommodate a minimum-width PSD channel stop with little or no increase in the spacing between adjacent wells. A thin ring of PSD material can then encircle each well (Figure 4.16). This ring not only stops any possible leakage caused by charge spreading but also allows complete freedom to route the leads in any pattern desired. If the PSD rings are drawn when the wells are placed, or if they are automatically produced during mask generation, then the designer can subsequently ignore NMOS channel formation. These PSD rings correspond to the base-over-isolation (BOI) scheme used for the same purpose in standard bipolar designs.

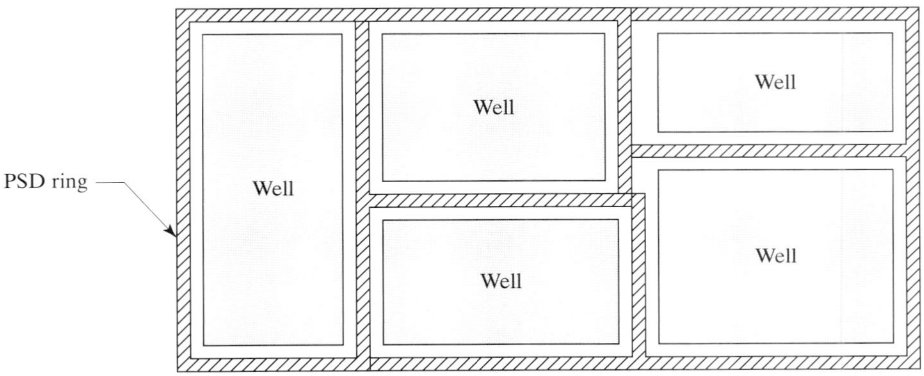

FIGURE 4.16 Sample layout showing the use of PSD rings to prevent NMOS channels.

4.4 PARASITICS

All integrated circuits contain electrical elements not required for their operation. These include reverse-biased isolation junctions, and resistances and capacitances between various diffusions and depositions. The circuit does not benefit from the presence of these *parasitic components*, but they can sometimes adversely affect its operation.

Parasitics are responsible for a number of different types of electrical failures. For example, capacitive coupling can inject noise into sensitive circuitry. The type of parasitics that will be discussed in this section concern the forward biasing of junctions that normally remain reverse-biased. When these junctions forward bias, current begins to flow between circuit nodes that normally remain isolated from one another. If these currents are small and the circuit is relatively insensitive to their presence, then these leakages may produce only subtle parametric shifts. Larger currents can catastrophically disrupt the operation of the circuit. The malfunctioning circuit may actually *latch up*, causing it to continue malfunctioning even after the removal of the triggering event. Latchup can cause physical destruction of an integrated circuit due to excessive power dissipation and consequent overheating. Even if the circuit does not self-destruct, normal operation can only be restored by interrupting the power.

Two important parasitic mechanisms involve currents flowing through the substrate. *Substrate debiasing* occurs when parasitic currents induce voltage drops in a resistive substrate. If these voltage drops become large enough, they can forward-bias one of the isolation junctions. The forward-biased junction then injects current into other circuit nodes, causing potentially catastrophic malfunctions. *Minority carrier injection* occurs when a forward-biased junction injects minority carriers into the isolation, a tank, or a well. Some of these carriers diffuse several hundred microns before recombining and can easily cross reverse-biased junctions that block majority carrier flow.

4.4.1. Substrate Debiasing

Substrate debiasing becomes a problem when currents flowing through the substrate generate voltage drops of a few tenths of a volt or more. This substrate current consists of majority carriers that cannot surmount reverse-biased isolation junctions, but sufficient debiasing may cause one or more isolation junctions to forward-bias and inject minority carriers into active circuitry.

Figure 4.17 shows a typical example of substrate debiasing in a standard bipolar process. Substrate PNP transistor Q_1 injects its collector current I_C directly into the substrate. This current then flows laterally to substrate contact SC_1. Because of the presence of substrate resistance R_s, the substrate voltage immediately under NPN transistor Q_2 rises. Only a few hundred millivolts of substrate debiasing are necessary to forward-bias the collector-substrate junction of a saturated common-emitter NPN.

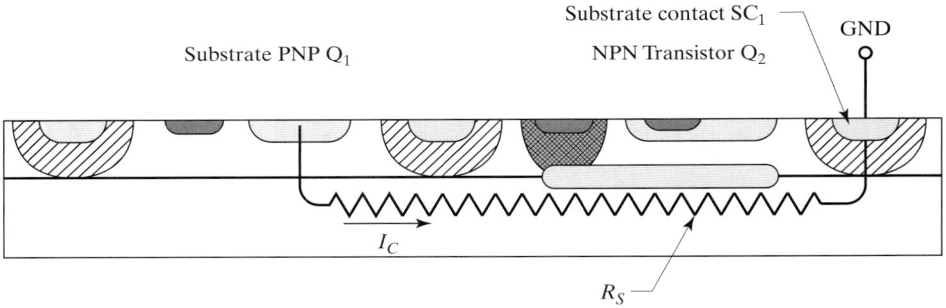

FIGURE 4.17 Cross section of a standard bipolar die showing potential substrate debiasing caused by substrate resistance R_s.

Effects

The voltage required to forward-bias a PN junction depends on both current density and temperature. Table 4.1 lists typical forward-bias voltages for the collector-substrate junction of a minimum-area NPN transistor constructed in a standard bipolar process. This table is useful for estimating susceptibility to substrate debiasing. For example, a circuit using 100μA minimum currents can probably tolerate 1μA of leakage. If it must operate at 125°C, then Table 4.1 indicates that substrate debias-

TABLE 4.1 Forward voltages for a typical collector-substrate junction of a minimum NPN transistor in standard bipolar, as a function of temperature and current.[31]

Current	25°C	85°C	125°C	150°C
10nA	0.43V	0.29V	0.19V	0.13V
100nA	0.49V	0.36V	0.27V	0.22V
1μA	0.55V	0.43V	0.35V	0.30V
10μA	0.61V	0.50V	0.43V	0.39V
100μA	0.67V	0.57V	0.51V	0.47V

[31] Based on $V_{BE}(150°C, 1μA) = 0.3V$.

ing must not exceed 0.35V. If the same circuit has to operate at 150°C, then it can tolerate no more than 0.30V of debiasing.

Figure 4.18 depicts the cross section of a standard bipolar wafer containing a single substrate current injector and a single substrate contact. R_1 models the lateral resistance through the substrate, while R_2 models the vertical resistance beneath the substrate contact. The total resistance of the substrate R_s equals the sum of the lateral and vertical components: $R_s = R_1 + R_2$.

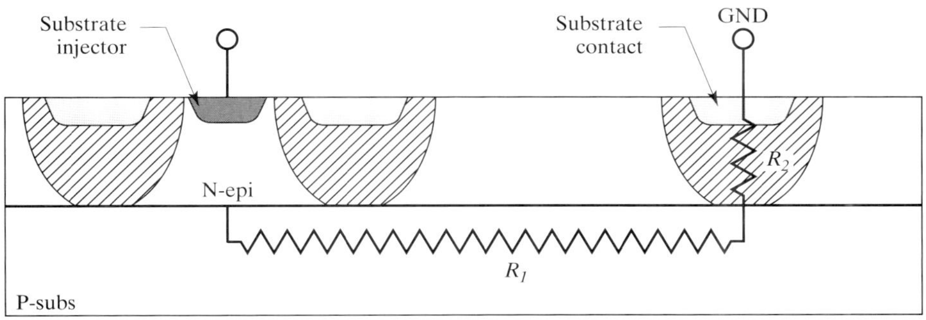

FIGURE 4.18 Simplified model of substrate debiasing in a standard bipolar process.

The relative magnitudes of R_1 and R_2 depend upon the process. Standard bipolar uses a lightly doped substrate and a heavily doped isolation diffusion, so $R_1 >> R_2$. The value of R_1 depends on various geometric factors, including the cross-sectional area of both the injector and the substrate contact as well as the distance between them. Typical values of R_1 range from hundreds to thousands of Ohms,[32] while R_2 rarely exceeds 10Ω.[33] The presence of a network of isolation diffusions criss-crossing the die complicates the computation of substrate resistance because the low sheet resistance of the isolation (usually about 10Ω/□) allows each substrate contact to extract current from a large area of isolation. This effect complicates the computations to such an extent that only empirical measurement or sophisticated computer simulation can yield accurate results.

Two points should be kept in mind when using lightly doped substrates. First, substrate resistance always increases with separation. A substrate contact placed adjacent to the injector will extract some of the current before it ever reaches the substrate. Contacts placed further away require the current either to flow through long stretches of isolation or to pass through the highly resistive substrate. Second, a substrate contact to a heavily doped isolation diffusion draws current not only from the substrate immediately beneath it but also from adjoining stretches of isolation. This effectively magnifies the area of substrate contacts, allowing even a minimum-size contact to have an effective area of many hundreds of square microns. Because of this effect, a scattering of minimum-size substrate contacts throughout the die will have a much lower effective resistance than a single large contact.

---

[32] An estimate of the lateral resistance R_1 can be obtained by examining the spreading resistance $R_{sp} = \rho/d$, where ρ is resistivity and d is the diameter of the points of contact. Spreading resistance assumes a semi-infinite slab of uniformly doped material and a probe separation much wider than the probe diameter; these conditions are only approximately met by typical substrate contacts. Assuming that the cross-sectional areas of injector and substrate contact are 1mil^2 each, this yields d = 28.7μm. A substrate with a resistivity of 10Ω·cm would have a spreading resistance of 3.48kΩ. More accurate results can be obtained by applying various correction factors: G. A. Gruber and R. F. Pfeifer, "The Evaluation of Thin Silicon Layers by Spreading Resistance Measurements," *National Bureau of Standards Special Publication 400–10,* Spreading Resistance Symposium, NBS, Gaithersburg, Maryland: June 1974.

[33] The vertical resistance through single-diffused isolation can be approximated by dividing the diffusion into multiple layers of constant doping. Computations for a diffusion with a surface doping of 10^{20}cm^{-3}, a minimum dopant concentration of 10^{17}cm^{-3}, and a depth of 5μm yield a resistance of about 4Ω/mil^2.

CMOS and BiCMOS processes usually employ a heavily doped substrate and a lightly doped epi, so $R_1 << R_2$. The value of R_1 is usually so small that it can safely be ignored. The value of R_2 depends on the thickness of the epi layer and its resistivity. A typical value is about 600kΩ/μm^2 (1kΩ/mil^2). This value can be used to compute the area of substrate contacts required for a CMOS or BiCMOS design, as explained in the following section.

Even on a heavily doped substrate, contacts placed immediately adjacent to a substrate injector will exhibit less resistance than ones placed far away. This *proximity effect* falls off rapidly with distance, and substrate contacts placed hundreds of microns away are no more effective than those placed on the opposite side of the die. The proximity effect occurs because carriers can flow directly to the adjacent contact, rather than having to flow down to the substrate, across, and up to a distant contact. Contacts placed immediately adjacent to a substrate injector can also help prevent localized debiasing of the highly resistive isolation, protecting adjacent tanks from injection from the isolation sidewalls.

Preventative Measures

Integrated circuits should inject as little current into the substrate as possible, as this not only minimizes substrate debiasing but also helps limit noise and cross-talk caused by modulation of the substrate potential. The collector current of substrate PNP transistors flows directly into the substrate, so these devices should be used sparingly, and no single device should conduct more than a milliamp or two. Lateral PNP and vertical NPN transistors can inject large substrate currents when they saturate, but techniques have been developed to minimize this problem (Sections 8.1.4–5). The exact requirements for substrate contacts depend on the nature of the substrate and isolation:

Heavily doped substrates. The contacts in the scribe seal can usually extract 5 to 10mA without undue debiasing. If higher substrate currents are anticipated, then the total area of contacts required can be computed using the following formula:[34]

$$A_c = 10\frac{\rho t_{epi} I_s}{V_d} \qquad [4.1]$$

This formula assumes a uniform lightly doped isolation, such as the P-epi of N-well CMOS and BiCMOS processes. A_c represents the required total area of substrate contacts in μm^2, ρ is the resistivity of the epi in Ω·cm, t_{epi} is the epi thickness in microns, I_s is the maximum substrate current in milliamps, and V_d is the maximum allowable debiasing in volts (from Table 4.1). The thickness of the epi is reduced by up-diffusion of dopants from the underlying substrate and from the presence of a heavily doped (if thin) contacting diffusion, such as PSD. Consider a die with a P-epi resistivity of 10Ω·cm and an effective epi thickness of 7μm. If the substrate must conduct 20mA without more than 0.3V of debiasing, then 47,000μm^2 (72mil^2) of substrate contacts are required. Subtracting the area of substrate contacts in the scribe seal yields the required area of additional contacts. These can be inserted wherever space exists in the layout. As a precaution against localized debiasing, substrate contacts should ring any device injecting more than 1mA.

---

[34] This formula is derived from the fundamental equation $R = \rho t/A$. It neglects fringing effects, which tend to reduce the effective resistance of small substrate contacts. The formula provides a first-order approximation of the worst-case substrate resistance.

Lightly doped substrates with heavily doped isolation. No simple formula exists for computing the area of substrate contacts required to protect a lightly doped substrate from debiasing. A scattering of ten or twenty substrate contacts across the die will, when combined with the scribe seal, handle at least 5 to 10mA. Any device that injects 100μA or more should have substrate contacts located nearby, and any device that injects 1mA or more should be ringed with as much substrate contact as possible. Sensitive low-current circuitry should reside at least 250μm (10mil) away from any substantial source of substrate injection, since debiasing on lightly doped substrates tends to localize around the point of injection. Once the layout has been completed, additional substrate contacts should be scattered throughout the layout wherever room exists. A large number of small substrate contacts scattered throughout the layout will prove more effective than a few large contacts. Even with all of these precautions, designs that inject more than 10mA into the substrate may experience debiasing. Apart from adding more substrate contacts or moving sensitive circuits away from substrate injectors, the only remedies for such problems are the addition of a heavily doped substrate or the use of backside contacting.

Lightly doped substrates with lightly doped isolation. A few processes use a lightly doped substrate in combination with a very resistive isolation. This situation can arise when a BiCMOS design is constructed on a lightly doped substrate to save costs. Such designs cannot rely on the scribe seal to extract more than a few milliamps of substrate current. Large numbers of substrate contacts scattered across the die will help extract substrate current, but some degree of localized substrate debiasing is almost inevitable. Sensitive circuits should be located far away from major sources of substrate injection. Since substrate modulation can inject substantial noise into high-impedance circuitry, consider placing wells under resistors and capacitors to isolate them from substrate noise coupling. Sensitive MOS circuitry may also employ NBL to isolate NMOS transistors from the substrate (Section 11.2.2). In some cases, it may be possible to add strips of heavily doped material to the isolation without increasing the well-to-well spacings (Figure 4.16). This strategy effectively converts the design into one that uses a lightly doped substrate in conjunction with a heavily doped isolation. This stratagem substantially reduces the number and area of substrate contacts required to extract large substrate currents. Backside contact can also provide a large reduction in substrate resistance, but it is difficult to obtain Ohmic contact to a lightly doped substrate unless a backside diffusion is performed to increase the surface doping concentration.

4.4.2. Minority-Carrier Injection

Junction isolation relies on reverse-biased junctions to block unwanted current flow. The electric fields set up by depletion regions repel majority carriers, but they cannot block the flow of minority carriers. If any isolation junction forward-biases, it will inject minority carriers into the isolation. Many of these carriers recombine, but some eventually find their way to the depletion regions isolating other devices.

Effects

Figure 4.19 shows a cross section of a standard bipolar circuit. Suppose that the collector of NPN transistor Q_1 connects to a pin of the integrated circuit, and that the external circuitry experiences occasional transient disturbances that pull current out of this pin. If transistor Q_1 is off, then these transients pull its tank below ground, forward-biasing the collector-substrate junction of Q_1 and injecting minority carriers (electrons) into the substrate. Most of these carriers recombine, but some diffuse across to other tanks, such as T_1.

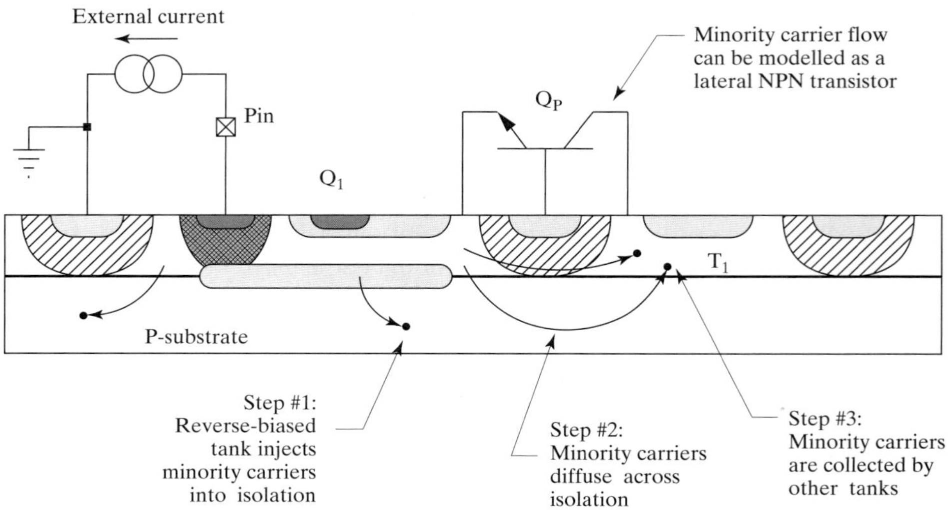

FIGURE 4.19 Example of minority-carrier injection into the substrate of a standard bipolar process. Lateral NPN transistor Q_P models the transit of minority-carriers across the isolation.

The transit of minority carriers across the isolation is analogous to the flow of minority carriers through a bipolar transistor. The tank pulled below ground acts as the emitter of lateral NPN transistor Q_P. The isolation and substrate act as the base of this transistor, and any other reverse-biased tank acts as a collector. Each reverse-biased tank forms a separate parasitic transistor corresponding to Q_P. The betas of these parasitic lateral NPN transistors are very low because most of the minority carriers recombine in transit. The parasitic bipolar between two adjacent tanks might have a beta of 10, but the beta between two widely separated tanks might not even reach 0.001. Even such low gains can cause circuit malfunctions. Suppose that a forward-biased tank injects a minority current of 10mA into the substrate. If the parasitic associated with another tank has a beta of 0.01, then this tank will collect 100μA of current—easily enough to disrupt the operation of a typical analog circuit.

Substrate contacts cannot, by themselves, stop minority-carrier injection since minority carriers travel by diffusion and not by drift. Minority carriers are best collected by reverse-biased junctions. However, substrate contacts still provide majority carriers to feed recombination. Since most minority carriers recombine in the isolation, substrate contacts remain necessary to prevent substrate debiasing.

In some cases, minority-carrier injection can cause a circuit to latch up. Early CMOS processes suffered from a form of this malady that has since come to be called *CMOS latchup.*[35] Figure 4.20A shows the cross section of a portion of a CMOS die consisting of an NMOS transistor M_1 and a PMOS transistor M_2. In addition to these two desired MOS transistors, this layout contains two parasitic bipolar transistors. Lateral NPN transistor Q_N's emitter is the source of M_1, its base is the isolation, and its collector is the N-well of M_2. Lateral PNP transistor Q_P's emitter is the source of M_2, its base is the N-well, and its collector is the isolation. Figure 4.20B shows the two parasitic bipolar transistors drawn in a more familiar fashion. In this schematic, R_1 represents the well resistance of M_2, and R_2 represents the substrate resistance. These two resistors normally ensure that both bipolar transistors remain off. As long as this remains the case, neither parasitic conducts any current and the integrated circuit works as intended. When a transient disturbance turns on either transistor, the current flowing through this device will turn on the other parasitic as well. Each transistor then supplies the other's base current. Once both tran-

[35] R. R. Troutman, "Recent Developments in CMOS Latchup," *IEDM Tech. Dig.*, 1984, pp. 296–299.

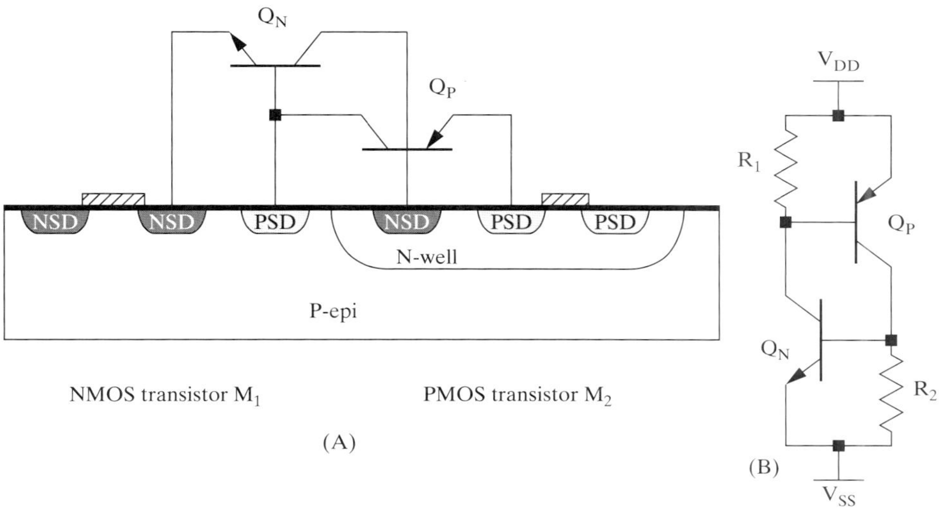

FIGURE 4.20 (A) Cross section of a CMOS die showing diffusions that form the two parasitic bipolar transistors Q_P and Q_N; (B) equivalent schematic showing these transistors along with well resistance R_1 and substrate resistance R_2.

sistors begin to conduct, they will continue to do so even if the transient disturbance that initiated conduction is removed. The circuit has *latched up* and it will remain in this state until power is removed. The integrated circuit can actually conduct so much current that it overheats and self-destructs. Even if this does not occur, latchup causes circuit malfunctions and excessive supply current consumption.

CMOS latchup can be triggered in one of two ways. If the source of NMOS transistor M_1 is pulled below ground, it will inject minority carriers (electrons) into the substrate, turning on parasitic transistor Q_N. This transistor will then turn on Q_P. Alternatively, the source of PMOS transistor M_2 may be pulled above the well. It will then inject minority carriers (holes) into the well and will turn on parasitic transistor Q_P. This transistor then turns on Q_N. Latchup can only occur if the product of the betas of transistors Q_N and Q_P exceeds unity. If the beta product is less than unity, then transient disturbances may occur, but the circuit cannot actually latch up (Section 11.2.7). CMOS latchup also requires the existence of four distinct semiconductor regions arranged in the sequence PNPN. A discrete device called a *silicon controlled rectifier* (SCR) has this same four-layer structure. CMOS latchup is sometimes described in terms of a parasitic SCR consisting of PSD, N-well, P-epi and NSD. This point of view is analogous to the transistor approach discussed above because the SCR operates in the same manner as the pair of coupled transistors.

The obvious way to stop CMOS latchup consists of reducing the beta of either or both parasitic transistors. If the product of these betas is less than unity, then latchup cannot occur. This is usually achieved by increasing layout spacings, which in turn increases the width of the neutral base regions of the parasitic lateral transistors. Alternatively, the amount of dopant present in the neutral base region of one (or both) parasitic transistors may be increased. Both of these approaches increase the Gummel number of one or both transistors and reduce the beta product.

Although many CMOS processes claim immunity to latchup, these claims are true only in a somewhat narrow sense. The PNPN structure inherent in the CMOS transistors of such a process lacks sufficient gain to establish regenerative feedback, but minority-carrier injection still occurs. The collected carriers can still cause circuit malfunctions, and if positive feedback exists in the circuit, these malfunctions can still cause a form of latchup. The significance of this observation is frequently underestimated. Any integrated circuit that experiences unanticipated minority-carrier injection can potentially latch up. Even if it does not actually do so, it is still likely to malfunction. Not only do electrons injected into the substrate pose a potential threat, but so do holes unintentionally injected into wells or tanks.

Preventative Measures (Substrate Injection)

Fundamentally, there are four ways to defeat minority-carrier injection: (1) eliminate the forward-biased junctions that cause the problem, (2) increase the spacing between components, (3) increase doping concentrations, and (4) provide alternate collectors to remove unwanted minority carriers. All of these techniques provide some benefit, and in combination they can correct almost any minority-carrier injection problem.

The simplest solution, at least in theory, consists of eliminating the forward-biased junctions that inject minority carriers. This goal is often very difficult to achieve. In a standard bipolar process, tanks must not go below substrate by more than about 0.3V or they will inject minority carriers into the substrate. In an N-well CMOS process, no well and no NSD region residing in the epi may go below substrate potential. If the voltage on a pin slews rapidly, parasitic inductance can cause transients that pull the pin above supply or below ground. The faster the node slews, the smaller the parasitic inductance required to cause such transients. Modern switching speeds have become so fast that the inductance of pin and bondwire alone often cause objectionable transients. Substrate injection has become difficult, if not impossible, to eliminate.

Minority-carrier injection into the substrate will cause fewer problems if potential injectors are separated from sensitive circuitry. In most designs, only a few devices connect to pins. With a little forethought, these devices can be placed far away from sensitive circuitry. In many cases, the layout will naturally favor this sort of separation. For example, power transistors inject minority carriers during transients. Since these transistors form part of the output circuitry, they will typically be placed far away from sensitive input circuitry to minimize electrical and thermal feedback. This same arrangement also minimizes the circuit's vulnerability to minority-carrier injection.

Additional dopant added to the isolation regions of the die will reduce the gain of the parasitic lateral bipolar. CMOS and BiCMOS processes often employ P+ substrates for just this reason. All other factors being equal, a process incorporating a heavily doped substrate will provide greater immunity to electrical upsets than one that uses a lightly doped substrate. However, a heavily doped substrate cannot, by itself, prevent minority carriers from moving laterally through isolation regions separating adjacent tanks or wells. In order to obtain the full benefits of the heavily doped substrate, the process must use a heavily doped isolation, or the designer must add suitable guard rings.

The isolation doping can also be increased by adding a deep-P+ diffusion. Most CMOS processes do not include any suitable diffusion. Some BiCMOS processes include one for constructing certain components (such as DMOS transistors)—usually as part of a process extension. Standard bipolar processes sometimes offer a deep-P+ process extension for constructing high-current lateral PNP transistors. If a suitable diffusion exists, it can be placed in the isolation regions of the die to help increase the isolation doping. This technique can help offset the very light doping of the PBL portion of an up-down isolation system, and can minimize lateral conduction of minority carriers between adjacent tanks or wells.

Minority carriers are collected in disproportionate numbers by reverse-biased junctions near the point of injection. Not only do carriers have less distance to travel to reach a nearby junction, but the nearer junctions also block the flow of carriers to more distant ones. Designers can take advantage of this *shadow effect* to erect deliberate barriers to the flow of minority carriers by placing reverse-biased junctions between the point of injection and vulnerable diffusions. A reverse-biased junction used in this manner is called a *minority-carrier guard ring.* Figure 4.21 shows a typical layout for such a guard ring in a standard bipolar process. Tanks T_1 and T_2 connect to pins that may experience voltage transients. These are surrounded by a third tank, T_3, that collects a significant fraction of the minority carriers injected by T_1 and T_2. Tanks T_1 and T_2 connect to pins, so it is quite natural to place them along one side of the die or even in a corner (as shown in Figure 4.21). This not only minimizes the length of interconnecting leads but also eliminates the need for guard rings along two edges.

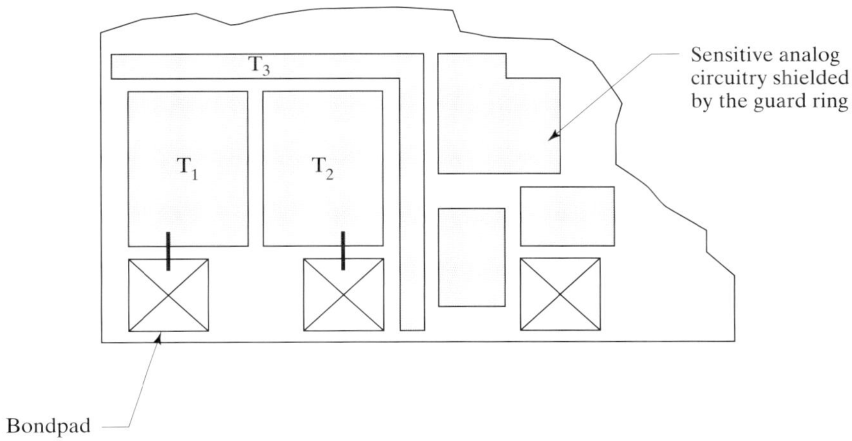

FIGURE 4.21 Sample electron-collecting minority-carrier guard ring (T_3) implemented in a standard bipolar process.

The key to designing efficient guard rings consists of making them deep, wide, and low-resistance. The deeper the guard ring, the larger the fraction of passing minority carriers it can collect. N-well makes a more effective electron-collecting guard ring than NSD, and an epi tank makes a better electron-collecting guard ring than an emitter diffusion. If the guard ring can be connected to produce a large reverse bias, then the depletion region surrounding it will widen and the collection surface will be forced even deeper into the silicon. Thus, electron-collecting guard rings connected to the positive supply become more effective than those connected to substrate potential. Since diffusing minority carriers move randomly, some will actually be collected by the bottom of the guard ring's depletion region. A wider guard ring will therefore collect more minority carriers than a narrow one will. Also, narrow diffusions do not penetrate as deeply as wide diffusions because dopants become diluted by lateral dispersion. A diffusion two or three times wider than minimum will contain enough dopant to obtain the maximum possible junction depth. Low resistance also helps improve the effectiveness of a guard ring, especially if it cannot be strongly reverse-biased. Collected carriers can forward-bias a high-resistance guard ring and cause it to re-inject minority carriers. The lower the vertical resistance of the guard ring, the larger the current it can collect before it saturates and re-injects.

Figure 4.22 shows cross sections of two minority-carrier guard rings designed to collect electrons injected into the substrate. Figure 4.22A shows a substrate guard ring for standard bipolar.[36] This guard ring includes all four available N-type materials: N-epi, deep-N+, NBL, and emitter. The NBL helps obtain the maximum possible junction depth, while deep-N+ and emitter reduce the vertical resistance. The wider this structure, the more effectively it will collect minority carriers. Most designers

---

[36] W. Davis, *Layout Considerations,* unpublished manuscript, 1981, p. 53.

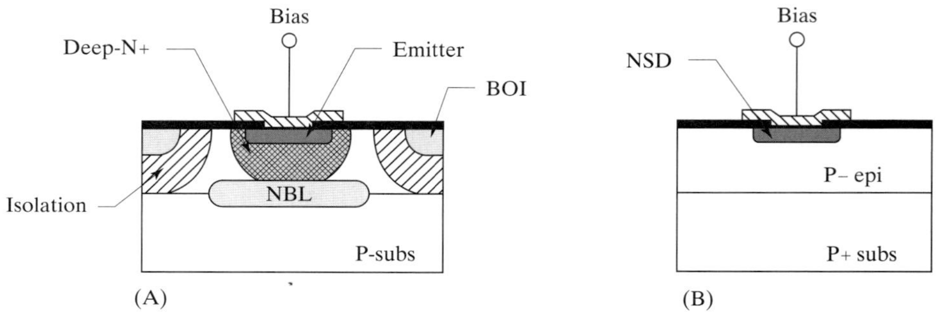

FIGURE 4.22 Cross sections of two representative electron-collecting guard rings: (A) standard bipolar[37] and (B) CMOS.

compromise between efficiency and area by making the deep-N+ strip no more than twice minimum width and by spacing the other layers accordingly. If possible, this guard ring should connect to the highest supply voltage available on the die. The guard ring will still function connected to substrate potential, but it may saturate unless all parts of the guard ring connect to the substrate terminal by a direct metal run. This is probably not practical in a single-level metal process since gaps must be left in the metallization to allow leads to pass through. Single-level-metal guard rings should connect to the positive supply and should contain as few gaps as possible.

BiCMOS layouts can produce electron-collecting guard rings similar to those in Figure 4.22A, although in this case N-well replaces the N-epi tank. Electron-collecting guard rings are considerably more difficult to construct in CMOS-only processes. The N-well becomes extremely resistive in the absence of deep-N+ and NBL, and most CMOS devices operate at relatively low voltages. Figure 4.22B shows an alternate CMOS minority-carrier guard ring. NSD has a relatively low resistance, but it is too shallow to capture more than a small percentage of the electrons in the substrate. The wider the NSD strip, the more effective the guard ring. A width of at least 8 to 10μm is recommended, although narrower guard rings do provide some benefit. The NSD guard ring should, if possible, connect to a supply pin. The reverse bias across the NSD-epi junction drives the depletion region deeper into the epi and increases the apparent depth of the guard ring. In low-voltage processes, a strongly reverse-biased NSD guard ring will often generate secondary carriers due to impact ionization. This problem can be minimized by connecting the low-voltage NSD guard ring to ground instead of to a power supply.

Guard rings of the type shown in Figure 4.22A can reduce substrate injection by a factor of 10 to 100 providing they are used in conjunction with a heavily doped substrate. The P–/P+ interface between the lightly doped epi and the heavily doped substrate repels minority carriers (Section 8.1.5), constraining them to remain within the relatively thin epi layer. This greatly improves the collection efficiency of the guard ring.[38] A simple modification to the electron-collecting guard ring of Figure 4.22A can further increase its attenuation. Instead of connecting the guard ring directly to a supply voltage, it is connected back to the substrate so that a majority-carrier current flows through the substrate beneath the guard ring (Figure 4.23). This type of guard ring is intended to protect against minority carriers originating on only one side of the ring—in this case, the right side. The deep-N+ sinker in the center of the guard ring collects most of these minority carriers. The resulting cur-

[37] C. Jones, "Bipolar Parasitics," unpublished report, 1988, p. 43.

[38] L. S. White, G. R. M. Rao, P. Linder, and M. Zivitz, "Improvement in MOS VLSI Device Characteristics Built on Epitaxial Silicon," in *Silicon Processing,* American Society for Testing and Materials STP 804, 1983, pp. 190–205.

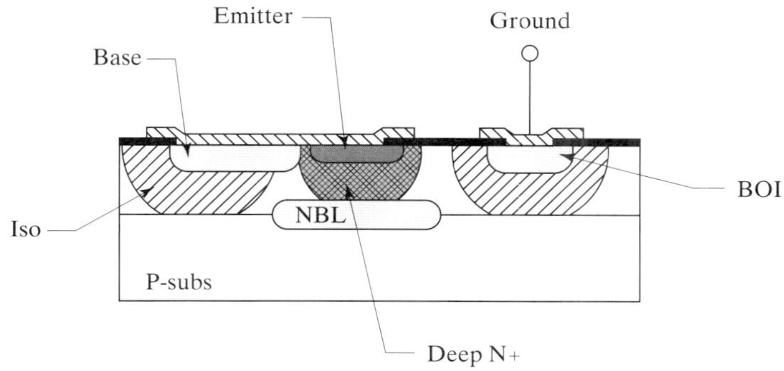

FIGURE 4.23 Cross section of an improved minority-carrier guard ring for collecting electrons injected into the substrate.

rent flows out of the sinker and into the attached metal plate. The base/iso diffusion at the left side of the structure re-injects this current into the substrate in the form of majority carriers. Since the nearest substrate contact lies on the other side of the structure, the majority carriers flow back underneath the guard ring. This current locally debiases the substrate and creates an electric field that opposes minority-carrier flow. The minority carriers are forced upward and toward the guard ring, where they are ultimately collected, or they are held in the substrate until they recombine. The inventor[39] claims an attenuation factor in excess of one million for this structure. While this degree of attenuation may not be achieved in every process, this guard ring will provide more attenuation than those shown in Figure 4.22, especially at higher currents.

The modified guard ring in Figure 4.23 suffers from several drawbacks. It provides enhanced attenuation of carriers flowing in only one direction—in this case from right to left. Substrate contacts cannot be placed near the guard ring on the side facing the injected carriers. This structure also relies upon deliberate debiasing of the substrate, which could potentially forward-bias adjacent junctions. The principle behind this style of guard ring also applies to the design of ordinary guard rings of the sort shown in Figure 4.22. In all cases, it is better to place the electron-collecting guard ring adjacent to the injector, and to place substrate contacts inside of the guard ring. The majority-carrier substrate current that flows under the guard ring generates an electric field opposing the flow of minority carriers, thereby enhancing the performance of the guard ring.

Minority-carrier guard rings can also prevent holes injected into a tank from reaching the substrate and debiasing it. This situation can occur whenever a bipolar transistor saturates, regardless of whether this transistor is an NPN or a PNP (Section 8.1.4–5). Power transistors can easily inject tens or even hundreds of milliamps into the substrate. A heavily doped layer such as NBL can reduce minority-carrier injection from a P-type region through an N-well or N-epi tank to substrate. NBL also helps to minimize tank or well resistance and therefore makes it more difficult to develop the debiasing required to trigger CMOS latchup. CMOS processes generally do not incorporate NBL due to the cost of the extra masking step and to manufacturing difficulties associated with the fabrication of buried layers. Standard bipolar and analog BiCMOS processes frequently use NBL to reduce NPN collector resistance. If NBL exists, it should be added to all tanks or wells that can tolerate its presence, in order to minimize substrate injection and to improve latchup immunity.

---

[39] F. Van Zanten, U.S. Patent # 4,466,011, 1984.

Figure 4.24 shows a hole-collecting guard ring constructed in a standard bipolar process. As a first line of defense against hole injection into the substrate, the tank is floored with NBL and ringed with deep-N+. Any hole attempting to reach the isolation must pass through one or the other of these heavily doped regions. The large population of electrons in these regions enhances recombination and prevents most holes from successfully crossing. Far more importantly, the N+/N– boundary exhibits a built-in potential gradient caused by the outdiffusion of majority carriers that helps confine holes inside the tank until they recombine or are collected (Section 8.1.5).

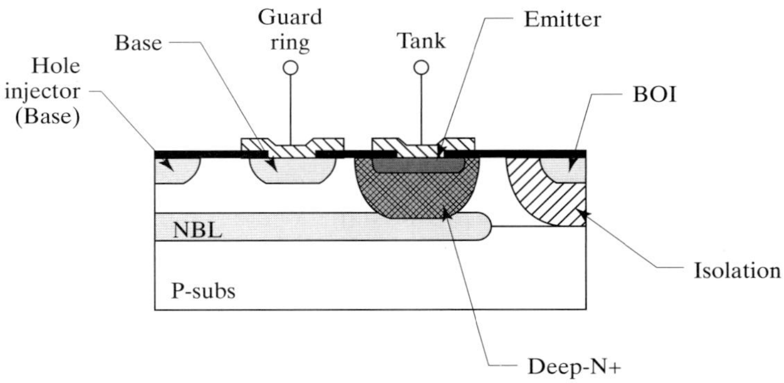

FIGURE 4.24 Cross section of a minority-carrier guard ring for collecting holes injected into a tank.[40]

The guard ring in Figure 4.24 includes a base ring placed just inside the deep N+. This ring usually connects to ground, but it will remain reasonably effective even if it is tied to the tank terminal. Any hole impinging on the depletion region surrounding the base ring will be drawn across by the electric field. Holes become majority carriers inside the base diffusion and can be removed through the contact. The base ring collects almost all of the holes as long as the tank contains NBL. Even without the deep-N+ ring around the outside edge of the tank, the base ring will collect at least 90% of the holes. This type of guard ring is largely ineffectual without NBL.

CMOS devices may experience hole injection into wells if a PSD region rises above the well potential, as might occur if a pin connected to a PMOS source or drain rises above supply. Effective hole collection rings cannot be constructed in a pure CMOS process due to the absence of NBL. The doping gradient of the well causes a downward drift of holes toward the underlying substrate, rendering PSD guard rings placed around the edges of the well ineffectual. CMOS processes must therefore rely upon low-resistance substrate contacts to extract any hole current injected into the substrate.

BiCMOS processes can construct hole-collection rings similar to those in Figure 4.24. These rings are not quite as effective on BiCMOS processes as on standard bipolar because the graded profile of the BiCMOS well opposes the potential barrier raised by the NBL. This problem is exacerbated by the relatively light doping of BiCMOS NBL regions required to avoid autodoping the P-epi. Despite these problems, an overall efficiency in excess of 95% is usually achievable by using base hole-collection rings in combination with NBL and deep N+. Section 13.2 discusses several additional types of hole guard rings and some of the difficulties associated with constructing hole guard rings in a BiCMOS process.

[40] Jones, p. 18ff.

Preventative Measures (Cross-injection)

Circuit upsets caused by minority-carrier injection into a tank or well can often be eliminated by placing each potential emitter of minority carriers in its own tank or well. As a rule, any PMOS transistor whose source connects to an external pin should occupy its own well. Similarly, any base resistor, HSR resistor, or lateral PNP collector connecting to a pin is best placed in its own tank. The small amount of extra space required to construct separate tanks or wells will be amply repaid by the elimination of even one parasitic. If, on the other hand, several devices connect to a common pin, then these can all occupy a common tank or well.

The hole-collection rings discussed previously have been designed to minimize injection of holes into the substrate. Another type of minority-carrier guard ring can prevent holes injected by one device from interfering with the operation of other devices in the same tank or well, a problem called *cross-injection.* Consider the case where two lateral PNP transistors occupy a common tank. If either transistor saturates, some fraction of the carriers it emits will be collected by the adjacent transistor. The resulting increase in collector current may disturb the operation of the circuit, particularly if the devices were intended to match one another. Cross-injection can be prevented by placing each transistor in its own tank, but this wastes area because of the large spacings associated with the isolation diffusion. A more compact solution employs a type of minority carrier guard ring called a *P-bar* (Figure 4.25).[41]

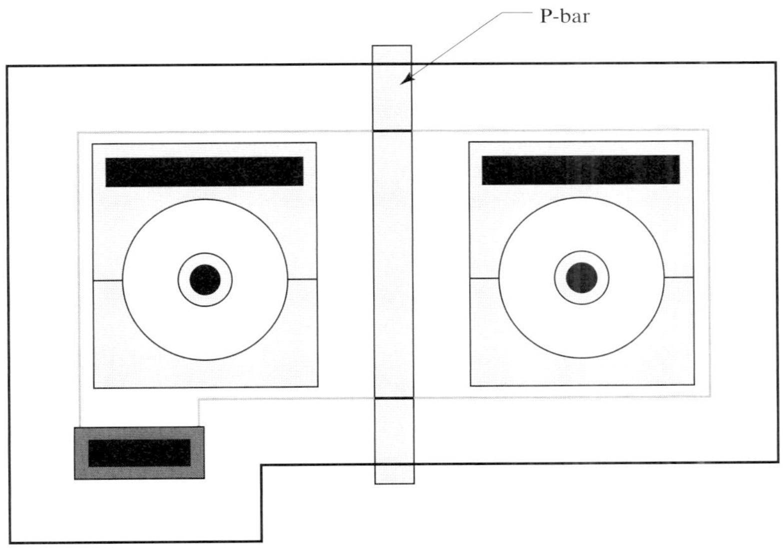

FIGURE 4.25 Example of a P-bar used to prevent cross-injection between two lateral PNPs.

A P-bar consists of a minimum-width strip of base diffusion placed between the two transistors.[42] Each end of the P-bar extends out into the isolation far enough to guarantee electrical contact. This arrangement ensures that the P-bar electrically connects to the isolation without requiring contacts. Now suppose that the lateral PNP on the left side of the P-bar saturates and begins to inject holes into the tank. In order for these holes to reach the lateral PNP transistor on the right, they must first pass underneath the P-bar. The base diffusion forming the bar reaches fairly deeply into the epi and leaves little room for carriers to pass underneath it. Most of

[41] Davis, p. 27; Jones, p. 10.

[42] Jones, p. 10.

the holes traveling from left to right will be collected by the P-bar and shunted to ground. This structure thus acts as a specialized type of minority-carrier guard ring. The presence of NBL beneath the P-bar ensures a low-impedance path for base current passing from the right-hand transistor to the tank contact at the lower left corner of the tank. The tank contact on the left side of this structure therefore suffices for both transistors.

Although the collection efficiency of P-bars can only be determined through empirical measurement, several observations are in order. As the tank bias increases, the depletion region surrounding the bar deepens and progressively pinches off the N-epi underneath it. Devices operating at high tank-to-substrate potentials therefore obtain a higher degree of isolation from a P-bar than devices operating near substrate potential. A wider P-bar also increases collection efficiency, in part because the pinched portion of the tank becomes wider and in part because the wider base region diffuses deeper into the epi. Even a minimum-width P-bar provides a high degree of isolation against minority-carrier cross-injection due to the up-diffusion of the underlying NBL and the formation of a depletion region beneath the P-bar.

The P-bar has many applications. Bipolar circuits often contain current mirrors composed of lateral PNP transistors with a common base connection. These transistors often occupy a common tank, but if one transistor saturates then the currents provided by the adjacent transistors increase. P-bars placed between the saturating transistor and the adjacent devices will prevent this effect without unduly enlarging the tank. Another common application consists of an NPN driving either a lateral or a substrate PNP transistor, in which the collector of the NPN connects to the base of the PNP. Minority-carrier conduction from the PNP to the NPN can initiate positive-feedback latchup by triggering the SCR inherent in this structure. A P-bar placed between the transistors may suppress the latchup, although this is not guaranteed unless the collection efficiency of the bar exceeds the reciprocal of the beta product of the two transistors.

P-bars also find use in CMOS processes, where they typically consist of PMoat. This type of P-bar exhibits a lower collection efficiency than its bipolar counterpart due to the shallowness of the PMoat diffusion and the absence of NBL. The lack of a buried layer greatly increases the well resistance beneath the P-bar, so prudence dictates the inclusion of well contacts on both sides of the bar. This structure can help increase a circuit's latchup immunity without requiring separate wells. If one PMOS transistor in the tank has a source or drain connecting to an outside terminal, then a transient can potentially forward-bias this PMoat into the well. The resulting minority-carrier injection can disturb adjacent transistors and can even lead to latchup. The strategic placement of a few minimum-width PMoat P-bars can provide considerable protection against this sort of cross-injection without consuming as much area as separate wells require.

Another type of minority-carrier guard ring called an *N-bar* can also protect against minority-carrier cross-injection. An N-bar consists of a strip of deep-N+ placed between two devices occupying a common tank (Figure 4.26).[43] The N-bar typically serves as a tank contact for the devices around it since the spacings surrounding the deep-N+ are large enough to allow room for both emitter diffusion and a contact. The doping gradient surrounding the N-bar repels minority carriers, and most of the carriers that overcome this gradient recombine inside the deep-N+ before they pass through it. Unfortunately, the N-bar generally stops short of the

[43] Davis, p. 31.

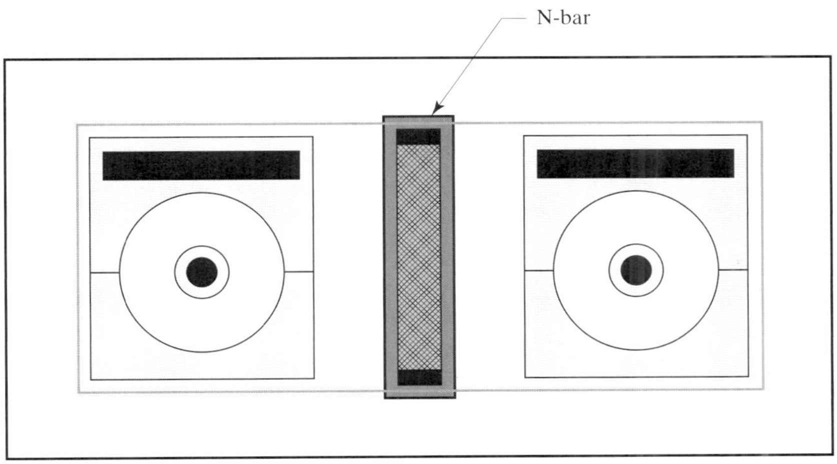

FIGURE 4.26 Example of an N-bar used to simultaneously provide tank contact and to minimize cross-injection between two lateral PNP transistors.

P-isolation on either side of the tank to avoid forming an N+/P+ junction that would break down at a relatively low voltage. These gaps allow minority carriers to bypass the N-bar, so an N-bar usually exhibits a lower collection efficiency than a P-bar. Still, the combination of a highly efficient collector contact with a moderately effective minority-carrier guard ring sometimes finds applications in high-current lateral PNP current mirrors and similar circuits.

4.5 SUMMARY

This chapter discusses a number of common failure mechanisms of integrated circuits. Table 4.2 summarizes these mechanisms, along with typical symptoms and suggested corrective actions. Even a cursory glance at the table reveals the interdisciplinary nature of the subject. Some mechanisms are primarily electrical, while others depend upon chemical or electrochemical processes. Some of these failure mechanisms require a knowledge of device physics to counteract them, while others require knowledge of processing and packaging technology. Only by amassing a working knowledge of many fields can one hope to design integrated circuits that will function reliably over a lifetime of use.

4.6 EXERCISES

Refer to Appendix C for layout rules and process specifications.

4.1. A certain copper-doped aluminum alloy can safely operate at current densities of $5 \cdot 10^5 \text{A/cm}^2$. If the metallization thickness equals 8kÅ, but thins by 50% when passing over oxide steps, then how much current can a 10μm-wide lead carry across an oxide step?

4.2. Propose a scribe seal structure for a single-level-metal standard bipolar process. Draw a cross section of this structure and explain the purpose of each of its components.

4.3. Lay out a 15kΩ, 8μm-wide HSR resistor. Field plate the resistor as well as possible, including flanges where necessary. The field plate should overhang HSR by at least 6μm and base by at least 8μm.

4.4. Modify the layout from Exercise 4.3 to include channel stops constructed from emitter diffusion. Assume that the channel stops must overlap the field plates by 4μm.

4.5. Construct a minimum-size, standard-bipolar, lateral PNP using a circular emitter geometry. Fully field-plate both the emitter and the collector, leaving space for base metallization. Assume the emitter field plate must overlap the collector by 2μm and the collector field plate must overhang the collector by 8μm.

TABLE 4.2 Summary of failure mechanisms.

Failure Mechanism	Symptoms	Corrective Actions*
Electrostatic discharge (ESD)	Gate oxide ruptures either immediately or after delay, junctions shorted or leaky.	***Add ESD protection devices, do not route leads over thin emitter oxide.***
Electromigration	Open or short circuits after long-term operation, usually at high temperature.	Use copper-doped aluminum, use refractory barrier metal, ***use adquate lead widths, use adequate bondwires.***
Antenna effect	Small gate oxides connected to large conductors suffer delayed failure.	***Reduce ratio of conductor area to gate oxide area, add diodes.***
Dry corrosion	Open circuit failures, moisture accelerates failure.	Use nitride PO, ***minimize PO openings.***
Mobile ions	Threshold shifts under high-temp biased operation, relaxes after unbiased bake.	Use phosphosilicate glass, use polysilicon gate MOS, ***minimize PO openings, use adequate scribe seals.***
Hot carriers (in MOS)	Threshold shifts under high-temp biased operation, relaxes after unbiased bake.	***Limit drain-source voltages, use LDD structures, use long-channel devices.***
Zener walkout	Breakdown voltage drifts, relaxes after unbiased bake.	***Use buried Zener*** (if available).
Parasitic channels & Charge spreading	Leakage currents at high voltage. If they appear after high-temp biased operation and relax after high-temp bake, charge spreading is responsible.	Use (111) silicon, add channel stop implants, ***add base-over-iso, use channel stops, use field plates.***
Substrate debiasing	Latchup, parametric shifts that occur under specific biasing conditions.	***Maximize substrate contact, place contacts near injectors.***
Minority-carrier injection into the substrate	Latchup, parametric shifts that occur under specific biasing conditions.	Use P+ substrate, ***maximize substrate contact, separate sensitive circuitry, add NBL to shared wells, use deep-P+ in isolation, add guard rings.***
Minority carrier cross-injection	Latchup, mismatches between merged devices.	***Use P-bar or N-bar, place devices in separate tanks or wells.***

* Possible solutions listed in ***bold italics*** are under the control of circuit and layout designers; the remaining solutions can only be implemented by process engineers.

4.6. Compute the area of substrate contacts necessary to extract 25mA from a die that uses an 8μm-thick, 10Ω·cm, P-type epi layer on top of a 0.01Ω·cm P-type substrate. Assume a maximum allowed debiasing of 0.3V.

4.7. Lay out a standard-bipolar NPN transistor with a 20μm by 40μm emitter. Arrange the transistor to minimize the distance between the emitter and collector contacts. The transistor should include deep-N+ in the collector to reduce collector resistance. Place an electron-collecting guard ring around this transistor, following the cross section shown in Figure 4.22.

4.8. Lay out a 2000/5 PMOS transistor. Divide the transistor into a sufficient number of fingers to obtain a roughly square aspect ratio. Construct a hole-collecting guard ring similar to that in Figure 4.24 that encircles the PMOS transistor. Make sure that the NBL overlaps the deep-N+ diffusion by at least 4μm to provide an adequate seal at the point of intersection.

4.9. Lay out an example of a P-bar separating two minimum-size standard bipolar lateral PNP transistors. The P-bar should extend at least 4μm into the isolation to ensure electrical continuity.

4.10. Several failed devices have been de-encapsulated (*decapped*) for microscopic examination. Suggest at least one failure mechanism consistent with each of the following observations:

a. A metal trace from a bond pad has melted open.
b. A greenish deposit covers the bond pads.
c. The gate oxide of a minimum-size NMOS has ruptured at one point, shorting the poly gate to the underlying epi.
d. A thin, dark filament appears across the base of a large NPN transistor. The transistor's base-collector junction appears shorted.

4.11. A new high-voltage, low-current operational amplifier has just completed burn-in testing. Sample units were operated under bias at 150°C for 1000 hours. Parametric testing reveals that the input offset voltages of the amplifier have drifted several millivolts during testing, and the supply currents have increased by 20%. What failure mechanisms might be responsible for these symptoms, and how can the designer determine what to fix?

5 Resistors

Resistors provide specific and controlled amounts of electrical resistance. They are useful in a variety of applications, ranging from current limiting to voltage division. Analog circuits usually include many resistors, so it is fortunate that they are relatively easy to integrate. Although the tolerance of any particular integrated resistor is relatively poor (±30%), the tracking between matched pairs of integrated resistors is excellent (±0.1%). Laser-trimmed thin-film resistors can achieve tolerances of better than ±0.1%, but only at the cost of additional processing steps.

Most processes offer a choice of several different resistor materials. Some are better suited to fabricating high-value resistors and others to fabricating low-value ones. The precision and temperature variation of different materials also varies widely. Circuit designers usually select appropriate materials for each resistor and mark their schematics accordingly. Sometimes different symbols identify the various types of resistors; sometimes the type of each resistor is printed beside it. The choice of resistor materials can have tremendous impact on circuit performance, so substitutions should not be made without careful consideration of the consequences.

5.1 RESISTIVITY AND SHEET RESISTANCE

The International System of Units[1] (SI) defines the *Ohm* (Ω) as the standard unit of resistance. As with other SI quantities, a system of prefixes allows this unit to be scaled upward or downward. Table 5.1 lists most of the prefixes used by engineers, along with their official SI abbreviations and those used by the simulation program SPICE.[2]

The value of a resistor can be computed given its dimensions and composition. Each material possesses a characteristic *resistivity,* usually measured in Ω·cm.

[1] The International System of Units, or *Système Internationale (SI),* is more commonly called the metric system.

[2] The acronym SPICE stands for *Simulation Program with Integrated Circuit Emphasis,* the most familiar and widely used circuit simulator. Developed by Larry Nagel and others under the supervision of D. O. Pederson at the University of California at Berkeley, SPICE was first released in 1972.

Name of Prefix	Value	SI Symbol	SPICE Symbol
atto-	10^{-18}	a	
femto-	10^{-15}	f	F
pico-	10^{-12}	p	P
nano-	10^{-9}	n	N
micro-	10^{-6}	μ	U
milli-	10^{-3}	m	M
kilo-	10^{3}	k	K
mega-	10^{6}	M	MEG
giga-	10^{9}	G	
tera-	10^{12}	T	

TABLE 5.1 Selected prefixes of the International System of Units (SI).

Resistivity is the inverse of *conductivity,* so if one of these two properties is known, then the other can be determined (a resistivity of 10Ω·cm implies a conductivity of $0.1(\Omega\cdot\text{cm})^{-1}$, and *vice versa*). Conductors have very low resistivities, while doped semiconductors have moderate resistivities (Table 5.2). The resistivity of a true insulator such as silicon dioxide is virtually infinite.

Material	Resistivity Ω·cm (25°C)
Copper, bulk	$1.7 \cdot 10^{-6}$
Gold, bulk	$2.4 \cdot 10^{-6}$
Aluminum, thin film	$2.7 \cdot 10^{-6}$
Aluminum (2% silicon)	$3.8 \cdot 10^{-6}$
Platinum silicide	$3.0 \cdot 10^{-5}$
Silicon, N-type ($N_d = 10^{18}$ cm^{-3})	0.25
Silicon, N-type ($N_d = 10^{15}$ cm^{-3})	48
Silicon, intrinsic	$2.5 \cdot 10^{5}$
Silicon dioxide (SiO_2)	$\sim 10^{14}$

TABLE 5.2 Resistivities of selected homogeneous materials.[3]

Figure 5.1 shows a simple resistor constructed from a rectangular slab of a homogeneous material having resistivity ρ. This resistor is contacted at either end by perfectly conductive plates. If the slab of resistive material has length *L,* width *W,* and thickness *t,* then its resistance *R* equals:

$$R = \rho \frac{L}{Wt} \qquad [5.1]$$

Integrated resistors consist of diffusions or depositions that can be modeled as films of constant thickness. It is therefore customary to combine resistivity and thickness into a single term called the *sheet resistance* R_s. In the case of a homogeneous material, $R_s = \rho/t$. The formula for resistance can now be rewritten as follows:

$$R = R_s\left(\frac{L}{W}\right) \qquad [5.2]$$

[3] Resistivities vary substantially depending on the conditions of preparation; for example, bulk resistivities of pure materials are considerably smaller than the thin-film values. Values for Cu, Au, Al, PtSi: W. R. Runyan and K. R. Bean, *Semiconductor Integrated Circuit Processing Technology* (Reading, PA: Addison-Wesley, 1994), pp. 535, 546, 548. Doped silicon: W. R. Thurber, R. L. Mattis and Y. M. Liu; *National Bureau of Standards Special Publication 400-64:* 1981, p. 42. Intrinsic silicon: B. G. Streetman, *Solid State Electronic Devices,* 2nd ed. (Englewood Cliffs, NJ: Prentice-Hall, 1980), p. 443. SiO_2: Runyan, *et al.,* p. 63.

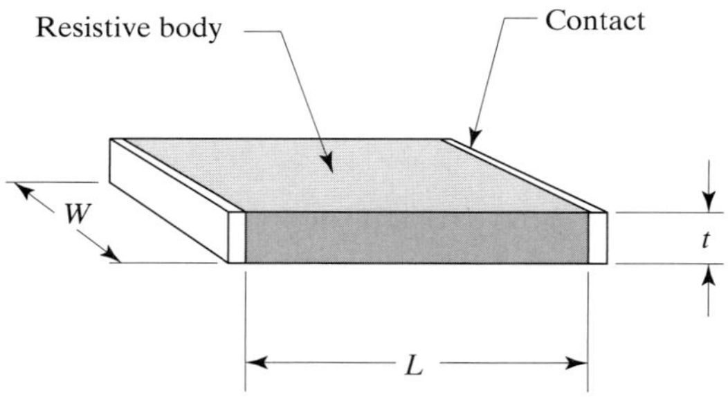

FIGURE 5.1 Layout of a simple resistor consisting of a rectangular slab of resistance material contacted by perfectly conductive terminations.

Resistors are often specified in terms of their (L/W) ratio, which, although technically dimensionless, is usually assigned fictitious units of *squares* (□). A resistor with equal length and width contains one square; a resistor with a length twice its width consists of two one-square resistors in series, or two squares, and so forth. The sheet resistance R_s is usually given in units of Ohms per square (Ω/□). The value of a resistor can be found by multiplying the number of squares it contains by its sheet resistance. For example, a resistor containing ten squares of 150Ω/□ material will have a value of 1.5kΩ.

Although the sheet resistance of a homogeneous film can be easily computed, many integrated resistors consist of inhomogeneous diffusions. No simple formula exists for determining the sheet resistance of such a diffusion. One can determine the sheet resistance of an ideal Gaussian diffusion using Irwin's graphs,[4] but real diffusions do not necessarily follow these idealized profiles. In practice, sheet resistances of diffusions are usually determined by empirical measurement instead of by computation.

5.2 RESISTOR LAYOUT

Figure 5.2 shows the layout of the simplest possible resistor, consisting of a simple rectangle of resistance material with contacts at either end. The low resistance of a contact effectively shorts out the material underneath it. Almost all of the current exits the contact along its inner edge, facing the main body of the resistor. The *drawn length* of the resistor L_d therefore equals the distance measured from the inner edge of one contact to the inner edge of the other. Similarly, the width of the strip of resistance material is called its *drawn width* W_d. The drawn length and width can be used to determine the approximate value of the resistor using equation 5.2.

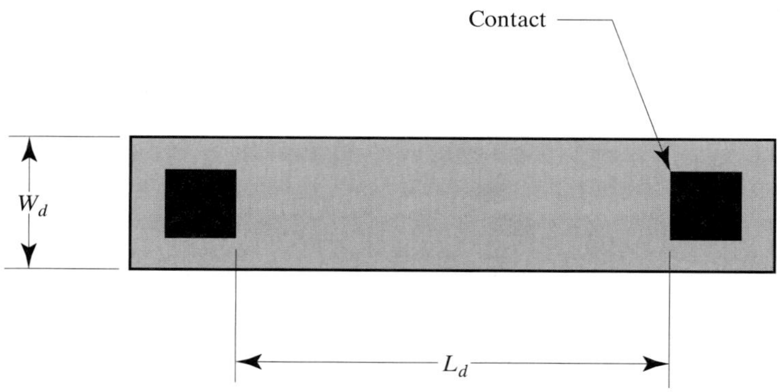

FIGURE 5.2 Layout of a simple strip resistor.

[4] J. C. Irvin, "Resistivity of Bulk Silicon and Diffused Layers in Silicon," *Bell Syst. Tech. J.,* Vol. 41, #2, 1962, pp. 387–410.

However, there are several factors at work in an integrated resistor that do not occur in the simple resistor in Figure 5.1. Photolithography and etching can cause oxide openings to grow or shrink slightly. Outdiffusion can widen a resistor and thus reduce its resistance. Nonuniform current flow near the contacts can increase its resistance. An examination of each of these terms will show which must be taken into account in order to accurately predict the value of a resistor.

The most significant corrections to the resistor equation are those associated with width rather than length, because most resistors are much narrower than they are long. The resistor equation can therefore be rewritten as follows:

$$R = R_s\left[\frac{L_d}{W_d + W_b}\right] \qquad [5.3]$$

The *width bias* W_b models the difference between the drawn width and the effective width. Outdiffusion adds about 20% of the junction depth to the drawn width of a diffused resistor.[5] For example, a base resistor with a junction depth of 1.25μm will have a width bias of about 0.25μm due to outdiffusion. This causes about a 5% error in the value of a 5μm-wide base resistor—enough to be worth attempting to correct. The width bias for a given diffusion can be determined experimentally by measuring a set of resistors of varying widths. Layout rules sometimes include tables of width biases. If available, these should be used when calculating resistor values.

Equation 5.3 implicitly assumes uniform current flow through the resistor. The layout in Figure 5.2 violates this assumption because the contacts do not extend entirely across the ends of the resistor. The current must crowd inward as it approaches the contacts, making the actual value of the resistor slightly larger than that estimated by width and length. The effect of lateral nonuniform current flow can be computed using the following formula:[6]

$$\Delta R = \frac{R_s}{\pi}\left[\frac{1}{k}\ln\left(\frac{k+1}{k-1}\right) + \ln\left(\frac{k^2-1}{k^2}\right)\right] \qquad [5.4]$$

where $k = W_e\,/\,(W_e - W_c)$, with W_e being the effective width of the resistor and W_c the width of the contact. The effective width of the resistor equals the sum of the drawn width W_d and the width bias W_b. The quantity ΔR represents the increase in resistance caused by nonuniform current flow at both ends of the resistor. For example, a resistor 5μm wide containing 3μm-wide contacts at either end will be 0.05 squares longer than equation 5.3 predicts. Since most resistors are at least ten squares long, this factor usually causes less than 1% error and is therefore inconsequential.

Current also flows nonuniformly in the vertical dimension as it enters and exits the resistor contacts. The current must bend upward to exit through the surface of the resistor, and it crowds toward the inside edges of the contacts as it does so. This current crowding produces a slight increase in overall resistance. This crowding effect is usually considered part of the *contact resistance* between the resistor and its metallization; see Section 5.3.4.

In summary, the width bias is usually important, and the effects of nonuniform current are usually not. Instead of applying corrections for nonuniform current flow,

---

[5] A. B. Glaser and G. E. Subak-Sharpe, *Integrated Circuit Engineering* (Reading, MA: Addison-Wesley, 1977), p. 127. See also P. R. Gray and R. G. Meyer, *Analysis and Design of Analog Integrated Circuits*, 3rd ed. (New York: John Wiley and Sons, 1993), p. 139; D. J. Hamilton and W. G. Howard, *Basic Integrated Circuit Engineering* (New York: McGraw-Hill, 1975), p. 150.

[6] C. Y. Ting and C. Y. Chen, "A Study of the Contacts of a Diffused Resistor," *Solid State Elect.*, Vol. 14, 1971, p. 434. This formula is strictly valid only for $W_c >> W_e - W_c$.

the designer should make the resistors long enough to avoid its influence. If the resistors are at least five squares long, then the effects of nonuniform current flow will probably total less than 5% and can usually be neglected.

Large resistors are often folded, producing so-called *serpentine* or *meander* resistors (Figure 5.3). These resistors usually employ rectangular turns (Figure 5.3A) rather than circular turns (Figure 5.3B). Rectangular turns are not only easier to draw but also allow the spacing between the turns of the resistor to be easily adjusted. The circular end segment can be split into two arcs to allow the insertion of an additional resistor segment, but this often requires redrawing the resistor.

Current does not flow uniformly around the bends in a serpentine resistor. Each square corner adds approximately 0.56 squares.[7] Neglecting process biases and end effects, the value of the resistor of Figure 5.3A is

$$R = R_s\left(\frac{2A + B}{W} + 1.12\right) \quad [5.5]$$

The contribution of a corner square is usually rounded to 1/2 square; thus the often-quoted rule, "a corner counts half a square." The slight errors implicit in this assumption rarely have any practical significance.

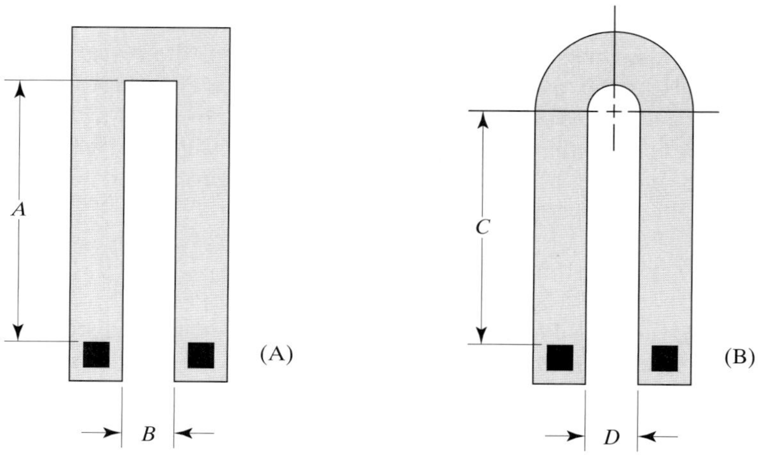

FIGURE 5.3 Layout of serpentine resistors with (A) rectangular turns and (B) circular turns. In the case of circular turns, spacing, *D*, is assumed to equal the width, *W*, of the resistor.

The 180° circular end segment of Figure 5.3B adds 2.96 squares to the resistor.[8] Neglecting process biases and end effects, the resistance of this structure equals

$$R = R_s\left(\frac{2C}{W} + 2.96\right) \quad [5.6]$$

The contribution of a circular end is usually rounded to 3 squares.

Sometimes a resistor becomes so narrow that contacts cannot reside inside it without violating design rules. This problem is usually overcome by enlarging the ends of the resistor to form heads around the contacts. The resulting structure is called a *dogbone* or *dumbbell resistor* because of its characteristic shape (Figure 5.4). The *drawn length* L_d of a dogbone resistor is measured from contact to contact, and the *drawn width* W_d is measured across the body of the resistor. The approximate value of the

[7] Glaser, *et al.*, p. 118. Grebene cites a value of 0.53: Grebene, p. 140. Reinhard cites one of 0.65: D. K. Reinhard, *Introduction to Integrated Circuit Engineering* (Boston: Houghton Mifflin, 1987), p. 191.

[8] Glaser, *et al.*, p. 118. Grebene's value appears to be in error: Grebene, p. 140.

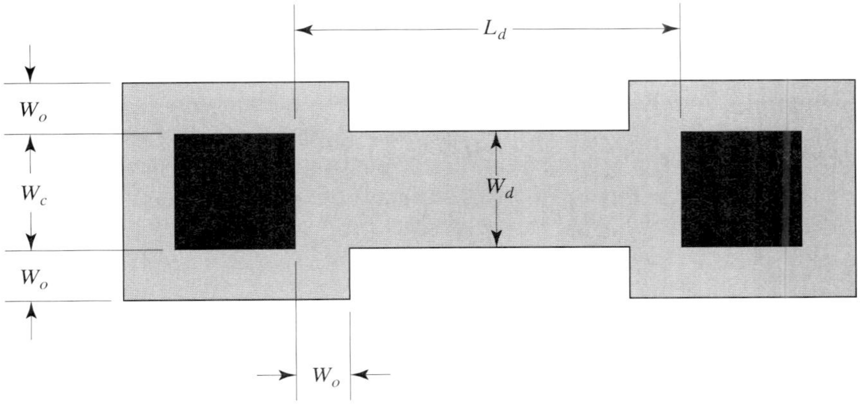

FIGURE 5.4 The dimensions of a dogbone resistor required for computing its value.

resistor can then be computed using equation 5.3. The effects of bends in the resistor can be handled in the same manner as for the strip resistor.

The effects of lateral nonuniform current flow differ for dogbone and strip resistors. In a strip resistor, such as that in Figure 5.2, the current crowds inward toward the contact, and the effective value of the resistor increases. In a dogbone resistor, the current spreads out as it enters the head, and the effective value of the resistor decreases. This effect can be minimized by making the width of the contact W_c equal to the width of the resistor W_d and by minimizing the overlap W_o of the resistor head over the contact. Table 5.3 lists published correction factors ΔR for two dogbone heads (one at either end of the resistor).

TABLE 5.3 Correction factors ΔR for dogbone resistors.[9]

$\mathbf{W}_o$	$\mathbf{W}_c$	ΔR
W_d	W_d	−0.7□
½W_d	W_d	−0.3□

The correction factor ΔR is usually less than 0.3 squares because most strip resistors use an overlap of less than half the drawn width. The contact resistance (Section 5.3.4) incorporates the effects of nonuniform current flow in the vertical plane. These corrections are negligible if the resistor is more than about five squares long.

Dogbone resistors do not pack as densely as strip or serpentine resistors (Figure 5.5). Many designers consider dogbone resistors to be more accurate than strip or serpentine resistors of the same width. True, the errors caused by nonuniform current flow are smaller for dogbones than for either strip or serpentine resistors, but this does not actually improve the accuracy of the resistor. Matched resistors should always be laid out in identical sections, and as long as each section has the same layout, it does not matter whether it uses dogbone heads or not.

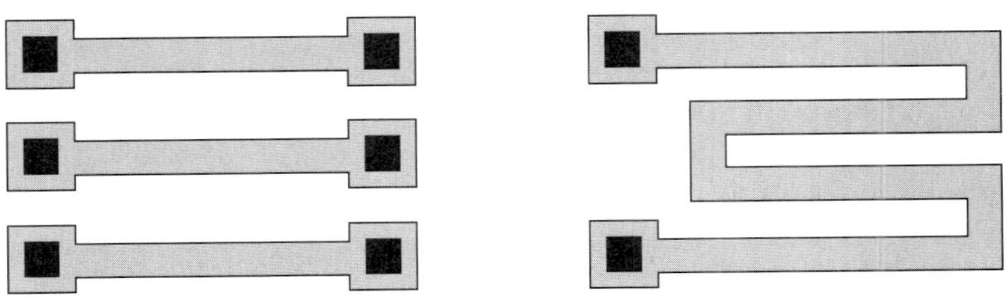

FIGURE 5.5 Sample dogbone and serpentine resistors, showing the poorer packing density caused by the presence of enlarged heads.

[9] Reinhard, p. 192.

5.3 RESISTOR VARIABILITY

The value of a resistor depends on numerous factors, including process variability, temperature, nonlinearity, and contact resistance. Other, less-significant factors that primarily affect resistor matching include orientation, stress and temperature gradients, thermoelectric effects, nonuniform etch rates, NBL push, voltage modulation, charge spreading, and PSG polarization. These less-significant factors will be discussed in Chapter 7.

5.3.1. Process Variation

The value of a resistor depends upon its sheet resistance and its dimensions. Sheet resistances vary because of fluctuations in film thickness, doping concentration, doping profiles, and annealing conditions. The dimensions of a resistor vary because of photolithographic inaccuracies and nonuniform etch rates.

Modern processes maintain sheet resistances within ±20 or ±25%,[10] except for pinch resistors. Base pinch resistors are formed from the region corresponding to the neutral base of an NPN transistor. Base pinch sheet resistance strongly correlates with NPN beta because both are functions of the doping and thickness of the neutral base region. A high NPN beta indicates a high base pinch sheet and *vice versa.* Most processes specify a 3:1 variation (±50%) in beta, and base pinch resistors vary by a similar amount.

Linewidth control is a measure of dimensional variation introduced by photolithography and processing. It depends only weakly on width for feature sizes of a micron or more.[11] In other words, if a 5μm feature can be held to a tolerance of ±1μm, then so can a 25μm feature. Consequently, linewidth control measured as a percentage of feature size improves with increasing feature size. Most processes can maintain linewidth control to within about ±20% of their minimum feature size. For example, a standard bipolar process with a minimum feature size of 5μm will probably have a linewidth control of about ±1μm.

If the sheet resistance variation and the linewidth control are known, then the actual tolerance for a resistor of effective width W_e can be computed using the following equation[12]

$$\delta R + \frac{C_L}{W_e} + \delta R_s \qquad [5.7]$$

where δR is the tolerance of the resistor, C_L is the linewidth control of the applicable layer, and δR_s is the variability of the sheet resistance. The equation assumes that the resistor is long enough to ignore length variations. Suppose that a resistor with an effective width of 2μm is drawn on a layer with a linewidth control of ±0.25μm using a material with a sheet resistance variation of ±25%. Equation 5.7 predicts that the resulting resistor will vary by ±38%. If the effective width were increased to 10μm, then the variability would be reduced to ±28%.

This information can be summarized as a set of design guidelines:

- Where tolerance does not matter, use minimum-width resistors and expect variations of about ±50%. Diffused resistors should not be made narrower

---

[10] The variability figures used in this section represent the three-sigma limits of a Gaussian distribution.

[11] Linewidth control of submicron features depends quite strongly on the precise dimensions of the geometry, but submicron resistors are rarely employed.

[12] Much of the variability in resistor sheets and linewidth control is due to unpublicized process biases; these add linearly rather than in quadrature. Glaser, *et al.* correctly use a linear sum but do not offer justification: Glaser, *et al.*, p. 121. C_L and δR_S are usually quoted as three-sigma values, in which case δR is also a three-sigma value.

than about 150% of their junction depth, or the amount of dopant may be inadequate to achieve the targeted depth.

- Where moderately precise tolerances are required, use resistors two or three times as wide as the minimum feature size and expect variations of ±35%.
- Where maximum precise tolerances are needed, use resistors approximately five times as wide as the minimum feature size and expect variations of ±30%.

These rules assume ±25% sheet variation and a linewidth control of ±20% of the minimum feature size. Base pinch resistors are exceptions to the above rules since their values depend primarily on their sheet variation, and increased widths offer little benefit. The width of a base pinch resistor should equal at least 150% of minimum to ensure that enough dopant diffuses down to form a pinched region. Resistor matching follows somewhat different rules than resistor tolerance (see Section 7.3).

These rules also require modification when dealing with deep diffusions such as N-well. The precision of deeply diffused resistors depends on outdiffusion as it does on linewidth control. In this case, the width required for a given accuracy should be computed using the junction depth or the minimum feature size, whichever is larger. Thus, an 8μm-deep N-well resistor should be made at least 16μm wide to obtain moderate accuracy.

5.3.2. Temperature Variation

Resistivity depends on temperature in a complex, nonlinear manner. Like any nonlinear function, this one can be expanded into a polynomial series. Unless considerable accuracy is required or the temperature varies widely, only the first two terms of the series are significant

$$R(T) = R(T_0)\,[1 + 10^{-6}TC_1(T - T_0)] \qquad [5.8]$$

where $R(T)$ is the resistance at the desired temperature, T; $R(T_0)$ is the resistance at some other temperature, T_0; and TC_1 is the *linear temperature coefficient of resistivity* (TCR) in parts per million per degree Celsius (ppm/°C). A factor of 10^{-6} has been inserted into equation 5.8 to balance the units involved. Table 5.4 lists typical linear temperature coefficients for several common integrated resistance materials. Most of these values are reasonably accurate between 0 and 50°C. The temperature coefficients of polysilicon are exceptions; they depend on annealing conditions and can vary significantly from the values listed in Table 5.4.

Matched resistors must consist of the same material to ensure that temperature variations do not upset their matching. In addition, the TCRs of different diffusions are sometimes used by circuit designers to temperature-compensate circuits. For these reasons, different resistance materials cannot be arbitrarily substituted for one another.

5.3.3. Nonlinearity

Ideal resistors exhibit a linear relationship between voltage and current. Practical resistors always exhibit some degree of nonlinearity; in other words, their resistance varies with applied voltage. Nonlinearity, or *voltage modulation,* arises from several sources, including self-heating, high-field velocity saturation, and depletion region encroachment.

A resistor dissipates power equal to the product of the voltage across it and the current through it. This dissipation causes internal heating. Plastic packages conduct heat rather poorly, so even small amounts of internal heating can cause substantial

TABLE 5.4 Typical linear temperature coefficients of resistivity for selected materials at 25°C.[13]

Material	TCR, ppm/°C
Aluminum, bulk	+3800
Copper, bulk	+4000
Gold, bulk	+3700
160Ω/□ Base diffusion	+1500
7Ω/□ Emitter diffusion	+600
5kΩ/□ Base pinch diffusion	+2500
2kΩ/□ HSR implant (P-type)	+3000
500Ω/□ Polysilicon (4kÅ N-type)	−1000
25Ω/□ Polysilicon (4kÅ N-type)	+1000
10kΩ/□ N-well	+6000

temperature increases inside the package. Most resistors have relatively large temperature coefficients, and even modest increases in temperature can cause significant resistance variations. Suppose 10mA flows through a 1kΩ HSR resistor, which consequently dissipates 100mW. Assuming a thermal impedance of 80°C/W and a temperature coefficient of 3000ppm/°C, this dissipation results in an 8°C temperature rise and a 2.4% increase in resistance.

Poly resistors are especially vulnerable to self-heating because they reside on top of the thick-field oxide, which acts as a thermal insulator between the poly resistors and the silicon substrate. Very little heat can escape upward through the protective overcoat and encapsulation, so the temperature rise ΔT between the resistor and the silicon substrate, in degrees Celsius, equals

$$\Delta T = 71\,\frac{V^2 t_{ox}}{R_s L} \qquad [5.9]$$

where R_s is the sheet resistance of the poly in Ω/□, t_{ox} is the thickness of the field oxide in Angstroms (Å), L is the length of the resistor in microns, and V is the voltage applied across the resistor. The temperature-induced nonlinearity equals the product of the temperature rise and the linear TCR of the resistor. This nonlinearity can usually be neglected for resistors that experience less than 1°C of self-heating.

The rate of carrier drift in weak electric fields is proportional to the electric-field intensity. As the field increases, the drift velocity of the carriers eventually becomes diffusion-limited, and the resistance begins to increase. The critical electric field intensity where this nonlinearity begins equals approximately 0.2V/μm for electrons and 0.6V/μm for holes.[14] The electric field should be kept well below these critical

[13] Temperature coefficients depend strongly on the conditions of fabrication and measurement, so the values given here are only approximations. Values for Al, Cu, and Au determined by linear interpolation of data from G. W. C. Kay and T. H. Laby, *Tables of Physical and Chemical Constants,* 15th ed. (Essex, England: Longman Scientific and Technical, 1986), pp. 117–118. Base, emitter, pinch, and HSR diffusions: Gray, *et al.,* p. 139. Base and HSR diffusions: Grebene, pp. 138, 153. Poly: W. A. Lane and G. T. Wrixon, "The Design of Thin-Film Polysilicon Resistors for Analog IC Applications," *IEEE Trans. Electron Devices,* Vol. 36, No. 4, 1989, pp. 738–744. See also discussion and curves for various diffusions in Hamilton, *et al.,* pp. 277–279; and P. Norton and J. Brandt, "Temperature Coefficient of Resistance for p- and n-type Silicon," *Solid State Electronics,* Vol. 21, 1978, pp. 969–974.

[14] Muller, *et al.* give curves showing high-field mobility saturation from which these values can be obtained by examination: R. S. Muller and T. I. Kamins, *Device Electronics for Integrated Circuits,* 2nd ed. (New York: John Wiley and Sons, 1986), p. 36.

intensities to minimize nonlinearity. Assuming a safety factor of two, the minimum resistor length L_{min} equals

$$L_{min} = (6.7\,\mu\text{m}/V) \cdot V_{max} \quad \text{for N-type silicon} \qquad [5.10A]$$

$$L_{min} = (3.3\,\mu\text{m}/V) \cdot V_{max} \quad \text{for P-type silicon} \qquad [5.10B]$$

where V_{max} is the maximum voltage applied across the resistor. Resistors shorter than L_{min} can be constructed, but their resistance will increase at high voltages.

Poly resistors may also exhibit nonlinearities if they are so short that appreciable voltage drops can appear across individual poly grains. Under these conditions, the resistance of the barrier regions between grains becomes a function of the voltage drop across the resistor. These nonlinearities can be neglected if the resistor length is at least a thousand times the diameter of an individual grain.[15] Since most polysilicon films have grain sizes of approximately 0.5 to 1μm, this means that precision poly resistors should be made at least 50 to 100μm long to avoid nonlinearity effects. The length of poly resistors should also satisfy equations 5.10A and 5.10B.

Additional voltage nonlinearities occur in lightly doped diffused resistors, especially base pinch resistors, due to modulation of the depletion regions around reverse-biased junctions. Figure 5.6 shows a cross section of a base pinch resistor. The depletion regions widen toward the high-voltage end of the resistor because the reverse biases are largest there. In other words, the resistor is pinched more severely at the high-voltage end than at the low-voltage end. This pinching effect becomes more pronounced as the voltage across the resistor increases. Voltage nonlinearities of up to 1%/V have been reported for HSR,[16] and similar (or larger) values can be expected for base pinch resistors.

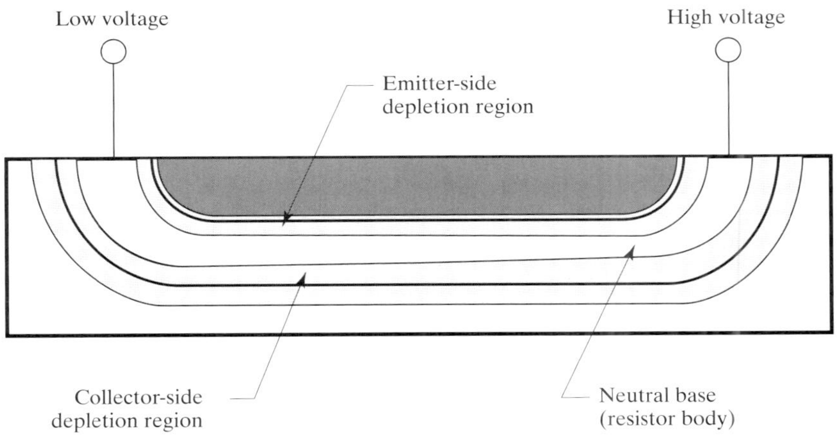

FIGURE 5.6 Cross section of a base pinch resistor showing the intrusion of the depletion regions into the neutral base. Notice that the high-voltage end of the resistor narrows slightly.

Heavily doped resistors do not experience appreciable voltage nonlinearities because the depletion regions extend only a minute distance into them. Emitter resistors, for example, show practically no voltage modulation.

Depletion regions also cause an increase in resistance when significant tank bias is applied. As the voltage difference between the resistor and the tank increases, the depletion regions widen and the resistance increases. This effect is called *tank modulation.* A 160Ω/□ base resistor experiences a tank modulation of about 0.1%/V.[17]

---

[15] Lane, *et al.*, p. 741.

[16] W. Bucksch, "Quality and Reliability in Linear Bipolar Design," *TI Tech. J.*, Vol. 4, #6, 1987, pp. 61–69. See also Grebene, p. 153.

[17] Value for 200Ω/N base: Hamilton, *et al.*, p. 155.

If the voltage between the base resistor and its tank were to increase by 10V, then the resistance would increase by 1%. The tank modulation for a 2kΩ/□ high-sheet resistor is about 1%/V, and base pinch resistors have tank modulations of at least 1%/V. Matched resistors either require carefully controlled tank biasing or relatively low-sheet materials.

Another form of voltage modulation occurs when leads cross a lightly doped resistor. The electric fields generated by the leads cause carriers to redistribute in the body of the resistor in much the same way that the field generated by an MOS gate redistributes carriers in the backgate. This *conductivity modulation* can cause several percent variation in the value of a 2kΩ/□ HSR resistor. HSR resistors are especially susceptible to this effect because they are both very thin and very lightly doped. Accurate HSR resistors should be field plated to minimize variations caused by conductivity modulation. Split field plates (Section 7.2.8) are recommended because they reduce nonlinearities caused by conductivity modulation from the field plates themselves. Poly resistors show less conductivity modulation than diffused resistors because they are usually much more heavily doped. Poly resistors with sheet resistances of 1kΩ/□ or less generally do not suffer from conductivity modulation.

5.3.4. Contact Resistance

Each resistor contains at least two contacts, and each of these adds some resistance to the overall structure. This resistance results from the presence of a potential barrier between the resistance material and the metallization. Although carriers can tunnel through this barrier, they lose some energy in the process. This energy loss varies with current and is therefore best described as a specific resistance of the contact interface, measured in Ω·μm². This *contact resistance* depends on the nature of the materials in contact and on processing conditions. Contact resistance can vary greatly from lot to lot, and designers should assume that it varies from essentially zero to the specified maximum unless the design rules state otherwise. The resistance R_c added by a single contact having width W_c and length L_c equals[18]

$$R_c = \frac{\sqrt{R_s \rho_c}}{W_c} \coth\left(L_c \sqrt{R_c/\rho_c}\right) \qquad [5.11]$$

where R_s is the sheet resistance of the resistor material, ρ_c is the specific contact resistance, and coth( ) represents the hyperbolic cotangent function. Equation 5.11 also accounts for current crowding in the vertical dimension (Section 5.2). Table 5.5 lists typical observed specific contact resistances for a number of contact systems.

TABLE 5.5 Typical contact resistances for various contact systems.[19]

Contact System	Contact Resistance, Ω·μm²
Al-Cu-Si to 160Ω/□ base	750
Al-Cu-Si to 5Ω/□ emitter	40
Al-Cu/Ti-W/PtSi to 160Ω/□ base	1250
Al-Cu/Al-Cu (Via)	5
Al-Cu/Ti-W/Al-Cu (Via)	5

[18] H. Murrmann and D. Widmann, "Current Crowding on Metal Contacts to Planar Devices," *IEEE Trans. on Elect. Dev.*, ED-16, #12, 1969, pp. 1022–1024.

[19] Murrmann, *et al.* give a value of 650Ω·μm² for 150Ω/□ base; D'Andrea, *et al.* give a value of 1000Ω·μm² for 180Ω/□ base and 900Ω·μm² for 140Ω/□ base: G. D'Andrea and H. Murrmann, "Correction Terms for Contacts to Diffused Resistors," *IEEE Trans. on Elect. Dev.*, ED-17, 1970, pp. 484–485. Emitter: Murrmann, *et al.* Schottky contact value: by extrapolation from base value using data in Bucksch. Via values are the author's estimate. For a theoretical analysis, see Runyan, *et al.*, pp. 522ff.

Because contact resistances are strongly process-dependent, these values are only indicative of what any given fab can actually guarantee.

The aluminum-copper-silicon (Al-Cu-Si) metal system employed by older processes exhibits significant contact resistance, especially for lightly doped materials such as 160Ω/□ base diffusion. The use of refractory barrier metal significantly increases contact resistance variability due to sintering problems. The addition of a silicide layer beneath the barrier metal (Al-Cu/Ti-W/PtSi) eliminates this problem and gives contact resistances only slightly larger than those of aluminum-copper-silicon. Modern CMOS and BiCMOS processes use silicided contacts in combination with heavily doped PSD and NSD diffusions to achieve contact resistances similar to those of emitter ($40\Omega \cdot \mu m^2$ or better).

Consider a 1kΩ base resistor constructed with 8×8μm contacts in a standard bipolar process using an Al-Cu-Si metal system. Equation 5.11 indicates that each contact adds 43Ω. The resistor therefore gains 86Ω, or about 9% of its value. Standard bipolar base resistors that total less than ten squares per segment may require oversized contacts to avoid excessive variability due to contact resistance.

5.4 RESISTOR PARASITICS

No practical resistor is ever completely isolated from its environment. Capacitive and inductive coupling inevitably occur at higher frequencies, and some types of resistors also experience junction leakage. Circuit designers can model these interactions by replacing each integrated resistor with a subcircuit containing several ideal components. One of these components is an ideal resistor, while the remainder are *parasitic components* that represent the undesired but unavoidable interactions of the resistor with the rest of the die. These *subcircuit models* are also of interest to layout designers because they illustrate the limitations of various types of resistors.

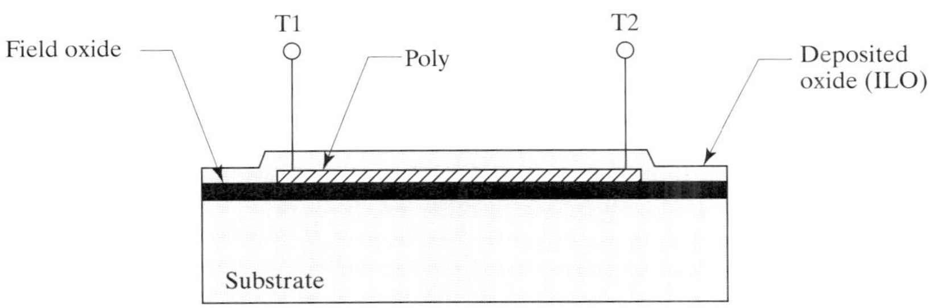

FIGURE 5.7 Cross section of a polysilicon resistor.

Figure 5.7 shows the cross section of a typical polysilicon resistor layout. The resistor is surrounded on all sides by oxide, an excellent insulator that exhibits virtually no leakage. The oxide also acts as a capacitive dielectric that couples the resistor to adjoining components. Most poly resistors are laid out on field oxide having a capacitance of about $0.05\text{fF}/\mu m^2$ (Section 6.1). Ignoring fringing effects, a resistor that is 5μm wide and contains 100 squares has a total substrate capacitance of about 0.125pF. This capacitance is distributed uniformly along the resistor, so it cannot be accurately modeled by a single capacitor. This *distributed capacitance* can be approximated using the *π-section* circuit of Figure 5.8A in which C_1 and C_2 are ideal capacitors, each representing half of the distributed capacitance. If a single π-section does not adequately model the distributed capacitance, then multiple π-sections can be used. Figure 5.8B shows a model incorporating two π-sections: R_1 and R_2 are ideal resistors with each being equal to half

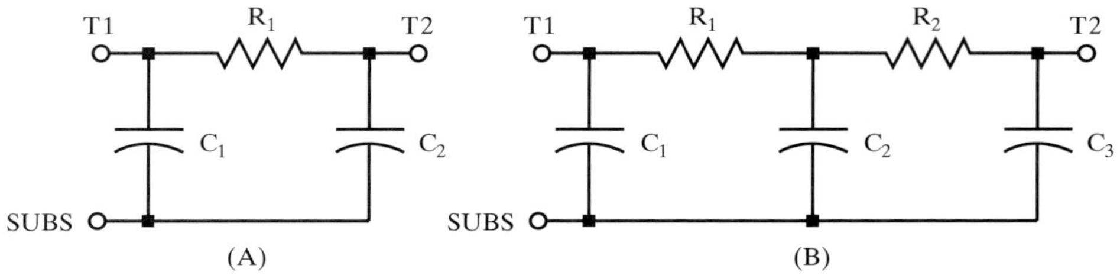

FIGURE 5.8 Subcircuit models for polysilicon resistors that approximate distributed substrate capacitance using π-sections: (A) single π-section and (B) dual π-section.

of the total resistance, and C_1, C_2, and C_3 are ideal capacitors; C_1 and C_3 each equal one-fourth of the distributed capacitance, and C_2 equals one-half of it.[20]

Leads routed across a polysilicon resistor introduce additional parasitic capacitances. The capacitance of the interlevel oxide (ILO) typically equals that of the field oxide, or about 0.5fF/μm^2. Thus, a 3μm lead crossing a 5μm-wide resistor at right angles produces about 7.5fF of coupling capacitance. This is an extremely small capacitance, but even so it can couple noise into high-impedance circuitry. Noisy signals should not be routed over polysilicon resistors in delicate analog circuits such as voltage references or low-noise amplifiers.

Figure 5.9 shows a simplified cross section of a diffused resistor. One or more reverse-biased junctions isolate this resistor from the remainder of the die. Tank contacts are required to maintain the necessary reverse bias across these junctions.

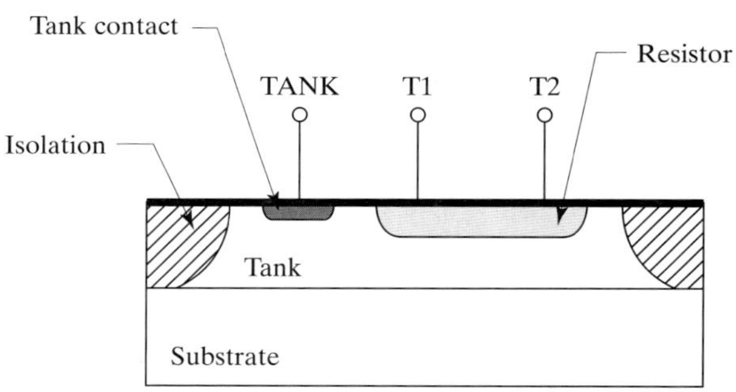

FIGURE 5.9 Cross section of a typical diffused resistor.

The parasitics of this diffused resistor consist chiefly of reverse-biased junctions: one between the resistor and the tank, and a second between the tank and the substrate. These junctions form distributed structures that are frequently modeled using π-sections. Figure 5.10 shows two single π-section subcircuit models for the resistor of Figure 5.9.[21]

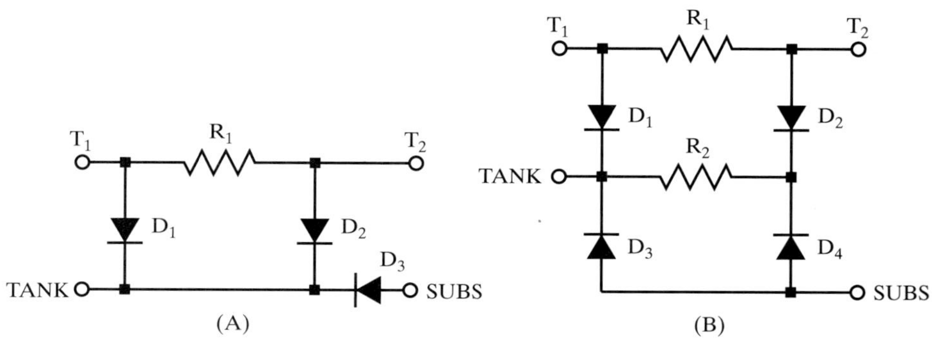

FIGURE 5.10 Subcircuit models for diffused resistors: (A) neglecting tank resistance and (B) including tank resistance.

[20] Y. Tsividis, *Mixed Analog-Digital VLSI Devices and Technology* (New York: McGraw-Hill, 1996), p. 166.

[21] For a similar discussion see Hamilton, *et al.*, pp. 160–182.

The subcircuit of Figure 5.10A includes an ideal resistor and three diodes. D_1 and D_2 each model half of the total area of the resistor-tank junction, while D_3 models the full area of the tank-substrate junction. This subcircuit remains reasonably accurate as long as the tank resistance is relatively small compared to the resistance of R_1. In the case of a more resistive tank (for example, a PSD resistor in N-well), the subcircuit of Figure 5.10B is preferable. This model includes a resistor, R_2, that models the effects of tank resistance, and diodes, D_3 and D_4, that model the distributed nature of the tank-substrate junction. Both of these subcircuits can incorporate additional π-sections to enhance their accuracy at higher frequencies.

The reverse-biased diodes associated with a diffused resistor cause several undesirable effects. If the tank voltage ever falls below the resistor voltage, the resistor-tank junction will forward-bias and inject minority carriers into the tank. This can trigger latchup (Section 4.4.2). Even if latchup does not occur, large currents may flow through the tank contact. The proper biasing of resistor tanks requires considerable thought. The three tank biasing schemes shown in Figure 5.11 illustrate several common arrangements.

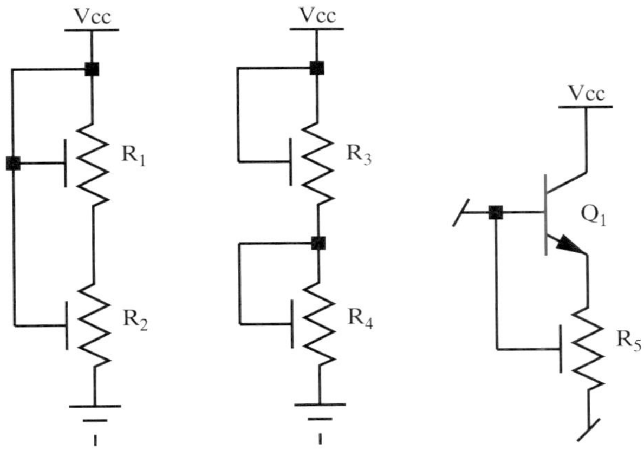

FIGURE 5.11 Several methods of biasing the tank connections of diffused resistors.

The resistor divider formed by R_1 and R_2 is biased as simply as possible: both tanks connect to a power supply that is more positive than either resistor. This reverse-biases the tank-substrate junctions, but each resistor sees a different relative tank bias and experiences a different amount of tank modulation. The resulting resistance variations cause the divider ratio to vary with supply voltage. The resistor divider formed by R_3 and R_4 illustrates a tank connection that is better suited for matched resistor dividers. Each tank connects to the positive end of its respective resistor. If the two resistors are equal in value, then the relative tank biases seen by each are also equal, and they will experience the same tank modulation. This technique can be extended to encompass resistors of unequal values by dividing them into multiple sections occupying separate tanks.

R_5 illustrates yet another method of biasing a resistor tank. Assuming that transistor Q_1 always conducts, the tank of R_5 will be biased one base-emitter drop above the positive terminal of the resistor. This connection illustrates one of the many possible configurations in which the tank connects neither to the resistor nor to a supply. Like all too many of these configurations, this one has the potential to forward-bias the resistor-tank junction. Suppose that the transistor is conducting and that R_5 is biased several volts above ground. If the base of Q_1 is suddenly pulled low, the resistor-tank junction will momentarily forward-bias. Connections that do not tie back to the positive end of the resistor nor to a supply should be carefully analyzed for sneak conduction paths similar to this one.

The reverse-biased junctions that isolate a diffused resistor can also suffer avalanche breakdown. This is of particular concern for emitter resistors and base pinch resistors since these usually avalanche at about 7V. If the bias across a resistor may exceed its breakdown voltage, then it should be constructed in multiple segments placed in separate tanks. The depletion regions associated with reverse-biased junctions also have considerable capacitance. This capacitance depends upon junction doping and reverse bias and is typically 1 to 5fF/μm^2. This value is substantially larger than the 0.5fF/μm^2 typically associated with poly resistors, so the high-frequency performance of diffused resistors cannot match that of poly resistors.

Most designers favor poly resistors because the absence of junctions makes parasitics less of a concern. However, the sheet resistance and temperature coefficient of polysilicon are not always as desirable as those of a particular type of diffused resistor. For example, N-well resistors are often more compact than poly resistors in processes with a polysilicon sheet resistance of 25Ω/□. Thus, diffused resistors still see occasional use even in processes that offer polysilicon resistors.

5.5 COMPARISON OF AVAILABLE RESISTORS

Most processes offer several types of resistors optimized for different applications. This section compares the various types of resistors presented in Chapter 3, and also presents several additional resistors that are useful for special applications.

5.5.1. Base Resistors

Base resistors are available in standard bipolar (Figure 3.13) and also in analog BiCMOS (Figure 3.51). Base sheets in standard bipolar typically range from 100 to 200Ω/□, allowing Ohmic contact to be made directly to the resistor. Processes that do not silicide the contacts often have rather high contact resistances. The base sheet in BiCMOS and advanced bipolar processes is usually somewhat higher than standard bipolar (300 to 600Ω/□), so reliable Ohmic contact requires the addition of a more heavily doped diffusion beneath the contacts. The resistance of the resulting composite structure can be computed using formulas similar to those used for HSR resistors (Section 5.5.4).

The base diffusion is best suited for laying out resistors in the range of 50Ω to 10kΩ. Larger resistors are usually constructed from HSR and smaller ones from emitter. Base sheet control is relatively precise, and base resistors are doped heavily enough to minimize tank modulation. These considerations often outweigh the area savings possible with HSR resistors, especially for precisely matched ones.

Field plates should be employed for high-voltage resistors to prevent charge spreading (Section 4.3.2). Leads can be routed across base resistors without causing significant conductivity modulation. Care should still be taken in routing noisy signals across base resistors because the oxide over the resistor is thinner than the field oxide, and capacitive coupling may inject noise into the circuitry attached to the resistor.

Base resistors must be placed in a suitable tank, consisting of either an N-epi region in standard bipolar or an N-well in analog BiCMOS. This tank should contain as much NBL as possible to reduce tank resistance. NBL can also help minimize noise coupling between resistors occupying a common tank by providing a low-impedance path from the tank to a clean supply node. NBL not only helps minimize debiasing, but it also acts as a barrier to minority carriers and consequently enhances latchup immunity. In the absence of NBL, the base diffusion also pushes somewhat deeper and therefore becomes more resistive. The NBL geometry should overlap the resistors sufficiently to ensure that all portions of the resistor experience the same amount of NBL push. An NBL overlap of 5 to 8μm will usually suffice.

Base resistors are often merged with other devices in a common tank. In order to prevent possible latchup problems, these devices should not inject minority carriers into the shared tank. For example, a resistor placed in the same tank as a lateral PNP can collect minority carriers emitted by the transistor if it saturates. Resistors can safely be merged with other resistors and with NPN transistors that do not saturate. If NPN transistors are included in a merged tank, a plug of deep-N+ should be used to contact the NBL to prevent tank debiasing and noise coupling (Section 13.1). If only resistors are present in a tank, then the deep-N+ plug can be omitted to save space.

Base resistors are probably the best general-purpose resistors available in standard bipolar. When offered, poly resistors are usually preferred over base resistors because they have smaller parasitic capacitances and because they have no tank junctions that might forward-bias.

5.5.2. Emitter Resistors

Emitter resistors are available in standard bipolar (Figure 3.14) and in some varieties of analog BiCMOS. Their sheet resistances typically range from 2 to 10Ω/□, so there is no difficulty in directly contacting the emitter diffusion. Because of its relatively low sheet, emitter is only suitable for relatively small resistors (0.5 to 100Ω). Larger resistors are usually constructed from base or HSR. Voltage modulation and conductivity modulation are both negligible in emitter resistors.

Capacitive coupling between an emitter resistor and an overlying lead can become significant. Processes that employ thin emitter oxides may exhibit oxide capacitances of up to $0.5\text{fF}/\mu\text{m}^2$ (0.3pF/mil^2), but processes that employ thick emitter oxide will have only a small fraction of this capacitance. Thin emitter oxides are also vulnerable to rupture during ESD events. Leads connecting to outside pins should not route across thin-oxide emitter diffusions unless they connect to them (Section 4.1.1). Thick emitter oxides are less vulnerable to rupture, so leads connecting to outside pins can safely cross them.

Emitter resistors must reside in a suitable tank. The usual arrangement consists of an emitter resistor enclosed within a base diffusion, which is in turn enclosed in a tank or an N-well region (Figure 3.14). When this configuration is employed, the base diffusion must connect to an equal or lower voltage than the emitter resistor. In turn, the tank enclosing the base diffusion must connect to a voltage equal to or higher than the base. This can be achieved by connecting the base diffusion to the low-voltage end of the resistor and connecting the tank to the high-voltage end. The emitter resistor must not be biased more than about 6V above the base region, or the emitter-base junction will avalanche. This limitation can be circumvented by dividing the resistor into several sections, each contained in its own base region.

Emitter resistors need not be enclosed in a base region to isolate them from the surrounding tank because the N-epi sheet resistance is much larger than the emitter sheet resistance. Even though the emitter resistor may electrically connect to the tank, the difference in resistivities ensures that most of the current flows through the resistor and not through the tank. Considerable space can be saved by eliminating the base diffusion (Figure 5.12). This layout is particularly suited for emitter resistors employed as tunnels. A *tunnel,* or *crossunder,* is a low-value resistor used to jumper leads on a die using only a single level of metal (Section 13.3.3). Tunnels should consume minimal area and achieve a very low resistance, something that the layout of Figure 5.12 does admirably.

In analog BiCMOS, emitter resistors can be placed directly into the P– epi. This tankless style of emitter resistor can also be constructed in standard bipolar by coding base entirely across the tank that contains an emitter resistor. Unfortunately, this structure has a breakdown voltage of about 7V that decreases to about 5V if the

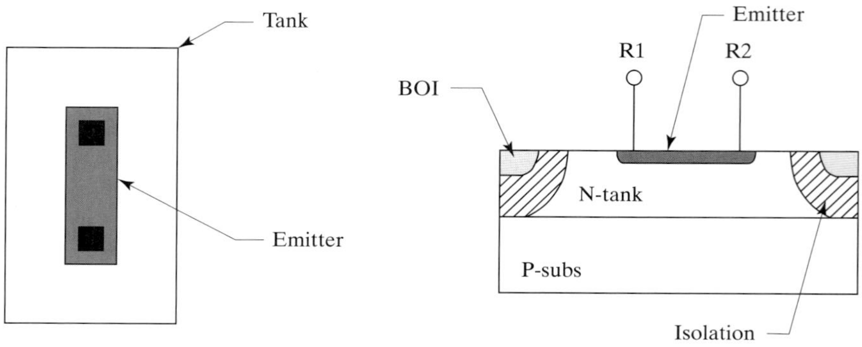

FIGURE 5.12 Layout and cross section of an alternate style of emitter resistor that eliminates the enclosing base diffusion to save space (compare to Figure 3.14).

tank is omitted. Emitter resistors formed directly into isolation also tend to leak. An emitter resistor can safely reside in isolation when it serves as a tunnel for a lead operating at substrate potential (Section 13.3.3). Leakage is obviously of no concern in this case, but care must still be taken to ensure that debiasing does not interfere with proper circuit operation.

Emitter resistors are frequently employed in standard bipolar to ballast power transistors and to serve as current sense resistors. They are also used extensively as tunnels in single-level-metal processes. Emitter resistors are uncommon in BiCMOS layouts since low-sheet polysilicon resistors are more compact and less susceptible to parasitics.

5.5.3. Base Pinch Resistors

Base pinch resistors can be constructed in standard bipolar (Figure 3.15) and in analog BiCMOS. Their effective sheet resistance is typically 2 to 10kΩ/□, allowing very compact layouts. Pinch resistors are typically used to construct high-value resistors that cannot be economically implemented using other diffusions. The sheet resistance of base pinch resistors is poorly controlled and a process variation of ±50% is typical. NBL should always be placed beneath a base pinch resistor because NBL push helps increase its sheet resistance.

The tank modulation of base pinch resistors is on the order of 1%/V or greater. Base pinch resistors can only be matched if they are of identical dimensions and if they see the same relative tank biasing. The biasing scheme of resistors R_3 and R_4 in Figure 5.11 must be used instead of the scheme of resistors R_1 and R_2. Failure to ensure equal tank biases may cause mismatches of up to ±20%. Even with perfectly matched layouts, errors of ±5% may result from the inherent lack of base pinch sheet control.[22] Superior matching can be achieved using even the narrowest high-sheet resistors, so if this implant is available, consider using it to replace base pinch resistors.

Like emitter resistors, base pinch resistors are limited to about 6V by the breakdown voltage of the base-emitter junction. Pinch resistors can withstand larger differential voltages if they are constructed in multiple segments. Since the emitter pinch plate connects to the tank, each segment of the resistor must occupy a separate tank. This is very wasteful of area, and constitutes yet another reason for considering alternative resistors such as epi-pinch resistors or narrow HSR resistors.

In summary, the pinch resistor is a marginal device used primarily to construct compact high-value resistors. These see frequent use in noncritical roles, as (for

[22] Grebene cites a value of ±6%, and Gray, *et al.* one of ±5%: Grebene, p. 147; Gray, *et al.*, p. 139.

example) base turn-off resistors for transistors. Occasionally designers will employ base pinch resistors to compensate for beta variation in NPN transistors because base pinch sheet resistance correlates with NPN beta. For most other applications, HSR resistors are superior to pinch resistors.

5.5.4. High-Sheet Resistors

High-sheet-resistance (HSR) implants are available as extensions for most standard bipolar processes. These implants exhibit sheet resistances of 1 to 10kΩ/□, depending upon implant dose, junction depth, and subsequent annealing conditions. The temperature coefficient of HSR resistors can be minimized by performing an incomplete anneal.[23] HSR resistors consist of light, shallow boron implants into an N-epi tank (Figure 3.19). Ohmic contact cannot be made directly to the HSR implant because it is both too lightly doped and too shallow, so the heads of the resistor consist of base diffusions. Figure 5.13 shows the layout and dimensions of a sample HSR resistor.

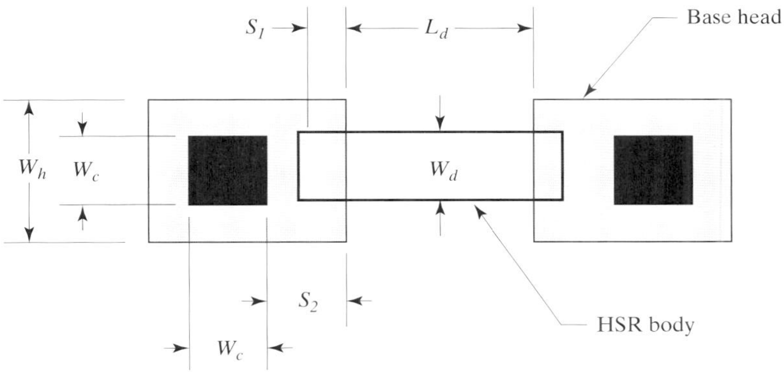

FIGURE 5.13 Layout of an HSR resistor, showing relevant dimensions.

The drawn length L_d of an HSR resistor is customarily measured between the inner edges of the two base heads. The value of the resistor equals

$$R = R_s\left[\frac{L_d - L_b}{W_d + W_b}\right] + 2R_h \qquad [5.12]$$

where R_s is the sheet resistance of the HSR implant, W_b is the width bias, and L_b is the length bias. The width bias applies to the HSR implant forming the resistor body, while the length bias applies to the separation between the base heads. The resistance of each base head, R_h, is computed separately since it does not depend on the HSR sheet resistance. The junction depth of the HSR implant generally does not exceed 0.5μm, so the major contributors to the width bias are usually photolithography and etching. The length-bias term accounts for the outdiffusion of the base heads and is about 20% of the junction depth of the base diffusion. This term can usually be ignored since it is so small. The resistance of the base heads is difficult to predict because of nonuniform current flow. The following equation forms a useful approximation

$$R_h = kR_{sb}\frac{S_2 + W_{bb}/2}{W_b + W_{bb}} \qquad [5.13]$$

---

[23] J. L. Stone and J. C. Plunkett, "Recent Advances in Ion Implantation—State of the Art Review," *Solid State Technol.*, Vol. 9, #6, 1976, pp. 35–44.

where R_{sb} is the base sheet resistance and W_{bb} is the width bias computed for base resistors. The arbitrary constant k accounts for nonuniform current flow in the resistor and typically equals about 0.7. This constant approaches one as the head is elongated, as is sometimes done to accommodate lead routing.

High-voltage HSR resistors are notoriously prone to charge spreading. The sheet resistance of HSR is far greater than that of base, so the currents involved are smaller and the effects of leakage become magnified. All HSR resistors operating at voltages approaching or exceeding the thick-field threshold require careful field plating. The thin, lightly doped HSR implant is also vulnerable to voltage modulation effects. For example, a merged HSR resistor divider connected to a 20V supply will experience severe mismatches. This same divider functions properly if the resistors are segmented and placed in separate tanks so that each segment experiences the same tank modulation. As this example suggests, the effect of tank connections is much more critical for HSR resistors than for deeper and more heavily doped base resistors. A merger that works well with base resistors may cause unacceptable errors if the resistors are converted to HSR.

Conductivity modulation also affects HSR resistors. Leads in single-level-metal designs must often route over HSR resistors for topological reasons. Conductivity modulation will occur if the voltage on such a lead differs significantly from the voltage on the resistor. This effect frequently magnifies noise coupling between leads and resistors. Since base diffusions experience much less conductivity modulation than HSR, noise coupling can often be minimized by elongating the base head and running the leads over base diffusion rather than over HSR implant (Figure 5.14). Elongated base heads are also useful when the HSR resistor is too short to accommodate all of the leads that must cross it. The resistance of an elongated base head can be approximated using equation 5.13 where the constant k is set equal to one.

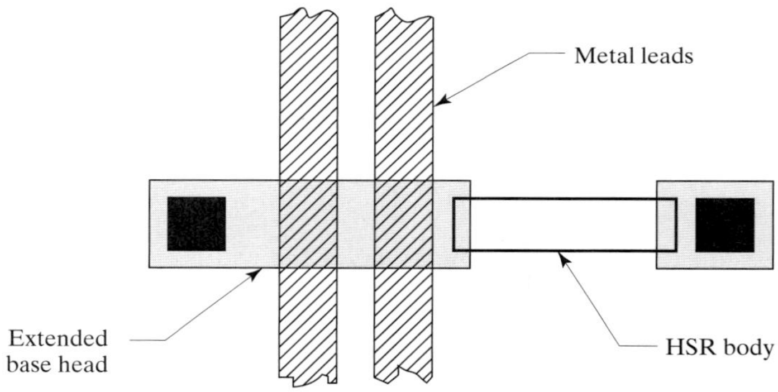

FIGURE 5.14 Example of an HSR resistor with an extended head.

The shallowness of the HSR implant limits the avalanche voltage of HSR resistors to a small fraction of the planar breakdown. Most HSR resistors can only withstand 20 to 30V. By comparison, base resistors can typically withstand 50 to 60V despite their heavier doping. The HSR breakdown can be increased a few volts by filleting all of the corners in order to reduce the electric field intensity at these points. Much higher operating voltages can be achieved by dividing the HSR resistor into multiple segments, placing each segment in a separate tank, and connecting each tank to the positive end of the enclosed segment. This arrangement not only enables the resistor to withstand operating voltages up to the tank-substrate breakdown but also reduces voltage modulation.

As with base resistors, HSR resistors should always have NBL placed underneath them. This not only helps prevent tank debiasing but also acts as a barrier to minority-carrier injection if a resistor momentarily forward-biases into the tank. The sheet resistance of the shallow HSR implant is not materially altered by the presence or absence of NBL, but HSR is more affected by NBL shadow than is base. HSR resistors can be merged with other devices in a common tank providing that none of the merged devices inject minority carriers into the shared tank.

HSR implants come in a range of sheet resistances. The smaller sheets (such as 1kΩ/□) are of little value because they cannot save enough die area to justify their cost. The larger sheets (such as 5kΩ/□) are extremely vulnerable to charge spreading and conductivity modulation. The optimal sheet resistance lies in the neighborhood of 2kΩ/□, which allows substantial area savings without having to worry too much about field plates or lead routing. If higher sheet resistances must be employed, then all of the resistors should be field plated to prevent surface charges from modulating them. These field plates also prevent the inadvertent routing of leads across the HSR resistors. For these higher sheet values, consider sectioning larger resistors and applying separate field plates to each section to lessen the conductivity modulation caused by the plates. Split-field plates (Section 7.2.8) may also prove useful for precisely matched resistors. Voltage modulation and tank modulation worsen as the sheet resistance increases, so tank connections must be carefully scrutinized.

HSR resistors are useful for packing large amounts of resistance into a limited die area. They are less variable than base pinch resistors, and they have a much larger sheet resistance than base. Most standard bipolar designs operating at supply currents of less than a milliamp make extensive use of HSR. CMOS and BiCMOS processes rarely implement HSR resistors because doped polysilicon offers a superior alternative. Although the poly sheet resistances commonly offered by these processes are lower than HSR sheet resistances (500Ω/□ versus 2kΩ/□), the narrower poly widths and spacings usually enable the construction of more compact layouts. Polysilicon does not experience tank modulation, and conductivity modulation is generally negligible for sheet resistances of less than 1kΩ/□.

5.5.5. Epi Pinch Resistors

An *epi pinch resistor* resembles a base pinch resistor in that it consists of a resistive region pinched between two junctions to increase its sheet resistance. In this case, the resistive layer consists of N-epi pinched between the underlying substrate and an overlying base diffusion (Figure 5.15). The substrate dopant diffuses upward during the isolation drive to produce an effective sheet resistance of 5 to 10kΩ/□.

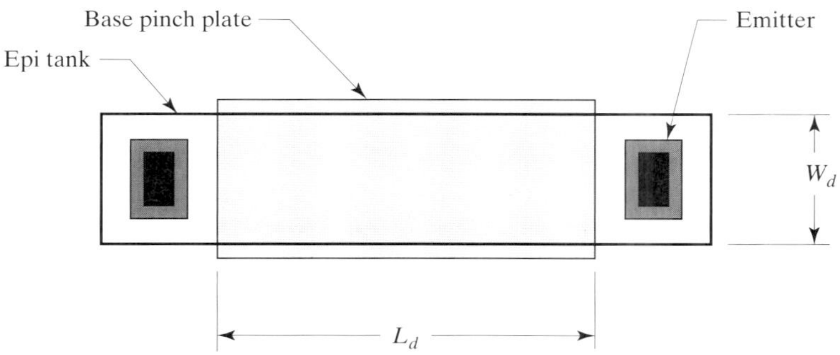

FIGURE 5.15 Layout of an epi pinch resistor (epi-FET).

As one would expect of a highly resistive material, the pinched epi suffers from severe voltage modulation effects. At higher voltages, the positive end of the resistor can deplete entirely through. Once the resistor pinches off, current flow ceases to increase and becomes largely independent of voltage bias. The *pinchoff voltage* at which this occurs depends upon epi thickness, epi doping, and base junction depth.

For a typical 40V standard bipolar process, the epi pinchoff voltage lies between 20 and 40V.

The epi pinch resistor is an example of a class of devices that pinch a lightly doped semiconductor region between closely spaced reverse-biased junctions. This device is actually a JFET (Section 1.5). The epi pinch resistor is also called an *epi-FET* (see Section 12.3.2).

The voltage modulation of epi pinch devices is so severe that they are rarely used as ordinary resistors. Instead, they find almost exclusive application in startup circuits, where they provide a trickle of current. The voltage modulation and eventual pinchoff are desirable in this application because they limit the increase of the startup current with voltage and thus reduce the current consumption of the part at higher voltages. Epi pinch resistors are not laid out with any particular precision because of their extremely large sheet variability (at least ±50%, not counting voltage modulation). The value of the resistor is usually computed using equation 5.3, using the drawn width and length shown in Figure 5.15. The width bias is negative because the out-diffusion of the isolation regions encroaches upon the pinched resistance layer.

Epi pinch resistors are often laid out in a serpentine pattern. Many of these layouts use the circular turns illustrated in Figure 5.3B. This practice is sometimes incorrectly described as an attempt to prevent premature avalanche breakdown of the sharp corners, although fillets have little effect on the breakdown voltages of deep diffusions. Actually, the rounding is so drastic that it transforms a rectangular turn into something resembling a circular turn. Circular turns are less affected by this rounding, and their values are therefore easier to compute.

Pinch resistors can also be constructed in a BiCMOS process by using base to pinch N-well. The layout appears identical to that in Figure 5.15, but narrower drawn widths are allowed because the width bias is now positive. The width of a well pinch resistor should still equal at least 150% of the nominal well junction depth, or the amount of dopant may not suffice to invert the P-epi beneath the base plate. The overlap of the base plate over the N-well must also be increased to account for the outdiffusion of the N-well. In the absence of specific layout rules, this overlap should be at least 150% of the epi thickness.

Epi and well pinch resistors are not used in large numbers. They provide a convenient means of obtaining small currents for startup circuits, but their inherent variability and nonlinearity usually preclude their application elsewhere.

5.5.6. Metal Resistors

Although the sheet resistance of aluminum metallization is small, it is by no means insignificant. Standard bipolar metallization is about 10 to 15kÅ thick and exhibits a sheet resistance of 20 to 40mΩ/□. The smaller feature sizes of CMOS processes usually dictate metal thicknesses of considerably less than 10kÅ, with correspondingly larger sheet resistances.

Metal resistors typically have values between 50mΩ and 5Ω. Resistors in this range are used for constructing current sense circuits and for ballasting large power bipolar transistors. A metal resistor can be laid out as either a straight run or as a serpentine pattern. The resistor should reside over field oxide to prevent oxide steps from causing variations in the metal sheet resistance. In a double-level-metal process, resistors can be constructed using either metal layer. Second-metal leads can route over a resistor placed on first metal, but first-metal or poly leads should not route beneath a resistor placed on second metal because they may introduce nonplanarities in the resistor.

Accurate voltage sensing across metal resistors requires special techniques. The leads of the resistor are only extensions of the resistance layer, so any excess lead length causes significant errors in the sense voltage. Therefore two sets of leads are employed: one pair to conduct current through the resistor and the other pair to sense the voltage developed across it. Voltage drops in the current leads do not alter the voltage difference across the sense leads, and no significant voltage drops occur in the sense leads because little or no current flows through them. The use of separate current-carrying and voltage-sensing leads is called a *Kelvin connection* (Section 14.3.2).

Figure 5.16A shows one means of providing Kelvin connections for a metal resistor. The resistor is implemented using a single level of metal, and the sense points are simply taps connecting into the side of the resistor. These sense leads should be as narrow as possible to ensure that the length of the resistor is determined by the spacing between the leads and not by their width. The body of the resistor should extend beyond either sense lead by a distance at least equal to the resistor width, and preferably by at least twice this width. These extensions promote uniform current flow in the vicinity of the sense leads and ensure accurate voltage sensing.[24]

Figure 5.16B shows an alternate layout that uses a second layer of metal to tap the center of the resistor. The sense leads should occupy the upper metal layer to preserve the planarity of the resistor. This layout is less sensitive to variations in current flow near the sense points. If the resistor is laid out to extend past both sense points for a sufficient distance, then the layouts of Figure 5.16A and 5.16B will have similar accuracies.

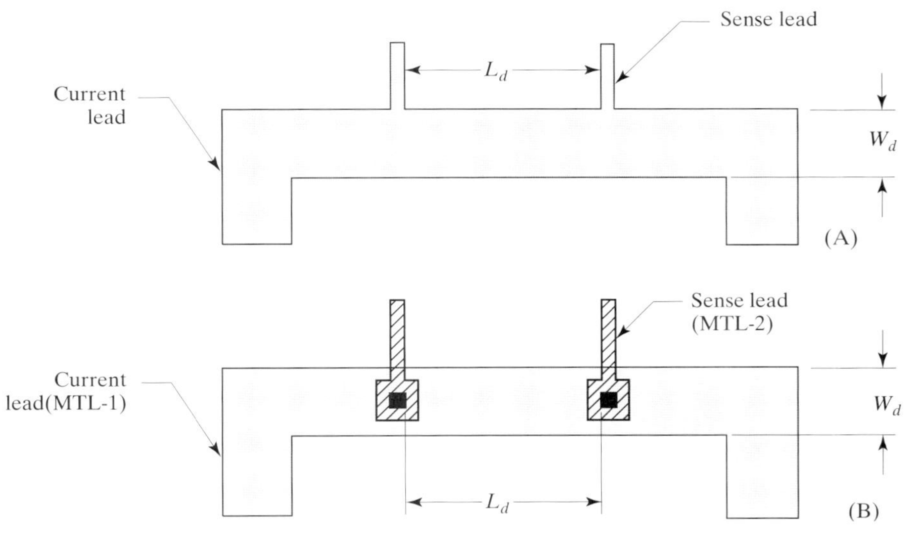

FIGURE 5.16 Two styles of Kelvin-connected metal sense resistors: (A) single-level-metal layout and (B) double-level-metal layout.

The values of metal resistors depend primarily upon the thickness of the metal layer, and only secondarily upon its composition. The sheet control of the resistor thus approximately equals the variation of metal thickness. Most processes experience some ±20% variation in metal thickness over field oxide, so a similar variation in metal sheet resistance can be expected.

5.5.7. Poly Resistors

Polysilicon resistors are available in CMOS (Figure 3.32) and BiCMOS processes. The poly used for constructing MOS gates is heavily doped to improve conductivity

---

[24] In practice, one must often settle for less-optimal layouts; see B. Murari, "Power Integrated Circuits: Problems, Tradeoffs, and Solutions," *IEEE J. Solid-State Circuits*, Vol. SC-13, #6, 1978, pp. 307–319.

and has a sheet resistance of about 25 to 50Ω/□. Lightly doped polysilicon can have sheet resistances of hundreds or even thousands of Ohms per square. Intrinsic or lightly doped polysilicon can be doped with NSD and PSD implants to provide additional choices of sheet resistance. The NMoat and PMoat coding layers cannot be used to dope polysilicon resistors because these layers generate moat geometries as well as implant regions. Instead, one must use coding layers that produce geometries only on the NSD and PSD masks.

The resistivity of polysilicon depends not only on doping but also on grain structure. The boundaries between grains interfere with the orderly flow of carriers and raise the resistivity of the material. Small-grained poly films therefore exhibit higher resistivities than large-grained films. These differences are most pronounced in lightly doped poly, which may exhibit a resistivity that is several orders of magnitude greater than similarly doped monocrystalline silicon.

The heterogeneous nature of polysilicon also affects its temperature coefficient of resistivity. Lightly doped poly exhibits a strongly negative TCR, while heavily doped poly has a positive TCR. For example, 4kÅ 500Ω/□ poly has a TCR of about –1000ppm/°C, while 4kÅ 70Ω/□ poly has a TCR of about 500ppm/°C. A suitably doped poly film will exhibit a linear TCR of zero. For 4kÅ poly, this point lies in the neighborhood of 200Ω/□.[25] Although this allows the construction of poly resistors having very small linear temperature coefficients, these devices still retain a significant parabolic temperature coefficient. Process variations will also affect the temperature coefficient of the poly. In practice, it is rarely possible to hold temperature coefficients to tolerances of better than ±250ppm/°C. Even so, this represents a significant advance over diffused resistors with temperature coefficients of several thousand ppm/°C.

Figure 5.17 shows a poly high-sheet resistor composed of lightly doped N-type polysilicon. NSD implants coded around either end of the resistor produce low-resistivity heads for the contacts. If the resistors must match accurately, these implants should overlap the poly so that the entire width of the resistor receives the dopant. Otherwise, the NSD implantation need only overlap the contacts sufficiently to ensure electrical continuity.

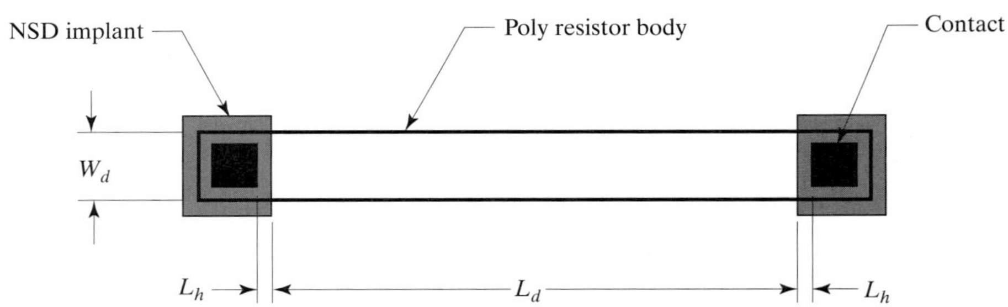

FIGURE 5.17 High-sheet poly resistor with implanted heads.

The total resistance of a high-sheet poly resistor can be computed by dividing it into sections and calculating the sheet resistance of each section

$$R = R_s\left[\frac{L_d}{W_d + W_b}\right] + 2R_h\left[\frac{L_h}{W_d + W_b}\right] \qquad [5.14]$$

where R_s is the sheet resistance of the poly used to construct the resistor body, R_h is the sheet resistance of the poly used to construct the heads, and L_h is the overlap of the implant over the contact. Unlike HSR resistors, poly resistors do not exhibit

---

[25] Values taken from Lane, *et al.*, p. 740.

nonuniform current flow at the boundaries between the body of the resistor and its heads because these have the same width. The effects of nonuniform current flow at the contacts can be analyzed using the same techniques used for diffused resistors (equations 5.4 and 5.11).

The width bias W_b models the oversizing or undersizing that occurs during photolithography and poly etching. This bias can equal as much as a micron, so it can have a significant effect upon narrow poly resistors. By the same token, narrow resistors will exhibit large process variations due to linewidth control. Most processes can control poly dimensions to within about 10% of the minimum width. For example, a process that can fabricate a 0.8μm gate will probably have a poly linewidth control of about 0.08μm. Although this is a remarkable degree of precision, narrow poly resistors still experience considerable variability. The effects of linewidth variation can be minimized by increasing the resistor width to several times minimum. Extremely narrow poly resistors may also experience increased resistance variability due to the growth of individual grains across the entire width of the resistor (an effect sometimes called *bamboo poly*).

Poly resistors should reside on top of field oxide. This not only reduces the parasitic capacitance between the resistor and the substrate but also ensures that oxide steps do not cause unexpected resistance variations. If the parasitic capacitance of the field oxide is still too large for a given application, consider using a second poly layer (if available) since the interlevel oxide will further reduce the parasitic capacitance. In some BiCMOS processes, deep-N+ can be coded underneath a resistor to thicken the field oxide by dopant-enhanced oxidation.[26] If this technique is employed, make sure to extend the drawn boundaries of the deep-N+ several microns beyond the resistor on all sides to ensure that it will reside on planar oxide.

Poly resistors do not tolerate transient overloading as well as monocrystalline silicon does. The oxide surrounding a poly resistor does not conduct heat well, and excessive power dissipation can cause localized overheating (Section 5.3.3). At temperatures beyond 250°C, the annealing process resumes. This can produce irreversible changes in resistance due to the movement of grain boundaries or the activation of incompletely incorporated dopants. In extreme cases, polysilicon resistors may experience a phenomenon similar to what occurs in Zener zaps.[27]

Not all CMOS processes provide for constructing resistors. If the dopant used to decrease the resistivity of the gate poly cannot be blocked, then the sheet resistance of the poly becomes so low that large-value resistors require too much room to construct. If the gate polysilicon is silicided, then its resistance drops even further, to around 2Ω/□. Silicided poly cannot provide enough resistance for most applications, so other structures, such as N-well resistors, must be used. Some processes provide a means for blocking silicide from selected regions using a *silicide block mask*. This layer should enclose the body of the resistor but not the ends where the contacts reside. The silicided portions at either end of the resistor perform the same function as the NSD-doped heads of the high-sheet poly resistor (Figure 5.17). The resistance of a silicided-head poly resistor can be computed using Equation 5.14, but R_h is so small in comparison to R_s that the second term of the equation can usually be ignored.

Poly forms the best resistors available on most processes. Even if the poly sheet is only half or a third of that of a diffusion, the narrower poly pitch will probably result in a smaller layout. Poly resistors do not experience tank modulation, and

---

[26] Not all processes allow poly resistors on top of deep-N+ because of planarization concerns.

[27] D. M. Petkovic' and N. D. Stojadinovic', "Polycrystalline Silicon Thin-film Resistors with Irreversible Resistance Transition," *Microelectronics Journal*, Vol. 23, #1, 1992, pp. 51–58.

conductivity modulation is generally a minor concern as long as the sheet resistance does not exceed 1kΩ/□.

5.5.8. NSD and PSD Resistors

Diffused resistors can be constructed using the NSD and PSD implants of a CMOS or BiCMOS process (Figure 3.33). These resistors usually exhibit sheet resistances of 20 to 50Ω/□, and they are almost immune to voltage modulation. NSD and PSD are relatively shallow diffusions that avalanche at relatively low voltages due to sidewall curvature. NSD resistors residing in the P-epi are limited by the NSD/epi breakdown voltage, but PSD resistors can operate at relatively high voltages when segmented and placed in separate wells.

Some processes silicide the moat regions to reduce their resistance. The use of such a *clad moat* technology prevents the formation of useful NSD and PSD resistors without using a silicide block mask. If a silicide block is used, the heads of the resistor must remain silicided to ensure proper contact. The value of such a resistor can be computed using Equation 5.14, but R_h is so low that the second term of the equation can be ignored.

NSD and PSD resistors are not often used because most CMOS and BiCMOS processes offer poly resistors with equal or greater sheet resistances. NSD and PSD resistors see occasional use in ESD devices because their parasitic junction diodes can serve as voltage clamps. Diffused resistors can also withstand larger power transients than poly resistors because the thermal conductivity of silicon far exceeds that of the field oxide. Again, this consideration primarily applies to resistors used in transient suppressors and ESD structures. High-sheet poly resistors are preferable to NSD and PSD resistors for most other applications.

5.5.9. N-well Resistors

Sometimes a large resistor must be fabricated in a CMOS process that lacks high-sheet poly. A high-value resistor can be created using a stretch of N-well contacted at either end by NMoat regions (Figure 5.18A). By itself, the N-well exhibits a sheet resistance of as much as 10kΩ/□. Even higher sheet resistances can be produced by pinching the well with PSD (Figure 5.18B). The PSD implant forms a reasonably effective pinch plate despite its shallow junction depth because most of the conduction normally occurs in the uppermost—and most heavily doped—portions of the well. In analog BiCMOS, base can be substituted for PSD to produce still higher

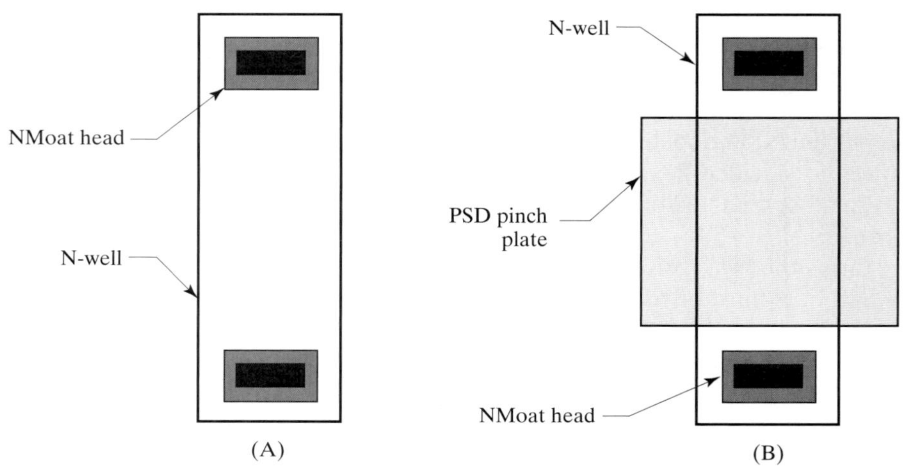

FIGURE 5.18 (A) N-well resistor and (B) N-well resistor with PSD pinch plate (field plates not shown).

sheet resistances. The resulting devices closely resemble epi-pinch resistors, but they usually possess much higher initial resistances (20kΩ/□ or more) and lower pinchoff voltages (20V or less). A base pinch plate can be made still more effective by coding base implant alone rather than in conjunction with moat. Oxidation-enhanced diffusion forces the base junction deeper into the silicon and produces an even thinner pinched channel. This structure can produce pinchoff voltages of 10V or less, depending on the depths and dopings of N-well and base.

N-well resistors without pinch plates are used for many of the same applications as base pinch resistors. They exhibit similar process variabilities and tank modulation effects, but N-well resistors usually have even larger temperature coefficients than base pinch resistors (6000ppm/°C versus 2500ppm/°C). N-well resistors without pinch plates may suffer from surface depletion and inversion unless they are properly field plated. The field plate should extend beyond the drawn outline of the N-well far enough to account for outdiffusion; in practice this means that the field plate must overlap the drawn N-well by a distance slightly greater than the junction depth.

Pinched N-well resistors are much less susceptible to conductivity modulation and charge spreading because the pinching material acts as a field plate and protects the underlying resistor from surface effects. The heads of the resistor protruding from beneath the pinch plate remain vulnerable. The metal leads contacting each head of the resistor should extend to cover the exposed portions of the N-well.

When laying out N-well resistors, remember that the well does not achieve its full junction depth unless the drawn geometry is at least twice as wide as the well is deep. N-well resistors with widths less than this will exhibit progressively higher sheet resistances. Base-pinched N-well resistors are especially vulnerable to this effect because the pinched well region is so thin. In extreme cases, the base may succeed in punching entirely through a narrow N-well resistor, causing it to appear as an open circuit.

5.5.10. Thin-film Resistors

Integrated resistors are usually fabricated from materials optimized for other uses, so they do not perform as well as discrete resistors fashioned from materials selected specifically for use in resistors. These specialized materials can also be deposited as thin films on the surface of an integrated circuit. The resulting *thin-film resistors* easily outperform diffused resistors and poly resistors. Thin-film resistors can achieve temperature coefficients of less than 100ppm/°C with nearly perfect linearity, and they can be adjusted to tolerances of better than ±0.1% by laser trimming. These resistors are desirable components for high-performance analog applications, but the specialized process technologies they require are costly and are not always available.

Thin-film resistors can be made from a wide variety of materials, the most common of which are *nichrome* (a nickel-chromium alloy) and *sichrome* (a silicon-chromium mixture). Resistors made from these materials have varying sheet resistances and temperature coefficients, but all utilize essentially the same layout (Figure 5.19). These resistors do not require contacts since the thin-film material is deposited immediately prior to the uppermost layer of metallization. Any top-level

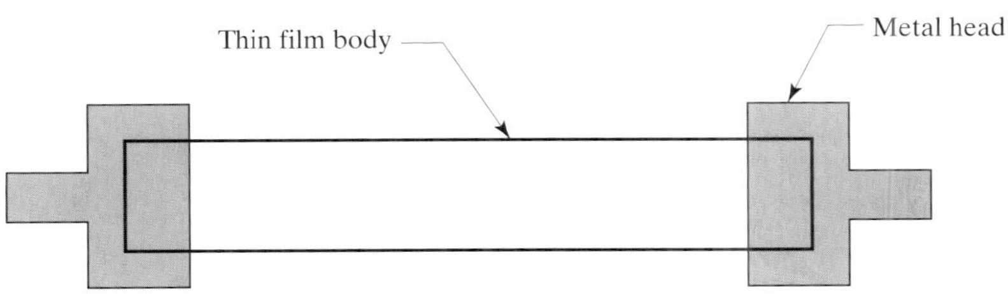

FIGURE 5.19 Layout of a thin-film resistor.

metal lead touching the resistor body will contact the resistor. Top-metal leads cannot cross thin-film resistors without shorting to them, but lower levels of metal are not so constrained. These resistors should be laid out on field oxide because any steps in the underlying surface may cause significant variability.

The value of a thin-film resistor can be computed using equation 5.3. The length of the resistor is measured between the inner edges of the two metal plates forming its heads. These should overhang the resistor body sufficiently to ensure that misalignment will not leave any portion of the resistor head uncovered. No corrections are required for nonuniform current flow at the ends of the resistor. Serpentine resistors use the correction factors previously discussed.

Thin-film resistors are preferable to all others, so designs that employ them will not require most other types of resistors. Diffused resistors do, however, have better power handling capabilities. Thin-film resistors subjected to overheating often exhibit permanent changes in sheet resistance due to annealing of their crystalline structure. Extreme overheating can even cause open-circuit failures. Diffused resistors have larger volumes and are in intimate contact with the silicon die, which serves to provide additional heat-sinking. They are therefore preferred to thin-film resistors for applications subject to severe transient overloads—for example, ESD protection circuitry.

5.6 ADJUSTING RESISTOR VALUES

One often finds small variable resistors called *trimmers* scattered around printed circuit boards. Manual or robotic adjustment of these resistors allows a circuit to be adjusted after assembly to achieve tighter tolerances than can be economically obtained using high-precision components. Analog integrated circuits also make considerable use of adjustable resistors to counter process variations and to provide for design uncertainties. The relatively large amounts of process variability inherent in high-volume wafer fabrication can prevent high-precision circuitry from meeting specification. In such cases, each integrated circuit must be adjusted after it is manufactured by *trimming* the value of one or more components. Trimming is usually performed during wafer-level testing, and it usually requires specialized test equipment and procedures.

Circuit design is fraught with uncertainties, so experienced designers often allow room for adjustments after the initial design is evaluated. A redesign of a circuit that affects only component values is called a *tweak.* Tweaks adjust the mean of a distribution, whereas trimming minimizes its variance. All tweaked units receive the same adjustment, whereas each trimmed unit receives individualized adjustment.

5.6.1. Tweaking Resistors

Integrated circuits are normally laid out with little regard for future modification. This approach is justifiable since most changes affect the entire mask set, and most designs require several passes to meet specifications. Resistor tweaks represent an exception to this rule because they need not affect the entire mask set. If the resistors are properly designed, their values can be tweaked by altering only one mask. Tweakable resistors can drastically reduce the time required to obtain fully parametric devices. Several wafers can be held out of the first wafer lot just prior to the step required to tweak the resistors. After sample wafers have been evaluated, a new mask can be made incorporating corrected resistor values. The time required to complete the processing of the tweaked wafers is usually much less than the time required to fabricate a new wafer lot.

There are four commonly used methods for tweaking resistors: sliding contacts, sliding heads, trombone slides, and metal options. Each technique involves a different mask, and no one method can adjust all types of resistors.

Sliding Contacts

Sliding contacts are the simplest type of resistor tweak.[28] Figure 5.20 shows two examples of resistors that incorporate sliding contacts.

Sliding contacts are easiest to implement for resistors without heads (Figure 5.20A). The body of the resistor is extended so that one contact can slide inward or outward. The metal plate for this contact is also elongated so that it covers the contact no matter where the latter is located. This precaution prevents having to purchase a new metal mask to perform the tweak. The sliding contact should initially reside at a point midway between the limits of travel (Figure 5.20A).

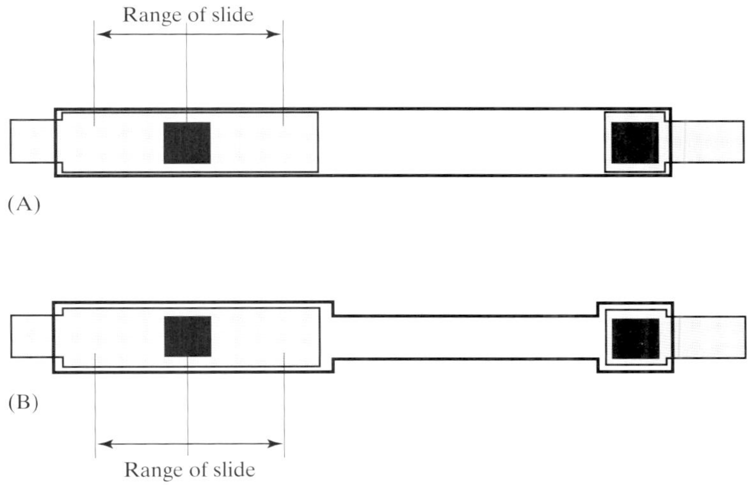

FIGURE 5.20 Two styles of sliding contacts: (A) without heads and (B) with heads.

Sliding contacts are somewhat more difficult to construct if the resistor uses enlarged heads. A sliding contact can still be made as long as the material of the head has the same resistivity as the material of the body (Figure 5.20B). The head must be enlarged to accommodate the sliding contact, and this complicates the computation of its resistance. An approximate value can be obtained by assuming that the resistor consists of two sections in series: one narrow and one wide. The total resistance equals the sum of the resistances of each segment. Nonuniform current flow at the juncture of the two segments introduces a small error, but this can be ignored because the resistor will probably be trimmed anyway. The sliding contact removes or adds a certain length to the wider segment. Movement of the contact does not significantly affect the nonuniformities of current flow, either at the joint where the head meets the body of the resistor or at the contacts.

Sliding contacts work well for resistors whose body and heads consist of the same material, as do standard bipolar base and emitter resistors. Sliding contacts are much less useful if the head is constructed of a low-sheet material, as is the case for most resistors with sheet resistances of more than about 200Ω/□. The sliding contact can only make trifling changes in such a resistor, as it can only move within the confines of the low-sheet material used to form the resistor heads. High-sheet resistors are usually tweaked using *sliding heads*.

---

[28] Grebene, p. 156.

Sliding Heads

Figure 5.21 illustrates the layout of a resistor incorporating a sliding head at its left end. The body of this resistor consists of a high-sheet material such as HSR implant or lightly doped poly. The heads consist of a lower-sheet material used to ensure Ohmic contact. The resistance can be reduced by extending the sliding head further into the resistor body. If sufficient room is provided to pull the head back, then the total resistance can also be increased by sliding the head toward its contact.

FIGURE 5.21 Layout of a resistor with a sliding head.[29]

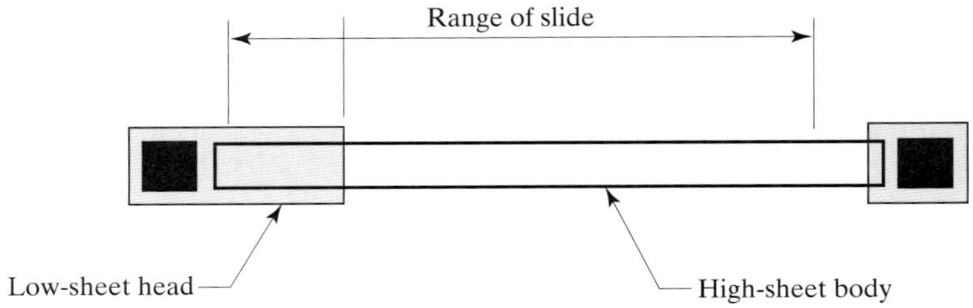

The sliding-head resistor can be modeled as a series combination of two separate resistors, one representing the body and the other representing the heads. Although nonuniform current flow causes slight errors in the calculation, these can be ignored because the resistor will probably require adjustment anyway.

Sliding heads are commonly used for adjusting HSR implant and high-sheet polysilicon resistors. The same concept works for pinch resistors, except that the pinch plate slides forward or backward instead of the heads. Poly resistors that use a silicide block mask can also be adjusted by sliding the silicide block toward or away from the contact.

Trombone Slides

Serpentine resistors can be adjusted by sliding turns inward or outward, a technique colorfully referred to as a *trombone slide*. This adjustment alters the total length of the resistor without changing the number of turns it contains (Figure 5.22). Room left adjacent to the resistor allows for its extension. If the resistor occupies a tank or is enclosed in an implant, then these geometries should also enclose the region into which the resistor can slide.

FIGURE 5.22 Resistor adjustment by means of a trombone slide.

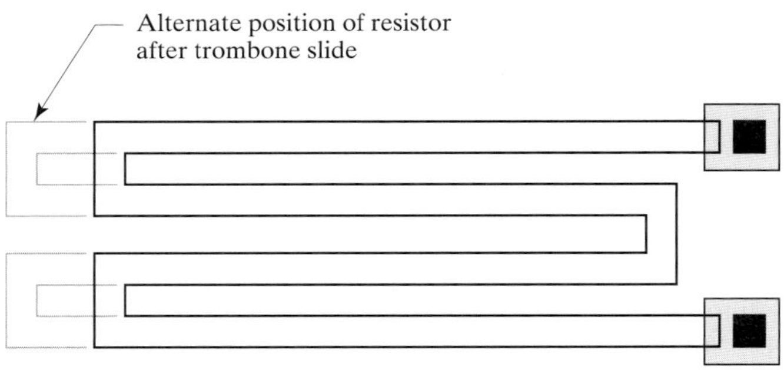

Metal Options

Another method that enables limited tweaking of resistors involves dividing the resistor into multiple sections. The majority of these sections are connected in series to form the final resistor, but several sections are left unconnected as spares. These

[29] *Ibid.*

spares include contacts, each covered by just enough metal to meet the minimum design rules. The resistance can be adjusted by adding or removing segments using a new metal mask. The flexibility of this tweaking scheme is limited by the number of possible combinations of the segments. Segments can be joined in parallel as well as in series, but some potential combinations do not properly balance the thermoelectric potentials generated by the contacts (Section 7.2.7).

5.6.2. Trimming Resistors

Resistors can be trimmed using fuses, Zener zaps, or laser trims. Fuses and Zener zaps act as programmable switches, allowing a network of resistors to be reconfigured in much the same way as a metal option. Laser trimming can provide incremental adjustment with a resolution of better than ±0.1% of the initial resistor value. However, this technique is only applicable to thin-film resistors, and it requires the assembly/test site to purchase automated laser trimmers.

Fuses

A *fuse* is simply a short section of minimum-width metal or polysilicon connected between two bondpads. It is programmed, or *blown*, by passing a large current between the bondpads, causing the fuse material to vaporize. After programming, the fuse becomes an open circuit.

Figure 5.23A shows a typical example of a metal fuse, which consists of nothing more than a constricted segment in a wide metal lead. A small opening in the protective overcoat above the fuse allows the vaporized metal to escape during programming. This opening represents a potential pathway for contaminants to enter the die, but if it is omitted, the vaporized metal cannot escape and the fuse sometimes reforms after programming. The explosive expansion of the metal vapor can also crack the protective overcoat. These cracks allow the metal vapor to escape, but they sometimes extend into adjacent circuitry. Most manufacturers agree that the inclusion of an opening in the protective overcoat is a small price to pay to prevent fuse regrowth and overcoat cracking.

Figure 5.23B shows a typical polysilicon fuse. The polysilicon is laid out as a small resistor connecting two bondpads. As with the metal fuse, a small opening allows the vaporized silicon to escape without rupturing the protective overcoat. The polysilicon fuse must be heavily doped to obtain a sufficiently low resistance

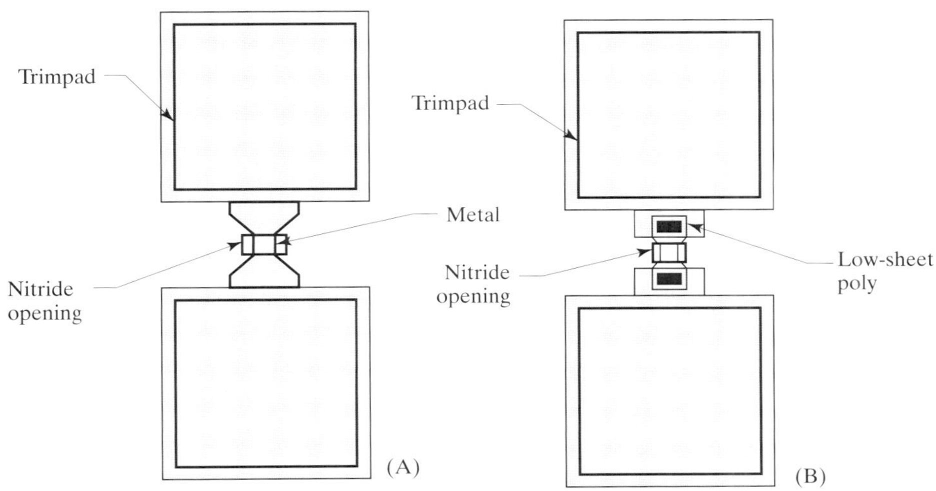

FIGURE 5.23 Layouts of (A) a typical metal fuse and (B) a polysilicon fuse.

to allow programming. The fuse must have a total resistance of no more than 200Ω to ensure reliable programming with a 10V pulse.

The pads used to program fuses are not normally bonded, so they need only be large enough to allow the probe needles to reliably land on them. These pads are sometimes called *trimpads* to distinguish them from true bondpads. Despite their reduced dimensions, trimpads still require substantial die area. Some processes allow active circuitry to reside underneath trim pads, in which case very little area is wasted. Trim pads must be placed so that the probe needles can reach them; this usually requires that they reside around the periphery of the die. Reducing the number of trimpads saves die area and reduces the cost of the integrated circuit. The number of trimpads required to blow the fuses can be minimized if several fuses are connected in series or in parallel.

Aluminum fuses are easy to program. Aluminum melts at a relatively low temperature (660°C) but boils at a much higher temperature (2470°C).[30] By the time vaporization begins, most of the metal is already molten and is readily ejected. A pulse of several hundred milliamps for a period of a few milliseconds will reliably blow the fuse. The ejected aluminum tends to splatter onto the probe needles and eventually fouls them so badly that shorts occur. The probe card may require occasional cleaning to alleviate this problem. Metal fuses incorporating refractory barrier metal can also be programmed reliably; the aluminum is ejected first, but the thin refractory metal soon follows.

Polysilicon fuses are more difficult to program. Silicon melts at a much higher temperature than aluminum (1410°C versus 660°C), and it is brittle and prone to fracturing from the stresses of rapid heating. The polysilicon may crack before it begins to melt unless the programming current pulse has an extremely fast rise time ($<$ 25nS). Such fractures cause the current flow to stop and prevent proper programming. Mechanical stresses can cause cracked fuses to reform at any time. This type of reliability failure can be prevented by ensuring that the programming pulse has a sufficiently fast rise time to blow the fuse before cracks appear.

Both polysilicon and aluminum fuses are difficult to program after encapsulation. The plastic seals the opening over the fuse and prevents the conductive material from being ejected. Not only does this impair the blowing of the fuses, but it also makes them prone to regrowth because conductive material remains near them. Some designs use circuits to detect changes in resistance between a programmed fuse and a reference fuse that is left unprogrammed. Although such circuitry is complex and unwieldy, it does allow post-package trimming of resistors. Modern CMOS and BiCMOS designs sometimes substitute *electrically programmable read-only memory* (EEPROM) elements for the fuses to save space and to simplify the programming circuitry.

The programming process can cause high voltages to appear on circuitry connected to the fuse. A large current flows through the fuse in the instant before it blows, and the sudden interruption of this current can cause voltage transients due to parasitic inductances. These transients can avalanche junctions or damage thin gate oxides. The magnitude of the transients can be reduced by minimizing the loop area of the programming circuit. Especially difficult cases may benefit from the addition of integrated Zener clamps. Circuit designers can help minimize the impact of fuse programming transients by placing the fuses on the least-vulnerable end of

---

[30] Melting and boiling point data: R. C. Weast, ed., *CRC Handbook of Chemistry and Physics,* 62nd ed. (Boca Raton, FL: CRC Press, 1981), pp. B-2 – B-48. Values are rounded to the nearest ten degrees.

a resistor. For example, the resistor normally trimmed in a Brokaw bandgap[31] connects between the emitter of an NPN transistor and ground. The trims should be placed on the grounded end of the resistor so that transients must pass through the rest of the resistor before reaching the base-emitter junction. This same connection also saves a trimpad because the lowest fuse connects to ground.

Several fuses in combination can provide additional trim resolution. The resistor segments should be *binary weighted* to ensure that the achievable values of resistance are uniformly spaced across the trim range. This not only provides the maximum possible trim range for a given precision but also simplifies the design of the test program by allowing the use of a binary search algorithm. Binary weighting can be implemented in either of two ways. If the voltage across a resistor requires trimming, then the resistors should be connected in series and weighted $R_{lsb} : 2R_{lsb} : 4R_{lsb} : 8R_{lsb} \ldots$ where R_{lsb} is the value of the *least significant bit* (LSB) of the trim network.[32] Figure 5.24A shows a 3-bit binary-weighted voltage trim scheme. If the current through a resistor requires trimming, then the resistors should be connected in parallel and weighted in the ratio $R_{msb} : R_{msb}/2 : R_{msb}/4 : R_{msb}/8 \ldots$ where R_{msb} is the value of the *most significant bit* (MSB) resistor (Figure 5.24B).

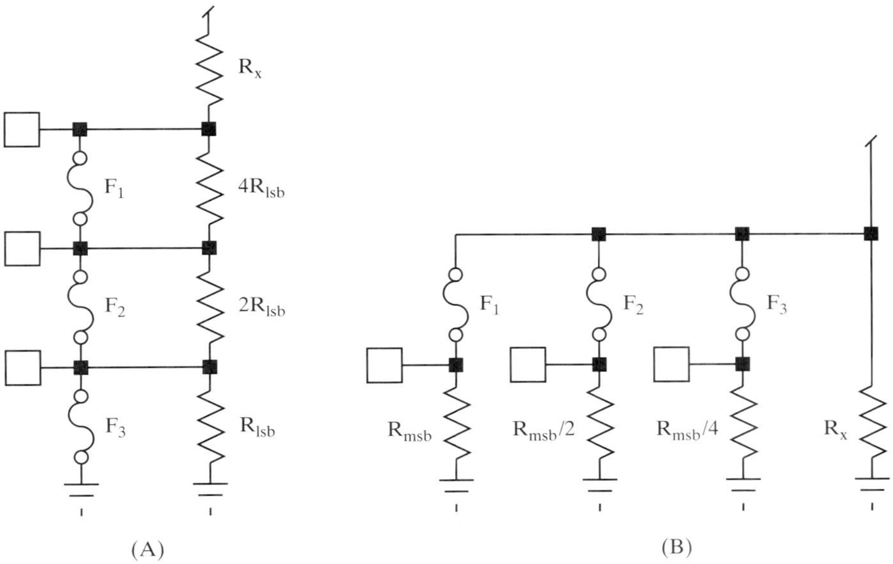

FIGURE 5.24 Two different binary-weighted resistor trim schemes using fuses: (A) series-connected and (B) parallel-connected. Both cases assume that the ground pad is used to program the fuses.

A precision trim scheme often requires a very small LSB resistance. Minimum-width resistors of less than about 1kΩ may overheat during trimming, causing resistance variations and, in extreme cases, outright failures. Smaller resistors can be implemented as parallel combinations of two or more resistors, each having a relatively large resistance, but this technique becomes impractical for resistors much smaller than 200Ω. A technique called *differential trimming* can implement arbitrarily small LSB resistances. This scheme requires two resistors per trim bit rather than one. The two resistors are connected in parallel while the fuse remains intact. Blowing the fuse disconnects one of the resistors and leaves the other to conduct the current alone. The effective LSB resistance equals the difference between the two

[31] A. P. Brokaw, "A Simple Three-Terminal IC Bandgap Reference," *IEEE J. Solid-State Circuits*, SC-9, 1974, pp. 388–393.

[32] See Grebene, pp. 156–158.

resistor values. Figure 5.25 shows an example of differential trim applied to the LSB of a two-bit, series-connected trim scheme. If R_A has a value of 1kΩ and R_B has a value of 250Ω, then the parallel combination of these resistors equals 200Ω. The difference between R_B acting alone and the two resistors acting in parallel therefore equals 50Ω.

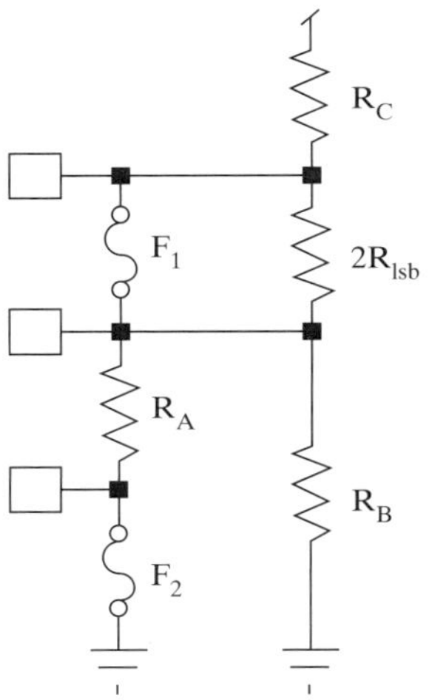

FIGURE 5.25 Differential trim scheme applied to the LSB fuse, F_2, of a series-connected binary-weighted trim network.

Differential trimming requires two trimpads per bit, while standard trimming requires only one pad per bit. The additional trimpads required for differential trim make this technique uneconomic for all but the smallest resistors. Resistors smaller than about 500Ω usually benefit from differential trimming, while larger resistors are better implemented as standard trims.

Fuse trim schemes require trim pads around the periphery of the die, but precision resistors normally reside in the interior to minimize mechanical stresses. Long leads are required to connect fuses at the edge of the die to resistors in the middle. These leads not only waste die area but can also pick up noise from other circuitry. If CMOS transistors are available, then these can be used as switches to reconfigure the resistor network. These transistors can, in turn, be controlled by fuses placed around the periphery of the die. Since the fuse leads no longer connect directly to the trim resistors, they are less susceptible to noise, and their length and routing become less critical. Care must be taken to ensure that the CMOS transistors have small on-resistances compared with the trim resistors. The gate drive voltage for these transistors must be derived from a well-regulated supply, since this voltage will modulate their on-resistance. The design of remotely programmed trim networks is beyond the scope of this text, but the foregoing comments should convey the general concept.

Some designers have attempted to use remotely programmed poly fuses to implement *look-ahead trimming.* A voltage sufficient to switch the transistor but inadequate to blow the fuse can be used to test the results of programming the fuse before committing to it. Unless the circuitry is specifically designed to switch at low voltages, the look-ahead process may overstress the poly fuse, causing it to increase in value to the point where it can no longer be reliably programmed.[33]

Poly fuses have been reliably programmed after encapsulation by using circuitry to compare the resistance of the blown fuse to an unprogrammed reference fuse.

---

[33] D. W. Greve, "Programming Mechanism of Polysilicon Resistor Fuses," *IEEE Trans. on Electron Devices,* Vol. ED-29, #4, 1982, pp. 719–724.

Even if programming does not result in a clean open circuit, it almost invariably causes resistance changes that can be detected in this manner.

Zener Zaps

Zener diodes short-circuit when severely overloaded, and this phenomenon forms the basis of the trim device called a *Zener zap*. The Zener diodes connect across segments of the resistor network in the same manner as the fuses shown in Figures 5.24 and 5.25. The Zeners must be oriented so that they remain reverse-biased during normal operation, and the voltage placed across each Zener must not exceed the base-emitter breakdown voltage. The Zeners appear as open circuits until they are programmed, after which they appear as shorts. The act of programming a Zener is called *zapping*, so these Zeners are called zap Zeners, or Zener zaps.

Figure 5.26A shows the layout of a Zener zap constructed in a standard bipolar process. The device has the same basic structure as a small NPN transistor. The collector and emitter terminals of the NPN together form the cathode of the Zener and the base terminal serves as the anode. Because the device is used as a Zener, little or no current flows through its collector contact. Deep-N+ is therefore unnecessary and can be omitted to save space. The emitter and base contacts should be placed as close as layout rules allow to facilitate the zapping process. The emitter should extend as close to the base contact as possible to minimize the series resistance of the Zener. Although the illustrated Zener zap structure uses a separate collector contact, an alternate layout involves stretching the emitter so that it extends beyond the base diffusion and thus shorts to the tank. This layout theoretically consumes less space, but since the tanks can extend underneath the adjacent trimpads, the two layouts actually consume similar amounts of area.

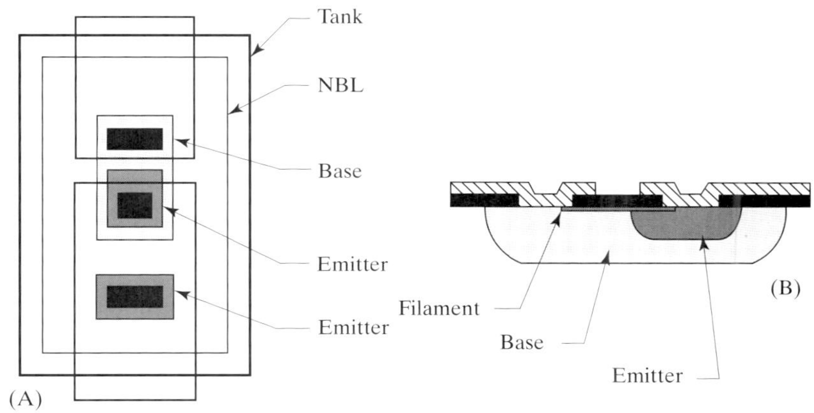

FIGURE 5.26 Zener zap constructed using base and emitter diffusions from a standard bipolar process: (A) layout of unprogrammed Zener and (B) cross section of programmed Zener.

Programming involves forcing a large reverse current through the diode to avalanche its base-emitter junction.[34] A programming current of about 250mA will result in a total drop of as much as 10 to 20V across the Zener, much of this due to internal series resistance. The resulting power dissipation in a very limited volume of silicon results in extreme localized heating. The aluminum metal contacting the silicon melts, and a molten filament of aluminum-silicon alloy flows underneath the oxide and bridges the gap between the contacts (Figure 5.26B). Once this filament forms, the resistance of the Zener zap drops to a few Ohms.

[34] G. Erdi, "A Precision Trim Technique for Monolithic Analog Circuits," *IEEE J. Solid-State Circuits,* SC-10, 1975, pp. 412–416.

Zener zaps require the same arrangements of pads as fuses, so the networks illustrated in Figures 5.24 and 5.25 are applicable to Zeners as well as to fuses. Because larger voltages are generally required to program Zeners than fuses, care must be taken to ensure that the circuitry connected to the trim network can withstand a momentary overvoltage condition. As long as a few kilohms of resistance lie between the zaps and the remainder of the circuit, programming will generally cause no harm. If necessary, diodes or Zeners can be used to clamp the voltage seen by delicate circuitry.

Zener zapping involves the formation of an aluminum-silicon alloy, and this in turn presumes that aluminum is directly touching the silicon. The presence of refractory barrier metal or silicide between the aluminum metal and the silicon interferes with the zapping process. Experimental fabrication of Zener zaps on a process using refractory barrier metal demonstrated that the Zeners could be zapped, but only with difficulty.[35] The programming current had to be nearly doubled, and some wafer lots of material resisted zapping. Based on this experiment, Zener zaps appear unsuitable for processes employing refractory barrier metal or silicided contacts.

Unlike fuses, Zener zaps do not require openings in the protective overcoat. This not only eliminates a potential pathway for contaminants to enter the die but also raises the possibility of trimming packaged units. While post-package trimming is certainly possible, in practice it has rarely been implemented because of the large number of pins (or alternatively, the large power devices) required for zapping.

Very short emitter resistors can be zapped by the same mechanism as Zeners. Attempts have been made to regulate the zapping process to provide infinite adjustability. Since the molten filament moves at a finite speed, it is theoretically possible to halt the programming process before the gap between the contacts is fully bridged.[36] In practice, the filament moves so quickly and so erratically that it is difficult to control, and this scheme cannot be recommended for production use.

Laser Trims

Another method of trimming uses a laser to alter the resistance of a thin conductive film. These films commonly consist of *nichrome* (a nickel-chromium alloy) or *sichrome* (a silicon-chromium mixture). The laser beam causes localized heating, which alters the grain structure or the chemical composition of the material to drastically increase its resistance. The protective overcoat remains intact since the material does not actually melt or vaporize. Each shot of the laser affects a circular zone about 3 to 10μm in diameter. By moving the laser incrementally while performing a series of shots, a continuous line of high-resistance material can be formed. It is possible to virtually sever a resistor by this process, allowing discrete adjustments to a network of resistor segments. Alternatively, the value of the resistor can be continuously monitored and the trim halted once the desired resistance has been achieved. Continuous trimming allows finer resolution, but it also alters the temperature coefficient of the resistor because current continues to flow through portions of the material that have been altered by heat from the laser beam. The change in temperature coefficient is proportional to the increase in resistance caused by trimming, but it rarely exceeds 100ppm/°C. Discrete trimming entirely avoids this problem because current only flows through links that have not been altered by exposure to the laser beam.

---

[35] F. W. Trafton, private communication.

[36] R. L. Vyne, W. F. Davis, and D. M. Susak, "A Monolithic P-channel JFET Quad Op Amp with In-Package Trim and Enhanced Gain-Bandwidth Product," *IEEE J. Solid-State Circuits,* Vol. SC-22, #6, 1987, pp. 1130–1138. Also see R. L. Vyne, "Method for resistor trimming by metal migration," US Patent 4 606 781, Aug. 1986.

Figure 5.27A shows one common style of continuously trimmed, thin-film resistor. The laser beam first moves laterally across the resistor. When the resistance has increased to about 90% of the target value, the laser beam begins to move down the length of the resistor. This longitudinal cut causes a more gradual increase in resistance and therefore allows finer trim resolution. This trim technique can produce tolerances of better than ±0.1%, but it requires resistors at least 30 to 50μm wide. Figure 5.27B shows another style of layout frequently employed for continuous trimming. Discrete trims usually employ loop or ladder networks similar to those of Figures 5.27C and D. The value of the resistor will depend on how many of its segments have been severed. The segments of a discrete network can be made arbitrarily narrow, but they must lie some 10μm apart in order to allow the laser to sever only one link at a time.

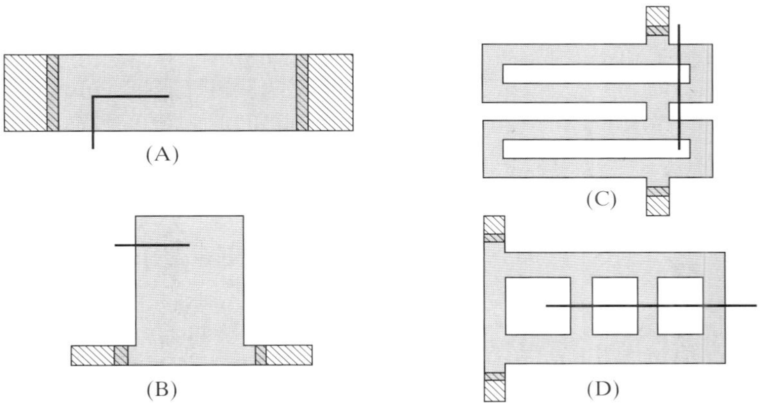

FIGURE 5.27 Four different schemes for laser-trimming thin-film resistors: (A) notched bar, (B) tophat, (C) looped layout, and (D) ladder layout.[37] The heavy black lines show the path of the laser beam through the resistor.

Lasers are also used to sever metal and poly links. Unlike electrical fuses, these laser-ablated links are normally covered by a thin layer of oxide or oxynitride. The laser penetrates this overcoat and heats the fuse to the point of vaporization. The resulting pressure shatters the overcoat and allows the fuse material to explosively erupt outward. The extreme temperatures and pressures generated by laser ablation allow a very clean cut with minimal splattering. The networks used for laser trimming are similar to those shown in Figure 5.24, but they require no trimpads. Differential trimming is unnecessary and there is no danger of damaging sensitive circuitry. The width of the laser-ablated link is quite critical. Excessively narrow links cannot be successfully ablated because insufficient material exists to produce the pressure necessary to rupture the overcoat. Very wide links require multiple laser shots and tend to splatter. Typical laser-ablated links are about 1μm wide by 15μm long. The links must reside far enough from adjacent circuitry to prevent the laser from interfering with other devices. A spacing of 10 to 15μm usually provides adequate clearance for the trimming process.

5.7 SUMMARY

Resistors are the most common type of passive component in analog integrated circuits, and many different types are available. For standard bipolar products, lower values of resistance are obtainable using base and emitter resistors while higher values usually require HSR implant resistors. Extremely low-value resistors can be fabricated using aluminum metallization, while pinched structures can provide high

[37] After Glaser, *et al.*, p. 358.

resistances as long as their nonlinearity is not a concern. CMOS and BiCMOS processes typically offer doped polysilicon resistors that are superior to diffused resistors. Metal resistors are still used for very low-value resistors, and well resistors are sometimes used to obtain large values of resistance when high-sheet poly is unavailable. Thin-film resistors offer superior temperature stability at a higher manufacturing cost.

The value of most resistors can be determined by means of a simple computation involving width and length. A width correction factor may be needed to account for process size adjusts and outdiffusion. Each rectangular 90° turn inserted into a serpentine resistor adds approximately a half square. Each circular 180° turn adds approximately three squares. The corrections for nonuniform current flow near the contacts can usually be ignored because these rarely amount to more than a half square.

Provision can be made for tweaking resistors by means of sliding contacts, sliding heads, trombone slides, and metal option jumpers. Trimming can be performed by means of fuses, Zener zaps, or laser adjustment. Laser trimming of thin-film resistors can produce extremely precise and stable resistors, while Zener zaps potentially allow post-package trimming for canceling package shifts.

5.8 EXERCISES

Refer to Appendix C for layout rules and process specifications.

5.1. What is the sheet resistance of a 7kÅ aluminum film if its resistivity equals 2.8μΩ·cm?

5.2. Suppose a standard bipolar base resistor has a value of 2kΩ, a sheet resistance of 160Ω/□, a uniform width of 8μm, and two 5×5μm contacts. Compute the change in resistance caused by each of the following:
- a. A width bias of 0.4μm.
- b. Nonuniform current flow.
- c. Contact resistance, assuming Al-Cu-Si metallization.

5.3. Lay out a 20kΩ minimum-width serpentine standard bipolar base resistor. Place the two contacts as close to one another as possible. Assume a width bias of 0.4μm and ignore all other sources of error except the contributions of turns. The resistor tank should have a roughly square aspect ratio.

5.4. Using the guidelines of Section 5.3, recommend a width for each of the following types of resistors, assuming a moderate degree of accuracy is required:
- a. Standard bipolar base.
- b. Standard bipolar HSR.
- c. Analog BiCMOS base.
- d. Analog BiCMOS poly-2.

5.5. If a 2kΩ/□ HSR resistor has a value of 34.4kΩ at 25°C, calculate its value at 125°C given a linear TCR of +2200ppm/°C. What percentage change does this represent?

5.6. Suppose that a 200Ω/□ poly resistor must support a voltage differential of 5V. How short could this resistor be made before voltage nonlinearities become a concern? Consider both self-heating and granularity.

5.7. Lay out a 30kΩ HSR resistor with a width of 8μm. Account for the effects of width bias, base head resistance, and contact resistance. Assume a width bias of 0.2μm, a length bias of 0.15μm, and a base width bias of 0.4μm. The metallization system is Al-Cu/Ti-W/PtSi. Compute the change in resistance caused by each of the factors considered when designing the resistor.

5.8. Lay out a serpentine high-sheet poly-2 resistor with a width of 4μm and a value of 150kΩ. Assume the body of the resistor has a sheet resistance of 600Ω/□ and the NSD-doped heads have a sheet resistance of 50Ω/□. To allow for misalignment, the overlap of NSD over contact must equal at least 2μm. Account for turns, but ignore all other correction factors.

5.9. Design a 5kΩ standard bipolar base resistor with a sliding contact allowing a $\pm$10% adjustment of its value. Use a drawn resistor width of 8μm and a width bias of 0.4μm. Account for turns, but ignore all other correction factors.

5.10. Design a 25kΩ standard bipolar HSR resistor incorporating a sliding head allowing a $\pm$20% adjustment of its value. Assume the resistor width equals 8μm and that the HSR width bias equals 0.2μm. Account for turns, but ignore all other correction factors.

5.11. Design a PSD-doped poly-2 resistor with a value of 50kΩ and a width of 4μm. Allow an adjustment of $\pm$25% in value by means of a trombone slide. Account for turns, but ignore all other correction factors.

5.12. Design a polysilicon fuse. Use a strip of minimum-width poly-1 as the fuse, and place a 5$\times$5μm opening in the protective overcoat over the fuse. Contact either end of the fuse with an array of at least four contacts. Keep metal a minimum of 2μm away from the opening in the protective overcoat. Assume the trimpads require 75$\times$75μm openings in the protective overcoat.

5.13. Lay out a series-connected binary-weighted resistor network consisting of four resistors, the smallest of which equals 1kΩ. Assume all of these resistors consist of PSD-doped poly-2 with a width of 4μm. Make all resistors from one or more 1kΩ segments to ensure precise matching. Assume a width bias of 0.2μm and ignore all other correction factors. Connect an array of four polysilicon fuses (designed in Exercise 5.12) to the resistor array to complete the trim network.

5.14. Lay out a standard bipolar Zener zap. Assume that the emitter contact has a width of 8μm and minimize all other dimensions. Assume the trimpads require 75$\times$75μm openings in the protective overcoat. The tank and NBL geometries can reside beneath the trimpads.

6 Capacitors

Capacitors are a class of passive elements useful for coupling AC signals and for constructing timing and phase shift networks. They are relatively bulky devices that store energy in electrostatic fields. The microscopic dimensions of integrated circuits preclude the fabrication of more than a few hundred picofarads of capacitance. Even this tiny amount suffices for certain crucial applications, particularly for compensating feedback loops. Most analog integrated circuits contain at least one capacitor.

In addition to capacitors, discrete circuits often contain inductors, transformers, and saturable reactors. These devices use magnetic fields to store and manipulate energy. Electromagnetic storage takes even more room than electrostatic storage, so only a few nanohenries of inductance can be economically integrated. These tiny inductors become useful only at frequencies beyond 100MHz. Because they find such limited application in integrated circuit design, this text does not further discuss inductors.

6.1 CAPACITANCE

The International System of Units (SI) defines the *Farad* (F) as the standard unit of capacitance. A Farad is an extremely large amount of capacitance. Most discrete circuits employ capacitors ranging from a few picofarads to a few thousand microfarads.[1] No more than a few hundred picofarads can be economically integrated, so larger capacitors must reside off-chip. Most systems use a number of discrete capacitors in conjunction with each analog integrated circuit.

All of the capacitors used in integrated circuits are *parallel-plate capacitors,* which consist of two conductive plates called *electrodes* attached to either side of a slab of insulating material called the *dielectric* (Figure 6.1). In the simple parallel-plate

[1] The correct abbreviations for picofarads and microfarads are pF and μF. Historically, a number of other abbreviations were used, including μμF for picofarads and mF and mFd for microfarads. The use of such nonstandard abbreviations should be avoided. At the same time, it is inadvisable to use the abbreviation mF for millifarads because of the historical meaning attached to this term.

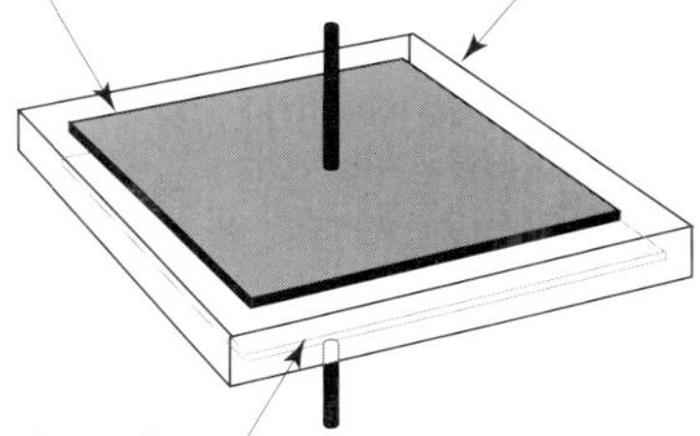

FIGURE 6.1 Construction of a simple parallel-plate capacitor.

capacitor, the two electrodes are assumed to have the same dimensions and to reside directly opposite one another.

The value of the simple parallel-plate capacitor can be computed using the following approximate equation

$$C \cong 0.0885 \frac{A\varepsilon_r}{t} \qquad [6.1]$$

where C is the capacitance in picofarads (pF), A is the area of either electrode in square microns (μm^2), t is the thickness of the dielectric in Angstroms (Å), and ϵ_r is a dimensionless constant called the *relative permittivity* or the *dielectric constant*. ϵ_r depends on the nature of the dielectric. Table 6.1 lists the relative permittivities of several materials that are frequently used in integrated circuits. The entries for oxide and nitride list a range of values because the properties of these materials depend on deposition conditions.

TABLE 6.1 Relative permittivities and dielectric strengths of selected materials.[2]

Material		Relative Permittivity (Vacuum = 1)	Dielectric Strength (MV/cm)
Silicon		11.8	30
Silicon dioxide (SiO_2)	Dry oxide	3.9	11
	Plasma	4.9	3–6
	TEOS	4.0	10
Silicon nitride (Si_3N_4)	LPCVD	6–7	10
	Plasma	6–9	5

Consider a capacitor with a plate area of $0.1mm^2$ constructed using a 200Å ($0.02\mu m$) dry oxide film. If the dielectric has a relative permittivity of about four (as is usually the case) then the capacitance will equal about 180pF. This example helps explain why it is so difficult to integrate capacitors of more than a few hundred picofarads.

Reducing the thickness of the dielectric increases its capacitance, but it also increases the electric field imposed across it. Sufficiently intense electric fields can break covalent bonds and can cause avalanche multiplication of the resulting carriers. The passage of these carriers through the dielectric gradually damages its molecular structure and can lead to long-term failures. If the electric field increases beyond a certain point, a positive-feedback mechanism occurs that quickly short-circuits the

[2] Values for SiO_2, Si_3N_4 are from A. C. Adams, "Dielectric and Polysilicon Film Deposition," in S. M. Sze, ed., *VLSI Technology*, 2nd ed. (New York: McGraw-Hill, 1983), pp. 259, 263. Critical field for silicon taken from D. J. Hamilton and W. G. Howard, *Basic Integrated Circuit Engineering* (New York: McGraw-Hill, 1975), p. 135. See also W. R. Runyan and K. R. Bean, *Semiconductor Integrated Circuit Processing Technology* (Reading MA: Addison-Wesley, 1994), pp. 67–68 for a discussion of the distribution of breakdown voltages.

capacitor. To prevent catastrophic failure, the electric field across the dielectric must never exceed a critical value called the *dielectric strength.* Table 6.1 lists the dielectric strengths of various materials in megavolts per centimeter (MV/cm). The maximum voltage V_{max} that a parallel plate capacitor can withstand equals

$$V_{max} = 0.01tE_{crit} \qquad [6.2]$$

where t is the dielectric thickness in Angstroms (Å) and E_{crit} is the dielectric strength in MV/cm. According to this formula, the maximum voltage that a 200Å dry oxide can withstand equals about 20V. Long-term reliability requires this value be derated by about 50%, so a 200Å oxide is usually rated for 10V operation. Oxide films grown on polysilicon may require further derating because of the presence of microscopic irregularities at the polysilicon/oxide interface. The presence of these *asperities* causes localized intensification of the electric field and reduces the dielectric strength of the oxide.[3] A 200Å dry oxide grown on polysilicon might therefore be rated for no more than 5V.

When the thickness of the dielectric has been reduced as far as the operating voltage allows, then only a high-permittivity dielectric can further increase the capacitance per unit area. Certain ceramics, such as barium strontium titanate, have relative permittivities of several thousand. Although these materials can be deposited on an integrated circuit, the costs involved render them economical for only a few applications. Designers must instead turn to other, more commonly available, materials. Silicon nitride is often chosen because it has a permittivity roughly twice that of oxide. Unfortunately, thin nitride films are prone to the formation of *pinholes*—small areas of inadequate thickness that compromise the dielectric strength of the film. Some processes sandwich a nitride layer between two oxide layers to obtain a composite dielectric less susceptible to pinhole formation.[4] The effective relative permittivity ϵ_{eff} of an oxide-nitride composite dielectric can be computed using the formula

$$\varepsilon_{eff} = \frac{t_{ox} + t_{nit}}{\left(\dfrac{t_{ox}}{\varepsilon_{ox}}\right) + \left(\dfrac{t_{nit}}{\varepsilon_{nit}}\right)} \qquad [6.3]$$

where t_{ox} and t_{nit} are the thicknesses of oxide and nitride, and ϵ_{ox} and ϵ_{nit} are their relative permittivities. For example, if 200Å of nitride with a relative permittivity of 7.5 is sandwiched between two 50Å oxide films with relative permittivities of 3.9, then the composite has an effective relative permittivity of 5.7. The resulting film has the dielectric strength of 300Å dry oxide, yet it has 50% more capacitance per unit area.

Capacitors that use oxide or oxide-nitride dielectrics are identified by a bewildering array of different names. An *oxide capacitor* employs silicon dioxide as its dielectric. This oxide is usually grown on a lower electrode comprising of either a silicon diffusion or a polysilicon deposition. The upper plate usually consists of metal or doped polysilicon. An *ONO capacitor* resembles an oxide capacitor except that it employs an oxide-nitride-oxide composite dielectric to obtain a higher capacitance per unit area. *Poly-poly capacitors* employ two polysilicon electrodes in combination with either an oxide or an ONO dielectric. *MOS capacitors* consist of a thin layer of grown oxide formed on a silicon diffusion that serves as one of the electrodes. The

[3] N. Klein and O. Nevanlinna, "Lowering of the Breakdown Voltage of Silicon Dioxide by Asperities and at Spherical Electrodes," *Solid-State Electronics,* Vol. 26, #9, 1983, pp. 883–892.

[4] The exact mechanism by which oxidation improves dielectric integrity is uncertain, but may involve charge trapping; see K. K. Young, C. Hu, and W. G. Oldham, "Charge Transport and Trapping Characteristics in Thin Nitride-Oxide Stacked Films," *IEEE Electron Device Letters,* Vol. 9, #11, 1988, pp. 616–618.

other electrode consists of either metal or doped polysilicon. If gate oxide is used to form a MOS capacitor, the resulting structure is often called a *gate oxide capacitor.* Despite their many names, all of these structures are variations upon a common theme: that of the *thin-film capacitor.*

The value of a thin-film capacitor may vary due to voltage modulation effects within its electrodes, but its maximum possible capacitance depends solely on the dielectric. This *dielectric capacitance* can be computed using equation 6.1. If the two electrodes are of different sizes, then the common area of the two plates is used in the equation. For example, in the hypothetical thin-film capacitor of Figure 6.2, only the cross-hatched area where the two electrodes overlap contributes to the capacitance; therefore the effective area of the electrodes equals 300μm^2.

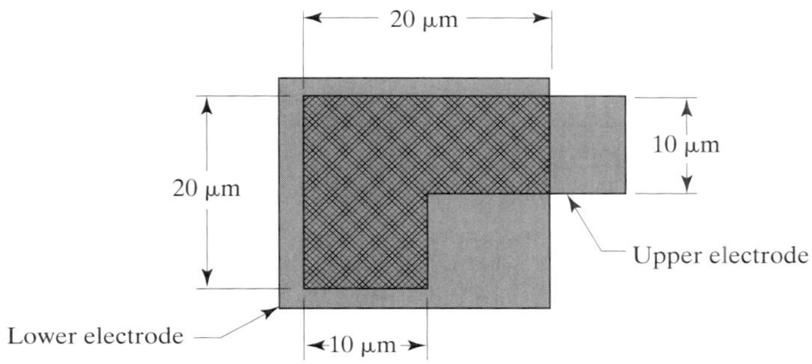

FIGURE 6.2 Hypothetical example of a thin-film capacitor. The crosshatched region where the two plates intersect forms the effective area of the capacitor plates, or in this case 300μm^2.

Equation 6.1 slightly underestimates the true capacitance because the electric field is not entirely confined to the region between the electrodes. The field actually flares out around the edges—an effect called *fringing* (Figure 6.3). The fringing field increases the apparent width of the capacitor plates by an amount proportional to the thickness of the dielectric. This effect is usually ignored because the thickness of the dielectric is much less than the dimensions of the electrodes. For example, consider a capacitor using a 500Å dielectric. Assuming that the capacitor has circular plates 25μm in diameter, the error caused by fringing fields equals about 0.7%.[5] For larger capacitors, or those that employ thinner dielectrics, the effects of fringing fields are even smaller.

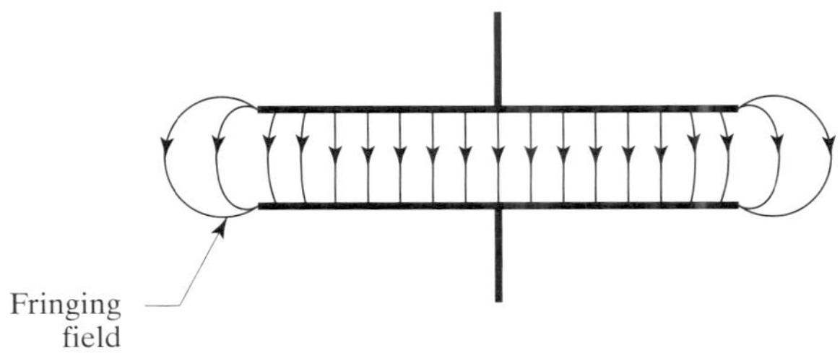

FIGURE 6.3 Illustration of the fringing field surrounding a parallel-plate capacitor embedded in a dielectric of constant permittivity.

Another type of integrated capacitor is the *junction capacitor,* which uses the depletion region surrounding a reverse-biased junction as a dielectric. The permittivity and dielectric strength of silicon are about three times those of oxide, so junction capacitors can obtain high capacitances per unit area. The benefits of compact size are offset by extreme voltage nonlinearity caused by variations in depletion region width with applied bias. The *zero-bias capacitance* C_{j0} serves as a measure of

---

[5] Computed from a formula in C. H. Séquin, "Fringe Field Corrections for Capacitors on Thin Dielectric Layers," *Solid State Electronics*, Vol. 14, 1971, pp. 417–420.

the value of the capacitor. As the reverse bias across the junction increases, its depletion region widens and its capacitance decreases.

The zero-bias capacitance can be computed using equation 6.1, given the thickness of the depletion region. In the case of an abrupt junction between a heavily doped region and a uniform lightly doped one, the zero-bias depletion region width W_o (in Angstroms) equals[6]

$$W_0 \cong 3.10^{11}\sqrt{1/N} \quad [6.4]$$

where N is the concentration of dopant atoms per cubic centimeter on the lighter side of the junction. In practice, this equation only applies to shallow, heavily doped diffusions formed in a lightly doped epi (for example, NSD in P-epi). Deeper junctions exhibit considerable grading of the doping profile, and equation 6.4 ceases to even approximate reality. No simple closed-form equation exists for computing the depletion region widths of diffused junctions, but Lawrence and Warner[7] have published junction capacitance curves for diffusions into a constant background concentration.

The area of junction capacitors is also relatively difficult to compute. Figure 6.4 shows the three-dimensional profile of a typical planar diffusion. As the dopant is driven down, it outdiffuses in all directions to form curved sidewalls. These sidewalls intersect each other to form rounded (*filleted*) corners.

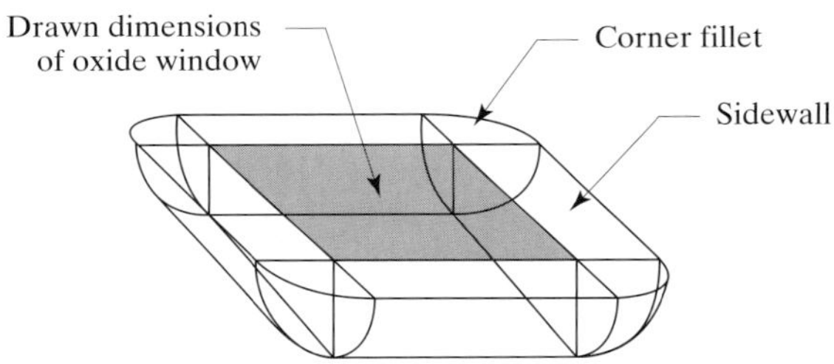

FIGURE 6.4 Three-dimensional view of a diffused junction, showing the sidewalls and filleted corners produced by outdiffusion beyond the drawn dimensions of the oxide window.

The area of a diffused junction consists of three components: the area of the bottom surface, which approximately equals the area of the oxide window shown in gray in Figure 6.4; the area of the sidewalls, which is proportional to the perimeter of the drawn geometry; and the area of the corner fillets. Approximating the sidewalls as cylindrical segments and neglecting the corner fillets, the total area of the junction equals

$$A_{total} = A_d + \frac{\pi}{2}x_j P_d \quad [6.5]$$

where A_d and P_d are the drawn area and perimeter of the oxide window, and x_j is the junction depth of the diffusion.[8]

While one can use the Lawrence-Warner curves to predict the capacitance of a diffused junction, the results are approximate and the process tedious. A simpler method of determining junction capacitances uses the following empirical equation

$$C_{total} = C_a A_d + C_p P_d \quad [6.6]$$

---

[6] This equation assumes that the built-in potential is 0.7V, which is a reasonable approximation for light-to-moderate doping levels.

[7] H. Lawrence and R. M. Warner, Jr. "Diffused Junction Depletion Layer Capacitance Calculations," *Bell System Tech. J.*, Vol. 34, 1955, pp. 105–128.

[8] This formula, and the technique associated with its use, is discussed at some length in Hamilton, *et al.*, pp. 129–135.

where the *areal capacitance* C_a represents the capacitance per unit area, and the *peripheral capacitance* C_p represents the capacitance per unit periphery. The values of these two constants are determined by measuring two or more junction capacitors with very different perimeter-to-area ratios. This technique is considerably more accurate than *a-priori* computations because the experimentally determined constants take into account the vast majority of nonidealities.

Junction capacitors are a staple of standard bipolar design because this process does not produce thin oxides suitable for use as capacitor dielectrics. The base-emitter junction usually provides the highest capacitance per unit area. Experimental measurements on a 40V standard bipolar process with a 160Ω/□ 2μm-deep base gave $C_a = 0.53\text{pF/mil}^2$ ($0.82\text{fF}/\mu\text{m}^2$) and $C_p = 0.072\text{pF/mil}$ ($2.8\text{fF}/\mu\text{m}$).[9]

Junction capacitors customarily employ one of two competing styles of layout. The *plate capacitor* (Figure 6.5A) maximizes junction area, while the *comb capacitor* (Figure 6.5B) maximizes junction periphery. The comb capacitor will have more capacitance per unit area than the plate capacitor if the spacing between the fingers S_f is sufficiently small. Quantitatively, the comb layout is superior whenever the following inequality is met:[10]

$$S_f < \frac{2C_p}{C_a} \qquad [6.7]$$

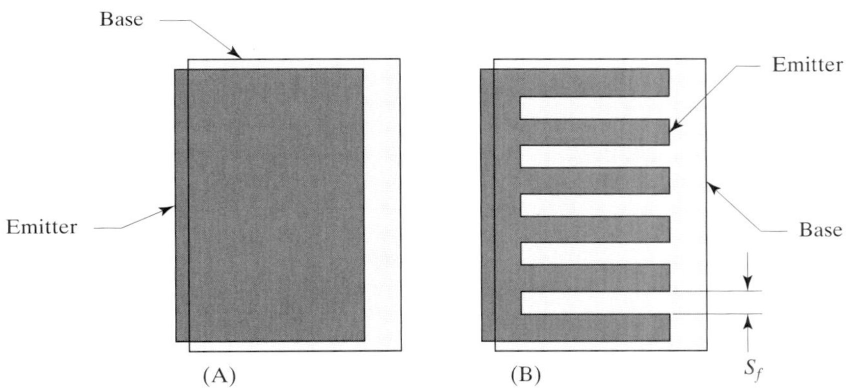

FIGURE 6.5 Two different styles of diffusion capacitor: (A) plate and (B) comb. The tank, NBL, contact, and metal layers are omitted for clarity.

In practice, most versions of standard bipolar can obtain slightly more capacitance from the comb capacitor. Comb capacitors are therefore a familiar sight on older analog layouts. They rarely appear on CMOS and BiCMOS designs because thin-film capacitors can provide equal or larger capacitances per unit area with fewer parasitics than junction capacitors.

A number of different symbols have been used to represent capacitors on schematics. Figure 6.6A shows the standard symbol for a generic capacitor. On schematics, this symbol is usually supplemented by annotations indicating the type of capacitor, its value, and the identity of each of its plates. Figure 6.6B shows a symbol originally used for tubular foil capacitors. The curved electrode represented the outside layer of foil, which was usually grounded to form an electrostatic shield for the rest of the capacitor. This symbol is frequently used to represent integrated capacitors because the two plates are easily distinguished from one another. The curved plate usually (but not always) represents the bottom electrode of an

---

9 F. W. Trafton and R. A. Hastings, "A Study of Emitter-Base Junction Capacitance," unpublished paper, 1989.

10 *Ibid.*

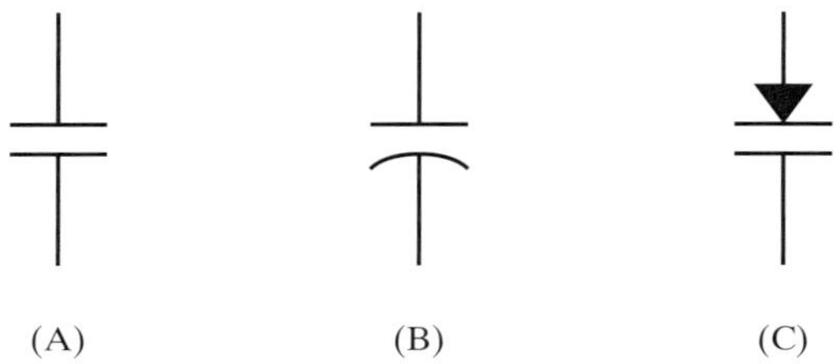

FIGURE 6.6 Typical schematic symbols for capacitors.

integrated capacitor.[11] The symbol of Figure 6.6C represents a junction capacitor. The arrowhead indicates the P-type electrode (*anode*), while the unadorned plate indicates the N-type electrode (*cathode*).

Junction capacitors are sometimes represented as PN junction diodes with values given in terms of junction areas. This practice is understandable from a simulation point of view, since the junction capacitor is modeled as a PN diode. On the other hand, the layouts used for diodes and junction capacitors differ because the capacitor always remains reverse-biased, while a true PN diode operates under forward as well as reverse bias. Many designers therefore use the symbol of Figure 6.6C to distinguish junction capacitors from diodes.

6.2 CAPACITOR VARIABILITY

Integrated capacitors display considerable variability, due mostly to process variation and voltage modulation. There are several lesser sources of variability that only become important for the construction of accurately matched capacitors. These include electrostatic fields and fringing effects, nonuniform etch rates and gradients in doping, film thickness, temperature, and stress. An analysis of these lesser effects appears in Chapter 7.

6.2.1. Process Variation

Both thin-film and junction capacitors experience significant process variations, the causes of which are unique to each type of capacitor. In MOS capacitors, the dielectric consists of a thin film of silicon dioxide grown on monocrystalline silicon. The thickness of this film rarely exceeds 500Å, and in low voltage processes it may be less than 100Å thick. Since the silicon-oxygen bond is approximately 1.5Å long,[12] these films consist of no more than a few hundred atomic monolayers. Much research and development has been directed toward achieving precise control of thin oxide dielectrics. Modern CMOS processes routinely control gate oxide capacitance to within ±20%, and some processes can maintain ±10%.[13]

Dielectrics deposited or grown on polysilicon or metal electrodes are less well controlled than gate oxide. The permittivity of the dielectric film depends not only on thickness, but also on composition, which can vary substantially depending on the conditions of growth or deposition. ONO dielectrics are particularly variable because they are formed by a three-step process consisting of initial oxide growth followed by nitride deposition and subsequent surface oxidization. Each of these steps introduces its own uncertainties, so ONO capacitors typically vary by at least ±20% over process.

[11] The curved plate of a capacitor is sometimes used to denote the upper electrode of an integrated capacitor because this plate is outermost and thus corresponds to the outside foil of a tubular capacitor. Because of the potential for confusion, the plates should always be explicitly labeled.

[12] R. C. Weast, ed., *Handbook of Chemistry and Physics*, 62nd ed. (Boca Raton, FL: CRC Press, 1981), p. F-178.

[13] The variability figures cited in this section represent the 3-sigma limits of a Gaussian distribution.

Junction capacitors are usually constructed from base and emitter diffusions. The emitter-base depletion region width depends on numerous factors, including the average base doping concentration, the base doping profile, and the emitter junction depth. These factors produce at least ±20% variation in a plate capacitor. A comb layout varies even more than a plate layout for several reasons. First, the peripheral capacitance depends more strongly on emitter junction depth than does the areal capacitance. Second, the peripheral capacitance is more susceptible to surface effects such as oxide charge modulation and boron suckup. Third, the intersecting tails of adjacent emitter diffusions modulate the base doping between the fingers and, consequently, vary the peripheral capacitance. Comb capacitors generally vary by at least ±30% over process. These tolerances do not include the effects of voltage modulation and temperature variation.

6.2.2. Voltage Modulation and Temperature Variation

Ideally, the value of a capacitor should not depend on the bias placed across it. Junction capacitors do not even approximate this ideal because the reverse bias placed across the junction modulates the width of its depletion region. Similar effects are observed in many thin-film capacitors because one (or both) electrodes consist of doped silicon subject to depletion effects. MOS capacitors are particularly vulnerable to depletion modulation because their lower electrode is lightly doped and therefore easily depleted, but even poly-poly capacitors with relatively heavily doped electrodes exhibit small voltage nonlinearities due to depletion of the polysilicon. These effects completely vanish if both electrodes consist of metal or silicide.

Figure 6.7 shows the general character of the voltage variation exhibited by a junction capacitor. The capacitance gradually decreases from the zero-bias value C_{j0} as the reverse-bias increases, because the depletion region gradually widens. Eventually the electric field across the depletion region becomes so intense that the junction avalanches, which occurs in standard bipolar base-emitter junctions at about –7V. The capacitance of a forward-biased junction actually increases, because the depletion region narrows as the external bias begins to counteract its built-in potential. As the forward-bias approaches the built-in potential, the depletion region collapses and the junction capacitance drops away rapidly.[14] The forward-bias enhancement of junction capacitance is not particularly useful because forward-biased diodes conduct current. Even a forward-bias of only 0.3V will cause

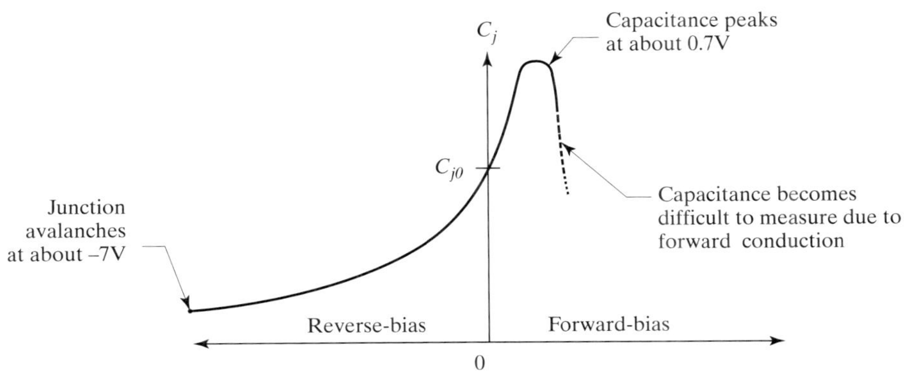

FIGURE 6.7 General behavior of base-emitter junction capacitance under bias. The minimum capacitance just prior to avalanche equals about 40 to 50% of the zero-bias value C_{j0}.

[14] The usual depletion-capacitance formulas indicate that capacitance asymptotically approaches infinity at the built-in potential. This is incorrect; actually capacitance peaks at the built-in potential at a large but finite value; see B. R. Chawla and H. K. Gummel, "Transition Region Capacitance of Diffused p-n Junctions," *IEEE Trans. Electron Devices*, ED-18, 1971, pp. 178–195.

noticeable conduction at higher temperatures, so most designers completely avoid forward-biasing junction capacitors.

Junction capacitors are often used as compensation capacitors because their voltage variability rarely impairs their usefulness in this role. The compensation capacitor is sized so that its absolute minimum value still stabilizes the circuit. This requires that the nominal zero-bias capacitance equal about three times the minimum value needed to stabilize the circuit. Given such a large safety factor, the designer need not worry too much about the exact size or shape of the capacitor.

MOS capacitors may also exhibit strong voltage modulation effects. Figure 6.8 shows the capacitance curve for an NMOS transistor configured as a MOS capacitor. Majority carriers drawn up from the bulk silicon accumulate beneath the gate oxide when the gate is biased negative with respect to the backgate. The capacitance of the device in accumulation is determined solely by the gate dielectric, as given by equation 6.1. When the gate is biased positively, majority carriers are repelled from the surface and a depletion region begins to form. As the bias increases, this depletion region widens and its capacitance decreases. Once the gate bias equals the threshold voltage, sufficient minority carriers will have been drawn up from the bulk to invert the surface. After inversion occurs, larger forward biases only increase the concentration of minority carriers and do not affect the width of the depletion region. Therefore the capacitance levels off at a new, lower value.[15] This minimum capacitance C_{min} may equal less than 20% of the gate oxide capacitance.[16]

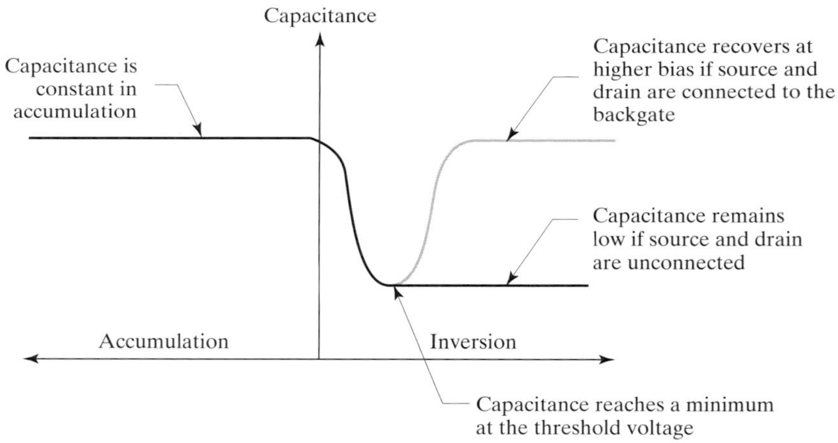

FIGURE 6.8 General behavior of a MOS transistor employed as a capacitor. Different curves are obtained depending on the connection of the source and the drain.

The above analysis applies as long as the source and drain diffusions are absent or unconnected. If these diffusions exist and are connected to the backgate, then the behavior of the MOS transistor capacitance is somewhat more complicated. Once strong inversion occurs, a conducting channel shorts the source and drain terminals. This channel then becomes the lower plate of the capacitor, and the capacitance rises to again equal the gate oxide capacitance (Figure 6.8).

MOS transistors used as capacitors are generally biased outside the capacitance dip centered on the threshold voltage. The source and drain diffusions are unneces-

[15] Actually, the capacitance may recover somewhat if the measurements are performed at very low frequencies. This phenomenon is caused by modulation of generation and recombination within the inversion region due to the electric field projected by the gate electrode. This effect is of no practical significance to the layout designer.

[16] It is rarely necessary to evaluate C_{min} because MOS capacitors are normally operated to retain the full gate oxide capacitance. Most device physics texts discuss the evaluation of C_{min}; for example see B. G. Streetman, *Solid State Electronic Devices*, 2nd ed. (Englewood Cliffs, NJ: Prentice-Hall, 1980), p. 296 ff.

sary if the device operates in accumulation, but they must be present and electrically connected to the backgate if it is to achieve full capacitance in inversion. An NMOS transistor operates in accumulation if the gate is biased negative to the backgate and in inversion if the gate is biased positive to the backgate. Similarly, a PMOS transistor operates in accumulation if the gate is biased positive to the backgate and in inversion if the gate is biased negative to the backgate. One can usually determine by examination whether or not source and drain diffusions are needed for a given MOS capacitor. If in doubt, the diffusions should always be included because they cause no harm even if the device operates in accumulation.

Junction and MOS capacitors both have one electrode formed of lightly doped silicon that is prone to depletion modulation, so these devices exhibit considerable voltage variation. Most other types of capacitors use highly conductive electrodes and exhibit much less voltage modulation. Heavily doped silicon electrodes still exhibit a small amount of voltage modulation, usually no more than 50 to 100ppm/V.[17] This small voltage modulation becomes significant only in precisely matched capacitors operating at different voltages, as is the case in charge-redistribution DACs. Capacitors with electrodes formed from metal or, in the case of the lower plate, silicided poly typically have voltage modulations of less than 5ppm/V.[18]

6.3 CAPACITOR PARASITICS

All integrated capacitors have significant parasitics. The desired capacitance results from the electrostatic interaction between two large-area electrodes. These same electrodes also electrostatically couple to the rest of the integrated circuit, producing unwanted parasitics. The parasitic capacitances associated with one plate usually outweigh those associated with the other, so the orientation of the capacitor becomes quite important.

Figure 6.9A shows a simple subcircuit model of the parasitics associated with a poly-poly capacitor. This model also applies to other types of capacitors whose electrodes are both deposited layers. Ideal capacitor C_1 represents the desired capacitance of the structure. C_2 represents the parasitic capacitance between the lower

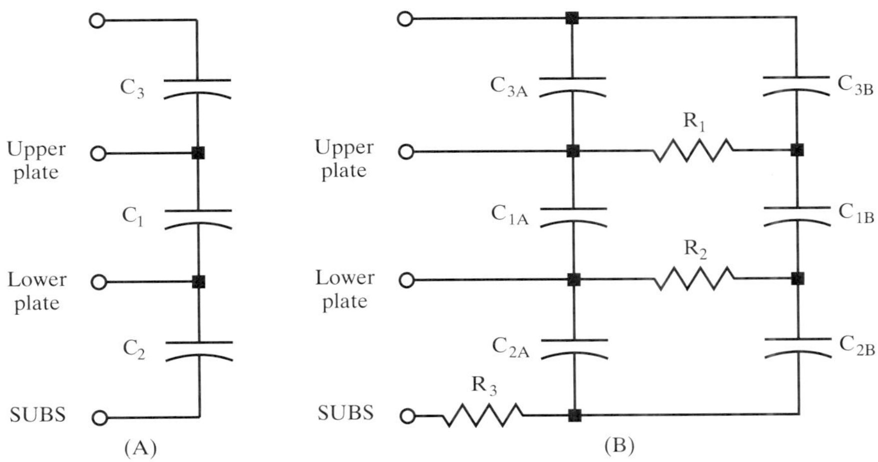

FIGURE 6.9 Subcircuit models for poly-poly capacitors: (A) a simple model without series resistance, and (B) a model incorporating series resistance using single π-sections.

---

[17] A linear coefficient of –148.3ppm/V and a quadratic coefficient of –9.1ppm/V^2 were reported by R. H. Eklund, R. A. Haken, R. H. Havemann and L. N. Hutter, "BiCMOS Process Technology," in A. R. Alvarez, ed., *BiCMOS Technology and Applications*, 2nd ed. (Boston: Kluwer Academic, 1993), p. 123.

[18] A linear coefficient of 1.74ppm/V and a quadratic coefficient of –0.4ppm/V^2 were reported by Eklund, *et al.*, p. 123.

electrode and the substrate. The value of this parasitic can be calculated using the area of the lower plate and the thickness of the field oxide. Capacitor C_3 represents the parasitic capacitance associated with the upper plate. This capacitance is usually much smaller than C_2, and it becomes significant only if another conductor overlies the capacitor. No leads should route across a capacitor unless they connect to it, not only because they add unwanted capacitance but also because of the potential for noise coupling. C_3 can become significant when a metal shield is placed over the capacitor to help improve matching (see Section 7.2.8). In this case, the capacitance C_3 coupling the upper plate of the capacitor to the shield can be computed using the area of the upper electrode and the thickness of the interlevel oxide (ILO).

The series resistance of the polysilicon electrodes may become significant at higher frequencies. The subcircuit model of Figure 6.9B incorporates this series resistance by dividing all of the capacitors into single π-sections. Resistor R_1 models the series resistance of the upper plate, while R_2 models that of the lower plate. Capacitors C_1, C_2, and C_3 are divided into equal sections C_{1A}/C_{1B}, C_{2A}/C_{2B}, and C_{3A}/C_{3B}. Resistor R_3 represents the finite resistance of the substrate, which is often as large as—or larger than—the series resistance of either plate.

Capacitors using diffused electrodes are modeled by replacing the parasitic capacitors with diodes. These diodes remain reverse-biased under normal operating conditions, but leakage currents still flow through them, and these may become significant at higher temperatures. Much larger currents will flow if the diodes even momentarily forward-bias. Since forward conduction involves significant minority carrier flow, inadvertent forward-biasing of junction capacitors can cause latchup.

Figure 6.10A shows a subcircuit model for a base-emitter junction capacitor. The emitter plate is usually connected to the tank, placing the base-emitter and base-collector junctions in parallel. Although the base-collector junction does not add much capacitance, every bit helps. Diodes D_1 and D_2 model the paralleled base-emitter and base-collector junctions. Diodes D_3 and D_4 model the collector-substrate junction. Resistor R_1 models the distributed resistance of the base plate, which is greatly increased by the pinching action of the emitter plate. The resistance of the emitter plate is negligible because the emitter sheet resistance is so much smaller than that of the pinched base. Resistor R_2 models the resistance of the substrate. This model does not include parasitic capacitances coupled to the emitter plate, but these can be modeled as ideal capacitances if desired.

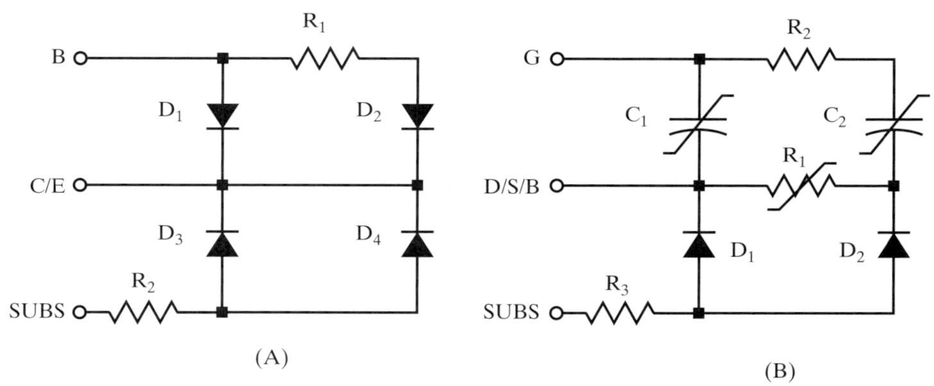

FIGURE 6.10 Subcircuit models for (A) a junction capacitor where C/E denotes the collector/emitter electrode and B denotes the base electrode; and (B) an MOS or gate oxide capacitor where G denotes the deposited gate electrode and D/S/B denotes the drain/source-backgate electrode.

Figure 6.10B shows a subcircuit model for a MOS capacitor. The capacitance of the structure is modeled using two voltage-dependent capacitors C_1 and C_2, each representing half of the capacitance of the MOS structure. R_1 models the distributed resistance of the lower plate of the capacitor, which, in accumulation, consists of well resistance and, in inversion, consists of channel resistance. In either case, this resistance depends strongly upon voltage. R_2 models the resistance of the top-plate electrode, which is often negligible in comparison to R_1. Diodes D_1 and D_2 represent the junction isolating the N-well from the P-epi. Each of these diodes has an area equal to half of the well-epi junction area. Resistor R_3 models the resistance of the substrate contact system that extracts current from the epi side of the well-epi junction. The diagonal slashes through R_1, C_1, and C_2 indicate that these components are voltage-dependent.

6.4 COMPARISON OF AVAILABLE CAPACITORS

Most processes offer only a limited selection of capacitors. Standard bipolar offers base-emitter junction capacitors, and (with the addition of one extra masking step) thin oxide capacitors. CMOS processes are usually limited to MOS capacitors, but some of these processes offer extensions to build either gate oxide capacitors or poly-poly capacitors. BiCMOS processes offer greater flexibility, as they can typically fabricate both MOS and base-NSD junction capacitors, and they may also provide poly-poly capacitors. This section analyzes the strengths and weaknesses of each of these types of capacitors.

6.4.1. Base-emitter Junction Capacitors

Base-emitter junction capacitors exist in both standard bipolar (Figure 3.16) and analog BiCMOS processes. They provide excellent capacitance per unit area at zero bias (typically 0.5pF/mil^2, or 0.8fF/μm^2), but this capacitance falls away with increasing reverse bias. A reverse bias of only –1V causes the capacitance to drop to 75% of its zero-bias value. The capacitance loss slows as the bias increases, so a reverse bias of –5V reduces the capacitance by 50%. Most circuits apply several volts of reverse bias to the capacitor, so the effective capacitance per unit area of base-emitter junction capacitors equals approximately 0.3pF/mil^2 (0.5fF/μm^2). The extreme variability of junction capacitors limits them to noncritical roles such as noise filters and compensation capacitors for closed-loop feedback circuits.

The layout designer must decide whether to use a plate layout (Figure 6.5A) or a comb layout (Figure 6.5B). If the areal and peripheral capacitances are known, then equation 6.7 will indicate which layout requires the least space. If the values of the areal and peripheral capacitances are not known, then the plate layout should be favored over the comb layout because the value of the latter depends strongly on its peripheral capacitance, which is difficult to compute on an *a-priori* basis.

The series resistance of a junction capacitor is greatly increased by the pinching action of the emitter plate. The best structure for minimizing this resistance consists of a series of minimum-width emitter fingers interdigitated with base contacts. This configuration minimizes the effects of base pinch resistance at the cost of increased area. The comb structure provides a reasonably low series resistance because the unpinched base regions between the emitter fingers are much less resistive than the pinched base regions underneath them. The emitter fingers can run either horizontally or vertically; the orientation that gives the shortest fingers also gives the minimum series resistance (the layout in Figure 6.5B follows this rule but the one in Figure 3.16 does not). The plate layout (Figure 6.5A) has by far the highest series resistance.

Junction capacitors are normally laid out inside a tank. Contact must be made to this tank to ensure that the base-collector junction remains reverse-biased. This contact also places the base-collector junction in parallel with the base-emitter junction and thus slightly increases the total capacitance. The tank contact can be made by simply extending the emitter plate beyond the base plate (Figure 3.16). The isolation junction acts as a parasitic capacitance tied to the cathode (emitter) plate, as represented by diodes D_1 and D_2 in Figure 6.10B. The addition of NBL to the tank substantially increases this capacitance. NBL serves no useful function in a junction capacitor, so it is often omitted to minimize the bottom-plate parasitic capacitance. If the anode (base) plate is connected to substrate potential, then this parasitic junction capacitor appears in parallel with the desired capacitance. In this case, NBL may be added to maximize the available capacitance per unit area.

If the anode connects to substrate potential, then the base diffusion can extend out into the isolation to save area (Figure 6.11). The base contact can occupy any portion of the base diffusion, even those portions over isolation. In most processes, the emitter cannot reside over isolation and must instead occupy a tank following the appropriate layout rules. The capacitor of Figure 6.11 consists of two sets of fingers branching from a common tank/emitter contact in the middle; this arrangement helps minimize finger lengths and parasitic resistances.

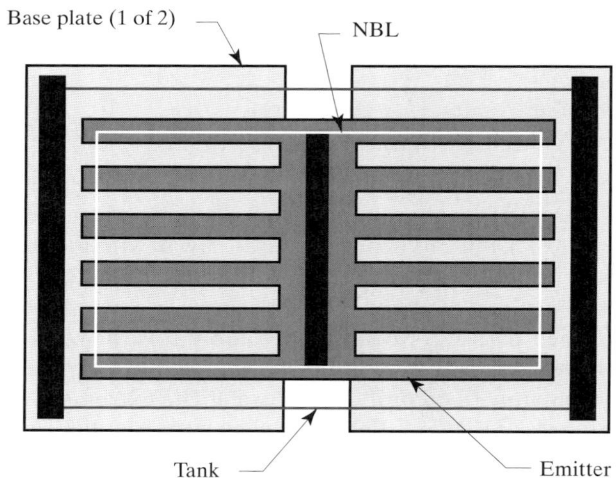

FIGURE 6.11 A junction capacitor with base plates extending into isolation.

A base-NSD junction capacitor can be constructed in the P-epi of a BiCMOS process. The P-epi is so lightly doped that it does not significantly affect the breakdown of the base-NSD junction. The well diffusion can be omitted as long as the anode plate of the junction capacitor connects to substrate potential. The capacitance of the structure increases slightly if N-well is coded beneath it due to the added capacitance of the base/N-well junction. This increase in capacitance rarely exceeds 20% of the total, making it debatable whether the N-well is worth the space it consumes.

While more capacitance can be obtained from a junction held under a slight forward-bias, it is very difficult to prevent forward conduction at higher temperatures. Forward-biasing a junction capacitor will cause substantial current flow, although this is not necessarily destructive to the device. Certain circuit configurations actually use forward conduction to clamp the voltage across the capacitor, but the vast majority of junction capacitors are intended to remain reverse-biased at all times.

Junction capacitors have relatively small breakdown voltages. The standard bipolar base-emitter junction typically avalanches at about 6.8V, so the reverse bias across such a junction capacitor should not exceed about 6V. Avalanche breakdown should be avoided because it increases junction leakage by generating surface trap sites that promote recombination in the depletion regions.

The value of a junction capacitor can be increased slightly by placing a metal plate over the emitter to produce a capacitor using the emitter oxide as its dielectric. If the process employs a thin emitter oxide, then a substantial amount of capacitance can be obtained in this manner, but yields may be reduced by the tendency of thin emitter oxides to form pinholes. Thick emitter oxide does not have this vulnerability, but it provides so little capacitance per unit area that the metal plate produces no noticeable benefit. Most modern processes use thick emitter oxides and thus do not significantly benefit from the addition of a metal plate to junction capacitors.

The die symbolization is often placed over junction capacitors because these devices contain some of the largest areas of unmetallized silicon in the layout. This practice is not harmful, although it does preclude the placement of a metal plate over the capacitor. The symbolization must meet all applicable design rules since it occupies an electrically active region of the die.

Processes that do not experience excessive emitter-isolation leakage can use junction capacitors formed by placing emitter diffusion over isolation regions. Since isolation outdiffuses much farther than emitter, an emitter geometry can be placed directly over an existing isolation region. This technique can produce 100 to 500pF of junction capacitance with little or no increase in die area. The emitter plate of the capacitor has a relatively low capacitance, so the entire capacitor need not be metallized to ensure proper operation. This is an important consideration because leads must cross the emitter-in-iso capacitor to connect adjacent components to one another.

6.4.2. MOS Capacitors

A MOS transistor can be pressed into service as a capacitor, but its lightly doped backgate increases its parasitic resistance. Better results are obtainable using a thin oxide dielectric formed on a heavily doped diffusion. MOS capacitors are sometimes constructed in standard bipolar using the emitter diffusion as the lower electrode. Unless the process forms an exceptionally thin emitter oxide, an additional masking step is required to produce a suitable dielectric oxide.

MOS transistors are ill-suited for use as capacitors, but on CMOS processes they are often the only choice. An MOS transistor used as a capacitor should be biased to avoid the capacitance dip near the threshold voltage (Figure 6.8). This places the device in either of two favorable biasing modes: *accumulation* or *strong inversion*. Accumulation requires biasing the gate of an NMOS negative to its backgate, or biasing the gate of a PMOS positive to its backgate. A constant bias of at least one volt will ensure that the transistor operates within a relatively linear portion of its capacitance curve, limiting voltage variation to about $\pm 10\%$. Source and drain electrodes serve no function and can be eliminated as long as the device operates solely in accumulation. Figure 3.34 shows a MOS capacitor of this sort.

An NMOS transistor enters strong inversion when its gate is biased positive to its backgate by the sum of its threshold voltage plus 1V. A PMOS transistor operates in strong inversion when its gate is biased negative to its backgate; again the bias should exceed the threshold voltage by at least a volt. A MOS capacitor operating in inversion requires source/drain electrodes to contact the channel. These electrodes normally connect to the backgate terminal. The layout of an inversion-mode capacitor is identical to that of a regular MOS transistor (Figures 3.29, 3.30).

A MOS transistor operated as a capacitor has substantial series resistance, most of which is associated with the lower electrode (resistor R_1 in Figure 6.10B). This resistance can be minimized by using a fairly short channel length, ideally 25μm or less. If the source and drain diffusions are omitted, then the backgate contact can run entirely around the gate (as illustrated in Figure 3.34).

Figure 6.12 shows one style of MOS capacitor compatible with standard bipolar processing. The capacitor dielectric consists of a thin oxide formed by an etch and regrowth process controlled by a special masking step. The lower electrode of the capacitor consists of an emitter diffusion enclosed in a tank. The upper electrode is formed from first-level metal. The sheet resistance of the emitter diffusion is so low that voltage modulation and series resistance can both be neglected.

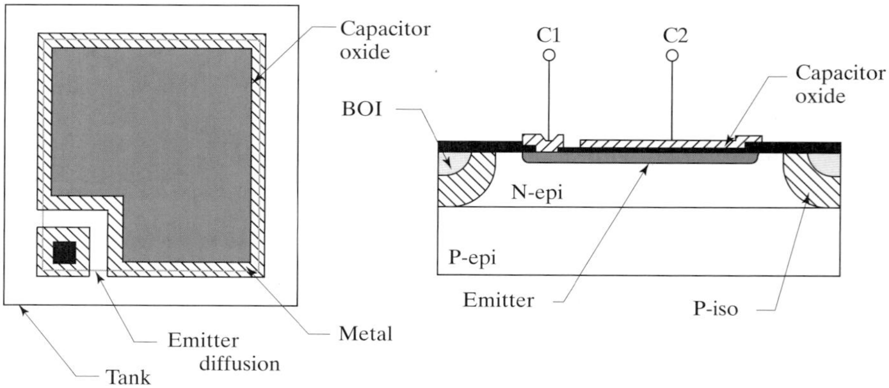

FIGURE 6.12 Layout and cross section of an MOS capacitor constructed in a standard bipolar process using a capacitor oxide mask.

The emitter plate of a MOS capacitor can be formed directly into the standard bipolar isolation, but the resulting N+/P+ junction has considerable parasitic capacitance, and it often exhibits excessive leakage. These difficulties can be circumvented by connecting the emitter plate to the same potential as the isolation. If the emitter plate must connect to a different potential, then it should be enclosed in a tank. This tank does not require the addition of NBL, which can be omitted to further reduce the parasitic capacitance between the emitter plate and the isolation.

Alternatively, the lower (emitter) plate of the MOS capacitor can reside inside a base region connected to the upper electrode. This configuration places the base-emitter junction capacitance in parallel with the thin oxide capacitance of the MOS capacitor to obtain a very high capacitance per unit area—often more than 1pF/mil^2 ($1.5\text{fF}/\mu\text{m}^2$). This type of structure is called a *sandwich capacitor* or a *stacked capacitor*. Like junction capacitors, sandwich capacitors exhibit extreme variability and low breakdown voltages. They are principally used for large compensation and supply line bypass capacitors.

MOS capacitors can also be formed in a BiCMOS process. Since the NSD implant follows gate oxide growth and poly deposition, the lower plate must consist of some other diffusion, usually deep-N+ (Figure 6.13). Deep-N+ has a higher sheet resistance than NSD (typically $100\Omega/\square$), so the lower plate parasitic resistance can be substantial. The heavily concentrated N-type doping often thickens the gate oxide by 10 to 30% through dopant-enhanced oxidation, resulting in higher working voltages but lower capacitance per unit area. The deep-N+ is often placed inside an N-well to reduce parasitic capacitance to the substrate. The N-well can be omitted if the larger parasitic capacitance and lower breakdown voltage of the deep-N+/P-epi junction are tolerable.

Regardless of how an MOS capacitor is constructed, its two electrodes are never wholly interchangeable. The lower plate always consists of a diffusion with substantial parasitic junction capacitance. This junction capacitance can only be eliminated by connecting the lower plate of the capacitor to substrate potential. The upper plate of the MOS capacitor consists of a deposited electrode that has relatively little

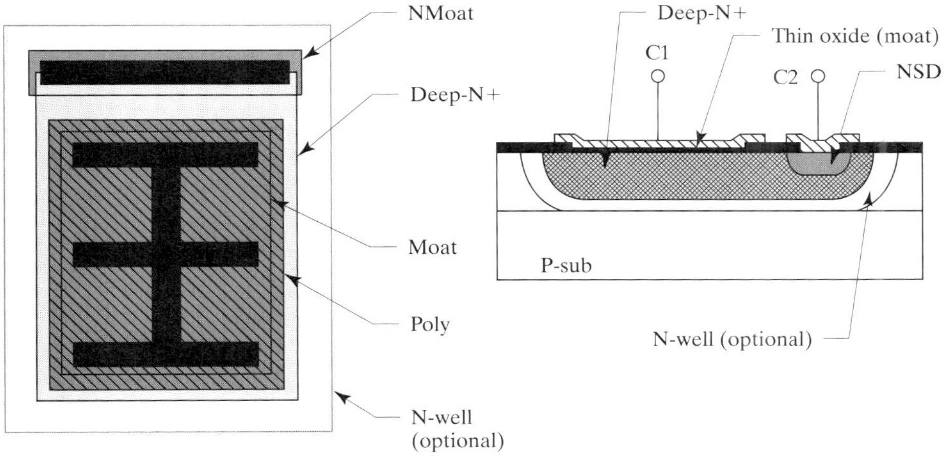

FIGURE 6.13 Layout and cross section of a deep-N+ MOS capacitor constructed in an analog BiCMOS process.

parasitic capacitance. A circuit designer usually attempts to connect an MOS capacitor so that the lower plate connects to the driven node (the one with the lower impedance). Swapping the two electrodes of an MOS capacitor may load a high-impedance node with unwanted parasitics and can potentially cause circuit malfunctions.

6.4.3. Poly-poly Capacitors

Both junction and MOS capacitors use diffusions as their lower electrodes. The junction isolating the diffused electrode exhibits substantial parasitic capacitance and restricts the voltages that can be applied to the capacitor. These limitations can be circumvented by using a deposited material such as polysilicon for both electrodes. Many CMOS and BiCMOS processes already incorporate multiple polysilicon layers, so poly-poly capacitors do not necessarily require any additional masking steps. For example, many processes blanket-dope the gate poly and add a second poly for constructing high-sheet resistors. The gate poly can serve as the lower electrode of a poly-poly capacitor, while the resistor poly (doped with a suitable implant) can form the upper electrode. The upper electrode can be doped using either the NSD or the PSD implant (Figure 6.14). The implant that yields the lowest sheet resistance will produce the best capacitor, since heavier doping not only reduces series resistance but also minimizes voltage modulation due to poly depletion.

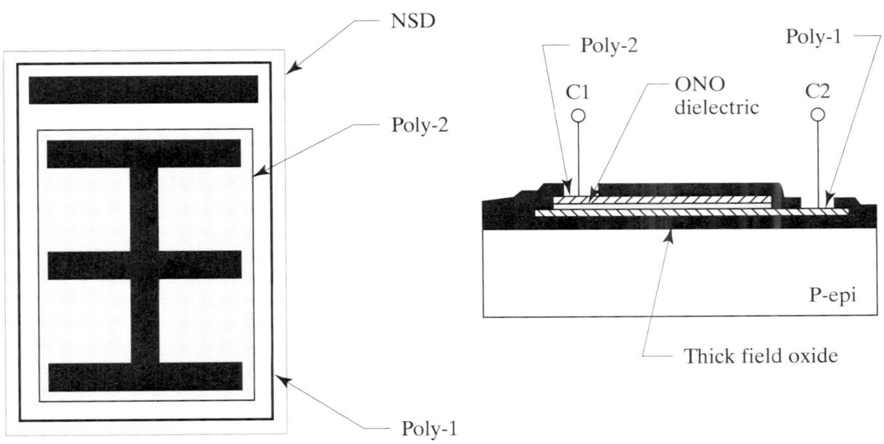

FIGURE 6.14 Layout and cross section of a poly-poly ONO capacitor. The entire capacitor has been enclosed in NSD because the gate poly is also N-type and the additional dopant only further reduces its sheet resistance.

Poly-poly capacitors always require at least one additional process step. Even if both of the electrodes consist of existing depositions, the capacitor dielectric is unique to this structure and consequently requires a process extension. The simplest way to form this dielectric is to eliminate the interlevel oxide (ILO) deposition that normally separates the two polysilicon layers, and in its place substitute a thin oxide grown on the lower polysilicon electrode. Using this technique, a capacitor forms wherever the two poly layers lie on top of one another. This is not a serious limitation as long as the second polysilicon layer is not used for interconnection.

Oxide has a relatively low permittivity. A higher permittivity, and therefore a higher capacitance per unit area, can be obtained using a stacked oxide-nitride-oxide (ONO) dielectric. The first step in forming the ONO dielectric consists of thermal oxidation of the lower polysilicon electrode. A chemical vapor deposited (CVD) nitride layer grown on top of the polysilicon is superficially oxidized to form the final composite dielectric. Unwanted regions of nitride can be removed using a suitable etch after patterning second poly.

The capacitors in Figures 6.13 and 6.14 use fingered contacts for the upper electrode. Alternatively, a sparse array of contacts can be speckled across the upper plate of the capacitor. A dense array of contacts unnecessarily slows down editing, verification, and reticle generation without providing any significant benefit. Some processes also allow the use of a single, large contact opening. All three of these styles of contacts provide a relatively low series resistance over the entire poly-2 plate. The lower electrode of the capacitor in Figure 6.14 is contacted only along one edge. Its series resistance can be reduced by ringing the entire structure with contacts. An even lower series resistance can be obtained by breaking the poly-2 plate into strips and interdigitating these strips with poly-1 contacts.

Poly-poly capacitors normally reside over field oxide. Oxide steps should not intersect the structure because they can cause surface irregularities in the lower capacitor electrode. Not only can these irregularities cause localized thinning of the dielectric, but they also concentrate the electric field. Both of these effects compromise the breakdown voltage of the capacitor.

Although oxide steps should not intersect a poly-poly capacitor, these capacitors are sometimes enclosed entirely within a deep-N+ diffusion. The heavy phosphorus doping accelerates LOCOS oxidation and produces a thicker field oxide that reduces the parasitic capacitance between the lower plate of the capacitor and the substrate. If the deep-N+ region connects to a quiet low-impedance node, such as analog ground, it will shield the lower capacitor electrode from substrate noise. A similar shield constructed from N-well is useful for situations where deep-N+ is not available, or where the layout rules do not allow capacitors to reside on top of deep-N+ due to planarization difficulties.

If there is a choice of dielectrics for use with poly-poly capacitors, several additional points should be considered. Composite dielectrics experience hysteresis effects at high frequencies (10MHz or above) due to the incomplete redistribution of static charges along oxide-nitride interfaces. If the value of the capacitor must remain constant regardless of frequency, then pure oxide dielectrics are preferable to ONO composite dielectrics. Oxide dielectrics typically have lower capacitance per unit area, but this is not always undesirable. Larger plate areas improve matching, so low-capacitance dielectrics are useful for improving the matching of small capacitors.

The voltage modulation of poly-poly capacitors is relatively small, as long as both electrodes are heavily doped. An unsilicided poly-poly capacitor typically exhibits a voltage modulation of 150ppm/V. The temperature coefficient of a poly-poly capacitor also depends on voltage modulation effects and is typically less than

250ppm/°C.[19] Both of these values will increase if either or both electrodes are lightly doped.

The two plates of a poly-poly capacitor are not wholly interchangeable. The upper electrode usually has less parasitic capacitance than the lower electrode. ONO capacitors may also exhibit asymmetric breakdown characteristics. The dielectric rupture process is triggered by field-aided emission of electrons from the surface of the negative electrode. The lower electrode of the ONO capacitor contains asperities produced by thermal oxidation, while the upper electrode lies on a smooth deposited oxide. The ONO capacitor therefore ruptures at a higher voltage if the lower electrode is biased positively with respect to the upper electrode. Proper orientation may be critical for certain applications, as an improperly oriented ONO capacitor may have its breakdown voltage reduced by 50% or more.

6.4.4. Miscellaneous Styles of Capacitors

Many other types of thin-film capacitors can be created by proper combinations of processing steps. For example, a capacitor can be fabricated using polysilicon for the lower electrode and metal for the upper electrode. The dielectric for this type of capacitor consists of grown oxide, deposited nitride, or an ONO stack. This structure usually requires an extra mask to remove ILO over the poly electrode to allow the formation of a capacitor dielectric. The high conductivity of the metal electrode reduces the voltage modulation of the capacitor, which in turn reduces its temperature coefficient.

The voltage modulation of a thin-film capacitor can be further reduced by using a highly conductive lower plate as well as a metallic upper plate. The dielectric must now be deposited rather than grown. CVD oxide is normally used, either alone or in combination with CVD nitride. The lower plate is not normally fashioned of aluminum because the oxide dielectric must reach a relatively high temperature to achieve proper densification. Incomplete densification may compromise the dielectric strength of the resulting film. Silicided poly can withstand the necessary temperatures, as can deposited refractory metals such as tungsten. Capacitors with two highly conductive electrodes experience very little voltage modulation. A capacitor employing a metal upper plate and a silicided lower plate has a reported voltage coefficient of 2ppm/V.[20]

Another type of thin-film capacitor can be constructed by forming a high-permittivity thin-film material over the lower electrode. For example, a tantalum lower plate can be superficially oxidized to form a tantalum pentoxide dielectric. This material has a relative permittivity of 22 and, even in very thin films, it exhibits a low incidence of pinhole defects.[21] Tantalum pentoxide capacitors are compatible with tantalum thin-film resistors, and these two types of specialized components are often manufactured using the same processing steps.

Special ceramic thin-film dielectrics such as barium strontium titanate exhibit permittivities of 1000 or more. These materials exhibit large process, temperature, and voltage variabilities. They may also experience capacitance shifts over time (*aging*) and severe hysteresis effects (*soakage*). Thus, while these materials allow the integration of nanofarads of capacitance, the resulting devices are useless for precision

---

[19] J. L. McCreary, "Matching Properties, and Voltage and Temperature Dependence of MOS Capacitors," *IEEE J. Solid-State Circuits*, SC-16, #6, 1981, pp. 608–616.

[20] Eklund, *et al.*, p. 123.

[21] Value for ϵ_r from A. B. Glaser and G. E. Subak-Sharpe, *Integrated Circuit Engineering* (Reading, PA: Addison Wesley, 1977), p. 355.

circuitry. Most wafer fabs are also not equipped to handle the deposition and etching of these specialized dielectric materials.

6.5 SUMMARY

Capacitance is not as readily integrated as resistance. Barring the use of exotic materials such as titanates, only a few hundred picofarads of capacitance can economically be integrated on a single die. Even this relatively small amount of capacitance suffices for many applications, including timers, capacitive dividers, and active filters.

Integrated capacitors generally fall into two categories: those using thin insulating films as dielectrics (*thin-film capacitors*) and those using reverse-biased junctions as dielectrics (*junction capacitors*). Properly constructed thin-film capacitors have fewer parasitics than junction capacitors, but they usually require additional processing steps. Most standard bipolar processes offer a base-emitter junction capacitor as part of the baseline process and an MOS capacitor as a process extension. CMOS processes always offer an MOS capacitor since an MOS transistor can be pressed into service in this role. This device exhibits an undesirable dip in capacitance near the threshold voltage and thus requires careful biasing. Many CMOS and BiCMOS processes also offer a poly-poly capacitor using a thin oxide or ONO stack dielectric specifically tailored for this application. Although the fabrication of this capacitor increases the process complexity and cost, its superior performance can often justify its inclusion.

The absolute accuracy of capacitors is relatively poor. Variations in doping and junction depth can cause junction capacitors to vary by up to ±30%. Dimensional variations cause thin-film capacitors typically to vary by at least ±10%. Capacitors are difficult to trim because the usual trim structures add excessive amounts of parasitic capacitance. If necessary, laser link trimming (Section 5.6.2) can be applied to most types of capacitors. The vast majority of circuits either trim resistors or current sources to compensate for capacitor variability, or they use topologies sensitive only to the matching of capacitors rather than to their absolute values.

6.6 EXERCISES

Refer to Appendix C for layout rules and process specifications.

6.1. Assume that a thermal oxide film with a relative permittivity of 3.9 can safely withstand a field of $5 \cdot 10^5$V/cm. How thick must the oxide be made to withstand an operating voltage of 15V? What is the capacitance of the resulting film in fF/μm^2?

6.2. What is the relative permittivity of a composite dielectric consisting of 60Å of dry oxide, 220Å of plasma nitride, and a further 50Å of dry oxide, assuming an oxide permittivity of 3.9 and a nitride permittivity of 6.8? What operating voltage can this dielectric withstand? What percentage improvement in capacitance will the composite dielectric provide over a pure oxide capacitor having the same working voltage?

6.3. Determine the approximate zero-bias junction capacitance of a device consisting of a square NSD region diffused into P-epi, given that the oxide window through which the NSD is implanted measures 10 × 20μm, the NSD junction depth equals 0.9μm, and the P-epi doping concentration equals $5 \cdot 10^{16}$ atoms/cm^3. Include the effects of sidewall capacitance.

6.4. A junction capacitor with a drawn area of 9mil^2 and a drawn periphery of 12mils has a zero-bias junction capacitance of 6.45pF. A second junction capacitor with a drawn area of 4.6mil^2 and a drawn periphery of 26.4mils has a zero-bias junction capacitance of 4.92pF. What are the areal and peripheral capacitances for this type of junction capacitor? What must the spacing between fingers equal in order for a comb capacitor of this type to obtain a higher capacitance per unit area than a plate capacitor?

6.5. Lay out a junction capacitor with a nominal zero bias capacitance of 10pF using the standard bipolar layout rules of Appendix C. Assume $C_a = 0.82\text{fF}/\mu\text{m}^2$ and $C_p = 2.8\text{fF}/\mu\text{m}$. Justify your choice of layout style (comb or plate).

6.6. Lay out a 5pF standard bipolar thin oxide capacitor. A special process extension will be used to produce a 450Å oxide with a relative permittivity of 3.9. The layout rules for the thin oxide layer TOX are as follows:

1. TOX width	10μm
2. EMIT overlap of TOX	4μm
3. METAL overlap TOX	4μm

6.7. Lay out a poly-poly capacitor having a minimum guaranteed capacitance of 20pF. Use a sparse array of contacts spaced 10μm apart to contact the poly-2 plate. Dope the poly-2 plate using NSD. Contact the poly-1 plate on at least three sides, and include all necessary metallization.

6.8. Construct a 5pF MOS capacitor from a PMOS transistor laid out according to the baseline CMOS process of Appendix C. Assume that the capacitor will operate in inversion, and use a channel length of no more than 20μm to minimize bottom-plate resistance. Use a sparse array of contacts spaced 10μm apart to contact the poly plate.

7 Matching of Resistors and Capacitors

Most integrated resistors and capacitors have tolerances of ±20 to 30%. These tolerances are much poorer than those of comparable discrete devices, but this does not prevent integrated circuits from achieving a high degree of precision matching. All of the devices in an integrated circuit occupy the same piece of silicon, so they all experience similar manufacturing conditions. If one component's value increases by 10%, then all similar components experience similar increases. The ratio between two similar components on the same integrated circuit can be controlled to better than ±1%, and in many cases, to better than ±0.1%. Devices specifically constructed to obtain a known, constant ratio are called *matched devices*.

Analog integrated circuits usually depend on matching to obtain much of their precision and performance. Any number of mechanisms can interfere with matching. Most of these mechanisms are known, and layout designers have devised ways to minimize their impact. This chapter covers the design of matched resistors and capacitors. Much of this information also applies to matching of other components, such as bipolar transistors (Section 9.2), diodes (Section 10.3), and MOS transistors (Section 12.4).

7.1 MEASURING MISMATCH

The *mismatch* between two components is usually expressed as a deviation of the measured device ratio from the intended device ratio. Suppose a designer lays out a pair of matched 10kΩ resistors. After fabrication, one pair of these resistors are found to equal 12.47kΩ and 12.34kΩ. The ratio between these resistors equals 1.0105, or approximately 1% more than the intended ratio of 1.0000. This pair of resistors therefore exhibits a mismatch of approximately 1%.

The concept of mismatch also applies to devices having ratios other than 1:1. The mismatch between any two devices equals the difference between the ratio of their measured values and the ratio of their intended values, divided by the ratio of their intended values. The final step, division by the ratio of intended values, normalizes the result so that it becomes independent of ratio. If the intended values are X_1 and X_2 and the measured values are x_1 and x_2, then the mismatch δ equals

$$\delta = \frac{(x_2/x_1)-(X_2/X_1)}{(X_2/X_1)} = \frac{X_1x_2}{X_2x_1} - 1 \qquad [7.1]$$

Equation 7.1 computes the mismatch of one specific pair of devices. The same measurements performed on a second pair of devices will yield a different mismatch. Measurements of a large number of device pairs will produce a random distribution of mismatches. An analysis of the mismatch distribution of a small sample of devices allows the designer to determine the percentages of units that are likely to fail specifications. In order for this analysis to yield valid results, the sample must fairly represent the capabilities of the process. Ideally, the sample should consist of 50 to 100 devices drawn from random locations on at least ten wafers that are drawn randomly from at least three wafer lots. In practice, one must often rely on a sample drawn from a single wafer lot. When selecting this sample, consider the following guidelines:

- The sample should include no fewer than twenty devices.
- The sample should include devices drawn from at least three wafers. Each wafer should contribute approximately the same number of devices to the sample.
- The wafers should be selected from various positions in the wafer lot. A three-wafer sample should include one wafer from the front of the lot, one from the middle, and one from the back.
- The sample devices should be selected from random locations on each wafer.
- If possible, the sample should include wafers from more than one wafer lot.
- Wafers that have been misprocessed or reworked do not fairly represent the process and should not be used for characterization.
- If at all possible, the sample units should be packaged using the same lead frames and encapsulation as production material.

Once a sample has been selected and all of the sample units have been measured, the resulting data must be statistically analyzed. The theory of such analyses is beyond the scope of this text,[1] but the simplified procedure given below provides a good example of the techniques and terminology involved.

Suppose the sample includes N units, the mismatches of which are $\delta_1, \delta_2, \delta_3 \ldots \delta_N$, as computed by equation 7.1. Mismatches can be either negative or positive quantities. The signs of these mismatches are significant and must be retained in order for the following computations to have meaning. Based on the computed mismatches, one can derive an average mismatch m_δ. This average, or *mean,* consists of the sum of all the mismatches divided by the number of sample units N

$$m_\delta = \frac{1}{N}\sum_{i=1}^{N}\delta_i \qquad [7.2]$$

where the sigma function $\Sigma(\,)$ represents the sum of all of the individual terms. Once the mean has been computed, one can determine the *standard deviation of the mismatches* s_δ:

$$s_\delta = \sqrt{\frac{1}{N-1}\sum_{i=1}^{N}(\delta_i - m_\delta)^2} \qquad [7.3]$$

---

1 Any good statistical text will cover this subject in some detail. A discussion of the propagation of error and its measurement is contained in J. Mandel, *The Statistical Analysis of Experimental Data* (New York: Dover, 1984).

The mean m_δ is a measure of the *systematic mismatch,* or *bias,* between the matched devices. Systematic mismatches are caused by mechanisms that influence all of the samples in the same manner. Consider the case of a pair of matched 2kΩ and 4kΩ base resistors laid out as single strips of base diffusion with contacts at either end. Both of these resistors have the same contact resistance, which will be assumed to equal 100Ω. This contact resistance causes a 5% increase in the 2kΩ resistor, but only a 2.5% increase in the 4kΩ resistor. Every pair of resistors shows the same imbalance, so the contact resistance represents a systematic mismatch. This mismatch could be eliminated by dividing both resistors into segments of 2kΩ. The 2kΩ resistor would contain one segment with 100Ω of contact resistance, which produces a 5% increase in its value. The 4kΩ resistor would contain two segments with a total of 200Ω of contact resistance, which produces a 5% increase in value. Since both segmented resistors experience the same percentage increase in value, they do not exhibit any systematic mismatch.

The standard deviation quantifies *random mismatch* caused by statistical fluctuations in processing conditions or material properties. These fluctuations are an unavoidable part of semiconductor manufacture, but their magnitude can sometimes be reduced if one can identify their underlying causes. Figure 7.1 shows a histogram of the mismatches of 30 units. The exact appearance of the histogram will change as additional units are added, but the values of the mean and the standard deviation fluctuate relatively little once the distribution contains 20 or 30 units.

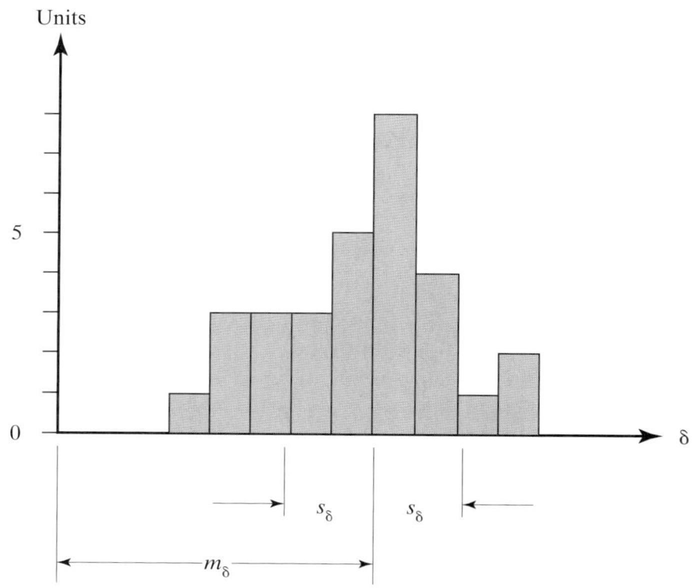

FIGURE 7.1 Histogram of the mismatch, δ, of 30 units, showing mean, m_δ, and standard deviation, s_δ.

Once the mean and standard deviation have been computed using equations 7.2 and 7.3, these can be used to predict worst-case mismatches using one of several indices. The *three-sigma mismatch* equals the sum of the absolute value of the mean plus three times the standard deviation. The *six-sigma mismatch* equals the sum of the absolute value of the mean plus six times the standard deviation. Suppose a pair of resistors shows a mean mismatch of –0.3% with a standard deviation of 0.1%. The three-sigma mismatch of these resistors equals 0.6%, while their six-sigma mismatch equals 0.9%. Less than 1% of all units should have mismatches greater than the three-sigma value, and virtually no units should have mismatches larger than the six-sigma value. The mismatch figures given in this text are all three-sigma values.

7.2 CAUSES OF MISMATCH

Random mismatches stem from microscopic fluctuations in dimensions, dopings, oxide thicknesses, and other parameters that influence component values. Although these statistical fluctuations cannot be entirely eliminated, their impact can be minimized through proper selection of component values and device dimensions. Systematic mismatches stem from process biases, contact resistances, nonuniform current flow, diffusion interactions, mechanical stresses, temperature gradients, and a host of other causes. A major goal of designing matched components consists of rendering them insensitive to various sources of systematic error. The following sections discuss the major known causes of mismatch and techniques for combating them.

7.2.1. Random Statistical Fluctuations

All components exhibit microscopic irregularities, or fluctuations. In the case of a polysilicon resistor, the edges of the poly exhibit microscopic irregularities that give them a slightly ragged appearance. Some of these irregularities stem from the granularity of the polysilicon, while others result from imperfections in the photoresist. The granularity of the polysilicon also causes variations in poly thickness and resistivity. Other types of devices exhibit different types of fluctuations, but all of these fall into one of two categories: fluctuations that occur only along the edges of the device and fluctuations that occur throughout the device. The former are called *peripheral fluctuations* because they scale with device periphery, while the latter are called *areal fluctuations* because they scale with device area. The nature of these scaling relationships can be deduced from statistical arguments.

Consider the case of a pair of matched capacitors, each having capacitance C. The random mismatch due to peripheral and areal fluctuations has a standard deviation s_C, that equals[2,3,4]

$$s_C = \frac{1}{\sqrt{C}}\sqrt{k_a + \frac{k_p}{\sqrt{C}}} \qquad [7.4]$$

where k_a and k_p are constants representing the contributions of areal and peripheral fluctuations, respectively. The contribution of the peripheral term decreases as the capacitance increases. For sufficiently large capacitors, the areal term dominates and the random mismatch becomes inversely proportional to the square root of capacitance. Most practical matched capacitors follow the inverse-square-root relationship fairly closely, so doubling the size of a pair of capacitors decreases their random mismatch by about 30%. The matching of capacitors of different values is dominated by the value of the smaller capacitor, not the larger one. In other words, a 5pF capacitor matches a 50pF capacitor about as well as it matches another 5pF capacitor.

---

[2] J. B. Shyu, G. C. Temes, and F. Krummenacher, "Random Error Effects in Matched MOS Capacitors and Current Sources," *IEEE J. Solid-State Circuits,* Vol. SC-19, #6, 1984, pp. 948–956.

[3] J. B. Shyu, G. C. Temes, and K. Yao, "Random Errors in MOS Capacitors," *IEEE J. Solid-State Circuits,* Vol. SC-17, #6, 1982, pp. 1070–1076.

[4] J. L. McCreary, "Matching Properties, and Voltage and Temperature Dependence of MOS Capacitors," *IEEE J. Solid-State Circuits,* Vol. SC-16, #6, 1981, pp. 608–616.

Now consider the case of a pair of matched resistors[5] having width W and resistance R. The random mismatch between these resistors has a standard deviation s_R that equals

$$s_R = \frac{1}{W\sqrt{R}}\sqrt{k_a + \frac{k_p}{W}} \qquad [7.5]$$

where k_a and k_p are constants representing the contributions of areal and peripheral fluctuations, respectively (Appendix D). This equation shows that random mismatches scale inversely with width. Doubling the width of a pair of matched resistors will at least halve their random offset. The mismatches also scale as the square root of resistance, so larger resistors match better than smaller ones. This leads to a very useful generalization concerning the widths of matched resistors. Suppose a resistance, R_1, requires a width, W_1, to obtain a certain degree of matching. The width, W_2, required to obtain the same degree of matching for a resistance R_2 equals the larger of the two following values:

$$W_2 = W_1\sqrt{\frac{R_1}{R_2}} \qquad [7.6A]$$

$$W_2 = W_1\sqrt[3]{\frac{R_1}{R_2}} \qquad [7.6B]$$

Equation 7.6A represents the extreme case where areal fluctuations dominate over peripheral fluctuations, while equation 7.6B represents the opposite extreme. The actual situation lies somewhere between these extremes, although areal effects generally predominate. As long as one takes the larger of the two widths given by the equations, the matching of the new resistor R_2 should always equal or exceed the matching of the original resistor R_1. In the case of matched resistors of different values, the smaller of the two resistances should be used in equations 7.6A and 7.6B.

An example will clarify the use of these equations. Suppose a pair of 6μm-wide 10kΩ resistors have a worst-case random mismatch of ±0.1%. What width is required to obtain the same degree of matching between 100kΩ resistors? Equation 7.6A predicts a minimum width of 1.90μm, while equation 7.6B predicts a minimum width of 2.78μm. The actual width required to obtain this degree of matching therefore lies somewhere between 1.90μm and 2.78μm. A conservative designer would probably make these resistors 3μm wide.

Equations 7.6A and 7.6B only apply to poly resistors in which the resistor is much wider than its largest poly grains. If this condition is not met, then the equations will underestimate the mismatch of the resistors. Most poly grains are less than 1μm across,[6] so matched poly resistors should be made at least 2 to 3μm wide.

N-type poly resistors seem to exhibit larger random mismatches than P-type poly resistors. On one advanced bipolar process, N-doped poly resistors exhibited approximately twice the random mismatch of P-doped poly resistors having similar dimensions and sheet resistance.[7] This effect may stem from dopant segregation

---

[5] Resistor matching is also treated in W. A. Lane and G. T. Wrixon, "The Design of Thin-Film Polysilicon Resistors for Analog IC Applications," *IEEE Trans. on Electron Devices,* Vol. 36, #4, 1989, pp. 738–744.

[6] A. C. Adams, "Dielectric and Polysilican Film Deposition," in S. M. Sze, ed., *VLSI Technology,* 2nd ed., (New York: McGraw-Hill, 1983), p. 244.

[7] M. Corsi, private communication, 1998.

at grain boundaries.[8] The exact explanation remains unclear, so it is not certain that P-type poly resistors will always exhibit less random mismatch than N-type poly resistors.

7.2.2. Process Biases

The dimensions of geometries fabricated in silicon never exactly match those in the layout database because the geometries shrink or expand during photolithography, etching, diffusion, and implantation. The difference between the drawn width of a geometry and its actual measured width constitutes the *process bias.* Process biases can introduce major systematic mismatches in poorly designed components. Consider the case of two matched poly resistors having widths of 2μm and 4μm, respectively. Suppose that poly etching introduces a process bias of 0.1μm. The ratio of the actual widths equals (2 + 0.1)/(4 + 0.1), or 0.512. This represents a systematic mismatch of no less than 2.4%! Since most processing steps have biases of at least 0.1μm, the layout designer must ensure that all matched devices are insensitive to process biases. In the case of resistors, process biases can be virtually eliminated by simply making both resistors the same width.

Process biases can also affect the length of a resistor. The length of most resistors is determined by the placement of their contacts. Suppose that these contacts have a process bias of 0.2μm. If one matched resistor was 20μm long and the other was 40μm long, then the mismatch due to this bias would equal (20 + 0.2)/(40 + 0.2), or 0.503. This represents a systematic mismatch of about 0.5%. The simplest way to avoid this bias consists of dividing both matched resistors into segments of the same size. If the resistors of the previous example were laid out in 20μm segments, then the ratio of the resistors would equal (20 + 0.2)/[2·(20 + 0.2)], or exactly 0.5. The same stratagem has already been shown to eliminate systematic mismatches due to contact resistances and nonlinear current flow at the ends of the resistors. Section 7.2.6 explains how to divide matched resistors into arrays of optimally sized segments.

Capacitors also experience systematic mismatches caused by process biases. Suppose a pair of poly-poly capacitors, one measuring 10×10μm and the other 10×20μm, both experience a poly etch bias of 0.1μm. The actual area of the 10×10μm capacitor equals $(10.1)^2$, or 102.1μm^2, while the actual area of the 10×20μm capacitor equals (10.1·20.1), or 203.01μm^2. The ratio of these two areas equals 0.5029, which represents a systematic mismatch of 0.6%.

In theory, matched capacitors become insensitive to process biases when their area-to-periphery ratios equal one another. In the case of two capacitors of the same value, this can be achieved by using the same geometry for both capacitors. Identical matched capacitors are usually laid out as squares because this reduces their area-to-periphery ratio, which in turn minimizes the contribution of peripheral fluctuations to their random mismatch. The problem becomes somewhat more difficult if the capacitors have values that are not in simple ratio. Although the smaller capacitor should still be laid out as a square, the larger capacitor must be laid out as a rectangle. Suppose the smaller capacitor, C_1, has dimensions L_1 by L_1

---

[8] J. C. C. Tsai, "Diffusion," in S. M. Sze, ed., p. 312.

(Figure 7.2). The larger capacitor, C_2, should have a length L_2 and a width W_2 equal to[9]

$$L_2 = \frac{C_2}{C_1}\left(1 + \sqrt{1 - \frac{C_1}{C_2}}\right) \qquad [7.7A]$$

$$W_2 = \frac{C_2}{C_1}\left(1 - \sqrt{1 - \frac{C_1}{C_2}}\right) \qquad [7.7B]$$

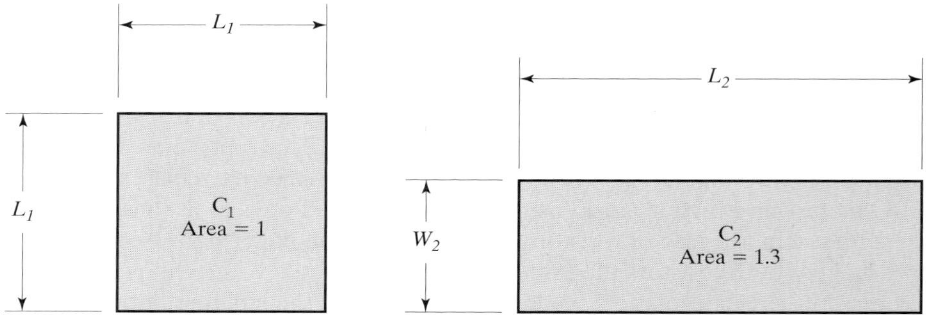

FIGURE 7.2 Matching capacitors using identical area-to-periphery ratios.

Although equations 7.7A and 7.7B theoretically eliminate systematic mismatches due to process bias, in practice things do not work out so nicely. Process biases are not constant quantities; they actually depend on the dimensions of the geometries in question. The process biases experienced by a rectangular capacitor do not precisely equal those experienced by a square capacitor. This problem becomes increasingly severe for larger ratios. Rectangular capacitors also increase the contribution of peripheral fluctuations to random mismatches. In practice, capacitors with ratios of more than 1.5:1 should not be constructed using equations 7.7A and 7.7B. In such cases, the designer should instead resort to using arrays of matched capacitor segments, or *unit capacitors.*[10]

Not all systematic mismatches are due to process biases. Other mechanisms that produce systematic mismatches include pattern shift, etch variations, diffusion interactions, mechanical stresses, thermal gradients, thermoelectrics, voltage modulation, charge spreading, and dielectric polarization. The following sections take up each of these topics in turn.

7.2.3. Pattern Shift

As discussed in Section 2.5.1, surface discontinuities left from the thermal annealing of the N-buried layer (NBL) propagate up through the monocrystalline silicon layer deposited during vapor-phase epitaxy. The resulting surface discontinuities become faintly visible under an optical microscope, particularly when lateral illumination is used. This image, called the *NBL shadow,* serves as a registration marker for the alignment of subsequent diffusions.

Process engineers have long been aware that surface discontinuities present in the substrate are not always faithfully reproduced in the final silicon surface. These discontinuities are frequently displaced laterally during epitaxial growth (Figure 7.3A). This effect is called *pattern shift.* Sometimes the various edges of

---

[9] Y. Tsividis, *Mixed Analog-Digital VLSI Devices and Technology* (New York: McGraw-Hill, 1996), pp. 220–223.

[10] M. J. McNutt, S. LeMarquis, and J. L. Dunkley, "Systematic Capacitance Matching Errors and Corrective Layout Procedures," *IEEE J. Solid-State Circuits,* Vol. 29, #5, 1994, pp. 611–616.

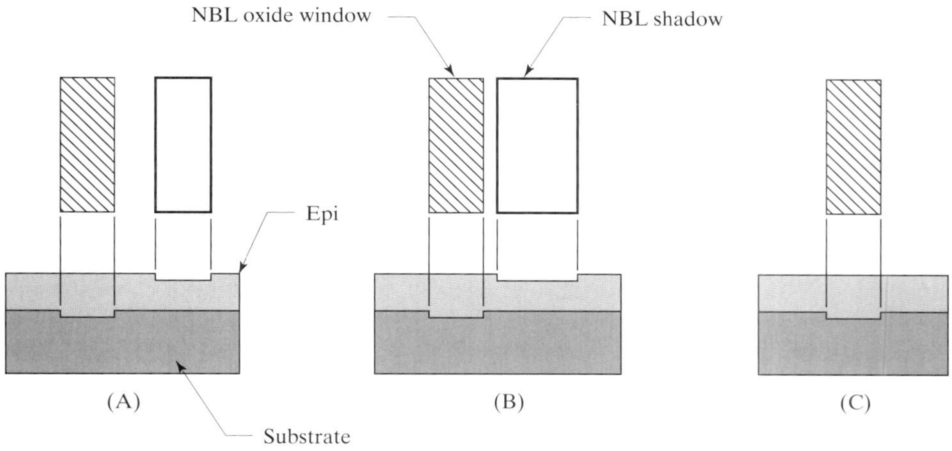

FIGURE 7.3 Effects of epitaxy on surface discontinuities: (A) pattern shift, (B) pattern distortion, and (C) pattern washout.

the discontinuity shift by different amounts, causing *pattern distortion* (Figure 7.3B). Occasionally the surface discontinuities completely vanish during the course of epitaxy, causing *pattern washout* (Figure 7.3C).

Pattern shift, distortion, and washout are all manifestations of the same underlying phenomenon. During vapor-phase epitaxy, reactant molecules adsorb on the silicon surface and move laterally until they find suitable locations where they can incorporate into the growing lattice. The exposed microsteps, formed by the intersection of the crystal lattice with the surface, encourage crystal growth in a specific direction and cause the surface topography to shift as epitaxy continues. Wafers that are (111)-oriented experience relatively severe pattern shift and distortion, which can be minimized by tilting the plane of the wafer by approximately 4° around a <110> axis.[11] Wafers that are (100)-oriented experience significant pattern distortion, but no pattern shift. The use of slightly tilted (100)-oriented wafers minimizes pattern distortion at the price of introducing pattern shift.

The magnitude of pattern shift depends upon the mobility of adsorbed reactants and the crystal orientation. Higher pressures, faster growth rates, and the presence of chlorine as a substituent in the reactant all favor increased pattern shift, while higher temperatures tend to reduce it. LPCVD deposition of dichlorosilane on (111) silicon tilted 4° will induce a pattern shift of 50 to 150% of the thickness of the epi along a <211> axis, while similar conditions using silicon tetrachloride will induce pattern shifts of 100 to 200% of the epi thickness.[12]

Due to the many variables involved, the direction and magnitude of pattern shift in a specific process can only be determined through experimental observation. A high-quality optical microscope with at least 100X magnification and a reflective illuminator are required. The shadow is usually clearest in the vicinity of minimum-geometry NPN transistors; it appears as a faint dark line not associated with any corresponding change in the oxide color. Once the shadow has been identified, the NBL shift can be estimated by comparison to features of known dimensions, such as contacts or narrow resistors. Metal lines are not recommended as a reference for comparisons because their process biases are often quite large. The deposition of interlevel oxide (ILO) obscures the NBL shadow, and planarization renders it

[11] W. R. Runyan and K. E. Bean, *Semiconductor Integrated Circuit Processing Technology* (Reading, MA: Addison-Wesley, 1994), p. 331.

[12] For further discussion, see Runyan, *et al.*, pp. 331–333.

completely invisible. Wafers used for measuring NBL shift must therefore be removed from the process prior to ILO deposition and planarization.

Pattern shift becomes a potential concern whenever matched devices are laid out in a process that employs a patterned buried layer, such as the NBL of standard bipolar and analog BiCMOS. If the NBL shadow intersects a component, it may interfere with diffusion and implantation and may cause subtle shifts in component value. Not all components are necessarily affected by pattern shift. Capacitors generally do not incorporate NBL, and polysilicon resistors usually reside over field oxide. On the other hand, diffused resistors are usually enclosed in tanks or wells containing NBL. HSR resistors are especially susceptible because the extremely thin high-sheet implant is less tolerant of slight surface discontinuities than are deeper diffusions.

Figure 7.4 depicts four matched base resistors laid out in a common tank containing NBL. Assuming this process exhibits a pattern shift to the right, then the NBL shadow intersects the leftmost resistor. There are several ways to avoid this intersection. The simplest approach consists of removing NBL from beneath the components. While this certainly eliminates the NBL shadow, it also needlessly increases the tank resistance and potentially leaves the circuit vulnerable to latchup. A better approach relies on knowledge of the direction of pattern shift. If the NBL shadow shifts to the right, then the overlap of the left edge of the NBL over the components can be increased so that the NBL shadow cannot fall across them, even with worst-case pattern shifts and misalignments. This usually requires that the NBL overlap the components by at least 120% of the nominal pattern shift. If no experimental data on the direction of the pattern shift exists, then the NBL must overlap all sides of the devices vulnerable to encroachment by the NBL shadow. In the layout in Figure 7.4, the left and right sides of the resistors are vulnerable, while the top and bottom are not. If no information on the magnitude of the pattern shift exists, then the NBL should overlap the components by at least 150% of the nominal epi thickness.

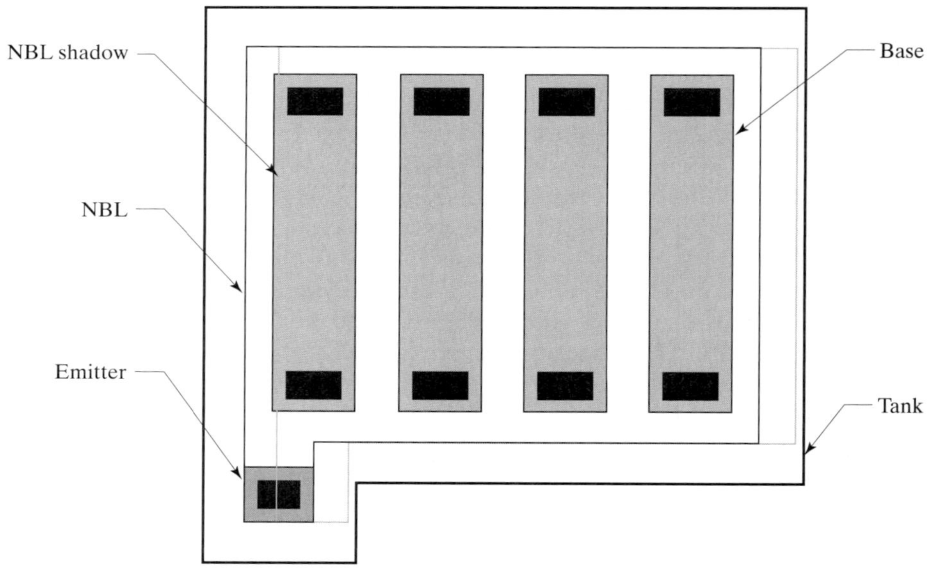

FIGURE 7.4 Layout showing an intersection of the NBL shadow with the leftmost base resistor.

7.2.4. Variations in Polysilicon Etch Rate

Poly resistors are created by etching a doped polysilicon film. The etching rate depends, at least to some extent, on the geometry of the poly openings. Larger openings grant more access to the etchant and thus clear more quickly than small openings. Consequently, sidewall erosion occurs to a greater degree around the edges of a large opening than around the edges of a small one. This effect causes widely separated poly geometries to have smaller widths than closely packed geometries do. Consider the case of three polysilicon resistors with no other nearby regions of poly (Figure 7.5). The edges of the resistors facing outward form the sidewalls of a vast opening that etches quickly and clears early. The edges of the resistors facing inward form sidewalls of narrow slits that etch more slowly and clear later. The middle resistor lacks outward-facing edges, so it has a slightly larger final width than either of the other resistors.

FIGURE 7.5 Variations in etch rate in an array of supposedly matched resistors. The exposed outer edges of the resistors experience overetching relative to the protected inner edges.

Although small, these variations in etch rate are sufficient to produce serious systematic mismatches. Suppose that a 10kÅ polysilicon film etches 90% anisotropically, resulting in an undercut of 0.1μm. The difference between the undercutting of the outward-facing and inward-facing edges is unlikely to be more than a small fraction of the total undercut, perhaps 0.02μm. As small as this value seems, it still represents 0.5% of the width of a 4μm resistor.

When a number of polysilicon strips are arrayed side-by-side, only the strips on the ends of the array experience etch rate variations. *Dummy resistors* (or *etch guards*) are often added to either end of an array of matched resistors to ensure uniform etching (Figure 7.6).[13] The dummy resistors may be constructed in one of two ways. *Unconnected dummies* are simply strips of polysilicon placed on either side of the array (Figure 7.6A). The spacing between the dummy segments and the adjacent resistors must match the spacing between the resistors of the array. The width of poly geometries has little effect on their etch rates, so the dummies can be made much narrower than the resistors they protect. This scheme has the slight disadvantage that the dummy segments remain electrically unconnected. Since the oxide isolating these segments is an exceptionally good insulator, a static electrical charge

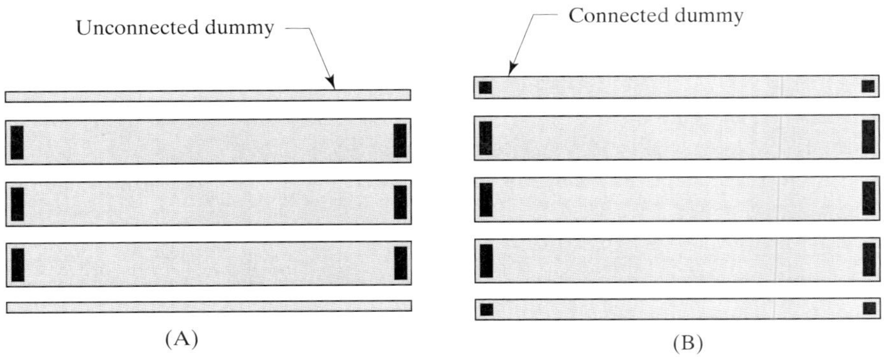

FIGURE 7.6 Examples of (A) unconnected dummy resistors and (B) connected dummy resistors.

[13] Y. Tsividis, pp. 229–231. Tsividis's structure includes dummies, but is not connected to neutralize thermoelectrics (Section 7.1.6).

can accumulate on the dummy segments. This charge might influence the behavior of adjacent resistors. Any possibility of electrostatic modulation can be eliminated by connecting the dummies to ground or to some other suitable low-impedance node (Figure 7.6B).

A less-desirable style of dummy resistor consists of a continuous ring of poly looped around the resistor array. Dry etching employs intense electromagnetic fields to generate and direct the reactive ions. These fields can interact with the loop of polysilicon to produce circulating currents that could affect etch rates during the final minutes of etching. Instead of an unbroken loop, consider using separate dummy segments like those in Figure 7.6. If a loop must be employed, then a gap should be left at some point to interrupt circulating currents.

Poly-poly capacitors experience the same etch rate variations as poly resistors. When matching arrays of capacitors, additional dummy capacitors should be placed around the edges of the array.[14] Figure 7.7 shows an array of six matched poly-poly capacitors with grounded dummies. Individual strips are again employed instead of a continuous ring. Notice that the poly-2 (shown in dark gray) has been drawn so that the spacing from dummy poly-2 to capacitor poly-2 matches the spacing between capacitor poly-2 regions. The dummies should always be electrically connected so that they can shield the matched capacitors from stray electrostatic fields (Section 7.2.8).

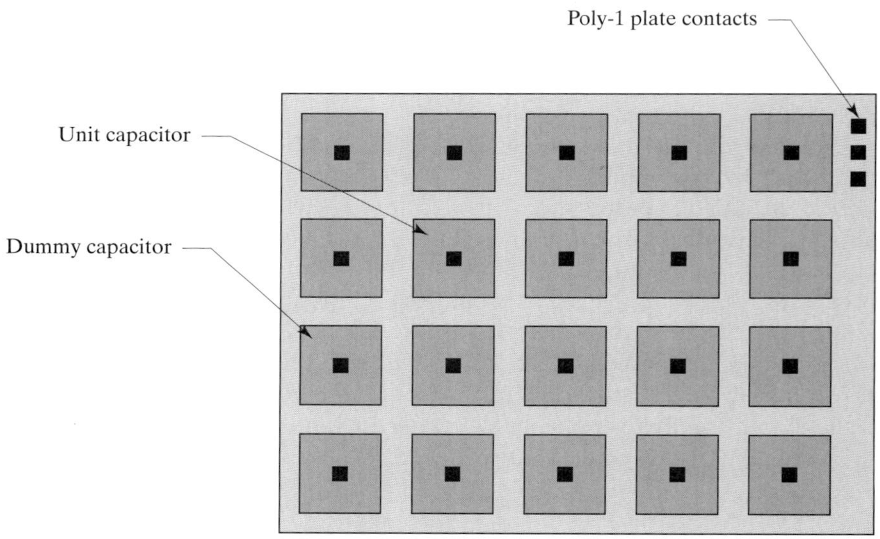

FIGURE 7.7 Matched capacitor array employing grounded dummies. The metal-2 electrostatic shield covering this array has been omitted for clarity.

Designers often use duplicates of the matched capacitors as dummy devices, presumably to provide better matching. The size of the dummies actually has no significant effect upon etch rates. As long as the array is covered by a metal plate to block fringing fields (Section 7.2.8), there is no need to use full-size dummy capacitors.

7.2.5. Diffusion Interactions

The dopants that form a diffusion do not all reside within the boundaries of its junction. Consider a P-type diffusion made into an N-type epi. The concentration of acceptors in the middle of the diffusion greatly exceeds the concentration of donors, so this region of silicon is P-type. The acceptor concentration decreases as one moves outward, but the donor concentration remains constant. At the *metallurgical*

---

[14] Y. Tsividis, p. 228.

junction, the acceptor concentration exactly equals the donor concentration. Beyond the junction, the acceptor concentration drops below the donor concentration and the silicon becomes N-type. The acceptor concentration falls to negligible levels some distance outside the junction. The portion of the dopant that lies outside the metallurgical junction is called the *tail* of the diffusion.

The tails of two adjacent diffusions will intersect one another. Assuming that both diffusions are of the same polarity, their tails add and each diffusion reinforces the other. Both diffusions will have slightly lower sheet resistances and slightly greater widths than they would have had in isolation from one another. Exactly the opposite situation occurs if the diffusions are of opposite polarities. The two intersecting tails counterdope one another, causing both diffusions to have slightly higher sheet resistances and slightly narrower widths.

The effects of diffusion interactions on matching resemble those discussed previously for variations in poly etch rate. The resistors occupying the ends of the array will have slightly different values than the resistors occupying the middle of the array. This source of systematic mismatch can be eliminated by adding dummy resistors to either end of the array. These diffused dummy resistors must have exactly the same width as the other resistors to ensure that their dopant profiles match. The dummy resistors should be connected to prevent the formation of floating diffusions that could exacerbate latchup sensitivity (Figure 7.8).

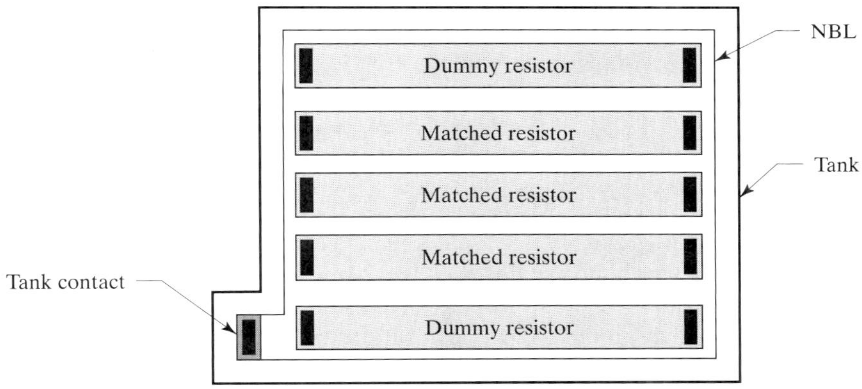

FIGURE 7.8 Matched array of diffused base resistors including grounded dummies.

Layouts, even those of ordinary unmatched resistors, can frequently be designed to eliminate diffusion interactions without significantly increasing die area. For example, Figure 7.9A shows a serpentine resistor that has been rather carelessly laid out. Not only are the spacings between the turns inconsistent, but the base heads also lie immediately adjacent to the resistor body. The layout of Figure 7.9B minimizes these potential sources of diffusion interactions by extending the heads of the resistor slightly outside the array. This modification requires little or no additional area because the tighter packing in the serpentine helps compensate for the room consumed by the extended base heads.

The layout of Figure 7.9C illustrates another type of diffusion interaction. An HSR resistor has been merged into a large tank containing a deep-N+ sinker. The deep-N+ diffusion is located immediately adjacent to the resistor body. Deep-N+ outdiffuses further than most diffusions and so has more opportunity to produce undesired interactions. The heavy phosphorus doping of the deep-N+ region can spawn lattice defects that can enhance diffusion rates in adjacent areas by a mechanism similar to emitter push (Section 2.4.2). Severe diffusion interactions may occur when these two mechanisms reinforce each other. Figure 7.9D shows a more prudent layout that places the deep-N+ sinker behind one head of the resistor, where it will have little or no effect on the resistance of the device.

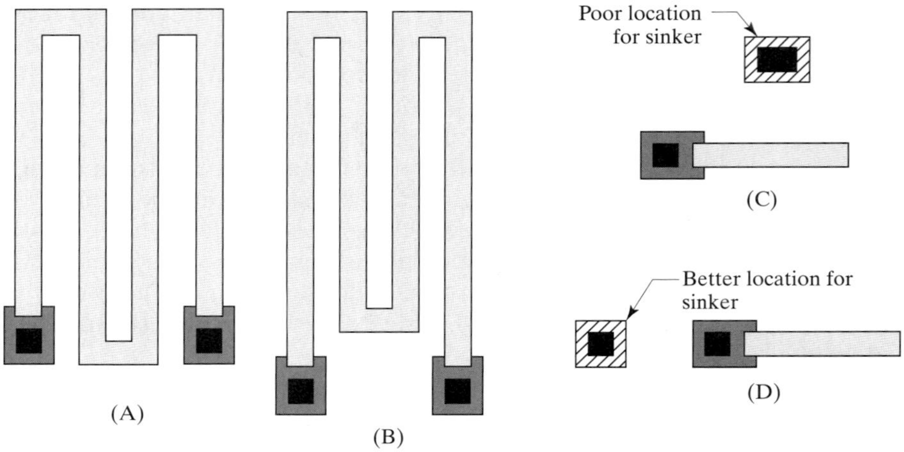

FIGURE 7.9 Examples of additional opportunities for reducing diffusion interactions: (A) poor serpentine resistor layout and (B) improved layout; (C) poor placement of deep-N+ sinker and (D) improved placement.

7.2.6. Stress Gradients and Package Shifts

Silicon is *piezoresistive* in that it exhibits changes in resistivity under stress. Variations in stress across the die will produce corresponding variations in resistor matching. Capacitors generally remain unaffected by stresses because the dimensions and permittivity of the dielectric are relatively insensitive to mechanical stress. Although well-matched capacitors usually exhibit less systematic mismatch than well-matched resistors, not all circuits can rely on matched capacitors. Layout techniques have been developed for minimizing the stress sensitivity of resistors. The severity of the problem varies from one design to the next because different forms of packaging produce higher or lower levels of stress.

Metal cans have been used to package semiconductors from the earliest days of the industry. Although they are expensive, they can provide a reliable low-stress means of packaging dice. In this type of package, the die rests on a metal plate called a *header.* After wirebonding, the package is hermetically sealed by welding a metal cap over the header. A metal can package induces little or no residual die stress as long as the die is mounted using epoxy rather than solder or gold eutectic. Although the metal header and the silicon die have vastly different coefficients of thermal expansion (Table 7.1), the epoxy absorbs the resulting strain and provides mechanical compliance. Precision integrated circuits that require extremely precise matching are often packaged in metal cans or in hermetically sealed ceramic packages that offer similar benefits.

TABLE 7.1 Coefficients of thermal expansion (CTE) for several materials used in packaging integrated circuits.[15]

Material	CTE ppm/°C
Epoxy encapsulation (typical)	24
Copper alloys	16–18
Alloy-42	4.5
Molybdenum	2.5
Silicon	2.5

[15] Values for epoxy, molybdenum, and silicon: R. E. Thomas, "Stress-Induced Deformation of Aluminum Metallization in Plastic Molded Semiconductor Devices," *IEEE Trans. on Components, Hybrids, and Manufacturing Technology,* Vol. CHMT-8, #4, 1985, pp. 427–434. Values for Alloy 42 and copper from "Leadframe Materials" *Semiconductor Reliability News,* Vol. 8, #9, September 1996, p. 5.

Plastic packages are much more widely used than either metal cans or ceramic packages. As Table 7.1 indicates, the coefficient of thermal expansion of plastic encapsulants is approximately ten times that of silicon. The epoxy resin flows into the mold at high temperature. As the encapsulated device cools, the difference between the coefficients of thermal expansion of silicon and epoxy generates residual stresses that remain permanently frozen into the packaged device. Measurements of electrical parameters before and after packaging will reveal differences called *package shifts* proportional to the level of residual stress. The input offset voltages of operational amplifiers and comparators and the output voltage of references often exhibit package shifts (Section 9.2.5). Package shifts set a lower limit on achievable accuracy that cannot be sidestepped through wafer-level trimming. Careful layout can usually reduce the sensitivity of a circuit to mechanical stress, which can in turn reduce the magnitude of the package shifts.

Power packaging requires an intimate thermal union between the die and its leadframe or its header in order to minimize heat build-up. The die attach for a power package consists of either silver-filled epoxy, solder, or gold eutectic. Silver-filled epoxies do not provide quite as good a thermal and electrical union as either of the other two alternatives, but epoxy mounting is desirable for precision devices because it produces lower residual stresses. Solder mounting is commonly employed in large metal tab or can packages. Since solder does not adhere to silicon, the back side of the die must be plated with a sputtered or evaporated metal film. Gold eutectic bonding uses a thin strip of gold foil called a *gold preform* placed between the die and the header. When heated, the gold preform alloys to both materials. Neither solder nor gold alloy has much mechanical compliance, so the thermal mismatch between the copper header and the silicon die generates extreme stresses.

Several methods exist for minimizing package stress. Although "low-stress" mold compounds exist, these do not seem to reduce packing stresses sufficiently to justify their use. A better procedure consists of coating each chip with polyimide resin prior to encapsulation. The polyimide overcoat provides mechanical compliance between the mold compound and the die. If used in conjunction with an epoxy die attach, polyimide overcoating can substantially reduce stress levels in plastic. The special equipment required and additional process time consumed make this an expensive proposition. Stresses due to solder or gold eutectic mounting can be minimized by constructing the header out of a material with a coefficient of thermal expansion similar to that of silicon, such as molybdenum or alloy-42 (a nickel-iron alloy containing 42% nickel). Molybdenum headers are expensive, while alloy-42 is brittle and exhibits poor thermal and electrical conductivity. The vast majority of products use plastic encapsulation in combination with copper alloy headers or leadframes, so the remainder of this section discusses the effects of this type of packaging.

Piezoresistivity

The piezoresistivity of (100)-oriented silicon varies with orientation and doping.[16] An N-type (100) silicon wafer exhibits maximum piezoresistivity along [100] directions and minimum piezoresistivity along <110> axes. N-type diffused and implanted resistors therefore exhibit minimum stress sensitivity if they lie along <110>

---

[16] Y. Kanda, "A Graphical Representation of the Piezoresistance Coefficients in Silicon," *IEEE Trans. on Electron Devices,* Vol. ED-29, #1, 1982, pp. 64–70.

axes. One of the <110> axes of the wafer lies parallel to the major wafer flat, while the other <110> axis lies perpendicular to it (Figure 7.10). Since dice are laid out in rows and columns relative to the wafer flat, the X- and Y-axes of the layout correspond to the desired <110> directions. The stress sensitivity of N-type monocrystalline resistors can therefore be minimized by laying them out either horizontally or vertically.

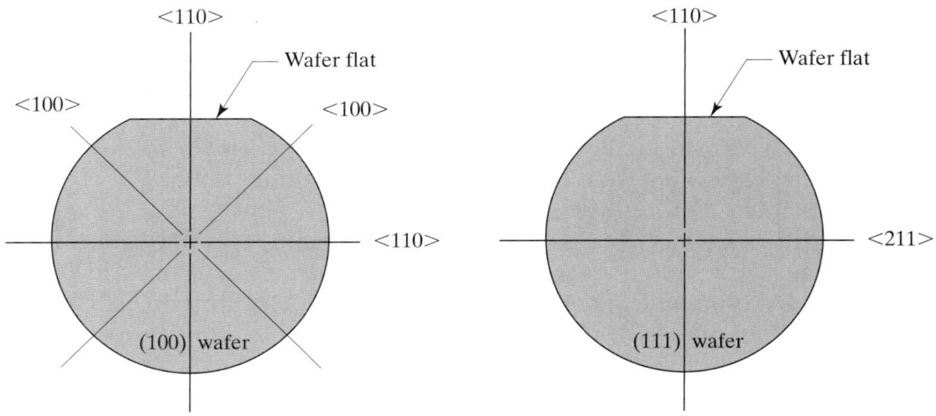

FIGURE 7.10 Identification of directions on (100) and (111) wafers.

A P-type (100) silicon wafer exhibits maximum piezoresistivity along <110> axes and minimum piezoresistivity along <100> axes. P-type diffused and implanted resistors therefore exhibit the least stress sensitivity if they lie along <100> axes. The <100> axes of a (100) wafer are rotated 45° to the wafer flat (Figure 7.10). P-type monocrystalline resistors therefore exhibit the least stress sensitivity if they are placed at 45° to the X- and Y-axes of the layout. When so oriented, the piezoresistivity of P-type monocrystalline resistors actually falls to zero. This does not occur with N-type resistors, which still retain some degree of piezoresistivity even when oriented in the optimum directions. This difference constitutes one reason for preferring P-type monocrystalline resistors to their N-type counterparts.

The piezoresistivity of a (111) wafer does not vary with direction. Although no reason exists to prefer one orientation over another, most resistors on (111) silicon are placed either vertically or horizontally to simplify packing and interconnection.

The piezoresistivity of monocrystalline silicon exhibits little dependence on doping concentrations as long as these do not exceed about 10^{18} atoms/cm^3. Almost all matched resistors use significantly lower dopant concentrations, so low-sheet and high-sheet materials exhibit no significant differences in piezoresistivity.

Polycrystalline silicon is an isotropic material, so its piezoresistivity is the same in all directions. The magnitude of this piezoresistivity drops as the resistivity of the poly increases.[17] Lightly doped poly of the sort normally used to make resistors has a relatively small (but nonzero) stress dependency. The [100]-oriented P-type diffused resistors on (100) silicon will have lower stress sensitivities than poly resistors, but the poly resistors may have better overall matching because they do not exhibit the voltage modulation problems that plague most types of diffused and implanted resistors (Section 7.2.8).

---

[17] H. Mikoshiba, "Stress-Sensitive Properties of Silicon-Gate MOS Devices," *Solid-State Electronics,* Vol. 24, 1981, pp. 221–232.

Gradients and Centroids

Figure 7.11 graphically illustrates the stress distribution across a typical integrated circuit. The drawing at the lower left is called an *isobaric contour plot.* The curved lines (called *isobars*) indicate the stress levels at various points on the die surface. Each isobar passes through a series of points of equal stress intensity. The stress intensity rises from a broad minimum in the middle of the die to maxima at the four corners. The graph above the contour plot shows the stress intensity along a line bisecting the die horizontally, while the graph at the right shows the stress intensity along a line bisecting the die vertically. By comparing the two graphs to the contour plot from which they were generated, the general nature of the latter should become apparent. Contour plots resembling this one are used in topographic mapping to display the three-dimensional shapes of hills and valleys. The distribution of stress on a die resembles a depression: it is lowest in the middle of the die and highest at the four corners.

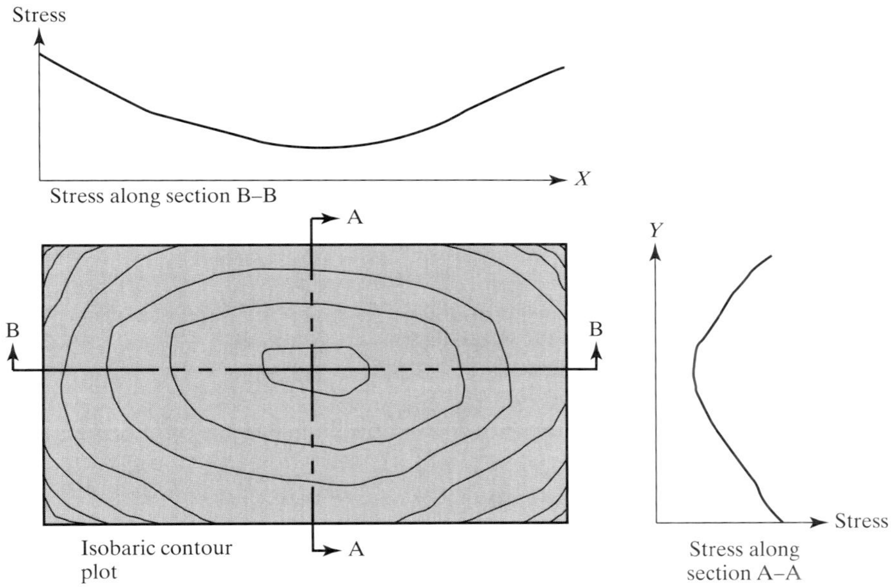

FIGURE 7.11 Isobaric contour plot of the stress distribution across the surface of a typical epoxy-mounted plastic-packaged (100) silicon die, together with two graphs showing the stress along section lines A–A and B–B.

The spacing of the isobars provides additional information about the stress distribution. The stress intensity changes rapidly where the isobars are spaced closely together, and slowly where they are spaced far apart. The rate of change of the stress intensity is called the *stress gradient.* This gradient is usually smallest in the middle of the die and slowly increases as one moves out toward the edges. The stress gradient is usually much greater at the extreme corners of the die than at any other point (Figure 7.12).

Matched devices should reside as close to one another as possible to minimize the difference in stresses between them. Although the finite size of the devices might seem to limit how closely they can be placed, certain layout techniques can produce remarkably small separations. The following analysis assumes that the stress gradient is approximately constant in the region between the matched devices. This is generally a reasonable assumption providing that the matched devices are placed to form as compact a structure as possible.

The stress difference between two matched devices is proportional to the product of the stress gradient and the separation between them. For the purposes of this

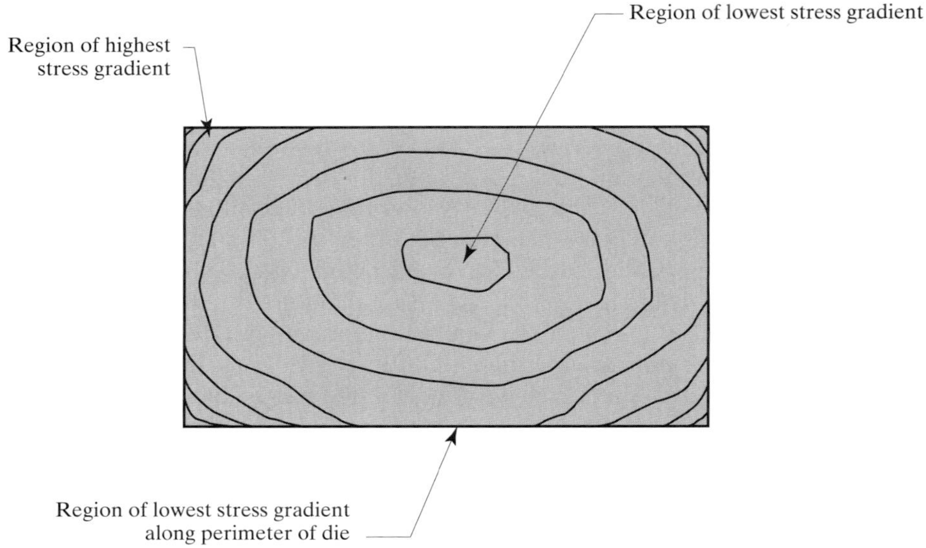

FIGURE 7.12 Isobaric contour plot showing regions of highest and lowest stress gradient.

calculation, the location of each device is computed by averaging the contribution of each portion of the device to the whole.[18] The resulting location is called the *centroid* of the device. The centroid of a rectangular device lies in its exact center. The centroids of other geometries can often be located by applying the *principle of centroidal symmetry,* which states that the centroid of a geometry must lie on any axis of symmetry of that geometry. Figure 7.13 shows how this principle can determine the centroids of a rectangle and a dogbone resistor. The centroid of practically any geometry used in layout can be determined in a similar manner.

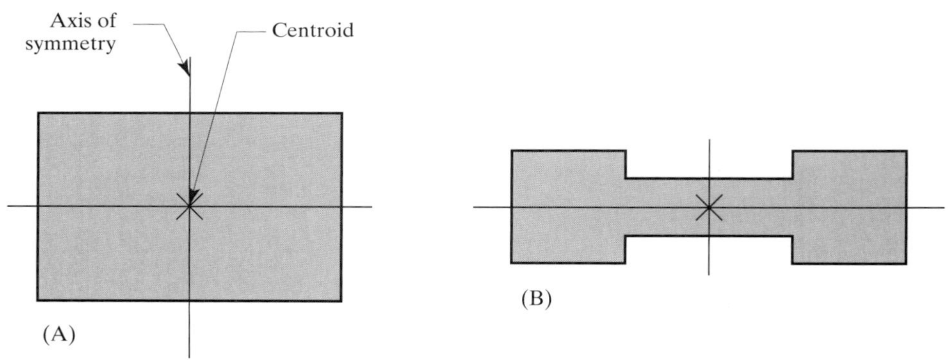

FIGURE 7.13 Locating (A) the centroids of a rectangle and (B) a dogbone resistor.

The effects of stress on resistors can be quantified in terms of piezoresistivity, centroid locations, and stress gradients. The magnitude of the stress-induced mismatch δ_s between two resistors equals

$$\delta_s = \pi_{cc} d_{cc} \nabla S_{cc} \qquad [7.8]$$

where π_{cc} is the piezoresistivity along a line connecting the centroids of the two matched devices, ∇S_{cc} is the stress gradient along this same line, and d_{cc} equals the

[18] Most texts on statics include a discussion of centroids and their relationship to the well-known principle of moments, *e.g.,* R. C. Hibbeler, *Engineering Mechanics: Statics,* 4th ed. (New York: Macmillian Publishing Co., 1998), p. 435.

distance between the centroids. This formula reveals several ways to minimize stress sensitivity. First, the designer can reduce the piezoresistivity, π_{cc}, by choosing a suitable resistance material or by orienting the resistors in the direction of minimum piezoresistivity. Second, the designer can reduce the magnitude of the stress gradient, ∇S_{cc}, by proper location of the devices and by selecting low-stress packaging materials. Third, the designer can reduce the separation between the centroids of the device, d_{cc}. The first two options have already been discussed, so we will now focus on reducing the separation of the centroids.

Common-centroid Layout

Suppose that a matched device is divided into sections. If these sections are all identical, and if they are arranged to form a symmetric pattern, then the centroid of the device will lie at the intersection of the axes of symmetry passing through the array. It is actually possible to arrange two arrayed devices so that they share common axes of symmetry. If this can be achieved, then the principle of centroidal symmetry ensures that the centroids of the two devices coincide. Figure 7.14A shows an example of such a *common-centroid* layout. The two devices are marked A and B, their axes of symmetry are shown as dotted lines, and their centroids are denoted by an "X" where the two axes of symmetry intersect.

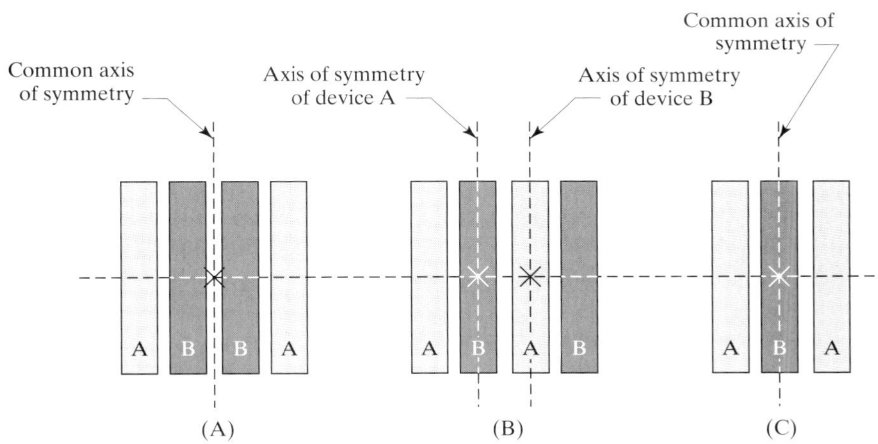

FIGURE 7.14 Examples of one-dimensional common-centroid arrays.

Equation 7.8 predicts that the stress-induced mismatch of a common-centroid layout equals zero because the separation between the centroids equals zero. This does not actually occur, because the stress gradient never remains precisely constant. Despite this limitation, common-centroid layout is still the single most powerful technique available for minimizing stress-induced mismatches. Well-designed common-centroid layouts can reduce the stress sensitivity of large devices by an order of magnitude or more.

Figure 7.14 shows three examples of common-centroid layouts produced by arraying segments of matched devices along one dimension. These types of layouts are usually called *interdigitated arrays* because the sections of one device interpenetrate the sections of the other like the intermeshed fingers of two hands. Figure 7.14A shows an interdigitated array consisting of two devices, each composed of two segments. If the devices are denoted by the letters A and B, then the arrangement of the segments follows the *interdigitation pattern* (or *weave*) ABBA. This pattern has an axis of symmetry that divides it into two mirror-image halves (AB and BA). A second axis of symmetry passes horizontally through the array, but this axis results from the symmetries of the individual segments rather than the symmetry of the interdigitation pattern.

Arrays that use the interdigitation pattern ABBA require dummies because the segments of one device occupy both ends of the array. Some designers prefer to use arrays that follow the pattern ABAB (Figure 7.14B) because they mistakenly believe that this pattern eliminates the need for dummies. This belief is founded on a misunderstanding of the function of the dummy devices. Etch variations and diffusion interactions depend on the arrangement of adjacent geometries. The addition of dummy devices ensures that each segment faces an identical arrangement of geometries. If the dummies are omitted, then the segments on the ends of the array face whatever geometries happen to reside nearby. The geometries adjacent to one end of the array are likely to differ from those next to the other. If this happens, then mismatches will result if dummies are omitted regardless of whether the array follows the pattern ABBA or ABAB. The pattern ABAB should be avoided because it does not completely align the centroids of the two devices, and the resulting separation of the centroids leaves the devices vulnerable to stress-induced mimatches.

Common-centroid layouts can also consist of devices of different sizes. Figure 7.14C shows an example that implements a 2:1 ratio using the pattern ABA. If the two devices on the end of the array are resistors, then they can connect either in series or in parallel. If they are capacitors, they can only connect in parallel because a series connection would introduce mismatches caused by the differences in parasitic capacitances between the upper and lower plates. More complicated patterns offer a larger number of possible ratios, especially if one considers that resistors can connect in series as well as in parallel. Table 7.2 lists a number of additional interdigitation patterns. The patterns marked by asterisks do not provide complete alignment of the centroids. In all such cases, one can achieve full alignment by using a larger number of segments.

A	**AA**	**AAA**	**AAAA**
AB*	**ABBA**	**ABBAAB***	**ABABBABA**
ABC*	**ABCCBA**	**ABCBACBCA***	**ABCABCCBACBA**
ABCD*	**ABCDDCBA**	**ABCDBCADBCDA***	**ABCDDCBAABCDDCBA**
ABA	**ABAABA**	**ABAABAABA**	**ABAABAABAABA**
ABABA	**ABABAABABA**	**ABABAABABAABABA**	**ABABAABABAABABAABABA**
AABA*	**AABAABAA**	**AABAAABAAABA***	**AABAABAAAABAABAA**
AABAA	**AABAAAABAA**	**AABAAAABAAAABAA**	**AABAAAABAAAABAAAABAA**

TABLE 7.2 Sample interdigitation patterns for arrays having one axis of symmetry.

The process of designing an interdigitated array begins with the identification of all of the components comprising the array. Matched devices occur in groups, and all of the devices in any one group must reside in the same array. A designer cannot identify the groups of matched components in a circuit without fully understanding how the circuit operates. The circuit designer must therefore identify the groups of matched components and pass this information to the layout designer.

Once the components comprising an array have been identified, they must be divided into segments. This process is not always a simple one. The designer should first check to see if all of the values have a *greatest common factor.* For example, two resistors having values of 10kΩ and 25kΩ have a greatest common factor of 5kΩ. The array can then consist of a number of segments each equal to the greatest common factor. For example, an array of a 10kΩ and a 25kΩ resistor could be laid out using seven 5kΩ segments.

In cases where a large common factor does not exist, try using the value of the smallest device as the value of a segment. Based on this value, determine the number of segments in the other devices. If any device requires a partial segment with a value less than about 70% of a complete segment, try dividing the smallest device value by increasingly large integer numbers (2, 3, 4 . . .) until a value is found that does not require a small partial segment. For example, suppose that a 39.7kΩ and a 144.5kΩ resistor must be arrayed. If we choose a segment resistance of 39.7kΩ, then the 144.5kΩ resistor would require 3.638 segments. This requires a 63.8% partial segment, so we divide the smaller resistance by two and try a segment resistance of 19.85kΩ. The larger resistor would need 7.280 segments, which would require a 28.0% partial segment. Dividing the smaller value by three produces a segment resistance of 13.233kΩ. The larger resistor would require 10.920 segments, so the array contains no partial segment that is less than 70% of a complete segment. This array could therefore consist of thirteen segments of 13.233kΩ and one segment of 12.174kΩ. In a few cases, this procedure produces a very small segment value. The designer should try larger segment values to see if one can be found that will allow all of the matched resistors to be constructed without using any partial segment that is less than about 70% of a complete segment. In certain cases, the designer may have to tolerate the presence of a small partial segment.

Once the segment value has been found, make sure that it is not so small that it precludes reasonable matching. Matched resistor segments should never contain less than five squares, and preferably they should contain at least ten. If the array seems to require short segments, consider using sections connected in parallel as well as in series. Capacitor segments (unit capacitors) should not have dimensions of much less than $100\mu m^2$. Capacitors are always connected in parallel because a series connection inserts parasitic capacitances that disturb matching. There are a few cases in which series-connected capacitors may form part of a set of matched capacitors, but such instances should stem only from the necessities of circuit design, not from the expediencies of layout.

Partial resistor segments are best implemented using sliding contacts to enable all of the segments to have exactly the same geometry. This precaution ensures that etch variations and diffusion interactions do not cause mismatches between the partial segment and the remainder of the array. The sliding contact can also be used to adjust the value of the partial segment should the initial setting prove incorrect. Partial unit capacitors should reside at one end of the array, where they will not unduly disturb the other unit capacitors. The partial unit capacitor should be sized using equations 7.7A and 7.7B to ensure that its area-to-periphery ratio equals that of the other unit capacitors.

Once the array has been segmented, a suitable interdigitation pattern must be chosen. The best interdigitation patterns obey all four rules of common-centroid layout listed in Table 7.3. The *rule of coincidence* states that the centroids of the matched devices should coincide at least approximately. Patterns that do not achieve coincidence exhibit larger stress sensitivities than those that do. The *rule of symmetry* states that the array should be symmetric around both axes. A one-dimensional array should derive one of its axes of symmetry from its interdigitation

1. ***Coincidence:*** The centroids of the matched devices should coincide at least approximately. Ideally, the centroids should exactly coincide.
2. ***Symmetry:*** The array should be symmetric around both the X- and Y-axes. Ideally, this symmetry should arise from the placement of segments in the array and not from the symmetry of the individual segments.
3. ***Dispersion:*** The array should exhibit the highest possible degree of dispersion; in other words, the segments of each device should be distributed throughout the array as uniformly as possible.
4. ***Compactness:*** The array should be as compact as possible. Ideally, it should be nearly square.

TABLE 7.3 The four rules of common-centroid layout.

pattern. For example, an array using the pattern ABBA will have an axis of symmetry dividing it into two mirror-image halves (AB and BA). A one-dimensional array must rely on the symmetries of the individual segments to produce its second axis of symmetry. In the case of resistors and capacitors, this should not present a problem because all of the segments should have symmetric shapes.

The *rule of dispersion* states that the segments of each device should be distributed throughout the array as uniformly as possible. The degree of dispersion is often evident to the eye, but it can be partially quantified by counting the number of repeated segments (*runs*). For example, the pattern ABBAABBA contains three runs of two segments each, while the pattern ABABBABA contains only one run of two segments. The latter pattern is therefore more disperse than the former. Dispersion helps reduce the sensitivity of a common-centroid array to higher-order gradients (nonlinearities). Dispersion is therefore especially important for arrays subject to large stress gradients.

The *rule of compactness* states that the array should be as compact as possible. Ideally, it should be square, but in practice it can have an aspect ratio of 2:1 or even 3:1 without introducing any significant vulnerability. If the aspect ratio of the array exceeds 2:1, then consider breaking the array into a larger or smaller number of segments. If the array consists of a few long segments, try doubling the number of segments and halving the value of each. Arrays consisting of many short (or small) segments are excellent candidates for the two-dimensional arrays discussed below.

All of the common-centroid layouts discussed so far array the devices in only one dimension. Such a *one-dimensional array* derives one of its axes of symmetry from its interdigitation pattern and one of its axes of symmetry from the symmetry of its segments. The segments can also be arranged to form a *two-dimensional array* deriving both of its axes of symmetry from its interdigitation pattern. This type of arrangement generally provides better cancellation of gradients than one-dimensional arrays, primarily because of the superior compactness and dispersion possible within a two-dimensional array.

Figure 7.15A shows two matched devices, each composed of two segments arranged in an array of two rows and two columns. This arrangement is often called a *cross-coupled pair.* Resistors are rarely laid out as cross-coupled pairs because the resulting arrays usually have unwieldy aspect ratios. Capacitors often produce very compact cross-coupled pairs, as do diodes and transistors. If the matched devices are large enough to segment into more than two pieces, then the cross-coupled pair can be further subdivided as shown in Figure 7.15B. This array exhibits more dispersion than a cross-coupled pair and is therefore less susceptible to higher-order gradients. This two-dimensional interdigitation pattern, or *tiling,* can be indefinitely extended in both dimensions.

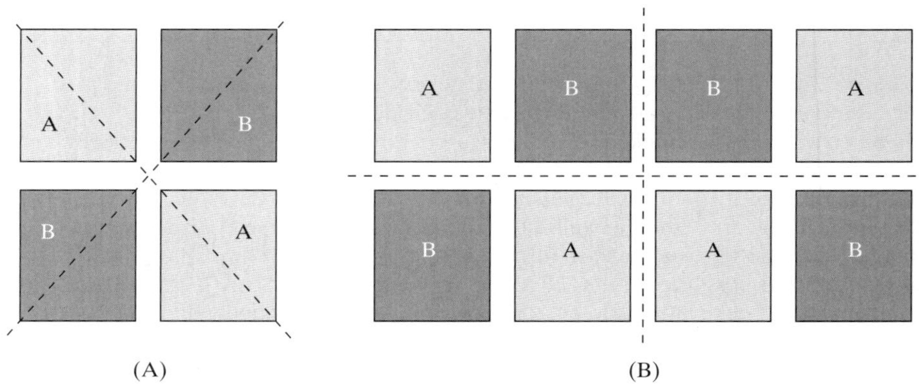

FIGURE 7.15 Examples of two-dimensional common centroid arrays.

The rules for creating one-dimensional arrays also apply to two-dimensional arrays. The sections should be arranged so that the array has two or more axes of symmetry intersecting at the point where the centroids of the matched devices coincide. Table 7.4 lists a number of sample interdigitation patterns for two-dimensional arrays. Each row in the table consists of four examples of a given pattern. These include the simplest possible array, an extension of the array in one dimension, and extensions of the array in two dimensions. Although more complex variations are possible, most two-dimensional arrays are relatively simple because capacitors and transistors—the devices most often arrayed in two dimensions—do not lend themselves to subdivision into a large number of segments.

TABLE 7.4 Sample interdigitation patterns for two-dimensional common-centroid arrays.

ABBA BAAB	**ABBAABBA BAABBAAB**	**ABBAABBA BAABBAAB ABBAABBA**	**ABBAABBA BAABBAAB BAABBAAB ABBAABBA**
ABA BAB	**ABAABA BABBAB**	**ABAABA BABBAB ABAABA**	**ABAABAABA BABBABBAB BABBABBAB ABAABAABA**
ABCCBA CBAABC	**ABCCBAABC CBAABCCBA**	**ABCCBAABC CBAABCCBA ABCCBAABC**	**ABCCBAABC CBAABCCBA CBAABCCBA ABCCBAABC**
AAB BAA	**AABBAA BAAAAB**	**AABBAA BAAAAB AABBAA**	**AABBAA BAAAAB BAAAAB AABBAA**

Location and Orientation

Any residual stress sensitivities not canceled by common-centroid layout remain proportional to the magnitude of the stress gradient. Matched devices should therefore occupy areas of the die where the stress gradients are smallest. As Figure 7.11 suggests, the stress gradient typically falls to a minimum in a broad region near the center of the die. The best locations for matched devices are therefore near the middle of the die. The stress gradient along the periphery of the die reaches a similar broad minimum in the middle of each side, with the lowest values usually appearing in the middle of the longer sides. If matched devices must reside along the die periphery, then they are best located in the middle of one of the longer sides of the die. Matched components should never be placed near corners because both stress intensity and stress gradients reach their maximum values here.

The stress distribution on the surface of a die exhibits symmetries that can be exploited to further improve matching. Most dice exhibit a symmetric stress distribution around at least one axis. In the case of (100) silicon, the stress distribution is usually symmetric around both the horizontal and vertical axes. Critically matched common-centroid arrays should be oriented so that one of their axes of symmetry aligns with either the horizontal or the vertical axis of the die (Figure 7.16A).

The picture is less clear in the case of (111) silicon because one axis of symmetry of the die lies along a <110> axis while the other lies along a <211> axis (Figure 7.16B). Some authors have suggested that the stress distribution is more symmetric

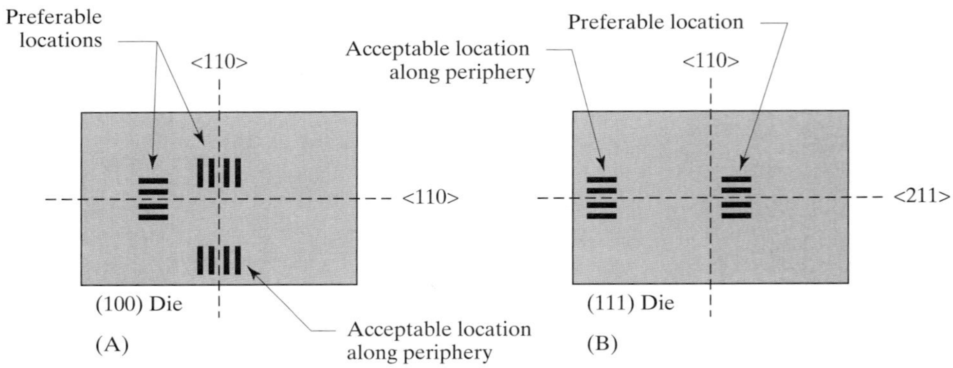

FIGURE 7.16 Locations for placing common centroid arrays on (100) and (111) dice, in the latter case assuming an axis of symmetry in the stress distribution around the <211> axis.

around the <211> axis than around the <110> axis,[19] so critically matched common-centroid arrays on (111) silicon should be oriented so that one of the axes of symmetry of the array aligns with the <211> axis of the die (Figure 7.16B).

The placement of the common-centroid array on an axis of symmetry of the stress distribution helps reduce residual mismatches by minimizing the stress gradient. If the stress distribution is symmetric around the chosen axis of symmetry, whatever stress-induced effects do occur will have opposite polarities on either side of this axis. Providing that the matched devices are also placed symmetrically around this same axis, the effects of stress on one half of the device will cancel the effects of stress on the other half. Whenever possible, critically matched devices should be placed to take advantage of this phenomenon.

The stress distribution on a die also depends on its size and shape. Larger dice generally exhibit higher levels of stress than small ones. Stress also tends to increase with aspect ratio, so elongated dice exhibit higher stress levels than square dice having similar areas. As previously mentioned, packaging also plays a major role in determining stress levels. Epoxy mounting provides mechanical compliance, which allows stresses to dissipate. Epoxy-mounted dice in metal cans or ceramic packages exhibit relatively little stress, regardless of die size or shape. The die area and aspect ratio become more important for parts encapsulated in plastic, or mounted with solder or gold eutectic. Table 7.5 offers some general guidelines for die aspect ratios in various types of packaging.

TABLE 7.5 Suggested die aspect ratios for analog layouts ($15\text{kmil}^2 \approx 9.7\text{mm}^2$).

Package Type	Die Size	Suggested Aspect Ratio	Maximum Aspect Ratio
Metal can/epoxy mount	Any	2:1 or less	Any
Plastic/epoxy mount	$<15\text{kmil}^2$	1.5:1 or less	3:1 or less
	$>15\text{kmil}^2$	1.5:1 or less	2:1 or less
Plastic/solder mount	$<15\text{kmil}^2$	1.5:1 or less	2:1 or less
	$>15\text{kmil}^2$	1.3:1 or less	1.5:1 or less

7.2.7. Temperature Gradients and Thermoelectrics

The electrical properties of many integrated components depend strongly on temperature. Most integrated resistors have temperature coefficients of 1000ppm/°C or more (Table 5.4). Assuming a temperature coefficient of 2500ppm/°C, a 1° temper-

[19] W. F. Davis, *Layout Considerations,* unpublished manuscript, 1981, pp. 66–67.

ature difference between two matched resistors produces a 0.25% mismatch. Thermal gradients of 1°C/mil (0.04°C/μm) can exist near a large power device.[20] To better understand how these thermal variations arise, we will briefly examine the concept of *thermal impedance.*

All electrical circuits dissipate some amount of power in the form of heat. This heat flows through the encapsulation out into the ambient environment. The average *junction temperature* of the die T_j equals

$$T_j = T_a + P_d\theta_{ja} \qquad [7.9]$$

where T_a is the ambient temperature of the environment, P_d is the power dissipated in the package, and θ_{ja} is a constant called the *junction-to-ambient thermal impedance.* The θ_{ja} for most plastic packages exceeds 100°C/W, limiting their power dissipation to about a watt. Specially constructed power packages exist that offer much lower thermal impedances (Table 7.6). These packages usually incorporate a metal tab or plate intended for mounting onto an external metal surface called a *heat sink.* Power packages are usually specified in terms of a *junction-to-case thermal impedance,* θ_{jc}. In this case, the average junction temperature, T_j, equals

$$T_j = T_c + P_d\theta_{jc} \qquad [7.10]$$

where T_c represents the case temperature of the package measured at a specified point on the heat tab or plate. Power packages often have remarkably low thermal impedances because of their special construction. Various manufacturers cite slightly different values due to variations in materials and manufacturing processes, but the values listed in Table 7.6 are representative of the industry.

TABLE 7.6 Typical thermal impedances for several common types of packages.[21]

Type of package	θ_{ja} (°C/W)	θ_{jc} (°C/W)
16-pin plastic dual in-line package (DIP)	110	
16-pin plastic surface-mount package (SOIC)	131	
3-lead plastic TO-220 power package		4.2
3-lead metal TO-3 can power package		2.7

One might expect the packages with the lowest thermal impedances to have the smallest thermal gradients, but in practice the exact opposite occurs. Power packages achieve their low thermal impedances by mounting the die on a heat sink. Heat flows vertically down to the heat sink and out of the package, rather than laterally across the die. Temperatures rise only where power is dissipated; other portions of the die remain at approximately the same temperature as the heat sink. Temperature differentials of up to 50° can appear across the surface of a die mounted in a power package; the thermal gradients are correspondingly large.

A package lacking a heat sink presents a very different picture. Silicon is a far better conductor of heat than epoxy, so heat flows laterally across the die until the entire die rises to a high temperature. Heat then percolates out from the die through

[20] See Figure 1 in R. J. Widlar and M. Yamatake, "Dynamic Safe-Area Protection for Power Transistors Employs Peak-Temperature Limiting," *IEEE J. Solid-State Circuits,* Vol. SC-22, #1, 1987, pp. 77–84.

[21] Values for 16-pin dip, 16-pin SOIC: *Power Supply Circuits Databook;* Texas Instruments #SLVD002, 1996, pp. 4–22. Values for TO-3, TO-220: *Power Products Data Book,* Texas Instruments #DB-029, 1990, pp. 4–72, 5–8. These values are broadly representative of the industry.

the plastic to the outside environment. The plastic package acts as a thermal insulation blanket that minimizes the magnitude of thermal gradients. Ordinary plastic packages rarely experience thermal differentials of more than a few degrees across the die unless the power dissipation of the circuit rapidly fluctuates.

Figure 7.17 shows an *isothermal contour plot* for a die containing one large heat source mounted in a power package. The curved lines on the surface of the die, called *isotherms,* represent adjacent points of equal temperature. Each isotherm represents a relatively large change in temperature, perhaps five degrees. The same general distribution of isotherms would also occur on a die in an ordinary plastic package, but the average die temperature would be much higher and the isotherms would represent smaller changes in temperature—perhaps one degree per isotherm.

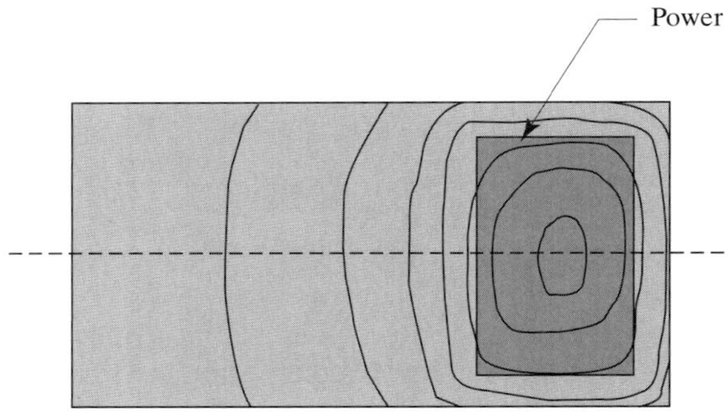

FIGURE 7.17 Isothermal contour plot of a die having only one major heat source. The axis of symmetry of the thermal distribution is marked by a dotted line.

The thermal gradients are largest around the perimeter of the power device and gradually decrease in magnitude as one moves away from it. Because the heat source has been placed symmetrically around the horizontal axis of the die, the heat distribution is also symmetric about this axis. The presence of this axis of symmetry can be used to improve the thermal matching of other components on the die.

Thermal Gradients

The relative spacing of the isotherms reflects the thermal gradient at each point on the die. The thermal gradient is large where the isotherms are spaced closely together, and it is small where they are spaced far apart. Thermal gradients are exactly analogous to the stress gradients discussed in the previous section. Assuming that the thermal gradient remains approximately constant in the vicinity of a pair of matched devices, then the thermally induced mismatch δ_T between the two devices equals

$$\delta_T = TC_1 d_{cc} \nabla T_{cc} \qquad [7.11]$$

where TC_1 is the linear temperature coefficient of the resistance material, d_{cc} is the distance between the centroids of the resistors, and ∇T_{cc} is the thermal gradient along a line connecting the centroids of the resistors.

Although common-centroid layouts are used to combat both stress and thermal gradients, their position and orientation differ depending on the application. The axes of symmetry of stress distributions are determined entirely by the packaging and therefore present rigid constraints on the layout. The axes of symmetry of thermal distributions are determined by the position and orientation of the power devices. The magnitude of thermally induced variations can be minimized by proper placement of the matched devices relative to the power devices.

Most dice contain only a few major heat sources, which are usually large bipolar or MOS power transistors. Whenever possible, these devices should lie upon an axis of the die in order to produce a symmetric thermal distribution. They should also lie as far as possible from critical matched devices. Consider the case of a die containing one power device (Figure 7.18A). This power device would ideally occupy one end of the die and align around an axis of symmetry of the die. This arrangement is preferable to placing the power device centrally because it allows more separation between the power device and critical matched devices. The placement of the matched devices is a compromise between stress effects—which favor a central location—and thermal effects—which favor placement as far away from the power device as possible. The optimal arrangement places the matched devices about halfway across the portion of the die not occupied by the power transistor (Figure 7.18A). A larger separation between the power device and the matched pair can sometimes be achieved by elongating the die to an aspect ratio of 1.3 or even 1.5. Although elongation increases stress levels, the improvement in matching due to increased separation may actually outweigh the effect of increased stresses. If matched devices must reside along the periphery of the die, then they should occupy the middle of the side opposite the power device. In this case, the aspect ratio of the die should be moderated to limit the effects of stress on the less-optimal location of the matched devices. Although the effects discussed above are difficult to quantify, the arrangements advocated here have been successfully used on many designs.

The layout of Figure 7.18A can easily be extended to incorporate two power transistors instead of one. Ideally the transistors would be placed one behind another, both at one end of the die and both on the same axis of symmetry (Figure 7.18B). This yields a layout resembling that of Figure 7.18A. Unfortunately, this arrangement often produces difficulties in routing the leads from the power devices to their respective pins. Many designs place the two power devices in adjacent corners of the die located on the opposite end from the matched devices. This arrangement has the advantage of separating the heat sources from the matched devices by the largest possible distance. However, the disadvantage is that the heat distribution is now asymmetric unless the devices operate at identical power levels. Another possible layout for two heat sources is shown in Figure 7.18C. This arrangement places one power device on each end of the die and the matched devices in the middle. This arrangement will probably prove satisfactory as long as a separation of at least 20 to 30mils can be achieved. It has the advantage of preserving a thermal axis of symmetry regardless of the dissipation levels in the individual power devices, and it locates the matched devices in the center of the die where the stress gradients are lowest.

The layout of Figure 7.18C can be extended to include additional devices. Figure 7.18D shows an example containing four power devices. This arrangement usually suffers from a lack of separation between the power devices and the matched devices. This problem can be partially remedied by increasing the aspect ratio of the die to 1.5:1 or even 2:1. Even a large aspect ratio will not necessarily affect matching because the power devices occupy the ends of the die where the stresses are greatest. Aspect ratios larger than 1.5:1 are somewhat risky for solder or gold eutectic mounting if the die's longer dimension exceeds 150 mils (3.8mm). The stresses accumulating in the corners of such a long die may actually cause mechanical damage to the metal system or the bond wires. Epoxy die mounting provides additional mechanical compliance and therefore allows larger aspect ratios.

A variety of considerations often constrain the placement of power devices. These include the location of bondpads, the routing of power buses and the placement of control circuitry. The compromises required to satisfy all of these constraints usually lead to a less-than-optimal layout. This does not necessarily constitute an insurmountable problem because common-centroid layout techniques can greatly reduce the impact of the remaining thermal mismatches.

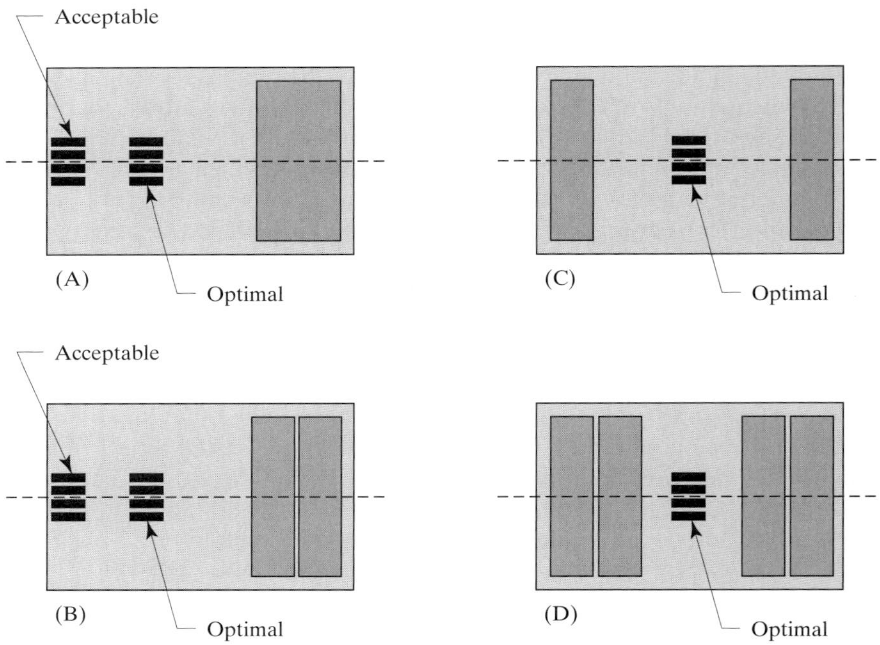

FIGURE 7.18 Various arrangements of one, two, and four power devices for optimal thermal matching. The power devices are shown as dark gray rectangles, and the axes of symmetry created by their placement are shown as dotted lines.

Thermoelectric Effects

Resistors display two distinct types of thermal variation. One is caused by the temperature coefficient of the resistance material. Common-centroid layout techniques can ensure that the average temperatures of two resistors track one another, so even materials with large temperature coefficients will match quite precisely. The other source of thermal variation is the *Seebeck effect,* also called the *thermoelectric effect.* As discussed in Section 1.2.5, a voltage differential called the *contact potential* arises whenever two dissimilar materials come in contact with one another. The contact potentials of metal/semiconductor junctions are strong functions of temperature, so if the contacts are held at different temperatures, a net voltage difference will appear across the resistor. This *thermoelectric potential* E_T equals

$$E_T = S\Delta T_c \qquad [7.12]$$

where S is the *Seebeck coefficient* (typically about 0.4mV/°C), and ΔT_c is the temperature difference between the two contacts of the resistor. A temperature difference of 1° across a resistor will thus generate a voltage differential of about 0.4mV between its contacts. This may appear inconsequentially small, but certain types of circuitry are extremely vulnerable to small voltage offsets. For example, a 0.4mV offset in a bipolar current mirror produces a 1.5% mismatch in the currents.

Common-centroid layout cannot eliminate thermoelectrics because they arise from differences in temperature between the ends of each resistor segment. Improperly arraying the device actually compounds the problem. The individual thermoelectric potentials generated by each segment of the resistor array of Figure 7.19A add to produce an overall thermoelectric potential far larger than that of any one segment. The thermoelectric potentials of the individual segments can be canceled by reconnecting them as shown in Figure 7.19B.

In order to obtain complete cancellation, the resistor should consist of an even number of segments, half connected in one direction and half connected in the other

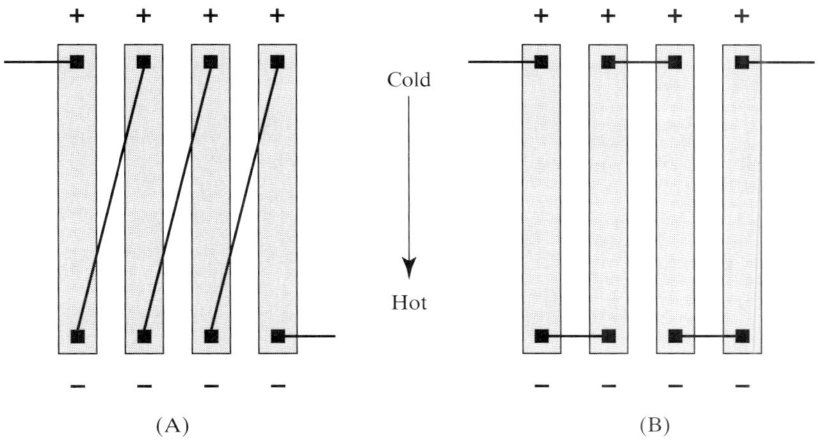

FIGURE 7.19 (A) Improper connection of resistor segments causes their thermoelectric potentials to add, while (B) proper connection of the segments cancels the thermoelectrics.

(Figure 7.19B). If the resistor has an odd number of segments, then one segment cannot be paired. Critically matched resistors should, if possible, consist of an even number of segments, but less-sensitive resistors can tolerate the presence of an unpaired segment.

The two contacts of a serpentine resistor should reside as close to one another as possible to minimize the impact of thermoelectrics. The serpentine resistor of Figure 7.20A will have unnecessarily large thermal variations due to an excessive separation between its contacts. The layout of Figure 7.20B reduces thermal variability and improves matching by bringing the resistor heads into closer proximity. However, this layout is vulnerable to misalignment errors. If the resistor body shifts downward relative to the resistor heads, then the length of the resistor increases by twice the misalignment. This vulnerability can be eliminated by orienting the resistor heads in opposite directions, so that any shift that increases the length of the resistor protruding from one head reduces the length protruding from the other. The layout of Figure 7.20C eliminates the misalignment vulnerability, but it places the base heads adjacent to stretches of the resistor body, which can lead to diffusion interactions. It is difficult to eliminate this minor defect without introducing more serious problems in the process.

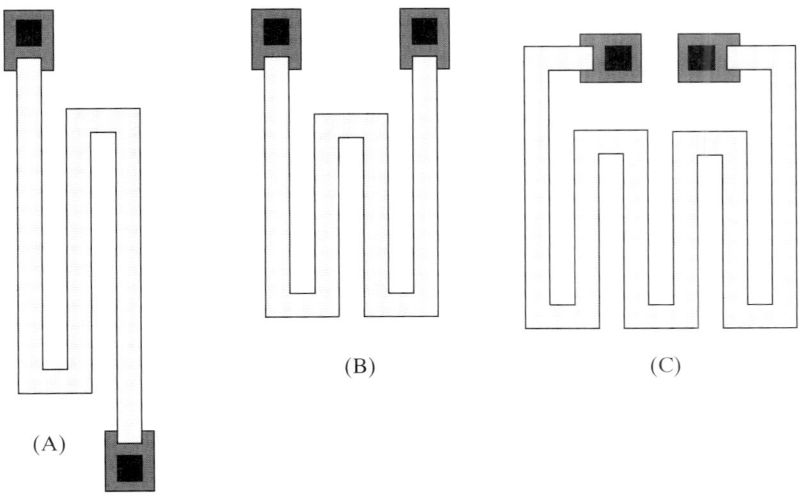

FIGURE 7.20 An HSR resistor with widely separated contacts (A) is prone to thermoelectric-induced offsets. Placing the contacts close together (B) minimizes thermoelectrics, but may increase variation due to misalignment. Placing the contacts close together and oriented in opposite directions (C) fixes both problems.

7.2.8. Electrostatic Interactions

Electrostatic fields can influence the value of both resistors and capacitors. An electrostatic field can cause depletion or accumulation of carriers within a resistive material. Most integrated resistors consist of lightly doped silicon and are therefore quite susceptible to voltage modulation. The electrostatic coupling of a capacitor to surrounding circuitry via fringing fields can cause unexpected variations in capacitance. Electrostatic fields can also couple noise into sensitive high-impedance nodes that often exist within arrays of matched resistors and capacitors.

The principal types of electrostatic interactions observed in resistors are voltage modulation, charge spreading, and dielectric polarization. The major electrostatic interactions in capacitors are fringing fields and dielectric relaxation. The following sections cover each of these mechanisms.

Voltage Modulation

The value of a resistor can be affected by the voltages present on adjacent nodes of the circuit. The value of a diffused resistor may vary with the voltage differential between the tank and the resistor body, an effect called *tank modulation* (Section 5.3.3). The mismatches caused by tank modulation in two or more diffused resistors cancel as long as the tank-to-body voltages of the resistors track one another. Matched resistors can safely reside in a common tank as long as they are of equal values and they experience the same bias. Otherwise, each resistor should occupy its own tank. The individual tanks must connect so that each resistor sees the same tank-to-body differential voltage. This is most easily accomplished by tying the positive end of each resistor to its respective tank. If the resistors are of different values, they all must be divided into segments of equal (or approximately equal) value, and each segment must reside in its own independently biased tank. Similarly, each section of an interdigitated resistor requires its own tank (Figure 7.21).

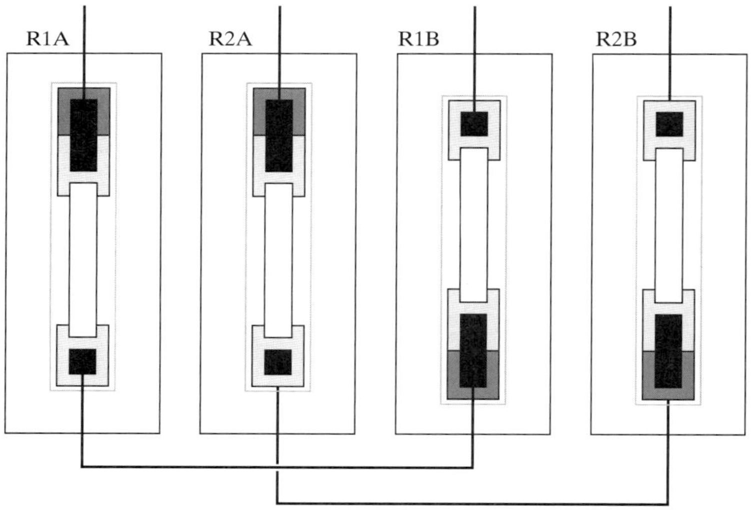

FIGURE 7.21 Two HSR resistors connected in order to cancel both thermoelectrics and tank modulation effects.

Separate tanks require enormous amounts of die area. Poly or thin-film resistors will surely provide a more compact solution. If deposited resistors are not available, consider whether the circuit can tolerate a small amount of voltage modulation. If the voltage across the matched resistors is directly (or even indirectly) trimmed, this may also compensate for voltage modulation. This is often the case for voltage references and regulators. The matched resistors should still be interdigitated to prevent thermal and stress gradients from causing package shifts and thermal drifts.

Many applications can tolerate a small amount of voltage modulation. The magnitude of the resulting systematic mismatches can be minimized by using relatively low-sheet diffusions, such as base or emitter, rather than high-sheet diffusions, such as HSR. For example, the voltage modulation of 160Ω/□ base equals about 0.1%/V while that of 2kΩ/□ HSR can approach 1%/V. Matched base resistors are usually merged into a common tank to save space, while matched HSR resistors frequently require separate tanks.

The proper determination of tank biasing requires a complete understanding of circuit specifications and design, so the circuit designer rather than the layout designer must determine the tank connections. These should appear on the schematic along with the type of resistance material required, the width of the resistor, and any special matching requirements. In cases where the connections of the resistors are not obvious, or appear to be in error, the layout designer should verify them before continuing.

Leads routed over resistors can also affect their operation. As a rule, leads that do not connect to matched resistors should not cross them. Not only may these leads capacitively couple noise into the resistor, but the electric field between the lead and the resistor can actually modulate the conductivity of the resistance material (a phenomenon called *conductivity modulation*). Emitter, base, and low-sheet poly (R_s< 200Ω/□) resistors rarely experience significant conductivity modulation. High-sheet resistors are more problematic; for example, metal-1 over 2kΩ/□ HSR can produce 0.1%/V of conductivity modulation. The impact of conductivity modulation depends on three factors: (1) the voltage difference between the lead and the underlying resistor, (2) the thickness of the intervening oxide, and (3) the area of intersection. A lead connected to an HSR resistor can safely route across the end of the resistor next to the head, while one routed entirely across the resistor array may cause problems because it has a larger area of intersection. Wires rarely need to route across resistors in a double-level metal process, but jumpers through resistor arrays are often unavoidable in single-level metal processes. For example, the interdigitated HSR resistor array shown in Figure 7.22 uses a jumper between segments R_{1A} and R_{1B} to allow a lead to exit from the left-hand terminal of R_2.

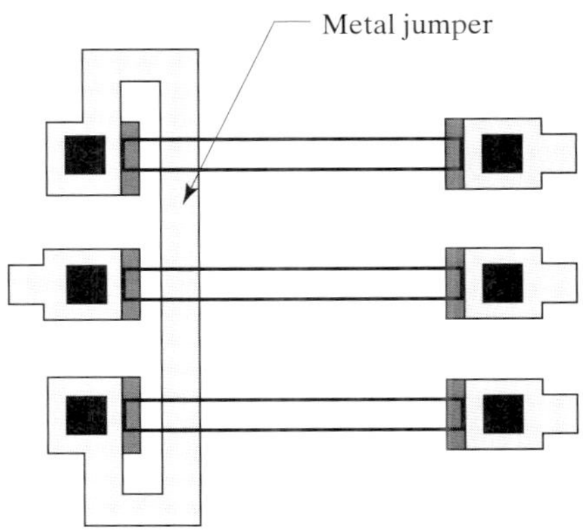

FIGURE 7.22 Portion of an interdigitated HSR array implemented in a single-level metal process showing the placement of a jumper between segments.

The jumper of Figure 7.22 is constructed so that it intersects each resistor segment in exactly the same manner. This precaution helps to minimize the mismatches produced by the stress of the aluminum lead on the underlying resistor.[22] Still,

[22] For a discussion of a similar stress-induced mismatch mechanism in MOS transistors, see H. Tuinhout, M. Pelgrom, R. P. de Vries, and M. Vertregt, "Effects of Metal Coverage on MOSFET Matching," *IEDM*, 1996, pp. 735–738.

the mere presence of a lead running over the resistor segments implies that stresses are imposed on the resistors, and therefore produces an inevitable degradation of matching. Whenever possible, one should not route any leads across critically matched resistors.

A more convenient or more compact layout often results if leads can cross resistors. A technique called *electrostatic shielding* (or *Faraday shielding*) can isolate a resistor from the influence of overlying leads. Electrostatic shielding not only prevents conductivity modulation but also provides considerable shielding against capacitive coupling.

Figure 7.23A illustrates the basic concept of an electrostatic shield. The shield is interposed between the two conductors—in this case the resistor and an overlying lead. The electrostatic interactions between the shield and the conductors on either side can be modeled as a pair of series-connected capacitors (Figure 7.23B). The AC voltage source, V_N, represents the time-varying voltage present on the overlying lead. Noise injected by V_N is attenuated by the RC filter formed by C_{P1} and R_1, where C_{P1} represents the capacitance between the overlying lead and the shield, and R_1 represents the resistance of the connection between the shield and AC ground. Whatever noise passes through C_{P1}–R_1 will be injected through C_{P2} onto R_2, where C_{P2} represents the capacitance between the shield and the sensitive node, and R_2 represents the resistance between the sensitive node and AC ground. The attenuation provided by the shield falls away with increasing frequency. Providing that the shield connects to a clean low-impedance node (such as signal ground), substantial attenuation may remain into the low RF region (1 to 10 MHz). At higher frequencies, it becomes increasingly difficult to guarantee a sufficiently low-impedance shield connection, so high-speed digital signals should not route across sensitive circuitry even if an electrostatic shield is present.

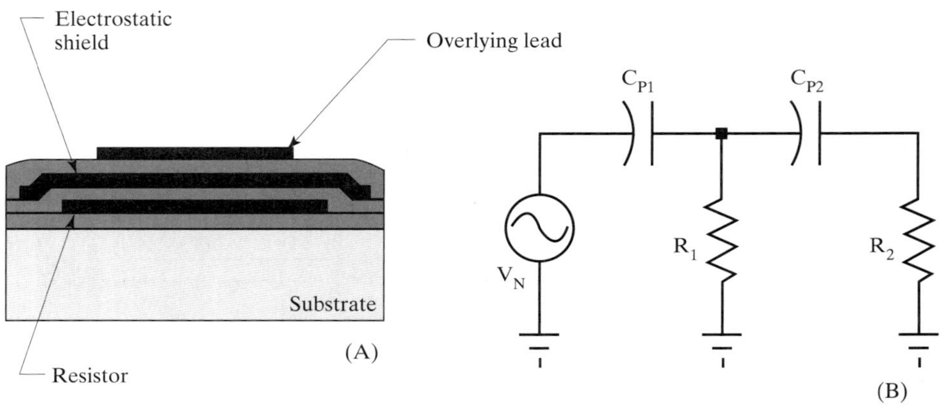

FIGURE 7.23 The concept of an electrostatic shield: (A) cross section and (B) equivalent circuit.

Figure 7.24 shows a practical example of an electrostatic shield. The poly resistor array includes dummies at either end, and both the shield and the dummies are connected to ground. Leads can cross the shielded resistors without modulating the conductivity of the poly resistors or injecting noise into them. Note that a fringing field exists around any conductor. In order to prevent this fringing field from coupling around the shielding, all leads should reside well away from the edges of the shield. An overhang of 5μm (shield-over-conductor) will intercept the majority of the fringing fields. Since metal can elastically deform, the electrostatic shield may also help minimize stresses generated in the poly resistors by the presence of the second-metal leads running across them.

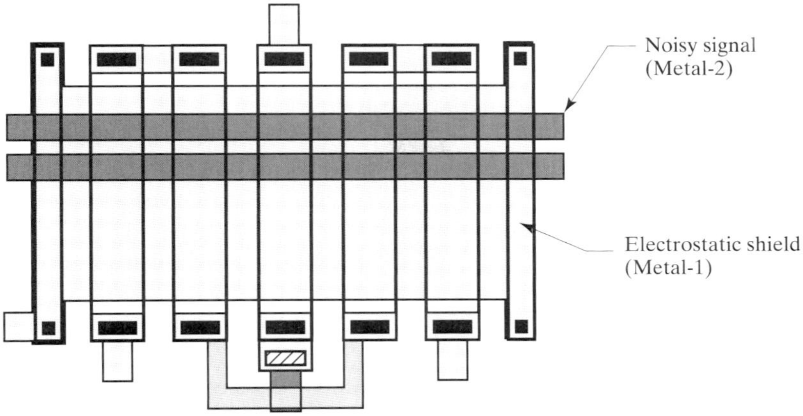

FIGURE 7.24 Example of electrostatic shielding applied to a matched poly resistor array.

Note that the electrostatic shield of Figure 7.24 covers the entire resistor array. Each resistor segment in the array sees a slightly different voltage. As long as the voltage difference across the array remains small and the sheet resistance of the resistors is less than about 500Ω/□, a common shield will suffice. If the resistors consist of high-sheet material, or if the voltage differential across the array exceeds a few volts, then the common electrostatic shield itself will cause objectionable conductivity modulation! In such cases, the shield should be divided into individual sections placed over each resistor segment. Each shield must overlap its respective resistor by several microns (after accounting for misalignment and outdiffusion) to ensure that fringing fields do not degrade its effectiveness. Sectioned shielding requires substantially more space than common shielding.

The substrate can also inject noise into deposited resistors and capacitors. One way to minimize this source of noise coupling consists of placing a well beneath the devices and connecting this well to an AC ground. Resistors and capacitors that are especially sensitive to noise can benefit from a combination of electrostatic shields above and below them.[23] A deep-N+ sinker placed beneath deposited devices and connected to an AC ground can provide electrostatic shielding while minimizing parasitic capacitance (by thickening the field oxide through dopant-enhanced oxidation).

Charge Spreading

The mechanisms behind charge spreading were discussed at length in Section 4.3.2. Briefly, circuit operation injects electrons into the oxide overlying the die. Although most of these electrons eventually return to the silicon, a few become trapped at the interface between the interlevel oxide and the protective overcoat, or between the protective overcoat and the mold compound. These electrons constitute a mobile charge capable of varying resistor values through conductivity modulation. The electric field required to cause a fractional-percent variation of a high-sheet resistor is an order of magnitude smaller than that required to invert the surface of the silicon. Matched high-sheet resistors are therefore extremely susceptible to long-term drifts caused by charge spreading. Base resistors are much less susceptible to charge

[23] K. Yamakido, T. Suzuki, H. Shirasu, M. Tanaka, K. Yasunari, J. Sakaguchi, and S. Hagawara, "A single-chip CMOS filter/CODEC," *IEEE J. Solid-State Circuits*, Vol. SC-16, 1981, pp. 302–307.

spreading due to their lower sheet resistance, and only precisely matched base resistors are significantly affected. High voltages, moisture, and mobile ion contamination amplify the effects of charge spreading. Designs that operate at voltages exceeding half of the thick-field threshold or that are fabricated on standard bipolar should be examined for potential charge spreading vulnerabilities.

Electrostatic shielding can minimize or even eliminate the effects of charge spreading on matching. The electrostatic shield also serves as a field plate, counteracting surface inversion across high-voltage tanks. The field plate must connect to a potential not greatly different from the tank bias, as described in Section 4.3.2. It is often connected to the more positive end of the resistor. Field plates may actually increase noise coupling into high-impedance resistors due to leads running above the field plate or due to fringing fields from adjacent components. Noisy signals must not be routed across field plates unless the field plates connect to low-impedance nodes that are relatively insensitive to capacitive coupling.

Figure 7.25 shows the same resistor array as Figure 7.21. Each resistor segment has been provided with an individual field plate protecting it against the effects of charge spreading. The field plates flange over the bodies of the resistors far enough to prevent channel formation. The gaps in the field plates have not been channel-stopped because the proximity of an adjacent diffusion could cause diffusion interactions. If channel stops are required, they should be carefully replicated on each section so that all of the resistors experience the same interactions. In most cases, channel stops are unnecessary as long as the field plates are properly flanged. Although diffused HSR resistors can rival the matching of deposited resistors, the area required for separate tanks and field plates makes them rather uneconomical.

FIGURE 7.25 HSR resistor array, field-plated to minimize charge spreading. With the exception of its metallization pattern, this array matches the one in Figure 7.21.

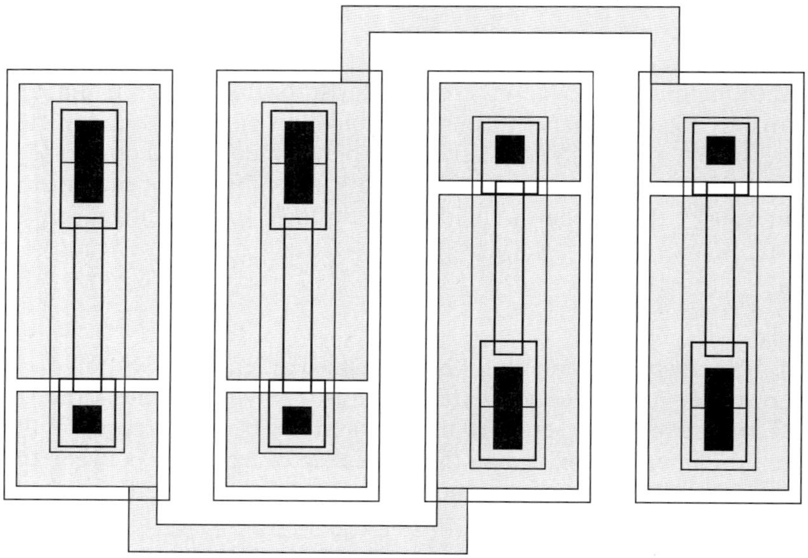

Dielectric Polarization

Electrostatic fields can also arise due to the movement of charges within an insulator, a phenomenon called *dielectric polarization.* Oxides contaminated by alkali metal ions such as sodium or potassium exhibit a large degree of polarizability due to the mobility of these ions within the oxide. As discussed in Section 4.2.2, these *mobile ions* slowly redistribute themselves under the influence of external electrical fields. The electric field seen at the surface of the silicon shifts over time as the mobile ions gradually assume a new configuration. If the external field suddenly vanishes, the new distribution of mobile ions generates a weak residual electric field oriented in the opposite direction to the original. This residual field gradually relaxes as the mobile ions return to their original distribution. The slow variation in field intensity seen beneath the oxide can modulate the value of a high-sheet resistor. The resulting *long-term drifts* are extremely undesirable.

The addition of phosphorus to silicon dioxide effectively immobilizes alkali metal ions, presumably through a sequestration mechanism involving phosphate groups. Phosphosilicate glass (PSG) thus exhibits much less polarization when contaminated with alkali metal ions than does pure oxide. Unfortunately, the phosphate groups themselves are slightly polarizable.[24] Phosphosilicate and borophosphosilicate glasses are subject to low levels of dielectric polarization in the presence of external electrical fields, and this in turn gives rise to hysteretic voltage modulation.

Dielectric polarization normally affects only high-sheet resistors. The polarization rates are usually far too slow to affect capacitors, and the resulting fields are too weak to affect resistors with sheets of less than about 500Ω/□. A dielectric polarization of approximately 0.1% has been observed in 2kΩ/□ high-sheet resistors constructed on a standard bipolar process using heavily phosphorus-doped BPSG.[25]

As the above example suggests, field plating is not a panacea for dielectric polarization. Field plates may actually intensify the phenomenon because they induce a deliberate electric field across the interlevel oxide. On the other hand, leaving the field plates off the resistors renders them vulnerable to charge spreading. This dilemma is best solved by avoiding the use of high-sheet resistors entirely. If this is impractical, then *split field plates* should be used. Figure 7.26 shows an example of a split field plate applied to a pair of matched high-sheet resistors similar to those shown in Figure 7.25.

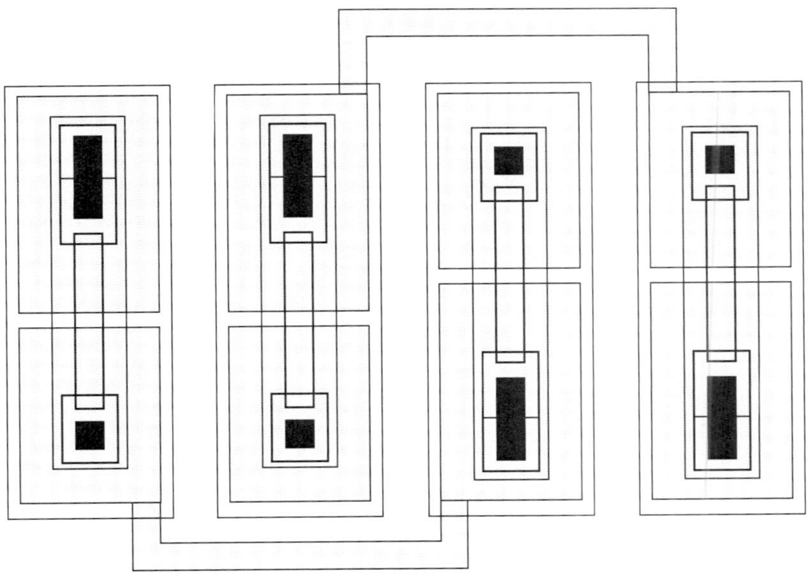

FIGURE 7.26 HSR resistor array field-plated to minimize dielectric polarization as well as to charge spreading and thermoelectrics (compare with Figure 7.25).

---

[24] E. H. Snow and B. E. Deal, "Polarization Phenomena and Other Properties of Phosphosilicate Glass Films on Silicon," *J. Electrochem. Soc.*, Vol. 113, #3, 1966, pp. 263–269.

[25] F. W. Trafton, private communication.

Split field plates differ from conventional field plates in the location of their gaps. In a conventional field plate, as much of the plate as possible connects to the positive end of the resistor to offer maximum protection against surface inversion. In a split field plate, each resistor segment requires two field plates, one connecting to either end of the resistor. The gap between the two plates falls in the exact center of the resistor. The split field plate subjects half of the resistor to an electric field equal and opposite to that seen by the other half. The dielectric polarization in each half of the resistor exactly cancels the polarization in the other. Split field plates also reduce voltage nonlinearity effects caused by the field plates, because the accumulation effects beneath one plate balance the depletion effects under the other.[26] Split field plates can be applied to resistors occupying a common tank, but each segment must have its own split field plate.

Split field plates are recommended for all applications where resistors having sheet resistances in excess of 1kΩ/□ must match to better than ±0.5%. Split field plates are not necessary for base resistors since the lower sheet resistance of this diffusion generally renders dielectric polarization negligible. They are likewise not necessary for the majority of polysilicon resistors because these usually exhibit sheet resistances less than or equal to 1kΩ/□.

Dielectric Relaxation

Capacitors are also susceptible to dielectric polarization in the form of a hysteretic effect called *dielectric relaxation* or *soakage.* Suppose that the capacitor is suddenly charged to produce a corresponding electric field between its electrodes. As the dielectric polarizes, the electric field intensity gradually decreases (or *relaxes*) to a lower value. The relaxation of the electric field causes a corresponding droop in capacitor voltage. After the capacitor is suddenly discharged and then disconnected, the polarization dissipates and a reverse bias gradually accumulates across the capacitor. Charge storage capacitors (such as those used in timers and sample-and-hold circuits) are especially intolerant of dielectric relaxation errors.

Charge spreading can affect capacitors in much the same manner as dielectric polarization. When the capacitor is charged, the shifting electrostatic fields cause a gradual redistribution of charges along insulating interfaces. This results in a change in the electric field intensity that is similar to what occurs in dielectric relaxation. Charge spreading can occur along the interfaces of a composite dielectric, such as the oxide-nitride-oxide sandwich used to fashion ONO capacitors. The movement of charges along the interfaces of the capacitor dielectric can occur very rapidly, so variations in capacitor value may occur at frequencies in excess of 1MHz.

High-quality oxide dielectrics are preferred for critically matched capacitors, particularly for those that operate at high frequencies. The use of a grown oxide dielectric usually reduces dielectric relaxation to negligible levels.[27] Charge spreading and polarization phenomena arising outside the capacitor structure can be eliminated using electrostatic shielding. Figure 7.27 shows one method of shielding a poly-poly

---

[26] J. Victory, C. C. McAndrew, J. Hall, and M. Zunino, "A Four-Terminal Compact Model for High-Voltage Diffused Resistors with Field Plates," *IEEE J. Solid-State Circuits,* Vol. 23, #9, 1998, pp. 1453–1458.

[27] LPCVD oxides also exhibit little dielectric relaxation, but the same is not true of TEOS oxides: J. W. Fattaruso, M. De Wit, G. Warwar, K. S. Tan, and R. K. Hester, "The Effect of Dielectric Relaxation on Charge-Redistribution A/D Converters," *IEEE J. Solid-State Circuits,* Vol. 25, #6, 1990, pp. 1550–1561.

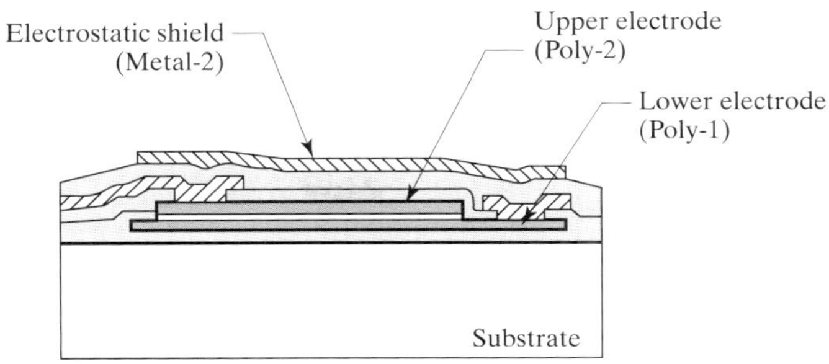

FIGURE 7.27 Cross section of a poly-poly capacitor incorporating an electrostatic shield plate. Note the overlap of the electrostatic plate over the upper electrode.

capacitor. The sensitive high-impedance node connects to the upper electrode of the capacitor. An electrostatic shield constructed from second-level metal completely covers the upper electrode. This electrostatic shield connects to the lower electrode of the capacitor to form a sandwich-capacitor structure. The sensitive node is now entirely enclosed within an electrostatic shield. Both the lower capacitor electrode and the electrostatic shield should overhang the upper capacitor electrode by at least 3 to 5μm to suppress fringing fields. This structure is relatively insensitive to dielectric polarization and charge spreading providing that both the capacitor oxide and the interlevel oxide above the upper electrode are not phosphorus-doped.

7.3 RULES FOR DEVICE MATCHING

The previous sections have discussed the various mechanisms responsible for causing mismatches. This information would ideally be used to formulate a set of quantitative rules for each process. The layout designer could then apply these rules to obtain matched devices of any desired degree of accuracy. In practice, the time and effort required to generate quantitative matching rules are usually prohibitive. This section therefore presents a set of qualitative rules applicable to many processes. In the absence of more precise matching data, these rules can be used to lay out structures with at least a modest degree of confidence.

The following rules employ the terms *minimal, moderate,* and *precise* to denote increasingly precise degrees of matching. These terms have the following meanings:

- **Minimal Matching:** Approximately ±1% three-sigma mismatch, or 6 to 7 bits of resolution. Suitable for general-purpose use, such as for degenerating current mirrors in biasing circuitry.
- **Moderate Matching:** Approximately ±0.1% three-sigma mismatch, or 9 to 10 bits of resolution. Suitable for ±1% bandgap references, op-amp and comparator input stages, and most other analog applications.
- **Precise Matching:** Approximately ±0.01% three-sigma mismatch, or 13 to 14 bits of resolution. Suitable for precision A/D and D/A converters and for all other applications requiring extreme precision. Capacitors can more easily obtain this level of matching than resistors.

7.3.1. Rules for Resistor Matching

Minimal matching can be obtained without much difficulty, and moderate matching can be reliably obtained using interdigitation. Precisely matched resistors are difficult to construct due to variations in contact resistance and the presence of thermal and stress gradients. The following rules summarize the most important principles of resistor design:[28]

1. Construct matched resistors from a single material.
   Resistors constructed from different materials do not even approximately match one another. Process variations will cause unpredictable shifts in the value of one resistor relative to the other, and the differing temperature coefficients of the two materials prevent them from tracking over temperature.
   Do not construct matched resistors from different materials!
2. Make matched resistors the same width.
   Uncorrected process biases will cause systematic mismatches in resistors of different widths. If for some reason one resistor must be made wider than another, consider constructing the wider resistor from a number of sections connected in parallel. Width effects are not entirely independent of temperature and stress, so even if moderately or precisely matched resistors are trimmed, they should still consist of sections of a uniform width.
3. Make matched resistors sufficiently wide.
   In the absence of experimental data, assume that minimal matching of resistors containing 30 or more squares requires 150% of the minimum allowed width of the deposition or diffusion, moderate matching requires 200%, and precise matching requires 400%. For example, if the minimum drawn width of a poly line equals 2μm, then minimal matching requires 3μm, moderate matching 4μm, and precise matching 6μm. If the smaller of the matched resistors contains less than 20 to 30 squares, consider increasing the width of the resistors (see Equations 7.6A and 7.6B). If the smallest resistor contains more than 100 squares, consider reducing the resistor width. In no case, however, should one reduce the width of matched deposited resistors below about 2μm because of the potential for excessive variability caused by granularity effects.
4. Where practical, use identical geometries for resistors.
   The existence of corner and end effects precludes precise matching of resistors having different geometries. Resistors having the same width, but different lengths or shapes, can easily experience mismatches of ±1% or more. Matched resistors should be divided into sections as discussed in Section 7.2.6. Sectioning is not necessary for minimal (and even moderate) matching if the parameters controlled by the resistors are trimmed at wafer probe. Wafer-level trimming compensates for most (but not all) variations in resistors caused by geometric factors. Allow an extra 1 to 2% of trim range for unexpected mismatches between resistors of different geometries.
5. Orient matched resistors in the same direction.
   Resistors oriented in different directions may vary by several percent. Diffused resistors show the largest orientation mismatches, but even polysilicon resistors are affected to some degree. Most resistors should be oriented vertically or horizontally. P-type diffused resistors on (100)-oriented silicon may experience less stress-induced variation if they are oriented at 45° to the X- and Y-axes.
6. Place matched resistors in close proximity.
   Mismatches increase with separation. Minimally matched resistors can be spaced a few mils apart, moderately matched resistors should be placed adja-

---

[28] Tsividis provides a list of ten general matching rules, entries in which correspond to numbers 1–3, 5, 6, and 13 in the list provided in the text: Tsividis, pp. 234–236.

cent to each other, and precisely matched resistors must be interdigitated. Dice larger than about 25kmil^2 (16mm^2), dice that dissipate more than 250mW, dice in heat-sunk packages, designs that experience junction-to-ambient temperature rises in excess of 20°C, and dice mounted using solder or gold eutectics will experience larger mismatches. If any of these conditions apply, then minimally matched resistors should be placed adjacent to each other, and moderately or precisely matched resistors should be arrayed and interdigitated. All resistors in high-stress locations (corners or edges) or within 10mils (250μm) of a power device dissipating more than 100mW should be interdigitated.

7. Interdigitate arrayed resistors.
Arrayed resistors should be interdigitated to produce a common centroid layout. The resulting array should have an aspect ratio no greater than 3:1, and each resistor segment should be at least five times (and preferably ten times) longer than it is wide. The interdigitation pattern should obey the rules of common-centroid layout (Table 7.3). Arrangements having an even number of segments per resistor are preferable to those having an odd number because even numbers of segments offer better rejection of thermoelectrics. Consider connecting some segments in parallel if this will reduce the total area of the array. If the array requires a large number of segments, consider arranging these into multiple banks connected to form a two-dimensional array.

8. Place dummies on either end of a resistor array.
Arrayed resistors should include dummies. Poly dummies need not have the same width as the other segments. Diffused dummies should always have the same geometry as adjacent segments. Remember to keep the spacing between adjacent segments constant, and equal to the spacing between the dummies and their neighboring resistor segments. Whenever possible, connect dummy resistors to a quiet low-impedance node.

9. Avoid short resistor segments.
Very short resistor segments may introduce considerable variation due to contact resistance. Moderately matched resistor segments should contain no fewer than five squares, and precisely matched resistors should contain no fewer than ten. Precisely matched poly resistors should have a total length of no less than 50μm to minimize nonlinearities caused by granularity.

10. Connect matched resistors in order to cancel thermoelectrics.
Arrayed resistors should always be connected so that equal numbers of segments are oriented in either direction. If the array contains an odd number of segments, then one must remain unpaired. A single unpaired segment does not produce much mismatch, but arrays should, if possible, contain no unpaired segments. Serpentine resistors should be constructed so that their heads lie near one another, because this improves thermoelectric cancellation.

11. If possible, place matched resistors in low stress areas.
The stress distribution reaches a broad minimum in the middle of the die. Any location ranging from the center halfway out to the edges will lie within this broad minimum. If precisely matched resistors must reside close to an edge, then they should be placed near the middle of one side of the die, preferably a longer side. The stress distribution reaches a maximum in the die corners, so avoid placing matched devices anywhere nearby.

12. Place matched resistors well away from power devices.
For purposes of discussion, any device dissipating more than 50mW is a power device, and any device dissipating more than 250mW is a major power device.

Precisely matched resistors should reside on an axis of symmetry of the major power devices using one of the optimal symmetry arrangements of Section 7.2.6. Such resistors should also reside no less than 200 to 300μm (8 to 12mils) away from the closest power device. P-type diffused resistors may exhibit less package shift if placed diagonally on (100)-oriented silicon, but this arrangement precludes placement on an axis of symmetry of the power device. If large thermal gradients are expected, then these resistors should be arrayed vertically or horizontally to allow symmetrical placement because the thermal gradients are likely to produce more mismatch than stress gradients. For moderate matching, the matched devices need not lie on an axis of symmetry of the major power devices. They should, however, reside no less than 200 to 300μm away from major power devices and at least 100μm away from smaller power devices. Minimal matching can be achieved anywhere on the die, but if the devices must reside next to a major power device, then they must be interdigitated.

13. Place precisely matched resistors on axes of symmetry of the die.
On (100) silicon, precisely matched resistors should be placed so that the axis of symmetry of the resistor array aligns with one of the two axes of symmetry of the die. P-type diffused resistors may benefit from a diagonal orientation that minimizes their stress sensitivity. On (111) silicon, the axis of symmetry of the array should align to the <211> axis of symmetry of the die. If large numbers of matched devices exist, reserve the optimal locations for the most critical devices.

14. Consider tank modulation effects.
Tank modulation becomes significant for precisely matched resistors having sheet resistances of 100Ω/□ or more, moderately matched resistors having sheet resistances of 500Ω/□ or more, and minimally matched resistors having sheet resistances of 1kΩ/□ or more. Substitute poly resistors for diffused resistors where possible. If diffused resistors must be employed, then consider whether use of a lower sheet material will allow merging the matched resistors into a common tank. For example, moderately matched resistors can be constructed from 160Ω/□ base diffusions placed in the same tank. Trimmed resistors subject to known and controlled voltage biases can usually occupy a common tank regardless of their sheet resistance because trimming largely compensates for the effects of tank modulation.

15. Sectioned resistors are superior to serpentines.
Serpentine resistors are suitable for constructing large, minimally matched resistors or for constructing minimally matched resistors that will be trimmed. All other resistors should use arrays of sections.

16. Use poly resistors in preference to diffused ones.
Polysilicon resistors can be made much narrower than most types of diffused resistors, and, providing that they are sufficiently long, their small widths will not cause any significant increase in mismatch. Also, poly resistors do not require tanks and are therefore immune to tank modulation.

17. Place deposited resistors over field oxide.
Deposited materials, including polysilicon, experience increased variation when crossing oxide steps. Even minimally matched deposited resistors should not cross oxide steps or other surface discontinuities.

18. Choose P-type poly resistors in preference to N-type poly resistors.
Empirical results suggest that boron-doped poly resistors match better than those doped with phosphorus or arsenic. Since the reasons for this difference

are not fully understood, it is possible that some poly resistors may not follow this rule.

19. Do not allow the NBL shadow to intersect matched diffused resistors.
The NBL shadow should not intersect any precisely matched diffused resistor, or any moderately matched shallow diffused resistor (such as HSR). If the direction of NBL shift is unknown, allow adequate overlap of NBL over the resistor on all applicable sides. If the magnitude of the NBL shift is unknown, then overlap NBL over the resistor by at least 150% of the maximum epi thickness.

20. Consider field plating and electrostatic shielding.
Field plate any matched resistor operating above 50% of the thick-field threshold. Field plate all moderately matched diffused resistors having sheet resistances of 500Ω/□ or more. Consider split field plates for precisely matched diffused resistors having sheet resistances of 500Ω/□ or more. Precisely matched polysilicon resistors with sheet resistances of 500Ω/□ or more should have electrostatic shields placed over them where possible.

21. Avoid routing unconnected leads over matched resistors.
Whenever possible, unconnected leads should not run over matched resistors. Unconnected leads can run across minimally matched resistors having sheet resistances of less than 500Ω/□ or moderately matched resistors having sheet resistances of less than 100Ω/□, but the designer should carefully scrutinize all such layouts for potential noise coupling. Leads can run over electrostatic shields and field plates, providing that the latter connect to low-impedance nodes that can absorb the anticipated levels of noise injection. Beware of running high-speed digital signals across matched resistors regardless of whether field plates or electrostatic shielding exists.

22. If leads cross resistors, they should cross all resistor segments in the same manner.
The presence of a lead above a resistor generates mechanical stresses that may alter the value of the resistor. The effects of stress on mismatch will be minimized as long as the lead crosses the resistor array orthogonally, in order to intersect each resistor segment at the same point and with the same area of intersection. Even with this precaution, the presence of the lead will inevitably impair matching, so critically matched resistors should remain uncovered, if possible. Note that this rule cannot always be obeyed, particularly in the case of single-level metal designs and for resistors with sliding contacts.

23. Avoid excessive power dissipation in matched resistors.
Power dissipated within matched resistors can generate thermal gradients that can degrade matching. As a guideline, one should avoid dissipating more than 1mW/mil^2 (1.5μW/μm^2) in precisely matched resistors. Moderately matched resistors can tolerate several times this level of power dissipation. Resistors that dissipate larger amounts of power should be interdigitated. High currents in narrow resistors may also induce velocity saturation nonlinearities (Section 5.3.3).

7.3.2. Rules for Capacitor Matching

Properly constructed capacitors can obtain a degree of matching unequaled by any other integrated component. Matched capacitors form the basis of most data conversion products such as analog-to-digital (A/D) and digital-to-analog (D/A) converters. Untrimmed oxide-dielectric capacitors packaged in plastic can achieve ±0.01% matching. This suffices to allow construction of 14-bit and perhaps even

15-bit converters. Beyond this, some type of wafer-level trimming is required to maintain accuracy. Matching of ±0.001% can be obtained using trimmed oxide-dielectric capacitors packaged in plastic, making possible 16 to 18-bit monolithic converters. Higher-precision products usually employ hybrid assemblies rather than single-die circuits.

Precisely matched capacitors usually employ a thick oxide dielectric in conjunction with deposited electrodes. Junction capacitors have difficulty maintaining even minimal matching due to their extreme temperature dependence and the effects of outdiffusion. Composite dielectrics are inferior to pure oxide dielectrics because the multiple steps required to produce the composite increase its variability. Dielectric relaxation can also degrade matching in composite dielectrics at higher frequencies. Thick oxide dielectrics are favored over thin ones because they are more tolerant of changes in oxide thickness. A ±10Å variation in a 100Å oxide represents ±10% of the oxide thickness, while the same variation in a 500Å oxide represents only ±2% of its thickness. Silicided polysilicon (*polycide*) is sometimes used for constructing the lower plates of matched capacitors because its low resistance minimizes depletion effects and because this material can withstand the high temperatures required to densify the deposited-oxide dielectric. Aluminum is the material of choice for the upper plate due to the absence of depletion effects in this material. Capacitors constructed using unsilicided poly electrodes experience significant surface depletion, even when the poly is heavily doped. These capacitors therefore exhibit temperature coefficients several times as large as metal-polycide capacitors. Despite these problems, poly-poly capacitors can still obtain adequate matching for all but the most precise applications.

The following rules summarize the most important principles of constructing matched deposited-electrode capacitors:[29]

1. Use identical geometries for matched capacitors.
   Capacitors of different sizes or shapes match poorly, so matched capacitors should always use identical geometries. If the capacitors are not the same size, then each should consist of a number of segments (or *unit capacitors*), all of which are copies of the same geometry. The larger capacitor should consist of multiple segments in parallel, while the smaller capacitor should have fewer segments in parallel. Unit capacitors should not connect in series because differences between the parasitic capacitances of the upper and lower plates will produce systematic mismatches. If the required ratio does not lend itself to division into an integer number of unit capacitors, then one nonunitary capacitor should be inserted into the larger of the matched capacitors. The aspect ratio of this nonunitary capacitor should not exceed 1.5:1 (Section 7.2.2).
2. Use square geometries for precisely matched capacitors.
   Peripheral variations are a major source of random mismatch in capacitors. The smaller the periphery-to-area ratio, the higher the obtainable degree of matching. The square has the lowest periphery-to-area ratio of any rectangular geometry and therefore yields the best matching. Rectangular capacitors with moderate aspect ratios (2:1 or 3:1) can be used to construct moderately matched capacitors, but precisely matched capacitors should always be square. Oddly shaped geometries should be avoided because it is difficult to predict the magnitude of their peripheral variations.

---

[29] Some of these rules follow guidelines proposed by M. J. McNutt, et. al., p. 615.

3. **Make matched capacitors as large as practical.**
Increasing the size of capacitors reduces random mismatch. An optimal capacitor size exists, beyond which gradient effects cause increasing variability. The optimum dimensions of square capacitors in several CMOS processes are reported to lie between 20 × 20μm and 50 × 50μm.[30] Capacitors larger than about 1000μm² should be divided into multiple unit capacitors, as proper cross-coupling will minimize gradient effects and improve overall matching.
4. **Place matched capacitors adjacent to one another.**
Matched capacitors should always reside next to one another. If a large number of capacitors are involved, they should be arranged to form a rectangular array having as small an aspect ratio as possible. For example, if thirty-two matched capacitors are required, then consider using a 4 × 8 array. Alternatively, a 5 × 7 array could be constructed and the three unused capacitors connected as dummies. Adjacent rows of unit capacitors should have equal spacings between them, as should adjacent columns of unit capacitors. The row and column spacings need not be the same.
5. **Place matched capacitors over field oxide.**
Any surface discontinuities in the thick-field oxide will cause corresponding variations in the topography of the capacitor dielectric. Matched capacitors should always reside over field oxide well away from the edges of moat regions and diffusions.
6. **Connect the upper electrode of a matched capacitor to the higher-impedance node.**
The higher-impedance node of the circuit usually connects to the upper electrode, because this generally exhibits less parasitic capacitance than the lower electrode. Substrate noise also couples more strongly to the lower electrode than to the upper electrode. Some arrays may require the high-impedance node of the circuit to connect to the lower plate in order to allow an array of unit capacitors to share a common lower plate. If substrate noise coupling is a concern, consider placing a well under the entire array. This well should connect to a clean analog reference voltage, such as signal ground, so that it can serve as an electrostatic shield for the lower electrode(s) of the capacitor array.
7. **Place dummy capacitors around the outer edge of the array.**
Dummy capacitors will shield the matched capacitors from lateral electrostatic fields and will eliminate variations in etch rates. The dummy capacitors need not have the same width as the capacitors of the array as long as an electrostatic shield covers the array. Otherwise, fringing fields can easily extend 30 to 50μm, and arrays of identical dummy capacitors must extend at least this far to ensure precise matching. Moderate matching generally requires only a minimum-width ring of dummy capacitors, and minimal matching does not require dummy capacitors at all. Both electrodes of each dummy capacitor must be connected to prevent static charges from accumulating on the dummy electrodes and interfering with the operation of adjacent devices. The spacing between the dummy capacitors and the adjacent unit capacitors should equal the spacing between rows of unit capacitors.
8. **Electrostatically shield matched capacitors.**
Electrostatic shielding provides several benefits. First, it contains fringing fields to the capacitor array, thereby eliminating the need for wide arrays of

---

[30] Shyu, Temes, and Yao, p. 1075.

dummy capacitors. Second, it allows leads to route over the capacitors without causing mismatch or noise injection. Third, it prevents electrostatic fields from adjacent circuitry from interfering with the matched capacitors. Fourth, it reduces the effects of packaging stress on the underlying capacitors. All precisely matched capacitors should be electrostatically shielded, and this shielding should extend over the dummies placed around the matched capacitors to seal the array against the entry of electrostatic fields. Even minimally matched capacitors can benefit from electrostatic shielding; if no dummy capacitors are used, then the shield should overlap the capacitors by at least 3 to 5μm.

9. Cross-couple arrayed capacitors.
Capacitor arrays lend themselves to cross-coupling because the unit capacitor forms a compact square rather than an elongated rectangle. The array typically consists of several rows and columns of capacitors. Even in the case of two matched capacitors of equal values, a very compact cross-coupled array can be constructed by dividing each capacitor into two halves. Cross-coupling minimizes the effects of oxide gradients on capacitor matching and provides protection against stress and thermal gradients. The centroids of the matched capacitors should precisely align. In practice, this is difficult to achieve with larger capacitor arrays, so the designer must often settle for less-than-optimal interdigitation patterns.

10. Consider the capacitance of leads connecting to the capacitor.
The leads that connect a matched capacitor into the circuit will contribute some capacitance of their own. This capacitance becomes of concern when one tries to construct moderately or precisely matched arrays of capacitors. Each unit capacitor should have two minimum-width leads connecting to its top electrode, so that each capacitor will have equal overall lead capacitance. If an array contains a nonunitary capacitor, ideally its number of leads should be equal to twice the ratio of its capacitance to the unit capacitance, but this is often difficult to achieve in practice. The total lead area on each capacitor should be computed, and additional leads should be inserted until the ratio of the lead capacitance equals the ratio of the intended capacitors.

11. Do not run leads over matched capacitors unless they are electrostatically shielded.
The capacitance between the overlying lead and the upper plate of the capacitor will induce a mismatch between the matched capacitors unless the area of the lead overlying each capacitor is identical. Even then, fringing fields and electrostatic noise coupling will degrade the performance of the matched capacitors. If leads must run over matched capacitors, an electrostatic shield should be inserted between the capacitors and the leads.

12. Use thick-oxide dielectrics in preference to thin-oxide or composite dielectrics.
Thick-oxide dielectrics exhibit less mismatch due to dimensional variations, so they are preferable to thin-oxide dielectrics. Composite dielectrics, such as oxide-nitride-oxide sandwich dielectrics, have more mismatch than homogeneous dielectrics because multiple processing steps all affect their final capacitance per unit area. While minimal and moderate matching can be obtained using thin-oxide or composite dielectrics, precise matching usually requires thick-oxide dielectrics.

13. If possible, place capacitors in areas of low stress gradients.
The stress distribution reaches a broad minimum in the middle of the die. Any location ranging from the center of the die halfway out to the edges lies within this broad minimum. Avoid placing capacitors along the edges of the die, and

especially in the corners of the die, as the stress in these regions is substantially greater than elsewhere.

14. Place matched capacitors well away from power devices.
The temperature coefficient of heavily doped poly-poly capacitors is relatively low (perhaps 50ppm/°C), and the temperature coefficient of metal-polycide or metal-metal capacitors is even lower. The direct impact of temperature on matched capacitors is much smaller than on resistors. Matched capacitors should still reside at least 200 to 300μm away from power devices that dissipate 250mW or more.

15. Place precisely matched capacitors on axes of symmetry of the die.
On (100) silicon, precisely matched capacitor arrays should be placed so that the axis of symmetry of the array aligns with one of the two axes of symmetry of the die. On (111) silicon, the axis of symmetry of the array should preferably lie on the <211> axis of symmetry of the die. If large numbers of matched devices are required, reserve the optimal locations for the most critically matched devices.

7.4 SUMMARY

Most integrated circuits contain numerous matched resistors and capacitors. The layout designer should determine which components must match, and with what degree of precision. Using this information, a die floor plan can be constructed showing the relative locations of power devices and matched components. The most critically matched devices should occupy locations near the middle of the die on an axis of symmetry of the power devices. Even if no power devices are present on the die, stress considerations still dictate the placement of the most sensitive matched components near the middle of the die. Preliminary die planning before layout begins will pay dividends in the form of easier circuit construction and interconnection as well as better performance.

Matched devices should never reside in the corners of the die or near a major heat source. Sometimes matched resistors must be placed along a die edge to facilitate the placement of trimpads for fuses or Zener zaps. In this case, the resistors are best placed near the middle of one of the sides of the die.

The matching of resistors and capacitors is very much a function of how well these components are laid out. Two resistors haphazardly laid some distance away from one another using different sizes and shapes oriented in different directions can easily mismatch by several percent. If the same resistors are constructed as an array of interdigitated sections, then they will certainly match to better than ±0.1%.

The exact degree of matching achievable from any given layout is difficult to determine. The hard data required to quantitatively evaluate matching performance is rarely available in a manufacturing environment, so the layout designer must make decisions based on limited information. Although this is a difficult and sometimes frustrating process, the designer can greatly improve the performance of many circuits by following the principles discussed in this chapter.

7.5 EXERCISES

Refer to Appendix C for layout rules and process specifications.

7.1. A pair of capacitors were designed to have values of 5pF and 2.5pF. Measurements on ten units provide the following pairs of values: (5.19pF, 2.66pF), (5.21pF, 2.67pF), (5.19pF, 2.65pF), (5.23pF, 2.66pF), (5.21pF, 2.68pF), (5.12pF, 2.67pF), (5.25pF, 2.68pF), (5.15pF, 2.63pF), (5.21pF, 2.61pF), (5.28pF, 2.61pF). What is the three-sigma worst-case mismatch between these two capacitors?

7.2. A twelve-wafer lot of wafers numbered 1, 2, 3 . . . 12 are available for use as samples in an experimental determination of mismatch. Provide detailed instructions for selecting a sample of 30 units from this lot.

7.3. A pair of 3pF capacitors have a measured standard deviation of mismatch of 0.17%. How large must the capacitors be made to ensure that they will achieve an estimated six-sigma worst-case mismatch of ±0.5%? Assume that systematic mismatches are negligible.

7.4. A certain design contains a pair of 3μm-wide resistors that have a measured standard deviation of mismatch of 0.32% and a measured systematic mismatch of +0.10%. Assuming that the systematic mismatch does not vary with width, how wide would the resistors have to be made to achieve an estimated three-sigma mismatch of ±0.5%?

7.5. Divide the following matched resistances into segments following the rules of Section 7.2.2, assuming a sheet resistance of 500Ω/□. In each case, state the number of segments in each resistor, and the segment resistances.

a. 10kΩ and 15kΩ.
b. 7.5kΩ and 11kΩ.
c. 3.66kΩ and 11.21kΩ.
d. 75.3kΩ and 116.7kΩ.

7.6. Divide the following matched capacitors into unit capacitances following the rules of Section 7.2.2, assuming an areal capacitance of 1.7fF/μm^2. State the number of unit capacitors in each device, the unit capacitor values, and their dimensions.

a. 4.0pF and 8.0pF.
b. 1.8pF and 4.2pF.
c. 3.7pF and 5.1pF.
d. 25pF and 25pF.

7.7. Choose an optimal orientation for each of the following types of resistors:

a. Standard bipolar HSR resistors.
b. Analog BiCMOS N-well resistors.
c. P-type polysilicon resistors.
d. Nichrome thin-film resistors (nichrome is a polycrystalline metal alloy).

7.8. Devise one-dimensional interdigitation patterns for each of the following cases:

a. Two resistors with a ratio of 4:5.
b. Two resistors with a ratio of 2:7.
c. Three resistors with a ratio of 1:3:5.
d. Four resistors with a ratio of 1:2:4:8.

7.9. Lay out a resistor divider consisting of two 3kΩ, 8μm-wide base resistors placed in a common tank. Include all necessary metallization, including a separate lead for the tank contact.

7.10. Construct a resistor divider consisting of two 25kΩ, 8μm-wide HSR resistors.

7.11. The divider of Exercise 7.10 forms part of a die having dimensions of 2150μm by 1760μm. These dimensions do not include scribe streets and seals, which may be ignored for the purposes of this exercise. Draw a rectangle having these dimensions and place the divider in the best possible location for optimal matching. Assume the design contains no major heat sources.

7.12. Repeat Exercise 5.13, laying out the resistors to obtain precise matching. One of the resistors would normally contain a single segment; replace this resistor with a series-parallel network containing an even number of 1kΩ segments.

7.13. Suppose the trim network in Exercise 7.12 forms part of a die having an active area of 5.3mm^2, of which a power transistor consumes 3.6mm^2. This area does not include scribe streets and seals, which need not be considered for this exercise. Choose an aspect ratio for the die and construct a rectangle having the requisite area. Place another rectangle in the layout to denote the location of the power device. Now locate the trim network in an optimal location along one side of the die. Place the trimpads as close to the edge of the die as possible. Space all metallization at least 8μm from the edge of the die.

7.14. Construct a capacitor array using analog CMOS poly-poly capacitors. The capacitors should have the following values: 0.5pF, 1pF, 2pF, 4pF, and 8pF. Assume that all capacitors share a common poly-1 plate, and bring a lead from each capacitor's poly-2 plate out to the edge of the array. Use copies of the unit capacitors as dummies and cover the array with a metal-2 shield. Take whatever other measures are required for precise matching.

8 Bipolar Transistors

The *bipolar junction transistor* (BJT) is among the most versatile of all semiconductor devices. In addition to its obvious applications as a voltage or current amplifier, it can also serve as the basis of voltage and current references, oscillators, timers, pulse shapers, amplitude limiters, nonlinear signal processors, power switches, transient protectors, and many other types of circuits. There are also certain applications for which bipolar transistors are ill-suited, the most important of these being low-power digital logic. Most logic is now constructed using *complementary metal-oxide-semiconductor* (CMOS) circuitry. Bipolar transistors remain important for constructing most analog circuits, although many of these circuits now contain CMOS elements as well.

Much of the information required to understand the operation and construction of bipolar transistors does not appear in elementary texts. This chapter opens with a review of several of these topics, including beta rolloff, avalanche breakdown, thermal runaway, and device saturation. The remainder of the chapter covers the design of small-signal bipolar transistors. This information lays the foundation for the more specialized topics covered in Chapter 9.

8.1 TOPICS IN BIPOLAR TRANSISTOR OPERATION

Figure 8.1 shows a simple model of an NPN transistor. Diode D_1 represents the base-emitter junction of the transistor. Current-controlled current source I_1 models the minority carrier current flowing across the reverse-biased base-collector junction. The current through I_1 equals the current through D_1 multiplied by the transistor's *forward active current gain* β_F. In terms of terminal currents, this relationship becomes

$$I_c = \beta_F I_B \qquad [8.1]$$

Unlike an MOS transistor, the BJT requires a constant base current to sustain the flow of collector current. This base current represents unavoidable losses due to recombination in the neutral base and carrier injection from base to emitter.

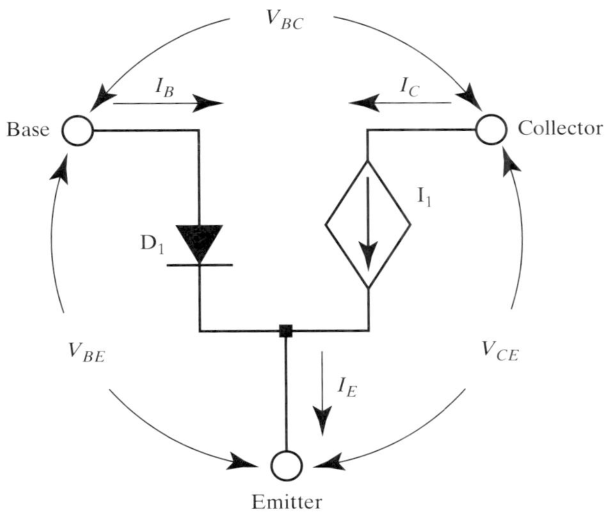

FIGURE 8.1 Simplified three-terminal model of an NPN transistor.

Because conduction requires constant base current, the bipolar transistor is often called a *current-controlled device.* This is somewhat misleading because the transistor can also be driven by a base-emitter voltage that forward-biases diode D_1 and provides base current to the transistor.

The base-emitter junction of a bipolar transistor is, for all intents and purposes, identical to the silicon junction diode discussed in Section 1.2.2. The base current of the transistor depends exponentially upon the base-emitter bias V_{BE}. If the forward beta remains constant and the collector-to-emitter bias V_{CE} suffices to maintain the transistor in the normal active region, then the collector current I_C also becomes an exponential function of V_{BE}. These relationships can be expressed by the following formulas:[1]

$$I_c = I_s e^{(V_{BE}/V_T)} \quad [8.2]$$

$$V_{BE} = V_T \ln\left(\frac{I_c}{I_s}\right) \quad [8.3]$$

The *emitter saturation current* I_S depends on several factors, including the doping profiles of the base and emitter diffusions and the effective area of the base-emitter junction. The *thermal voltage* V_T scales linearly with absolute temperature, and at 298K (25°C) it equals 26mV. The base-emitter voltage V_{BE} exhibits a negative temperature coefficient[2] of about –2mV/°C. This may seem small, but since collector current depends exponentially on base-emitter voltage, an increase of only 18mV in V_{BE} doubles the collector current. A 1°C temperature difference between two bipolar transistors will cause an 8% mismatch between their collector currents, corresponding to a temperature coefficient of 80,000ppm/°C! This enormous temperature coefficient has profound implications for the design of matched devices and power transistors.

---

[1] These formulas are slightly simplified; the actual Ebers-Moll equations include a term accounting for reverse conduction: $I_C = A_e J_S\,[\exp(V_{BE}\,/\,kT) - 1]$. The presence of the "–1" term has no significant effect upon conduction in the forward active region at any reasonable bias level and thus has been omitted. This simplified equation also neglects the ideality factor (or emission coefficient) η, which is very near unity for NPN transistor operating at moderate currents.

[2] Most of this temperature coefficient stems from the presence of V_T in the equation for V_{BE}, but the temperature coefficient of I_S also has some impact on V_{BE}.

8.1.1. Beta Rolloff

Elementary textbooks often assume that beta remains constant, but it actually varies considerably depending on collector current. Figure 8.2 shows typical beta curves for small signal NPN and lateral PNP transistors constructed in a standard bipolar process. The NPN beta remains relatively constant over a wide range of collector currents, which to some extent justifies the assumption of constant beta. At high current levels, typically beyond 5mA/mil^2 of emitter area (8μA/μm^2), the NPN beta begins to roll off. A similar but more gradual rolloff occurs at very low current levels, usually below 10nA/mil^2 (15pA/μm^2). High-current beta rolloff is caused by high-level injection, while low-current beta rolloff results from several mechanisms, including recombination within the depletion regions and at the oxide-silicon interface, and shallow emitter effects (Section 8.3.3).

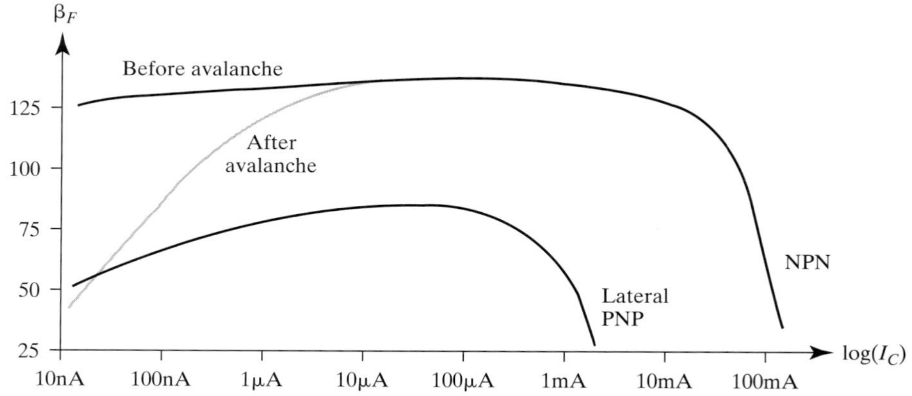

FIGURE 8.2 Beta versus collector current plots for small-signal NPN and lateral PNP transistors. The curve marked in gray shows the effect of emitter-base avalanche on NPN beta.

The beta curve of the lateral PNP differs considerably from that of the NPN. Not only does the lateral PNP have a lower peak beta, but it also exhibits more pronounced high-current and low-current beta rolloffs. Several factors account for the differences between the beta curves of the vertical NPN and the lateral PNP. The emitter of the PNP is much more lightly doped than that of the NPN, so the PNP exhibits a lower emitter injection efficiency that reduces its peak beta. The flow of carriers near the surface of the lateral transistor increases surface recombination and exacerbates low-current beta rolloff. The lightly doped base of the lateral PNP causes high-level injection to begin at relatively low current levels and thus accentuates high-current beta rolloff. The regions of high-current beta rolloff and low-current beta rolloff often overlap, causing a pronounced peaking of the beta curve.[3] Transistors that exhibit such peaking are actually operating in high-level injection at the point of peak beta, complicating the design of certain types of circuits.

8.1.2. Avalanche Breakdown

The maximum operating voltages of a bipolar transistor are determined by the breakdown voltages of the base-emitter and base-collector junctions. Depending upon biasing conditions, several different breakdown voltages may be observed. The three most important of these are denoted V_{EBO}, V_{CBO}, and V_{CEO}. Each of these is measured between two of the transistor terminals with the third terminal left unconnected (*open*).

[3] I. Getreu, *Modeling the Bipolar Transistor* (Beaverton, Oregon: Tektronix, 1976), p. 48ff.

The breakdown voltage of the transistor base-to-emitter with collector held open is represented by V_{EBO}. For an NPN transistor constructed on a standard bipolar process, V_{EBO} equals approximately 7V. This breakdown voltage remains relatively constant over process and temperature, so NPN transistors biased into emitter-base avalanche form useful Zener diodes. Base-emitter avalanche rapidly degrades the beta of NPN transistors because hot carriers generated by the avalanche process induce the formation of recombination centers along the oxide-silicon interface.[4] These recombination centers increase the recombination current within the depletion regions. This current adds to the base current of the transistor and drastically decreases the low-current beta of the device (Figure 8.2). Although high-current betas are not affected to the same degree, one should avoid avalanching any NPN transistor not specifically intended to operate as a Zener diode.

The breakdown of the transistor collector-to-base with emitter left open is represented by V_{CBO}. This breakdown voltage depends on several factors relating to the base-collector junction, all of which are discussed in greater detail in Section 8.2. Both the base and the collector of most bipolar transistors are lightly doped, so V_{CBO} is usually quite large. For NPN transistors constructed on standard bipolar processes, it can range from 20V to 120V or more. Since collector-base breakdown is largely subsurface, it does not generate surface recombination centers and therefore does not affect beta. Both the V_{EBO} and the V_{CBO} of the lateral PNP transistor depend on the breakdown of the base-epi junction, so the beta of lateral transistors is largely unaffected by any form of avalanche.

The breakdown of the transistor collector-to-emitter with base held open is represented by V_{CEO}. This is substantially smaller than V_{CBO} due to an effect called *beta multiplication*. Low levels of impact ionization begin to occur at voltages well below the nominal breakdown voltage. Since the base terminal of the transistor is left unconnected, any avalanche injection into the base produces a corresponding (and larger) increase in collector current. The additional carriers transiting across the base-collector junction increase impact ionization and generate additional base drive. At a voltage equal to V_{CEO}, this positive feedback mechanism becomes self-sustaining and the transistor avalanches. The V_{CEO} of an NPN transistor usually equals about 60% of its V_{CBO} (Section 8.2.4).

Beta multiplication can be suppressed by connecting the base terminal to the emitter, holding the transistor in cutoff and preventing amplification of the collector-base leakage. The breakdown voltage collector-to-emitter with base shorted to emitter is denoted V_{CES}. This breakdown voltage can approach V_{CBO} as long as the base resistance of the transistor is relatively small and the extrinsic base terminal connects to a potential equal to or less than the extrinsic emitter potential. If for any reason current begins to pass through the transistor, the breakdown voltage will suddenly decrease to nearly V_{CEO}. This phenomenon, called *snapback,* occurs even if the extrinsic base and emitter terminals are shorted, since the necessary bias develops across the internal base resistance of the transistor. Figure 8.3 shows idealized curve tracer plots illustrating this phenomenon. As soon as the trace labeled V_{CES} exceeds 60V, it immediately snaps back to 43V. As the current through the transistor increases, the avalanche voltage begins increasing again. This results partially from extrinsic collector resistance and partially from high-current beta rolloff decreasing the beta multiplication effect. The V_{CEO} of many transistors also shows a small amount of snapback due to low-current beta rolloff. For example, the transistor in Figure 8.3 registers an initial V_{CEO} breakdown of 43V and snaps back to a

---

4 B. A. McDonald, "Avalanche Degradation of h_{FE}," *IEEE Trans. on Electron Dev.*, Vol. ED-17, #10, 1970, pp. 871–878.

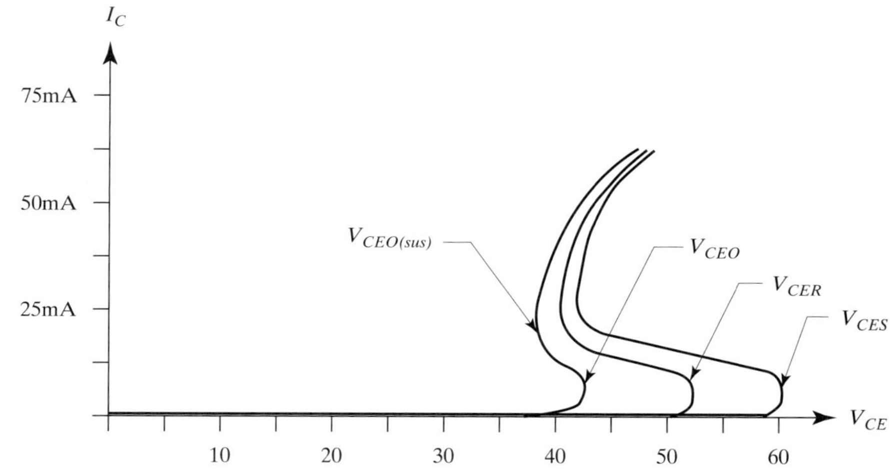

FIGURE 8.3 Idealized curve tracer plots of V_{CEO}, V_{CER}, and V_{CES} in an NPN transistor.

sustained V_{CEO} (sometimes called $V_{CEO(sus)}$) of 38V. Allowing for a small safety margin, this device would rate a V_{CEO} of approximately 36V.

The middle trace of Figure 8.3 represents the collector-to-emitter breakdown voltage of the transistor with a resistor connected between the base and emitter terminals (V_{CER}).[5] This condition lies between V_{CEO} and V_{CES} and shows the expected intermediate breakdown voltage along with a pronounced snapback.

8.1.3. Thermal Runaway and Secondary Breakdown

Bipolar transistors operating at relatively high power levels fall prey to a failure mechanism called *thermal runaway*. To illustrate how thermal runaway occurs, suppose that a large bipolar transistor suddenly begins to dissipate significant power. The center of the transistor quickly becomes warmer than its outer edges, causing the V_{BE} of the center to drop slightly. Due to the exponential character of the current-voltage relationship, a small change in base-emitter voltage produces a large change in collector current. Increased power dissipation occurs in the hotter portions of the transistor, leading to a further decline in V_{BE}. The region of the transistor that conducts current steadily shrinks as it grows hotter, until practically all of the current funnels through a very small, very hot area called a *hot spot*.

If the temperature in the hot spot reaches some 350 to 450°C, junction leakages become so large that the transistor essentially shorts out. Catastrophic failure occurs either when metallization is drawn through the contacts (as in a Zener zap structure) or when the silicon melts, cracks, or vaporizes. This type of self-destructive runaway does not always occur. Sometimes the increased current density causes beta to roll off far enough to stabilize the hot spot at a high, but not immediately destructive, temperature.[6] The presence of a "stable" hot spot dangerously overstresses a transistor and renders it vulnerable to electromigration, thermally accelerated corrosion, and various other long-term failure mechanisms.

A transistor that contains a stable hot spot often self-destructs during turn-off. Failure occurs due to an apparent avalanche of the collector-base junction, often at a voltage substantially below the rated V_{CEO} of the transistor. This unexpected

---

[5] V_{CER} has also been used to refer to the collector-to-emitter breakdown voltage with a reverse bias applied to the base-emitter junction. This mode of biasing is sometimes used in power circuits to raise the V_{CE} breakdown of the transistor above V_{CES}.

[6] P. L. Hower, D. L. Blackburn, F. F. Oettinger, and S. Rubin, "Stable Hot Spots and Second Breakdown in Power Transistors," *National Bureau of Standards*, PB-259 746, Oct. 1976.

reduction in avalanche voltage is called *secondary breakdown.*[7] It is a consequence of extremely high current densities within the transistor, due in this case to the presence of a stable hot spot. The velocity of carriers in the lightly doped collector increases in order to support the increasing current flow through this zone. Eventually the carrier velocity reaches its maximum limit (the *carrier saturation velocity*). Once this occurs, the electric field across the neutral collector intensifies, and the avalanche voltage of the transistor snaps back to a lower value, V_{CEO2}. If the voltage across the collector exceeds V_{CEO2}, the transistor avalanches and the resulting power dissipation quickly destroys the device.[8]

Secondary breakdown can also occur in transistors that have not experienced thermal runaway. During turnoff, the base lead withdraws charge from the neutral base. Charge removal begins in the portion of the base adjacent to the emitter periphery, and progresses inward toward the center of the transistor. As turnoff proceeds, conduction in the transistor collapses into an ever-shrinking area. This *emitter current focusing* causes the momentary appearance of extremely high current densities in small portions of the transistor. These current densities may become large enough to trigger secondary breakdown, especially if the transistor is conducting a large current at the time it is turned off.[9,10]

Thermal runaway and secondary breakdown can be avoided by restricting the operating conditions of the transistor. Figure 8.4 shows a graph illustrating the *forward-bias safe operating area* (FBSOA) of a typical bipolar transistor. The safe operating region is bounded by four separate curves.[11] The horizontal line represents the maximum current that the metallization and bondwires can safely carry without eventual electromigration failure. The vertical line represents the maximum voltage that can be placed across the transistor without fear of avalanche (usually assumed to equal $V_{CEO(sus)}$). A curve passing diagonally across the plot represents the maximum power that the device can dissipate without producing excessive temperatures

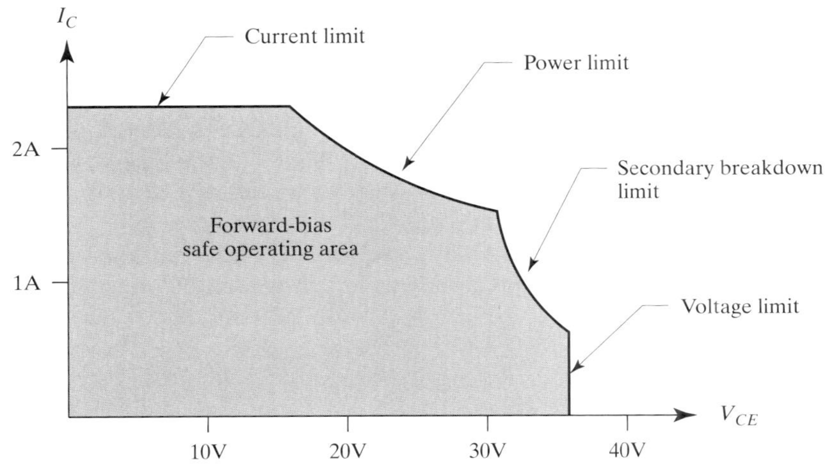

FIGURE 8.4 Forward-bias safe operating area (FBSOA) plot of a typical NPN power transistor.

[7] B. A. Beatty, S. Krishna, and M. S. Adler, "Second Breakdown in Power Transistors Due to Avalanche Injection," *Trans. on Electron Dev.*, Vol ED-23, #8, 1976, pp. 851–857.

[8] J. G. Kassakian, M. F. Schlect, and G. C. Verghese, *Principles of Power Electronics* (Reading, MA: Addison-Wesley, 1992), pp. 522–525.

[9] Kassakian, *et al.*, pp. 521–522.

[10] P. L. Hower and W. G. Einthoven, "Emitter Current-Crowding in High-Voltage Transistors," *IEEE Trans. on Electron Devices*, Vol. ED-25, #4, 1978, pp. 465–471.

[11] F. F. Oettinger, D. L. Blackburn, and S. Rubin, "Thermal Characterization of Power Transistors," *IEEE Trans. on Electron Devices*, Vol. ED-23, #8, 1976, pp. 831–838.

within the package. A fourth and final curve clips off the portion of the safe operating area where secondary breakdown may occur. A very robust transistor may not exhibit any FBSOA reduction due to secondary breakdown. A properly heat-sunk but poorly designed power transistor may lose a substantial fraction of its potential FBSOA to secondary breakdown. Section 9.1.2 discusses ways to increase the safe operating area of bipolar transistors.

8.1.4. Saturation in NPN Transistors

An NPN transistor enters saturation when both its base-emitter and base-collector junctions simultaneously forward-bias. Power transistors are often intentionally operated in saturation to reduce the collector-to-emitter saturation voltage $V_{CE(sat)}$ and to minimize power dissipation. Unfortunately, saturation also produces a whole host of problems. The unintentional saturation of bipolar transistors has probably caused more circuit malfunctions than any other device-related design flaw.

Saturation affects discrete and integrated transistors in different ways. The saturation of a discrete transistor merely prolongs its *turnoff time* (also called its *reverse recovery time*). As soon as the base-collector junction begins to forward-bias, minority carriers flow across it in both directions. In an NPN, holes flow into the collector and electrons into the base. A substantial population of excess minority carriers soon accumulates on both sides of the collector-base junction. Now suppose that the external circuit reduces the external base-emitter bias to zero. The transistor does not turn off immediately because the resistance of the neutral base impedes the withdrawal of the stored charge. The transistor continues to conduct until this charge is withdrawn or until it recombines. Saturation therefore increases the turnoff time by at least an order of magnitude. Fast bipolar logic usually incorporates anti-saturation clamps (as in LSTTL logic) or uses circuit topologies that are immune to saturation (as in ECL and DCML logic).

Saturation increases the reverse recovery time of an integrated bipolar transistor, but it also has other deleterious effects due to the presence of an additional junction between the collector and the isolation. Figure 8.5 shows a cross section of a typical NPN transistor fabricated on a standard bipolar process. The arguments presented for this structure apply equally to any other junction-isolated, vertical NPN transistor, including the CDI NPN of analog BiCMOS (Figure 3.48). These arguments do not apply to fully oxide-isolated transistors, which behave as if they were discrete devices.

The presence of junction isolation introduces a fourth terminal consisting of the P– substrate and the P+ isolation. This *substrate* terminal must always be biased to

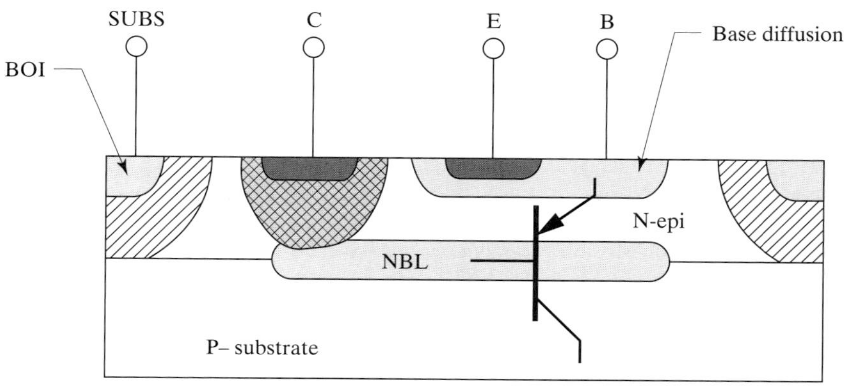

FIGURE 8.5 Cross section of an NPN transistor fabricated on a standard bipolar process, showing the parasitic PNP transistor.

a lower voltage than the collector terminal to avoid forward biasing the isolation junction. This reverse-biased junction isolates the transistor from the remainder of the integrated circuit as long as there are few minority carriers in the neutral collector. Junction isolation fails as soon as the base-collector junction begins to forward-bias. Many of the holes injected across the base-collector junction eventually diffuse across the collector to the collector-substrate junction. The electric field across this reverse-biased junction draws these holes into the substrate, where they become majority carriers.

Saturation can also be explained by imagining a PNP transistor superimposed upon the cross section of the NPN transistor (Figure 8.5). The collector-base junction of the vertical NPN also forms the base-emitter junction of this *parasitic PNP.* When the NPN saturates and forward-biases its collector-base junction, the parasitic PNP transistor turns on and diverts excess base drive into the substrate. This unexpected diversion of base drive has several unpleasant consequences. For one, an integrated NPN transistor cannot be driven as deeply into saturation as a discrete transistor can because the parasitic PNP steals its base drive as soon as it begins to saturate. Integrated NPN transistors therefore tend to have higher $V_{CE(sat)}$ voltages than equivalently constructed discrete transistors.

A saturating NPN also represents an unexpected source of substrate current that can potentially lead to substrate debiasing (Section 4.4.1). If the base drive to the transistor exceeds a few milliamps (as is often the case for power transistors), then special provisions may become necessary to prevent substrate debiasing. Substrate debiasing can be averted by adding guard rings to prevent the holes in the collector from reaching the substrate, or by designing the base-drive circuit to reduce the base-drive once the transistor begins to saturate (Section 9.1.3).

Saturation also causes a failure mechanism called *current hogging.* When a transistor saturates, some of its base current flows through the base-collector junction rather than the base-emitter junction. This diversion of current reduces the base-emitter voltage.[12] Many circuits connect the base-emitter junctions of several transistors in parallel and expect the collector currents of these devices to track the areas of the respective base-emitter junctions. This relationship ceases to apply if one of the transistors saturates because this transistor experiences a drop in V_{BE} relative to the remaining transistors. The base current of the saturating transistor therefore increases at the expense of the other transistors. In more colloquial terms, the saturating transistor *hogs* the base drive.

Figure 8.6 shows a simple current mirror constructed from three NPN transistors. Current source I_{BIAS} feeds diode-connected transistor Q_1. All three transistors see the same base-emitter voltage and therefore draw the same collector currents, *providing that all three transistors operate in the normal active region.* To illustrate why this is the case, suppose transistor Q_3 saturates. Much of its base current diverts into parasitic PNP transistor Q_P, causing the base-emitter voltage of Q_3 to decrease. Eventually Q_3's V_{BE} drops to a point just sufficient to satisfy the *intrinsic collector current* I_3, restoring equilibrium. In this case, the intrinsic collector current equals the sum of the extrinsic collector current, I_{t3}, and the base current of Q_P. Transistor Q_2 sees the same base-emitter bias as Q_3, so its collector current, I_2, equals the intrinsic collector current, I_3. In summary, when one transistor in a mirror saturates, the extrinsic collector currents of all the other transistors decrease to equal the

[12] A vertical transistor is particularly susceptible to current hogging, because its heavily doped emitter region raises the base-emitter voltage at a given bias level a few tens of millivolts above the corresponding base-collector voltage, so current flows preferentially through the base-collector junction.

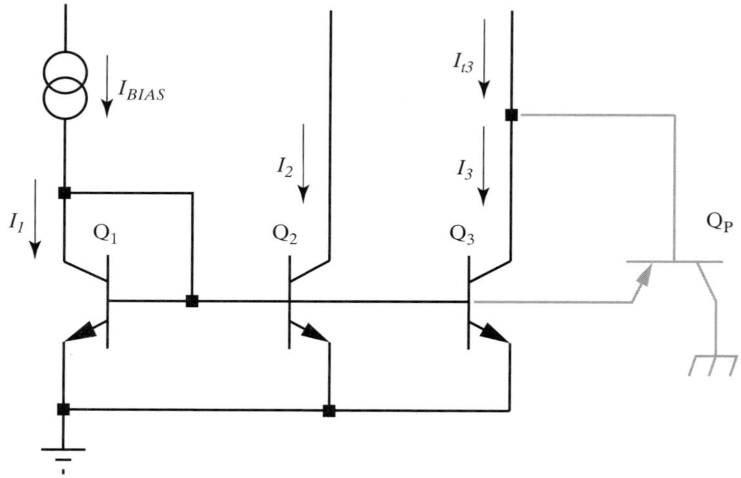

FIGURE 8.6 An example of a circuit that exhibits current hogging. Q_P represents the parasitic PNP transistor present in the structure of vertical NPN Q_3.

intrinsic collector current of the saturated transistor. This disturbs the balance of the circuit and often leads to serious malfunctions.

Circuit designers have developed several cures for current hogging, including base-side ballasting and Schottky clamps. *Base-side ballasting* requires the insertion of matched resistors into the base leads of each transistor (Figure 8.7).

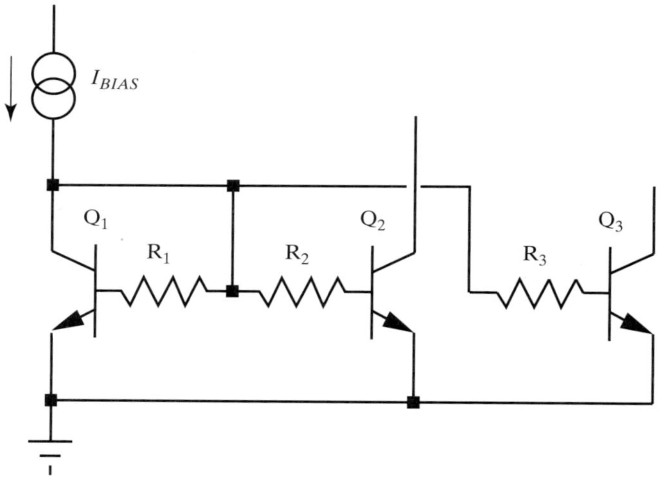

FIGURE 8.7 Base-side ballasting applied to the circuit shown in Figure 8.6.

The base ballasting resistors must ratio inversely to the emitter areas of their respective transistors. For example, if Q_2 has twice the emitter area of Q_1, then R_2 must equal half the resistance of R_1. The base ballasting resistors must match in order to maintain collector current matching. The base-side ballasting prevents current hogging by introducing localized negative feedback. If any one of the transistors begins to saturate, its base current will increase slightly. This causes a corresponding increase in the voltage drop across its base ballasting resistor, forcing a drop in the V_{BE} of this transistor. Typically, no more than 50 to 100mV appears across the base ballasting resistor of a saturating transistor, producing a correspondingly small disturbance in the base-bias currents.

A diffused base ballasting resistor must not occupy the tank of the NPN transistor it protects because then the ballasting resistor can forward-bias into the tank

(Figure 8.8). In effect, the parasitic PNP transistor simply moves from one point in the structure to another.

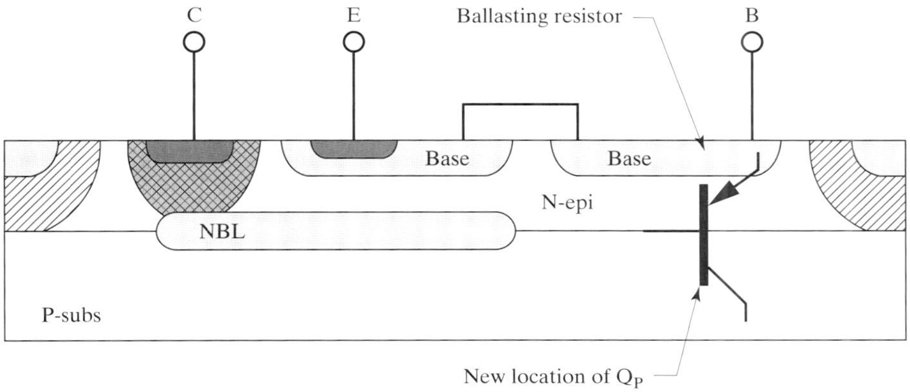

FIGURE 8.8 A base-side ballasting resistor becomes ineffective when merged into the same tank as the NPN it protects.

A clamping diode connected across the base-collector junction of a transistor will prevent it from entering saturation. In order for the clamping diode to function properly, it must have a lower forward voltage than the base-collector junction. Only a few types of Schottky diodes, most notably those constructed using platinum or palladium silicides, have the necessary characteristics. Figure 8.9A shows the connection of a Schottky clamp diode in parallel with the base-collector junction of an NPN transistor. The resulting *Schottky-clamped NPN* is often represented by the symbol of Figure 8.9B. The Schottky clamp works by providing an alternate path for current that would otherwise flow through the forward-biased base-collector junction. Because the diode prevents the base-collector junction from conducting, a Schottky-clamped NPN neither injects appreciable minority carriers nor experiences the prolonged turn-off times characteristic of saturation. The Schottky clamp does not prevent the base current from increasing, but it does prevent it from exceeding the collector current the transistor would normally conduct. Schottky-clamped transistors are used extensively in switching circuitry and bipolar logic to eliminate saturation-induced propagation delays. Section 10.1.3 discusses the layout of Schottky-clamped NPN transistors in further detail.

FIGURE 8.9 (A) Schottky-clamped NPN transistor and (B) its conventional schematic symbol.

Saturating NPN transistors can also cause problems when merged with other devices. When a transistor saturates, its forward-biased base-collector junction injects minority carriers into its collector. These carriers can be collected by the reverse-biased junctions of other devices merged into the same tank. Sneak currents associated with minority conduction may cause circuit malfunctions or even catastrophic latchup. Section 13.2.1 discusses these difficulties in greater detail.

8.1.5. Saturation in Lateral PNP Transistors

Figure 8.10 shows a cross section of a lateral PNP transistor constructed in a standard bipolar process. An epi tank forms the base of the device while a small plug of base diffusion placed in the center of the tank acts as its emitter. The collector consists of an annular base diffusion surrounding the emitter. A substantial fraction of the total emitter-base junction area consists of sidewalls in close proximity to the surrounding collector. Most of the holes travel laterally from the sidewalls of the emitter to the sidewalls of the collector. Some holes are either injected from the bottom surface of the emitter, or diffuse away from the surface. These errant minority carriers can flow across the reverse-biased junctions isolating the transistor from the isolation and substrate. Parasitic PNP transistor Q_{P1} represents the undesired flow of holes down to the substrate, while Q_{P2} represents the flow of holes to the sidewalls of the tank. Figure 8.10 shows both of these parasitics superimposed upon the cross section of the transistor, along with the desired lateral PNP transistor Q_L.

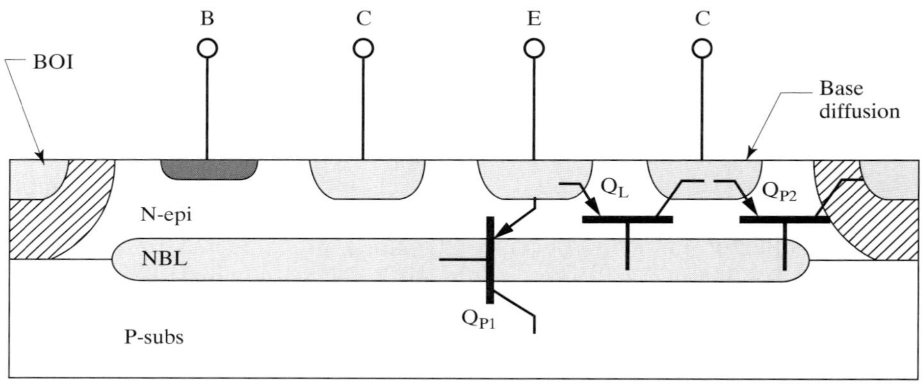

FIGURE 8.10 Cross section of a lateral PNP transistor constructed on a standard bipolar process showing parasitic substrate PNP transistors Q_{P1} and Q_{P2}.

A significant fraction of the total emitter current will be lost if measures are not taken to block the flow of holes to the substrate. The *collector efficiency* η_C of a lateral PNP equals the ratio of the current collected by lateral transistor Q_L to the sum of the collector currents of Q_L, Q_{P1}, and Q_{P2}. In terms of collector, emitter, and base terminal currents I_C, I_E, and I_B, the collector efficiency η_C equals

$$\eta_c = \frac{I_c}{I_E - I_B} \qquad [8.4]$$

For reasons explained below, the addition of NBL causes the collector efficiency of a lateral PNP to rise from less than 0.1 to near unity.[13] While lateral PNP transistors can be constructed in CMOS processes, the absence of NBL makes them extremely inefficient. Standard bipolar and analog BiCMOS processes both incorporate NBL and generally provide excellent lateral PNP transistors.

NBL improves the collection efficiency of a lateral PNP by repelling minority carriers moving toward it. This repulsion occurs because of the presence of an opposing electric field at the interface between the heavily doped NBL and the lightly doped N-epi. The presence of this electric field can be explained as fol-

[13] Analog BiCMOS processes can create lateral PNP transistors with very narrow base widths, producing collector efficiencies of 0.3 or more without using NBL.

lows: The difference in doping concentrations between N-epi and NBL causes a similar difference in majority carrier concentrations. A diffusion current consisting of electrons flows from the NBL to the N-epi. This diffusion current must be balanced by an equal and opposite drift current. In order to set up this drift current, the NBL must become slightly more positive than the N-epi. The resulting electric field attracts electrons from the N-epi back into the NBL. This same electric field also causes holes to scatter off the N+/N– interface and to rebound back into the overlying epi.[14] The scattered carriers continue to wander randomly through the epi tank. Those moving downward are again repelled from the N+/N– interface, while those moving upward eventually reach the collector-base depletion layer of the lateral PNP. Most of these carriers are eventually collected by the lateral transistor, leading to a large increase in collector efficiency. A small fraction of the downward-moving holes have sufficiently large instantaneous velocities to surmount the opposing electric field and to actually enter the NBL. Most of the holes that enter the NBL recombine within it due to the large population of available majority carriers.

A few of the holes scattered from the NBL interface move laterally through the epi until they reach the isolation sidewalls. Although this may seem a serious problem, the dimensions of the transistor actually preclude any significant loss of current through lateral parasitic conduction. The cross section in Figure 8.10 has been exaggerated vertically to create a relatively compact drawing. The lateral distance to the sidewall is roughly an order of magnitude larger than the vertical separation between the upper edge of the NBL interface and the lower edge of the collector-base depletion region. At higher collector voltages the depletion region extends down to the N+/N– interface, closing off the path for lateral parasitic conduction to the sidewalls. Even at low collector voltages, this path is so long and so narrow that few minority carriers can traverse its entire length without coming into contact with the collector-base depletion region.

Lateral parasitic conduction increases dramatically during saturation. The collector possesses a large sidewall area facing the isolation across a relatively narrow gap. This geometry forms an efficient PNP transistor that activates as soon as the collector forward-biases, as happens in saturated transistors and those operated in reverse-active mode. When a lateral PNP transistor saturates, the emitter current remains unchanged. Whatever injected holes are not collected by the lateral transistor travel to the substrate instead. Therefore lateral PNP current mirrors continue to operate properly even if one or more of their transistors in the mirror saturate. All that happens is that the unused collector current flows to the substrate. This does not present a problem as long as the collector currents do not exceed a few milliamps. If substrate injection cannot be tolerated, then a lateral PNP transistor can be fitted with a Schottky clamp. Section 9.1.3 discusses two alternate methods of limiting saturation in lateral PNP transistors.

Split-collector lateral PNP transistors are seriously affected by the saturation of any one of their several collectors. The saturating collector segment re-injects holes from all of its junction surfaces. Some of these are collected by the isolation sidewall, but significant portions of the re-injected carriers are collected by the adjacent collector segments. Consequently, the saturation of one collector in a split-collector lateral PNP causes the current drawn by the adjacent collectors to increase.

---

[14] NBL is often said to "reflect" minority carriers. This choice of wording is somewhat misleading because the mean free path of the carriers is usually quite short relative to the dimensions of the transistor. Therefore the "reflected" carriers do not move far before their motion is randomized through lattice interactions.

8.1.6. Parasitics of Bipolar Transistors

An integrated bipolar transistor behaves quite differently from an idealized textbook device due to the many parasitic elements it contains. Perhaps the most important of these is the PN junction which isolates the transistor from the rest of the die. If this junction forward-biases, it will inject current into the substrate. Leakage currents and capacitive coupling can still occur even if the junction remains reverse-biased. The substrate connection must therefore be counted as one of the terminals of the integrated bipolar transistor. Figure 8.11B shows one conventional method of representing an NPN transistor with an explicit substrate connection. This symbol is often called a *four-terminal* NPN to distinguish it from the *three-terminal* NPN in Figure 8.11A. PNP transistors can also have both three-terminal and four-terminal symbols (Figure 8.11C and D). Although the three-terminal symbols do not show a substrate connection, one still exists as long as the process employs junction isolation. This implicit substrate connection can cause considerable trouble if the designer forgets its presence and assumes that the transistor is truly isolated from the remainder of the die. Many designers routinely use four-terminal symbols rather than three-terminal ones to avoid problems of this sort.

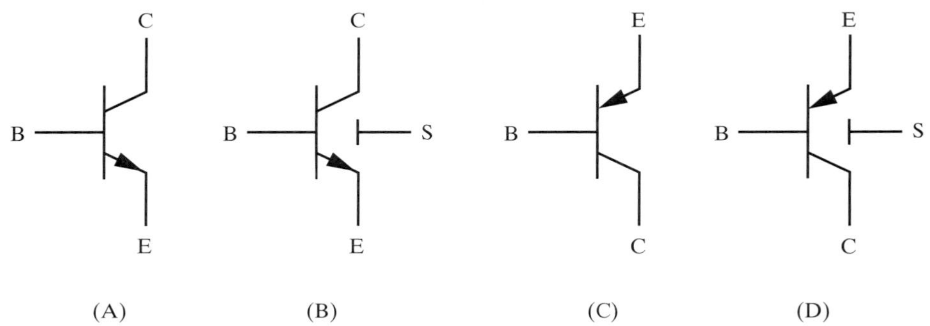

FIGURE 8.11 Symbols for three-terminal and four-terminal transistors (E: emitter, B: base, C: collector, S: substrate).

A complete model of the parasitics of a bipolar transistor includes a number of distributed effects, the discussion of which lies beyond the scope of this text. Figure 8.12 shows simplified parasitic models for a vertical NPN and a lateral PNP formed in a standard bipolar process. These same models also apply to most other junction-isolated transistors, including the vertical NPN and lateral PNP of analog BiCMOS.

The vertical NPN transistor model contains an ideal three-terminal NPN transistor Q_1 that models the intended functionality of the device. Transistor Q_2 represents the parasitic substrate PNP transistor that forward-biases when the vertical NPN transistor saturates (Section 8.1.4). Zener diodes D_{BE}, D_{BC}, and D_{CS} are not intended to model the behavior of the forward-biased junctions, as these are already subsumed into the behavior of transistors Q_1 and Q_2. The Zener diodes instead model the avalanche breakdown, leakage, and capacitance associated with the respective junctions. For example, diode D_{BE} models the base-emitter capacitance C_{BE} of the transistor as well as the base-emitter breakdown voltage V_{EBO}. Diode D_{BC} models the base-collector capacitance C_{BC} as well as the base-collector breakdown voltage V_{CBO}. Diode D_{CS} models the collector-substrate capacitance C_{CS} and the collector-substrate breakdown voltage. The collector-base capacitance C_{BC} and the collector-substrate capacitance C_{CS} are of concern because they limit the operating frequency of the transistor. The transistor switches faster if these junctions are made smaller. The avalanche voltages of the diodes D_{BE}, D_{BC}, and D_{CS} also set the operating voltage of the transistor, as discussed in Section 8.1.2.

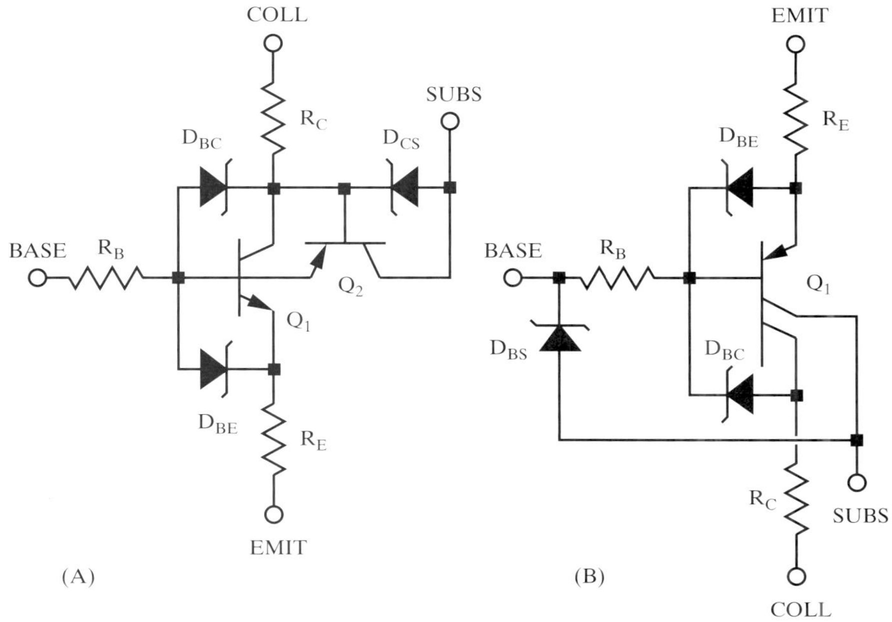

FIGURE 8.12 Subcircuit models for (A) vertical NPN transistor and (B) lateral PNP transistor.

Resistances R_E, R_B, and R_C model the Ohmic resistances of the emitter, base, and collector diffusions, respectively. Resistor R_E is usually quite small and is frequently neglected. Resistor R_B has a profound effect upon the switching speed of the transistor, as the base current must flow through a low-pass RC filter formed by R_B in series with capacitance C_{BE} of diode D_{BE}. The resulting low-frequency rolloff limits the maximum operating frequency of the transistor. Transistors with lower base resistances switch faster than those with high base resistances. The base resistance of a typical vertical NPN consists of the sum of the pinched base resistance beneath the emitter and the resistance of the extrinsic base underneath the contact. The presence of the pinched resistance causes the behavior of the base resistance to vary with bias in a very complex manner.[15] The collector resistance R_C limits the *saturation voltage* $V_{CE(sat)}$ that the transistor can achieve in saturation. Power transistors must have low collector resistances in order to minimize their saturation voltages at high currents. Also, the collector resistance of a transistor must not become too great or the transistor may intrinsically saturate even when the terminal voltages appear to indicate operation in the normal active region.

The lateral PNP transistor model of Figure 8.12B uses an ideal dual-collector PNP Q_1 to model the desired PNP transistor and its substrate PNP parasitic (Section 8.2.3). Some portion of the collector current flows through each of the two collectors. If the collector of the four-terminal transistor saturates, then most of the current instead flows to the substrate terminal (Section 8.1.5). Zener diodes D_{BE}, D_{BC}, and D_{BS} model the avalanche breakdown and capacitance effects of the three junctions of the lateral PNP transistor. Diode D_{BE} models the base-emitter junction's capacitance C_{BE} and its breakdown voltage V_{EBO}. Diode D_{BC} models the

[15] J. R. Hauser, "The Effects of Distributed Base Potential on Emitter-Current Injection Density and Effective Base Resistance for Stripe Transistor Geometries," *IEEE. Trans. on Electron Devices*, Vol. ED-11, #5, 1964, pp. 238–242.

base-collector junction's capacitance C_{BC} and its breakdown voltage V_{CBO}. The two breakdown voltages V_{EBO} and V_{CBO} are about equal because both junctions consist of the same diffusions. Capacitance C_{BC} is usually quite large due to the construction of the lateral PNP transistor, partly accounting for its poor frequency response. Diode D_{BS} models the base-substrate junction's avalanche voltage and its capacitance C_{BS}. This capacitance, which is also rather large, does not exist in a vertical transistor. The base-substrate capacitance of a lateral PNP also accounts for some of this transistor's slow frequency response.

Resistors R_E, R_B, and R_C model the Ohmic resistances of the emitter, base, and collector, respectively. Although both the emitter and the collector resistances may equal a few hundred Ohms, neither greatly affects the operation of the transistor. Resistor R_B is quite large because of the light doping of the N-epi tank. The presence of NBL does not completely counteract the effect of light doping because conduction in the lateral PNP transistor occurs near the surface, and the base current must consequently flow through virtually the entire thickness of the epi. This large base resistance forms an RC filter with capacitances C_{BC} and C_{BS}, accounting for the notoriously sluggish frequency response of lateral transistors.

8.2 STANDARD BIPOLAR SMALL-SIGNAL TRANSISTORS

Standard bipolar was originally used to construct digital logic circuits. Designers soon realized that this process could also fabricate analog integrated circuits such as voltage references, operational amplifiers, and comparators. None of these circuits operate at particularly high voltages or currents. Most of their transistors conduct, at most, a couple of milliamps. The design of these *small-signal transistors* emphasizes small size, high gain, and high speed at the expense of power-handling capability. Although designers may differ on exact definitions, most would probably agree that small-signal transistors handle currents of less than 10mA and power levels of less than 100mW. Transistors that exceed these limits begin to resemble power transistors more than small-signal devices (Section 9.1.2).

The standard bipolar process was optimized to fabricate NPN transistors, but its various diffusions can also create substrate and lateral PNP transistors. The design principles of small-signal transistors remain much the same regardless of the details of the process. The structure of any optimized bipolar transistor resembles that of a standard bipolar NPN. Most nonoptimized transistors resemble either the substrate or the lateral PNP. By studying the transistors available in standard bipolar, one can gain insight into how transistors are implemented in other processes.

8.2.1. The Standard Bipolar NPN Transistor

The standard bipolar NPN contains several features intended to optimize its performance. These include a heavily doped emitter; a precisely tailored base profile; a thick, lightly doped N-epi; a heavily doped NBL; and a deep-N+ sinker (Figure 8.13).

The emitter diffusion is heavily doped with phosphorus to maximize its emitter injection efficiency. The solid solubility of phosphorus in silicon allows doping levels exceeding 10^{20} atoms/cm^3, allowing the construction of a highly efficient emitter that takes full advantage of a carefully tailored base profile. Arsenic is sometimes added to the emitter diffusion to compensate for lattice strains induced by the heavy phosphorus doping. This precaution eliminates defects that might otherwise migrate into the neutral base and degrade the beta of the transistor.

The standard bipolar base diffusion has been tailored to provide a combination of high beta, high Early voltage, and moderate V_{CEO}. Light doping aids in main-

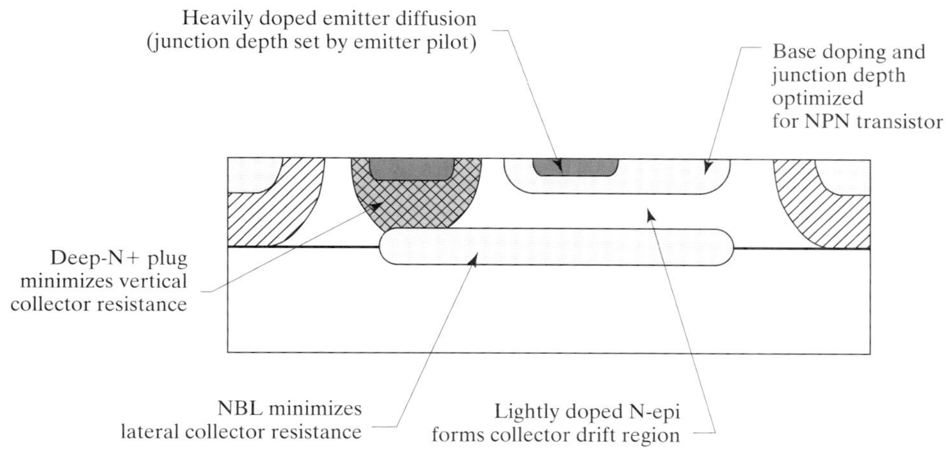

FIGURE 8.13 Key features of the standard bipolar vertical NPN transistor.

taining beta control because it allows the use of a wider, and therefore more controllable, neutral base region. A lightly doped base also increases the planar V_{CBO}, and by extension, the V_{CEO} of the vertical NPN. The standard bipolar base diffusion must still contain sufficient dopant to allow direct Ohmic contact since no shallow P+ diffusions exist in this process. Too low a surface doping concentration can also cause surface inversion and parasitic channel formation that can reduce the low-current beta of the transistor. Compromises between these conflicting requirements usually result in a base sheet resistance of 100 to 200Ω/□.

The NPN collector consists of three separate regions: a lightly doped N-epi, an N+ buried layer, and a deep-N+ sinker. The inclusion of a lightly doped layer adjacent to the collector-base junction increases V_{CEO} and Early voltage by allowing the formation of a wide depletion region extending primarily into the collector. This lightly doped *drift region* lies sandwiched between the base and a heavily doped *extrinsic collector.* In standard bipolar, the drift region consists of the remaining lightly doped N-epi beneath the base diffusion and above the NBL, while the extrinsic collector consists of the NBL and the deep-N+ sinker. The drift region, although lightly doped, does not impede the flow of collector current as long as it remains entirely depleted. The drift region depletes through at higher currents due to velocity saturation and at higher collector-to-emitter voltages due to the extension of the base-collector depletion region. Between these two extremes lies a range of collector voltages and currents that cannot entirely deplete the drift region, causing the effective resistance of the neutral collector to increase (an effect sometimes called *quasisaturation*[16]). Quasisaturation can be minimized by keeping the drift region as thin as possible consistent with the V_{CEO} rating of the process.

NBL creates a low-resistance path across the bottom of the transistor, but the current must still flow upward to reach the collector contact. The lightly doped epi separating the contact from the underlying NBL is highly resistive, so the inclusion of a deep-N+ sinker can reduce the total collector resistance by as much as an order of magnitude. For example, a typical minimum-size NPN transistor with NBL but without deep-N+ exhibits a collector resistance of about 1kΩ, while the same structure with the addition of a deep-N+ sinker has a collector resistance of about 100Ω. Power transistors always include deep-N+, but small-signal devices conducting no more than a few hundred microamps often omit it to save space.

---

[16] G. Massobrio and P. Antognetti, *Semiconductor Device Modeling with SPICE,* 2nd ed. (New York: McGraw-Hill, 1993), pp. 111–115.

Construction of Small-signal NPN Transistors

Small-signal NPN transistors usually employ square or rectangular emitters. Figure 8.14 shows two examples, both of which include the features discussed in the previous section. These transistors differ only in the placement of their base and emitter contacts. The structure of Figure 8.14A places the emitter between the collector and base contacts, forming a *collector-emitter-base* (CEB) layout. The structure of Figure 8.14B places the base contact between the collector contact and the emitter, forming a *collector-base-emitter* (CBE) layout. The CEB layout places the emitter and collector contacts closer together and therefore slightly reduces the collector resistance. All other factors remaining equal, the CEB layout will outperform the CBE layout. The difference is so small, however, that many designers use the two layouts interchangeably. The substitution of one style of transistor for the other often simplifies the lead routing of single-level-metal designs.

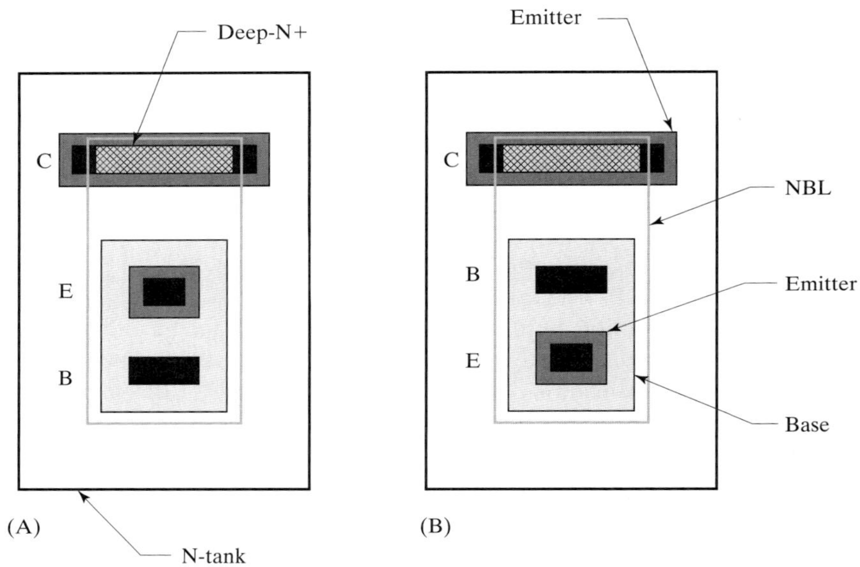

FIGURE 8.14 Two styles of NPN transistors: (A) collector-emitter-base (CEB), and (B) collector-base-emitter (CBE).

The size of an NPN transistor, or more precisely, the magnitude of its saturation current I_S, scales approximately linearly with drawn emitter area. Other factors that exert a lesser degree of influence on scaling include lateral conduction, current crowding, and emitter push. Carriers injected laterally from the emitter sidewalls suffer more recombination than those injected vertically because the laterally injected carriers must cross a wider and more heavily doped base region than the vertically injected ones. Surface states along the oxide-silicon interface also increase lateral recombination. Because of these effects, transistors with large area-to-periphery ratios usually have higher betas than those with smaller ratios. Measurements of two NPN transistors fabricated on a standard bipolar op-amp, one with a 1.0 × 1.0mil emitter and the other with a 1.5 × 3.5mil emitter,[17] showed that the smaller transistor exhibited a peak beta of 290 and the larger one a beta of 520. Most small NPN transistors employ either square or slightly elongated rectangular emitters to make efficient use of the available space while retaining a high area-to-

[17] B. A. Wooley, S.-Y. J. Wong, and D. O. Pederson, "A Computer-Aided Evaluation of the 741 Amplifier," *IEEE J. Solid-State Circuits,* Vol. SC-6, 1971, pp. 357–365.

periphery ratio (Section 9.2.1). Larger emitters also drive deeper into the base diffusion and thus reduce the neutral base width.

All of the other geometries that form the NPN transistor are placed relative to the emitter geometry, beginning with the emitter contact. This contact usually occupies as much of the emitter area as possible in order to minimize emitter resistance. The emitter diffusion should overlap the emitter contact equally on all sides to ensure uniform lateral current flow. The base diffusion must overlap the emitter on all sides sufficiently to prevent lateral punchthrough. To save space, the base is usually contacted along only one side of the emitter. The base contact can be elongated without enlarging the transistor, significantly reducing base resistance.

The outdiffusion of the P+ isolation determines the minimum tank overlap of base diffusion, which is among the largest spacings of the process. Up-down isolation can reduce this spacing by perhaps a third (Section 3.1.4). The base region occupies a tank contacted at one end by a deep-N+ sinker. Deep-N+ outdiffuses and must therefore reside some distance away from both the base and the isolation diffusions. Emitter coded over the deep-N+ increases the surface doping to ensure reliable Ohmic contact. Both the deep-N+ sinker and the emitter diffusion used to contact it can be elongated without increasing the size of the transistor. The deep-N+ sinker is often omitted from low-current devices to save space. Regardless of whether a sinker is used, the NBL should fill as much of the tank as possible to minimize the overall collector resistance. Minimum-geometry transistors often exhibit little overlap of NBL and deep-N+. As long as the drawn geometries touch one another, the collector resistance will be reduced sufficiently to allow low-current operation without significant collector voltage drops. If the transistor must conduct more than a few milliamps, then the transistor should be enlarged to allow the NBL to completely enclose the deep-N+ sinker.

Many variations of the standard bipolar NPN layout exist, especially for single-level-metal processes. The lack of second metal forces the designer to route leads through the transistor. Selective use of CEB and CBE layouts allows some rearrangement of the terminal ordering and often eliminates the need to jumper one or more signals. Transistors can also be stretched to allow one or more leads to route between their terminals. Stretched transistors exhibit increased base and collector resistances and capacitances that could interfere with proper circuit operation, so designers should avoid stretching devices whenever possible.

Figure 8.15A shows a typical *stretched-collector transistor*. The collector and base contacts have been moved apart to allow two leads to pass between them. This modification increases collector resistance and the collector-to-substrate capacitance.

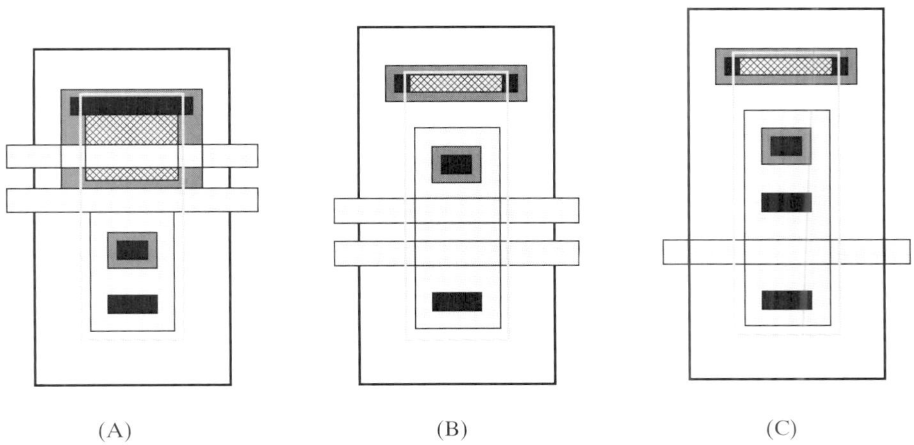

FIGURE 8.15 Three examples of stretched NPN transistors.

Nothing can be done to eliminate the added capacitance, but the increased resistance can be partially offset by elongating the deep-N+ and emitter diffusions surrounding the collector contact. In processes that employ a thin emitter oxide, leads routed over an extended emitter diffusion may represent an ESD vulnerability (Section 4.1.1), and only the deep-N+ and not the enclosing emitter diffusion should be elongated.

Figure 8.15B shows a *stretched-base transistor.* The base region of this transistor has been elongated to allow two leads to pass between the base and emitter contacts. The elongated base geometry causes an increase in base resistance and collector-base capacitance and a corresponding reduction in the frequency response. Stretched bases usually have a greater effect on circuit performance than stretched collectors, so they should be employed only when absolutely necessary, especially in high-speed signal paths. Figure 8.15C shows a different type of stretched-base transistor in which a lead tunnels through the base diffusion. This arrangement not only increases collector-base capacitance, but also inserts considerable resistance between the two base contacts. Assuming a base sheet of 160Ω/□, the resistance between the two base contacts of the illustrated transistor equals about 200Ω.

Double-level metal (DLM) virtually eliminates the need for stretched transistors. Not only does it eliminate stretched devices and tunnels, but it also reduces the die area because metal jumpers and vias can reside on top of other devices. DLM also allows the use of a small number of standardized layouts that can be fully characterized and modeled to enable more accurate simulation. The widespread implementation of multilevel metal systems has almost eliminated the use of stretched devices, but the technique still has merit for certain applications, for example in photodevices where metal-2 must act as a light shield, or in power devices where the upper metal layers are devoted to power routing.

Several approaches exist for scaling up NPN transistors, all of which seek to increase emitter area without reducing device performance. This turns out to be an elusive goal that requires different strategies for different applications. The naïve approach consists of enlarging the emitter while retaining the same overall geometry (Figure 8.16A). This yields a high beta due to an increased emitter area-to-periphery ratio, but it also entails several serious disadvantages. The lightly doped pinched base region beneath the emitter has a sheet resistance of about 2 to 10kΩ/□. This resistance introduces unwanted phase shifts and greatly slows transistor switching. More subtly, it causes nonuniform current flow at higher current levels. The portions of the emitter lying closest to the base contact experience the highest base-emitter bias and, consequently, inject more carriers than portions of the emitter far from the base contact. Figure 8.16B illustrates this effect, called *current*

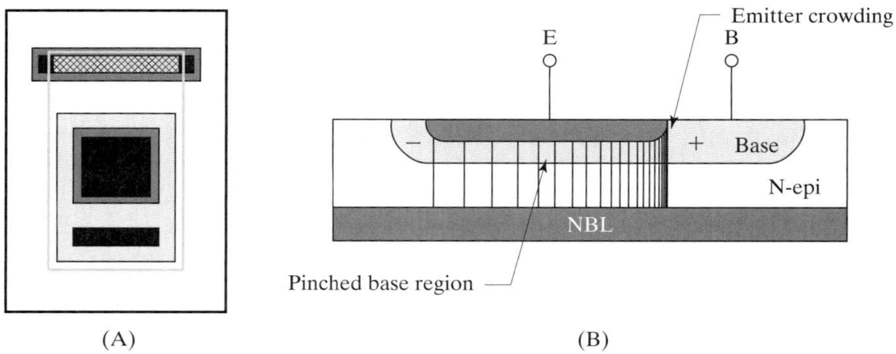

FIGURE 8.16 (A) Layout of a compact emitter NPN transistor and (B) a cross section of the active region of this device that illustrates the effects of emitter crowding.

crowding or *emitter crowding.* Its severity can be appreciated by remembering that a mere 18mV of debiasing doubles the emitter current.[18]

Current crowding reduces the active area of the emitter by forcing most of the conduction to occur near the base contact. This effect complicates device matching, reduces beta, and makes the transistor more susceptible to secondary breakdown. While current crowding is undesirable in small-signal transistors, it can actually enhance the robustness of properly designed power devices (Section 9.1.2).

An alternative style of layout uses long, narrow emitter stripes, or *fingers.*[19] Base contacts placed along both sides of each emitter finger help minimize base resistance, increasing switching speed and enhancing immunity to secondary breakdown (Figure 8.17A). The relatively small area-to-periphery ratio of this transistor yields a lower beta than the structure of Figure 8.16. This *narrow-emitter transistor* also becomes vulnerable to thermal runaway at emitter current densities of more than a few mA/mil^2 of emitter (Section 9.1.2).

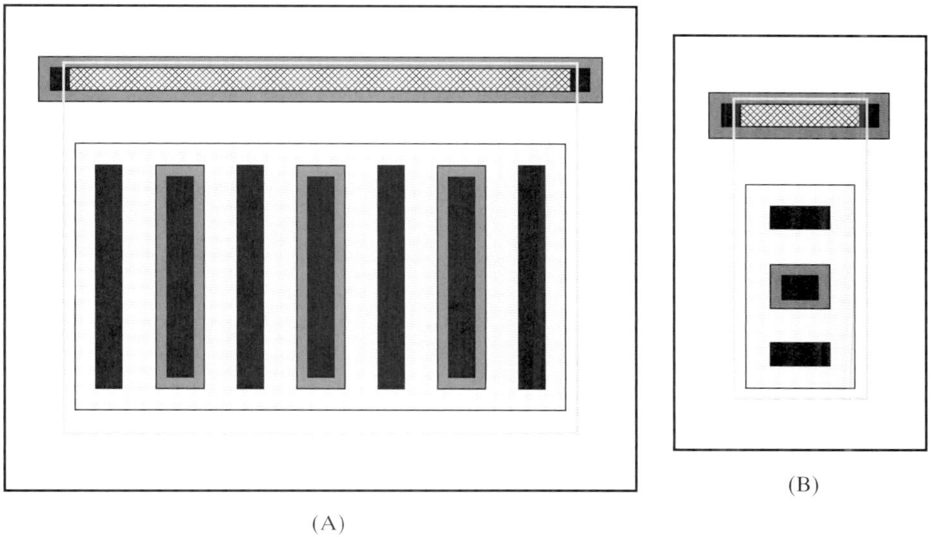

FIGURE 8.17 (A) The narrow-emitter transistor and (B) its equivalent minimum-size layout, the double-base transistor.

The compact emitter transistor in Figure 8.16 functions best in applications requiring high beta and moderate speeds. The narrow emitter transistor in Figure 8.17 provides superior frequency response, but inferior beta. A minimum-geometry transistor can also employ an emitter stripe paralleled by base contacts on either side, producing the somewhat misnamed *double-base transistor* (Figure 8.17B). This structure reduces the base resistance to approximately one-quarter that of the single-base layout of Figure 8.15.[20] Both the large-emitter and the narrow-emitter transistor function poorly at high emitter current densities, so neither is suitable for use as a power device (Section 9.1.1).

8.2.2. The Standard Bipolar Substrate PNP Transistor

Since NPN and PNP transistors differ only in doping polarities, one could theoretically create a PNP transistor by inverting all of the doping polarities of the standard

[18] R. J. Whittier and D. A. Tremere, "Current Gain and Cutoff Frequency Falloff at High Currents," *IEEE Trans. on Electron Devices*, Vol. ED-16, #1, 1969, pp. 39–57.

[19] A. B. Grebene, *Bipolar and MOS Analog Integrated Circuit Design* (New York: John Wiley and Sons, 1984), p. 76.

[20] A. B. Glaser and G. E. Subak-Sharpe, *Integrated Circuit Engineering* (Reading, MA: Addison-Wesley, 1977), p. 52.

bipolar process. The collector would then consist of a lightly doped P-type epi with the addition of a *P-buried layer* (PBL) and a deep-P+ sinker. The base would consist of a lightly doped N-type diffusion and the emitter of a heavily doped P-type one. The limitations of boron doping make it difficult to realize this structure. Buried layers should consist of slow-diffusing dopants such as arsenic or antimony. Boron diffuses relatively rapidly, so any attempt to fabricate a P-buried layer requires a radical process redesign to eliminate subsequent high-temperature drives. Standard bipolar NPN transistors also employ a heavily doped emitter to maximize emitter injection efficiency. The solid solubility of boron in silicon is only one-third that of phosphorus,[21] so the high-current performance of the PNP suffers accordingly. Even if these problems are somehow overcome, the mobility of holes in silicon is only about one-third that of electrons.

Standard bipolar cannot fabricate a fully isolated vertical PNP transistor. Although some processes do offer both vertical NPN and vertical PNP transistors, these *complementary bipolar* processes require several additional processing steps. As a compromise, standard bipolar offers a type of vertical PNP transistor called a *substrate PNP*. This transistor lacks full isolation because it employs the P-type substrate as its collector. Its base consists of N-epi and its emitter of base diffusion (Figure 8.18). The performance of the transistor suffers because these materials are not tailored to their respective tasks. On the other hand, the substrate PNP requires no additional process steps, and it costs nothing to add one to a standard bipolar design.

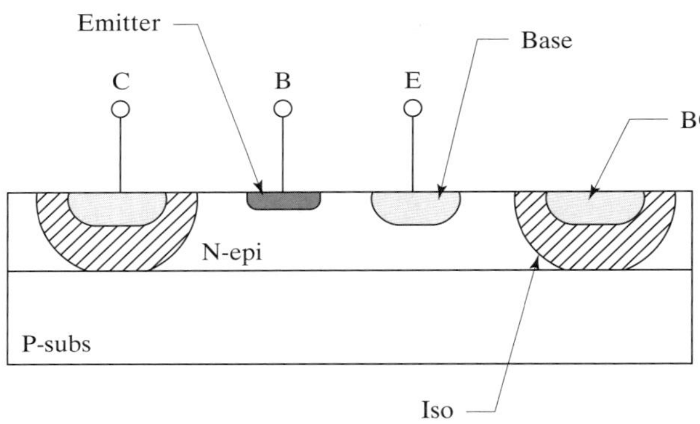

FIGURE 8.18 Cross section of a typical substrate PNP transistor fabricated in standard bipolar.

The various diffusions of the standard bipolar process take their names from the functions they perform in a vertical NPN transistor. These same diffusions play very different roles in a substrate PNP transistor. The *emitter* of a substrate PNP consists of *base* diffusion and the *base* consists of an N-tank contacted by means of *emitter* diffusion. As this example shows, one must pay close attention to the difference between the names of diffusions and the roles they play in a given structure.

The standard bipolar emitter diffusion typically has a dopant concentration in excess of 10^{20} atoms/cm^3, while the base diffusion rarely exceeds 10^{17} atoms/cm^3. The lighter doping greatly reduces the emitter injection efficiency of the substrate PNP. The still-lighter doping of the N-epi causes premature beta rolloff due to high-level injection. A substrate PNP typically retains a beta of 30 out to about 0.5mA/mil^2 (0.8μA/μm^2), while a vertical NPN transistor retains a beta of 150 out to perhaps 20mA/mil^2 (30μA/μm^2). The vertical resistance of the base diffusion exceeds that of the emitter diffusion by an order of magnitude, so a minimum-emitter

[21] F. A. Trumbore, "Solid Solubilities of Impurity Elements in Si and Ge," *Bell System Technical Journal,* Vol. 39, No. 1: 1960, pp. 205–233. Datapoints are taken at 1100°C.

substrate PNP typically exhibits 100Ω of emitter resistance, compared to the 10Ω of a comparable NPN transistor. The emitter resistance causes few problems at the low current densities characteristic of substrate PNP transistors. The increased emitter resistance even provides emitter ballasting, and this combined with high-current beta rolloff ensures a high degree of immunity to thermal runaway.

Standard bipolar employs a relatively lightly doped N-epi. In the absence of NBL, the epi-substrate junction diffuses upward during the long isolation drive. The base region of the substrate PNP is therefore both thin and lightly doped—much more so than the epi thickness might suggest. Most of the carriers flow vertically from emitter to substrate rather than laterally from emitter to isolation, so this transistor does not exhibit the difficulties associated with lateral current flow (Section 8.2.3). Substrate PNP transistors often have peak betas of 100 or more, but high-level injection causes their betas to peak at current densities of less than 1mA/mil^2 (1.5μA/μm^2).

The collector of the substrate PNP consists of a series combination of the P-substrate and the P+ isolation diffusion. The lightly doped substrate increases both the Early voltage and the V_{CEO} rating of the transistor, but strictly speaking it does not act as a drift region because it is not bounded by an adjacent P+ layer. Collector voltage drops rarely have much effect upon the substrate PNP itself, but they can cause debiasing of adjacent circuitry at higher current levels. Standard bipolar designs rarely experience objectionable levels of debiasing as long as the collector current in each substrate PNP does not exceed 1 to 2mA and the total collector current of all substrate PNP transistors does not exceed 10mA. A standard P+ isolation diffusion supplemented by base-over-iso can cope with current levels of this magnitude as long as substrate contacts are located near the substrate transistors. Any substrate PNP transistor that injects 1mA or more should have additional substrate contacts surrounding it (Section 4.4.1). Substrate PNP transistors become progressively more impractical as current levels increase; circuit designers may wish to consider substituting lateral PNP transistors for high-current substrate devices to minimize substrate debiasing. Alternatively, backside substrate contact can be employed to remove almost any amount of substrate current without debiasing.

Substrate PNP transistors operate primarily by vertical conduction and therefore scale with drawn emitter area. Several other factors influence the precise scaling of these devices, including outdiffusion, lateral conduction, and surface effects. As with vertical NPN transistors, precisely matched substrate PNP transistors must use identical emitter geometries.

Construction of Small-signal Substrate PNP Transistors

The simplest style of substrate PNP consists of an N-tank containing emitter and base regions (Figure 8.19A). The tank serves as the transistor's base, and the base diffusion as its emitter. The emitter diffusion beneath the base terminal of the transistor allows Ohmic contact to the tank. The gain of this transistor depends somewhat on the area-to-periphery ratio of the emitter. This occurs in part because laterally emitted carriers must travel further to reach the isolation than vertically emitted carriers must travel to reach the substrate. Laterally emitted carriers experience additional recombination due to defect sites along the oxide-silicon interface. Large base diffusions also drive deeper into the epi and thus narrow the base width of the substrate PNP. In summary, larger and more compact emitter structures produce higher gains.

The transistor in Figure 8.19A is constructed by first laying out an emitter of the desired size and shape using base diffusion. The base contact consists of a rectangle of emitter diffusion placed on one side of the emitter geometry and spaced away

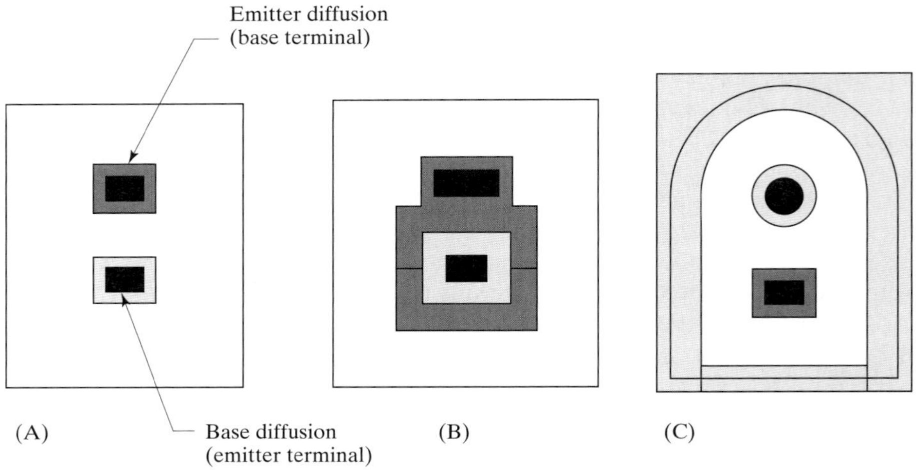

FIGURE 8.19 Examples of (A) standard, (B) emitter-ringed, and (C) verti-lat styles of substrate PNP transistor.

from it by the minimum allowed base-to-emitter spacing. The emitter diffusion need only be wide enough to contain a minimum-width contact. At least one substrate contact should reside near each substrate PNP. Ideally, this contact should abut the substrate PNP, but wiring constraints frequently force it to reside several mils away. Some separation between the substrate contact and the transistor can be tolerated as long as the collector current of the substrate PNP does not exceed a milliamp or two. Higher collector currents will probably debias the substrate unless substrate contacts are placed adjacent to the transistor.

Figure 8.19B illustrates another type of substrate PNP layout where the emitter consists of a rectangle of base diffusion surrounded on all sides by a thin ring of emitter diffusion. This emitter ring adjoins but does not overlap the drawn base diffusion. The emitter ring helps discourage lateral conduction by raising the voltage required to forward-bias the emitter sidewalls. The emitter-ringed transistor exhibits a higher beta at low current densities than the transistor in Figure 8.19A does. This advantage disappears at higher current densities because the base forward-biases into the emitter ring. Although this structure has some merit, most layouts do not employ it because it requires more room than the simple layout of Figure 8.19A, and because the latter structure has sufficient gain for most applications. If an emitter ring is used, then the spacing between the ring and the base diffusion's contact must equal the spacing between the emitter of a vertical NPN and its base contact. The other dimensions of this transistor follow much the same rules as apply to the structure of Figure 8.19A.

The device of Figure 8.19C is sometimes called a *tombstone PNP* or a *cathedral PNP* because of its distinctive tank outline. The same transistor is also called a *verti-lat PNP* because it attempts to capitalize upon both vertical and lateral conduction in the same device. The emitter of this transistor consists of a circular plug of base diffusion. The tank geometry passes around the emitter in a semicircular arc digitized around the same center as the emitter. A ring of base diffusion also surrounds the tank, overlapping into it as far as layout rules allow. This base ring helps counteract the high sheet resistance of the periphery of the isolation caused by its extensive outdiffusion. Lateral conduction in the transistor occurs outward from the plug of base diffusion serving as the emitter, to the ring of base diffusion serving as the collector. Conduction also occurs vertically downward from the emitter plug to the underlying substrate. This style of transistor requires much the same field plating as does a lateral PNP. The field plate should entirely cover the exposed surface of the N-epi at the semicircular end of the transistor, and it should extend toward the base contact as far as the metal spacing rules allow. Failure to properly field-plate the transistor may cause unexpected leakage phenomena as well as a degradation of low-current beta (Section 8.2.3).

A verti-lat PNP should theoretically have a higher beta than a standard substrate PNP because its lateral conduction pathway has been optimized by minimization of the neutral base width and by the incorporation of a base field plate. In practice, the structure of Figure 8.19B generally outperforms the verti-lat transistor because it suppresses lateral conduction and relies instead on the more efficient vertical conduction.

FIGURE 8.20 Higher-current substrate PNP transistor employing two wide emitter stripes and a single narrow base stripe.

Figure 8.20 shows the layout of a larger substrate PNP transistor. Like the large vertical NPN transistor in Figure 8.17A, this device employs interdigitated emitter and base stripes. Each emitter consists of a fairly wide (~25μm) strip of emitter diffusion. Since PNP transistors exhibit beta rolloff at relatively low current densities, hot spotting and secondary breakdown generally do not occur and the emitter width is limited only by the resistance of the pinched region beneath the base diffusion. Because this pinched sheet may exceed 10kΩ/□, emitter stripes more than 25 to 50μm wide are not advisable. The base contacts consist of thin stripes of emitter diffusion placed between adjacent emitter fingers. If desired, two additional stripes of emitter diffusion can be placed on the ends of the transistor to further reduce base resistance. The large substrate contact encircling the transistor helps limit substrate debiasing. The illustrated contact cannot deal with more than a few milliamps of substrate current without saturating; a much wider strip of contact would be required to deal with the maximum collector current of which this structure is capable. The substrate contact has been interrupted along the top of the transistor to allow the emitter and base leads to emerge on first-level metal. If double-level metallization is available, then the contact should form an unbroken ring around the transistor to minimize collector resistance.

8.2.3. The Standard Bipolar Lateral PNP Transistor

Although standard bipolar cannot fabricate an isolated vertical PNP, it does offer an isolated *lateral PNP* consisting of two separate base diffusions placed in a common tank. One of these diffusions serves as the emitter and the other as the collector. When the emitter forward-biases, holes flow into the tank and travel laterally to the collector. Lateral transistors generally have slower switching speeds and lower betas than vertical devices. Although little can be done to boost switching speeds, proper design can substantially improve the beta. The layout designer can also vary the base width of a lateral PNP by moving the emitter and collector nearer together or further apart. A narrower basewidth produces a higher beta and a lower Early voltage, while a wider basewidth produces the opposite effect. The product of beta and Early voltage remains approximately constant regardless of basewidth.

The beta of a lateral PNP depends on at least five different factors: emitter injection efficiency, base doping, base recombination rate, base width, and collector efficiency. All but the last two of these are beyond the control of the layout designer. The emitter injection efficiency of the lateral PNP suffers from the use of a relatively lightly doped base diffusion as the emitter of the transistor. The even lighter doping of the N-epi causes the early onset of high-level injection and a corresponding beta rolloff beyond a current density of about 100μA per minimum emitter. Some processes incorporate a deep-P+ diffusion intended specifically to construct lateral PNP transistors having better high-current characteristics, but even with this modification, beta rolloff still begins at or below 500μA per minimum emitter.

The standard bipolar lateral PNP transistor suffers from elevated recombination rates caused by the use of (111) silicon. The same surface states that increase the thick-field threshold also act as recombination centers and decrease the lifetime of minority carriers traveling near the surface. The field oxidation also spawns oxidation-induced lattice defects that migrate a short distance into the silicon. Both of these factors reduce the beta of the lateral PNP to a fraction of that of the corresponding substrate PNP. Annealing in a reducing atmosphere (such as the nitrogen-hydrogen mixture called *forming gas*) reduces the surface state concentration and increases the beta of the lateral PNP transistor at the cost of reducing the thick-field threshold. Historically, the introduction of compressive nitride protective overcoats produced an increase in lateral PNP beta, presumably because of the reducing conditions prevailing during nitride deposition. Together with smaller feature sizes, reducing anneals increased the peak beta of standard bipolar lateral PNP transistors from less than 10 to more than 50.

The *drawn* basewidth of a lateral PNP equals the separation between the drawn base diffusions serving as its emitter and collector (Figure 8.21, dimension W_{B1}). The *actual* base width of the transistor is more difficult to determine because it depends on two-dimensional current flow. Near the surface, the actual base width W_{B2} is considerably less than the drawn base width W_{B1} due to outdiffusion. As one proceeds deeper into the silicon, the sidewalls of the emitter and collector curve away from one another. Carriers do not move in straight lines, and the greater distance traveled along alternative pathways also increases the *effective* basewidth of the transistor (W_{B3}). The effective base width of the transistor consists of a weighted average of all possible paths by which carriers might traverse the neutral base. The complexity of this problem precludes simple closed-form solution, and the solutions that exist provide little insight into transistor design.[22,23] Three general observations can still be made. First, the effective base width for purposes of computing punchthrough consists of the drawn basewidth minus twice the outdiffusion distance, or W_{B1}. Only

---

[22] D. E. Fulkerson, "A Two-Dimensional Model for the Calculation of Common-Emitter Current Gains of Lateral p-n-p Transistors," *Solid-State Electronics,* Vol. 11, 1968, pp. 821–826.

[23] K. N. Bhat and M. K. Achuthan, "Current-Gain Enhancement in Lateral p-n-p Transistors by an Optimized Gap in the n+ Buried Layer," *IEEE Trans. on Electron Devices,* Vol. ED-24, #3, 1977, pp. 205–213.

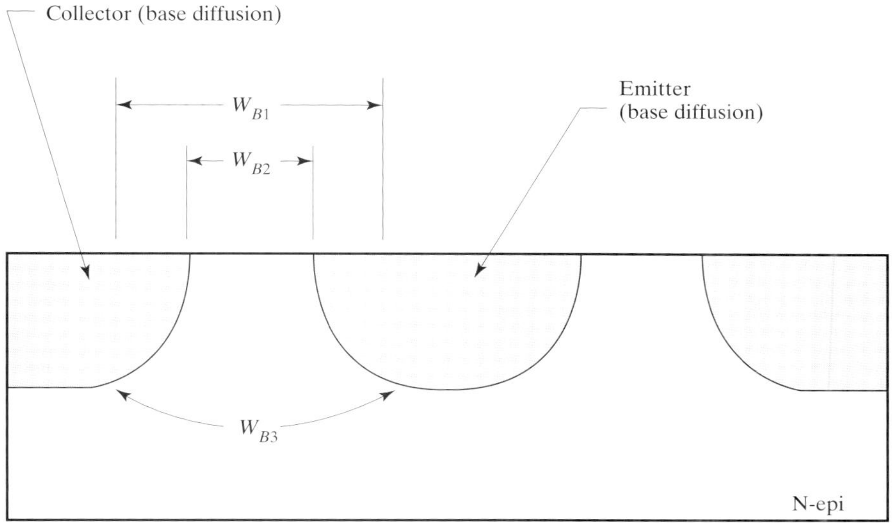

FIGURE 8.21 Cross section of a lateral PNP transistor depicting three different measures of neutral base width discussed in the text (drawn base width, W_{B1}, actual base width at the surface, W_{B2}, and effective base width beneath the surface, W_{B3}).

the smallest drawn basewidths exhibit any significant decrease in operating voltage due to punchthrough. Second, the effective base width for purposes of computing beta (W_{B3}) substantially exceeds the actual base width at the surface (W_{B2}) due to the contribution of subsurface conduction. This causes beta to scale more weakly with drawn base width than one might expect. If a transistor with a drawn basewidth of 8μm has a peak beta of 80, then one with a drawn basewidth of 16μm will have a peak beta greater than 40. Third, the Early voltage of a lateral PNP is inversely proportional to peak beta. Suppose a transistor with a peak beta of 80 has an Early voltage of 70V; a wider-base transistor exhibiting a beta of 60 should have an Early voltage of approximately $(80/60) \cdot 70 = 93$V.

Designers have long known that the gain of a lateral PNP appears to increase when NBL is placed underneath the emitter. This increase occurs because the presence of the NBL restricts the motion of holes to a smaller region of the N-epi and therefore reduces recombination. NBL also greatly increases collection efficiency by blocking the injection of holes into the substrate (Section 8.1.5). Because only a few of the carriers manage to surmount the built-in potential surrounding the NBL, the increase in base current caused by recombination in the NBL is quite small.

One might also expect some percentage of the emitted carriers to escape by traveling laterally beneath the collector diffusion. Experience shows that this does not occur to any appreciable extent while the transistor remains in the normal active region. The NBL updiffuses through the epi, establishing a retrograde doping profile that tends to drive the minority carriers upward along the built-in potential gradient. At the same time, the depletion region surrounding the collector reaches deep into the lightly doped N-epi and therefore intercepts a large fraction of minority carriers attempting to pass underneath it. Only a small percentage (usually less than 1%) of emitted carriers successfully escape the collector to reach the isolation sidewall. Transistors with shallow collector regions may experience greater sidewall losses (Section 8.3.2).

Construction of Small-signal Lateral PNP Transistors

The lateral PNP transistor traditionally consists of a small plug of base diffusion surrounded by a larger annular base region (Figure 8.22). The central base plug serves as the emitter of the transistor, while the surrounding base ring serves as its collector. This structure ensures that almost all of the carriers injected by the emitter are intercepted

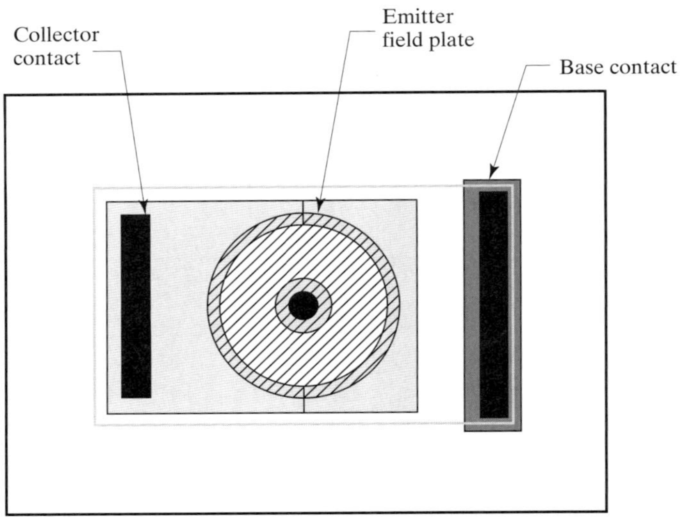

FIGURE 8.22 Layout of a lateral PNP transistor showing emitter field plate.

by the intended collector before they can reach the isolation. The apparent reverse beta of this style of lateral PNP transistor is always far smaller than its forward beta. In the reverse mode, the outer ring of base becomes the emitter and the majority of the injected carriers are injected toward the isolation sidewalls rather than to the small base plug in the center of the transistor. Thus, despite identical doping levels in emitter and collector, the lateral PNP transistor remains a profoundly asymmetrical device.

The circles and arcs used in constructing lateral PNP transistors are actually polygonal approximations. The number of segments in these polygons determines how closely they resemble ideal circles and arcs. The number of segments in the complete circle should be evenly divisible by four to ensure symmetry around both axes. Circular approximations with sixty-four sides are recommended for most applications. Annular geometries, such as the collector in Figure 8.22, usually consist of two matched halves to eliminate certain difficulties sometimes encountered during photomask generation.[24]

As stated previously, the peak beta of a lateral PNP depends inversely upon its effective base width. Carriers injected from the emitter sidewalls travel a shorter path than carriers injected from the bottom surface. Smaller emitters therefore have shorter effective base widths. Some designers mistakenly assume that the periphery-to-area ratio determines the beta of a lateral PNP, but the actual determining factor is the distance from the middle of the emitter to its periphery.

Practical lateral PNP transistors usually employ a circular emitter geometry just large enough to contain a minimum-size contact (Figure 8.22). The emitter contact should be made circular and concentric with the emitter to ensure that all portions of the emitter periphery are equidistant from the contact edge. If they are not, some portions of the emitter periphery would inject an undue share of the carriers. This nonuniform current flow can cause subtle mismatches, especially in split-collector transistors. The collector of the lateral PNP generally consists of a square of base diffusion having a circular hole in its center. The emitter occupies the center of this hole. The drawn base width equals the difference between the radius of the emitter and the radius of the collector opening. One end of the collector extends sufficiently to allow placement of a contact inside it. Although debiasing occurs along the collector periphery, this only results in a slight increase in the effective saturation voltage of the transistor due to premature saturation of the debiased portion of the col-

[24] The boundary of a so-called *semisimple figure* becomes coincident with itself at one or more points, while the boundary of a *simple figure* does not. Some pattern generation algorithms have problems with semisimple figures, so they are often avoided.

lector. If this increase in saturation voltage causes concern, then additional collector contacts can be added to reduce collector debiasing. For most applications, the additional contacts are not worth the space they consume.

The collector of the lateral PNP resides in the tank that forms its base. A strip of emitter diffusion added to one end of the tank serves as a base contact. A deep-N+ sinker is not normally required because the base current in the lateral PNP rarely exceeds a few tens of microamps. If space is at a premium, even a minimum-size base contact will suffice. On the other hand, a lateral PNP should always contain as much NBL as possible to minimize unwanted substrate injection.

Figure 8.22 shows a metal plate covering the exposed portions of the N-tank between the emitter and the collector. This *field plate* prevents unwanted surface inversion or accumulation from occurring due to charge spreading and field interactions with the adjacent collector. Without the field plate, the beta of the lateral PNP fluctuates depending on surface potentials.[25] If a negative charge accumulates at the surface, then the minority carriers move toward the oxide interface and the beta of the transistor decreases. Similarly, a positive potential repels the carriers from the surface and increases the beta. The presence of mobile ions greatly increases the magnitude of the instabilities caused by these surface charges. The use of a field plate connected to the emitter not only stabilizes the beta of the transistor, but also helps to increase it by repelling carriers from the oxide-silicon interface.

Lateral PNP transistors that lack field plates often exhibit collector-to-emitter leakage at voltages well below the thick-field threshold. These leakages result from parasitic channel formation. The geometry of the lateral PNP ensures that the parasitic channel will have a large W/L ratio, while the voltage differential present between collector and emitter attracts mobile ions and surface charges. Leakages have been observed in standard bipolar lateral PNP transistors at collector-to-emitter voltages of only 5 to 10V on a process having a thick-field threshold in excess of 40V. Properly designed field plates will completely eliminate these leakage currents. The field plate should connect to the emitter of the transistor and should entirely cover the exposed N-epi between the emitter and the collector. The field plate should overlap the periphery of the collector so that misalignment between the metal and the base diffusion cannot expose the epi surface. An overlap of 2 to 3μm (about 0.1mil) is more than enough, because the outdiffusion of the base helps shrink the size of the collector opening that the field plate must cover.

A minimum-emitter lateral PNP occupies considerably more space than a minimum-emitter NPN. A single lateral PNP can, however, be subdivided to form several smaller transistors sharing a common base and emitter. Figure 8.23A shows a simple example of such a *split-collector lateral PNP*. This transistor contains two partial collector segments, each occupying half of the emitter periphery. Since the emitter injects carriers uniformly in all directions, each collector receives half of the total injected current. This device therefore behaves as if it were actually two separate transistors, each having an effective emitter size one-half that of a normal lateral PNP. The transistor in Figure 8.23B carries the process of splitting the collector still further. Instead of two half-sized collectors, this transistor contains four quarter-sized ones. One can even construct a split-collector transistor containing segments of different sizes. For example, the transistor in Figure 8.23C includes three, one-sixth collectors and two, one-quarter collectors.

The multiple collectors will match one another within about ±1% as long as they all possess identical geometries placed symmetrically.[26] The split collectors in

---

[25] R. O. Jones, "P-N-P Transistor Stability," *Microelectronics and Reliability*, Vol. 6, 1967, pp. 277–283.

[26] B. Gilbert, "Bipolar Current Mirrors," in C. Toumazou, F. J. Lidgy, and D. G. Haigh, *Analog IC Design: The Current-Mode Approach* (London: Peter Peregrinus, 1990), p. 250.

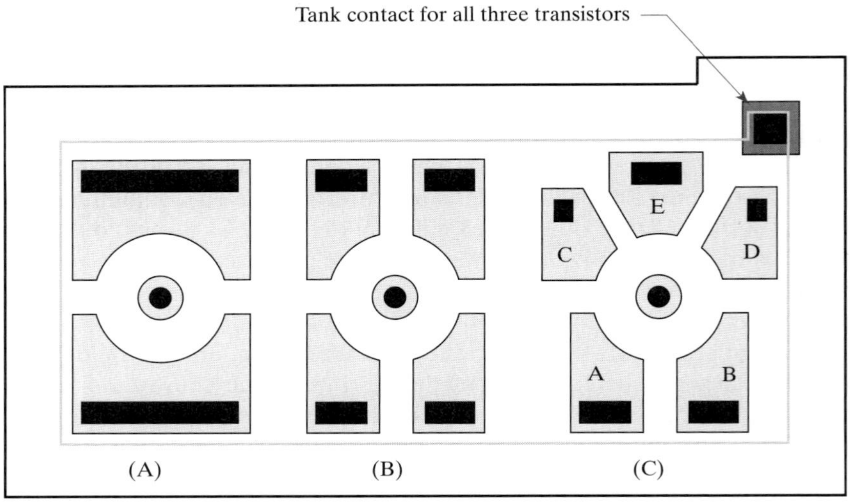

FIGURE 8.23 Examples of split collector transistors: (A) 1/2–1/2, (B) 1/4–1/4–1/4–1/4, (C) 1/6–1/6–1/6–1/4–1/4. The field plates have been omitted for clarity.

Figures 8.23A and 8.23B meet these criteria and therefore match fairly precisely. The split collectors in Figure 8.23C do not all have the same geometries and therefore do not all precisely match one another. Collectors A and B will match, as will collectors C and D. Collector E will not match collectors C and D because the former possesses a different geometry than the latter. Similarly, the ratio between collectors A and C will not exactly equal 2:3 because these segments have different geometries.

Split-collector laterals are frequently employed to construct current mirrors. A simple 1:1 current mirror can be constructed from a split collector transistor with two half-sized collectors. One collector connects back to the common base to form reference transistor Q_{1A} (Figure 8.24A). The other collector serves as the output transistor Q_{1B}. This arrangement saves considerable space, but it probably does not match as precisely as two separate lateral PNP transistors placed next to one another. Emitter degeneration cannot improve the matching of the split-collector mirror because the two transistors share a common emitter. Many schematic diagrams denote split collectors by placing multiple collector leads on a single base bar (Figure 8.24B). The ratio of the split collectors is indicated by values placed next to each collector lead. The passage of a base lead through the transistor does not indi-

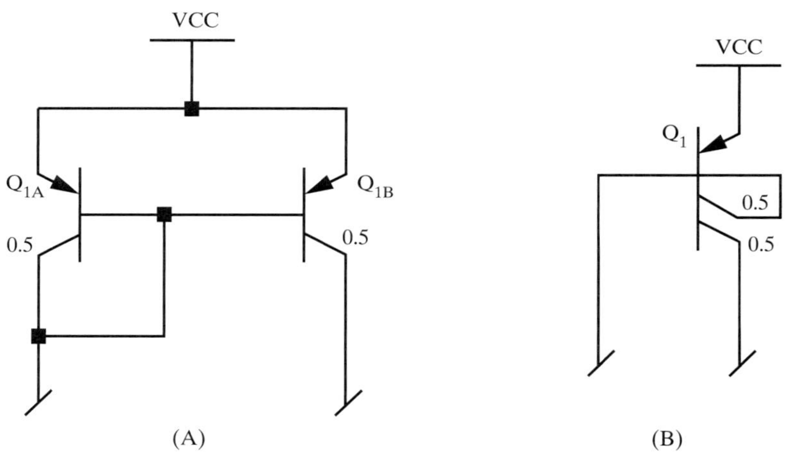

FIGURE 8.24 Schematic diagrams for 1:1 current mirrors constructed using split collector lateral PNP transistors: (A) conventional schematic diagram and (B) simplified schematic diagram.

cate the presence of multiple base terminals, but rather indicates two separate connections to the same base terminal. Using this style of schematic representation, the circuit of Figure 8.24A reduces to the considerably more compact, if less familiar, schematic of Figure 8.24B.

Some designers prefer to digitize lateral PNP transistors using a square emitter surrounded by a square collector ring. This style of lateral PNP is easier to digitize than the circular geometries previously discussed, but its base width increases slightly due to the greater length of the diagonal conduction paths and the poorer area-to-periphery ratio of the square emitter. Some designers fillet the four corners of the opening in the annular collector geometry so that the base width does not increase at the corners. Figure 8.25 shows examples of two types of square lateral PNP transistors..

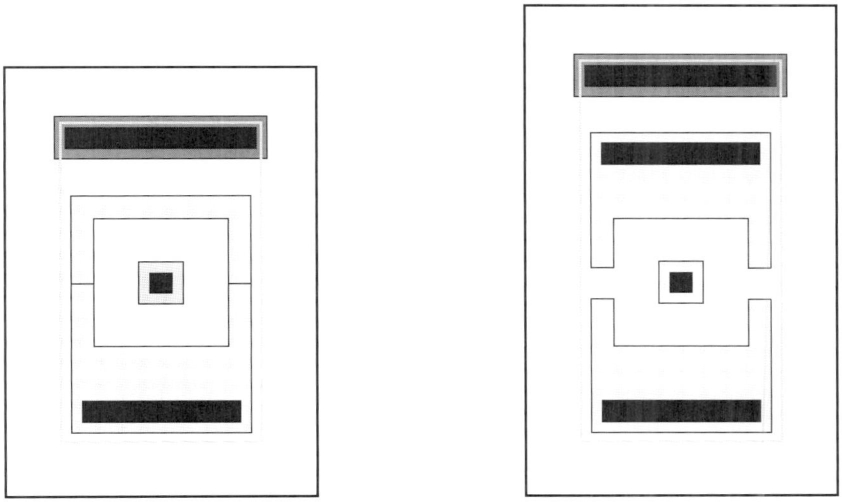

(A) (B)

FIGURE 8.25 Square geometry lateral PNP transistors: (A) minimum-emitter and (B) 1/2–1/2 split collector. The field plates have been omitted for clarity.

The lack of radial symmetry in a square split-collector transistor prevents the use of anything except half and quarter collectors. Apart from this limitation, the same rules apply to the construction of square split-collector PNP transistors as to circular ones. In all cases, the emitter should be kept as small as possible and a field plate should entirely cover the exposed N-epi between the emitter and the collector.

Some designers claim that lateral PNP transistors scale with drawn emitter area, while others believe that they scale with drawn emitter periphery. The geometry of the emitter actually plays a more crucial role than either its area or its periphery. As pointed out previously, the effective base width is proportional to the distance from the center of the emitter to the collector periphery. The use of large square or circular emitters drastically reduces the beta of the transistor. An emitter elongated into a thin stripe can possess a very large area without requiring a proportionate increase in base width (Figure 8.26A). The *elongated-emitter lateral PNP* has an emitter geometry reminiscent of the outline of a hot dog, so it is sometimes called a *hot-dog transistor.*

The area and periphery of the elongated-emitter transistor both increase at about the same rate, so arguments about whether the transistor scales with area or periphery are largely academic. An elongated-emitter transistor will not match

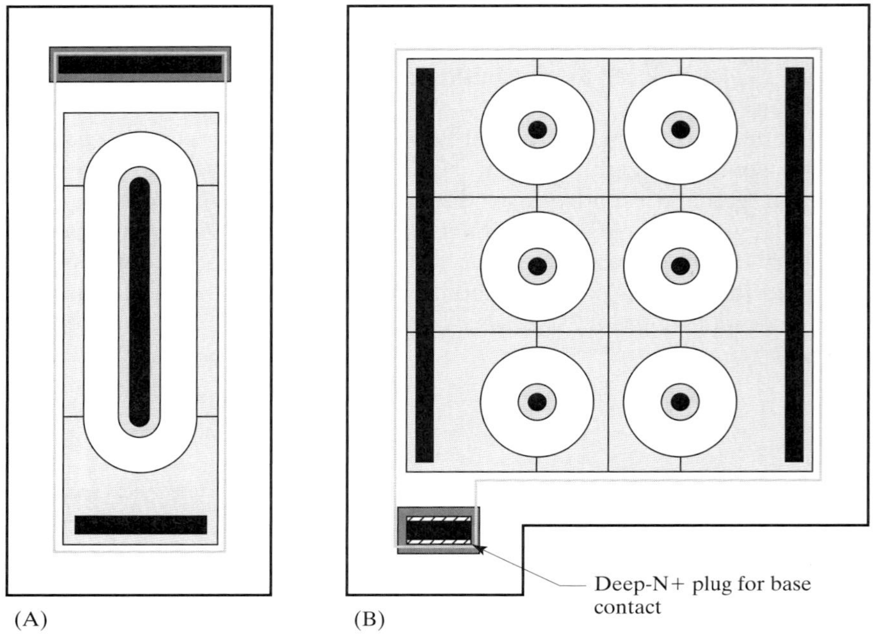

FIGURE 8.26 Higher-current lateral PNP transistors: (A) elongated-emitter or hot-dog transistor and (B) a small arrayed-emitter transistor.

a circular-emitter transistor because of junction sidewall effects, current crowding, high-level injection, and various other sources of mismatch. The beta of the elongated-emitter transistor decreases slightly as its emitter lengthens, although this decrease is much smaller than would occur if the emitter geometry was, for example, an enlarged square. This decrease results from the contribution to the effective base width from carriers moving down the length of the emitter stripe. The transistor in Figure 8.26A has a relatively large collector resistance that may disturb transistor matching and reduce beta at low collector-to-emitter voltages. If this causes concern, then the collector contacts can be placed along the longer sides of the collector to reduce the collector resistance.

Larger lateral PNP transistors can also be produced by arraying a multitude of minimum-size emitters (Figure 8.26B). Because the geometry of each emitter cell is exactly the same as that of a minimum-emitter device, the current-handling capability of this *arrayed-emitter transistor* is directly proportional to the number of emitters it contains. Large devices of this type often stagger alternating columns of emitters to achieve a tighter hexagonal packing arrangement.

Although scaling by area and scaling by periphery both give approximately the same results when applied to the lateral PNP transistors in Figure 8.26, scaling by periphery extends more naturally to the case of split-collector laterals. Each split collector functions as a separate transistor whose size depends on the fraction of the emitter periphery it subtends. Split-collector transistors employing elongated emitters also follow the same principle. Most designers scale lateral PNP transistors by drawn-emitter periphery, and split-collector transistors are assigned the appropriate fractions of the periphery of their shared emitter.

8.2.4. High-voltage Bipolar Transistors

The maximum operating voltage of a process cannot exceed that of its weakest device. In standard bipolar processes, the vertical NPN transistor usually breaks down before either the lateral or the substrate PNP. The NPN V_{CEO} thus determines the maximum operating voltage. The V_{CEO} of a vertical NPN transistor is related to the avalanche voltage of the planar base-collector junction V_{CBOP} by the equation[27]

$$V_{CEO} = \frac{V_{CBOP}}{\sqrt[n]{\beta_{\max}}} \qquad [8.5]$$

where β_{max} represents the peak beta of the device and *avalanche multiplication factor n* usually lies in the range $3 < n < 6$. Low-beta devices have higher V_{CEO} ratings, but nominal betas below about fifty begin to restrict the usefulness of the device. Most processes therefore rely on a high base-collector planar breakdown voltage to obtain adequate V_{CEO} ratings. The width of the drift region determines V_{CBOP}, and thicker epi layers produce correspondingly higher breakdown voltages. The depth of the isolation diffusion must increase to keep pace with the epi thickness, but the other steps in the process remain the same. Manufacturers of standard bipolar processes usually offer several epi thicknesses corresponding to convenient operating voltages, such as 20V, 40V, and 60V. If a process offers a choice of voltage ratings, use the lowest one possible because the higher voltages require larger isolation spacings.

Parasitic channel formation and charge spreading become increasingly serious concerns at higher voltages. Charge spreading may cause low levels of leakage at operating voltages somewhat below the thick-field threshold V_{TF}. Circuits operating at low currents or employing precision matching are especially prone to leakage problems and therefore require careful field plating and channel stopping. At voltages exceeding the thick-field threshold, metallization can directly induce parasitic channel formation and complete field plates and channel stops become mandatory for proper circuit operation (Section 4.3.2). The exposed surface of the base region between the emitter and the collector of a lateral PNP transistor should always be field-plated regardless of operating voltages.

The breakdown voltage of any junction depends on its curvature: the sharper the curvature, the lower the breakdown voltage. All diffused and implanted junctions have a characteristic sidewall curvature. Deeper junctions have less sidewall curvature and therefore exhibit higher breakdown voltages. The effects of sidewall curvature can be quite dramatic. The observed breakdown voltage of a base-collector junction may equal only 60V, even though the planar avalanche voltage exceeds 120V. The observed breakdown voltages of shallow junctions often depend on geometry because the curvature of the corners of the diffusion exceeds the curvature of the sidewalls.[28] The observed breakdown voltages of such diffusions decrease slightly when the geometries contain acute angles. Deeper diffusions are less prone to this effect because outdiffusion rounds the corners off. Junctions with depths greater than 3μm rarely show any significant reduction in operating voltage due to the presence of 90° vertices. Base diffusions with a junction depth of about

---

[27] A. Grove, *Physics and Technology of Semiconductor Devices* (New York: John Wiley and Sons, 1967), pp. 230–234.

[28] C. Basavanagoud and K. N. Bhat, "Effect of Lateral Curvature on the Breakdown Voltage of Planar Diodes," *IEEE Electron Device Letters*, Vol. EDL-6, #6, 1985, pp. 276–278.

2μm may experience a reduction of 5 to 10V in breakdown voltage due to 90° vertices, and relatively shallow HSR diffusions may experience even larger reductions.

Designers must sometimes push the operating voltage of base or HSR diffusions close to their respective limits. Under such circumstances, the 90° vertices of rectangular geometries become liabilities. The designer must round off each such corner with a small arc, or *fillet,* to achieve the full operating voltage. The radius of these fillets should exceed the junction depth of the diffusion. Fillets having a radius of 5μm (0.2 mil) serve admirably for both base and HSR diffusions. Both outside and inside corners should receive fillets. Circular geometries, such as fillets, sometimes produce false diagnostics during verification, so some designers use chamfers instead. A *chamfer* is a small diagonal facet drawn perpendicular to a line bisecting the vertex. Chamfers are not quite as effective as fillets because they still contain discernible vertices, although these are less acute than the original corner. If chamfers are used, their lengths should exceed the junction depth of the diffusion.

Figure 8.27 shows two examples of NPN transistors that incorporate fillets. The transistor of Figure 8.27A incorporates fillets only on the corners of the base diffusion. These suffice to obtain the full V_{CBO} and are all that are necessary. Some designers also fillet the transistor tanks, although the isolation diffusion is so deep that these fillets have little or no effect. Up-down isolation may occasionally benefit from fillets because it uses shallower isolation diffusions. Figure 8.27B shows an NPN transistor incorporating fillets on base, NBL, and tank geometries. The larger fillets are customarily drawn concentric with the base fillets to maintain constant spacings.

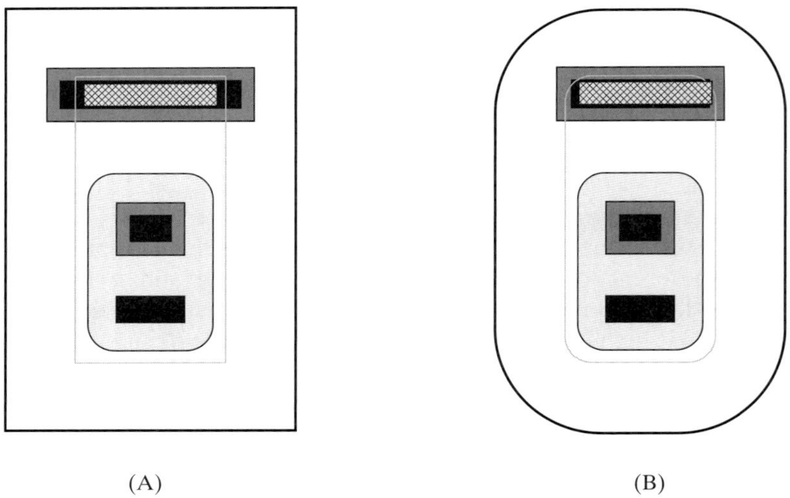

FIGURE 8.27 NPN transistors incorporating high-voltage fillets (A) on base only and (B) on base, NBL, and isolation.

High-voltage layouts often require increased spacings between certain diffusions to accommodate wider depletion region widths. Spacings that frequently require modification include base–base, HSR–HSR, base–HSR, base–iso, and HSR–iso. Other spacings that might require adjustment include collector–base, collector–iso, NBL–iso, deep-N+–iso, and deep-N+–base (where the collector is defined as the emitter region around the collector contact). The need for increased spacings for higher-voltage diffusions causes some difficulties in verification. The simplest procedure, and one that will certainly produce a functional design, consists of applying the larger spacing rules to all diffusions. On the other hand, a considerable amount of space can be saved by applying the larger rules only to devices operating at higher voltages. This requires some means of distinguishing between high- and low-voltage geometries during design rule verification. One technique consists of drawing a geometry on a special layer around low-voltage devices to distinguish them from high-voltage ones. Some designers prefer to code a figure around high-voltage devices rather than around low-voltage ones, but this practice is not recommended. The inadvertent omission of a low-voltage marker will produce errors that become apparent during verification, while the omission of a high-voltage marker may go undetected because it does not produce any design rule violations.

Designers sometimes attempt to push the voltage ratings of devices beyond their specified limits by using special circuit topologies. The most common example of this practice consists of pushing the V_{CEO} of a vertical NPN beyond its rated maximum by ensuring the device's base terminal never sees a high-impedance state. In effect, the circuit designer relies on the V_{CER} rating of the transistor rather than the V_{CEO}. This practice can cause yield problems because V_{CER} ratings are rarely specified or controlled by the wafer fab. Transistors that are pushed beyond their rated V_{CEO} may also latch up during turn-off due to the snap-back of V_{CER} to $V_{CEO(sus)}$ during conduction. Whenever possible, one should use a higher-voltage process rather than attempting to push the ratings of a lower-voltage one.

8.3. ALTERNATIVE SMALL-SIGNAL BIPOLAR TRANSISTORS

Many additional types of bipolar transistors exist. Process extensions allow standard bipolar to fabricate NPN transistors with extremely high betas and lateral PNP transistors with improved high-current performance. Analog BiCMOS offers bipolar transistors with reduced feature sizes that can equal or even surpass the performance of standard bipolar. Advanced bipolar and BiCMOS processes offer extremely high-speed transistors suitable for fast digital logic. These advanced transistors are also useful for constructing ultra-fast amplifiers and comparators. Even an N-well CMOS process can produce a low-gain substrate PNP useful for constructing voltage and current references for CMOS circuits. This section takes a look at all of these alternative devices.

8.3.1. Extensions to Standard Bipolar

The standard bipolar process was perfected in the late 1960s. Since that time, many other processes have been developed that offer superior device performance. Standard bipolar's only real advantages over these other processes are its simplicity and its low cost. Most process extensions proposed for standard bipolar have proven either too costly or too complex to justify their widespread adoption. Two options that have enjoyed some measure of popularity are the *super-beta NPN* and the *deep-P+ lateral PNP.*

Figure 8.28 depicts cross sections of a standard NPN transistor and a super-beta NPN. The super-beta device employs a deeper emitter diffusion that decreases the width of the neutral base to less than 0.1μm. Betas in excess of 5000 are possible,[29] but the thin, lightly doped base punches through at collector-to-emitter voltages of only 1 to 3V, and Early voltages lie in the same range. These limitations restrict super-beta transistors to a few specialized applications. They are, for example, useful for constructing the input differential stages of low-input-current operational amplifiers. Circuits that use super-beta transistors invariably employ regular NPNs as well, so the regular emitter diffusion must remain part of the process flow. Super-beta circuits

---

[29] W. M. Gegg, J. L. Saltich, R. M. Roop, and W. L. George, "Ion-Implanted Super-Gain Transistors," *IEEE J. Solid-State Circuits,* Vol. SC-11, #4, 1976, pp. 485–491.

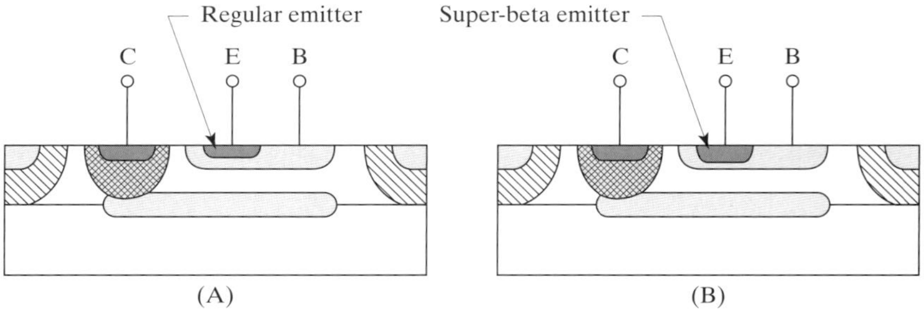

FIGURE 8.28 Comparison of representative cross sections of (A) a standard bipolar NPN and (B) a super-beta NPN.

have now largely been supplanted by BiFET and BiCMOS alternatives, but a few of the older products remain in production due to the low cost of this mature process technology.

Figure 8.29 shows cross sections of a standard bipolar lateral PNP and a corresponding deep-P+ lateral. The latter device employs a special *deep-P+* diffusion that is more heavily doped and deeper than the regular base diffusion. The increase in dopant concentration improves the emitter injection efficiency of the lateral PNP, while the deeper junction ensures that a larger percentage of emitter injection occurs from the sidewalls.[30] The high-current beta of a deep-P+ lateral does not roll off as quickly as that of a base lateral. Deep-P+ laterals can operate at current densities two or three times as large as base laterals. The beta for a typical emitter (10μm diameter) falls to half its peak value at around 200 to 500μA, as compared to 100 to 200μA for a base lateral. Although this increase in performance may seem relatively small, it can shrink the area required for a deep-P+ lateral PNP transistor to half that required by a base lateral, and it can be implemented at modest cost. Any design that contains a power lateral PNP transistor may benefit from the incorporation of deep-P+ lateral PNP transistors.

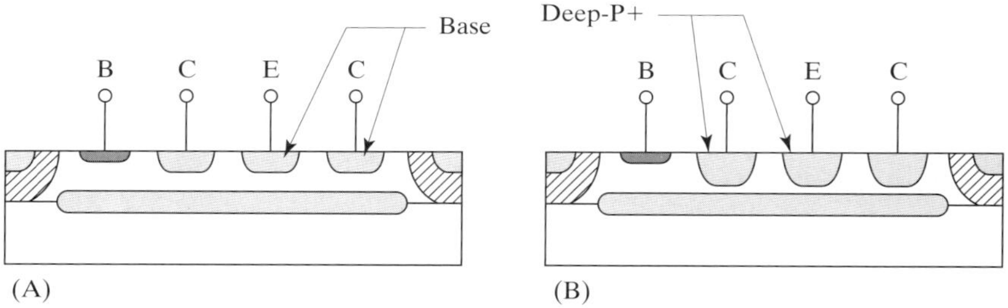

FIGURE 8.29 Comparison of representative cross sections of (A) a standard bipolar lateral PNP and (B) a deep-P+ lateral PNP.

8.3.2. Analog BiCMOS Bipolar Transistors

The analog BiCMOS process discussed in Section 3.3 employs a P-type (100) epi instead of the N-type (111) epi favored by standard bipolar. Analog BiCMOS components must therefore occupy N-wells isolated from one another by regions of P-epi. Except for the substitution of N-wells for N-tanks, the construction of bipolar devices in analog BiCMOS parallels that in standard bipolar.

An unexpected difficulty may arise from the use of N-well to form the collector of the NPN. The graded nature of the well causes the resistivity of its lowest portions to greatly exceed the resistivity of standard bipolar N-epi. The vertical resistance through the analog BiCMOS N-well is therefore much greater than the vertical

[30] B. Murari, "Power Integrated Circuits: Problems, Tradeoffs, and Solutions," *IEEE J. Solid-State Circuits,* Vol. SC-13, #3, 1978, pp. 307–319.

resistance through the standard bipolar N-epi. Unless the transistor includes a deep-N+ sinker, the vertical collector resistance will cause a soft transition from saturation to normal active operation similar to that caused by quasisaturation (Figure 8.30). This soft transition not only causes excessively high saturation voltages, but it also makes the transistor very difficult to accurately model. The inclusion of deep-N+ in the transistor eliminates the soft transition, but only at the cost of significantly increasing device area.

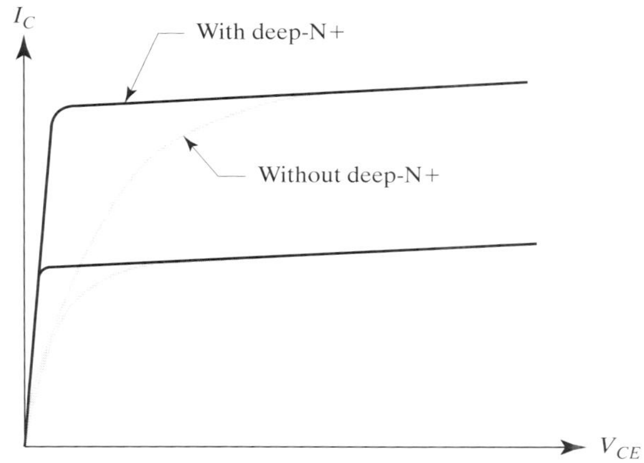

FIGURE 8.30 Comparison of saturation characteristics of CDI NPN transistors with and without the addition of deep-N+ sinkers.

The relatively shallow junction depth of the N-well limits the operating voltage of the CDI NPN transistor to a typical value of 15 to 20V. An alternative structure can provide higher voltages at the cost of reduced safe operating area and poorer Early voltage. Figure 8.31 illustrates the layout and cross section of this *extended-base NPN transistor.*

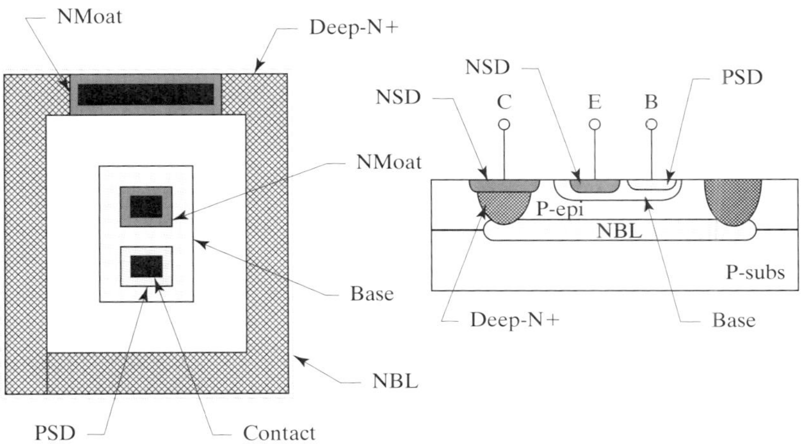

FIGURE 8.31 Layout and cross section of an extended-base NPN transistor.

Unlike the CDI NPN, the extended-base transistor does not employ N-well in its construction. The base of this transistor consists of a combination of a regular base diffusion and an isolated P-epi region, the latter lying sandwiched between the base diffusion and the underlying NBL. A ring of deep-N+ isolates the base from the surrounding P-epi and simultaneously allows contact to the NBL. The collector of this transistor consists of the NBL beneath its extended base and the deep-N+ ring surrounding it. The P-epi portion of the extended base acts as a drift region that greatly increases the operating voltage of this structure. The collector-base depletion region cannot penetrate far into the heavily doped NBL and instead moves upward through the lightly doped P-epi. Because the depletion region intrudes primarily into the neutral base rather than into the collector, the Early voltage of this structure is somewhat lower than that of the CDI NPN. Although the additional dopant in the P-epi increases the Gummel number of the extended base, the increase is smaller than one might expect because the P-epi is very lightly doped in comparison with the base diffusion. The base diffusion terminates the drift region and restricts the upward growth of the collector-base depletion region, preventing base punchthrough. Multiple-well processes (Section 11.2.2) often use P-well as a substitute for the base diffusion in these devices.

The extended-base structure has a higher planar base-collector breakdown voltage than the CDI NPN and therefore also exhibits a higher V_{CEO} rating, typically ranging from 40 to 60V. The base diffusion can be eliminated entirely, resulting in an *epi-base transistor.* The elimination of the base diffusion greatly reduces the Gummel number of the device. The epi-base transistor offers a beta of several hundred at the cost of reduced operating voltage (due to punchthrough) and increased base resistance. One advantage of the epi-base transistor is that it does not require a separate base diffusion.

If the deep-N+ isolation ring is replaced by a similar N-well ring, the resulting device requires only one mask step that does not normally appear in the analog CMOS process flow. The resulting transistors can handle only relatively small currents due to their high collector resistances, but they still offer much better low-current beta characteristics and device matching than any other bipolar transistors compatible with CMOS processing (Section 8.3.3). Unfortunately, pure CMOS processes rarely offer an NBL layer suitable for the construction of an epi-base device.

The analog BiCMOS substrate PNP (Figure 3.50) employs an emitter constructed of PSD rather than base, because the larger junction depth of the base diffusion reduces the punchthrough voltage of the transistor. The performance of the PSD substrate PNP roughly equals that of a substrate constructed in standard bipolar. Since substrate PNP transistors inject current into the substrate, the designer must take precautions to avoid substrate debiasing (Section 4.4.1).

Lateral PNP transistors constructed in analog BiCMOS exhibit surprisingly high peak betas. The relative shallowness of the base diffusion allows the emitter and collector of the transistor to be placed in close proximity to one another. Simultaneously, the graded nature of the well (aided by the presence of a phosphorus channel stop implant) helps increase the punchthrough voltage near the surface where the base is narrowest. Most of the minority-carrier injection occurs deeper in the transistor, where the built-in potential of the base-emitter junction decreases due to the graded nature of both the well and the base diffusion, so the presence of higher surface doping levels does not unduly increase the Gummel number of the transistor. The graded nature of the well—again aided by the phosphorus channel stop—generates an electric field that forces minority carriers down and away from the oxide-silicon interface. Carriers overcoming this electric field still experience relatively low levels of surface recombination due to the low surface state charge of (100) silicon. Finally, the small feature sizes possible with the shallow diffusions and superior photolithography of analog BiCMOS allow the construction of very small emitters (typically 5μm in diameter). This increases the proportion of minority carriers injected from the emitter sidewall, while at the same time reducing the distance traveled by those carriers. All of these factors act in concert to raise the peak beta of the lateral PNP so much that it may exceed that of the CDI NPN. This high peak beta also helps extend the usable current range, allowing each minimum emitter to conduct as much as 100μA while retaining a beta of twenty. The smaller cell size of the analog BiCMOS lateral PNP enables the construction of very area-efficient power lateral PNP transistors even without the aid of a deep-P+ diffusion.

The size of the emitter has a strong effect on the beta of an analog BiCMOS lateral PNP, primarily because carriers emitted from the bottom surface of the emitter must travel farther than those emitted from the sidewalls. Additionally, the graded well doping generates a weak electric field that causes minority carriers to drift toward the NBL/N-well interface. This field prevents the interface from reflecting minority carriers toward the collector as efficiently as would otherwise occur. Worse yet, the NBL layers used in analog BiCMOS processes are often relatively lightly doped to minimize lateral autodoping, while the wells in low-voltage processes are more heavily doped to prevent punchthrough. The reduced doping difference decreases the built-in field at the NBL/N-well interface, allowing carriers to penetrate into the NBL, where they recombine or travel to the substrate. Analog BiCMOS lateral PNP transistors should always employ minimum-size emitters to ensure the highest possible gain. Larger transistors should use arrayed emitters rather than elongated ones for the same reason.

PSD implants can also form the emitter and collector of a lateral PNP transistor. The shallowness of the PSD implant diminishes the size of the transistor, but it also reduces the collector efficiency. The source/drain implants are so shallow that a substantial percentage of the minority carriers travel underneath them and escape to the sidewalls. This problem is exacerbated by the recessed thick-field oxide and N-type channel stop implants that shadow the PSD collector, and by the graded nature of the well that imposes a downward drift on minority carriers. If PSD laterals must be used, their collector efficiency can be increased by widening the collector or by ringing the transistor with deep-N+.

Lateral PNP transistors constructed in analog BiCMOS do not require base field-plating unless the operating voltage of the transistor exceeds the thick-field threshold. The presence of the phosphorus channel stop implant produces a built-in potential that repels minority carriers from the surface, and the use of (100) silicon minimizes surface recombination experienced by minority carriers that do reach the surface. Leakages and beta variations are therefore much reduced in analog BiCMOS. The elimination of base field plates helps to reduce the overall size of the transistor, making the lateral PNP a more attractive component.

8.3.3. Bipolar Transistors in a CMOS Process

Straight CMOS processes offer few options for constructing bipolar transistors. The only bipolar device available in an N-well CMOS process is a substrate PNP. This transistor uses the same diffusions as the corresponding analog BiCMOS device, consisting of PSD for the emitter, N-well for the base, and the P-substrate for the collector. Unfortunately, this device rarely performs as well as its analog BiCMOS counterpart. The main reasons for the differences in performance are differences in N-well doping and the use of shallow clad moats.

The N-well doping profile of analog BiCMOS represents a compromise between the profiles best suited to bipolar transistors and those best suited to MOS devices. The CDI NPN requires a lightly doped well to act as its drift region. This same well forms a suitable base region for lateral and substrate PNP transistors, and it can serve as the backgate for relatively long-channel, high-voltage PMOS transistors. Modern CMOS processes usually have channel lengths that are much shorter than one micron. Higher well dopings are required to prevent punchthrough of these short channels. The heavier well doping increases the Gummel number of the substrate PNP and therefore reduces its gain. Many CMOS processes also alter the well dopant profile to increase the doping concentration beneath the surface in order to prevent vertical punchthrough and to increase latchup immunity. The increased subsurface doping associated with such *retrograde* well profiles increases the Gummel number of the substrate PNP, reducing its beta still further.

The extremely shallow source/drain implants required to produce short-channel MOS transistors cause a decrease in the emitter injection efficiency of bipolar transistors, probably because of recombination at the contact and increased surface recombination effects. Clad moats are particularly troublesome because silicidation consumes some of the underlying silicon and further thins the already-shallow source/drain diffusions. The effects of reduced emitter injection efficiency include reduced peak beta and a more pronounced low-current beta rolloff, often resulting in a peaked beta curve reminiscent of that of a standard bipolar lateral PNP transistor.

The gain of substrate PNP transistors constructed in submicron CMOS can vary from less than one to more than fifty. Transistors with gains of less than five exhibit extremely poor matching. Small variations in beta produce corresponding variations in base current. Since the base resistance of these transistors is relatively large and the base current is comparable to the collector current, any variation in base current produces a corresponding variation in base-emitter voltage, V_{BE}. Collector current mismatches of ±3 to 5% are not uncommon in low-gain CMOS substrate transistors. This problem becomes most severe in devices whose beta lies near unity; devices with extremely small betas behave as diodes and therefore largely avoid this problem.

If at all possible, a silicide block mask should be coded over the emitter regions of the transistor. A small silicided area underneath each emitter contact is still necessary, but this does not degrade the gain of the transistor nearly as much as full moat cladding. The transistor should consist of minimum-width strips of NSD interdigitated with similar strips of PSD (Figure 8.32). This arrangement reduces the base resistance of the transistor and helps minimize the influence of base current variations on device matching. The collector current of this transistor should be kept as small as possible to further reduce voltage drops in the neutral base. Current densities of 1 to 10μA/mil^2 (1.5–15nA/μm^2) are generally advisable for transistors used as part of voltage or current references.

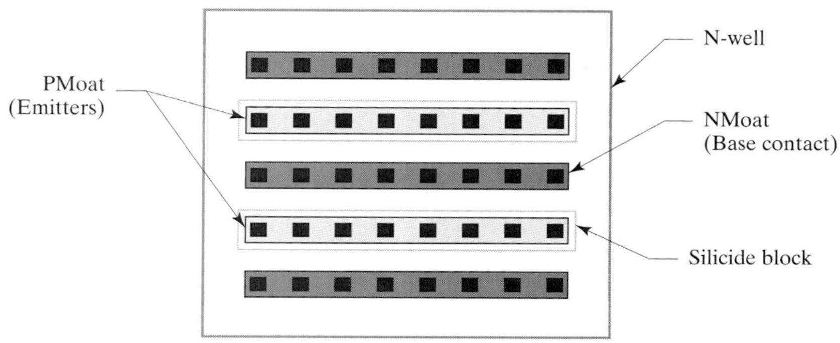

FIGURE 8.32 Layout of a substrate PNP transistor compatible with an N-well CMOS process.

Some designers have attempted to construct lateral PNP transistors in straight CMOS processes. These devices have generally proved unsatisfactory because of a combination of low gain and low collector efficiency. The effective beta of many of these devices does not exceed unity, and the effects of beta variation become so severe that even minimal matching cannot be maintained. The use of polysilicon to generate a self-aligned base region has enabled the fabrication of devices with usable betas, but even these devices exhibit excessive beta variation.[31] A superior alternative to substrate and lateral PNP transistors can be fabricated with the addition of one process step. The presence of NBL allows the construction of epi-base transistors that use N-well isolation rings. These transistors cannot handle large currents because of their high collector resistance, but they greatly outperform the other transistors available in CMOS processing. Unfortunately, few pure CMOS processes offer any form of buried layer, and epi-base transistors are usually found only in stripped-down analog BiCMOS processes. Low-voltage, multiple-well processes can sometimes construct a *retrograde-well NPN transistor.* This device uses a retrograde N-well (or an N-well incorporating a punchthrough stop implant), as discussed in Section 11.1.2. The retrograde N-well is counterdoped with a shallower P-well to form the base region of the transistor. This device is otherwise similar to a P-well epi-base transistor (Section 8.3.2).

8.3.4. Advanced-technology Bipolar Transistors

All of the bipolar transistors discussed up to this point are relatively slow. Three factors limit switching speed: junction capacitance, base resistance, and neutral base width. The base-emitter, base-collector, and collector-substrate junction capacitances must be charged and discharged in order to switch the transistor. The resistance of the neutral base limits the charging and discharging rates of these capacitors and thereby restricts the maximum switching speed of the device. Assuming that these limitations can somehow be overcome, the transit time of minority carriers across the neutral base still sets a fundamental limit on the maximum switching speed of the transistor.

The simplest way to reduce junction capacitances consists of minimizing junction areas. Improved photolithography allows smaller drawn geometries, which translate directly into smaller junction capacitances and faster switching speeds. Further improvements can be achieved by redesigning the transistor to eliminate unnecessary overlaps and spacings. For example, the conventional transistor of Figure 8.33A requires the emitter diffusion to overlap the emitter contact to allow for misalignment. The *washed-emitter transistor* of Figure 8.33B eliminates this overlap, substantially reducing the size of the emitter. This in turn allows the use of a smaller base, leading to additional performance improvements.

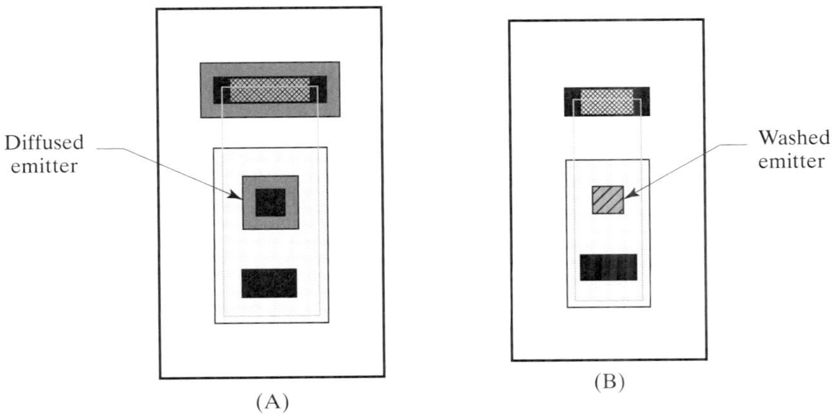

FIGURE 8.33 Comparison between (A) a conventional diffused emitter and (B) a washed emitter.

[31] E. A. Vittoz, "MOS Transistors Operated in the Lateral Bipolar Mode and Their Application in CMOS Technology," *IEEE J. Solid-State Circuits,* Vol. SC-18, #3, 1983, pp. 273–279.

A *washed emitter* consists of an emitter diffusion self-aligned to a contact opening by means of a special etching technique. The emitter is deposited using a conventional deposition and drive during which a thin emitter oxide forms. The emitter regions remain covered by photoresist during the contact oxide removal. After all of the other contacts have been opened and the photoresist has been removed, a brief, carefully timed etch strips the thin emitter oxide and forms emitter contacts. Outdiffusion of the emitter under the original oxide opening provides just sufficient overlap of emitter over contact to prevent base-emitter shorts.[32]

Even better results can be obtained by using a *polysilicon emitter.* Figure 8.34 shows an example of a CDI NPN fitted with a simple polysilicon emitter. This transistor is processed normally up through the completion of the base drive. Next, an oxide removal defines the extent of the drawn emitter. A layer of arsenic-doped polysilicon is deposited and patterned over the exposed emitter opening. A brief period of heating causes arsenic to diffuse from the polysilicon into the exposed monocrystalline silicon, producing an extremely thin and heavily doped emitter diffusion that self-aligns to the emitter oxide removal. This process has the same effect as using a washed emitter does, but it allows the formation of much thinner and more precisely controlled emitter diffusions.

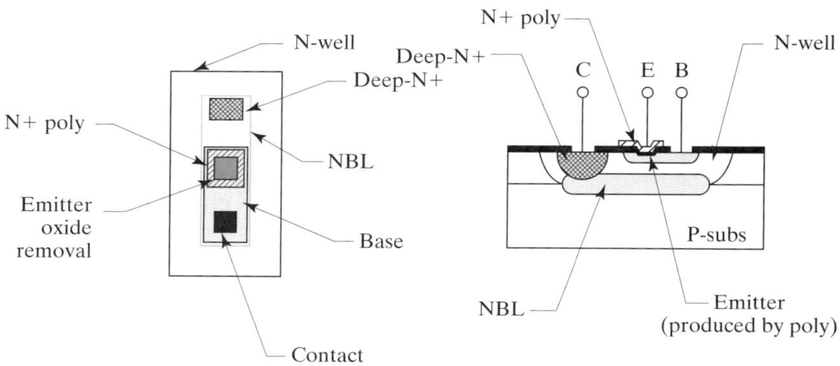

FIGURE 8.34 Layout and cross section of a CDI NPN transistor with a polysilicon emitter.

Polysilicon emitter transistors exhibit betas up to six times greater than those of equivalent diffused-emitter structures. The polysilicon emitter structure is believed to increase the emitter injection efficiency of the transistor by somehow preventing majority carriers in the base from flowing into the emitter. Various mechanisms have been proposed to explain this phenomenon. One theory credits the presence of grain boundaries in the polysilicon with reducing minority carrier mobility. Another theory emphasizes the role of the thin interfacial oxide that often separates the poly from the monocrystalline silicon.[33] Interestingly, polysilicon emitters also improve the beta of PNP transistors.[34]

Because the polysilicon emitter structure inherently possesses a higher beta, a more heavily doped base region can be employed to reduce base resistance. The polysilicon emitter structure also allows the creation of an extraordinarily thin emitter diffusion. Emitter punchthrough cannot occur because of the existence of a layer of polysilicon between the true emitter of the transistor and the emitter contact.

---

[32] R. S. Muller and T. I. Kamins, *Device Electronics for Integrated Circuits,* 2nd ed. (New York: John Wiley and Sons, 1986), pp. 306–307.

[33] Z. Yu, B. Ricco, and R. Dutton, "A Comprehensive Analytical and Numerical Model of Polysilicon Emitter Contacts in Bipolar Transistors," *IEEE Trans. Electron Devices,* Vol. ED-31, 1984, pp. 773–784.

[34] C. M. Maritan and N. G. Tarr, "Polysilicon Emitter p-n-p Transistors," *IEEE Trans. on Electron Devices,* Vol. ED-36, #6, 1989, pp. 1139–1143.

Furthermore, the depth of the emitter junction can be controlled with great precision. These factors allow the use of a much thinner base than might otherwise be possible, and the resulting reduction in neutral base width translates into a smaller minority-carrier transit time and a much faster transistor. The thinner base also allows the use of a thinner epi, drastically reducing outdiffusion of deep-N+ and N-well and therefore greatly shrinking the size of the transistor. The benefits of polysilicon emitters become readily apparent if one compares the transistors of Figures 8.33 and 8.34.

Additional reductions in device area and junction capacitance can be achieved through the use of partial oxide isolation. LOCOS processing can oxidize entirely through a thin epitaxial layer to separate adjacent tanks without using isolation diffusions. The elimination of sidewall junctions not only reduces collector-substrate capacitance, but also shrinks tank dimensions. Figure 8.35A illustrates the layout and cross section of an NPN transistor in which the base diffusion abuts the oxide isolation. The elimination of base-isolation spacings produces a corresponding reduction in device area. Figure 8.35B shows a more radical structure in which the emitter also abuts the oxide isolation. Such *walled-emitter* structures generally require a more abrupt termination of the oxide region than conventional LOCOS processing can provide. Various modifications of the LOCOS technique can reduce the width of the bird's beak,[35] and special anisotropic etch techniques can entirely eliminate it.

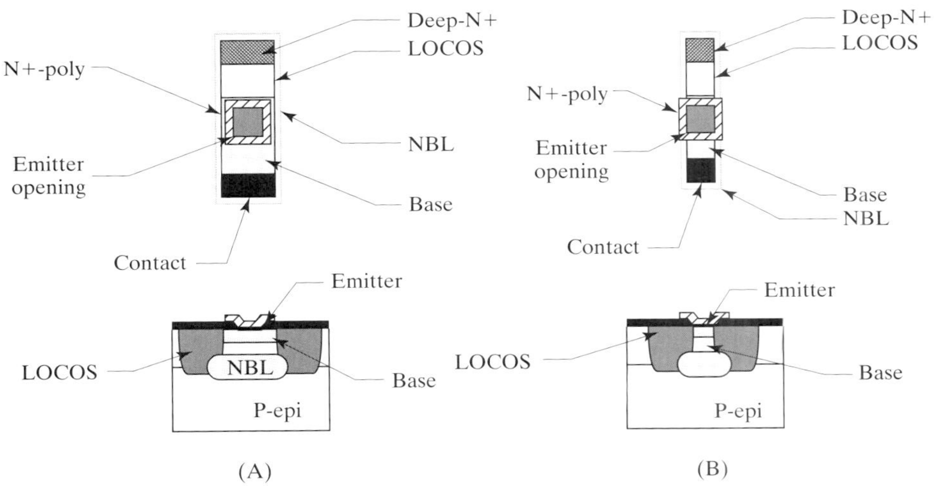

FIGURE 8.35 Partial oxide-isolated NPN transistors: (A) conventional, and (B) walled-emitter.

Modern, high-speed NPN transistors usually employ a combination of partial oxide isolation and polysilicon emitter technology. A variety of clever techniques have been proposed to allow the base contacts to be placed in extremely close proximity to a polysilicon emitter. The *super self-aligned* (SSA) structure of Figure 8.36 represents one such approach. Both the base and the emitter are contacted through doped polysilicon. The base polysilicon is deposited first and aligns to a contact opening. The emitter polysilicon forms after the base polysilicon and thus aligns to it. Both base and emitter ultimately self-align to the same contact opening. This structure can achieve switching times of less than 250pS, allowing the construction

[35] K. Y. Chiu, J. L. Moll, and J. Manoliu, "A Bird's Beak Free Local Oxidation Technology Feasible for VSLI Circuits Fabrication," *IEEE Trans. on Electron Devices*, Vol. ED-29, #4, 1982, pp. 536–540.

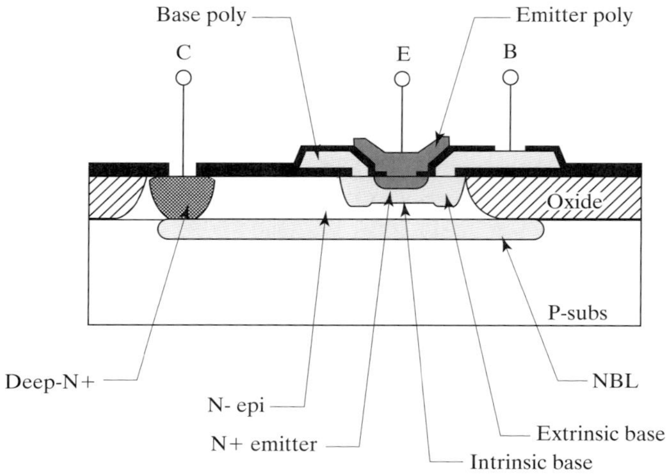

FIGURE 8.36 Cross section of a super self-aligned transistor.

of very fast logic circuits.[36] On the other hand, the complexity of this process limits its use to applications that demand speed at any price. Most analog BiCMOS processes target lower levels of performance to minimize processing costs. For example, the polysilicon-emitter CDI transistor in Figure 8.34 requires the same number of masks as a conventional CDI NPN, but it offers much smaller device sizes and lower junction capacitances.

8.4 SUMMARY

Bipolar transistors are extremely versatile devices, but they have their limitations. Improperly designed bipolar transistors may suddenly fail under heavy loads due to thermal runaway or secondary breakdown. Saturating bipolar transistors can also inject current into the substrate, debiasing adjacent circuitry and causing catastrophic latchup failures. These problems have caused circuit designers much anguish, but they can almost always be overcome by proper circuit design and device layout.

Bipolar transistors fall into one of two general categories: small-signal transistors and power transistors. Small-signal transistors are optimized for dense packing rather than for power handling capability. These devices are primarily used in analog signal processing and control circuitry, where they are particularly prized for their high transconductance and superior device matching. Standard bipolar processes offer relatively high-performance vertical NPN transistors plus several varieties of somewhat less useful PNP transistors. BiCMOS processes generally incorporate a similar selection of bipolar devices. The increasing popularity of analog BiCMOS ensures that bipolar transistors will remain an important part of analog circuit design for the foreseeable future.

The layout of bipolar transistors can be tailored to suit specific applications. Bipolar power transistors are frequently designed for improved immunity to thermal runaway and secondary breakdown. Small-signal transistors are frequently laid

---

[36] S. Konaka, Y. Yamamoto, and T. Sakai, "A 30-ps Si Bipolar IC Using Super Self-Aligned Process Technology," *IEEE. Trans. on Electron Devices,* Vol. ED-33, 1986, pp. 526–531. Also see T. Y. Chiu, G. M. Chin, M. Y. Lau, R. C. Hanson, M. D. Morris, K. F. Lee, M. T. Y. Liu, A. M. Voschenkov, R. G. Swartz, V. D. Archer, S. N. Finegan, and M. D. Feuer, "The Design and Characterization of Nonoverlapping Super Self-Aligned BiCMOS Technology," *IEEE Trans. Electron Devices,* Vol. 38, #1, 1991, pp. 141–150.

out to minimize device mismatches. The following chapter examines the techniques used to create these optimized bipolar transistor layouts.

8.4. EXERCISES

Refer to Appendix C for layout rules and process specifications.

8.1. What is the magnitude of the thermal voltage V_T at –55°C? At 125°C?

8.2. A lateral PNP transistor exhibits an emitter current of 110μA, a collector current of 98μA, and a base current of 7μA. What is the transistor's current gain and its collection efficiency?

8.3. Select base-ballasting resistors for the circuit in Figure 8.7. Assume $I_{BIAS} = 100\mu A$, and design the base-ballasting resistors so that a transistor in deep saturation will not consume more than 10% of I_{BIAS}. Assume that the V_{BE} of an NPN drops by a maximum of 100mV when it enters deep saturation.

8.4. Lay out the circuit in Figure 8.7 using the resistor values computed in Exercise 8.3 and the standard bipolar layout rules in Appendix C. Assume that all three transistors have minimum emitter areas. Use 6μm HSR resistors for R_1, R_2, and R_3. Place Q_1, Q_2, and Q_3 alongside one another and interdigitate the base ballasting resistors for proper matching. Justify your choice of tank biasing for R_1, R_2, and R_3.

8.5. Lay out a minimum-size standard-bipolar NPN transistor, including a deep-N+ sinker. Construct a similar device, but omit the sinker. Allow space for all necessary metallization. Assuming that the areas of the two devices equal the areas of their respective tanks, what percentage area reduction does the omission of deep-N+ produce?

8.6. Lay out a standard-bipolar, stretched-collector CEB transistor that allows one minimum-width lead to pass between its collector and emitter. The transistor should have a minimum-size emitter. Minimize the collector resistance, assuming the process uses a thick emitter oxide.

8.7. Lay out a standard-bipolar, narrow-emitter transistor having four minimum-width emitter fingers, each 100μm long. Place a deep-N+ sinker along one side of the transistor, and make sure that NBL fully encloses the sinker to minimize collector resistance. Include all necessary metallization to allow connection of the transistor into a circuit.

8.8. Construct minimum-size, standard-bipolar substrate PNP transistors using the standard, emitter-ringed, and verti-lat layout styles. Allow room for all necessary metallization. For the verti-lat transistor only, provide an emitter field plate overlapping the collector by 2μm.

8.9. Construct a split-collector lateral PNP transistor containing four quarter-sized collectors arranged around a minimum circular emitter. Allow room for all necessary metallization, including an emitter field plate overlapping the collectors by 2μm.

8.10. Construct a power PNP transistor consisting of sixteen minimum-size circular-emitter cells arranged in a 4 × 4 pattern. Include sufficient collector contacts to ensure that each cell resides adjacent to at least one collector contact. Include at least a minimum-size deep-N+ sinker in the base contact. Allow room for all necessary metallization, including emitter field plates overlapping the collectors by 2μm.

8.11. Construct a set of merged PNP transistors occupying the same tank. One transistor should have an elongated emitter 25μm long, while the second transistor should consist of two half-size collectors arranged around a minimum-size circular emitter. Allow room for all necessary metallization, including emitter field plates overlapping the collectors by 2μm.

8.12. Modify the layout in Exercise 8.11 to field plate the collector of the elongated transistor and one of the two collectors of the split-collector transistor. Overlap the field plate over the collectors by at least 6μm. Use flanges to elongate all channels as far as possible without enlarging the tank.

8.13. Construct a high-voltage NPN transistor using the standard bipolar layout rules. Assume the transistor uses a deep-N+ sinker and that it has a minimum-size emitter. Use 4μm fillets on the base diffusion and include all necessary field plates and channel stops.

8.14. Lay out an extended-base NPN transistor using the analog BiCMOS layout rules. Overlap the NBL 5μm into the deep-N+ isolation ring to ensure an adequate seal between the two. Construct a 64μm^2 emitter and allow for all necessary metallization.

8.15. Construct a CMOS substrate PNP transistor containing two emitter fingers that are each 30μm long. Assume that the process only silicides contacts and therefore no silicide block mask is required.

9 Applications of Bipolar Transistors

The bipolar transistor possesses two key advantages over its MOS counterpart: higher transconductance and superior device matching. These advantages translate into faster circuits that consume less power and offer higher precision. Many high-performance operational amplifiers and comparators use bipolar circuitry to minimize input offset and maximize output drive. Some families of high-speed logic also employ bipolar transistors as output drivers. Voltage regulators and references almost always use bipolar circuitry to obtain precise temperature-invariant voltages. Almost all of the highest-speed and highest-accuracy integrated circuits rely on bipolar circuitry in one form or another.

The *transconductance* of a bipolar transistor equals the ratio of the change in collector current to the change in base-emitter voltage. A high transconductance produces a large change in collector current for a small change in base-emitter voltage. The transconductance of a bipolar transistor is directly proportional to emitter current and is independent of emitter area, so even a small bipolar transistor can provide a large transconductance if it receives enough current. MOS circuitry dominates low-power design because MOS transistors retain moderate transconductances at very low currents. As the current levels increase, bipolar transistors become increasingly attractive. A micropower amplifier probably uses all-CMOS circuitry to conserve power, but a high-drive amplifier probably incorporates a bipolar output stage to reduce output impedance and minimize standby currents. The bipolar transistors in this output stage must handle high currents while dissipating large amounts of power. Small-signal transistors, even ones with enlarged emitters, perform poorly in power applications, so a variety of specialized layouts have been developed for this purpose.

The high transconductance of bipolar transistors also improves their base-emitter voltage matching. An untrimmed differential input stage constructed using bipolar transistors can routinely achieve three-sigma input offset voltages of less than ±1mV over temperature. Only exceptionally well constructed (and very large)

MOS input stages can hope to rival this performance.[1] Ratioed bipolar transistors can also generate very accurate voltage differentials, which form the basis of most voltage and current references. MOS references, even carefully constructed ones, rarely perform as well as their bipolar counterparts.

Although bipolar transistors offer distinct advantages over MOS, many designers remain reluctant to use them. Bipolar transistors can fall prey to a number of failure mechanisms that rarely affect MOS designs. The problem of saturation in bipolar transistors has no direct equivalent in MOS design. Improperly constructed bipolar transistors frequently self-destruct under heavy loads, while MOS transistors rarely do so. Carelessly matched bipolar transistors are far more vulnerable to thermal gradients than similar MOS devices. This chapter explains how to retain the unique advantages of bipolar transistors while avoiding their many pitfalls.

9.1 POWER BIPOLAR TRANSISTORS

The previous chapter discussed the layout of small-signal transistors. These devices usually employ minimum-area emitters to conserve space. These small emitters are acceptable because small-signal transistors rarely conduct more than a fraction of a milliamp. Transistors conducting larger currents experience beta rolloff unless their emitter areas increase in proportion to their emitter currents in order to maintain a constant emitter current density. The beta of a typical vertical NPN transistor begins to roll off at current densities of about 1mA/mil^2 (1.5 μA/μm^2) of emitter. To conserve space, power transistors generally operate at lower betas than their small-signal counterparts. A beta of ten is often chosen as a minimum acceptable limit for high-current operation. Power NPN transistors can usually handle 5 to 10mA/mil^2 (8 to 15μA/μm^2) before their beta drops below ten. PNP transistors cannot handle more than a small fraction of this current density. Although substrate PNPs may retain a beta of ten up to current densities of 1mA/mil^2, substrate injection usually limits them to maximum currents of a few milliamps. Lateral PNP transistors rarely achieve more than 250μA/minimum emitter. Most high-current circuits avoid the use of PNP transistors entirely, even if this eliminates otherwise-attractive circuit topologies.

Small-signal transistors can handle up to about 10mA and 100mW without any precautions. Beyond this point they become increasingly vulnerable to failure mechanisms caused by high currents and high power dissipation. These problems become especially acute in transistors handling currents in excess of 100mA or dissipating power in excess of 500mW. Such transistors require specialized layouts to protect them from thermal runaway and secondary breakdown. With careful layout, one can successfully integrate transistors capable of conducting 10A and dissipating 100W. Power transistors of this magnitude require so much die area that they completely dominate the layout of the integrated circuit. The cost of constructing a large integrated power device greatly exceeds the cost of purchasing an equivalent discrete device. Very high-power or high-current devices also require special packages that do not readily accommodate large numbers of pins. Most integrated power transistors conduct less than 2A and dissipate less than 10W. Power lateral PNP transistors require enormous amounts of die area, and few designs incorporate PNP transistors conducting more than 500mA. The vast majority of power bipolar transistors are therefore power NPN devices.

Many different power NPN layouts have been proposed. Each offers its own unique combination of advantages and disadvantages. No single structure outper-

---

[1] H. C. Lin, "Comparison of Input Offset Voltage of Differential Amplifiers Using Bipolar Transistors and Field-Effect Transistors," *IEEE J. of Solid-State Circuits,* Vol. SC-5, #6, 1970, pp. 126–129.

forms all the others in all applications. In order to make an intelligent choice, the designer must understand the mechanisms that cause power transistors to fail.

9.1.1. Failure Mechanisms of NPN Power Transistors

The three most common problems encountered in the design of power bipolar transistors are emitter debiasing, thermal runaway, and secondary breakdown. These three problems all result from the large currents and the high power dissipations typical of power devices. None of these mechanisms cause much trouble in small signal transistors, but all impose significant constraints on power transistor design.

Emitter Debiasing

The term *emitter debiasing* refers to a nonuniform current distribution that may develop in a power bipolar transistor due to voltage drops in the extrinsic base and emitter, and in their respective leads. The high transconductance of bipolar transistors makes these devices very susceptible to changes in base-emitter bias. Small voltage drops down the base or emitter leads can radically redistribute current flow through the transistor. Some portions of the transistor may conduct little or no current, while others conduct far more current than they were designed to handle. The overloaded portions of the transistor become vulnerable to thermal runaway and secondary breakdown.

Figure 9.1 shows an example of emitter debiasing occurring between the separate emitter fingers of a power transistor. In the accompanying schematic, transistors Q_1 to Q_4 represent the four emitter fingers and resistors R_1 to R_3 represent the resistance of the metal leads connecting the fingers together. Assume that each emitter

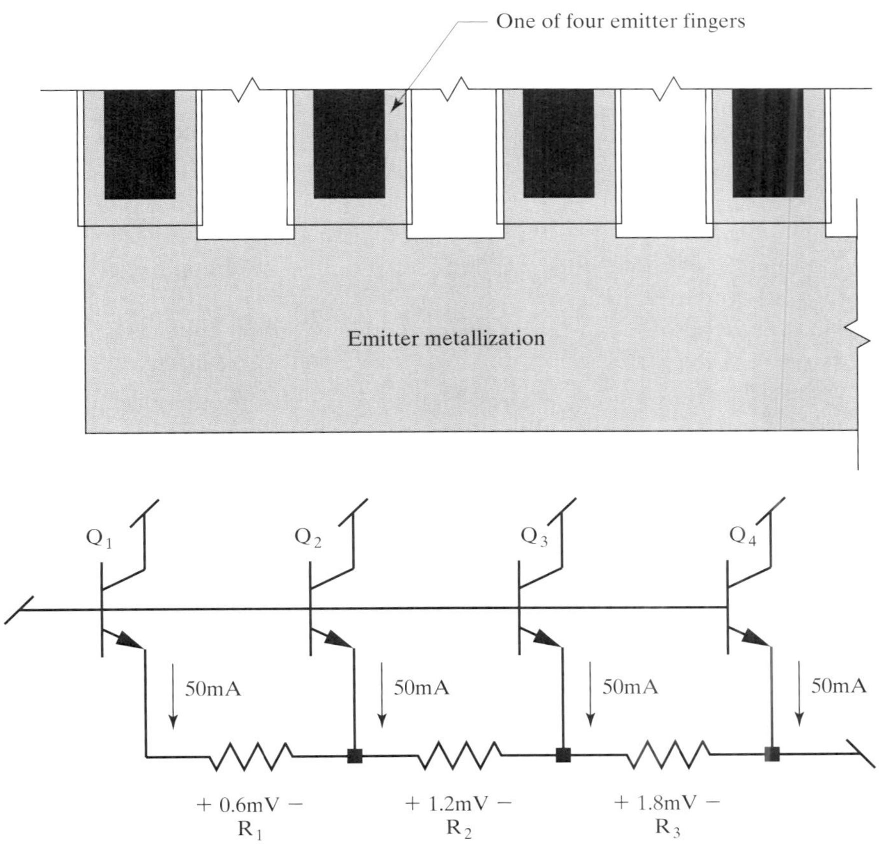

FIGURE 9.1 The layout and equivalent schematic of a power transistor having four emitter fingers. The values listed on the schematic follow the computations described in the text.

finger conducts 50mA, and that each resistor consists of one square of 20kÅ aluminum with a sheet resistance of 12mΩ/□. The total drop across the three resistors equals 3.6mV. The ratio of emitter currents η between two transistors whose base-emitter voltages differ by a voltage ΔV_{BE} equals

$$\eta = e^{\Delta V_{BE}/V_T} \quad [9.1]$$

where V_T represents the thermal voltage of silicon (approximately 26mV at room temperature). In this example, the ratio of currents equals 1.15, so the rightmost finger Q_4 would conduct about 15% more current than the leftmost finger Q_1. Analog BiCMOS processes encounter even more severe debiasing problems because they use thinner metallization (typically 10kÅ).

The previous example illustrates the severity of emitter debiasing—relatively small currents flowing through short, wide leads still cause 3.6mV of debiasing. A technique called *emitter ballasting* can greatly reduce the impact of debiasing. Emitter ballasting requires the insertion of resistors into each emitter lead (Figure 9.2). These resistors are typically sized to provide a voltage drop of 50 to 75mV at full rated current. For example, emitter fingers conducting 50mA each might employ 1Ω ballasting resistors. The addition of these ballasting resistors forces the emitter current to redistribute about equally between the emitter fingers. If any emitter finger attempts to draw more than its fair share of current, then the voltage drop across its ballasting resistor increases. This limits the amount of current that can flow through this emitter finger. Voltage drops between ballasted emitters appear primarily across the ballasting resistors rather than across the base-emitter junctions of the transistors. Thus 3.6mV of debiasing between two emitters ballasted with 1Ω resistors would result in a 1.8mA current increase in one emitter and a 1.8mA current decrease in the other. These numbers are only approximations, but they serve to show how much benefit emitter ballasting provides. A ballasted transistor can generally tolerate debiasing equal to 25% of the voltage drop across the ballasting resistors. A transistor that drops 50mV across each ballasting resistor can easily tolerate 10mV of debiasing in its emitter leads. If the layout would produce more debiasing, the size of the ballasting resistors can be increased to compensate. Remember, however, that the voltage drop across the ballasting resistors adds to the saturation voltage of the transistor, lowers its effective transconductance, and increases its power dissipation. If the design requires more than about 100mV of ballasting, consider altering the metallization pattern or the aspect ratio of the transistor.

Emitter debiasing can also develop within a single emitter finger (*intrafinger debiasing*). Voltage drops accumulate as the current flows along the finger. One end of the emitter finger sees a larger base-emitter voltage and therefore conducts more

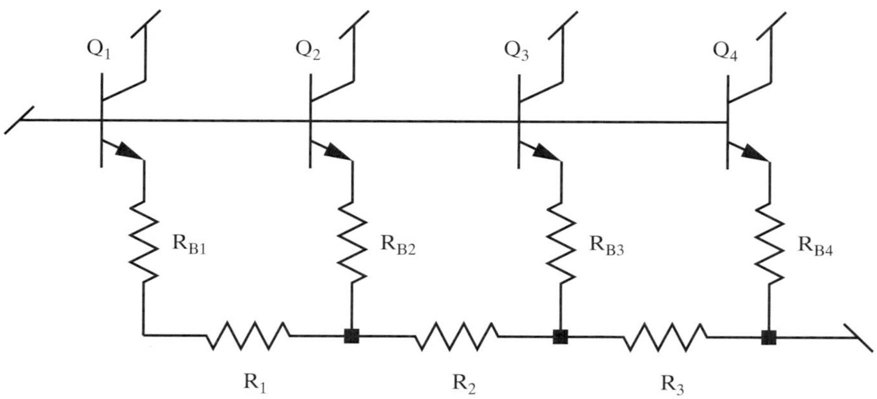

FIGURE 9.2 Connection of ballasting resistors to the segmented power transistor in Figure 9.1; R_{B1} to R_{B4} are the ballasting resistors for the emitter fingers Q_1 to Q_4.

current than the other. Debiasing along a long emitter finger can actually become a more serious problem than debiasing between separate fingers. For a narrow emitter finger like that in Figure 9.3, the voltage drop from one end to the other should not exceed 5mV. Assuming that an emitter lead of constant width runs down the emitter finger, and assuming that equal currents flow into the emitter lead along each increment of its length, then the total voltage drop from one end of the emitter contact to the other end equals

$$\Delta V_{BE} = \frac{LR_s I_E}{2W} \quad [9.2]$$

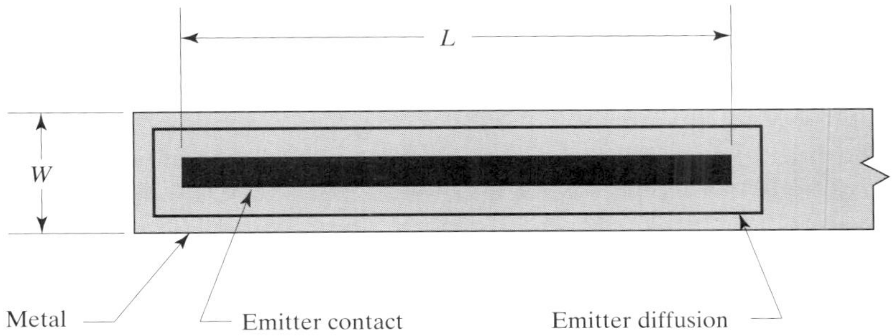

FIGURE 9.3 A sample emitter finger layout showing measurements L and W used in Equation 9.2.

where R_s represents the sheet resistance of the metallization, W is the width of the emitter lead, L is the length of the emitter contact, and I_E equals the total current flowing out of the entire emitter finger (Figure 9.3). For example, suppose an emitter finger conducts 50mA along a lead 300μm long by 30μm wide constructed from 12mΩ/□ aluminum. Equation 9.2 indicates that the debiasing along this emitter finger equals 9mV, which exceeds the maximum suggested debiasing of 5mV. Although this computation does not consider the redistribution of emitter current in response to debiasing, it still demonstrates the severity of the debiasing problem.

Several options exist for reducing intrafinger debiasing. The emitter fingers may be shortened and widened. This not only minimizes the finger length but also allows the use of wider metal leads. Alternatively, the transistor may employ a larger number of shorter emitter fingers of the same width as the originals. A ballasting technique also exists that applies to individual fingers (Section 9.1.2), but it can only provide a limited amount of ballasting, which may not suffice to compensate for poor emitter finger design.

Thermal Runaway and Secondary Breakdown

Both thermal runaway and secondary breakdown result from an intensification of current flow through portions of the power transistor. In the case of thermal runaway, the current flow localizes in response to increasing temperature. Suppose that one portion of the power transistor becomes slightly warmer than the rest. The V_{BE} required to maintain constant collector current drops by 2mV/°C, so a temperature rise of only a few degrees results in significant emitter debiasing. Almost all of the current flows through the hottest portion of the transistor, raising its temperature still further. In a matter of milliseconds, the region of conduction collapses to a tiny *hot spot* comprising only a few percent of the transistor's area. Perhaps beta rolloff can limit the collapse of the hot spot sufficiently to prevent catastrophic device failure, or perhaps not. Even if the hot spot stabilizes, the transistor will be so severely overstressed that it will become vulnerable to other failure mechanisms such as secondary breakdown, electromigration, and thermally accelerated corrosion.

Since thermal runaway involves emitter debiasing, ballasting resistors can provide some measure of protection against it. If each finger of a multi-emitter transistor has its own ballasting resistor, then a hot spot that develops in one finger cannot steal current from the other fingers. Even in a worst-case scenario in which hot spots develop in all of the fingers, each hot spot absorbs only a fraction of the total current. Usually 50mV of ballasting suffices to control thermal runaway, but more ballasting is sometimes necessary to offset voltage drops in the emitter metallization system.

Hot spots can still develop in individual emitter fingers even if all of the fingers have ballasting resistors. If each finger has its own ballasting resistor, then the current drawn by any one hot spot will decrease as the number of emitter fingers increases. Distributed emitter ballasting (Section 9.1.2) also provides some measure of protection against hot spot formation. Extremely demanding applications may require a combination of distributed emitter ballasting and individual ballasting resistors for each emitter finger.

Secondary breakdown occurs when the emitter current density in a transistor exceeds a critical threshold value J_{crit}. Beyond this point, the sustained collector-to-emitter breakdown voltage $V_{CEO(sus)}$ snaps back to a new, lower value called the *secondary breakdown voltage* V_{CEO2} (Section 8.1.3). A transistor is most vulnerable to secondary breakdown when it is in the process of turning off. The collector-to-emitter voltage across the transistor rises as the emitter current through the transistor decreases. Secondary breakdown occurs if the collector-to-emitter voltage exceeds V_{CEO2} while the emitter current density exceeds J_{crit}. Once avalanche begins, the base drive circuit can no longer turn the transistor off, and the transistor is soon destroyed by overheating or metallization failure.

Transistors driving inductive loads are extremely vulnerable to secondary breakdown. Consider a power transistor Q_1 driving a high-side inductive load L_1 (Figure 9.4A). As soon as Q_1 begins to turn off, the inductive kick-back of L_1 drives the collector voltage V_{CE} upward until recirculation diode D_1 begins to conduct (Figure 9.4B). The collector voltage reaches its maximum value almost immediately, long before the emitter current drops to zero. Secondary breakdown will occur if the collector voltage exceeds V_{CEO2} while the emitter current density exceeds J_{crit}.

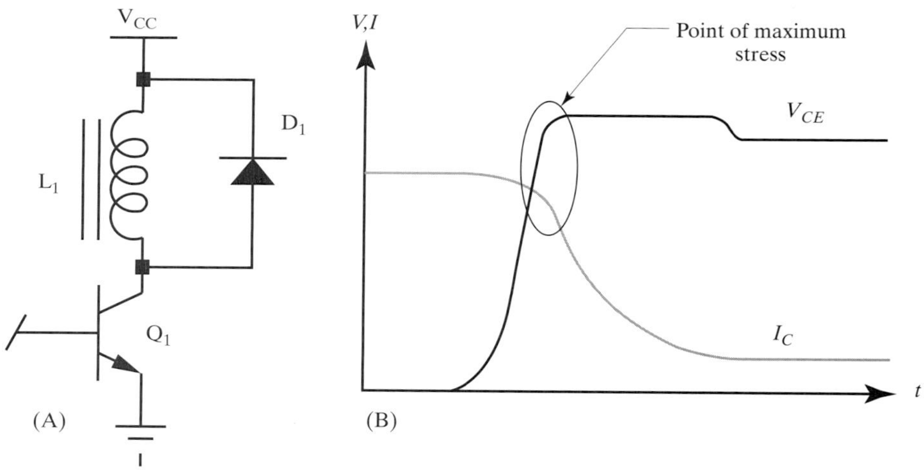

FIGURE 9.4 An example of (A) a bipolar transistor driving an inductive load and (B) the waveforms associated with the turn-off interval of the transistor.

Conservative design rules dictate that power transistors operate at an emitter current density of no more than 5 to 10mA/mil^2 (8 to 15μA/μm^2). These current densities lie well below those required to trigger secondary breakdown in order to provide a safety margin for emitter debiasing, emitter current focusing, and thermal gradient formation.

9.1.2. Layout of Power NPN Transistors[2]

Over the years, a number of alternative layouts have been proposed for NPN power transistors. Each layout has certain strengths and weaknesses, so knowledge of several different types of layouts will aid the designer in choosing the best style for any given application. Any layout can be scaled by adding or removing emitter sections or by connecting several power devices in parallel.

A transistor used in a *linear-mode* application remains in the forward active region for long periods of time. Linear transistors must withstand large collector-to-emitter differentials while simultaneously conducting large collector currents. Such a transistor must cover enough area to allow for heat dissipation. As a general rule, linear-mode transistors should not dissipate more than 100mW/mil^2 (150μW/μm^2) of emitter nor conduct more than 5mA/mil^2 (8μA/μm^2). These are conservative guidelines and, with sufficient ballasting and heatsinking, it is possible to successfully operate transistors at several times these stress levels. Still, one should generally follow these conservative guidelines unless empirical measurements show that a transistor can safely operate at higher stress levels.

Transistors used in *switched-mode* applications operate either in cutoff where no current flows, or in saturation where collector-to-emitter differentials remain small. Switching transistors dissipate power only during brief switching intervals. The average power dissipation of switching transistors remains relatively small, so they rarely experience hot spot formation or thermal runaway. On the other hand, switching applications generate many opportunities for emitter current focusing during turn-off. Conservative designs should not exceed an emitter current density of 10mA/mil^2 (16μA/μm^2) to ensure that emitter current focusing does not trigger secondary breakdown.

Transistors that drive purely capacitive loads such as MOS gates conduct current only as infrequent short-duration pulses. Such *pulsed-mode* transistors typically conduct large currents for a few hundred nanoseconds, then rest for a few microseconds before conducting again. Pulsed-mode transistors are unharmed by emitter-current focusing because the external capacitive load will quench conduction regardless of whether secondary breakdown occurs. Pulsed-mode transistors are also immune to thermal runaway because hot spots cannot form and localize in less than a few microseconds.[3] Most pulsed-mode applications rely on high-current beta rolloff and collector resistance to limit conduction. This practice is acceptable as long as the pulse duration does not exceed 1μS, the intervals between pulses are no shorter than 250nS, and the *average* emitter current density does not exceed 10mA/mil^2 (16μA/μm^2). The metallization for pulsed-power transistors should be designed following the electromigration rules for intermittent currents described in Section 14.3.3.

The Interdigitated-emitter Transistor

The oldest style of power transistor, the *interdigitated-emitter transistor,* remains in use because it can operate at higher speeds than any other style of bipolar power transistor. Figure 9.5 shows an interdigitated-emitter transistor constructed using a single-level-metal standard-bipolar process.

This transistor consists of a number of emitter fingers, each having its own dedicated emitter ballasting resistor. The ballasting resistors are all formed from a single

---

2 F. W. Trafton, "High Current Transistor Layout," unpublished manuscript, 1988.

3 H. Melchior and M. J. O. Strutt, "Secondary Breakdown in Transistors," *Proc. IEEE,* Vol. 52, 1964, p. 439.

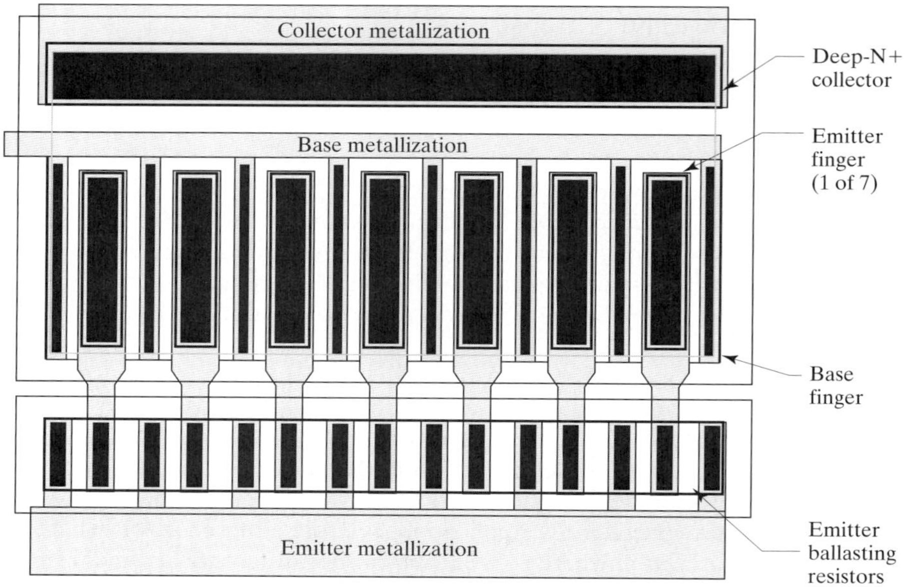

FIGURE 9.5 An example of an interdigitated-emitter power transistor. Each emitter finger has a separate ballasting resistor. Metallization is shown in gray for emphasis.

strip of emitter diffusion placed in a separate tank. The emitter diffusion is not isolated from the tank because small current leakages from one finger to another cause no harm. Each emitter finger connects to two ballasting resistors placed in parallel, each consisting of about one square of emitter. Assuming a minimum emitter sheet resistance of 5Ω/□, this provides a ballasting resistance of 2.5Ω per finger. This resistance will provide 50mV of ballasting at 20mA of emitter current.

The interdigitated-emitter transistor is extremely vulnerable to intrafinger debiasing. The voltage drop down each emitter finger computed using equation 9.2 should not exceed 5mV. A large number of short emitter fingers are preferable to a small number of long fingers. The width of the emitter fingers also affects performance. Widening the emitter fingers also widens the pinched base regions underneath them. The resulting increase in base resistance causes the transistor to exhibit slower switching and increased emitter focusing. The fastest and most robust designs incorporate minimum-width emitter fingers, but it is difficult to place enough metal on narrow emitter fingers to prevent them from debiasing. Double-level metal helps, but narrow fingers still do not make efficient use of available area. Most designers compromise on an emitter width of 0.3 to 1.0mil (8 to 25μm). In this style of transistor, the emitter contacts are always made as large as possible to reduce emitter resistance.

Base contacts along either side of each emitter finger reduce the base resistance and enable faster switching. Base contacts placed on either end of the emitter array ensure that the end fingers turn off as quickly as the others. If these end contacts were omitted, the end fingers would turn off more slowly than the others. This could lead to emitter current focusing and secondary breakdown during turnoff. Minimum-width base contacts help conserve space, and relatively few designs require more base metallization. Power transistors operating at high current densities may, however, experience enough beta rolloff to necessitate wider base metallization. Computation can show whether or not any particular design experiences significant base-lead debiasing. Designs exhibiting more than 2 to 4mV of base debiasing should be redesigned to reduce base metallization resistance. The comb-style base metallization in Figure 9.5 exhibits much less metallization resistance than the

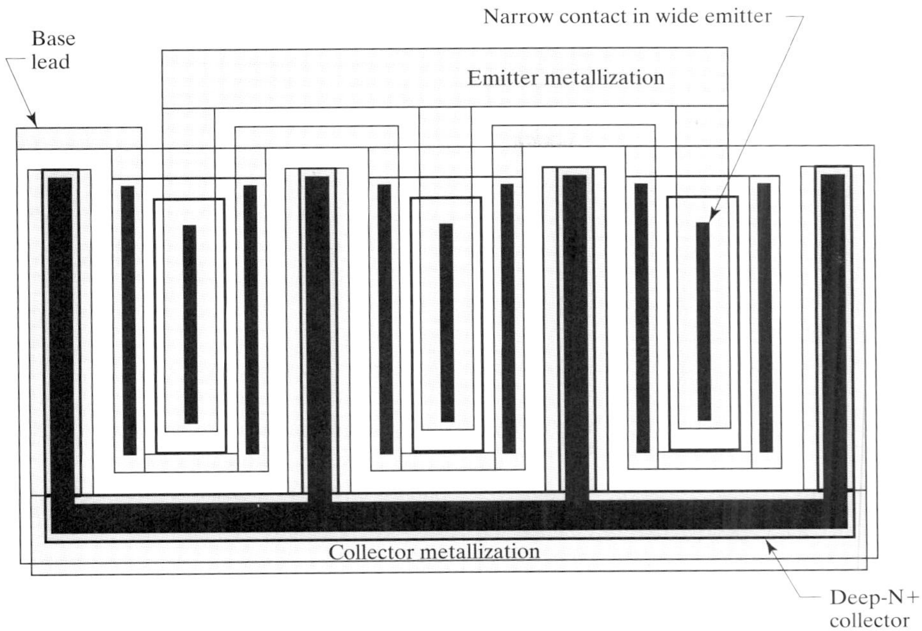

FIGURE 9.6 An example of wide-emitter narrow-contact power transistor. Emitter ballasting resistors, although not shown, can easily be added. Metallization is shown in gray for emphasis.

serpentine metallization in Figure 9.6. Unfortunately, many single-level-metal designs do not lend themselves to the use of comb-style base metallization.

The transistor in Figure 9.5 contains deep-N+ only along one side. This may suffice for a linear-mode device operating at a collector-to-emitter voltage differential of a half-volt or more. A switching transistor is a different matter, as their efficiency is determined by their collector-to-emitter voltage drop in saturation (their *saturation voltage*). If the saturation voltage is too large, then the transistor dissipates too much power. At high currents, the collector resistance of a switching transistor equals the sum of its vertical deep-N+ resistance and its lateral NBL resistance.[4] The vertical resistance of the deep-N+ sinker can be reduced by increasing its area. The sinker should not be less than 10μm wide to ensure that outdiffusion does not dilute its doping and increase its vertical resistance. The lateral NBL resistance can be reduced by contacting the NBL along a longer periphery or by decreasing the distance between the active portions of the device and the sinkers. Placing sinkers along both sides of the transistor reduces the NBL resistance by a factor of four, and an unbroken ring of deep-N+ around the transistor reduces it still further. The NBL should extend to the outer edge of the deep-N+ sinker to ensure a low-resistance connection between the two.

An unbroken ring of deep-N+ around a power transistor also forms a hole-blocking guard ring that helps control substrate injection during saturation. When an NPN transistor saturates, all of its unused base drive flows to the substrate. The guard ring does not reduce the amount of base drive consumed by the transistor, but it does prevent the majority of it from flowing to the substrate. Section 9.1.3 discusses several techniques for limiting the base current consumed during saturation.

[4] The drift region usually contributes little or no resistance since it depletes through under the influence of reverse bias or velocity saturation.

The Wide-emitter Narrow-contact Transistor

The interdigitated-emitter transistor uses relatively narrow emitter fingers to reduce base resistance and to control emitter crowding. This structure's low base resistance allows it to operate at higher frequencies than any other. Unfortunately, the narrow emitters are quite prone to emitter crowding. Emitter debiasing causes conduction to concentrate at the exit end of each finger, while thermal gradients focus conduction into the middle of the transistor. In either case, current tends to localize at one point in each emitter finger. Ballasting resistors can help ensure that the fingers conduct equal currents, but they cannot prevent intrafinger debiasing. Even well-ballasted interdigitated-emitter transistors tend to develop hot spots at higher current densities.

If each emitter finger is divided into a large number of individually ballasted sections, then no one portion of the emitter finger can contact more current than any other. Although it is generally not feasible to segment an emitter finger in this manner, placing a narrow emitter contact in a wide emitter finger provides similar benefits.[5] Figure 9.6 shows the resulting *wide-emitter narrow-contact transistor.*

The use of a wide emitter finger and a narrow contact produces the equivalent of a distributed network of ballasting resistors. This network consists partly of emitter resistance and partly of pinched base resistance. The emitter resistance is largest at the periphery of the emitter and smallest in the center directly beneath the narrow contact. Conversely, the base resistance is smallest at the periphery and is largest in the center directly under the emitter contact. These two forms of ballasting complement one another. At low currents, the base resistance is relatively insignificant, and current distributes uniformly across the width of the emitter finger. As the current increases, debiasing in the pinched base region causes conduction to move out toward the periphery of the emitter finger. The current must now flow through a larger emitter resistance. The resulting emitter voltage drops counteract the movement of current toward the emitter periphery. Together, the base-side and emitter-side distributed ballasting ensure that conduction occurs relatively uniformly across the entire width of the emitter finger. This type of emitter ballasting is distributed along the length of the emitter finger, so it protects all portions of the device against emitter debiasing and the formation of hot spots.

The emitter must overlap the contact by a distance sufficient to provide adequate ballasting. Typical wide-emitter narrow-contact structures employ emitter overlaps of 0.5 to 1.0mils (12 to 25μm). Larger overlaps unnecessarily slow the frequency response of the transistor, while smaller overlaps may not provide enough distributed ballasting to fully protect against thermal runaway and secondary breakdown. Transistors operating under extreme conditions often benefit from additional ballasting resistors inserted into the leads of each emitter finger, as illustrated in the interdigitated-emitter transistor in Figure 9.5.

Some designers employ a trapezoidal emitter contact tapering from a wide low-current end to a narrow high-current end. This design provides additional ballasting at the high-current end of the emitter finger to offset the effect of voltage drops in the emitter metallization. The lack of matching between metal and emitter sheet resistances makes this type of design problematic, and most wide-emitter narrow-contact transistors employ minimum-width contacts instead. One must not stretch the contact to the ends of the emitter finger, because this eliminates the ballasting at these points and renders them vulnerable to hot spot formation.

---

[5] A. B. Grebene, *Bipolar and MOS Analog Integrated Circuit Design* (New York: John Wiley and Sons, 1984), p. 510.

The transistor in Figure 9.6 employs multiple base regions with fingers of deep-N+ interdigitated between them. This structure minimizes collector resistance at the cost of increasing area and complicating lead routing. In single-level-metal layouts, the base lead must serpentine through the transistor between the collector and emitter metallization. The added length of the serpentined base lead can cause significant debiasing in the base metallization. Even with the distributed ballasting, the transistor should not contain metallization drops of more than a few millivolts. Base debiasing can be reduced by a factor of roughly four by connecting both ends of the serpentined emitter lead. Layouts employing double-level metal frequently use comb or grid arrangements to combat base metallization debiasing.

The wide-emitter narrow-contact structure is remarkably robust. The distributed emitter ballasting helps prevent thermal runaway and secondary breakdown within individual fingers, allowing the device to operate at higher current densities than comparable interdigitated structures do. Wide-emitter narrow-contact transistors that must operate under especially harsh conditions may benefit from the insertion of an additional 50 to 75mV of emitter ballasting in the leads of the individual emitter fingers. This structure does not switch as quickly as the interdigitated-emitter transistor, but the degradation is not as large as one might expect since considerable high-current conduction occurs along the emitter periphery.

The Christmas-tree Device

Another type of power transistor layout is nicknamed the *Christmas-tree device* because of the peculiar shape of its emitter geometry (shown in dark gray in Figure 9.7). Historically, this structure was widely used in linear applications because of its exceptional resistance to thermal runaway. It is rarely used for switching applications, because the same features that improve its immunity to thermal runaway degrade its ability to withstand emitter current focusing during turnoff.

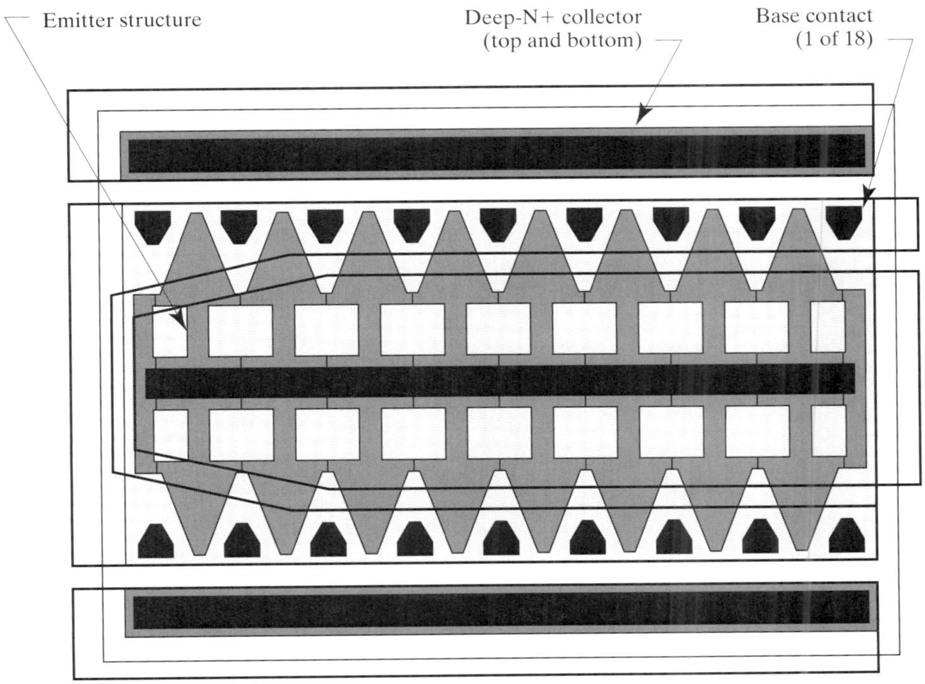

FIGURE 9.7 An example of a Christmas-tree power transistor. The base is shown in light gray and the emitter in dark gray in order to highlight the peculiar structure of the emitter.

The emitter of this transistor consists of a central spine surrounded by a complex branching structure of triangular prongs that give the transistor its picturesque name. Most of the conduction occurs in the triangular prongs along the emitter periphery. These connect to the central spine of the emitter through narrow emitter strips that act as ballasting resistors. At low currents, all portions of the emitter conduct. As the current increases, emitter crowding forces conduction out toward the periphery, causing current to flow through the ballasting resistors incorporated into the emitter structure. This device gains its resistance to thermal runaway from a large amount of distributed ballasting. Unfortunately, the great width of the emitter structure renders it vulnerable to emitter current focusing. As the transistor begins to turn off, the area of conduction retreats from the periphery toward the central spine. Because the spine represents only a small portion of the total emitter area, the emitter current density increases dramatically during the final stages of turnoff. This concentration of current flow can (and often does) trigger secondary breakdown. The wide-emitter narrow-contact structure exhibits superior immunity to secondary breakdown because the distance from the periphery to the center of the emitter is not as great and the effects of emitter focusing are not as dramatic.

The Christmas-tree device serves best in applications that dissipate large amounts of power, but where abrupt turn-off transitions never occur. Historically, this style of device was frequently chosen for the series-pass devices of linear voltage regulators and the output stages of audio power amplifiers. A number of variations on the Christmas-tree device have been developed in an attempt to minimize its vulnerability to emitter focusing while retaining its immunity to hot spot formation. None of these variants are as robust as the wide-emitter narrow-contact transistor, or its descendent, the cruciform-emitter transistor.

The Cruciform-emitter Transistor

The *cruciform-emitter transistor* represents an evolutionary development of the wide-emitter narrow-contact structure that seeks to incorporate additional emitter ballasting without rendering the device vulnerable to secondary breakdown. The emitter of this device consists of a series of cross-shaped (*cruciform*) sections stacked end-to-end to form a continuous emitter finger (Figure 9.8). The base contacts occupy the small notches between the arms of the crosses.

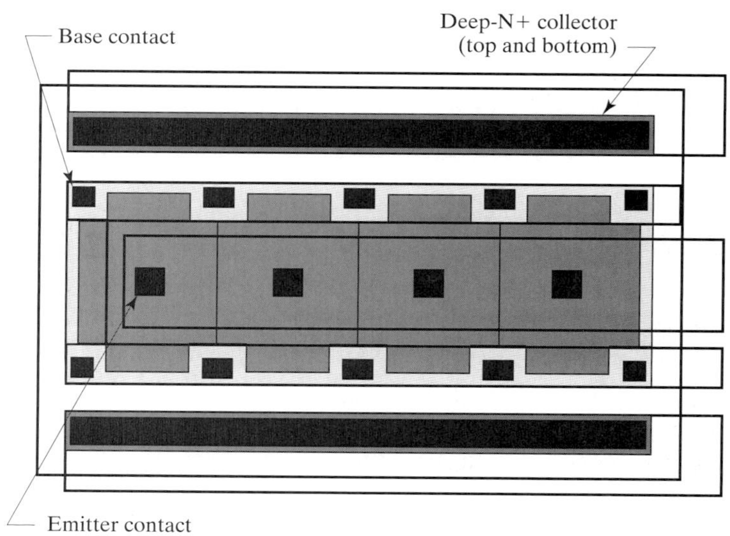

FIGURE 9.8 An example of a cruciform-emitter transistor. The base is shown in light gray and the emitter in dark gray for emphasis.

The width of the cruciform emitter has been increased to 3 to 5mils (75 to 125μm) to obtain additional ballasting. The narrow emitter contact has also been replaced by a series of small, square or circular contacts occupying the center of each cross. All of the emitter current must flow through these contacts, producing a distributed three-dimensional ballasting effect considerably more efficient than the two-dimensional ballasting generated by the wide-emitter narrow-contact structure. Consequently, the cruciform emitter combines the best features of the wide-emitter narrow-contact transistor and the Christmas-tree device. The cruciform transistor does not have quite the immunity to secondary breakdown that the wide-emitter narrow-contact transistor does, but it vastly outperforms the Christmas-tree device in this respect. The cruciform emitter transistor also makes extremely efficient use of space.

The cruciform structure suffers from two drawbacks. First, the small size of the emitter contacts renders them vulnerable to electromigration. All of the emitter current must cross the sidewalls of these contacts, and this produces very high localized current densities in the metallization. Even refractory barrier metal has its limits, which this transistor may exceed. Some designers replace the single contact in the center of each cruciform with an array of minimum contacts to increase the sidewall perimeter. Second, the compact design of the cruciform emitter can cause extreme localized heating at high power levels. Less area-efficient transistors are actually preferable to more compact ones from the standpoint of heat dissipation. If the heat produced by the transistor spreads over a wider area, then the thermal impedance between the transistor and the package decreases and the transistor can handle more power before it overheats. The cruciform structure is best suited for switching applications, as these are more strongly constrained by current-handling capability than by power dissipation.[6]

Power Transistor Layout in Analog BiCMOS

Any of the power transistors discussed above can also be implemented in analog BiCMOS. Figure 9.9 shows a BiCMOS version of a wide-emitter narrow-contact transistor. Double-level metallization allows the base contacts to completely encircle each emitter finger, whereas in a single-level-metal design they can only reach two or three sides of each finger. The complete ring of base contact helps ensure that all portions of the emitter periphery are equally active. An unbroken ring of deep-N+ sinker minimizes collector resistance and blocks substrate injection during saturation.

Figure 9.9B shows further details of the metallization system. Emitter current flows from the narrow emitter contacts to vias placed parallel to them. Passing up through these vias, the current reaches a metal-2 plate covering the top of the transistor. This plate minimizes emitter debiasing by reducing the metal-2 resistance to an absolute minimum. The resistance in the metal-1 plates actually serves as emitter ballasting and therefore is unobjectionable. The base metallization consists of a grid of first-level metal covering the base contacts. The base current exits the transistor through a metal-2 jumper placed between the emitter metal-2 plate and the encircling collector metal-2. If necessary, a second base lead can exit on the other side of the emitter plate. The collector metallization consists of a complete ring of metal-1 covering the collector contact and a U-shaped metal-2 plate covering the collector on three sides of the transistor. Vias along the inner edges of the collector contact allow current to flow through both levels of metallization. The collector lead can exit any side of the transistor except the side where the emitter lead exits. The

---

[6] A related structure called the *H-emitter transistor* is described in F. F. Villa, "Improved Second Breakdown of Integrated Bipolar Power Transistors," *IEEE. Trans. on Electron Devices*, Vol. ED-33, #12, 1986.

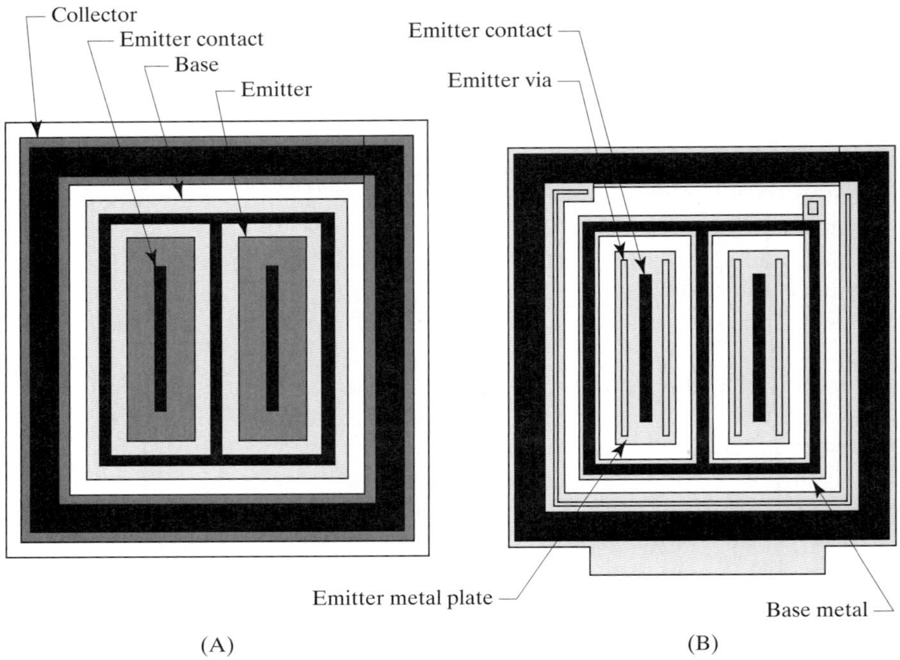

FIGURE 9.9 A wide-emitter narrow-contact transistor constructed in analog BiCMOS using double-level metal: (A) diffusions and (B) metal-1 pattern. The metal-2 pattern is not shown.

best arrangement places the collector lead diametrically opposite of the emitter lead. This minimizes the resistance of the collector metallization by ensuring that half of the current flows through the metal on either side of the transistor.

The structure in Figure 9.9 has been used to fabricate pulse-power transistors capable of operating at emitter current densities of more than 100mA/mil^2 (160μA/μm^2). This structure uses an overlap of emitter over contact of 8 to 12μm and a continuous ring of deep-N+ sinker at least 8μm wide. The NBL should completely overlap the deep-N+ sinker to minimize resistance and to ensure that no minority carriers can escape through a lightly doped portion of the extrinsic collector. This structure can continuously conduct in excess of 10mA/mil^2 (16μA/μm^2) and can operate as either a linear dissipative device or as a switching element. The distributed emitter ballasting inherent in the wide emitter fingers prevents hot spots from forming even at very high power levels. The use of a solid metal-2 plate to terminate the emitters helps minimize emitter debiasing, making individual emitter ballasting resistors unnecessary for all but the most demanding applications.

Selecting a Power Transistor Layout

All of the power transistor layouts presented in this section have their advantages and their disadvantages. The Christmas-tree device is best suited for linear applications that do not experience rapid switching transients. The interdigitated emitter transistor provides the best switching speeds and frequency response, but it requires individual ballasting resistors on each finger to avoid thermal runaway. The wide-emitter narrow-contact and cruciform transistors excel in switching applications. A wide-emitter narrow-contact transistor with ballasting resistors in each emitter finger is virtually immune to secondary breakdown at the voltages normally encountered in integrated circuit applications (10 to 40V). Some applications require that a small portion of the transistor's emitter be brought out independently to act as a sensing element. The interdigitated emitter structure offers the easiest insertion of a sense emitter and the best matching of the sense emitter to the remainder of the transistor. Table 9.1 summarizes these advantages and disadvantages.

	Interdigitated Emitter	Wide-emitter Narrow-contact	Christmas-tree Device	Cruciform Transistor
Thermal runaway	Good*	Good	Excellent	Excellent
Secondary breakdown	Fair	Excellent	Poor	Good
Frequency response	Excellent	Good	Fair	Fair
Compactness of layout	Poor	Good	Good	Excellent
Ease of emitter sensing	Excellent	Fair	Poor	Poor

* Assumes individually ballasted emitter fingers; otherwise *poor.*

TABLE 9.1 Comparison of four types of power NPN layouts.

9.1.3. Saturation Detection and Limiting

Both lateral PNP and vertical NPN transistors inject current into the substrate when they saturate. Substrate injection wastes supply current and may cause substrate debiasing and device latchup. Several techniques have been developed to suppress substrate injection, either by intercepting minority carriers before they reach the substrate or by preventing the transistor from saturating in the first place. Most of these techniques require specialized layouts.

The emitter of a lateral PNP continuously injects minority carriers into its tank. When the transistor saturates, most of this current flows to the substrate. Small-signal transistors rarely inject enough current to warrant concern, but a few designs incorporate large lateral PNP transistors that conduct tens or even hundreds of milliamps. Currents of this magnitude can easily produce enough debiasing to trigger latchup.

Figure 9.10A shows one way to prevent minority carriers from reaching the substrate. This transistor incorporates a continuous, unbroken ring of deep-N+ around the outside edge of its tank. This ring merges with the underlying NBL and completely

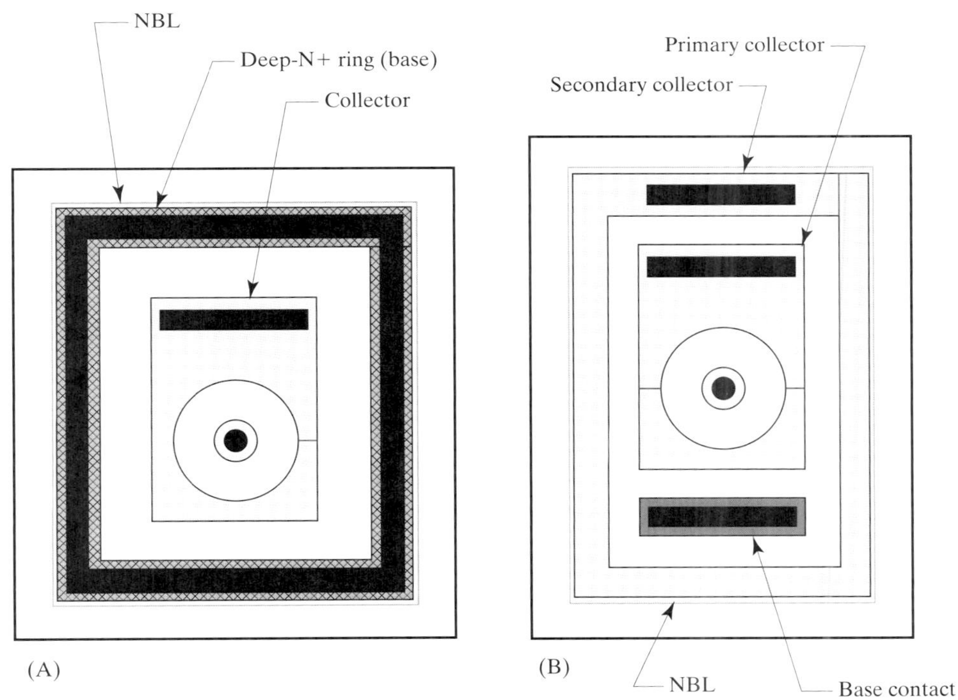

FIGURE 9.10 Two examples of lateral PNP transistors modified to minimize saturation: (A) transistor ringed with deep-N+ and (B) transistor with secondary collector.

encloses the base region of the lateral PNP within a wall of heavily doped N-type silicon. The built-in potential caused by the N+/N− doping gradient repels minority carriers. The few that actually penetrate the deep-N+ or the NBL usually recombine before reaching the isolation. Although this structure is compatible with BiCMOS processing, it may prove less effective in CDI processes. Most of these processes use relatively lightly doped NBL layers to reduce lateral autodoping during epitaxial deposition. The well doping gradient causes minority carriers to drift down toward the NBL/N-well interface, and a substantial fraction of these carriers may penetrate the NBL and enter the substrate. Measurements on a 7V analog BiCMOS process that uses relatively shallow, heavily doped wells yielded NBL penetration figures in excess of 10%.[7]

Figure 9.10B shows another method for preventing substrate injection. The illustrated device incorporates a ring of base diffusion completely encircling its primary collector. This ring acts as a *secondary collector.* As long as the primary collector does not saturate, few carriers can reach the secondary collector and it conducts little current. When the primary collector saturates, the carriers begin to flow to the secondary collector. So long as the secondary collector does not simultaneously saturate, it collects most of the carriers and prevents them from reaching the isolation sidewalls. The secondary collector is sometimes called a *ring collector* because it often takes the form of an unbroken ring enclosing the primary collector.

The secondary collector can perform one of several functions depending on how it is connected. If it connects to ground, then it returns any carriers it collects to the ground return line. When the transistor saturates, the emitter current flows to ground, but it does not pass through the substrate. This provides approximately the same electrical functionality as an unprotected lateral transistor without the risk of substrate debiasing. Alternatively, the secondary collector can connect to the base lead. When the transistor saturates, the carriers collected by the secondary collector add to the base current and cause the apparent beta to rapidly decline. This connection provides the same functionality as a deep-N+ ring, while consuming considerably less space. If the designer wishes to increase the efficiency of the ringed collector still further, then the base ring can be supplemented by a deep-N+ ring placed outside it.

Secondary collectors can also be constructed in analog BiCMOS processes. These should consist of base diffusion (or shallow P-well) rather than PSD, because the source/drain implant is frequently too shallow to act as an efficient collector. Analog BiCMOS secondary collectors are generally less effective than their standard bipolar counterparts due to the downward drift of minority carriers produced by the well doping gradient.

A secondary collector can also function as a saturation detector. Current begins to flow through the secondary collector as soon as the primary collector saturates, and the current stops as soon as the primary collector ceases to saturate. The secondary collector can be used to dynamically control the base drive to prevent the primary collector from saturating. Instead of dumping the unwanted current to ground, a *dynamic antisaturation circuit* throttles the base drive back to reduce the emitter current. The negative feedback loop required to control the base drive may become unstable unless properly compensated, and the phase shift across the secondary collector is difficult to model. Secondary collectors used for saturation detection do not have to encircle the entire transistor because they need only intercept a small fraction of the minority carriers to generate the necessary control sig-

---

[7] N. Gibson, unpublished report, 1998.

nal. Since the dynamic antisaturation circuit must incorporate a relatively large signal delay to ensure stability, a deep-N+ ring may be required to suppress transient substrate injection.

Saturating NPN transistors also inject current into the substrate. Small-signal transistors rarely inject enough substrate current to necessitate antisaturation rings, but power transistors are another matter entirely. Any saturating NPN transistor that conducts more than a few milliamps of base drive requires some form of protection against minority carrier injection. The first (and best) solution consists of a deep-N+ ring surrounding the periphery of the collector. This ring not only contains minority carriers inside the tank, but it also reduces the collector resistance of the transistor at the same time. Minority carriers that recombine within the tank or the deep-N+ guardring become majority carriers in the collector, and from there they pass through the transistor to the emitter.

A base diffusion placed within the collector of an NPN transistor collects minority carriers and can also act as a saturation detector. Power switching transistors often include a small base diffusion in the collector tank connected to a dynamic antisaturation circuit (Figure 9.11). Antisaturation circuits for grounded-emitter transistors are difficult to construct because the secondary collector must operate at or near ground potential. A knowledgeable circuit designer can usually find ways around this difficulty. NPN transistors fitted with antisaturation circuitry should still employ a complete deep-N+ ring around the periphery of the collector to contain minority carriers injected during transients, as well as to reduce the collector resistance of the power transistor.

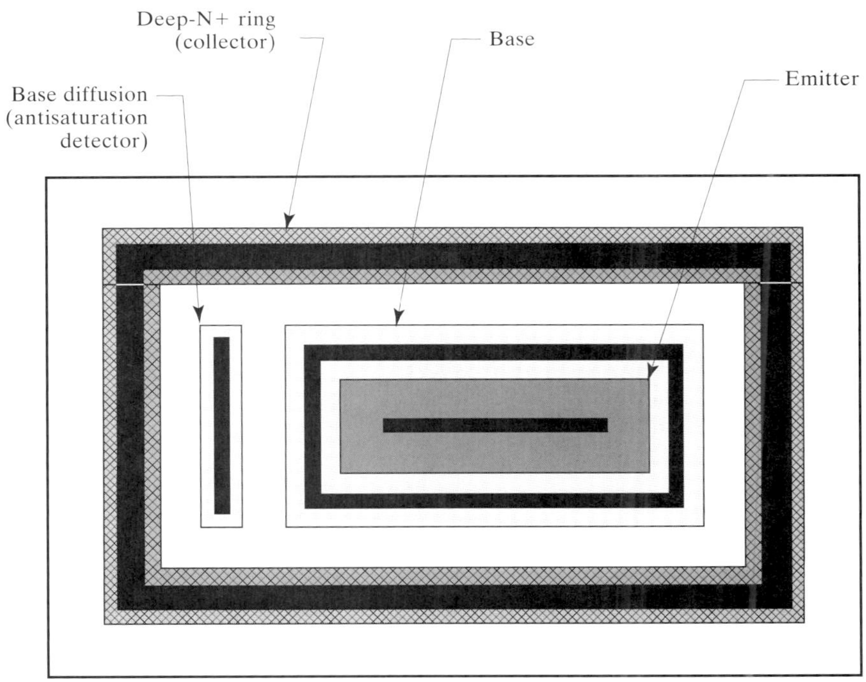

FIGURE 9.11 An NPN switching transistor incorporating both a complete deep-N+ ring and an outer collector that functions as a saturation detector.

No generally accepted symbols exist for transistors with secondary collectors or deep-N+ guardrings. Figure 9.12 shows a set of symbols that have achieved some degree of industry recognition. A thick base bar denotes a power transistor, or more generally, any transistor requiring special layout (A). A diagonal slash across the collector lead of an NPN indicates the presence of a deep-N+ sinker (B), as does

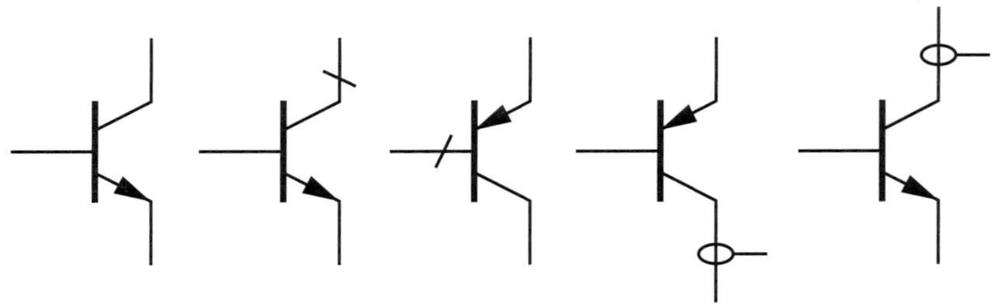

FIGURE 9.12 Proposed symbols for (A) a power NPN, (B) a power NPN with a deep-N+ collector, (C) a lateral PNP with deep-N+ in base, (D) a lateral PNP with a secondary collector, and (E) an NPN transistor with saturation detection.

the presence of a similar slash across the base lead of a lateral PNP (C). The addition of a small ring encircling the collector lead of a lateral PNP denotes the addition of a secondary collector[8] (D), while the addition of a similar ring around the collector lead of an NPN transistor denotes a base diffusion placed in the tank as a saturation detector (E). The majority of these circuits use NPN transistors because of their superior device characteristics.

9.2 MATCHING BIPOLAR TRANSISTORS

Many analog circuits require matched bipolar transistors. Current mirrors and current conveyors use them to replicate currents; amplifiers and comparators use them to construct differential input stages; references use them to produce known voltages and currents; and Gilbert translinear circuits use them to perform analog computations. All of these applications depend on precise matching of collector currents and base-emitter voltages, sometimes from transistors of the same size, and sometimes from ones of different sizes.

NPN collector currents scale approximately with drawn emitter area, but no model can precisely predict the influence of emitter geometry on matching. It is therefore very difficult to match transistors with different sizes and shapes of emitters. Most bipolar circuits employ simple integer ratios such as 1:1, 2:1, 4:1, or 8:1. These ratios are easily obtained by assembling multiple copies of an identical unit device. The same techniques can obtain virtually any ratio of small integer numbers, such as 2:3, 3:4, and 2:5. Ratios requiring more than eight or ten unit devices become increasingly impractical due to area requirements and the sensitivity of large devices to temperature gradients.

Two transistors having identical dimensions and operating at equal collector currents should theoretically develop exactly the same base-emitter voltage. In practice, small differences in the emitter saturation currents cause the two base-emitter voltages to vary slightly. The difference between the base-emitter voltages of two transistors operating at equal current densities is called the *offset voltage* ΔV_{BE} and can be computed from the following equation

$$\Delta V_{BE} = V_T \ln\left(\frac{I_{S1}}{I_{S2}}\right) \qquad [9.3]$$

where I_{S1} and I_{S2} are the emitter saturation currents of the two transistors. Given that the thermal voltage V_T equals 26mV at room temperature, a 1% mismatch in emitter saturation currents produces an offset voltage of 0.25mV. Saturation currents scale approximately with drawn emitter area, so a 1% variation in emitter area also produces a 0.25mV of offset.

---

[8] A different symbol for the ringed-collector PNP, as well as a unique application for this device, are found in H. Lehning, "Current Hogging Logic (CHL)—A New Bipolar Logic for LSI," *IEEE J. Solid-State Circuits*, Vol. SC-9, #5, 1974, pp. 228–233.

9.2.1. Random Variations

Random fluctuations in base doping and emitter junction area set the ultimate limits of vertical bipolar transistor matching. Other significant sources of random variation include recombination in the emitter-base depletion region and lateral injection across the base diffusion, both of which scale inversely with the emitter area-to-periphery ratio. Matched bipolar transistors therefore employ relatively compact emitter geometries. The three geometries favored for constructing matched vertical NPN transistors are squares, octagons, and circles (Figure 9.13). Each style of emitter has its proponents. The circle has the largest area-to-periphery ratio and therefore theoretically provides the best possible matching. However, circles are approximated as many-sided polygons during pattern generation. Squares require no such approximations, so many designers believe that they are rendered more precisely on the photomask. Octagons also require no approximations, and they possess slightly larger area-to-periphery ratios than do squares.

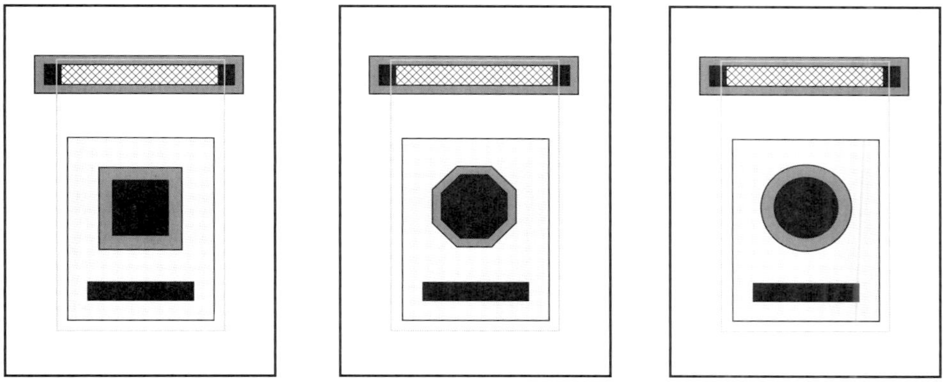

FIGURE 9.13 Examples of NPN transistors designed with square, octagonal, and circular emitters.

In practice, all three styles of emitters provide excellent matching. Although circles are approximated as polygons, identical circles produce identical polygons. The approximations involved in generating circular emitters therefore have little impact on their matching. Furthermore, the differences in area-to-periphery ratios among squares, octagons, and circles are relatively insignificant. The area-to-periphery ratio R_{AP} of any geometry can be determined using the following equation

$$R_{AP} = k_r \sqrt{A_e} \qquad [9.4]$$

where A_e represents the emitter area and k_r is a dimensionless constant equal to 0.250 for squares, 0.274 for octagons, and 0.282 for circles. Note that the area-to-periphery ratio is not itself a dimensionless quantity. For example, if the emitter area is measured in square microns, then R_{AP} will have dimensions of microns. Equation 9.4 shows that the reduction in peripheral effects gained by using a circular emitter can be equaled by simply increasing the area of a square emitter by 25%.

The mismatch between a pair of transistors due to peripheral and areal fluctuations has a standard deviation s that equals

$$s = \frac{1}{\sqrt{A_e}}\sqrt{k_a + \frac{k_r k_p}{\sqrt{A_e}}} \qquad [9.5]$$

where k_a and k_p are constants representing the contributions of areal and peripheral fluctuations, respectively. The contribution of the peripheral term decreases as the area increases. For sufficiently large emitters, the areal term dominates and the random mismatch becomes inversely proportional to the square root of emitter area

A_e. Most transistors with emitters that are more than two or three times the minimum diameter exhibit relatively little dependence on peripheral variation, again demonstrating the relative indifference of matching to geometric considerations (in particular, to the value of k_r).[9] Therefore, for most practical purposes, the standard deviation of the mismatch can be assumed to equal

$$s \cong \sqrt{\frac{k_a}{A_e}} \qquad [9.6]$$

Although large emitters exhibit less random mismatch than small ones, there are other factors to consider. Any increase in emitter size increases the spacing between the devices and therefore renders them more vulnerable to thermal and stress gradients. Large emitters also exhibit increased base pinch resistance. Because of these problems, one must avoid making matched emitters either too large or too small. As a general rule, the diameter of the emitter of a matched NPN transistor should not be less than twice nor more than ten times the minimum possible diameter. For example, a minimum contact width of 2μm and a minimum overlap of emitter over contact of 1μm produce a minimum emitter diameter of 4μm. Matched emitters in this process should have diameters of no less than 8μm and no greater than 40μm. More accurate guidelines require actual data that rarely exists for a production process.[10]

The choices between circular, square, and octagonal emitters are usually of little consequence, but there are exceptional cases where one type of emitter may confer a specific advantage. In the case of lateral PNP transistors, the emitter area must remain small to conserve beta. Circular emitters increase the area-to-periphery ratio of the emitter without increasing its diameter, and thus help not only to improve matching but also to raise beta. Therefore matched lateral PNP transistors often employ minimum-diameter circular emitters such as those in Figures 8.22 and 8.26B. The emitter should also overlap its contact equally on all sides to ensure an even distribution of emitter current. Therefore circular emitters should contain circular contacts, and octagonal emitters should contain octagonal contacts. These arrangements are not possible in processes that allow only minimum-dimension square contacts. In such cases, square emitters should be used rather than circular or octagonal ones.

Many circuits require matched transistors having unequal device areas. Although it is possible to connect identical unit transistors in parallel, the large amounts of die area required by collector isolation actually degrade matching by exacerbating the effect of thermal and stress gradients. Matched NPN transistors can occupy a common tank because the geometry of the collector has almost no effect on their matching (Figure 9.14A). The geometry of the base-collector junction also has relatively little impact on matching since most of the conduction occurs either directly underneath the emitter or immediately adjacent to it. Several emitters can therefore occupy the same base region (Figure 9.14B). The emitters must be placed far enough apart to prevent minority carriers that are injected by one from being collected by another.

---

[9] One study has shown that the area parameter alone provides a good fit to as small as 16μm^2: H.-Y. To and M. Ismail, "Mismatch Modeling and Characterization of Bipolar Transistors for Statistical CAD," *IEEE Trans. on Circuits and Systems - I: Fundamental Theory and Applications,* Vol. 43, #7, 1996, pp. 608–610.

[10] The real problem lies not in finding experimental data, but in finding data collected from a properly designed experiment that has undergone proper statistical analysis. A few measurements made on a couple of devices constitute neither a properly designed experiment nor a properly analyzed set of data. Decisions made on such basis are unlikely to be any better than those based on the admittedly *ad hoc* rule given in the text.

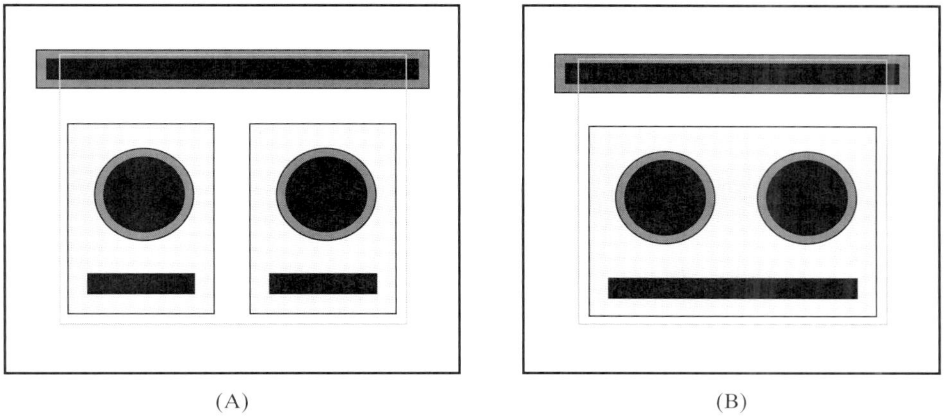

FIGURE 9.14 Two styles of multiple-emitter NPN transistors: (A) separate base regions in a common tank and (B) separate emitters in a common base region.

Similarly, the emitters should reside far enough inside the base diffusion to minimize lateral conduction that would otherwise produce mismatches between transistors having different numbers of emitters. These requirements can be met by increasing both the emitter-to-emitter spacing and the base overlap of emitter by 1 to 2μm. These increased spacings ensure that the individual emitters will not interact with one another or with the collector-base junction.

Matched lateral PNP transistors can also occupy a common tank to save area. Multiple emitters cannot occupy a single opening in a collector geometry because each emitter would interfere with the flow of minority carriers from the others. Instead, each emitter must occupy its own collector opening, and all of these openings must have identical dimensions. The outside dimensions of the collector geometry and the size and shape of the tank have little impact on matching. Therefore matched lateral transistors usually consist of rectangular arrays of minimum emitters placed in a common collector region (Figure 8.26B).

9.2.2. Emitter Degeneration

Regardless of the care taken in their construction, some types of bipolar transistors simply do not match very well. A technique called *emitter degeneration* can transfer the burden of matching from a set of bipolar transistors to a set of associated resistors. This technique will improve the overall matching of the circuit as long as the resistors match more precisely than the bipolar transistors. Emitter degeneration also increases the output resistance of bipolar transistors and therefore reduces the systematic errors due to finite Early voltages. The systematic mismatch in collector currents between two matched bipolar transistors operating at different base-collector voltages equals

$$\frac{I_{C1}}{I_{C2}} \cong 1 + \left(\frac{\Delta V_{BC}}{V_A}\right)\left(\frac{V_T}{V_T + V_d}\right) \qquad [9.7]$$

where $\Delta V_{BC} = V_{BC1} - V_{BC2}$, V_A is the Early voltage of the transistors, V_T is the thermal voltage (26mV at 25°C), and V_d is the voltage developed across the degeneration resistors. Equation 9.7 is only valid as long as V_{BC1} and V_{BC2} are both much smaller than V_A. This equation indicates that 50mV of degeneration reduces the Early error by approximately a factor of three.

Figure 9.15 shows a lateral PNP current mirror consisting of three lateral PNP transistors Q_1 to Q_3. Each of these transistors has an associated emitter degeneration

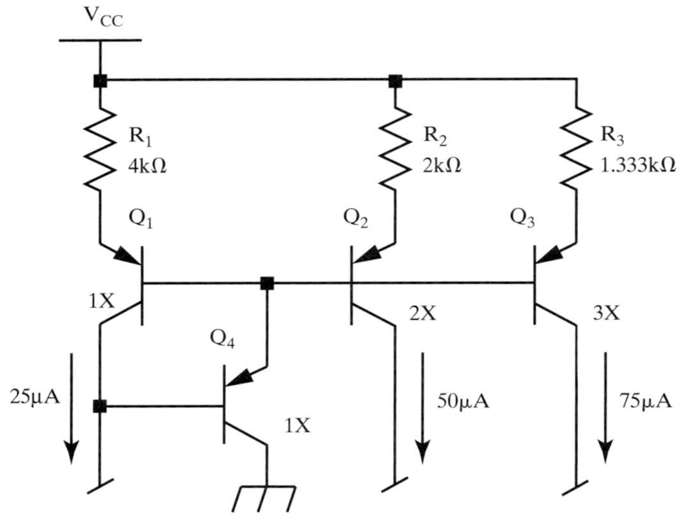

FIGURE 9.15 Lateral PNP current mirror that incorporates emitter degeneration resistors.

resistor R_1 to R_3. The mirror also uses a *beta helper* transistor Q_4 to minimize the effect of low betas on matching. Transistor Q_4 does not need to match any of the other transistors, and it does not normally require emitter degeneration.[11]

In this example, transistors Q_1 to Q_3 have device sizes of one, two, and three, respectively. These sizes represent the number of unit emitters in each transistor. This dimensionless notation avoids any possible confusion between emitter periphery and emitter area, and simultaneously frees the circuit designer from worrying about exact layout dimensions. The values of resistors R_1 to R_3 are inversely proportional to the sizes of transistors Q_1 to Q_3. Each resistor therefore generates the same voltage differential, which in this case equals 100mV. This voltage represents the amount of degeneration applied to the transistors. Approximately 50 to 75mV of degeneration suffices to ensure that the resistors determine the matching of the mirror rather than the transistors. Few circuits require more than 100mV of degeneration as long as the ratio of the emitter areas of the transistors lies within $\pm 10\%$ of the desired value.

The improvement in matching obtained through emitter degeneration depends on how well the resistors match and on the nature of the mismatches between the bipolar transistors. Well-matched resistors vary no more than $\pm 0.1\%$, while the currents of well-matched minimum-area emitters typically vary by $\pm 1\%$ or more. The matching of the degenerated transistors will approximately equal the matching of the emitter degeneration resistors. On the other hand, the area consumed by the resistors could also be used to increase the emitter areas. Increasing the emitter areas of matched NPN transistors usually proves more area-effective than adding emitter degeneration resistors, since well-matched resistors are not small. Emitter degeneration may sometimes provide better matching in the presence of large thermal gradients because resistors are less susceptible to these gradients than bipolar transistors.

Lateral PNP transistors might seem to benefit less from emitter degeneration than vertical NPN transistors do since much of the mismatch between lateral transistors stems from beta variations that remain unaffected by degeneration. Despite

[11] Sometimes a small resistor is added in series with the emitter of the beta helper as part of a frequency compensation network; this resistor does not take part in the matching of the mirror and its value is noncritical.

this, lateral PNP transistors are frequently degenerated because these devices use minimum emitters to maintain acceptable betas. Lateral PNP transistors also benefit from increased output resistance caused by emitter degeneration because these devices often have rather low Early voltages.[12] Emitter degeneration cannot improve the matching of split collector transistors because it is not possible to provide a separate emitter degeneration resistor for each split collector.

Large amounts of emitter degeneration are sometimes used to obtain noninteger ratios between transistors. The size of the transistors becomes relatively unimportant in the presence of 250 to 500mV of degeneration. For example, a 3.4:1 ratio can be obtained by ratioing a 3X transistor and a 10kΩ resistor with a 1X transistor and a 34kΩ resistor. This technique works equally well with both NPN and PNP transistors.

9.2.3. NBL Shadow

Surface discontinuities caused by oxidation during the NBL anneal propagate upward during epitaxial deposition to produce a surface discontinuity called the *NBL shadow*. A mechanism called *pattern shift* can displace the NBL shadow laterally by a distance of up to twice the epi thickness (Section 7.2.3). Mismatches can occur if the NBL shadow intersects the emitter of a vertical NPN transistor. Arrays of multiple-emitter NPN transistors are particularly vulnerable to pattern shifts perpendicular to their axis of symmetry. Consider the two transistors in Figure 9.16A. Pattern shift has displaced the NBL shadow toward the right, causing it to intersect the leftmost emitter of each device. Suppose that the affected emitters experience a 1% reduction in emitter area. Only one of Q_A's two emitters is affected, so its emitter area becomes 1.99. The sole emitter of Q_B is also affected, so its emitter area becomes 0.99. The new ratio between the two devices is 1.99:0.99, or about 2.01:1. This represents a mismatch of 0.5%, or an offset voltage of about 0.13mV.

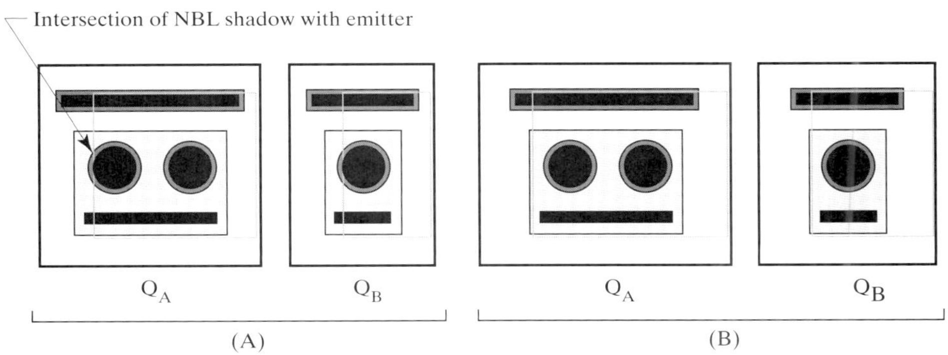

FIGURE 9.16 (A) NBL shadow causes mismatch between two transistors. (B) This mismatch is eliminated by oversizing NBL to prevent intersection of NBL and emitter.

Several ways exist to prevent pattern shift from causing mismatches. One approach consists of replacing the multiple-emitter transistor Q_A with two single-emitter transistors that are identical to Q_B. The NBL shadow will now intersect all three emitters in exactly the same manner, and the resulting systematic variations should cancel one another. Unfortunately, pattern distortion can also produce random variations in the NBL shadow that do not cancel one another. These can only be avoided by ensuring that the NBL shadow does not intersect the active area of the transistor, which, in the case of NPN transistors, is defined by the emitter diffusion.

---

[12] Gray, *et al.* conclude that NPN transistors derive more benefit from emitter degeneration than PNP transistors, but their arguments are flawed because they ignore the other factors discussed in the text; P. R. Gray and R. G. Meyer, *Analysis and Design of Analog Integrated Circuits*, 3rd ed. (John Wiley and Sons, New York: 1993), pp. 317–320.

If the direction of pattern shift is known, then the transistors can be laid out in a CEB array in which the main axis of symmetry lies parallel to the direction of pattern shift. In this way, the shift will displace the NBL shadow into either the collector contact or the base contact. The space required by these contacts usually suffices to prevent the NBL shadow from reaching the emitter. Pattern shift in (111) silicon usually occurs along the <211> axis. The direction of pattern shift in tilted (100) silicon depends on the direction of the tilt, which may vary from one manufacturer to the next. In order to properly orient the transistor array, the designer must determine the relationship between the X–Y coordinates of the layout and the wafer orientation. This information can usually be obtained by microscopic examination of a die. Planarization will obscure the NBL shadow, but a wafer can be removed prior to planarization for examination. Remember that the reticle array may differ from one device to another depending on the choices made during pattern generation, and that rotated or reflected reticle arrays will alter the apparent direction of pattern shift.

Sometimes the direction of pattern shift is not known, or layout considerations preclude the use of a specific orientation. In such cases, the overlap of NBL over emitter can be increased to prevent the NBL shadow from intersecting the emitter (Figure 9.16B). If no data exists on the magnitude of the pattern shift, then the designer should overlap the NBL over the emitter by 150% of the epi thickness.

Some designers eliminate the NBL shadow by omitting NBL from the matched transistors. Without NBL, the collector resistance of an NPN transistor can reach several kilohms. The V_{CEO} of the transistor may also diminish due to punchthrough of the lightly doped collector. CDI NPN transistors are particularly vulnerable to punchthrough due to the extremely light doping of the lowest portions of the N-well. One should not remove NBL from transistors unless characterization data indicates that the resulting device will function properly without it.

Lateral PNP transistors generally do not suffer from mismatches caused by the NBL shadow. The width of the collector usually suffices to prevent the shadow from intruding into the base region. If the NBL shadow does fall across the base region, it can cause mismatches by disturbing the flow of carriers from emitter to collector. These mismatches are easily eliminated by reorienting the transistors or by enlarging the NBL region as discussed above. One must never eliminate NBL from a lateral PNP, because this would greatly reduce its collection efficiency.

9.2.4. Thermal Gradients

Bipolar transistors are extremely sensitive to thermal gradients. The base-emitter voltage V_{BE} exhibits a temperature coefficient of about –2mV/°C, corresponding to a collector current temperature coefficient of about 80,000ppm/°C. Matched bipolar transistors are routinely expected to achieve offset voltages of less than ±1mV, corresponding to a temperature difference of only ±0.5°C. Temperature variations of this magnitude can easily occur in almost any integrated circuit.

Matched bipolar transistors are often used to construct differential pairs, ratioed pairs, and ratioed quads. A *differential pair* (also called a *diff* pair, an *emitter-coupled pair*, or a *long-tailed pair*) consists of two matched bipolar transistors whose emitters are connected as in Figure 9.17A. The input stages of amplifiers and comparators often consist of differential pairs whose collectors terminate into matched resistors or current mirrors. The input offset voltage of a bipolar amplifier or comparator depends largely upon the matching of the input differential pair. Various trimming schemes can reduce the random component of the offset voltage to a fraction of a millivolt. Trimming also minimizes the temperature coefficient of the offset

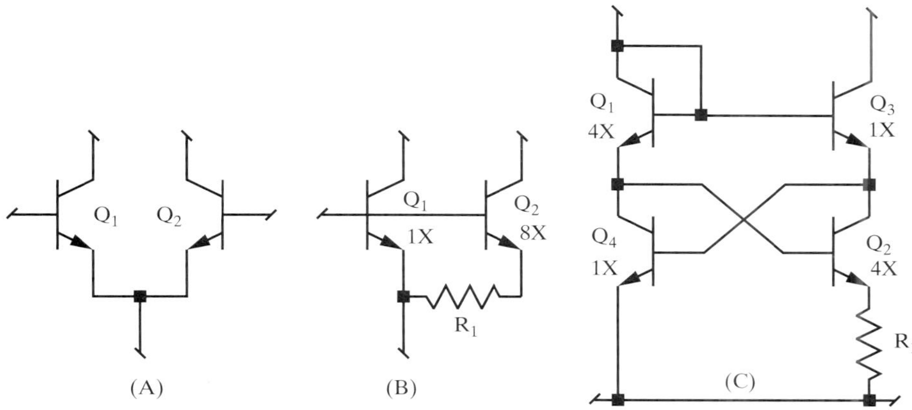

FIGURE 9.17 Three circuits containing matched NPN transistors: (A) differential pair, (B) ratioed pair, and (C) ratioed quad.

voltage as well as its absolute value,[13] so it is often used to minimize the vulnerability of high-gain amplifiers to a phenomenon called *thermal feedback*.

Thermal feedback occurs when one portion of a circuit influences another through thermal interactions rather than electrical ones. Changes in voltage or current within relatively high-power circuits (such as the output stages of an amplifier) produce localized temperature fluctuations that, in turn, generate small offsets between devices in the input stages of the circuit. The circuit then amplifies these offsets as if they were an electrical signal. The amplified offsets can produce further temperature variations, possibly even leading to oscillations. Since many amplifiers have voltage gains in excess of 10,000, even a very weak thermal interaction can cause significant thermal feedback. The frequency response of many commercial operational amplifiers contain low-frequency poles and zeros caused by this mechanism.[14] Thermal feedback can be minimized by increasing the separation of the input and output stages and by reducing the thermal sensitivity of the input stage. Many operational amplifiers place the input circuit on one side of the die and the output circuit on the other. Even so, thermal coupling remains a serious problem. The input differential pair of a high-gain amplifier should always be located and constructed to achieve the highest possible degree of matching in order to minimize its sensitivity to thermal variations.

A *ratioed pair* consists of two bipolar transistors whose emitter areas are in integer ratio. Assuming that the two transistors conduct equal currents, then their base-emitter voltages will differ by an amount ΔV_{BE} equal to

$$\Delta V_{BE} = V_T \ln\left(\frac{A_1}{A_2}\right) \qquad [9.8]$$

where V_T equals the thermal voltage (26mV at 25°C) and A_1 and A_2 represent the emitter areas of transistors Q_1 and Q_2.[15] The thermal voltage scales linearly with absolute temperature,[16] therefore ΔV_{BE} is a *voltage proportional to absolute*

---

[13] Mismatch in a differential pair has the same impact as deliberate ratioing of emitter areas; the ΔV_{BE} voltage so developed has a large positive temperature coefficient. Therefore minimizing the offset at one temperature also tends to minimize temperature variability.

[14] J. E. Solomon, "The Monolithic Op Amp: A Tutorial Study," *IEEE J. Solid-State Circuits,* Vol. SC-9, #6, 1974, pp. 314–332.

[15] This derivation ignores the ideality factor (or emission coefficient) η, which is usually very near unity for NPN transistors operating at moderate current levels.

[16] An *absolute temperature* is one measured with respect to absolute zero. The SI unit of absolute temperature is the Kelvin degree (K), which has the same magnitude as the Celsius degree (°C); 0°C ≅ 273K and 25°C ≅ 298K. The thermal voltage V_T equals kT/q, where k is Boltzmann's constant ($1.38 \cdot 10^{23}$J/K), T is the absolute temperature (in K) and q is the charge on the electron ($1.60 \cdot 10^{-19}$C).

temperature (VPTAT). The VPTAT produced by a ratioed pair of NPN transistors remains linear with temperature and independent of current over a remarkably wide range of operating conditions. A resistor connected between the emitters of a ratioed pair (as in Figure 9.17B) can transform this VPTAT into a *current proportional to absolute temperature,* or IPTAT.[17] IPTAT circuits form the basis of many precision voltage and current references.

In order for VPTAT and IPTAT circuits to operate properly, the ratioed pair must match very precisely. The most common ratio used in VPTAT and IPTAT circuits is probably 8:1, which produces a ΔV_{BE} of 54mV. A 1mV mismatch in such a circuit would produce approximately a 2% error in the voltage or current. A typical trimmed voltage reference must produce a voltage that varies no more than $\pm 1\%$ over all possible operating conditions. Poor layout often leads to excessive output voltage variation with input voltage (poor *line regulation*) or to excessive output voltage variation with output current (poor *load regulation*). Both of these problems usually stem, at least in part, from thermal feedback.

A *ratioed quad* is essentially a variation on the ratioed pair. The four transistors of the quad produce a VPTAT voltage that is usually imposed across a resistor to produce an IPTAT (Figure 9.17C). The VPTAT voltage ΔV_{BE} equals

$$\Delta V_{BE} = V_T \ln\left(\frac{A_1 A_2}{A_3 A_4}\right) \qquad [9.9]$$

where A_1 to A_4 are the emitter areas of transistors Q_1 to Q_4, respectively. The ratioed quad can provide a much larger ΔV_{BE} than a simple ratioed pair because the sizes of Q_1 and Q_2 multiply together. Two 4X transistors can produce a ΔV_{BE} of 72mV, while two 8X transistors can generate 108mV.

The extreme thermal sensitivity of bipolar transistors requires that matched devices be laid out to cancel thermal gradients. Critical matched devices almost always employ common-centroid layout techniques similar to those discussed in Section 7.2.6. Differential pairs usually employ the two-dimensional common-centroid layout shown in Figure 9.18. This configuration is popularly called a *cross-coupled quad.*[18] Common-centroid layouts help minimize the impact of thermal variations, but they cannot completely cancel nonlinear thermal variations. The exponential nature of the I_C vs. V_{BE} relationship sharply limits the cancellation achievable from common-centroidal layouts. Compactness is thus a highly desirable property of matched bipolar arrays. Most of the more complex common-centroidal arrangements lack compactness and therefore serve less well than the simple cross-coupled quad.

The VPTAT voltage developed by a ratioed pair increases as the logarithm of the area ratio, while offsets produced by thermal and stress gradients increase roughly as the square root of the ratio. As the area ratio increases, a point is eventually reached beyond which mismatch increases more rapidly than VPTAT. For any given design, there exists a ratio that will provide optimal matching. This optimal ratio depends on many factors, but in most cases it probably lies somewhere between 8:1 and 16:1. Even-number ratios greatly simplify the task of constructing common-centroid layouts, and smaller ratios consume less space, so 8:1 is probably the most popular choice of ratio. Similar arguments lead to ratioed quads of 4:1:1:4.

---

[17] In actuality, the temperature coefficient of the resistor will distort the linearity of the IPTAT. Most circuits use the IPTAT current to regenerate a VPTAT across another resistor. If the two resistors have the same temperature coefficient, then this can be (and is) neglected.

[18] Grebene, pp. 348–349, 365.

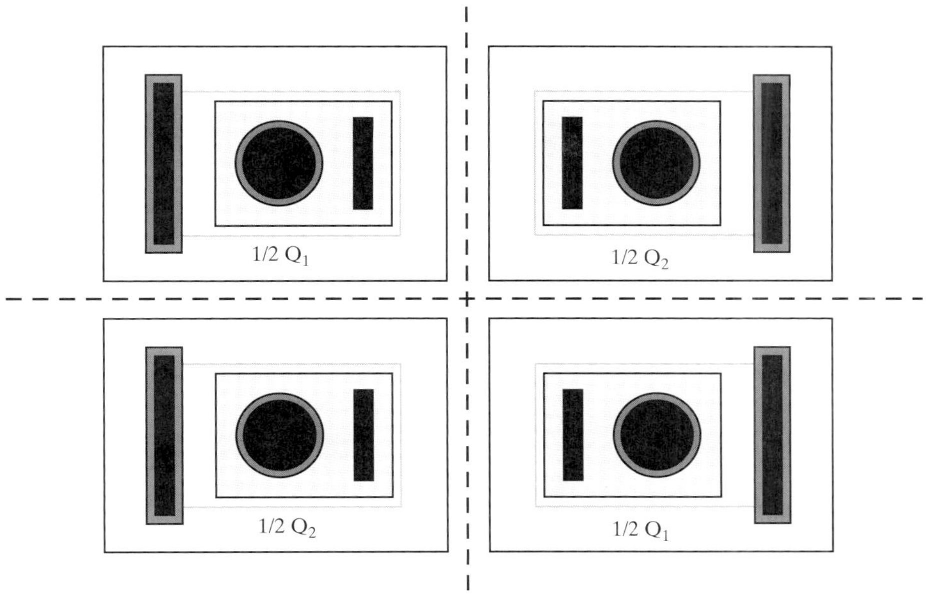

FIGURE 9.18 Example of cross-coupled bipolar transistors.

The layouts of ratioed pairs and ratioed quads must provide a high degree of symmetry in a compact layout. In a ratioed pair, one device usually possesses a single emitter and the other multiple emitters. One simple layout for a ratioed pair places the single-emitter device in the middle of two halves of the multiple-emitter device. The simplest structure embodying this arrangement is a one-dimensional common-centroid layout following the pattern ABA (Figure 9.19A). The placement of all of the emitters in a line generates a secondary axis of symmetry S_2 that enables the structure to reject thermal gradients perpendicular to the row of transistors, while the primary axis of symmetry S_1 rejects thermal gradients parallel to the line. The elongated shape of the array makes it more difficult to cancel gradients around S_1. Therefore arrays of this sort should be oriented so that the secondary axis S_2 lies parallel to the expected isotherms.

A more compact arrangement can be achieved for ratios that are multiples of 4:1. The emitters of the larger transistor Q_2 can then be arranged in two rows around an axis of symmetry S_2 (Figure 9.19B). Axis S_2 should also pass through the center of the single-emitter transistor Q_1. This arrangement is particularly beneficial for large ratios, such as 16:1, that would otherwise produce very elongated layouts. As before, the array should be oriented so that the secondary axis of symmetry S_2 lies parallel to the expected isotherms.

Ratioed quads are laid out as if they consisted of a pair of ratioed mirrors. Both the upper pair (Q_1 and Q_3 in Figure 9.17C) and the lower pair (Q_2 and Q_4) can employ layouts similar to those in Figure 9.19. Ideally, the two pairs should lie one above the other so that their primary axes of symmetry (S_1) coincide. This converts the entire arrangement into a two-dimensional common-centroid array. If for some reason this arrangement is not feasible, then each of the two pairs can be treated as an independent ratioed pair. A reasonable degree of matching will be achieved even if the two ratioed pairs reside some distance apart.

Some designers advocate laying out ratioed pairs using a circular array in which the smaller device occupies the center of a ring-shaped array of emitters forming the larger device. If the larger device contains a multiple of four emitters, then this arrangement will possess both horizontal and vertical axes of symmetry as well as a

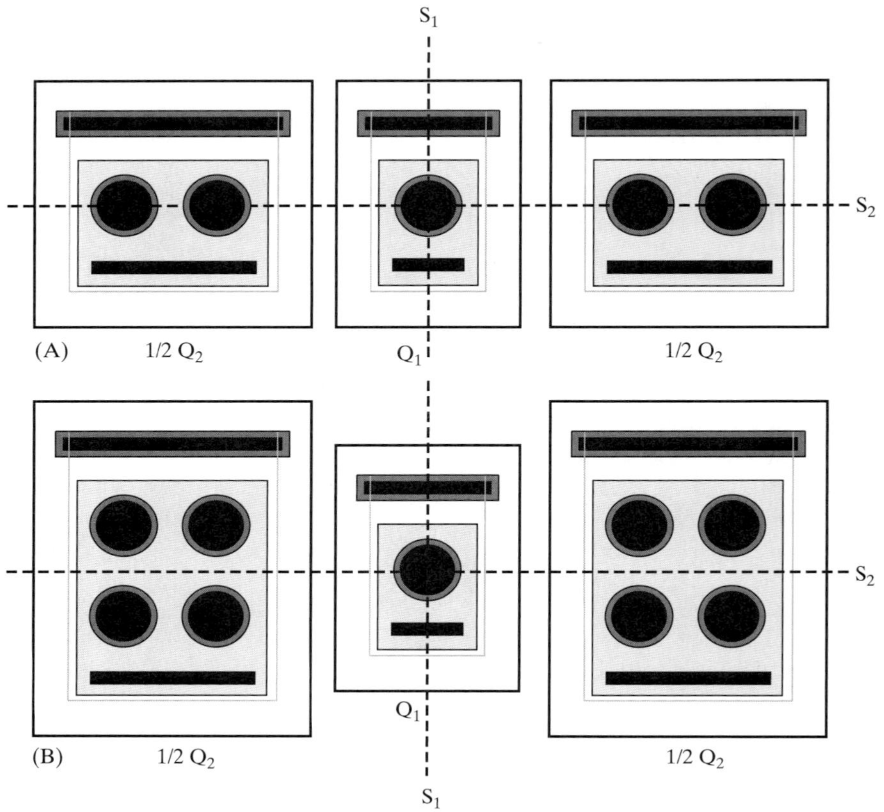

FIGURE 9.19 Two common-centroid layouts frequently used for constructing ratioed pairs.

number of subsidiary axes dependent on the total number of emitters. Although this arrangement has a high degree of symmetry, it consumes so much area that it is doubtful whether it actually matches as well as the simpler structures in Figure 9.19.

9.2.5. Stress Gradients

Mechanical stress can induce mismatches between bipolar transistors by altering their base-emitter voltages or by reducing their betas. The base-emitter voltage of a transistor depends on the bandgap voltage of silicon, which varies slightly under stress.[19] The degradation in beta under mechanical stress is principally due to mobility variations induced by piezoresistivity.[20] Together these effects can easily produce several millivolts of offset.

Assembled integrated circuits almost always experience a certain amount of stress. Plastic-molded devices are cured at elevated temperatures, and the mold compound elastically deforms as the assembled units cool. The resulting stresses become frozen into the assembled unit and do not dissipate over time. These residual stresses cause the base-emitter voltages of bipolar transistors to shift and can produce offsets between matched pairs of devices. These *package shifts* cannot be entirely trimmed out at wafer probe because they only appear after packaging. The residual stresses vary with temperature, so even post-package trimming cannot

[19] J. J. Wortman, J. R. Hauser, and R. M. Burger, "Effects of Mechanical Stress on p-n Junction Device Characteristics," *J. Applied Physics,* Vol. 35, #7, 1964, pp. 2122–2131.

[20] H. Mikoshiba and Y. Tomita, "Piezoresistance as the Source of Stress-induced Changes of Current Gain in Bipolar Transistors," *Solid State Electronics,* Vol. 25, #3, 1982, pp. 197–199.

entirely counteract package shifts. Package shifts therefore limit the minimum offset voltages obtainable between matched transistors.

Common-centroid layout techniques can greatly reduce the impact of stress on matched transistors. These techniques effectively move the matched transistors closer together and therefore equalize the stresses on them. Unfortunately, even the best common-centroid arrangements cannot cancel the higher-order components of the stress gradient. Matched transistors should therefore reside in low-stress regions of the die. Figure 9.20A shows the best locations for matched transistor arrays on a die fabricated in (100) silicon. These layouts assume that no significant sources of heat exist in the vicinity of the matched transistors. The best locations lie near the center of the die where the magnitude of the stresses reaches a broad minimum. The major axis of symmetry of the array S_1 should lie along one of the axes of symmetry of the die. This helps ensure that the isobars lie parallel to the secondary axis of the array S_2. If the matched transistors must reside along the side of the die, then they should be placed in the center of one side so that the primary axis of the array S_1 aligns to one axis of symmetry of the die. If the die is not square, then a location in the middle of the longer side is preferable to one in the middle of the shorter side. If possible, the transistors should reside at least 10mils (250μm) inside the edge of the die. Under no circumstances should critical matched bipolar transistors be placed near the corners of a die, as these experience excessively large stress gradients. In summary, the principles that guide the placement of matched bipolar transistors are analogous to the ones that apply to matched resistors, as discussed in Section 7.2.6.

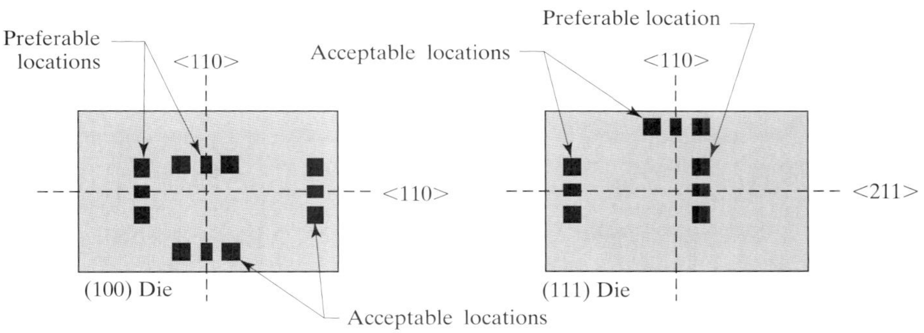

FIGURE 9.20 Locations for placing common centroid bipolar transistor arrays on (100) and (111) dice, in the latter case assuming an axis of symmetry exists in the stress distribution around the <211> axis (compare with Figure 7.16).

Figure 9.20B shows the best locations for critical matched transistors on a die fabricated in (111) silicon. Since isobars tend to lie symmetrically around the <211> axis, matched transistors ideally should reside around this axis instead of around the <110> axis. Locations near the center of the die will again give the lowest overall stress, although locations near either end of the die will provide acceptable matching. Less-critical matched transistors can also be placed on the <110> axis of symmetry. Again, matched transistors should reside at least 10 mils inside the die edge to avoid the increased stress gradients along the edges of the die, and matched transistors should never reside near the corners of the die because of the high stresses there.

Figure 9.21 shows the compromises required when the die contains both a significant heat source and critical matched transistors. Bipolar transistors are more susceptible to thermal gradients than to stress gradients, so the larger the distance between the heat source and the matched transistors, the better. The best compromise usually involves placing the heat source at one end of the die and the matched transistors at the opposite end. Consider increasing the aspect ratio of the die to 1.5:1, or even 2:1, since this helps increase the separation between the heat source

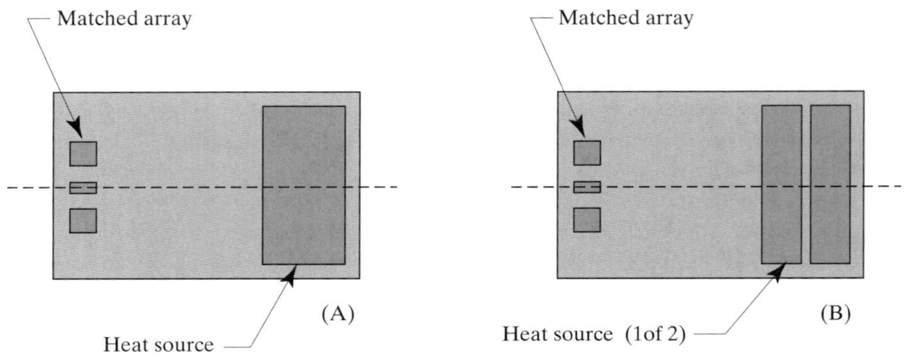

FIGURE 9.21 Layouts for matched transistor arrays on dice that contain one or more power devices.

and the matched transistors. The matched transistors should reside at least 5 to 10mils (125 to 250μm) away from the edge of the die opposite the heat source. This location should provide better overall matching than a location near the center of the die, even though the stresses are greater near the edges than at the center.

High-power integrated circuits may benefit from judicious application of emitter degeneration to transistors that would normally have sufficient area to match quite precisely on their own. High levels of power dissipation produce correspondingly large thermal gradients, especially in dice mounted on heat-sunk leadframes. These gradients have a much stronger effect on bipolar transistors than on passive components, so the matching of degenerated transistors usually proves superior to the matching of undegenerated ones. Emitter degeneration is particularly beneficial for matched transistors residing next to a large heat source; without degeneration such transistors often exhibit mismatches of many millivolts. Whenever emitter degeneration resistors are added to critically matched bipolar transistors, the resistors must be carefully laid out to ensure that they actually improve matching rather than degrade it. Poorly constructed degeneration resistors can easily ruin the matching of an otherwise well laid-out circuit.

9.3 RULES FOR BIPOLAR TRANSISTOR MATCHING

The previous section explained the mechanisms that cause mismatch in bipolar transistors. This section attempts to condense this information into a set of qualitative rules that will allow a designer to construct matched bipolar transistors with some degree of confidence, even if (as is usually the case) quantitative matching data is not available. For bipolar transistors constructed in a pure CMOS process, see Section 10.3.1.

The following rules use the terms *minimal, moderate,* and *precise* to denote increasingly precise degrees of matching. These terms should be interpreted as follows:

- **Minimal matching:** Typical three-sigma offset voltages of ±1mV or collector current mismatches of ±4%. This is suitable for constructing input stages of op-amps and comparators that must achieve three-sigma offsets of ±3 to 5mV without trim. It is also suitable for use in current mirrors for biasing noncritical circuitry.
- **Moderate matching:** Typical three-sigma offset voltages of ±0.25mV or collector current mismatches of ±1%. This level is suitable for use in ±1% bandgap references and in op-amps and comparators that must achieve ±1 to 2mV without trim. Since lateral transistors have difficulty maintaining this degree of matching, most untrimmed, moderately matched circuits use vertical NPN transistors instead.

- **Precise matching:** Typical three-sigma offset voltages of ±0.1mV or collector current mismatches of ±0.5%. This level of matching usually requires trimming or the addition of precisely matched degeneration resistors. Proper layout is still important because degeneration and trimming cannot entirely eliminate the effects of thermal gradients or package shifts. Lateral transistors cannot obtain this degree of matching unless they are heavily degenerated and the circuitry incorporates some means of base current cancellation. Circuits requiring precise matching usually employ heavily degenerated vertical NPN transistors.

9.3.1. Rules for Matching NPN Transistors

Vertical transistors inherently match better than lateral transistors because they are not subject to the vagaries of surface conduction. Most processes optimize the performance of their vertical NPN transistors at the expense of their lateral PNP transistors, which only strengthens the case for using NPN transistors. The following rules summarize the principles of designing matched NPN transistors:

1. Use identical emitter geometries.
   Transistors with different sizes or shapes of emitters match *very* poorly. Even minimal matching requires the use of identical emitter geometries. Matched transistors are therefore restricted to ratios of small integer numbers. The geometry of the base and collector regions matters much less than the geometry of the emitter region. Multiple emitters can thus reside in a common base region.
2. The emitter diameter should equal 2 to 10 times the minimum allowed diameter.
   The minimum diameter of the emitter equals the minimum contact width plus twice the minimum emitter overlap of contact. For example, a process having a minimum contact width of 2μm and a minimum overlap of 1μm has a minimum emitter diameter of 4μm. Matched emitters in this process should have diameters of 8 to 40μm. Emitter areas at the lower end of this range suffice for minimal matching. Moderate and precise matching generally require the use of larger emitters, but the presence of power devices may justify the use of smaller emitters to produce a compact structure that is less susceptible to thermal gradients.
3. Maximize the emitter area-to-periphery ratio.
   For a given emitter area, the transistor with the largest area-to-periphery ratio produces the best possible matching. Circular geometries provide the highest area-to-periphery ratios, but octagonal and square emitters are almost as good.
4. Place matched transistors in close proximity.
   Bipolar transistors are very sensitive to thermal gradients. Even minimally matched transistors should reside within a few hundred microns of one another. Moderately or precisely matched transistors should use common-centroid layout techniques to minimize the separation between the transistors.
5. Keep the layout of matched transistors as compact as possible.
   The use of common base and collector regions may cause slight mismatches, but the increase in compactness usually more than compensates. Layouts that arrange the emitters in tight clusters generally provide better matching than layouts that place them in a line. A pair of matched transistors of equal sizes should employ a cross-coupled layout.
6. Construct ratioed pairs and quads using even integer ratios between 4:1 and 16:1.
   Ratios that are too small or too large will match less well than those that lie within a certain range, typically between 4:1 and 16:1 for ratioed pairs and

between 4:1:1:4 and 8:1:1:8 for ratioed quads. The ratios used for quads tend to be smaller than those used for pairs, because quads develop larger VPTAT voltages for a given number of unit emitters.

7. Place matched transistors far away from power devices.
   Power devices represent a significant threat to bipolar transistor matching. Minimally matched transistors should lie at least 10mils (250μm) away from major power devices (those dissipating 250mW or more) and should not reside adjacent to any power device dissipating more than 50mW. Moderately matched devices should lie at least 5 to 10mils (125 to 250μm) from any device dissipating more than 50mW and should be placed at the opposite end of the die from major power devices. Precisely matched devices should be separated as far as possible from any power device. Consider elongating the die to a 1.5:1 or even a 2:1 aspect ratio to increase the separation between precisely matched transistors and major power devices. Power dissipations of a watt or more generally preclude precise matching unless the transistors are heavily degenerated.
8. Place matched transistors in low-stress areas.
   The presence of any significant heat source on the die precludes placing the matched transistors in the center, because they would lie too close to the heat source. In this case, moderately matched transistors should occupy the middle of the opposite end of the die from the heat source. Moderately matched transistors should not reside within about 10mils (250μm) of an edge of the die because stress levels increase near edges. Similarly, moderately matched transistors should be kept well away from the corners of the die where the stresses are greatest. Precise matching is very difficult to maintain in the presence of large thermal gradients.
9. Place moderately or precisely matched transistors on axes of symmetry of the die.
   Moderately or precisely matched transistor arrays should be oriented so that their major axis of symmetry, S_1, coincides with one of the axes of symmetry of the die. If possible, matched arrays should reside around the <211> axis of a (111) die rather than the <110> axis.
10. Do not allow the NBL shadow to intersect matched emitters.
    The NBL region of a moderately or precisely matched transistor should overlap its emitters by a distance sufficient to ensure that it does not intersect them. If the direction of NBL shift is unknown, allow adequate overlap of NBL on all sides of the emitter. If the magnitude of the shift is unknown, then overlap NBL over the emitter by at least 150% of the maximum epi thickness. Minimally matched transistors can forego this precaution because the impact of the NBL shadow is relatively small.
11. Place emitters far enough apart to avoid interactions.
    If multiple emitters must occupy a common base region, then space them far enough apart to prevent their depletion regions from intersecting. If the layout rules specify the spacing between unconnected emitters, use this rule for matched emitters regardless of how they are connected. If no such rule exists, then the spacing between the matched emitters should exceed the minimum spacing by 2 to 3μm.
12. Increase the base overlap of moderately or precisely matched emitters.
    If the base barely overlaps the emitter, misalignment can cause the lateral beta of a portion of the emitter periphery to increase enough to produce minor mis-

matches. The base region of a moderately or precisely matched transistor should overlap its emitter by 1 to 2μm more than minimum.

13. Operate matched transistors on the flat portion of the beta curve.
The β-*vs.*-I_C plot of an NPN transistor usually exhibits a plateau across a relatively broad range of currents. Whenever possible, matched transistors should operate on this plateau. Transistors operating at higher currents will suffer from high-level injection that may induce mismatches. In ratioed pairs and quads, high-level injection can cause deviations from the theoretical VPTAT voltages. Most NPN transistors do not enter high-level injection except at relatively high currents that can also cause undesirable self-heating. Transistors operated at a current level that is too low are prone to variations in beta due to surface effects. The low-current beta rolloff of most vertical NPN transistors occurs only at extremely low current densities, so this effect rarely interferes with device matching.
14. The contact geometry should match the emitter geometry.
A circular emitter should contain a concentric circular contact. Similarly, an octagonal emitter should contain an octagonal contact and a square emitter should contain a square contact. If the process allows only minimum-size square contacts, then use square emitters and square arrays of minimum contacts. The emitter contacts should fill as much of the emitter area as possible, except in cases where silicidation must be minimized to prevent beta degradation. These precautions help prevent interactions between the contact and the edge of the emitter from distorting the flow of emitter current.
15. Consider using emitter degeneration.
Minimally matched transistors will not normally benefit from emitter degeneration. Moderately matched transistors may benefit from degeneration in the presence of large thermal gradients. Precisely matched transistors are often degenerated, if for no other reason than to allow adjustment of their offset voltage by trimming the degeneration resistors. The degenerating resistors should develop at least 50mV for moderate matching and 100mV for precise matching. Emitter degeneration can also be used to match transistors with different emitter sizes or geometries. In this case, at least 200mV of degeneration should be employed for minimal matching and 500mV for moderate matching. This technique can achieve noninteger ratios between matched transistors. For example, a 1.64:1 ratio can be constructed from two transistors having equal emitter areas and a pair of emitter degeneration resistors with a ratio of 1:1.64.

9.3.2. Rules for Matching Lateral PNP Transistors

Lateral PNP transistors generally do not match as well as vertical NPN transistors. Their poorer matching is due partly to surface effects and partly to an inability to use large emitters. Emitter degeneration is frequently used to improve the matching of PNP current mirrors and whatever other circuits can tolerate its presence. The following rules summarize the principles of designing matched lateral PNP transistors:

1. Use identical emitter and collector geometries.
Both the emitter and the collector geometries affect conduction in lateral PNP transistors. Transistors with different emitter or collector geometries match very poorly. For minimal matching, only the size and shape of the inner periphery of the collector facing the emitter matters. For higher degrees of precision the entire collector geometry should be duplicated. The shape and size of the

tank (or well) are unimportant as long as none of the transistors saturate. If a transistor can saturate, it is safest to place it in its own tank. P-bar or N-bar isolation schemes (Section 4.4.2) should not be counted on to ensure complete isolation between matched devices. Shallow-collector transistors (such as analog BiCMOS devices constructed from PSD implants) should be placed in separate tanks or wells to minimize cross-injection caused by carriers passing underneath the shallow collectors.

2. Use minimum-size emitters for matched transistors.
   Larger emitters will degrade the beta of the transistor, and this effect probably hurts matching more than the increased area helps. Ratioed transistors should employ multiple copies of a minimum-emitter cell (Figure 8.26B).
3. Field-plate the base region of matched lateral PNP transistors.
   Field-plating ensures that electrostatic charges do not interfere with the flow of current across the neutral base. Improperly field-plated transistors are susceptible to long-term drifts that can play havoc with matching. Lateral PNP transistors constructed in analog BiCMOS processes that incorporate a channel stop implant across the neutral base do not require field-plating, because the channel stop performs this function. Still, the addition of field plates never hurts.
4. Split-collector lateral PNP transistors can achieve moderate matching.
   Moderate matching can be achieved only as long as all of the split collectors are identical copies of one another, and none of the collectors saturates. The presence of gaps between the collectors makes it impossible to accurately predict the division of current between split collectors of different sizes. The saturation of any split collector destroys the matching between the remaining split collectors. Split-collector laterals can be used to form very compact cross-coupled transistors that exhibit surprisingly precise matching.[21]
5. Place matched transistors in close proximity.
   Even minimally matched lateral PNP transistors should reside near one another to minimize the impact of thermal gradients. Moderately or precisely matched transistors may benefit from placement in a common base tank. If this is done, make sure that none of the transistors in the tank can saturate.
6. If possible, avoid constructing VPTAT circuits from ratioed lateral PNP transistors.
   An ideality factor ignored in the derivation of equations 9.8 and 9.9 becomes significant in high-level injection, where lateral PNP transistors usually operate. The VPTAT voltages developed by ratioed mirrors and quads often exhibit significant deviations from the values predicted by the equations due to the contribution of the ideality factor.
7. Place matched transistors far away from power devices.
   Minimally matched transistors should reside at least 10mils (250μm) away from major power devices and should not be placed adjacent to any device dissipating more than 50mW. Moderately matched devices should reside at least 5 to 10mils (125 to 250μm) away from any device dissipating more than 50mW, and they should be placed at the opposite end of the die from major power de-

---

[21] Gilbert reports ±0.1% typical matching from cross-coupled split-collector lateral PNP transistors; this is presumably a one-sigma value. See B. Gilbert, "Bipolar Current Mirrors," in C. Toumazou, F. J. Lidgy, and D. G. Haigh, *Analog IC Design: The Current-Mode Approach* (London: Peter Perigrinus, 1990), pp. 249–250.

vices. Precisely matched devices should be separated as far as possible from any power device. Devices that dissipate a watt or more generally preclude precise matching unless the matched transistors are heavily degenerated. Consider elongating the die to a 1.5:1 or even a 2:1 aspect ratio to increase the separation between precisely matched transistors and major power devices.

8. Place matched transistors in low-stress areas.
   Precisely matched transistors should occupy the center of the die, but the presence of any significant heat source generally precludes placing the matched transistors in the center of the die. Moderately matched transistors should instead occupy the middle of the end of the die opposite the heat source. They should not reside within about 10mils (250μm) of an edge of the die, and they should be kept well away from the corners of the die.
9. Place moderately or precisely matched transistors on axes of symmetry of the die.
   Moderately or precisely matched transistor arrays should be oriented so that their major axis of symmetry, S_1, coincides with one of the axes of symmetry of the die. If possible, matched arrays should be placed on the <211> axis of a (111)-oriented die.
10. Do not allow the NBL shadow to intersect the base region of a lateral PNP.
    The presence of the surface discontinuity that causes the NBL shadow distorts the flow of current across the neutral base of the transistor. If the direction of NBL shift is unknown, allow adequate overlap of NBL on all sides of the base region. If the magnitude of the shift is unknown, then overlap NBL over the base region by at least 150% of the maximum epi thickness. The NBL shadow will have little or no effect on matching if it merely intersects the collector of the transistor.
11. Operate matched lateral PNP transistors near peak beta.
    The β *vs.* I_C of a lateral PNP transistor usually exhibits a pronounced peak. Matched transistors should operate at or near this peak in order to minimize base current errors. Operating the transistor at either lower or higher current densities causes the beta to roll off and increases base current errors. Also, the nonidealities mentioned in rule 6 become increasingly important away from the point of maximum beta.
12. The contact geometry should match the emitter geometry.
    A circular emitter should contain a concentric circular contact. Similarly, an octagonal emitter should contain an octagonal contact and a square emitter should contain a square contact. These precautions help prevent interactions between the contact and the edge of the emitter from distorting the flow of emitter current.
13. Consider using emitter degeneration.
    Lateral PNP transistors usually benefit more from emitter degeneration than do vertical NPN transistors because of their lower Early voltages and the inadvisability of increasing their emitter areas. The degenerating resistors should develop at least 50mV for moderate matching and 100mV for precise matching. Emitter degeneration can also be used to match transistors with different emitter sizes or geometries. In this case, 200mV of degeneration should be employed for minimal matching and 500mV for moderate matching. This technique can also achieve noninteger ratios between matched transistors. Split collectors cannot be degenerated relative to one another because they share a common emitter.

9.4 SUMMARY

Bipolar power transistors are substantially more difficult to design than MOS power transistors. The negative temperature coefficient of V_{BE} makes bipolar transistors susceptible to thermal runaway, and current focusing during turnoff can destroy an otherwise robust transistor through secondary breakdown. Proper design can minimize these vulnerabilities. Bipolar transistors offer several unique advantages over MOS transistors: their high transconductance does not depend on large device areas or small channel lengths, and they also exhibit superior transient power handling capability due to the larger volume of silicon available to dissipate heat. Bipolar transistors perform exceptionally well as MOS gate drivers and as ESD protection devices (Section 13.4.3). Large lateral PNP transistors are incredibly robust. The large die area required to construct such a device spreads the heat dissipation over a corresponding volume of silicon, and the pronounced beta rolloff of the lateral PNP makes it almost impossible to destroy the transistor by excessive current conduction.

Bipolar transistors also exhibit better voltage matching characteristics than MOS transistors. The high transconductance of the bipolar transistor allows a single stage to generate higher gains and thus minimizes the number of matching transistors required. The emitter area of a vertical NPN transistor can be increased without impairing transconductance, while increasing the channel length of a MOS transistor rapidly decreases its transconductance and increases the required area. Matched MOS transistors almost always consume more area than matched bipolar transistors of similar precision. Properly ratioed bipolar transistors develop extremely accurate VPTAT voltages that form the basis of many voltage and current references. MOS transistors only develop VPTAT voltages when operated in subthreshold, a mode incompatible with high-temperature operation.

Although MOS transistors have supplanted their bipolar counterparts in many applications, bipolar transistors still have their advantages. The future of analog design appears to lie with analog BiCMOS processes that merge high-density CMOS with high-performance bipolar. These processes can provide high-density submicron CMOS logic in combination with precision analog functions currently achievable only through the use of bipolar transistors.

9.5 EXERCISES

Refer to Appendix C for layout rules and process specifications. For all power transistors, assume a minimum beta at full rated current of ten. Do not exceed an emitter current density of 5mA/mil^2 ($8\mu\text{A}/\mu\text{m}^2$) for linear-mode devices and 10mA/mil^2 ($16\mu\text{A}/\mu\text{m}^2$) for switched-mode devices.

9.1. What is the maximum current that can flow through a 100μm-long emitter finger with a metallization width of 12μm? Assume the metallization consists of 10kÅ of aluminum/copper/silicon alloy and that emitter debiasing must not exceed 5mV.

9.2. Construct an interdigitated-emitter power transistor using standard bipolar layout rules. The transistor is intended as a 500mA series-pass transistor for a linear regulator. Construct the transistor around a central spine of deep-N+ 20μm wide, and place banks of emitter fingers on both sides of this spine. Make the emitter fingers 20μm wide. Do not allow intrafinger debiasing to exceed 5mV, or base debiasing to exceed 3mV. Use emitter ballasting resistors that develop 50mV at full rated current. Include all necessary metallization.

9.3. Construct a wide-emitter narrow-contact power transistor for a lamp driver using standard bipolar layout rules. This switching transistor must handle 150mA of collector current and should contain as much deep-N+ as possible. Make all deep-N+ sinkers 16μm wide and all emitter fingers 24μm wide. Assume only one end of the base serpentine can be connected. Do not allow the sum of base metallization debiasing and emitter metallization debiasing to exceed 10mV.

9.4. Construct a cruciform-emitter power transistor for a relay driver using standard bipolar layout rules. This switching transistor must handle 700mA of collector current and should be ringed with a deep-N+ sinker 20μm wide. Maximize connection to the collector. Make the cruciform emitter sections 75μm wide and contact them with 10μm-diameter circular emitter contacts. Assume only one end of the base lead can be connected. Do not allow the sum of base metallization debiasing and emitter metallization debiasing to exceed 10mV.

9.5. Construct a wide-emitter narrow-contact transistor using analog BiCMOS layout rules. This gate driver transistor must conduct 500mA pulses. Assume that the transistor operates at a peak emitter current density of $75mA/mil^2$. Use an emitter overlap of emitter contact of 8μm and completely ring the transistor with a deep-N+ sinker no less than 10μm wide. Maximize metallization of both emitter and collector. Since the layout rules do not allow strip contacts, use rows of minimum-width contacts instead. Make the narrow contact for the emitter out of two rows of minimum contacts. Include all necessary metallization.

9.6. Construct the relay driver transistor from Exercise 9.4 using analog BiCMOS layout rules. Maximize metallization to both emitter and collector. Devise a suitable replacement for the circular emitter contacts.

9.7. Construct the circuit shown in Figure 9.15 using standard bipolar layout rules. Transistors Q_1, Q_2, and Q_3 are minimum-area lateral PNP transistors having one, two, and three emitters, respectively, and transistor Q_4 is a minimum-area substrate PNP. Resistors R_1, R_2, and R_3 consist of 6μm-wide base resistors placed in a tank connected to V_{CC}.

9.8. Construct the circuit shown in Figure 9.17C using analog BiCMOS layout rules. Transistors Q_1, Q_2, Q_3, and Q_4 should use square emitters having a width of 10μm. Include as many emitter contacts as possible. Compute the value of resistor R_1 necessary to produce a current of 10μA, and lay this resistor out using PSD-doped poly-2 6μm wide. Take all necessary precautions to obtain optimal matching.

9.9. Lay out the Brokaw bandgap cell shown in Figure 9.22A using standard bipolar layout rules. Transistors Q_1 and Q_2 should employ circular emitters with diameters of 10μm. Resistors R_1 and R_2 should be constructed as an interdigitated array of base resistors in a common tank. This tank should connect to the base of transistors Q_1 and Q_2. Include all necessary interconnection and label all devices.

9.10. Lay out the simple operational amplifier shown in Figure 9.22B using analog BiCMOS layout rules. Transistors Q_1 to Q_5 should employ 5×5μm square emitters. Transistors Q_6, Q_7, and Q_8 should use 8×8μm square emitters. Cross-couple Q_4 and Q_5. Include all necessary interconnection and label all devices. The values on this schematic represent the areas, in μm^2, of the respective emitters.

9.11. Lay out the Gilbert multiplier core shown in Figure 9.23 using analog BiCMOS layout rules. Transistors Q_1 to Q_4, Q_6, Q_7, Q_9, and Q_{10} use 8×8μm emitters. Transistors Q_5, Q_8, and Q_{11} use 6×6μm emitters. Lay out all transistors for optimal matching. Transistors sharing a common collector connection can occupy the same tank. Include all necessary interconnection and label all devices. The values on this schematic represent the areas, in μm^2, of the respective emitters.

9.12. Suppose the Gilbert multiplier core in Exercise 9.11 forms part of a die with an area of $7.6mm^2$ (not including scribe streets and seals), which also includes a power NPN transistor with an area of $4.3mm^2$. Select an aspect ratio for the die and place rectangles representing the outline of the die and power transistor. Place the multiplier core at an optimal location for best matching.

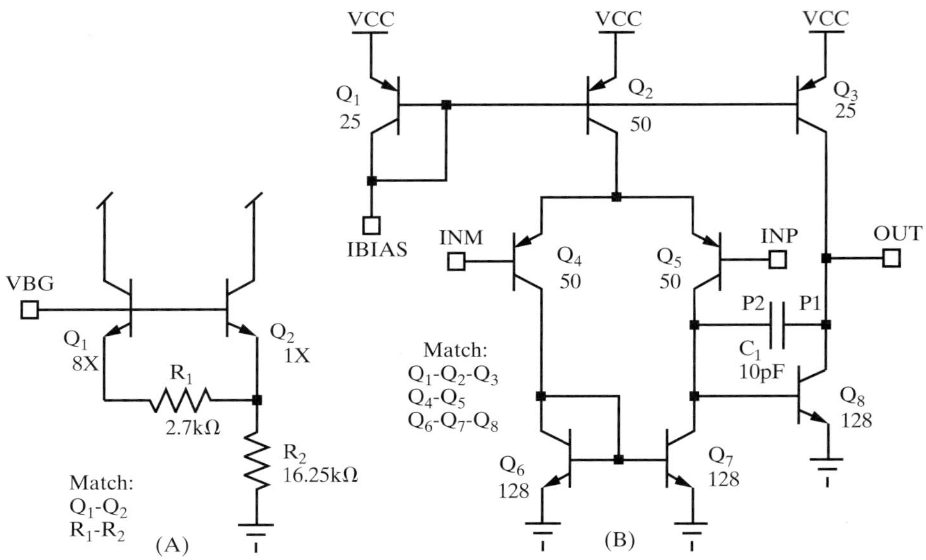

FIGURE 9.22 (A) Brokaw bandgap cell and (B) simple operational amplifier (B) for Exercises 9.9 and 9.10.

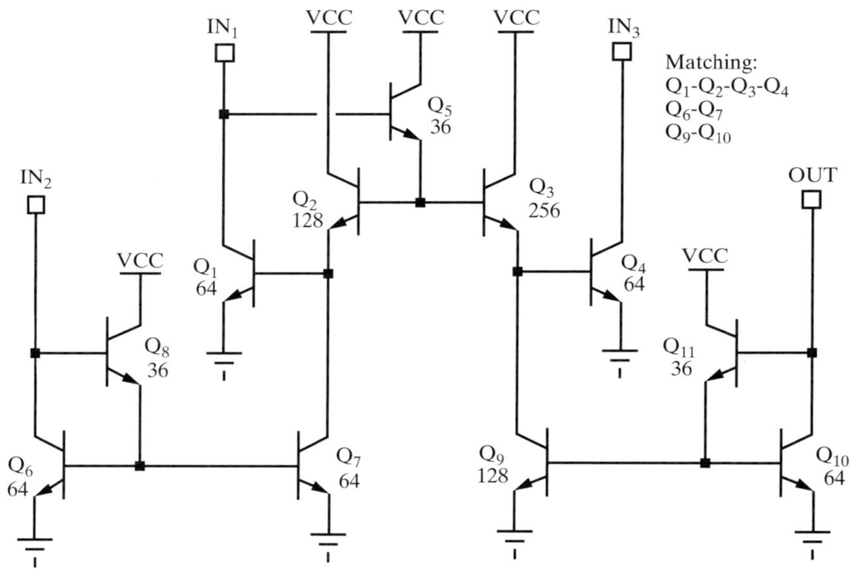

FIGURE 9.23 Gilbert multiplier core for Exercise 9.11.

10 Diodes

The device now called a *diode* was invented in the late nineteenth century, but it first saw widespread use in the galena crystal detector of 1907. This device was actually a Schottky diode formed between a metallic cat's whisker and semiconducting lead sulfide (galena). The copper oxide rectifiers and selenium stacks of the vacuum-tube era were also primitive Schottkies. Modern semiconductor diodes emerged from a different line of development that began with germanium point-contact diodes developed for military and computer applications. These were replaced in the mid-1960s by silicon PN-junction diodes similar to those in use today.

Diodes have found a number of applications in modern integrated circuits. Schottky diodes are often used as anti-saturation clamps for the collector-base junctions of NPN transistors. PN junction diodes form part of current mirrors and biasing networks. Junction diodes operated in reverse breakdown can also serve as voltage references and clamping devices. This chapter examines these and other applications of integrated diodes.

10.1 DIODES IN STANDARD BIPOLAR

The standard bipolar process can construct a wide variety of diodes. Of these, the most popular are the diode-connected transistor, the base-emitter Zener, and the Schottky diode. The first two are both variations of the bipolar NPN transistor, while the Schottky diode relies on the formation of a rectifying contact to lightly doped silicon. Not all versions of standard bipolar offer Schottky diodes, because they require the formation of platinum or palladium silicides and the addition of a special masking step to allow contact through the thick-field oxide. This section also discusses several additional types of Zener diodes sometimes available in standard bipolar.

10.1.1. Diode-connected Transistors

An NPN transistor consists of two back-to-back PN junctions, either of which could theoretically serve as a PN junction diode. In practice, the parasitic transistors associated with these junctions render them unsuitable for use as diodes. The collector-base diode loses most of its current to substrate due to parasitic PNP action. NBL does not stop this parasitic conduction, because the carriers can still flow to the isolation sidewalls. Ringing the tank with deep-N+ contains the minority carriers and minimizes the current loss, but only at the price of greatly enlarging the structure.[1] The collector-base diode also exhibits a relatively high series resistance due to the light doping of the tank.

The base-emitter diode loses the vast majority of its current to the enclosing tank due to parasitic NPN action. Most of the carriers injected by the emitter travel across the base and into the tank. The accumulation of electrons in the collector causes the base-collector junction to forward-bias, turning on the parasitic PNP and diverting current to the substrate.

A very useful type of diode is created by tying the collector and the base of an NPN transistor together (Figure 10.1). The resulting device is often called a *diode-connected transistor.* Most of the current flows through a diode-connected transistor from collector to emitter by means of transistor action. Only a small current flows through the base terminal, so the base resistance has little effect on the device's forward voltage. The forward voltage also remains independent of collector resistance as long as the transistor does not fully saturate. A typical diode-connected transistor can tolerate about 400mV of debiasing at 25°C, or about 200mV at 150°C. If the collector debiasing exceeds these limits, then the diode begins losing current to the substrate due to parasitic PNP action. Diode-connected transistors often incorporate NBL to minimize collector series resistance. Diodes that must conduct more than a few hundred microamps should also contain a deep-N+ sinker. Diodes conducting 10mA or more should be laid out as power devices and should incorporate deep-N+ rings to minimize substrate injection during transients.

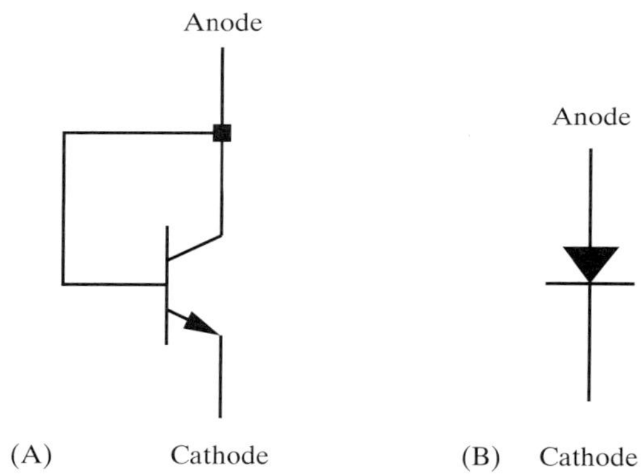

FIGURE 10.1 (A) Schematic and (B) symbol for a diode-connected transistor.

The diode-connected transistor will not suffer any loss of current to the substrate as long as collector debiasing does not exceed the limits given above. It also has much less series resistance than either the base-emitter diode or the base-collector diode. A minimum-size diode-connected transistor typically exhibits no more than 10 to 20Ω of series resistance. Its only serious drawback lies in its relatively low breakdown voltage, which is limited by the V_{EBO} of the NPN transistor to 6 to 8V.

Diode-connected transistors usually employ the CBE configuration rather than the CEB configuration (Figure 8.14). Although the CBE configuration has slightly more collector resistance, it allows first-level metal to connect the collector and base

[1] Some rather elaborate structures offer reasonable performance at the cost of large areas. See B. Murari, "Power Integrated Circuits: Problems, Tradeoffs, and Solutions," *IEEE J. Solid-State Circuits,* Vol. SC-13, #6, p. 307–319.

contacts. Many processes allow merged collector-base contacts, which save additional space (Figure 10.2). In this structure, the emitter diffusion surrounding the collector contact overlaps the base diffusion so that a single contact can touch both.[2] This contact must extend far enough into the base diffusion to account for misalignment and outdiffusion while still allowing sufficient base contact area for conduction. The contact must also extend into the collector far enough to counter misalignment while allowing adequate collector contact. Even with these overlaps, the merged structure is still considerably smaller than a traditional NPN layout.

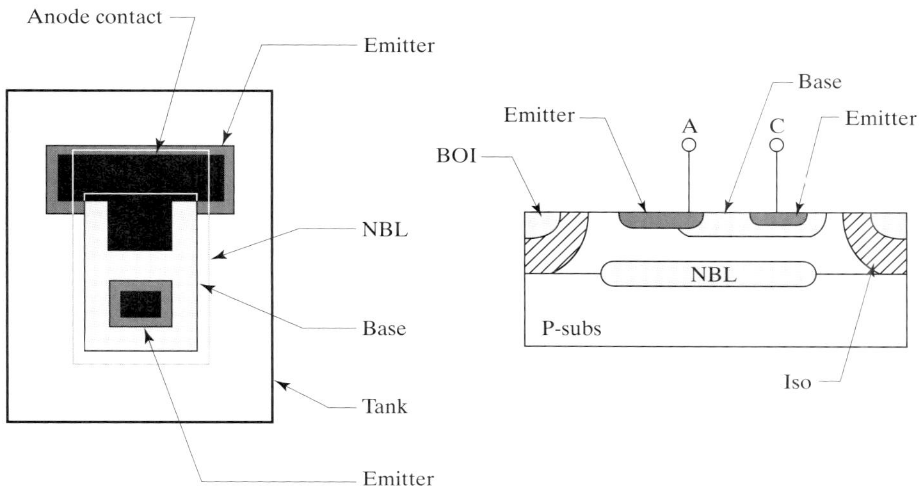

FIGURE 10.2 Layout and cross section of a standard bipolar diode-connected transistor.

The diode-connected transistor can serve as a convenient voltage reference. A base-emitter junction exhibits a forward voltage of about 0.65V at a current density of 1mA/mil^2 and a temperature of 25°C. A typical layout has an emitter junction area of 0.1mil^2 (about 60μm^2) and requires a current of about 100μA to develop a forward voltage of 0.65V. The forward voltage of a diode is relatively insensitive to small fluctuations in current. Even if the current through the diode were to double, the forward voltage would increase by only 18mV. The forward voltage also exhibits a temperature coefficient of about −2mV/°C. A stack of several diodes connected in series can develop larger voltages, but temperature and current variability increase proportionally.

A substrate PNP transistor can also serve as a diode, but the collector current of this device flows directly into the substrate. Currents that are much in excess of 1mA may debias the substrate enough to saturate the transistor. Diode-connected substrate transistors are sometimes used in CMOS processes that cannot fabricate other bipolar components (Section 10.3.1). They rarely see much use in processes that can fabricate isolated bipolar transistors.

Lateral PNP transistors make relatively poor diodes. They require large tanks that not only consume die area, but also contribute unwanted parasitic capacitance. Some portion of the collector current always flows to the substrate regardless of how thoroughly the device has been guard-ringed. The cathode current of a diode-connected lateral PNP is always less than its anode current. This loss prevents the use of lateral PNP transistors in applications in which current matching is critical. Still, diode-connected lateral PNP transistors are occasionally inserted into circuits to balance other PNP base-emitter voltage drops. An NPN transistor would not

2 "Diodes," *Semiconductor Reliability News,* Vol. III, #7, July 1991, p. 9.

serve as well because its base-emitter voltage does not exactly match that of a PNP. The base and collector of the lateral PNP form the cathode of a diode-connected PNP, while the emitter acts as its anode. The layout sometimes uses a merged cathode contact that is analogous to the merged anode contact of the diode-connected NPN transistor shown in Figure 10.2.

10.1.2. Zener Diodes

A reverse-biased diode conducts very little current until the voltage across it exceeds a certain value. Beyond this point, the current through the diode increases exponentially until it eventually approaches an asymptote defined by the series resistance of the diode (Figure 10.3). The breakdown curve usually shows a fairly definite inflection point or *knee* corresponding to the *breakdown voltage* of the diode. The magnitude of the breakdown voltage depends on the width and curvature of the depletion region of the diode (Section 1.2.4). Carriers can tunnel across a very thin depletion region by a quantum process discovered by Zener.[3] Diodes with breakdown voltages of less than 6V are called *Zener diodes* since they conduct primarily by Zener tunneling. Diodes with breakdown voltages in excess of 6V are properly called *avalanche diodes* because they conduct primarily by avalanche multiplication instead of by tunneling. Designers frequently use the term *Zener diode* to describe all junction diodes operated in reverse breakdown regardless of conduction mechanism.[4]

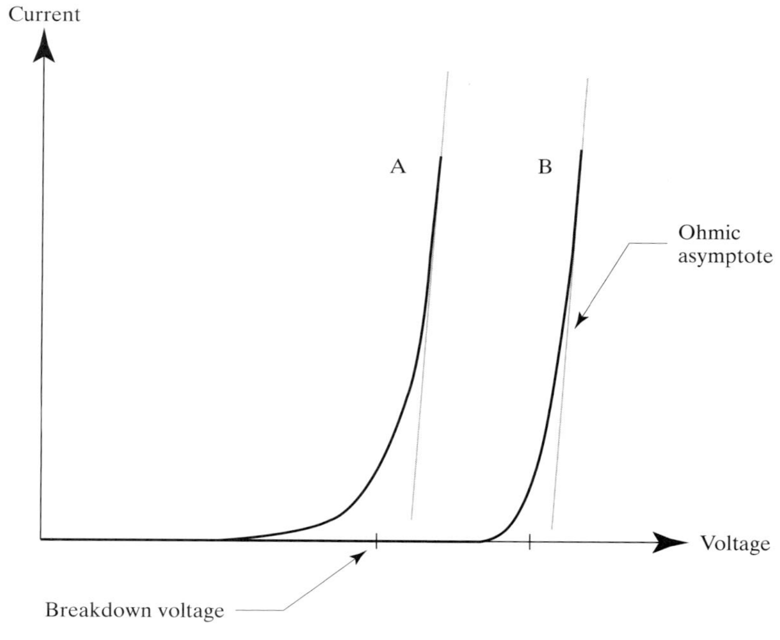

FIGURE 10.3 A comparison of the reverse breakdown characteristics of (A) a Zener diode and (B) an avalanche diode.

Zener tunneling exhibits a negative temperature coefficient, while avalanche breakdown exhibits a positive one. Diodes that break down at less than about 5.6V have negative temperature coefficients that increase in magnitude as the breakdown voltage diminishes. Diodes with breakdowns in excess of 5.6V exhibit increasingly positive temperature coefficients. The familiar emitter-base Zener has a break-

[3] C. Zener, *Proc. Roy. Soc.* A145, London: 1934, p. 523.

[4] Some authors use the term *breakdown diode* to refer to both Zener and avalanche diodes, but the engineering community has not widely adopted this practice.

down voltage of 6 to 8V with a positive temperature coefficient of 2 to 4mV/°C. A 40V base-collector Zener exhibits a much larger temperature coefficient, perhaps 35 to 40mV/°C.

Zener diodes with breakdown voltages of 5 to 6V have very small temperature coefficients. These diodes are sometimes used to construct temperature-independent voltage references, but they have several drawbacks that severely restrict their usefulness. Zener walkout (Section 4.3.1) causes the reference voltage to drift over time unless the device is specifically constructed to ensure subsurface breakdown. Zener references also require a supply voltage of at least 6V. Because of these disadvantages, Zener references have largely been replaced with lower-voltage alternatives such as bandgap references.

Zener diodes with breakdown voltages below 5V exhibit a rather gradual onset of reverse conduction (Figure 10.3A). This *soft breakdown* becomes more pronounced in lower-voltage devices. Diodes with breakdowns below about four volts have extremely soft breakdown characteristics. Designers often blame soft breakdown on leakage, but it is actually an unavoidable characteristic of Zener tunneling. In any event, Zener diodes with breakdown voltages of much less than 5V have little practical application because they do not exhibit well-defined reverse characteristics.

Surface Zener Diodes

The emitter-base junction of an NPN transistor forms a convenient Zener diode. Its breakdown voltage, V_{EBO}, depends on base doping and emitter junction depth. Most standard bipolar processes provide an emitter-base breakdown of about 6.8V. Advanced bipolar and BiCMOS processes often use lightly doped bases that exhibit emitter-base breakdowns of as much as 10V. Breakdown proceeds primarily by avalanche rather than by tunneling, so the temperature coefficient of the breakdown voltage is positive. A typical 6.8V emitter-base Zener exhibits a temperature coefficient of +3 to 4mV/°C.

Emitter-base Zeners have historically exhibited large amounts of process variation and long-term drift. Older bipolar processes used an emitter pilot step to control NPN beta, allowing the use of poorly controlled dopant sources such as boron nitride disks. The resulting variation in base doping and the compensatory changes in emitter junction depth caused V_{EBO} to vary by as much as ±1V. Ion implantation has dramatically improved base doping control, and most modern processes guarantee no more than ±0.25V initial variation. Process improvements have also reduced long-term drift due to Zener walkout, which now rarely exceeds 0.1V.

Emitter-base Zeners use essentially the same layout as NPN transistors (Figure 10.4). The emitter acts as the *cathode* of the Zener and the base acts as the *anode*. The tank serves only to isolate the Zener from the surrounding isolation. It should connect either to the cathode of the Zener or to some equal or higher voltage. The tank must never connect to the anode of the Zener lest the transistor become biased into the reverse-active region. If this occurs, then V_{ECO} breakdown will produce a snapback phenomenon similar to that exhibited by V_{CEO} breakdown (Section 8.1.2). Some designers leave the tank unconnected, but this practice is not recommended because it can amplify leakage currents. The floating tank acts as the base of a parasitic substrate PNP. The leakage current across the tank-isolation junction exceeds that across the smaller base-tank junction, and the difference between these currents forms the base drive of the substrate PNP. This mechanism can cause substantial loss of anode current to the substrate at elevated temperatures. Connecting the tank to the cathode prevents the parasitic PNP from amplifying leakage currents in this manner. The tank contact does not require a deep-N+ sinker because it only conducts

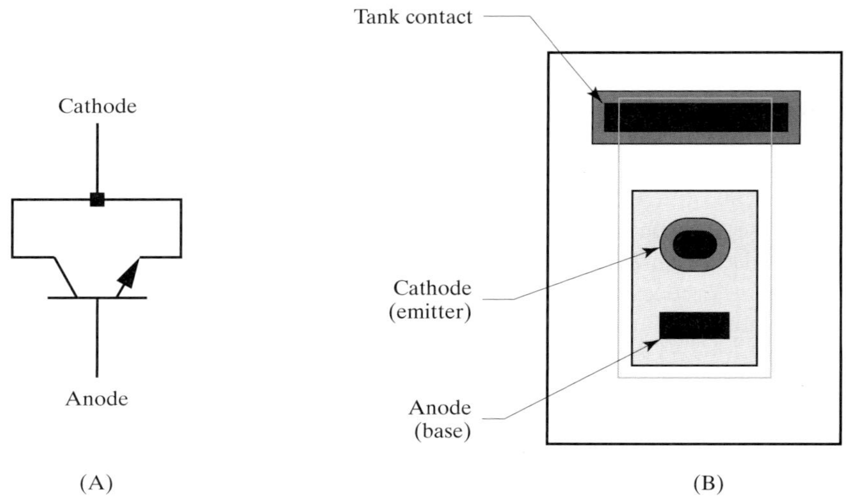

FIGURE 10.4 (A) Schematic and (B) layout of a typical base-emitter Zener diode.

leakage currents. NBL has little effect on a Zener and can therefore be omitted if the designer so chooses.

Emitter-base Zeners usually have circular or oval emitters like the one shown in Figure 10.4. These round geometries are intended to prevent electric field intensification at the corners of the emitter. Not all processes exhibit this phenomenon, but if it occurs it increases the variability of the breakdown voltage. One can determine whether a given process exhibits this effect by examining an avalanching rectangular emitter under a microscope in a darkened room. The avalanching junction will emit a faint white light.[5] If this light appears brightest at the corners, then electric field intensification is occurring at these points, and the device will benefit from the use of round emitters. Many designers routinely use round emitters because, even if they do not provide any benefit, they do no harm.

The base doping profile causes the base-emitter depletion region to narrow near the surface. The intensification of the electric field across the narrowed depletion region ensures that avalanche breakdown occurs at this point. Hot carriers produced by the intense electric field sometimes penetrate into the overlying oxide, where a small fraction becomes trapped. The gradual accumulation of this trapped oxide charge causes *Zener walkout* (Section 4.3.1). Some designers have attempted to suppress surface breakdown by flanging metal over the base-emitter junction to form a field plate. The thickness of the emitter oxide and the relatively low voltages placed across it prevent this field plate from having much effect. The avalanche process continues to occur near the surface and the field plate cannot prevent Zener walkout.

Emitter-base Zeners are relatively fragile devices because they dissipate energy in the relatively small volume of the emitter-base depletion region. The heat generated by this process can damage the junction or the adjacent contacts. Extreme overloads usually induce metal migration that causes permanent short-circuit failures. This mechanism has been employed as a replacement for fuses (Section 5.6.2). Base-emitter Zeners used as voltage references or clamp devices should not conduct more than about 10μA per micron of emitter periphery. For example, a 5×5μm emitter can safely conduct some 200μA. Zeners can tolerate much higher currents for brief periods of time, but long-term operation at elevated currents can cause

[5] "Junction Breakdown Characteristics," *Semiconductor Reliability News*, Vol. II, #1, January 1990, p. 8.

shifts in breakdown voltage due to junction damage. If an application requires more current than a small Zener diode can safely handle, consider using a power transistor to amplify the current conducted by the Zener, since this circuit takes much less area than a large Zener (Figure 13.22).

Some circuits connect the anode of the base-emitter Zener to the substrate. Since the base diffusion of the Zener operates at the same potential as the surrounding isolation, these two diffusions can overlap one another. This practice saves considerable area because it eliminates the large amount of spacing required to isolate the base diffusion from the isolation (Figure 10.5). A tank is placed beneath the emitter to prevent the isolation diffusion from reducing the Zener voltage. Although this tank remains unconnected, the diffusions around it are all biased to the same potential. The parasitic PNP transistor inherent in this structure cannot amplify leakage currents and is therefore harmless.

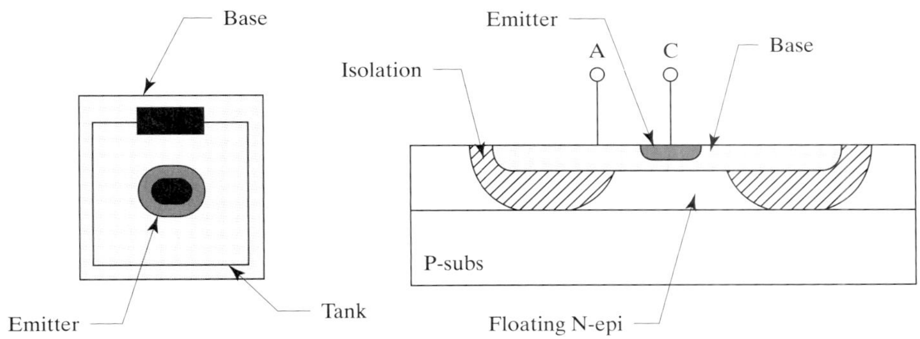

FIGURE 10.5 Layout and cross section of a nonisolated base-emitter Zener diode.

The nonisolated base-emitter Zener is vulnerable to both substrate debiasing and noise coupling. Nonisolated Zeners should never reside near structures that inject substrate currents of more than a few hundred microamps. Conservative designers often avoid using nonisolated Zeners, preferring to accept the extra die area of a conventional base-emitter Zener rather than risk unexpected debiasing or noise coupling.

Buried Zeners

If the avalanche region lies several microns beneath the oxide, hot carriers will scatter off the lattice and lose energy before they can reach the oxide interface. Zeners that avalanche beneath the surface are called *subsurface Zeners*, or more colloquially, *buried Zeners*. The breakdown voltages of buried Zeners remain constant throughout their operational lifetime, making these devices ideal for use as voltage references. This section presents several common varieties of buried Zeners that are compatible with standard bipolar processing.

A buried Zener can be constructed from the emitter and isolation diffusions of certain standard bipolar processes. This diode consists of a plug of P+ isolation covered by an emitter diffusion that overlaps into a surrounding tank (Figure 10.6).[6] The emitter counterdopes the surface of the isolation to form the cathode of the Zener. The anode consists of the portion of the isolation plug beneath the emitter. The N-epi surrounding the isolation plug prevents the emitter sidewall radius from further reducing the already-low breakdown voltage of the structure. The *emitter-in-iso* buried Zener usually exhibits a breakdown voltage of 5 to 6V and a temperature coefficient of about +1mV/°C.

---

[6] A. B. Grebene, *Bipolar and MOS Analog Integrated Circuit Design* (New York: John Wiley and Sons, 1984), pp. 133–134.

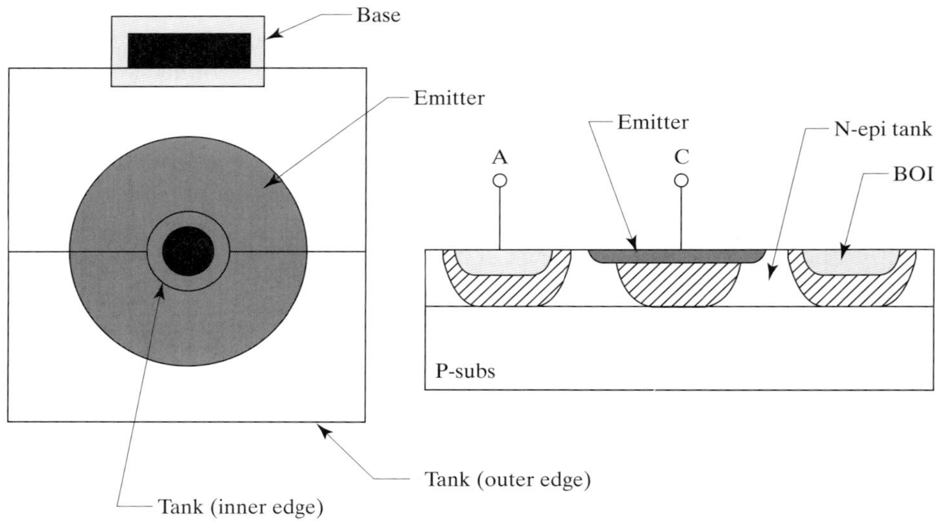

FIGURE 10.6 Layout and cross section of an emitter-in-isolation Zener diode.

Some processes use base over isolation to prevent channel formation. The presence of base in the isolation plug further reduces the breakdown voltage of the emitter-in-iso Zener. Although this may help minimize the residual temperature coefficient of the Zener, it also produces a softer breakdown characteristic. Breakdown voltages of less than 5V become rather indefinite and are therefore unsuited for use as voltage references. If the addition of base to the emitter-in-iso Zener causes its breakdown voltage to drop below 5V, then the designer should omit it from the area around the emitter plug.

The emitter-in-iso Zener shares most of the same drawbacks as the nonisolated base-emitter Zener. Its anode is electrically common to the substrate. This severely limits its range of applications and raises the possibility of substrate debiasing and noise coupling. Emitter-in-iso Zeners do not experience any appreciable long-term drift, but their initial voltage varies considerably because of the heavy deposition and long drive time required to fabricate the P+ isolation. A typical emitter-in-iso Zener exhibits a breakdown voltage of 5.4 ± 0.4V.

Several alternative styles of buried Zeners eliminate the drawbacks of the emitter-in-iso device at the cost of introducing additional processing steps. Figure 10.7 shows a buried Zener that requires the use of a deep-P+ diffusion inserted after isolation and before base.[7] The doping concentration in this deep-P+ diffusion signif-

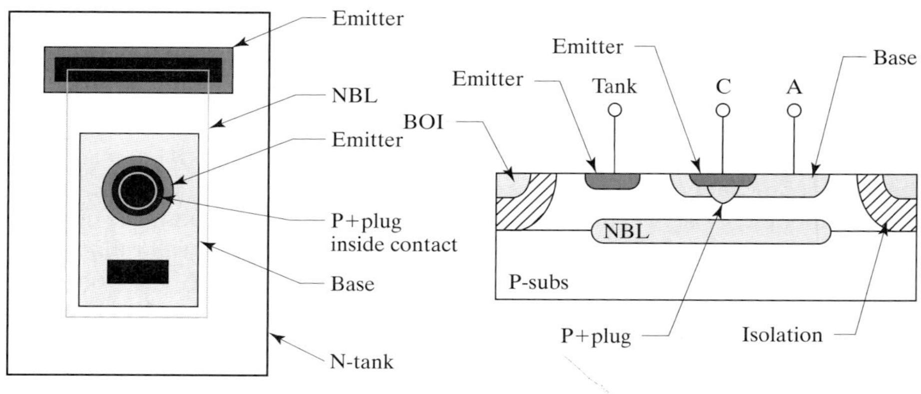

FIGURE 10.7 Buried Zener using a special deep-P+ diffusion in combination with emitter diffusion.

[7] Grebene, p. 134.

icantly exceeds that of the base diffusion but falls short of that of the emitter. The active region of the Zener consists of a plug of deep-P+ diffusion covered by emitter. The emitter diffusion is in turn enclosed by a base diffusion. The anode of the diode consists of the deep-P+ plug and is contacted by means of the surrounding base diffusion. The cathode of the diode consists of the emitter diffusion. The tank of this Zener is usually connected to the cathode to prevent beta multiplication of leakage currents. The structure in Figure 10.7 contains NBL, but it actually plays no role in the operation of the device. It can be omitted as long as the deep-P+ is sufficiently shallow to prevent punchthrough breakdown to the substrate.

The breakdown voltage of this structure can be tailored to suit a specific application by adjusting the profile of the deep-P+ diffusion. A lightly doped diffusion will have a higher breakdown voltage, while a heavily doped diffusion will have a lower one. The breakdown voltage cannot exceed that of the base-emitter junction, nor should it drop so low as to produce a soft breakdown characteristic. In practice, the breakdown voltage ranges from 5 to 6.5V. A typical device has a breakdown voltage of 6.3 ± 0.2V with a temperature coefficient of about +2mV/°C.

Another style of buried Zener substitutes a high-energy implant for the emitter diffusion of the standard base-emitter structure (Figure 10.8).[8] At high implant energies, the peak of the dopant distribution actually lies beneath the surface of the silicon. The high-energy implant therefore creates a shallow N-buried layer. The intersection of this layer with a plug of base diffusion forms a buried Zener diode. The breakdown voltage is set by adjusting the implant energy and dose used to fabricate the implanted NBL.

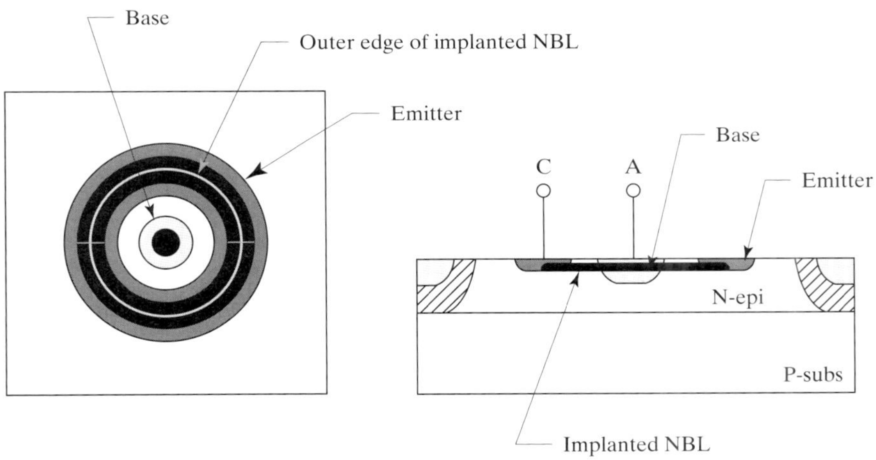

FIGURE 10.8 Buried Zener using an implanted NBL in combination with base diffusion.

All three of the buried Zeners just discussed provide breakdown voltages of 5 to 7V with relatively low temperature coefficients. Certain applications, such as ESD protection, require higher-voltage Zeners. These are usually constructed from stacks of series-connected base-emitter Zeners, possibly supplemented by diode-connected transistors. In addition to helping adjust the voltage of the stack, the negative temperature coefficient of the diode-connected transistors can offset part or all of the temperature coefficient of the Zeners. A few specialized high-voltage Zener structures take advantage of the large breakdown voltages between deep, lightly doped diffusions. Of these, the isolation-NBL Zener deserves special mention. This Zener

---

[8] Grebene, p. 134.

consists of a plug of isolation diffused into a plate of NBL. The NBL plate is enclosed in a tank contacted by means of a deep-N+ sinker. The voltage of the isolation-NBL Zener can vary by a few volts because it depends on a number of factors, including NBL and isolation doping, isolation drive time, and epi thickness. The isolation-NBL junction lies far beneath the surface and is therefore immune to Zener walkout. The relatively large area of the depletion regions and their depth beneath the surface also allow this structure to safely dissipate several times the power density of a base-emitter Zener.

10.1.3. Schottky Diodes

Schottky diodes depend on the formation of a rectifying Schottky barrier between a conductor and a semiconductor. Both P-type and N-type silicon form rectifying Schottky barriers with numerous metals and metal silicides. The forward voltages of the resulting diodes depend on the composition of the conductor and the polarity of the silicon (Table 10.1). Schottky diodes with forward voltages of less than 0.5V usually exhibit large junction leakages, especially at higher temperatures. This limitation generally prevents the use of P-type silicon in Schottky diodes.

TABLE 10.1 Typical forward voltages of Schottky diodes constructed from selected metals and silicides[9] (25°C, 1μA/μm²).

Material	N-type Silicon	P-type Silicon
Aluminum	0.54V	0.40V
Gold	0.62V	0.16V
Molybdenum	0.50V	0.24V
Palladium silicide (Pd_2Si)	0.57V	
Platinum silicide (PtSi)	0.66V	
Titanium silicide ($TiSi_2$)	0.42V	

The concentration of acceptors or donors in the silicon determines whether a given Schottky barrier exhibits rectifying or Ohmic behavior. Dopant concentrations exceeding 10^{17} atoms/cm^3 reduce the width of the depletion region to the point at which carriers can successfully surmount it by quantum tunneling. The resulting tunneling current increases almost exponentially with doping.[10] When the tunneling current exceeds the forward conduction current, the Schottky barrier begins to resemble a resistor. Practical Schottky diodes normally employ surface doping concentrations of no more than 10^{16} atoms/cm^3 to minimize tunneling effects.

Many applications require Schottky diodes with forward voltages significantly lower than those of PN junction diodes. Aluminum would appear to be an ideal material for constructing such diodes. Its forward voltage lies comfortably below that of a PN junction diode, yet not so low as to cause excessive junction leakage. Unfortunately, sintering can drastically alter the properties of an aluminum-silicon Schottky. The aluminum dissolves a small amount of silicon during sintering. As the wafer cools, some of the dissolved silicon redeposits at the metal-semiconductor interface as an aluminum-doped

---

[9] These values are derived from the barrier potential ϕ_B, assuming current density of 1μA/μm² and a Richardson constant of 120A/cm²/K². Under these circumstances, the forward voltage is 180mV less than ϕ_B. Barrier potentials from S. M. Sze, *Physics of Semiconductor Devices,* 2nd ed. (New York: John Wiley and Sons, 1981), pp. 290–291.

[10] Actually, theory predicts that the tunneling current varies exponentially with respect to the square root of the dopant concentration and linearly with respect to the applied voltage; See W. R. Runyan and K. E. Bean, *Semiconductor Integrated Circuit Processing Technology* (Reading, MA: Addison-Wesley, 1994), p. 524.

P-type semiconductor. This deposit constricts the Schottky contact area and, in extreme cases, may entirely cover the contact opening. This mechanism can cause the forward voltage of a Schottky to increase by several hundred millivolts.[11]

Few other metals exhibit the properties required to construct practical Schottky diodes. Most are difficult to sinter and exhibit excessive forward voltage variation. Modern integrated Schottky diodes employ the silicides of certain noble metals, most notably platinum and palladium. These *noble silicides* provide extremely stable and repeatable forward voltages lying in the desired range of 0.5 to 0.7V. The inability of noble silicides to withstand the temperatures required for source/drain annealing limits their application in CMOS processes. The forward voltages of the *refractory silicides* (such as titanium silicide) are rarely sufficient to prevent leakage, so processes that use these silicides rarely offer Schottky diodes.

Most processes indiscriminately silicide all contact openings. Those lying over heavily doped silicon become Ohmic contacts, while those residing over lightly doped N-type silicon become Schottky diodes. These processes generally prohibit the opening of contacts over lightly doped P-type silicon, because the resulting Schottky barrier has a contact resistance that is too high to function as an Ohmic contact and a forward voltage that is too low to function as a Schottky diode.

Figure 10.9 shows the layout and cross section of a Schottky diode constructed in a standard bipolar process. The metal system consists of a sandwich of platinum silicide, refractory barrier metal, and copper-doped aluminum. The Schottky barrier forms between the platinum silicide and the lightly doped N-type epi. In order to reach the epi, the Schottky contact must penetrate the thick-field oxide. If this contact opening were etched simultaneously with the base and emitter contacts, the latter would suffer severe overetching before the former cleared. Most standard bipolar processes include an additional etching step to thin the oxide over the Schottky contact openings prior to the regular contact oxide removal. The process extension required to form Schottky contacts therefore consists of a single masking step and a single oxide removal.

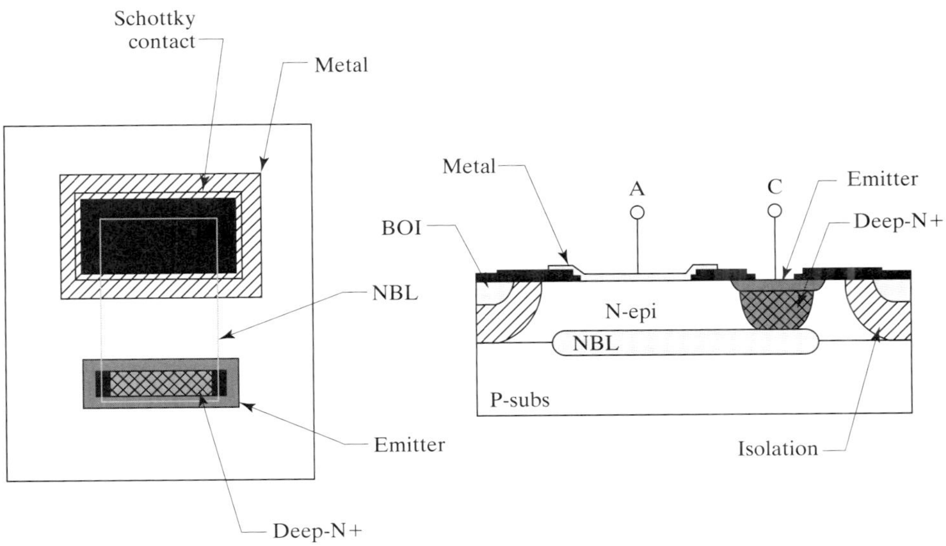

FIGURE 10.9 Layout and cross section of a field-plated Schottky diode.

[11] M. Mori, "Resistance Increase in Small-Area Si-Doped Al-n-Si Contacts," *IEEE Trans. on Electron Devices*, Vol. ED-30, #2, 1983, pp. 81–86.

The cathode of the Schottky includes NBL and deep-N+ to minimize its series resistance. A small plug of deep-N+ suffices to extract the cathode current as long as this does not exceed a milliamp or two. High-current Schottky diodes usually employ rings of deep-N+ around the periphery of their tanks to further reduce the series resistance. Large diodes can occupy empty spaces between other components, because the shape of the Schottky contact has little effect on its performance.

The planar breakdown voltage of a Schottky diode usually exceeds the V_{CEO} rating of the corresponding NPN transistor by at least a factor of two. Practical Schottky diodes obtain only a fraction of this theoretical breakdown voltage because the sharp edges of the Schottky contact greatly intensify the electric field. An unprotected Schottky usually begins to avalanche at a reverse bias of only a few volts. Several techniques have been developed that diminish the electric field intensity at the edges of the Schottky contact. The simplest of these consists of flanging the metallization over the Schottky contact to form a field plate. The full reverse-bias voltage appears between the field plate and the underlying silicon. The resulting vertical electrical field repels electrons from the surface of the silicon and makes it appear to be more lightly doped. This delays the onset of avalanche and increases the reverse-bias voltage rating by a few volts. The field plate need only overlap the contact by 3 to 5μm. This relatively simple and compact arrangement suffices for many low-voltage applications.

Figure 10.10 shows an alternative style of Schottky diode that can withstand much higher reverse voltages. This structure encloses the edge of the Schottky contact within a thin strip of base diffusion called a *field relief guard ring.*[12] The presence of the guard ring completely eliminates the lateral electric field at the edge of the Schottky contact and raises the breakdown voltage of the structure to equal the V_{CBO} of the base diffusion. These field-relief guard rings are completely unrelated to the minority carrier guard rings of Section 4.4.2.

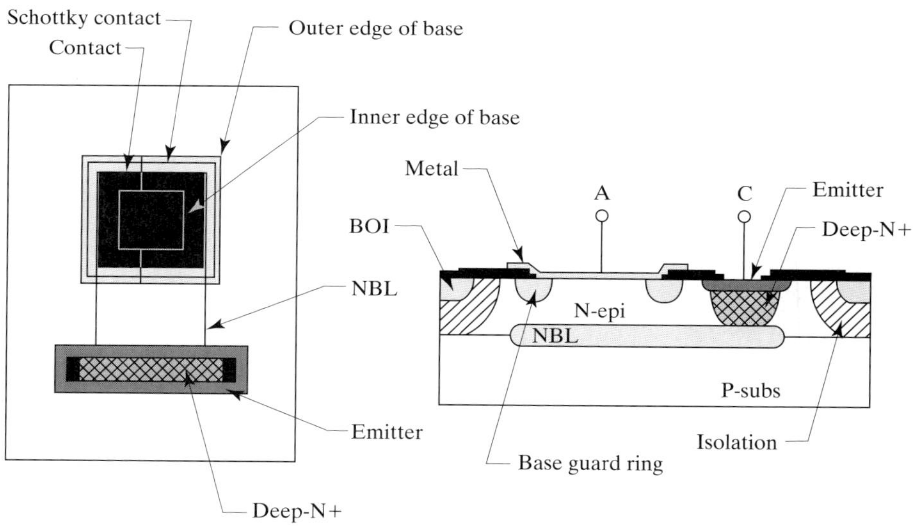

FIGURE 10.10 Layout and cross section of a base guard-ring Schottky diode.

The inclusion of a field-relief guard ring inevitably enlarges the Schottky diode. The spacing between the tank and the Schottky contact must increase to avoid punchthrough between the guard ring and the isolation. Outdiffusion also constricts

[12] M. P. Lepselter and S. M. Sze, "Silicon Schottky Barrier Diode with Near-Ideal I-V Characteristics," *Bell Sys. Tech. J.*, Vol. 47, #2 1968, p. 195–208.

the contact opening by several microns on all sides. These considerations affect the area of small Schottky diodes much more severely than they affect the area of large ones. Many designers routinely add guard rings to large Schottkies because they render the breakdown characteristics of the diode much more predictable and repeatable. On the other hand, these same designers frequently use field plates on small Schottkies to conserve area.

Schottky diodes are often used as antisaturation clamps for NPN transistors (Section 8.1.4). An antisaturation diode can occupy the same tank as the transistor it protects. A field-plated diode is easily created by extending the base contact out into the surrounding tank (Figure 10.11A). A guard-ringed diode requires the addition of a narrow strip of base around the periphery of the Schottky contact opening (Figure 10.11B). Minimum-size devices usually forego the addition of the guard ring to conserve area. Devices that must tolerate higher collector-base voltages may require guard rings, as may those that cannot tolerate any base-collector leakage.

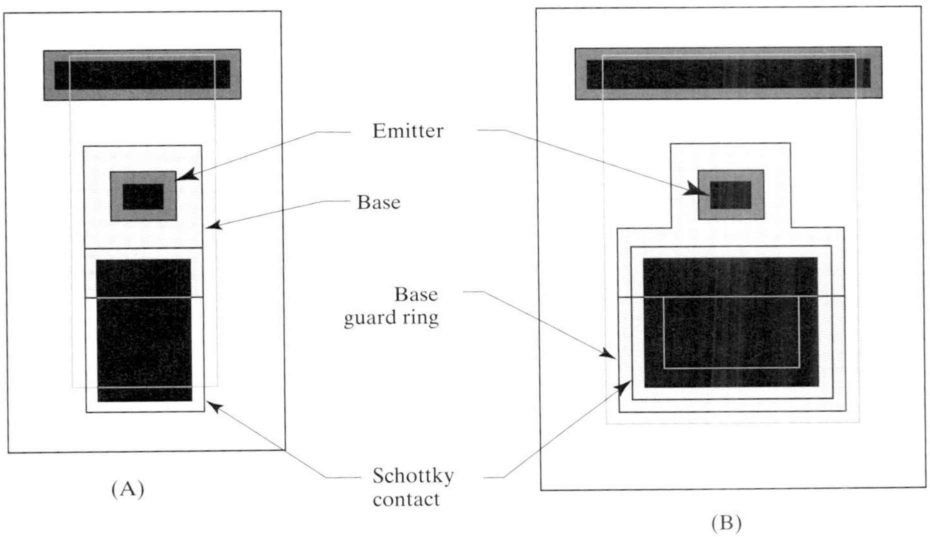

FIGURE 10.11 Schottky-clamped NPN transistors using (A) field plates and (B) base guard rings.

In order for a Schottky diode to serve as an antisaturation clamp, it must have a lower forward voltage than the base-collector junction of the protected transistor. The size of the required contact opening depends on three factors: the materials used to fabricate the Schottky, the maximum operating temperature, and the current it must conduct. Palladium silicide Schottky diodes make excellent antisaturation clamps because they have relatively small forward voltages. Platinum silicide Schottky diodes have larger forward voltages and therefore require careful attention to contact areas. Devices that must operate at high temperatures require larger contact areas because the Schottky forward voltage has a smaller temperature coefficient than the base-collector forward voltage. This phenomenon prevents platinum silicide Schottky antisaturation clamps from operating at temperatures much higher than 150°C. The same effect can cause guard-ringed Schottky diodes to begin to inject minority carriers to the substrate at temperatures in excess of 130 to 140°C. Junction leakage imposes similar temperature limitations on palladium silicide Schottky diodes.

The area of the Schottky contact required to protect a transistor from saturation must be determined by empirical measurement. In order to perform this measurement, current is forced through the base-emitter junction of a transistor whose collector remains unconnected. The base drive is increased until substrate injection exceeds a predetermined percentage of the base drive—perhaps 5%.

This measurement should be performed at the maximum operating junction temperature. The amount of base drive that the Schottky clamp can successfully absorb scales with the ratio of Schottky area divided by base-collector junction area. Based on a single experiment, one can determine approximately how much Schottky contact area is required to protect any given transistor. Outdiffusion may also affect the Schottky contact area of a guard-ringed device. Assuming the Schottky contact is rectangular, then the effective Schottky area, A_S, can be computed from the drawn dimensions of the base opening X_D and Y_D using the following equation

$$A_s = (X_D - 2\delta)(Y_D - 2\delta) \qquad [10.1]$$

where δ is a correction factor that accounts for base outdiffusion, photolithographic size adjusts, and so forth. Since forward current scales linearly with Schottky area, the correction factor can be determined by measuring the currents required to obtain a specified forward voltage for two Schottky diodes of different dimensions.

Schottky diodes also see occasional use as power devices. High current densities tend to activate parasitic mechanisms that would otherwise remain dormant. Figure 10.12 shows the parasitics associated with a typical Schottky diode. D_1 represents the Schottky diode itself, R_1 models the resistance of the NBL and deep-N+, and R_2 models the resistance of the N-epi beneath the Schottky contact. The epi contributes more resistance to a Schottky diode than to an NPN transistor because current in the Schottky must cross the entire thickness of the epi. This resistance is not necessarily undesirable because it acts as distributed ballasting and prevents thermal runaway under all but the most extreme conditions. Power Schottky diodes can therefore consist of a single contact opening of any desired shape.

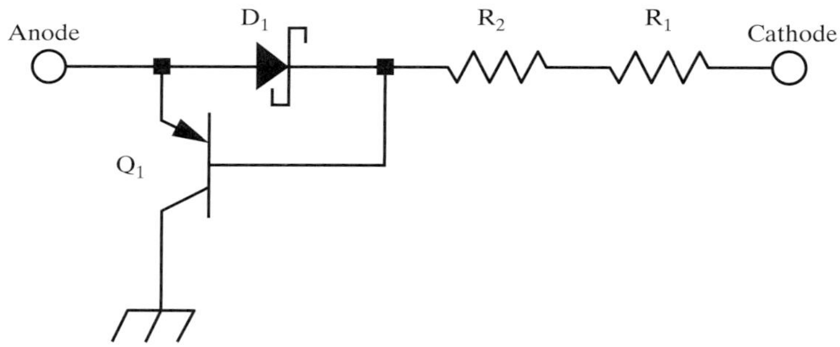

FIGURE 10.12 A simplified parasitic model of a Schottky diode.

Transistor Q_P in Figure 10.12 models minority carrier injection within the Schottky diode. Although Schottky diodes are majority carrier devices, they do experience some minority carrier injection at high current. Schottky field-relief guard rings will inject minority carriers into the cathode tank when the voltage across the diode exceeds the forward voltage of the junction between the guard ring and the N-epi. Field-plated devices lack this parasitic PN junction, but the Schottky barrier itself emits small numbers of minority carriers at high current densities. Unless these minority carriers are blocked, they will flow across the N-epi into the substrate. The presence of NBL in the Schottky helps minimize substrate injection, but power Schottky diodes should also contain a continuous ring of deep-N+ around their perimeter. This ring serves as both a cathode contact and as a minority carrier guard ring.

10.2 DIODES IN CMOS AND BICMOS PROCESSES

Every semiconductor process can produce at least one type of diode. Analog BiCMOS processes can fabricate most of the diodes available in standard bipolar. Pure CMOS processes offer fewer options and therefore deserve closer consideration.

Although one could theoretically construct PN junction diodes using any of the three junctions available in a CMOS process, only one can be biased into conduction under normal operating conditions. In an N-well CMOS process, the anodes of the NSD/P-epi diode and the N-well/P-epi diode both connect to the substrate. These diodes can only be forward-biased by pulling their cathode below the substrate. Not only do these diodes require a negative power supply, but they also pose a latchup hazard due to minority carrier injection into the substrate. The PSD/N-well junction does not exhibit these problems, but it contains a parasitic PNP that diverts a significant fraction of the diode current to the substrate. The PSD/N-well diode also has a large internal series resistance (Section 10.3.1).

Zener diodes can be constructed using either the PSD/N-well junction or the NSD/P-epi junction. The parasitic PNP transistor inherent in the PSD/N-well diode does not conduct as long as the junction remains reverse-biased. Similarly, the NSD/P-epi junction does not inject minority carriers into the substrate as long as it remains reverse-biased. Although these Zeners do function, their breakdown voltages always exceed the operating voltage of the process, and they find application only as ESD protection devices and gate protection clamp diodes.

Figure 10.13 shows a typical layout for a PSD/N-well Zener diode. Avalanche breakdown occurs across the depletion region surrounding the PSD implant. The high sheet resistance of the well tends to focus conduction at the edges of the PSD facing the NSD contact. The effective area of conduction can be increased by using long narrow stripes of PSD interdigitated with strips of NSD (Figure 10.13). Even with these improvements, the current-handling capability of the PSD/N-well Zener is substantially inferior to that of a base-emitter Zener.

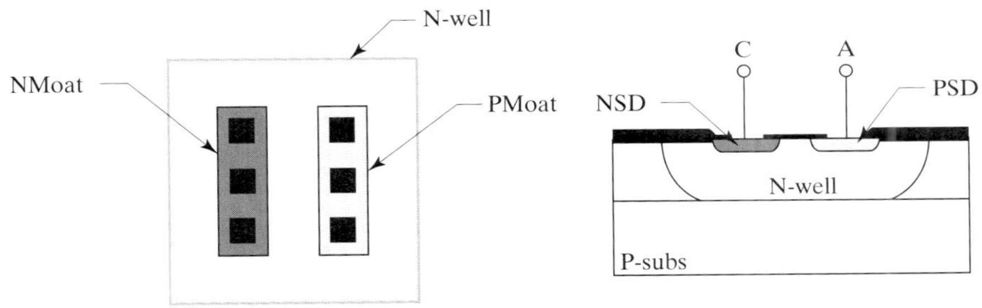

FIGURE 10.13 Layout and cross section of a PSD/N-well Zener diode.

Figure 10.14 shows a typical Schottky diode formed in a CMOS process between a noble silicide and the N-well. The moat geometry coded across the contact opening ensures that the contact resides over thin oxide. A ring of PSD diffusion placed around the edge of the contact opening serves as a field-relief guard ring. The reverse breakdown voltage of the Schottky is limited by the avalanche voltage of the PSD/N-well junction rather than by the planar breakdown voltage of the Schottky. A simple field plate can replace the PSD guard ring, but since field-plated Schottky diodes exhibit larger leakages than guard-ringed Schottkies, most designers opt to use guard rings whenever possible.

Schottky diodes fabricated in CMOS processes have relatively high series resistances due to the absence of NBL and deep-N+. This resistance can be minimized by elongating the Schottky contact and by surrounding it on all sides with cathode contacts. Larger Schottky diodes may employ interdigitated anode and cathode contacts. These measures cannot reduce the series resistance as effectively as NBL and deep-N+, so the resulting Schottky diodes are useful only for relatively low-current applications.

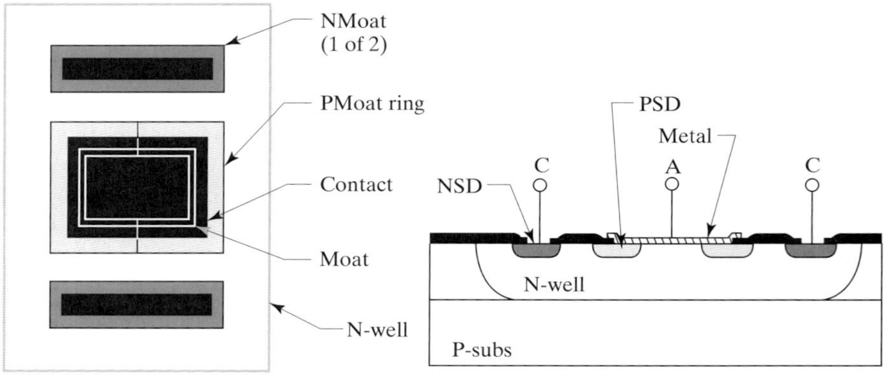

FIGURE 10.14 Layout and cross section of a PSD guard-ring Schottky diode in a CMOS process.

One other type of CMOS diode deserves mention. Some processes fabricate both P-type and N-type gate polysilicon. A polysilicon diode could reside over active circuitry and would therefore save considerable die area. Unfortunately, poly diodes exhibit much higher leakage currents than ordinary diffused diodes due to the presence of defects along the grain boundaries intersecting the junction. The properties of poly diodes also depend on annealing conditions, because the grain structure of the polysilicon changes during high-temperature processing. These concerns have prevented widespread utilization of poly diodes.

The selection of diodes available in analog BiCMOS closely parallels that offered by standard bipolar. Every BiCMOS process offers some variant of the diode-connected transistor, and most can also fabricate Zener diodes analogous to the base-emitter Zener of standard bipolar. The BiCMOS Zener often exhibits a higher breakdown voltage (7 to 10V) due to the use of a more lightly doped base region than standard bipolar. The shallowness of the diffusions in modern BiCMOS processes reduces the conduction volume in these Zeners, making them somewhat more fragile than their standard bipolar counterparts.

Analog BiCMOS processes that employ platinum or palladium silicides can fabricate Schottky diodes. The structure of these diodes resembles that of their standard bipolar counterparts. A moat geometry coded around the anode contact takes the place of the Schottky contact geometry (Figure 10.15). This substitution allows Schottky diodes to be constructed without requiring any additional masking steps. Many BiCMOS designers use these devices as replacements for diode-connected transistors.

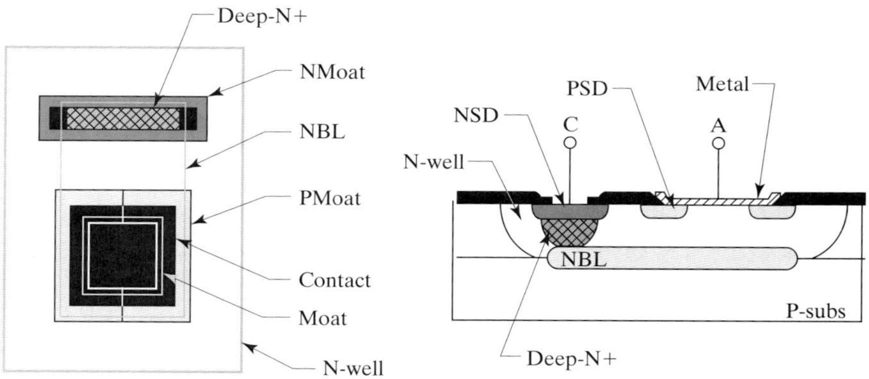

FIGURE 10.15 Layout and cross section of a PSD guard-ring Schottky diode in a BiCMOS process.

10.3 MATCHING DIODES

The devices discussed in this chapter fall into three broad categories: PN junction diodes, Zener diodes, and Schottky diodes. Different types of diodes do not match one another because they obey different principles of operation. Diodes of the same type can match, but only if they are properly constructed. This section briefly discusses the advantages and disadvantages of each type of matched diode and presents some guidelines to aid the designer in constructing matched diodes.

10.3.1. Matching PN Junction Diodes

Most junction diodes used in bipolar and BiCMOS processes are actually diode-connected transistors. As such, their layouts closely resemble those of conventional bipolar transistors. The only unique feature found in diode-connected transistors is the merged collector-base contact (Figure 10.2). The techniques used to match diode-connected transistors are otherwise identical to those used for other bipolar transistors (Sections 9.2 and 9.3).

Pure CMOS processes do not support the construction of diode-connected transistors, but they can still produce PN junction diodes. An N-well CMOS process can construct a PSD/N-well diode, and a P-well process can construct an NSD/P-well diode. Both of these devices contain parasitic vertical bipolar transistors that divert a significant fraction of their forward current to the substrate. The beta of these bipolar transistors can range from less than 0.1 to more than 10. A typical device has a beta of 2. The resulting base current generates a voltage drop across the well resistance that adds to the forward voltage of the diode. Any variation in beta produces a corresponding variation in the voltage drop across the well resistance. The matching of junction diodes therefore depends on the matching of parasitic bipolar betas and well resistances as well as forward voltages.

The layout of a CMOS PN junction diode is identical to that of a CMOS substrate transistor (Section 8.3.3). This device may be viewed either as a PN junction diode containing a parasitic bipolar transistor, or as a bipolar transistor having so low a gain that it resembles a diode. Either way, one must maximize beta and minimize well resistance to obtain any semblance of matching. These factors have such a strong impact that a device with a small junction area-to-periphery ratio will actually exhibit better matching than one that has a large area-to-periphery ratio. The best layout consists of an array of minimum-width junctions interdigitated with well contacts. Matched devices should be interdigitated with one another to produce a common-centroid array (Section 7.2.6). Figure 10.16 shows a pair of matched PSD/N-well diodes.

Many CMOS processes use silicided moats (*clad moats*), but the emitter of a bipolar transistor should not be silicided because this reduces its beta. The matching of PN junction diodes actually improves when they incorporate high-gain parasitic bipolar transistors, because the high gain reduces the currents that must flow through the well resistance. If a silicide block mask is available, then the designer should code this layer around the fingers of the diodes that form its PN junctions (Figure 10.16).

The matching of PN junction diodes also improves at low current densities, because the voltage drop across the well resistance decreases. The current density can be decreased either by increasing the area of the devices or by operating them at lower currents. CMOS diodes typically operate at current densities of 1 to $10\mu A/mil^2$ of junction area (4 to $40 nA/\mu m^2$).

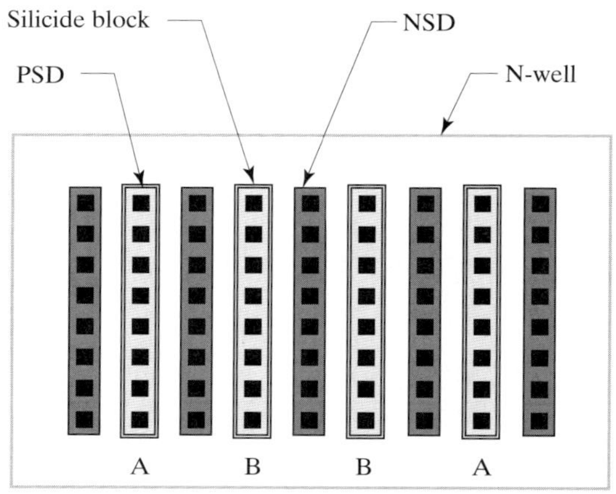

FIGURE 10.16 Layout of a pair of matched PSD/N-well diodes. The anodes of the two diodes are marked A and B.

Regardless of the care taken in their construction, PSD/N-well and NSD/P-well diodes will always exhibit residual mismatches of several millivolts because of the large emitter periphery-to-area ratios and the influence of well resistance and beta variation. Diode-connected transistors and Schottky diodes will almost certainly provide better matching.

10.3.2. Matching Zener Diodes

Zener diodes are difficult to match because their breakdown voltages depend so strongly on electric field intensity. Any curvature in the junction geometry intensifies the electric field and produces a localized reduction in breakdown voltage. The portion of the junction that has the largest curvature conducts the majority of the current and therefore sets the breakdown voltage of the device. Localized breakdown degrades matching because it reduces the effective area of the junction and magnifies the effects of outdiffusion. Matched Zeners should employ circular junction geometries to avoid introducing unnecessary corner curvature. Unfortunately, even circular junctions seldom break down uniformly. Linewidth variations produce minute irregularities that have larger curvatures than the remainder of the junction sidewall. The resulting variations in conduction are often visible under a microscope in a darkened room. The faint glow of the avalanching junction usually appears at only a few spots around the perimeter of the junction. Almost all of the current flows through these spots, which represent only a small fraction of the junction periphery. Devices that have identical layouts often show radically different patterns of conduction. These variations highlight the essentially random distribution of defects and linewidth variations.

Large devices generally exhibit a more uniform distribution of defects. Higher current densities also minimize variability by increasing the voltage drop across the lightly doped diffusion adjacent to the junction. The resulting voltage drop provides ballasting and helps distribute conduction across a larger area, but it also represents another source of variability.

Matched Zeners usually employ large circular geometries to minimize random variations and to eliminate edge effects. Both the anode and the cathode contacts should be circularly symmetric, or nearly so. If necessary, the amount of resistive ballasting in the device can be increased by spacing the contacts further away from the junction. Figure 10.17 shows a cross-coupled pair of matched base-emitter Zeners constructed following these guidelines. The shape of the anode contacts resembles a four-leaf clover, or *quatrefoil,* to allow leads to connect the cathodes to one another without stacking vias over contacts. The four individual Zeners occupy a common tank to minimize the separation between them. The tank isolates the

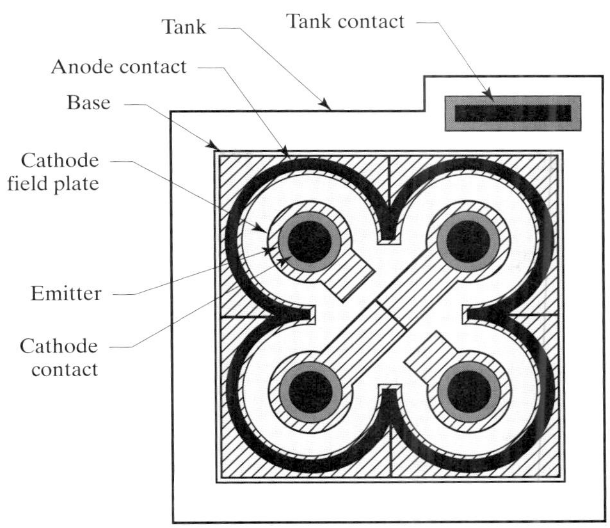

FIGURE 10.17 Quatrefoil layout of cross-coupled base-emitter Zeners. This layout assumes the use of double-level metal.

Zeners from the substrate, but it does not include NBL and deep-N+ because it does not have to conduct any significant current.

Even the elaborate quatrefoil layout of Figure 10.17 may not provide precise matching between surface Zeners, because these devices are susceptible to Zener walkout. The magnitude of this walkout depends on the total charge that has passed through the device, the magnitude and direction of the electric fields through the overlying oxide, and the concentration of mobile ions in the oxide. The layout in Figure 10.17 flanges the emitter metallization over the junction. This flange functions as a field plate that ensures that each Zener sees exactly the same voltage across its junction. The overlap of the field plate over the base-emitter junction should account for misalignment, outdiffusion, and fringing fields. An overlap of 5 to 8μm usually suffices. The field plate cannot stop Zener walkout, but it may help minimize its variability. Buried Zeners provide much better matching because they do not exhibit Zener walkout.

10.3.3. Matching Schottky Diodes

Like Zeners, Schottky diodes are inherently difficult to match. The characteristics of the Schottky barrier depend on several factors, including metal composition, silicon doping, edge effects, annealing conditions, and the presence or absence of surface contaminants. Most of these factors are difficult to control, so Schottky diodes usually exhibit larger mismatches than diode-connected transistors.

Matched Schottky diodes should always incorporate diffused guard rings rather than field plates because field-plated structures often exhibit leakages that could interfere with low-current matching. The contact opening should have a large area-to-periphery ratio to minimize mismatches due to linewidth variations. If possible, the diodes should incorporate NBL and deep-N+ to minimize the portion of the cathode resistance not directly associated with the Schottky contact. If NBL and deep-N+ are used, then several matched Schottky diodes can reside in the same tank or well. If deep-N+ and NBL are not used, then each Schottky should occupy its own well or tank, all of which should have exactly the same dimensions to ensure that the cathode resistances of the diodes match. Ratioed Schottky diodes should always consist of arrays of identical unit contacts, in much the same way (and for much the same reasons) as ratioed NPN transistors use arrays of identical unit emitters. Because Schottky diodes are very sensitive to thermal gradients, they should always use interdigitated or cross-coupled layouts similar to those employed for bipolar transistors.

Schottky diodes will not reliably match PN junction diodes or bipolar transistors. The voltage differential between a diode forward voltage and a Schottky forward voltage may vary by a few percent because of variations in surface conditions, anneal times, and other factors. Even the voltage differential between two Schottky diodes operated at different current densities may depend on processing conditions due to the nonideality factor in the diode equation, which typically plays a larger role in Schottky diodes than it does in junction diodes and bipolar transistors.

10.4 SUMMARY

Every semiconductor process offers one or more types of diodes. Standard bipolar and analog BiCMOS processes can fabricate excellent diode-connected NPN transistors that exhibit mismatches of only a few millivolts and can handle large amounts of current. These transistors can also function as emitter-base Zener diodes. Pure CMOS processes offer fewer options, but they may still be able to construct Schottky diodes if their metallization system includes a noble silicide.

10.5 EXERCISES

Refer to Appendix C for layout rules and process specifications.

10.1. Lay out a standard-bipolar diode-connected transistor using a minimum-size emitter. Compare the area of this device with the area of a minimum-size NPN transistor that does not contain deep-N+. Additional rules for constructing the collector-base contact of the diode are as follows:

1. **CONT** extends into **EMIT** 4μm
2. **CONT** overhang **EMIT** 6μm
3. **BASE** overhang **EMIT** 2μm

10.2. Lay out a minimum-size emitter-in-isolation Zener using standard bipolar rules. The emitter should overlap the isolation plug by 18μm.

10.3. What is the approximate temperature coefficient of a series combination of a 6.8V emitter-base Zener and two diode-connected NPN transistors?

10.4. Construct a standard bipolar field-plated Schottky diode with an area of 200μm^2. Overlap the field plate over the contact by 4μm. Include a deep-N+ sinker along one end of the device. Include all necessary metallization.

10.5. Construct a standard bipolar base guard-ringed Schottky diode having an effective area of 200μm^2. Assume that the correction factor δ equals 2.0μm. The layout rules for the base guard ring are as follows:

1. **BASE** overhang **CONT** 4μm
2. **BASE** extends into **CONT** 4μm

10.6. Construct a standard bipolar power Schottky diode with an effective area of 15000μm^2. Include a base guard ring constructed according to the rules given in Exercise 10.5. Ring the cathode tank with deep-N+ and provide as much metallization as possible, leaving space for a 12μm-wide cathode lead. The anode metallization should be at least 10μm wide at all points and should exit the device through a 12μm-wide anode lead opposite the cathode lead.

10.7. Why does the analog BiCMOS process of Appendix C not support Schottky diodes? What modification would allow their construction?

10.8. Lay out a pair of CMOS PSD/N-well diodes for optimal matching. The diodes should have drawn areas of 60 and 120μm^2, respectively. Draw silicide block (SBLOCK) around the anodes so that SBLOCK overlaps PMOAT by 1μm.

10.9. Lay out a quatrefoil Zener structure consisting of four individual diodes sharing a common anode, using standard bipolar rules. Use a diameter of 12μm for each of the four emitters. Include all necessary metallization, showing how the cathode lead exits from the device and how the tank contact is connected.

11 MOS Transistors

The *field-effect transistor,* or FET, has a long and complicated history. Its initial invention preceded that of the bipolar transistor by some seventeen years, but all of the early attempts to manufacture field-effect transistors failed because of processing problems.[1] Many of these problems were associated with the growth of thin, high-quality dielectric films. By the time these problems were finally overcome, Bardeen and Brattain had already developed the bipolar transistor.

Since the growth of thin dielectric films proved to be so difficult, the first practical field-effect transistors used reverse-biased junctions in place of dielectrics. The resulting devices were called *junction field-effect transistors* (JFETs). Although JFETs were relatively cumbersome devices, they offered much lower input currents than bipolar transistors could achieve. Certain types of operational amplifiers were designed with JFET input stages to reduce their input currents. These devices have become quite successful and are still being produced today.

Thin insulating films suitable for gate dielectrics were finally produced in 1960.[2] This achievement made possible the manufacture of the *metal-oxide-semiconductor field-effect transistor* (MOSFET), often simply called the *MOS transistor.* The early MOS devices still had their share of problems. Their threshold voltages were notoriously unstable, and their thin gate oxides were exceedingly vulnerable to electrostatic discharge (ESD). Once these problems were overcome, MOS transistors began to seriously challenge established bipolar technologies. MOS integrated circuits proved especially useful for low-power digital devices such as digital watches and pocket calculators.

The earliest MOS processes only offered PMOS transistors. These were soon superseded by processes that could produce both enhancement and depletion NMOS transistors. Demands for lower current consumption and greater design flexibility led to the introduction of processes that could simultaneously fabricate both NMOS and PMOS transistors. Although originally intended for digital applications, these *complimentary metal-oxide-semiconductor* (CMOS) processes could also be

---

1 D. Kahng, "A Historical Perspective on the Development of MOS Transistors and Related Devices," *IEEE Trans. on Electron Devices,* Vol. ED-23, #7, 1976, pp. 655–657.

2 D. Kahng and M. M. Atalla, "Silicon-silicon dioxide field-induced devices," Solid-State Device Research Conference, Pittsburgh, June 1960.

used to design a variety of analog integrated circuits. These soon began to replace bipolar integrated circuits in selected applications, but CMOS transistors were not able to duplicate all of the capabilities of bipolar. Many newer processes now merge both bipolar and CMOS transistors onto a common substrate.

This chapter describes the operation and construction of MOS transistors, particularly those employing self-aligned polysilicon gate technology. Chapter 12 covers a variety of more specialized devices, including high-voltage transistors, power transistors, and JFETs.

11.1 TOPICS IN MOS TRANSISTOR OPERATION

MOS transistors are relatively easy to understand and apply. Unlike their bipolar counterparts, they do not experience thermal runaway or secondary breakdown, and they have relatively few troublesome parasitics. Although most textbooks offer a relatively complete discussion of MOS transistor operation, a few topics are slighted because they rarely concern circuit designers. This section discusses several of these neglected topics, including the effects of process and layout on device parameters and the behavior of MOS transistors operating in breakdown.

11.1.1. Modeling the MOS Transistor

Figure 11.1 shows a simplified three-terminal circuit model of an NMOS transistor. No DC current flows through the gate terminal because of the insulating layer between it and the rest of the transistor. Capacitors C_{GS} and C_{GD} represent the gate-to-source and gate-to-drain capacitances produced by this *gate dielectric*. The slashes through these capacitors signify that their values depend on biasing. Voltage-controlled current source I_1 models the current flowing from drain to source through the channel beneath the gate.

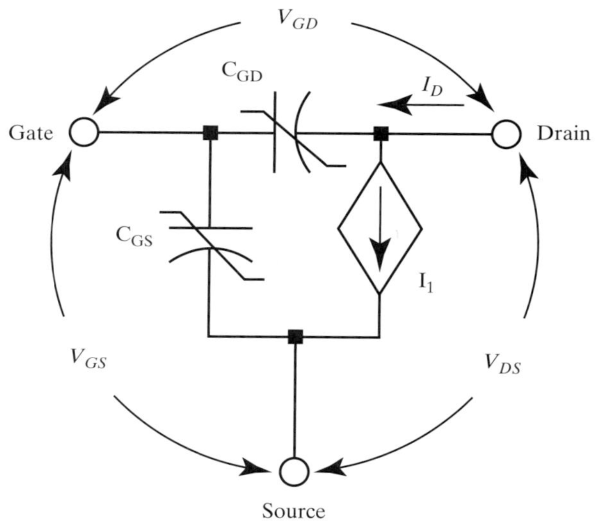

FIGURE 11.1 Simplified three-terminal model of an NMOS transistor.

The magnitude of the drain current I_D depends on the gate-to-source voltage V_{GS} and the gate-to-drain voltage V_{DS}. If the gate-to-source voltage is less than the threshold voltage V_t then no channel forms and the transistor remains in *cutoff* and conducts little or no current. A channel begins to form as soon as the gate-to-source voltage V_{GS} exceeds the *threshold voltage* V_t. The difference between these two quantities is sometimes called the *effective gate voltage* V_{gst}:

$$V_{gst} = V_{GS} - V_t \qquad [11.1]$$

A larger V_{gst} generates a stronger channel that can conduct more current. The drain current also depends on the drain-to-source voltage V_{DS}. If the drain-to-source voltage is less than the effective gate voltage, then the drain current varies linearly with drain-to-source voltage and the transistor is said to operate in the *linear region* (also called the *triode region*). If the drain-to-source voltage exceeds the threshold voltage, then the drain current becomes essentially independent of drain-to-source voltage and the transistor is said to operate in the *saturation region*. The relationship between I_D, V_{GS}, and V_{DS} can be described by a pair of equations—one for the linear region and the other for saturation:

$$If\ 0 \leq V_{DS} < V_{gst},\ I_D = k\left(V_{gst} - \frac{V_{DS}}{2}\right)V_{DS} \qquad [11.2A]$$

$$If\ V_{DS} \geq V_{gst},\ I_D = \frac{k}{2}V_{gst}^2 \qquad [11.2B]$$

These are the *Shichman-Hodges equations* for the NMOS transistor.[3] The parameter, *k*, is called the *device transconductance*, which is something of a misnomer because it has units of A/V^2 rather than A/V. The equations for the PMOS transistor differ in only one respect: equation 11.2A applies when $0 \geq V_{DS} > V_{gst}$, and equation 11.2B applies when $V_{DS} \leq V_{gst}$. For an enhancement-mode PMOS, V_{DS}, V_{GS}, V_t, k, and I_D must all be negative quantities for the equations to yield the proper results.

The Shichman-Hodges equations for the NMOS do not cover the case where $V_{DS} < 0$. Strictly speaking, this condition cannot occur because the source and drain of a MOS transistor are determined by electrical biasing rather than by arbitrary terminal markings. If one attempts to make the drain-to-source voltage less than zero, then the source and drain simply swap roles. The drain terminal now plays the role of the source, and *vice versa*. If one insists on retaining terminal names that no longer correspond to the roles that the terminals actually play, then the terminal conditions must be transformed into actual biasing conditions before applying the Shichman-Hodges equations (see Exercise 11.2).

Device Transconductance

The device transconductance, *k*, determines the amount of drain current that flows through a MOS transistor in response to a given V_{gst}. The device transconductance thus specifies the size of a MOS transistor in much the same way that the emitter saturation current I_S quantifies the size of a bipolar transistor. The device transconductance has units of A/V^2, or (more commonly) μA/V^2. It is related to the layout dimensions of the transistor by the following equation

$$k = k'\left(\frac{W}{L}\right) \qquad [11.3]$$

where *W* and *L* represent the width and length of the MOS channel and *k′* is a constant called the *process transconductance*, which equals

$$k' = \frac{\mu \varepsilon_o \varepsilon_r}{t_{ox}} \qquad [11.4]$$

where the quantity μ represents the *effective mobility* of the carriers (electrons in an NMOS, holes in a PMOS). Surface scattering reduces the mobility of carriers

---

[3] The equations given in the text ignore channel length modulation and the body effect. For further details see H. Shichman and D. A. Hodges, "Modeling and Simulation of Insulated-Gate Field-Effect Transistor Switching Circuits," *IEEE J. Solid-State Circuits*, SC-3, 1968.

confined within a MOS channel, so the effective mobilities appearing in equation 11.4 are considerably smaller than the bulk mobilities discussed in Section 1.1.1. The effective mobility of electrons and holes in silicon are about $675\text{cm}^2/\text{V}\cdot\text{s}$ and $240\text{cm}^2/\text{V}\cdot\text{s}$, respectively. The constant ε_o denotes a universal physical constant called the *permittivity of free space,* which equals $8.85\cdot 10^{-12}$ F/m. The constant ; ε_r represents the relative permittivity of the gate dielectric, which for pure silicon dioxide equals about 3.9. The actual permittivity of oxide may vary slightly from this theoretical value (see Table 6.1). The quantity t_{ox} represents the thickness of the gate dielectric. Substituting these values into the previous equation yields the following simplified formulas for the process transconductance of an NMOS transistor k'_n and of a PMOS transistor k'_p

$$k'_n \cong \frac{23000}{t_{ox}} \mu A/V^2 \quad [11.5\text{A}]$$

$$k'_p \cong \frac{8200}{t_{ox}} \mu A/V^2 \quad [11.5\text{B}]$$

where t_{ox} is measured in Angstroms (Å). MOS processes use the thinnest possible gate oxides to produce the largest possible device transconductances. The dielectric strength of gate oxide equals about 10^7V/cm, or about 0.1V/Å. In practice, the gate dielectric is restricted to substantially lower field intensities to prevent a delayed breakdown mechanism called *time-dependent dielectric breakdown* (TDDB). Transistors with gate oxides thicker than 500Å are generally restricted to field intensities of no more than $3\cdot 10^6$V/cm (30mV/Å). Thinner gate oxides can withstand somewhat higher electric field intensities,[4] so transistors with a gate oxide only 100 to 200Å thick can safely operate at fields in excess of $5\cdot 10^6$V/cm (50mV/Å). Assuming a conservative limit of 30mV/Å, then the maximum possible process transconductance for an operating voltage V_{op} equals

$$k'_n \cong \frac{690}{V_{op}} \mu A/V^2 \quad [11.6\text{A}]$$

$$k'_p \cong \frac{240}{V_{op}} \mu A/V^2 \quad [11.6\text{B}]$$

These formulas indicate that a 5V CMOS process can achieve an NMOS transconductance of about $138\mu\text{A/V}^2$ and a PMOS transconductance of about $48\mu\text{A/V}^2$. In practice, short-channel transistors often experience additional transconductance reductions caused by velocity saturation and other high-field effects. In order to construct a PMOS transistor with the same transconductance as a given NMOS, the W/L ratio of the PMOS must equal almost three times the W/L ratio of the NMOS. The PMOS transistor will thus require nearly three times the area of the NMOS. This disparity is most noticeable in power transistors, but even minimum-size logic gates often increase the size of PMOS transistors to compensate for their low process transconductances.

The device transconductance of MOS transistors decreases with temperature. This variation is primarily due to the temperature coefficient of the carrier mobility. As the temperature rises, lattice vibrations become more energetic and cause increased carrier scattering. Consequently, carrier mobilities are approximately pro-

---

4 C. M. Osburn and D. W. Ormond, "Dielectric Breakdown in Silicon Dioxide Films on Silicon, II. Influence of Processing and Materials," *J. Electrochem. Soc.*, Vol. 119, #5, 1972, pp. 597–603.

portional to the inverse square of absolute temperature.[5] The device transconductance at 150°C equals about half of its value at 25°C. The drain current for a given gate-to-source voltage scales somewhat similarly. Since an increase in temperature causes a decrease in drain current, MOS transistors generally do not exhibit thermal runaway. Power MOS transistors do not require ballasting, which not only simplifies their construction but also enables them to achieve extremely low on-resistances (Section 12.2.1).

The device transconductance given in Equation 11.4 corresponds to that found in most engineering texts,[6] as well as that used in the level-1 MOS model of the simulation program SPICE. Since SPICE was written at the University of California at Berkeley, this definition of device transconductance is often called the *Berkeley k.* A few authors use an alternative definition equal to one-half of the Berkeley k and adjust the Shichman-Hodges equations accordingly.

Threshold Voltage

The *threshold voltage* V_t equals the gate-to-source voltage required to just establish a channel beneath the gate dielectric when the backgate is connected to the source. An *enhancement-mode* MOS transistor requires the application of a non-zero gate-to-source voltage in order to form a channel. The channel of an enhancement-mode NMOS consists of electrons attracted to the surface of the P-type backgate by the positively charged gate electrode (Figure 11.2A). The threshold voltage of the enhancement NMOS is therefore positive. The channel of an enhancement PMOS consists of holes attracted to the surface of an N-type backgate by the negatively charged gate electrode (Figure 11.2B). The threshold voltage of an enhancement PMOS is therefore negative. Another type of MOS transistor exhibits a channel even at a gate-to-source voltage of zero. These *depletion-mode* transistors are normally conducting and require the application of an external gate-to-source voltage in order to turn them off (Figures 11.2C, D). The threshold voltage of a depletion NMOS is negative and the threshold of a depletion PMOS is positive.

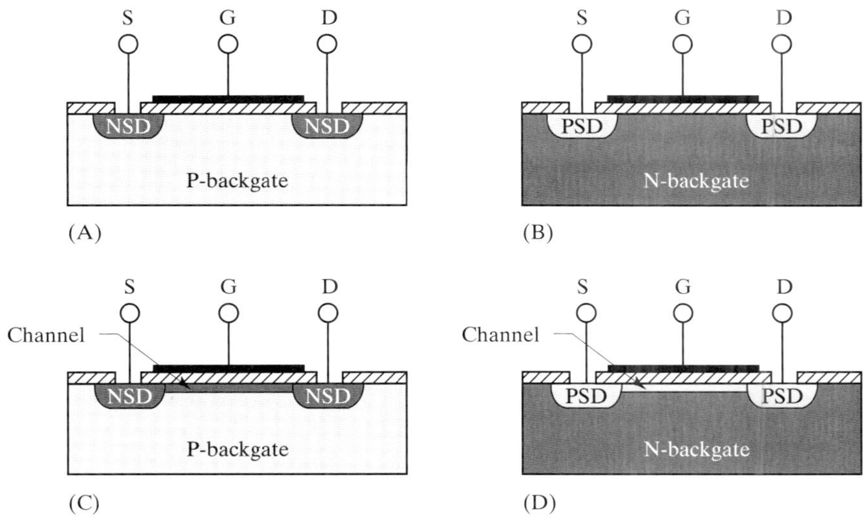

FIGURE 11.2 The four types of MOS transistors: (A) enhancement NMOS, (B) enhancement PMOS, (C) depletion NMOS, and (D) depletion PMOS.

---

5 Electron mobility varies as $T^{-2.42}$ and hole mobility varies as $T^{-2.20}$; see S.M. Sze, *Physics of Semiconductor Devices,* 2nd ed. (New York: John Wiley and Sons, 1981), p. 29.

6 For instance, see R. S. Muller and T. I. Kamins, *Device Electronics for Integrated Circuits,* 2nd ed. (New York: John Wiley & Sons, 1986), p. 430.

MOS transistors are often described as electrically actuated switches. An enhancement MOS resembles a normally open switch because it is normally off and it requires the application of an external gate bias to turn it on. A depletion MOS resembles a normally closed switch because it is on by default and it requires the application of an external gate bias to turn it off. Most processes are optimized to fabricate enhancement-mode transistors because these devices are much more convenient to use than their depletion-mode counterparts. Some processes also offer depletion-mode devices as process extensions.

The threshold voltage of a MOS transistor depends on several factors, including the gate electrode material, the doping of the backgate, the thickness of the gate oxide, the surface state charge density, and the oxide charge density (both fixed and mobile). Each of these factors will be considered in turn.

If the materials used to form the gate and the backgate are not identical, then a nonzero contact potential develops between them. This contact potential represents a net voltage difference between the two materials even if they are separated by an insulating layer. Any change in the contact potential produces a corresponding change in the threshold voltage. Modern MOS transistors almost always use heavily doped polysilicon gate electrodes. There are two choices of gate material: N+ poly or P+ poly. The substitution of a P+ gate for an N+ gate causes the contact potential to increase by about 1.2V, regardless of backgate doping (Table 11.1). Similarly, the substitution of an N+ gate for a P+ gate causes the contact potential to decrease by about 1.2V. These variations in contact potential will produce equal variations in threshold voltage.

TABLE 11.1 Computed poly gate-to-backgate contact potentials (10^{20}cm^{-3} poly).

Backgate Material	N+ Poly Gate	P+ Poly Gate
N-type, $N_D = 10^{14}$cm^{-3}	–0.36	0.82
N-type, $N_D = 10^{16}$cm^{-3}	–0.24	0.94
N-type, $N_D = 10^{18}$cm^{-3}	–0.12	1.06
P-type, $N_A = 10^{14}$cm^{-3}	–0.82	0.36
P-type, $N_A = 10^{16}$cm^{-3}	–0.94	0.24
P-type, $N_A = 10^{18}$cm^{-3}	–1.06	0.12

A few examples may help to explain the effects of swapping gate materials. Suppose that a certain NMOS transistor develops a threshold voltage of 0.7V using an N+ gate. If this same transistor were to use a P+ gate, it would have a threshold of about 1.9V. Suppose that the corresponding enhancement PMOS develops a threshold voltage of –0.7V using a P+ gate. If this transistor were to use an N+ gate, it would become a depletion device with a threshold of about +0.5V.

The backgate doping concentration also has a strong impact on the threshold voltage. In order to form a channel, enough carriers must be attracted to the surface to invert the silicon. A heavily doped backgate is difficult to invert and the gate must therefore exert a stronger electric field to muster enough carriers to create a channel. The magnitude of the threshold voltage therefore increases with backgate doping. The effect is small at low doping concentrations, but at higher doping levels it actually dominates the expression for the threshold voltage.

The thickness of the gate oxide can also play an important role in determining the threshold voltage of a transistor. A given gate-to-source voltage produces a weaker electric field across a thick gate oxide than across a thin one. Transistors with thick gate oxides are more difficult to invert than ones with thin gate oxides, so increasing the thickness of the gate oxide increases the magnitude of the threshold voltage. For example, the oxide in the field regions is made as thick as possible to raise the thick-field threshold. Unfortunately, thickness alone does not guarantee an adequate thick-field threshold for most processes. Consider the entries in Table 11.2 for a 10kÅ oxide. If the background doping equals $10^{15}cm^{-3}$, then the thick-field threshold only equals about 4V. This is obviously inadequate. Most processes use channel stop implants to raise the doping in the field regions.[8] If the channel stop implant raises the doping under a 10kÅ field oxide to $10^{17}cm^{-3}$, then the thick-field threshold will approach fifty volts. Since both the NMOS and the PMOS backgate doping levels are relatively low, most processes must employ a combination of boron and phosphorus channel stops to ensure that both thick-field thresholds lie well above the nominal operating voltages.

Backgate Doping	100Å	250Å	10kÅ
$10^{14}cm^{-3}$	–0.23	–0.21	0.89
$10^{15}cm^{-3}$	–0.08	–0.02	3.91
$10^{16}cm^{-3}$	0.14	0.35	13.9
$10^{17}cm^{-3}$	0.60	1.31	47.9
$10^{18}cm^{-3}$	1.86	4.28	162
$10^{20}cm^{-3}$	7.94	19.1	747

TABLE 11.2 Typical NMOS threshold voltages as a function of backgate doping and gate oxide thickness (N-type poly gate).[7]

The threshold voltage is also affected by the presence of residual charges within the gate oxide and along the oxide-silicon interface. These residual charges can be divided into three types: fixed oxide charge, mobile oxide charge, and surface state charge. The *fixed oxide charge* Q_f consists of defect sites scattered randomly throughout the oxide film. Gate oxides grown in dry oxygen at relatively low temperatures have very small fixed oxide charges. The fixed oxide charge can drastically increase if holes are injected into the oxide, as happens during oxide breakdown, hot carrier injection, and exposure to ionizing radiation.

The *mobile oxide charge* Q_m consists of positively charged mobile ions such as sodium and potassium.[9] The threshold voltage shift that they produce also depends on their location within the oxide film, and this in turn depends on gate biasing. Consider the case of an NMOS transistor whose gate oxide is contaminated by sodium. When a positive gate-to-source voltage is applied, the mobile ions move away from the positively charged gate electrode and move closer to the negatively charged backgate. As the mobile ions shift closer to the channel region, they exert an increasing effect on it. Thus the movement of the mobile ions causes the NMOS threshold to decrease. Any shift in the threshold voltage of MOS transistors can cause offsets. In order to obtain accurate matching, mobile ions must either be excluded from the process, or they must somehow be immobilized (Section 4.2.2). Modern processing techniques have reduced the magnitude of the mobile oxide charge to negligible proportions.

The *surface state charge* Q_{ss} is concentrated in a thin layer near the oxide/silicon interface. It is generally positive, but its magnitude depends on silicon crystal orientation and annealing conditions. The precise mechanisms responsible for generating the surface state charge are not fully understood, but they are believed to involve mismatches between the molecular structure of the silicon lattice and that of the

[7] The values in this table assume $Q_{ss} = 0$ and $\Phi_{MS} = -0.7V$.

[8] J. D. Sansbury, "MOS Field Threshold Increase by Phosphorus-Implanted Field," *IEEE Trans. on Electron Devices*, Vol. ED-20, #5, 1973, pp. 473–476.

[9] Sodium is the primary mobile ion encountered in silicon processing; lesser contributors include potassium and hydrogen ions. See B. E. Deal, "The Current Understanding of Charges in the Thermally Oxidized Silicon Structure," *J. Electrochem. Soc.*, Vol. 121, # 6, 1974, pp. 198C–205C.

oxide macromolecule.[10] Oxides grown on (100) silicon have less than half the surface state charge density of oxides grown on (111) silicon.[11] All modern MOS processes employ (100) silicon to minimize the impact of surface state charge on the threshold voltages of MOS transistors. Standard bipolar uses (111) silicon to deliberately raise the NMOS thick-field threshold. The surface state charge can also be reduced by annealing the oxide in a reducing atmosphere such as hydrogen or forming gas (a mixture of nitrogen and hydrogen). This anneal presumably offers opportunities for the molecular structure of the oxide to adjust to match that of the silicon. Although it is not possible to entirely eliminate the surface state charge, the use of (100) silicon in combination with proper annealing can minimize threshold voltage variation.

Each of the mechanisms discussed above contributes a small amount of variability to the threshold voltage. With care, threshold voltages can be held to within ±15%. This translates into a variation of ±0.1V in a threshold voltage of 0.7V. Threshold voltages also vary with temperature. The magnitude of this variation depends on the backgate doping and oxide thickness, but it typically ranges from –2mV/°C to –4mV/°C.[12,13] A temperature variation of –2mV/°C over a temperature range of –55 to 125°C produces a threshold variation of about ±0.2V. Combining this with the process variation yields a total variation of about ±0.3V. A transistor with a nominal threshold voltage of 0.7V might actually have a V_t as low as 0.4V or as high as 1.0V. Although it might seem that the minimum threshold could safely be decreased to 0.3V or less, this is actually not the case. MOS transistors continue to conduct small amounts of current when the gate-to-source voltage is less than the threshold voltage due to a mechanism called *subthreshold conduction.* The magnitude of the subthreshold current decreases exponentially, and the gate-to-source voltage must drop at least 0.3 to 0.4V below the threshold in order to reduce the drain current to negligible levels. The magnitude of the nominal threshold voltage must therefore equal at least 0.6 to 0.7V. Transistors with smaller threshold voltages are useful in certain applications, but they cannot be used as switching devices. Transistors used in very low current applications may require nominal threshold voltages of 0.8 to 1.0V to prevent objectionable subthreshold conduction. Alternatively, special circuit design techniques may be used to apply a reverse gate-to-source voltage to ensure proper cutoff.

11.1.2. Parasitics of MOS Transistors

A real-world MOS transistor contains a number of parasitic elements that affect its operation. Perhaps the most important of these are the junctions that isolate the source and drain regions from the backgate. These junctions remain reverse-biased during normal operation, but either or both of them may begin to conduct under certain circumstances. The forward-biased junctions will inject minority carriers into the backgate, at best causing unexpected leakage and at worst triggering latchup.

A complete model of all of the parasitics contained in an MOS transistor would include a number of distributed effects, the discussion of which lies beyond the

---

[10] This discussion is somewhat oversimplified, as several types of charges reside along the interface, including a component of the fixed oxide charge and a variety of interface traps (the so-called *fast surface states* and *slow surface states*). The component due to fixed oxide charge is always positive, while the interface traps may contribute either positive or negative charges (see Muller and Kamins, p. 152ff).

[11] Deal's data shows Q_{SS} values on (100) silicon equal to 20 to 30% of those on (111) silicon: Deal, *ibid.*

[12] R. Wang, J. Dunkley, T. A. DeMassa, and L. F. Jelsma, "Threshold Voltage Variations with Temperature in MOS Transistors," *IEEE Trans. on Electron Devices,* Vol. ED-18, #6, 1971, pp. 386–388.

[13] F. M. Klaasen and W. Hes, "On the Temperature Coefficient of the MOSFET Threshold Voltage," *Solid-state Electronics,* Vol. 29, #8, 1986, pp. 787–789.

scope of this text. Figure 11.3 shows a simplified parasitic model of an NMOS transistor constructed in an N-well CMOS process. The circuit contains a three-terminal NMOS transistor, M_1, that models the intended functionality of the device. Capacitors C_{GD}, C_{GS}, and C_{GB} represent the gate-to-drain, gate-to-source, and gate-to-backgate capacitances, respectively. The gate-to-backgate capacitance, C_{GB}, models the capacitance across the thin gate oxide separating the gate electrode from the backgate diffusion. This capacitance decreases as the transistor approaches inversion, and drops to zero when a channel forms. The gate-to-drain and gate-to-source capacitances, C_{GD} and C_{GS}, consist primarily of the overlap capacitances between the gate electrode and the respective diffusions. These have been greatly reduced by the introduction of self-aligned polysilicon gates, and they now consist mainly of fringing capacitances. The gate-to-source and gate-to-drain capacitances abruptly increase when a channel forms because of the sudden addition of the capacitance between the gate electrode and the channel.[14] The slashes drawn through C_{GD}, C_{GS}, and C_{GB} denote the voltage dependence of these devices.

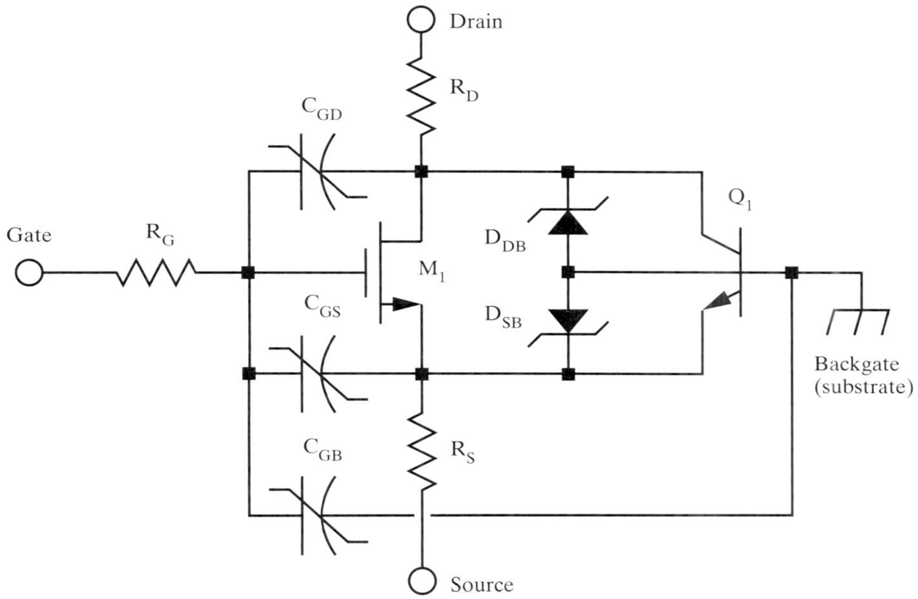

FIGURE 11.3 Simplified parasitic model of an NMOS transistor constructed in N-well CMOS.

Resistors R_G, R_S, and R_D represent the resistance of the gate, source, and drain terminals, respectively. The gate resistance actually forms a distributed network with the three gate capacitances, but this simplified model depicts it as a lumped quantity. The gate resistance has no effect on the DC performance of the transistor, but it does slow the switching speed because it limits the current that is available to charge and discharge capacitances C_{GS}, C_{GD}, and C_{GB}. The gate poly is often silicided to minimize R_G. The drain and source resistances R_D and R_S consist of the Ohmic resistances between the contact and the edge of the source/drain diffusions abutting the channel. These resistances can be minimized by siliciding the surfaces of the source/drain diffusions. A silicided poly gate electrode is sometimes called a *clad gate,* and silicided source/drain regions are also called *clad moats.* Many processes use clad gates, but generally only submicron processes have sufficient transconductance to merit clad moats.

[14] J. E. Meyer, MOS Models and Circuit Simulation, *RCA Rev.,* Vol. 32, March 1971, pp. 42–63.

Diodes D_{DB} and D_{SB} represent the drain-backgate and source-backgate junctions, respectively. The diodes model the junction capacitance added to the drain and source terminals by these junctions, and they also model their avalanche breakdown characteristics. Lateral NPN transistor Q_1 represents one possible path for minority carriers to travel from drain to source (or *vice versa*). Since the NMOS transistor resides in the epi, the minority carriers can also flow to adjacent NMOS source/drain regions, or to adjacent wells. The flow of carriers to adjacent wells can lead to CMOS latchup, as discussed below.

Figure 11.4 shows a simplified parasitic model for a PMOS transistor constructed in an N-well process. Capacitors C_{GS}, C_{GD}, and C_{GB} play the same roles in this device as their counterparts do in an NMOS. Similarly, resistors R_G, R_S, and R_D represent the same terminal resistances in the PMOS as they do in the NMOS. Diodes D_{SB} and D_{DB} represent the junctions between the source/drain regions and the backgate, in this case consisting of an N-well diffusion contacted through a fourth terminal. If either of these junctions forward-biases into the well, then the minority carriers will travel either to the other source/drain diffusion or to the substrate. Lateral PNP Q_1 represents the minority carrier conduction path from drain to source (or *vice versa*). Lateral PNP transistors Q_2 and Q_3 represent the minority carrier paths from the source/drain diffusions to the substrate.

FIGURE 11.4 Simplified parasitic model of a PMOS transistor constructed in N-well CMOS.

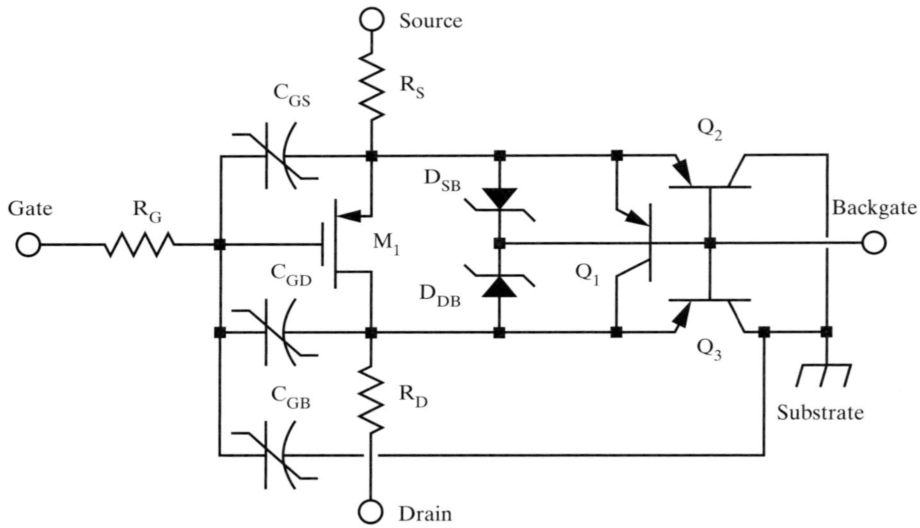

Breakdown Mechanisms

Several different mechanisms limit the operating voltage of MOS transistors. One of these corresponds to the V_{CER} breakdown mechanism seen in bipolar transistors (Section 8.1.2). For purposes of discussion, assume that the gate and backgate electrodes both connect to the source. As the drain-to-source voltage rises, it eventually reaches a point where the drain-backgate junction begins to break down. Avalanche multiplication injects large numbers of majority carriers into the lightly doped backgate, causing it to debias. The source-backgate junction begins to inject minority carriers into the backgate as soon as it forward-biases. Most of these minority carriers flow across to the drain, where they stimulate further avalanche multiplication. This beta multiplication process causes the breakdown voltage of the MOS transistor to snap back from the initial *trigger voltage* to a lower *sustain voltage* (Figure 11.5). Short-channel transistors have somewhat lower breakdown volt-

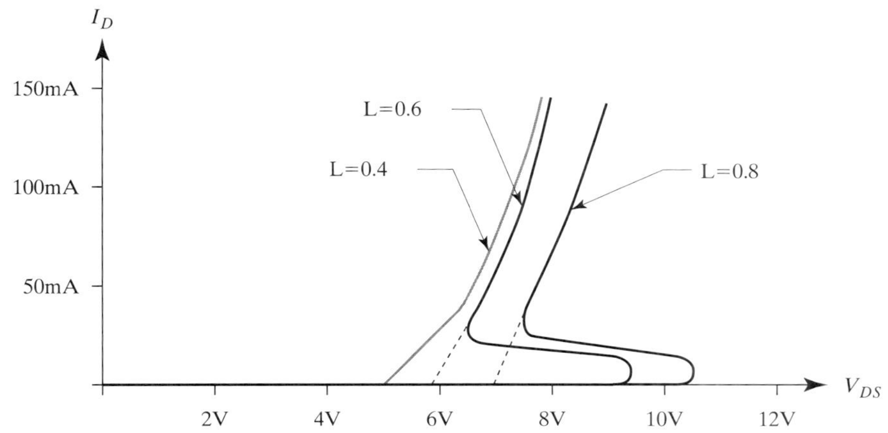

FIGURE 11.5 Breakdown characteristics of short-channel NMOS transistors. The solid curves show the breakdown characteristics in cutoff, and the dotted lines show the change in characteristics when in conduction.

ages because the narrower basewidth of their parasitic lateral bipolar transistor raises its gain and enhances the beta multiplication process.

Once an MOS transistor avalanches, most of its drain current flows through the parasitic bipolar transistor rather than through the MOS channel.[15] The device therefore becomes vulnerable to hot spot formation and current focusing. These mechanisms can severely restrict the transistor's ability to withstand transient overloading. Localized conduction is often triggered by the presence of sharp corners in the channel that cause electric field intensification and premature avalanche breakdown. The designer should attempt to eliminate as many of these corners as possible. Circular annular MOS devices (Section 11.2.6) are particularly robust because their channels have no corners at all. The conventional finger-style transistor has corners at either end of each channel, but these are not as vulnerable as one might expect because they lie along the periphery of the device where the temperatures are not as extreme. Switching transistors should not incorporate bends into their channels, because the resulting corners lie inside the device where localized conduction is more likely to result in thermal runaway.

The resistances of the source and drain diffusions help ballast the avalanching MOS transistor. High source/drain sheet resistances and large spacings between the contacts and the adjacent channel enhance the effectiveness of the ballasting. Silicidation of the source/drain regions eliminates most of the ballasting and renders the transistors more vulnerable to transient overloads. One can sometimes employ a silicide block mask to prevent the silicide from abutting the gate in order to introduce at least a small amount of ballasting. Additional ballasting can be generated by spacing the source/drain contacts further away from the channel.

Short-channel transistors sometimes experience another form of breakdown called *punchthrough*. The source/drain regions are heavily doped, so the depletion regions surrounding them extend primarily into the lightly doped backgate. The drain depletion region widens as the drain-to-source voltage increases. If the drain/backgate junction does not avalanche first, the drain depletion region will eventually extend entirely across the channel to touch the source depletion region. Punchthrough breakdown does not exhibit snapback because it does not activate the parasitic lateral bipolar transistor. The curves of Figure 11.5 highlight the differences between punchthrough (in the 0.4μm device) and avalanche breakdown (in

---

[15] F.-C. Hsu, P.-K. Ko, S. Tam, C. Hu, and R. S. Muller, "An Analytical Breakdown Model for Short-Channel MOSFETs," *IEEE Trans. on Electron Devices,* ED-29, #11, 1982, pp. 1735–1740.

the 0.6 and 0.8μm devices). The snapback characteristic of avalanche breakdown only appears if the transistor is in cutoff. Even low levels of subthreshold conduction produce enough beta multiplication to obscure the snapback characteristic (as shown by the dotted lines in Figure 11.5). Low-voltage isolated MOS transistors may also exhibit punchthrough breakdown between the drain and the substrate. In N-well CMOS processes, only the PMOS transistor normally experiences this form of punchthrough. This problem can be solved by using a deeper well diffusion, but this will result in larger spacings between components. Some low-voltage processes use high-energy implants called *punchthrough stops* to increase the doping at the bottom of the well to prevent vertical punchthrough. Alternatively, the well can be formed using a high-energy implant to produce a peak doping concentration deep beneath the surface of the silicon. Such a *retrograde well* has characteristics similar to those of a regular well augmented by a punchthrough stop.[16]

The thin gate dielectric is also vulnerable to a third form of breakdown. If the voltage across the gate oxide rises beyond a certain point, then avalanche generation begins to occur within the oxide itself. Holes generated by the avalanche process become trapped in the oxide, producing a positive charge that increases the electric field across the oxide. Below a certain field intensity, electron-hole recombination stabilizes the magnitude of the positive charge. Above this critical field intensity, the positive charge continues to grow and runaway conduction overheats and destroys the gate oxide.[17] This mechanism is called *dielectric breakdown,* or more commonly, *oxide rupture.* Unlike avalanche and punchthrough, oxide rupture is catastrophic and irreversible. The oxide rupture voltage places an effective limit on the gate-to-source voltage that an MOS transistor can withstand. Unless special precautions are taken in the design of the transistor, it also limits the achievable gate-to-drain voltage rating (see Section 12.1).

Oxide dielectrics may eventually be destroyed by electric field intensities that are somewhat lower than those required to trigger immediate rupture. Avalanche generation produces hot carriers that damage the integrity of the gate oxide. The damaged gate oxide allows larger currents to flow, producing an ever-increasing flow of hot carriers. At some point, the generation rate becomes so great that electron-hole recombination cannot keep pace, and catastrophic oxide rupture quickly ensues. The total amount of charge required to induce catastrophic failure remains roughly constant regardless of the magnitude of the currents involved, although it decreases at higher temperatures.[18] This mechanism is called *time-dependent dielectric breakdown* (TDDB). If the avalanche currents are relatively high, TDDB can occur in a matter of seconds. On the other hand, if the avalanche currents are very low, TDDB may require months or even years. The existence of TDDB explains why the voltage across the gate oxide must not exceed a small fraction of the actual rupture voltage.

MOS transistors are also subject to a fourth breakdown mechanism. The electric field across the pinched-off portion of an MOS channel can become very intense. The carriers flowing across the pinched-off region accelerate to very high velocities and become so-called *hot carriers.* Some of these carriers collide with the lattice and recoil out of the channel. Most of them travel into the backgate, and eventually contribute to the backgate current. Some of the hot carriers also travel into the gate

---

[16] R. D. Rung, C. J. Dell'oca, and L. G. Walker, "A Retrograde p-well for Higher Density CMOS," *IEEE Trans. on Electron Devices,* Vol. ED-28, #10, 1981, pp. 1115–1119.

[17] P. Soloman, "Breakdown in silicon oxide—A review," *J. Vac. Sci. Technol.,* Vol. 14, #5, 1977, pp. 1122–1130.

[18] G. A. Swartz, "Gate Oxide Integrity of NMOS Transistor Arrays," *IEEE Trans. on Electron Devices,* Vol. ED-33, #11, 1986, pp. 1826–1829.

oxide. Although most of these eventually return to the silicon, a few become permanently trapped and contribute to a slowly increasing fixed oxide charge (Section 4.3.1). As the transistor continues to operate, the accumulating fixed oxide charge causes the threshold voltage to gradually shift. This effect can easily disrupt matching between MOS transistors operating under different biases. If the threshold voltage shifts too far, the transistor may not even be able to switch on and off. Short-channel transistors are especially susceptible to hot-carrier generation because the high backgate doping levels required to prevent punchthrough shorten the pinched-off portion of the channel. The resulting electric fields produce hot carriers at lower voltages than would occur in a transistor with a lightly doped backgate. Transistors acting as switches are less susceptible to hot carrier generation because they normally operate either in the linear region or in cutoff. MOS transistors used in analog circuitry are more vulnerable, as these devices often operate continuously in the saturation region.

Any given MOS transistor will not necessarily experience all of these breakdown mechanisms. Long-channel transistors are usually limited by avalanche breakdown, oxide rupture, and TDDB. Short-channel transistors are usually limited by punchthrough, oxide rupture, and TDDB. Transistors operating for long periods of time under high drain-to-source voltages may also experience hot carrier-induced threshold voltage shifts.

CMOS Latchup

When a source/drain diffusion forward-biases into the backgate, it injects minority carriers that can flow to the reverse-biased junctions of adjacent devices. The exchange of minority carriers between adjacent NMOS and PMOS transistors can trigger *CMOS latchup* (Section 4.4.2). Minority-carrier guard rings can prevent latchup, but they are not necessarily easy to construct in CMOS processes.

The absence of NBL and deep-N+ makes it impossible to construct effective blocking guard rings in a pure CMOS process. One can still construct hole- and electron-collecting guard rings using PMoat and NMoat, although the collection efficiency of these shallow diffusions usually leaves much to be desired. Any PMOS transistor that can inject minority carriers into its well should be surrounded by a hole-collecting guard ring constructed of PMoat (Figure 11.6A). This guard ring should connect to substrate potential to reverse-bias the PMoat/N-well junction as

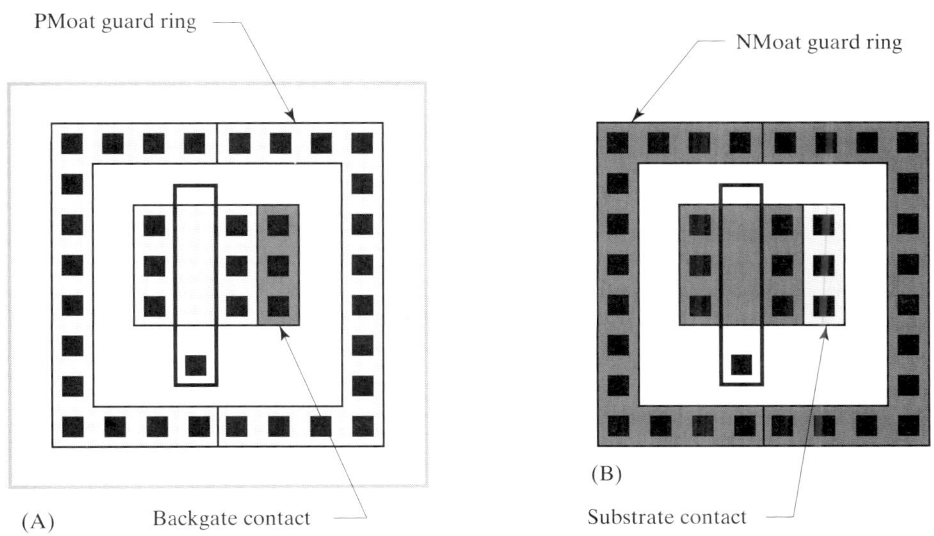

FIGURE 11.6 Examples of guard rings for protecting N-well CMOS transistors: (A) PMOS with a hole-collecting guard ring and (B) NMOS with an electron-collecting guard ring.

strongly as possible. The guard ring collects a percentage of the minority carriers injected laterally by the enclosed PMOS transistor. Although this reduces the lateral flow of carriers toward adjacent devices, it does not stop holes from traveling downward to the substrate. CMOS processes usually employ a P+ substrate to minimize debiasing.

Any NMOS transistor that can forward-bias into the substrate should be surrounded by an electron-collecting guard ring. These guard rings are usually constructed from NMoat rather than from N-well (Figure 11.6B). The deeper well diffusion intercepts a larger fraction of the carriers, but its large vertical resistance makes it prone to debiasing unless it connects to a relatively high-voltage supply. NMoat has a lower collection efficiency, but it is virtually immune to debiasing. NMoat guard rings are sometimes connected to a positive power supply to help drive the depletion region deeper into the substrate to enhance collection efficiency. In low-voltage processes, NMoat guard rings should be connected to substrate potential to minimize hot-carrier generation in their depletion regions (Section 13.2.3).

Guard rings, by themselves, cannot provide total latchup immunity. The flow of even a few minority carriers around the guard rings will trigger parasitic bipolar conduction if not for the presence of backgate contacts. Backgate contacts remove the collected carriers and prevent them from biasing the parasitic lateral bipolar transistors into conduction. Section 11.2.7 discusses the design of backgate contacts in greater detail.

11.2 SELF-ALIGNED POLY-GATE CMOS TRANSISTORS

Most modern CMOS and BiCMOS processes are designed to produce self-aligned poly-gate transistors. Figure 11.7 shows a layout and cross section of a simple self-aligned poly-gate NMOS. The backgate of this transistor consists of a P– epitaxial layer grown on a P+ substrate. The areas between adjacent transistors are called *field regions.* LOCOS oxidation covers these with a *thick-field oxide* that helps suppress parasitic channel formation. The nitride oxidation mask prevents thick oxide from growing in the *moat regions* where transistors will eventually reside. After the removal of the nitride, the moat regions are re-oxidized to form the *thin gate oxide* of the MOS transistors. Doped polysilicon is then deposited on top of the gate oxide to form the gate electrodes of the MOS transistors. After the poly has been patterned, a low-energy arsenic implant produces the source and drain regions of the transistors. This *N-source/drain* (NSD) implant does not possess enough energy to penetrate the polysilicon or the thick-field oxide. The NSD only penetrates the thin gate oxide regions bounded by the poly gate and the thick-field oxide. The resulting source/drain regions are said to *self-align* to the poly gate and the thick-field oxide.

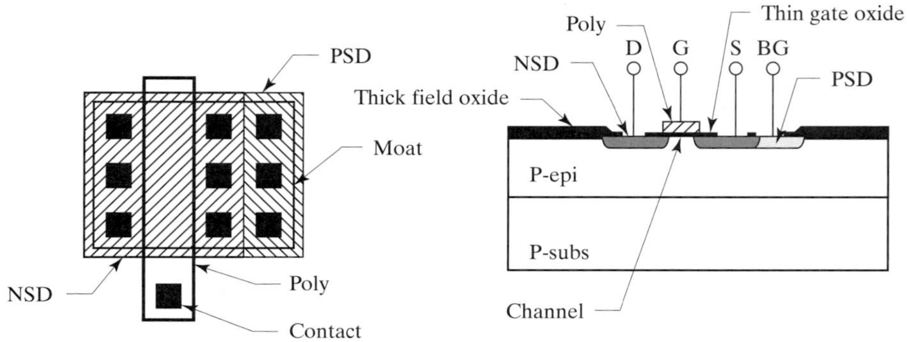

FIGURE 11.7 Layout and cross section of a simple self-aligned poly-gate NMOS transistor.

Next, a *P-source/drain* (PSD) implant is performed. This implant gains its name from the role it plays in the construction of PMOS transistors. The NMOS transistor uses PSD to contact the lightly doped P-epi backgate. A brief anneal activates the source/drain implants and completes the formation of the transistors.

Early MOS processes used aluminum to form the gate electrodes. Aluminum-gate processes are inferior to polysilicon-gate processes in several respects. Aluminum cannot withstand the temperatures required to anneal the source/drain implants, so it must be deposited after implantation. This precludes the self-alignment of the source/drain diffusions, so these implants must overlap the gate by an amount sufficient to account for misalignment. These overlaps greatly increase the gate-to-source capacitance, C_{GS}, and the gate-to-drain capacitance, C_{GD}, which in turn greatly reduces the switching speed of the transistor. The overlap capacitances of a poly-gate transistor are much smaller because the source and drain regions self-align to the gate.

Some attempts have been made to construct gate electrodes from refractory metals such as tungsten. These materials facilitate self-alignment because they can withstand the temperatures required to anneal the source/drain implants. Despite this advantage, refractory-metal gates have not enjoyed widespread success because they still exhibit threshold voltage variations caused by mobile ions and by variations in contact potential. Poly gates are preferred because they provide much more stable and reproducible threshold voltages (Section 4.2.2).

11.2.1. Coding the MOS Transistor

A simple N-well CMOS process requires a total of seven masks: N-well, moat, poly, NSD, PSD, contact, metal, and protective overcoat. The layout database contains the geometric information required to construct each of these seven masks. In the simplest database, the geometries for each mask are drawn upon a different layer. Figure 11.8A shows the layout of an NMOS transistor following this approach.

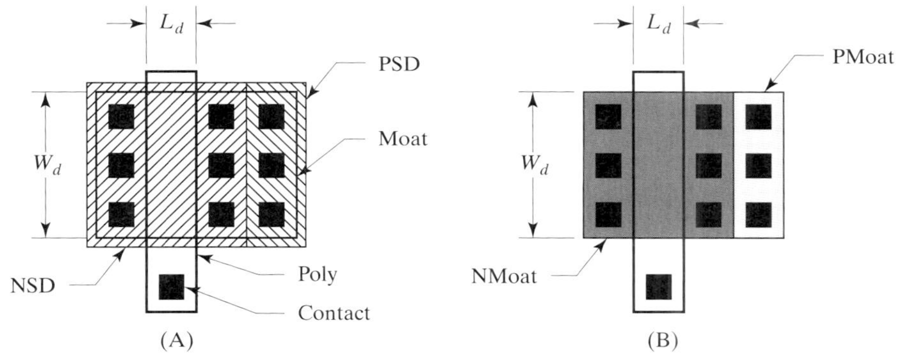

FIGURE 11.8 An NMOS transistor can be coded (A) using NSD, PSD, and moat mask layers or (B) using NMoat and PMoat coding layers.

Figure 11.8B shows another way to draw the same transistor. Two new layers called *NMoat* and *PMoat* are used to generate the NSD, PSD, and moat masks. A geometry placed on the NMoat layer produces corresponding geometries on both the moat and the NSD masks. The NSD geometry is automatically oversized to account for misalignment.[19] A geometry drawn on the PMoat layer creates corresponding geometries on the moat and PSD masks, and the PSD geometries are again automatically oversized.

---

[19] Although the moat geometries could be undersized instead, this would affect the drawn width of the devices and is therefore a less intuitive option.

The NSD, PSD, and moat layers used to construct the layout in Figure 11.8A are called *mask layers* because the information they contain is transferred directly to the corresponding masks without any intermediate processing. NMoat and PMoat are called *drawing layers* or *coding layers* because they are used only during the drawing (or *coding*) of the layout. The data on the coding layers must pass through a series of geometric transformations to generate the actual mask data.

Coding layers simplify the layout in a number of ways. Data entry takes less time because the layout contains fewer geometries, displays and plots become less cluttered, verification programs run more quickly, and databases consume less space. Each of these advantages may seem relatively insignificant, but together they make a strong argument for the use of coding layers.

The process of transforming coding data into mask data sometimes produces unexpected results on complicated geometries. After the NSD and PSD geometries are oversized, they must be trimmed so that the two implants do not overlap along abutting edges (Figure 11.8A). The trimming algorithm is relatively simple and straightforward as long as the NMoat and PMoat geometries do not contain bends or notches. The algorithm becomes much more complex if it must handle these special cases. The more complex the algorithm becomes, the more likely it is to produce unexpected results under circumstances that are not anticipated by its designer. The only way to entirely eliminate problems of this sort is to avoid the use of coding layers.

The choice between mask layers and coding layers is by no means an easy one. Many designers appreciate the simplifications introduced by coding layers, but they do not always understand the accompanying problems. The people responsible for writing verification and pattern generation programs must ultimately decide whether to use coding layers or to avoid them. This text uses NMoat and PMoat coding layers because they significantly simplify the illustrations.

Width and Length

The length of the transistor equals the distance between the source diffusion and the drain diffusion. The *drawn length* L_d of a self-aligned transistor equals the distance across the poly gate from source to drain as measured in the layout database. The *effective length* L_{eff} of the transistor may be slightly larger or smaller than the drawn length due to overetching, underetching, straggle, outdiffusion, and other factors.[20] These corrections remain relatively constant regardless of the dimensions of the gate, so L_{eff} can be approximated by

$$L_{eff} \cong L_d + \delta L \qquad [11.7]$$

where δL is constant for any given process. The value of δL is usually less than 1 μm, so it primarily affects short-channel devices. Submicron transistors, in particular, exhibit substantial differences between drawn and effective channel lengths. In such cases, the channel length, L, used in the Shichman-Hodges equations 11.2 must equal the effective channel length L_{eff} and not the drawn channel length L_d.

The poly gate must overhang both ends of the source/drain region to prevent the source and drain from shorting together. The width of a self-aligned MOS transistor is therefore set by the moat mask rather than the poly mask. The *drawn width* W_d equals the width of the moat geometry in the layout database or, equivalently, the width of the NMoat or PMoat geometry (Figure 11.8). The *effective width* W_{eff} varies slightly due to straggle, outdiffusion, the presence of the bird's beak, and

[20] G. Massobrio and P. Antognetti, *Semiconductor Device Modeling with SPICE,* 2nd ed. (New York: McGraw-Hill, 1993), pp. 279–283.

other factors. These corrections remain relatively constant regardless of the moat dimensions, so the effective width W_{eff} can be approximated by

$$W_{eff} \cong W_d + \delta W \qquad [11.8]$$

where δW is a constant for any given process. The value of δW is also usually less than 1μm.

11.2.2. N-well and P-well Processes

The NMOS transistors in Figure 11.8 are fabricated in a P– epitaxial layer deposited on a P+ substrate. The heavily doped substrate improves latchup immunity, but it introduces an additional process step. Providing that other measures have been taken to prevent latchup, the epitaxial layer can be eliminated and the transistors built directly into the substrate. Many early processes used this approach to minimize fabrication costs, and some digital processes continue to do so today. Most analog CMOS processes use an epitaxial layer because the epi doping can be controlled very accurately, and therefore the threshold voltages of transistors constructed in the epi vary less than the thresholds of those constructed in the substrate.

A P-epi allows the construction of NMOS transistors, and an N-epi allows the construction of PMOS transistors, but neither allows the construction of both simultaneously. In order to build complementary transistors, another diffusion must be added to counterdope the backgate region of one transistor or the other. If a P-epi is used, then a deep, lightly doped N-type diffusion must be added for PMOS transistors (Figure 11.9A). If an N-epi is used, then a deep, lightly doped P-type diffusion must be added for NMOS transistors (Figure 11.9B). These deep diffusions are commonly called *wells*. An N-type well is called an *N-well* and a P-type well is called a *P-well*. Most processes use either an N-well or a P-well, but not both. In these *single-well* processes, one type of transistor or the other resides in the epi. In an N-well process, the NMOS occupies the epi and the PMOS the N-well. In a P-well process,

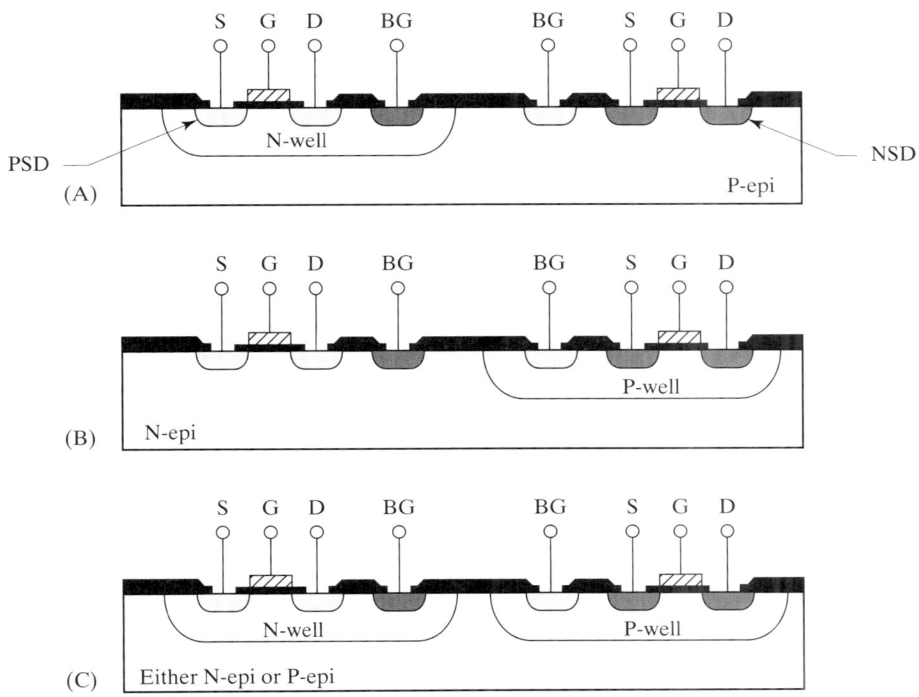

FIGURE 11.9 Three types of CMOS processes: (A) N-well, (B) P-well, and (C) twin-well.

the PMOS occupies the epi and the NMOS the P-well. Some processes include both an N-well and a P-well (Figure 11.9C). In such a *twin-well process,* the NMOS is formed in the P-well and the PMOS is formed in the N-well.

Single-well processes are simpler and cheaper than twin-well processes, but submicron processes sometimes require two wells. As the channel length of the transistor decreases, the backgate doping must increase to prevent punchthrough breakdown. The counterdoping mechanism that creates the well becomes difficult to control on heavily doped substrates. Heavy counterdoping also causes a slight reduction in carrier mobility and a more significant reduction in well-substrate breakdown voltages. These considerations force most submicron processes to use twin wells driven into a lightly doped epi.

The choice of epi also has several consequences. In a single-well process, transistors formed in the epi share a common backgate connection, while transistors formed in wells can be isolated from one another. Although the separate wells consume additional die area, isolation offers an extra degree of design flexibility. An N-well process produces isolated PMOS transistors, while a P-well process produces isolated PMOS transistors. A similar consideration affects the choice of the epi type for a twin-well process. If a P-epi is chosen, then this epitaxial layer shorts all of the P-wells on the die together, and all of the NMOS transistors share a common backgate connection. Similarly, if an N-epi is chosen, the epi shorts the N-wells together and all of the PMOS transistors share a common backgate connection.

N-well processes are favored over P-well processes for several reasons. Most schematics reference their power supplies to a common ground potential. If all of the power supplies deliver positive voltages with respect to ground, as is often the case, then ground becomes the most negative node in the circuit. The substrate of an N-well process can connect to this common ground, but the substrate of a P-well process must connect to the highest-voltage power supply. In multiple-supply systems, it is difficult to ensure that one power supply will always generate a higher voltage than the others, especially during start-up and shut-down. P-well processes are thus poorly suited to multiple-supply applications. One could theoretically reference multiple negative voltages to a positive ground, but this is rarely done in practice.

The mobility of carriers in the counterdoped well will be slightly less than the mobility of carriers in the epi. Since electrons are more mobile than holes, the NMOS transistor has a higher transconductance than the PMOS transistor. Many circuit designers prefer to degrade the performance of the already-inferior PMOS rather than reduce the superior transconductance of the NMOS. This consideration also favors the use of an N-well process.

BiCMOS processes generally employ a P-epi on a P-substrate because this combination simplifies the isolation of the bipolar transistors. The NPN transistor uses the lightly doped N-well as a collector region and the P-epi as isolation, a practice called *collector-diffused isolation* (CDI). Most analog BiCMOS processes are either N-well processes or twin-well processes built on a P-type epi.

N-well BiCMOS processes can construct isolated NMOS transistors as well as isolated PMOS transistors. The isolated NMOS uses a combination of NBL and deep-N+ (or N-well) to isolate the section of P-epi forming the backgate of the transistor (Figure 11.10). The NBL severs the isolated P-epi tank from the P-substrate beneath, and the deep-N+ (or N-well) ring isolates it from adjacent P-epi regions.[21] In order to ensure complete isolation, the ring must contain no gaps and the NBL

---

[21] E. Bayer, W. Bucksch, K. Scoones, K. Wagensohner, J. Erdeljac, and L. Hutter, "A 1.0μm Linear BiCMOS Technology with Power DMOS Capability," *BCTM Proceedings,* 1995, pp. 137–141.

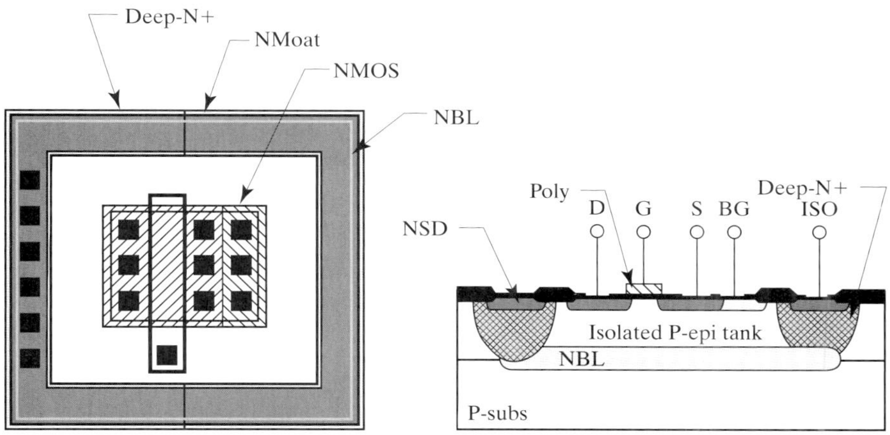

FIGURE 11.10 Layout and cross section of an isolated NMOS in an N-well analog BiCMOS process.

must overlap it sufficiently to allow for misalignment. If deep-N+ is available, then it is usually used instead of N-well because it produces a lower-resistance connection to the NBL. The N-well/epi junction usually has a higher breakdown voltage than the deep-N+/epi junction, so devices that must operate at a high voltage relative to the substrate will normally use N-well isolation rings, either alone or surrounding a deep-N+ isolation ring.

The isolation ring must connect to a voltage equal to or greater than that applied to the isolated P-epi tank. The source/drain regions easily punch through the lightly doped tank, so most isolated NMOS transistors cannot withstand the application of more than a few volts drain-to-isolation or source-to-isolation. These operating voltages increase slightly if the isolation ring connects to a potential midway between that of the backgate and that of the source/drain diffusions. This configuration allows a portion of the depletion region surrounding the isolation/NBL region to intrude into the lightly doped outer fringes of the NBL. Because the NBL dopant diffuses very slowly, the degree of field relief offered is slight and the improvement in operating voltage amounts to only a few volts.

The backgate region of the isolated NMOS consists of a thin, lightly doped P-epi layer. This layer has much more lateral resistance than the P-epi/substrate sandwich comprising the backgate of the nonisolated NMOS. Rapidly slewing signals can momentarily forward-bias the source/drain regions of an isolated NMOS into its backgate. Most of the injected minority carriers flow to the isolation ring, but a few travel from source to drain (or *vice versa*). If the transistor operates at a relatively large drain-to-source potential, then minority carrier injection can trigger V_{CER} breakdown and snapback. Adequate backgate contact minimizes the magnitude of the snapback and the likelihood of triggering it (Section 11.2.7).

11.2.3. Channel Stops

Self-aligned poly-gate MOS transistors form wherever poly intersects PMoat or NMoat geometries. Under certain circumstances, MOS transistors can also form underneath the thick-field oxide. These unwanted *parasitic transistors* interfere with the operation of the integrated circuit unless they are somehow suppressed.

The threshold voltages of the parasitic transistors can be raised by implanting the field regions with a suitable dopant before growing the thick-field oxide. A doping concentration of 10^{17} atoms/cm^3 beneath a 10kÅ field oxide will provide a thick-field threshold of nearly 50V (Table 11.2). This thick-field threshold will provide a comfortable safety margin for a 30V process. Implants used to raise the doping of field regions are called *channel-stop implants*.

The thick-field thresholds can also be raised by deliberately introducing surface state charges. Standard bipolar processes produce large surface state charges by using (111)-oriented silicon in combination with a final oxidizing anneal. The resulting positive charge raises the magnitude of the PMOS thick-field threshold and lowers the magnitude of the NMOS thick-field threshold. Standard bipolar uses a heavily doped P+ isolation system to suppress NMOS parasitic channel formation, and relies on the surface state charge to elevate the PMOS thick-field threshold. Thick-field thresholds of 40V can routinely be achieved by this means.

CMOS processes cannot tolerate the introduction of excess surface state charge because its effects are not limited to field regions. The magnitude of the surface state charge varies with processing conditions, and this, in turn, causes threshold voltage fluctuations. CMOS processes use (100)-oriented silicon and conduct an inert anneal to minimize the residual surface state charge. This anneal usually occurs in conjunction with the deposition of the protective overcoat. Many designers do not appreciate the importance of proper annealing. If a wafer is removed from processing before nitride deposition, then an anneal must be conducted in order to stabilize the threshold voltages and to sinter the contacts. Because this anneal does not necessarily duplicate the conditions of nitride deposition, the threshold voltages of no-nitride wafers do not always correspond to those of the finished product.

Most CMOS processes use two complementary channel-stop implants to suppress both NMOS and PMOS parasitic channels. All P-type field regions receive the P-type channel-stop implant to increase the magnitude of the PMOS thick-field threshold. Similarly, all N-type field regions receive the N-type channel stop implant to increase the magnitude of the NMOS thick-field threshold. Several methods have been devised to ensure the proper alignment of these channel-stop implants. The most common techniques involve either a blanket boron channel-stop implant and a patterned phosphorus channel-stop implant, or *vice versa*. Figure 3.23 shows the steps required to produce a blanket boron channel-stop implant and a patterned phosphorus channel-stop implant in an N-well CMOS process. Figure 11.11 shows the results.

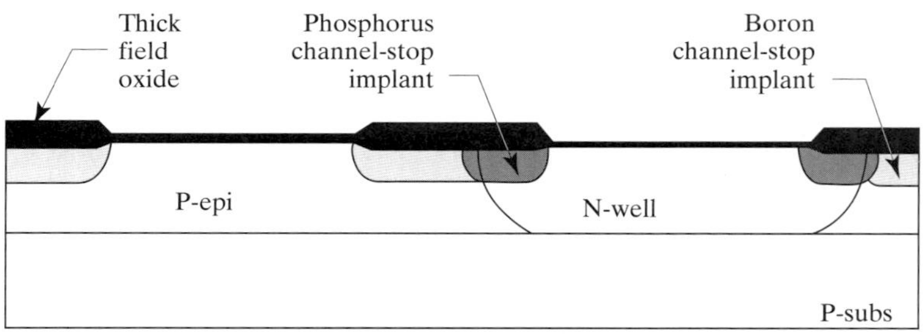

FIGURE 11.11 N-well CMOS wafer with boron and phosphorus channel-stop implants.

The channel-stop implants diffuse downwards during the long, high-temperature field oxidation. Lateral outdiffusion causes the two implants to intersect along the edges of the N-well. This zone of intersection limits the breakdown voltage of the N-well/P-epi junction. Fortunately, the channel-stop implants are sufficiently deep and lightly doped that the breakdown voltage lies well above normal operating voltages, and a 15V CMOS process can generally obtain an N-well/P-epi breakdown voltage in excess of 30V.

The patterned channel-stop implant requires a masking step. The locations receiving it depend on the type of process selected and which of the two implants is patterned. A patterned phosphorus channel stop in N-well CMOS goes into all N-well regions not inside moat. Although it is possible to draw the geometries for the channel-stop mask, it is much easier to generate the mask from existing coding layers. For example, the phosphorus channel-stop mask can be generated from the data present on the N-well, PMoat, and NMoat coding layers.

Submicron processes can sometimes dispense with one or both channel-stop implants. As the channel length shrinks, the backgate doping concentration rises and the operating voltage drops. The wells of a submicron process may therefore contain enough dopant to raise the thick-field threshold above the relatively low operating voltage of the process.

The channel-stop implants are designed to raise the NMOS and PMOS thick-field thresholds above the maximum operating voltage of the process. Many layout designers assume that this provides unconditional protection against parasitic channel formation, but in practice it does not. The maximum operating voltage of a process is usually set by either the gate oxide rupture voltage or by the breakdown voltage of the source/drain implants into their respective backgates. Certain components can operate at much higher voltages; for example a poly resistor is limited only by the breakdown of the thick-field oxide, which can easily reach several hundred volts. Similarly, the well-epi junctions can usually withstand several times the operating voltage of the process. Section 4.3.2 discusses the techniques used to suppress parasitic channel formation in circuits operating at or above the thick-field threshold.

11.2.4. Threshold Adjust Implants

Ideally, the threshold voltages of enhancement transistors should lie between 0.6 and 0.8V. The *native,* or *natural,* thresholds are determined by the doping of the gate and backgate and by the thickness of the gate oxide. Most processes dope the gate poly with phosphorus, reducing the magnitude of the NMOS threshold and increasing that of the PMOS. The natural NMOS threshold usually lies well below 0.6V and the magnitude of the natural PMOS threshold well above 0.8V. Over extremes of process and temperature, the NMOS goes into depletion, and the magnitude of the PMOS threshold exceeds 1.5V (Table 11.3). These thresholds are completely unacceptable for most applications.

Worst-case Corner	Natural NMOS	Adjusted NMOS	Natural PMOS	Adjusted PMOS
Minimum	–0.10	0.50	–1.75	–1.15
Nominal	0.20	0.80	–1.40	–0.80
Maximum	0.55	1.15	–1.10	–0.50

TABLE 11.3 Worst-case natural and adjusted threshold voltages for a typical 10V N-well CMOS process.[22]

The threshold voltage of an MOS transistor can be altered by implanting its channel region. A P-type implant produces a positive threshold shift and an N-type implant a negative one. NMOS and PMOS transistors using phosphorus-doped gate poly both require a positive threshold shift. Providing that the initial backgate dopant concentrations have been properly chosen, a single boron implant can adjust the thresholds of both types of transistors. This boron implant is called the *threshold adjust implant,* or simply the *threshold adjust.* Transistors receiving this implant are called *adjusted* transistors, while those not receiving it are called *native,* or *natural,*

[22] These values assume the listed natural V_t targets, ±0.15V threshold control and a –2mV/°C temperature coefficient.

transistors. The threshold adjust implant does not necessarily require a photomask. If the implant is performed across the entire wafer immediately after stripping the LOCOS nitride, then it appears in every moat region. This blanket implant simultaneously adjusts the threshold voltage of every MOS transistor to the targeted value. This practice precludes the fabrication of natural devices.

Circuit designers can often improve the performance of their circuits if they have access to both natural and adjusted transistors. Many processes therefore offer natural transistors as a process option. This option requires a single mask, properly called the *threshold adjust implant mask,* but more often referred to as a *natural* V_t *mask*. The associated coding layer has been given many names; in this text it is called *NatVT*.[23] This layer must be coded around the gate region of each natural transistor (Figure 11.12). The NatVT figure should slightly overlap the channel region to allow for misalignment and lateral outdiffusion. If a design does not use any natural transistors, then the NatVT mask can usually be omitted. A few processes may use the NatVT mask to fabricate certain other devices, such as Schottky diodes.

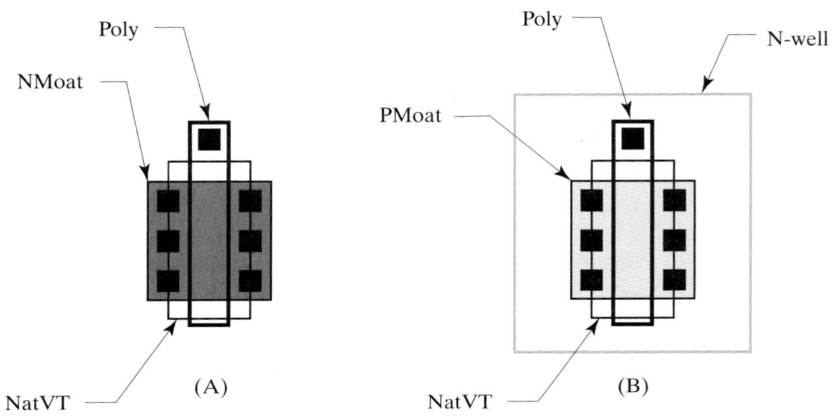

FIGURE 11.12 Layout of natural transistors using NatVT: (A) natural NMOS and (B) natural PMOS.

Although many processes have successfully used a single boron threshold adjust implant, submicron processes often require a different strategy. The boron implant reduces the magnitude of the PMOS threshold, but it also has the undesired effect of producing a *buried channel.* In order to obtain a large threshold shift, so much boron must be implanted that it actually inverts a thin layer of the backgate. The inversion region appears beneath the surface because this is where the peak doping concentration from the implant occurs. The buried channel lies so close to the surface that the electric field produced by the contact potential of the gate electrode inverts it, and it does not interfere with the normal operation of the transistor. This situation changes in a submicron transistor because the backgate doping increases as the channel length decreases. The increased doping partially shields the buried channel from the influence of the gate electrode. The gate can no longer fully invert the channel, so the buried channel begins to conduct current. Submicron buried-channel PMOS transistors are therefore somewhat leaky.

The buried channel can be eliminated by using a phosphorus channel-stop implant for the PMOS transistors. Since phosphorus induces a negative threshold shift, the PMOS transistor must begin with a relatively low threshold voltage, which can be achieved by using a boron-doped gate poly (Figure 11.13).

---

[23] The natural V_t mask is also called *NVT,* but this name is also used for the N-type threshold adjust mask used in a dual-doped poly process.

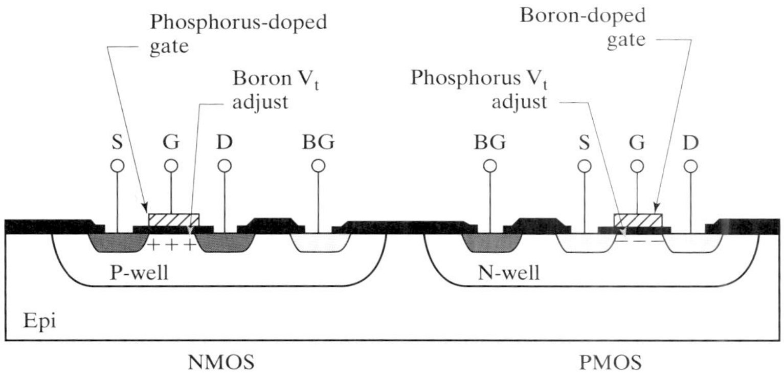

FIGURE 11.13 Cross section of dual-doped poly CMOS transistors.

The processing required to produce *dual-doped poly* CMOS transistors is as follows. After the LOCOS nitride has been stripped, the wafer is patterned using the mask for the *P-type V_t adjust* (PVT). This low-energy boron implant adjusts the NMOS threshold. Next, the wafer is patterned using the mask for the *N-type V_t adjust* (NVT). This low-energy phosphorus implant adjusts the PMOS threshold. After gate oxidation, the gate poly is deposited in a near-intrinsic state. Before etching, it is doped with a patterned boron implant using the *P-type gate poly* (PPoly) mask, followed by a patterned phosphorus implant using the *N-type gate poly* (NPoly) mask.

The most elaborate version of this process requires four new masks (PVT, NVT, PPoly, and NPoly). As long as no natural transistors are required, the P-well mask can be reused for the PVT step, and the N-well mask can be reused for the NVT step. If natural transistors are required, then the well masks cannot be reused in this manner and separate PVT and NVT masks become necessary. A well mask can also be used to define either the PPoly or the NPoly, but not both. Suppose that the N-well mask is used to pattern the NPoly. Both the P-well and the epi regions must receive the PPoly implant, and a special mask must be produced for this purpose. This process therefore requires at least one new mask, and possibly as many as four.

The number of masking steps can be reduced at the cost of compromising performance. For example, the gate poly can be doped with a blanket boron implant and a patterned phosphorus implant. The resulting poly is not quite as heavily doped as that formed by using two separate masking steps. The process can also use a blanket V_t implant followed by a patterned V_t implant of the opposite polarity in order to save one masking step. The transistor that receives both implants has somewhat more threshold voltage variation than it would have had using one implant. If both of these modifications are adopted, then only two masking steps are required instead of four. Even so, the additional steps add cost and complexity, so they are only used for submicron processes that would otherwise exhibit unacceptable leakage due to buried channel formation. In general, operating voltages of 5V or more require well dopings compatible with single-doped poly, while lower operating voltages require dual-doped gate poly.

Selective gate doping produces unwanted poly diodes unless the gate poly is silicided. PN junctions appear wherever the PPoly and the NPoly abut one another. They can be shorted by metal jumpers or, more conveniently, by silicidation. As long as the designer takes care not to block the silicide from locations where PPoly and NPoly abut one another, silicidation automatically shorts all of the poly diodes. Silicidation greatly increases the rate at which dopants diffuse through the poly, so the intersections between PPoly and NPoly must be spaced well away from the gate regions of adjacent MOS transistors.[24] Although the PPoly-NPoly junction does exhibit rectification, poly diodes are not recommended as circuit components because the presence of grain boundaries within the depletion regions causes substantial leakage.

Almost all CMOS processes adjust the NMOS and PMOS threshold voltages. The majority of analog processes offer natural NMOS and PMOS transistors either as part of the baseline process or as process extensions. A few processes offer additional threshold voltage options, such as depletion-mode transistors or low-V_t PMOS transistors. Each such option requires its own threshold adjust implant, formed through an additional masking step. Transistors using these special implants are coded much like natural transistors, except that NatVT is replaced by the layer that codes for the special implant.

11.2.5. Scaling the Transistor

Integrated circuits have become vastly more complex over the past thirty years. The first digital integrated circuits contained ten or twenty transistors; their modern equivalents contain tens of millions. This remarkable increase in complexity has largely been made possible by corresponding reductions in the size of individual transistors. From 1973 to 2000, minimum channel lengths went from 8μm to about 0.2μm.[25] These reductions in size have also improved the performance of the transistors. A set of guidelines called *scaling laws* have been developed that dictate how the various dimensions of an MOS transistor should be reduced to obtain the best performance.

Scaling laws fall into two general categories, both of which presume that width and length are multiplied by a *scaling factor S*. *Constant-voltage scaling* holds the operating voltage of the transistor constant while scaling its dimensions. As the transistor shrinks further and further, it becomes increasingly difficult to avoid hot-carrier generation and punchthrough breakdown. *Constant-field scaling* avoids these problems by reducing the supply voltage to keep the electric fields in the transistor constant regardless of scale. Most modern processes use some variant of constant-field scaling. Table 11.3 shows simplified rules for both constant-voltage and constant-field scaling laws.[26]

As an example, consider a process producing 5V transistors with minimum dimensions of 1μm long by 2.5μm wide using a 250Å gate oxide and a backgate doping concentration of $10^{16}cm^{-3}$. Suppose the channel length of this process is reduced to 0.8μm using constant-field scaling. The scaling factor S equals 0.8μm/1.0μm, or 80%. According to Table 11.4, the scaled transistor should have a minimum width of 2.0μm, a 200Å gate oxide, and a backgate doping of $1.25 \cdot 10^{16}cm^{-3}$. Since processes are generally identified by gate length, the original (100%) process would be considered a 1μm process, while the 80% shrink would be an 0.8μm process.

Shrinking a transistor actually improves its performance. The smaller dimensions reduce parasitic capacitances and increase switching speeds. The *gate delay* of a CMOS process equals the time required for a digital signal to propagate through a representative CMOS gate. As the transistors scale down, the gate delay decreases and the circuit can handle faster switching speeds. Early microprocessors operated

---

[24] Y. P. Tsividis, *Operation and Modeling of the MOS Transistor* (New York: McGraw-Hill, 1988), p. 439.

[25] The 8μm figure is from D. A. Pucknell and K. Eshraghian, *Basic VLSI Design,* 3rd ed. (Sydney: Prentice-Hall Australia, 1994), p. 7. Both figures are approximations of industry practice; much smaller dimensions are possible in a research environment.

[26] These laws have been adapted from Pucknell *et al.,* p. 129.

TABLE 11.4 Constant-voltage and constant-field scaling laws.

Quantity	Constant-voltage	Constant-field
Supply voltage	1	S
Minimum channel width	S	S
Minimum channel length	S	S
Gate oxide thickness	1	S
Backgate doping	$1/S^2$	$1/S$
Gate delay	S^2	S
Power-delay product	S^2	S^3

at clock speeds of 1 to 10MHz; their modern equivalents operate at 100 to 1000MHz.

Not only does a smaller transistor switch faster, but it requires less power to do so. CMOS logic gates require pulses of power to charge and discharge gate capacitances each time they switch. The faster the gate switches, the more transitions occur per second, and the larger the current consumption becomes. The supply current required by a gate can be reduced at the expense of increasing its gate delay. The product of gate delay and power consumption remains approximately constant for any given process. This *power-delay product* decreases as the size of the transistor shrinks. For example, 80% constant-field scaling reduces the power-delay product to about half of its initial value. As this example suggests, even relatively minor decreases in size significantly reduce power consumption. This is fortuitous, since otherwise a microprocessor running at several hundred megahertz would literally melt from its own waste heat.

Scaling laws are frequently applied to existing digital layouts to convert them for use with newer processes. Rather than laboriously recoding the layout, the designer simply runs a program that scales all of the data by a specified amount. This type of scaling is called an *optical shrink* because it produces the same results as photoreducing the existing mask set. Optical shrinks are denoted by the percentage scaling factor used to transform the data from their original, or *drawn*, dimensions to their final, or *shrunk*, dimensions. A 100% shrink indicates that the final dimensions equal the drawn dimensions, while an 80% shrink indicates that they equal 4/5 of the drawn dimensions. Figure 11.14A shows a 1μm transistor drawn at 100%. Figure 11.14B shows the same transistor optically shrunk to 80%. The optical shrink scales both the width and the length of the device in a manner consistent with either constant-voltage or constant-field scaling. The process engineers will adjust gate oxide thickness, backgate doping, and other parameters according to the desired type of scaling.

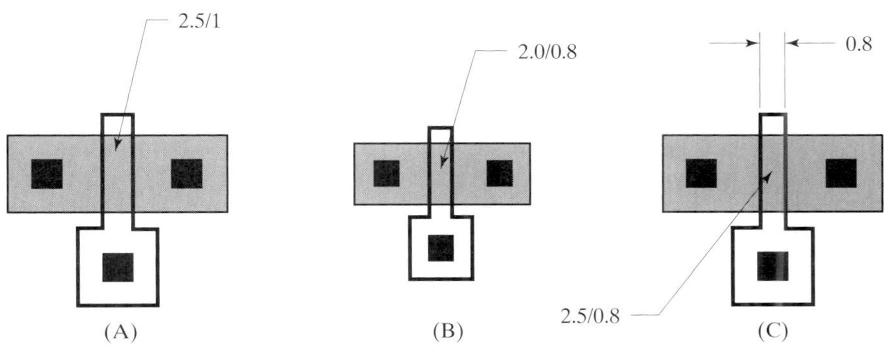

FIGURE 11.14 Examples of scaled MOS transistors: (A) drawn at 100%, (B) optically shrunk to 80%, (C) selectively shrunk to 80% of drawn gate length. The wells have been omitted for clarity.

An optical shrink affects all dimensions equally, but some are more difficult to scale than others. Multilayer metal systems have proved especially difficult to scale to submicron dimensions. Although fine-line metal systems certainly exist, they are very expensive. Many processes selectively scale channel length while retaining the previous dimensions for all other layout rules. Figure 11.14C shows a 1μm transistor whose gate has been shrunk to 0.8μm. The selective gate shrink requires a more complicated set of geometric transformations than a simple optical shrink, but it is still far simpler and quicker than a full relayout. The benefits of a selective gate shrink are somewhat less than for a full optical shrink (Table 11.5), but they are still sufficient to justify selective gate shrinks for many processes.

TABLE 11.5 Scaling laws for selective gate shrinks.

Quantity	Constant-voltage	Constant-field
Supply voltage	1	S
Minimum channel width	1	1
Minimum channel length	S	S
Gate oxide thickness	1	S
Backgate doping	$1/S^2$	$1/S$
Gate delay	S	1
Power-delay product	S	S^2

The scaling laws were originally developed for digital processes. CMOS logic circuits respond quite predictably to scaling, but the same is not true of analog or mixed-signal circuits. No set of predetermined scaling laws can comprehend the full complexity of analog circuit design. Indiscriminate scaling usually causes analog circuits to fail parametric specifications, and in some cases it may cause outright malfunctions. For example, an 80% optical shrink reduces all capacitors to 64% of their former values. Since analog designs rely on capacitors to stabilize feedback loops, a reduction in capacitance can actually destabilize the circuit. Constant-field scaling only makes matters worse because it simultaneously increases transconductance and reduces capacitance. Selective gate shrinks do not change capacitance values, but they are still risky because they can introduce unforeseen parametric changes in short-channel transistors that frequently prove more significant for analog circuits than for digital ones. To summarize, analog and mixed-signal circuits should not be scaled without re-evaluating the performance of the resulting circuit to ensure that it still meets functional and parametric specifications.

11.2.6. Variant Structures

The simplest type of self-aligned poly-gate transistor consists of a rectangle of NMoat or PMoat bisected by a strip of poly. This type of structure serves admirably for width-to-length ratios of less than ten. Transistors with larger W/L ratios become increasingly unwieldy unless they are divided into multiple identical sections connected in parallel. Figure 11.15A shows the layout of a three-section transistor. The paralleled fingers not only produce a more convenient aspect ratio, but they also save area because adjacent sections share source and drain fingers. The merger of adjacent source/drain fingers can also reduce parasitic junction capacitance by up to 50%.

The division of a transistor into sections can affect its matching, so circuit designers often specify the number of sections for critical transistors. The most common notation for a sectioned transistor is $N(W/L)$, which denotes N sections, each with a drawn width of W and a drawn length of L. Transistors specified in this manner should be laid out exactly as requested. If the transistor is specified as having dimen-

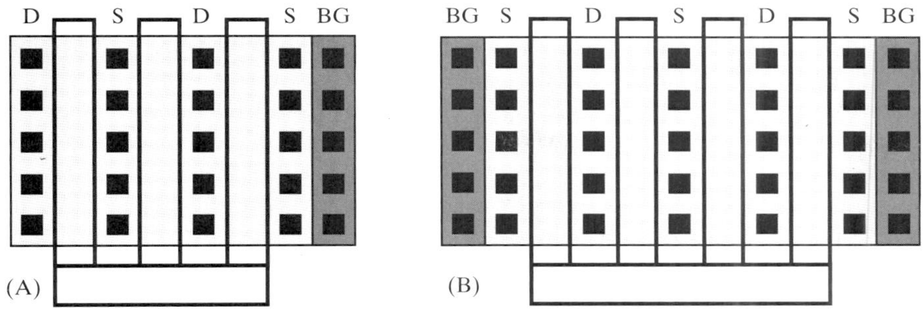

FIGURE 11.15 Sectioned transistors of (A) three and (B) four sections. The fingers are marked S (source), D (drain), and BG (backgate). The wells have been omitted for clarity.

sions *W/L*, this usually indicates that it can have any number of segments desired. If this transistor must match one having dimensions *N(W/L)*, then the former device should be laid out as a single section.

Transistors with even numbers of sections always contain odd numbers of source/drain fingers (Figure 11.15B). Such transistors are usually constructed with source fingers at either end. Not only does this allow the use of abutting backgate contacts on either or both ends, but it also reduces the number of drain fingers by one. This arrangement minimizes parasitic drain junction capacitance at the expense of source capacitance. Drain capacitance usually has more effect upon circuit performance than source capacitance, so a reduction in drain capacitance at the expense of source capacitance usually improves circuit performance.

Transistors sharing common source or drain connections are frequently merged to save space or to minimize parasitic junction capacitance. The merger is a relatively simple matter so long as both transistors contain sections of the same width. Differing widths require the use of a notched moat (Figure 11.16). The layout rules usually prohibit the placement of polysilicon immediately adjacent to a moat edge due to the large oxide step present at this location. The spacing between poly and moat S_{PM} forces a slight increase in the area of the shared source/drain finger, but one shared finger still consumes less area than two separate fingers.

Transistors M_1 and M_2 in Figure 11.16 share a common source region, so the drain fingers must occupy the ends of the array. This arrangement precludes the use of an abutting backgate contact, so the backgate contact is placed some distance away from the devices. The spacing between the backgate contact and the merged transistors may seem to eliminate any area benefit produced by the merger, but this backgate contact can also serve several other devices. Transistors sharing a common drain have source fingers on either end of the array and can therefore use abutting backgate contacts.

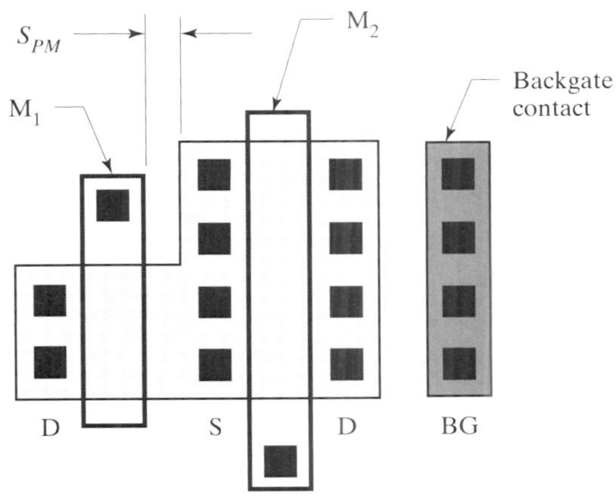

FIGURE 11.16 Merged transistors M_1 and M_2 share a common source (well omitted for clarity).

CMOS layout makes extensive use of merged devices to save space and to minimize capacitance. Figure 11.17 shows a simple layout of a two-input NAND gate that illustrates many of the techniques in common use. PMOS transistors M_1 and M_2 occupy a common well placed at the top of the layout. These transistors share a common drain region that not only reduces the width of the cell but also minimizes the drain capacitance on the output node Z. The two PMOS transistors also share a single backgate contact at the right end of the well. NMOS transistors M_3 and M_4 reside next to one another near the bottom of the layout. These transistors have been placed in series—the drain of M_3 simultaneously acts as the source of M_4. No contacts are necessary, because the current simply flows from one channel to the next. One strip of poly forms the gates of transistors M_2 and M_3, and a second strip of poly forms the gates of M_1 and M_4. The spacing between M_3 and M_4 is slightly larger than minimum. If desired, this spacing can be minimized by angling the gate leads toward one another.

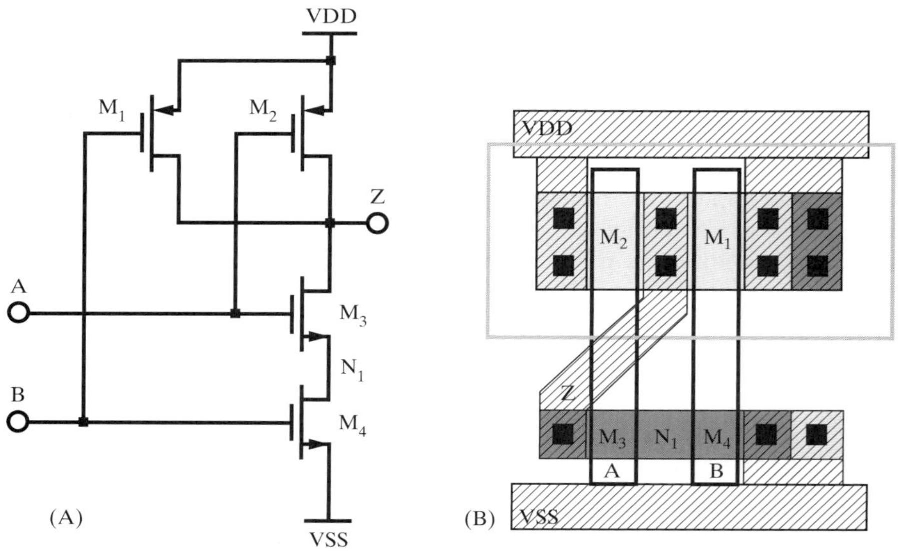

FIGURE 11.17 (A) Schematic and (B) layout of a two-input NAND gate.

The layout in Figure 11.17 follows the general guidelines of digital standard cell design. The power and ground rails run across the top and bottom of the cell, respectively. The width and spacing of these leads should be the same for all logic cells so they can stack end-to-end. The PMOS transistors occupy a common well spanning the top of the cell. When multiple logic cells stack end-to-end, their wells overlap to form a single contiguous region running the entire length of the assembly. This arrangement avoids the well-to-well spacings that would otherwise appear between adjacent cells. The NMOS transistors reside near the bottom of the cell, either in the epi or in another common well. Each cell contains at least one substrate and one backgate contact. Larger cells should contain additional substrate and backgate contacts wherever possible. The input and output connections exit from either the top or the bottom of the cell, whichever is more convenient for a given layout. Digital standard cells frequently contain special elements called *ports* and *prels* required by autorouting software. Since most autorouters cannot handle analog routing, there is little point in adding ports and prels to analog cells. The designer may still wish to employ concepts such as standard cell heights and consistent power and ground rail placement to allow analog cells to stack together. The height of analog cells is usually much greater than that of digital cells to accommodate their larger components and greater interconnection complexity. Additional rails are sometimes necessary to accommodate separate analog and digital supplies or to distribute several different supplies throughout a multisupply system.

Serpentine Transistors

Some designs require transistors with very long channels. The most convenient layout for such devices consists of a strip of NMoat or PMoat placed underneath a plate of polysilicon. A very compact layout results if one folds the moat into a serpentine pattern (Figure 11.18). The total channel length is computed by a procedure analogous to that used for serpentine resistors. Each 90° bend in the channel adds one-half of the transistor's width to its total length. The channel length of the transistor in Figure 11.18 therefore equals $2L_X + L_Y + W$. Serpentine transistors will not match precisely unless they have identical geometries, but most designs do not require especially precise matching from long-channel devices.

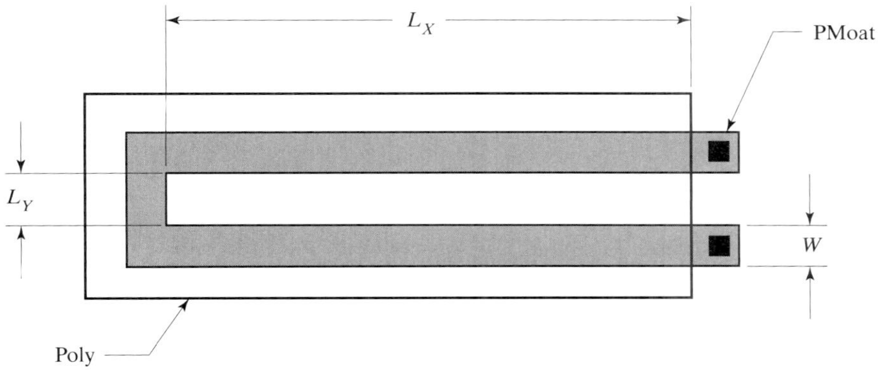

FIGURE 11.18 Serpentine PMOS transistor (well omitted for clarity).

Annular Transistors

The drain capacitance of an MOS transistor limits its switching speed and frequency response. Many circuits, analog as well as digital, can benefit from reduced drain capacitances. A smaller transistor has less capacitance, but it also provides less transconductance. These factors offset one another, so the smaller transistor is generally no faster than its larger counterparts. In order to actually increase switching speed, one must reduce the ratio of drain capacitance to transistor width C_D/W. Interdigitation reduces the C_D/W ratio by half because it surrounds each drain with two gates. This same principle can be carried still further by surrounding the drain on all four sides by an annular gate (Figure 11.19). An annular transistor will provide the smallest possible C_D/W ratio, but the decreased drain capacitance comes at the expense of increased source capacitance. The increased source capacitance is not necessarily injurious because the source often connects to a low-impedance node such as a power supply rail.

Two basic types of annular transistors exist: those that use a square gate geometry (Figure 11.19A) and those that use a circular gate (Figure 11.19B). The circular gate theoretically provides the highest C_D/W ratio because it minimizes the area-to-periphery ratio of the drain. The current flow through a circular gate is quite symmetric and the width of the transistor is easily computed. The current flow through square gates is less uniform and the effective width of the transistor is less easily computed. Square gates also have sharp corners that can induce premature avalanche breakdown due to electric field intensification.

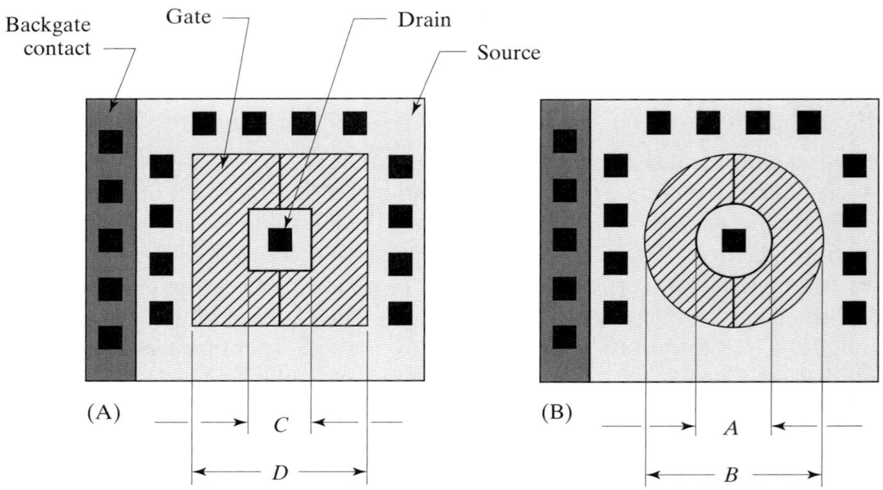

FIGURE 11.19 Annular MOS structures: (A) square and (B) circular.

The following equations (derived in Appendix D) give the width and length of a circular annular transistor in terms of the inner gate diameter A and the outer gate diameter B:

$$W = \frac{\pi\,(B - A)}{\ln\,(B/A)} \qquad [11.9A]$$

$$L = \frac{B - A}{2} \qquad [11.9B]$$

Some designers approximate the width of a circular annular transistor as the perimeter of a circle drawn halfway between the source and drain, giving $W \cong 1/2\ \pi\,(A + B)$. This approximation slightly overestimates the true width of the transistor. The errors caused by the approximation have little impact because precision circuits always rely on matching between identical devices rather than the properties of any one device.

The width and length of a square annular transistor are given by the following approximations, which do not correct for corner effects:

$$W \cong 2(C + D) \qquad [11.10A]$$

$$L \cong \frac{D - C}{2} \qquad [11.10B]$$

Annular transistors are often elongated to produce gate geometries similar to those in Figure 11.20. The C_D/W ratio of the elongated annular transistor is not much smaller than that of a conventional interdigitated transistor, so elongated structures are not recommended for minimizing drain capacitance. They are still sometimes used to produce an enclosed channel (Section 12.1.2). The W and L of an elongated circular annular transistor (Figure 11.20A) are approximately

$$W = \pi\,\frac{(B - A)}{\ln(B/A)} + 2U \qquad [11.11A]$$

$$L = \frac{B - A}{2} \qquad [11.11B]$$

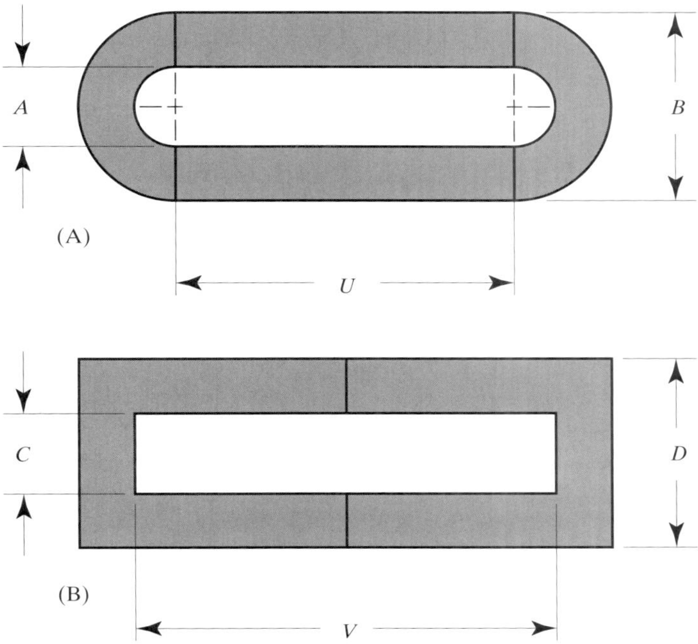

FIGURE 11.20 Gate geometries for elongated annular transistors: (A) circular and (B) square.

Similarly, the W and L of an elongated square annular transistor (Figure 11.20B) are approximately

$$W \cong 2V + C + D \quad [11.12A]$$

$$L \cong \frac{D - C}{2} \quad [11.12B]$$

11.2.7. Backgate Contacts

All MOS transistors require electrical connection to their backgates even though no current normally flows through them. MOS transistors that lack backgate contacts or have excessive backgate resistance are particularly prone to latchup. Every PMOS transistor contains a parasitic lateral PNP and every NMOS contains a parasitic lateral NPN. Together these form a parasitic SCR (Figure 4.20). The backgate contacts short the base-emitter junctions of these parasitic transistors, and the associated backgate resistances become base turn-off resistors (R_1 and R_2 in Figure 4.20). The SCR will remain off as long as the voltage across both of these resistors remains less than the forward voltage of the respective base-emitter junctions. The voltage necessary to trigger the SCR into conduction equals about 0.65 to 0.7V at 25°C, but this falls to 0.4 to 0.45V at 150°C due to the temperature coefficient of V_{BE}. Not only does the trigger voltage drop at high temperatures, but the betas of the parasitic transistors actually increase. Thus, CMOS latchup is most likely to occur at high temperature.

Most CMOS products must pass a standardized test that measures their latchup susceptibility. Positive and negative test current pulses are applied to each pin, while power is applied to the part. Depending on specifications, the magnitude of these test pulses may range from as little as ±100mA to as much as ±250mA. The supply current is measured both before and after the application of each test pulse. If these two currents are not approximately the same, then the part fails the test.

This latchup test can be modeled mathematically. Suppose that a test current I_T flows through the source/drain region of an MOS transistor M_1. In order to prevent latchup from occurring between M_1 and a complementary MOS transistor M_2, at least one of the following inequalities must be true:

$$\beta_{12}\beta_{21}(1-\eta_{c12})(1-\eta_{c21}) < 1 \qquad [11.13A]$$

$$I_T R_{B2}(1-\eta_{c12})\left(\frac{\beta_{12}}{\beta_{12}+1}\right) < V_{trig} \qquad [11.13B]$$

β_{12} represents the beta of the parasitic bipolar formed by minority carriers flowing from the source/drain region of M_1 to the backgate of M_2 in the absence of guard rings. β_{21} represents the beta of the parasitic bipolar formed by minority carriers flowing from the source/drain region of M_2 to the backgate of M_1, again in the absence of guard rings. η_{c12} represents the fraction of minority carriers flowing from M_1 to M_2 intercepted by guard rings. Similarly, η_{c21} represents the fraction of minority carriers flowing from M_2 to M_1 intercepted by guard rings. I_T equals the test current, R_{B2} equals the backgate resistance of M_2, and V_{trig} equals the trigger voltage of the SCR (about 0.4V at 150°C).

These equations provide some insight into the roles of guard rings and backgate contacts in suppressing latchup. Equation 11.13A represents the condition required to avoid sustained feedback. Minimizing parasitic betas β_{12} and β_{21} and adding guard rings to improve collector efficiencies η_{c12} and η_{c21} can help prevent sustained conduction. Any device meeting this criterion is invulnerable to CMOS latchup regardless of the magnitude of the test currents applied. Unfortunately, few CMOS processes can satisfy equation 11.13A because their transistors lie too close together and their guard rings are too inefficient. CMOS devices can still achieve conditional latchup immunity by satisfying the conditions of equation 11.13B. The four terms in this inequality represent the magnitude of the test current and the backgate resistance, the effectiveness of the guard rings, and the magnitude of the parasitic beta, respectively. The contributions of guard rings and backgate contacts multiply one another, producing a synergistic relationship between the two. Even if neither guard rings nor backgate contacts alone can stop latchup, a combination of the two frequently can. Guard rings require so much room that they can only be placed around a few devices—usually those that may potentially inject minority carriers into the die. Backgate contacts require much less area, so each transistor can have its own backgate contact or can at least share a backgate contact with another transistor.

The backgate of an NMOS must connect to a voltage less than or equal to its source, and the backgate of a PMOS must connect to a voltage greater than or equal to its source. In many applications, the backgate can be connected to the source. A few transistors operate under conditions in which it is difficult or impossible to distinguish source from drain. A few circuits connect the backgate to a voltage that is different from the source to increase the threshold voltage using the body effect. Some high-speed circuits also avoid connecting the source and backgate in order to minimize the capacitance appearing at the source node. All of these circuits require an independent backgate contact like that in Figure 11.21B. Transistors whose source and backgate operate at the same potential can use an abutting backgate contact (Figure 11.21A). This contact saves considerable area by eliminating the spacing between source and backgate diffusions. The intersection of two heavily doped diffusions produces a leaky and unreliable junction, but this defect can be tolerated as long as the two diffusions are connected together by metal or silicide.

Relatively small transistors such as those in Figure 11.21 require only one backgate contact. The distance across the transistor grows larger as additional segments

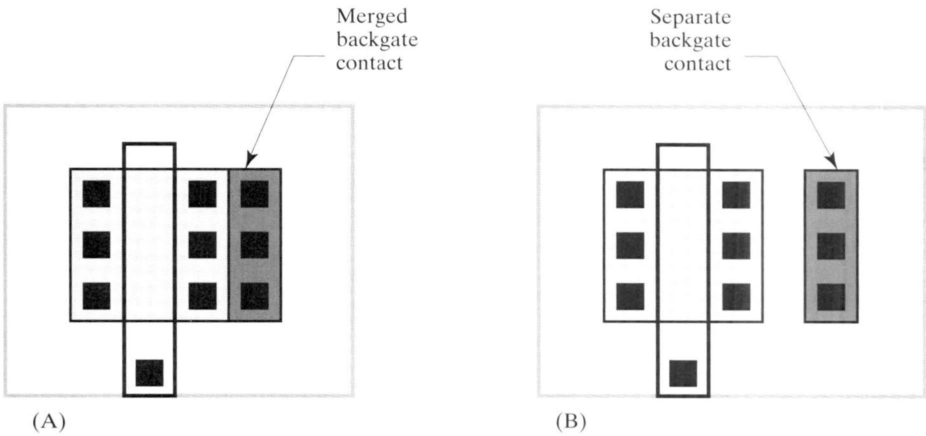

FIGURE 11.21 Examples of (A) abutting and (B) separate backgate contacts.

are added, and at some point this distance becomes so great that it produces an unacceptably large backgate resistance. A second backgate contact on the opposite side of the transistor reduces the distance to the backgate contacts at the cost of slightly increasing the device area (Figure 11.15B). The point at which a second backgate contact becomes necessary varies depending on the sheet resistance of the backgate. An NMOS transistor constructed in the P-epi above a P+ substrate would have a lower backgate resistance than a PMOS constructed in a shallow, lightly doped N-well. Adding a buried layer to a well likewise decreases the backgate resistance of transistors occupying it. Factors of this sort make it difficult to provide quantitative rules for backgate contact spacing. Some processes specify a maximum distance between any portion of a transistor and the nearest backgate contact. The maximum allowed distance becomes shorter as the backgate resistance increases. Typical spacings range from 25μm to 250μm. Transistors subject to large transients should use a smaller distance to provide additional latchup immunity. These include ones whose source/drain regions connect to pins, and those residing next to transistors whose source/drain regions connect to pins.

Large transistors with many fingers may require substrate contacts embedded within the body of the transistor itself. This is usually achieved by placing strips of backgate contact through the transistor at regular intervals (Figure 11.22A). Although these *interdigitated backgate contacts* reduce the distance to the nearest

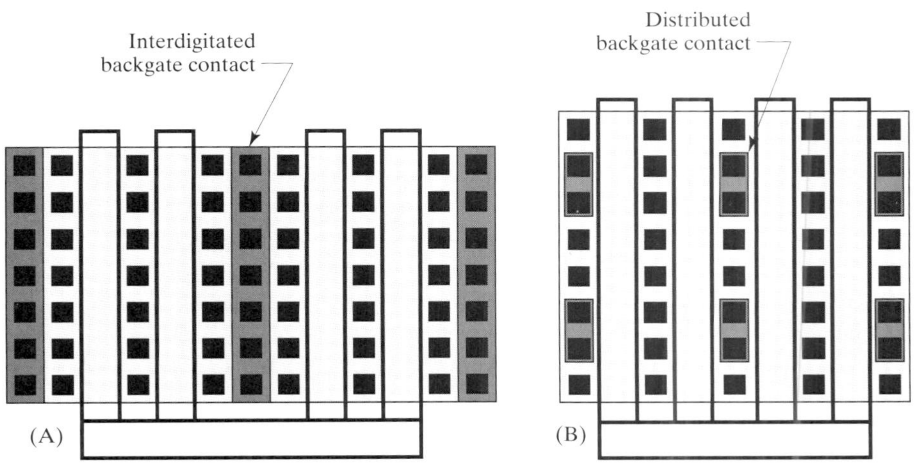

FIGURE 11.22 Additional styles of backgate contacts: (A) interdigitated backgate contact and (B) distributed backgate contact.

substrate contact, they also substantially increase the size of the transistor. Some processes allow another type of backgate contact, consisting of small plugs of backgate diffusion placed in holes within the source fingers of the transistor (Figure 11.22B). These *distributed backgate contacts* slightly increase the source resistance of the overall transistor, but they greatly reduce the area required by the backgate contacts. A substantial area savings can be obtained even if the transistor must be enlarged to compensate for increased source resistance. Distributed backgate contacts can be placed on every source finger, as shown, or they may be placed on only a few source fingers distributed at regular intervals across the transistor. A larger number of distributed backgate contacts further reduces backgate resistance, but not all applications necessarily require the same degree of latchup protection. Distributed backgate contacts always connect to the source of the transistor, so some applications may still require the use of interdigitated contacts.

Analog BiCMOS processes often contain additional diffusions that can help reduce the latchup susceptibility of MOS transistors. For example, many analog BiCMOS processes fabricate NPN transistors in the same N-well as PMOS transistors. The NBL that is used to reduce the collector resistance of the CDI NPN can also reduce the backgate resistance of the PMOS. If NBL is available, it should be placed in all MOS transistors that use the same well as the CDI NPN. Very efficient hole-blocking guard rings can be produced in many processes that offer deep-N+ and NBL. Some transistors may actually operate under conditions in which their source/drain regions regularly forward-bias into the backgate. Deep-N+ guard rings can help ensure that substrate injection from these PMOS transistors does not disrupt the operation of the rest of the circuit.

NMOS transistors are somewhat more difficult to protect from latchup than PMOS transistors. An isolated NMOS structure (Section 11.2.2) offers total immunity to CMOS latchup, but at the price of greatly increased backgate resistance. Although these transistors do not require a low backgate resistance in order to suppress CMOS latchup, parasitic lateral NPN action remains a concern. If the transistors must operate at relatively high voltages, they may become susceptible to snapback breakdown due to parasitic NPN action. Even if they operate well below the $V_{CEO(sus)}$ of the NPN, minority carrier injection still causes sluggish switching due to charge storage. A system of distributed backgate contacts can minimize these problems. An NMOS transistor fabricated above a P+ substrate and surrounded by a deep-N+ electron-collecting guard ring will also prove quite resistant to latchup. This style of transistor is suited to applications where the transistor must withstand severe transients, but where its source/drain regions do not routinely forward-bias into the backgate. Transistors of the latter sort are best constructed as isolated NMOS transistors.

11.3 SUMMARY

This chapter has covered the construction of conventional small-signal poly-gate CMOS transistors. The next chapter covers a variety of more specialized types of transistors, including extended-voltage transistors, power transistors, DMOS transistors, and JFETs. These transistors can fill a very wide range of applications, including many that are traditionally filled by bipolar transistors.

11.4 EXERCISES

Refer to Appendix C for layout rules and process specifications.

11.1. Suppose an enhancement NMOS has a threshold voltage of 0.7V and a transconductance of 220μA/V^2. Determine the region of operation and compute the drain current for each of the following biasing conditions:

a. $V_{GS} = 1.2\text{V}, V_{DS} = 2.3\text{V}$.
b. $V_{GS} = 1.2\text{V}, V_{DS} = 0.2\text{V}$.
c. $V_{GS} = -1.0\text{V}, V_{DS} = 4.4\text{V}$.

11.2. Suppose the enhancement NMOS in Exercise 11.1 is subjected to the following terminal voltages: $V_{GS} = 1.2\text{V}, V_{DS} = -2.3\text{V}$. Recognizing that the source and drain have swapped roles, determine the true electrical biasing conditions, the mode of operation, and the drain terminal current.

11.3. What is the process transconductance of an NMOS transistor having a composite gate dielectric consisting of 150Å of nitride ($\varepsilon_r = 6.8$) sandwiched between two layers of oxide, each 50Å thick ($\varepsilon_r = 3.9$)? *Hint:* See Section 6.1.

11.4. Estimate the process transconductances of NMOS and PMOS transistors having a maximum operating voltage of 15V.

11.5. Suppose an enhancement PMOS transistor with an N+ poly gate electrode has a nominal threshold voltage of –0.95V. What would the nominal threshold voltage become if the transistor used a P+ poly gate?

11.6. Would an enhancement PMOS transistor with a nominal threshold voltage of –0.4V serve as a useful device for constructing digital logic gates? Explain.

11.7. Lay out the inverter shown in Figure 11.23 using the poly-gate CMOS rules listed in Appendix C. Place a hole-collecting guard ring around the PMOS transistor and an electron-collecting guard ring around the NMOS. Connect the guard rings to provide the best possible protection.

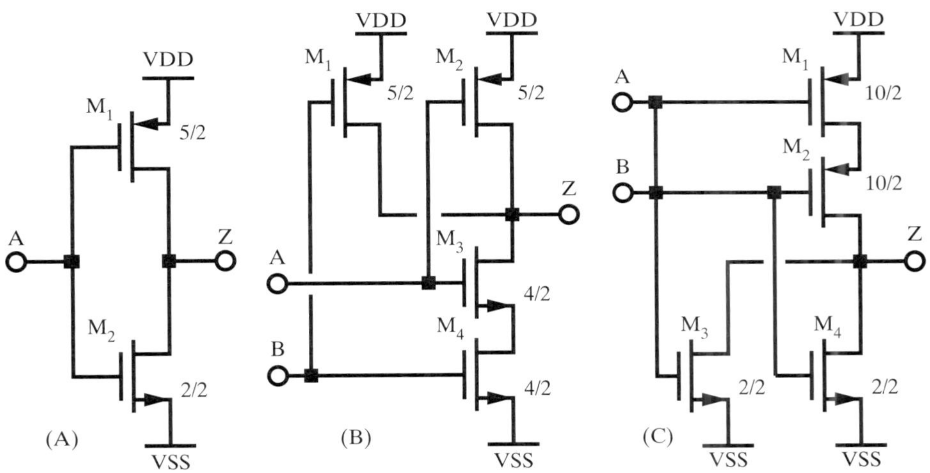

FIGURE 11.23 Three standard-cell logic gates: (A) Inverter, (B) NAND, and (C) NOR.

11.8. Construct an isolated NMOS transistor using the analog BiCMOS rules in Appendix C. The transistor should have a W/L ratio of 10/5. Include a single abutting substrate contact.

11.9. A 3μm CMOS process can withstand a maximum operating voltage of 12V and offers a gate delay of 2.3nS. Suppose constant-field scaling is applied to produce a 2μm process. Predict the maximum operating voltage and gate delay of the scaled process.

11.10. Lay out the following transistors using poly-gate CMOS rules. Include abutting backgate contacts, gate interconnections, and well geometries as necessary.
a. NMOS, 3(5/15)
b. NMOS, 12(20/5)
c. PMOS, 7(10/25)
d. PMOS, 4(10/3)

11.11. Lay out a natural NMOS transistor with dimensions of 3(10/3) and a natural PMOS transistor with dimensions of 25/25. Include abutting backgate contacts, gate interconnections, and well geometries as necessary. The layout rules for the NATVT layer are as follows:

1. **NATVT** width 4μm
2. **NATVT** overlap of **GATE** 2μm
3. **NATVT** spacing to **TOX** 4μm
4. **NATVT** spacing to **POLY** 4μm

Note: GATE is defined as the intersection of POLY with either NMOAT or PMOAT.

11.12. Construct standard-cell layouts of each of the three logic gates in Figure 11.23. The NAND gate should resemble the layout in Figure 11.17B. The VDD and VSS leads should be 4μm wide, and each cell should have at least one substrate contact and one well contact. Design the cells so they can be stacked together by abutting their VDD and VSS leads. No violations of layout rules should occur regardless of the order in which the cells are stacked.

11.13. Using the poly-gate CMOS rules of Appendix C, construct a serpentine PMOS with a nominal device transconductance of $0.01\mu A/V^2$. Fold the gate as many times as necessary to produce an approximately square layout.

11.14. Lay out the following annular transistors using poly-gate CMOS rules. Include abutting backgate contacts, gate interconnections, and well geometries as necessary.

a. Circular NMOS, 31.4/4
b. Square PMOS, 48/4
c. Elongated circular PMOS, 51.4/4

11.15. Construct a 5000/2 PMOS transistor using poly-gate CMOS rules. Divide the transistor into as many sections as required to produce an approximately square layout. Include enough interdigitated backgate contacts to ensure that no part of the transistor is more than 50μm from the nearest backgate contact.

12 Applications of MOS Transistors

Some designs require MOS transistors to operate at high voltages without experiencing breakdown or parametric shifts. Although breakdown voltages can be increased by using lower dopant concentrations and thicker gate oxides, hot-carrier generation remains a concern. A variety of specialized transistor structures have been developed to minimize hot-carrier generation at high voltages. This chapter examines several structures that are compatible with ordinary CMOS processing, as well as several that require process extensions.

MOS power transistors are particularly useful for high-current, low-resistance switching. Conventional CMOS transistors can conduct large currents, but have operating voltage limitations. Many processes offer *double-diffused MOS,* or DMOS, transistors as a process extension. These devices allow the construction of compact, high-voltage, high-current transistors.

MOS transistors are only one example of a wider category of devices called *field-effect transistors,* or FETs. Some processes also offer *junction field-effect transistors,* or JFETs. This chapter summarizes the properties of the JFET and presents a selection of typical layouts.

Finally, this chapter also discusses the matching of MOS transistors. MOS transistors used in analog circuits frequently require a high degree of matching. The techniques used to match MOS transistors differ quite markedly from those used for bipolar transistors.

12.1 EXTENDED-VOLTAGE TRANSISTORS

Early MOS processes produced relatively long-channel transistors with lightly doped backgates. Avalanche breakdown of the source/drain regions limited these transistors to operating voltages of 10 to 15V. Processing improvements have enabled channel lengths to decrease from about 8μm to less than 0.3μm. If backgate doping and operating voltages remained constant, the pinched-off portion of the channel would represent an ever-increasing percentage of the channel length. The resulting transistors would exhibit increased channel length modulation and premature punchthrough breakdown.

Channel length modulation can be minimized and punchthrough averted either by reducing the operating voltage or by increasing the backgate doping. Since the operating voltages of analog circuits are not easily reduced below 3 to 5V, most processes have opted to increase backgate doping. This minimizes the width of the pinched-off region at the cost of intensifying the lateral electric field across it. Intense electric fields generate hot carriers that in turn produce undesirable long-term parametric drifts (Section 4.3.1). Holes are more difficult to accelerate than electrons, so PMOS transistors are less prone to hot-carrier generation than are NMOS transistors.

The *operating voltage* of a MOS transistor equals the maximum drain-to-source voltage allowed during saturation, while the *blocking voltage* equals the maximum drain-to-source voltage allowed during cutoff. Since the pinched-off region disappears in cutoff, hot-carrier generation ceases and the blocking voltage is limited only by drain-backgate avalanche and gate oxide rupture. Hot-carrier generation may restrict the operating voltage to a lower value than the blocking voltage. Higher voltages do not necessarily cause instant failure, but they produce gradual shifts in both threshold voltage and device transconductance.[1] Analog circuits are particularly sensitive to hot carrier degradation because they frequently contain matched devices operating in saturation. MOS transistors in digital circuits enter saturation only during brief switching transients. Thus, digital circuits operating at low clock speeds can often ignore the voltage limitations imposed by hot-carrier generation.

Hot carriers become an increasingly serious problem as channel lengths diminish. Specialized transistor structures offer extended operating voltage ranges at the cost of additional process complexity and larger spacings. Virtually all submicron CMOS processes use some form of these *extended-voltage transistors*.

12.1.1. LDD and DDD Transistors

All extended-voltage transistors incorporate some form of specialized drain structure that absorbs a portion of the electric field that would otherwise appear across their channel. Figure 12.1A shows a plot of the lateral electric field intensity across the drain end of a saturated MOS transistor. This example assumes constant backgate and drain doping concentrations and abrupt junctions. The field intensity rises linearly across the pinched-off region and reaches a maximum at the drain metallurgical junction. The field then drops linearly across the depletion region inside the drain. The widths of the pinched-off region x_p and the drain depletion region x_d are proportional to the inverse square root of the doping of the respective regions. The voltages sustained across the pinched-off region V_p and across the drain depletion region V_d equal the areas of the respective triangles (Figure 12.1A).

The total drain-to-source voltage V_{DS} equals the sum of the areas of both triangles:

$$V_{DS} = \frac{E_{max}}{2}(x_p + x_d) \qquad [12.1]$$

The maximum electric field intensity E_{max} and the width of the pinched-off region x_p are limited by hot-carrier generation and channel length modulation, respectively. No such hard-and-fast limit exists on the width of the drain depletion region x_d. The

---

[1] C. Duvvury and S. Aur, "Hot-Carrier Degradation Effects in CMOS Technologies," *TI Technical Journal,* Vol. 8, #1, 1991, pp. 56–66.

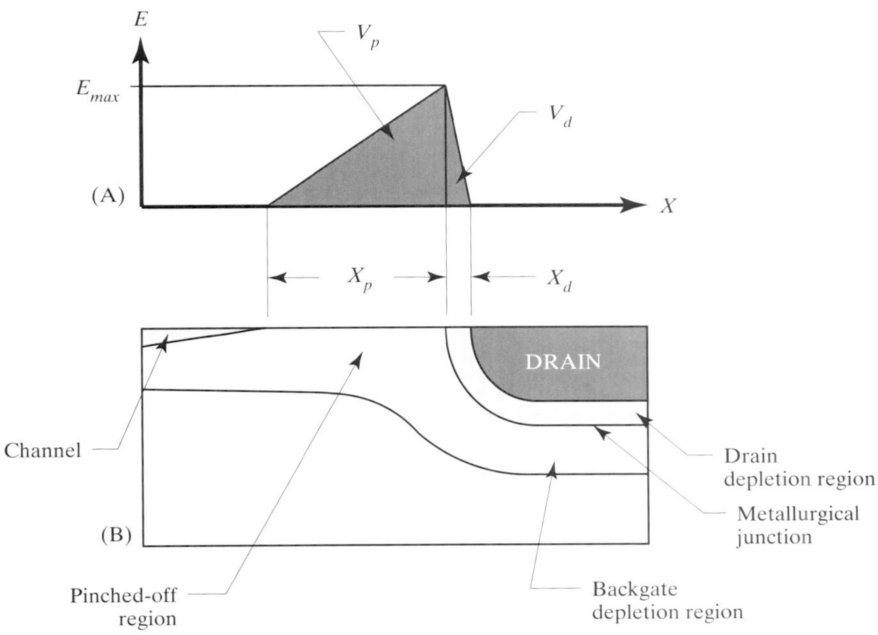

FIGURE 12.1 Diagram showing the lateral electric field across the drain end of a saturated MOS transistor.

only way to increase the operating voltage V_{DS} is to increase the width of the depletion region x_d. This in turn requires a more lightly doped drain.

In order to provide significant benefit, the width of the drain depletion region x_d must equal a significant fraction of the width of the pinched-off region x_p. This dictates that the drain doping not greatly exceed the backgate doping. Unfortunately, a lightly doped drain is also a highly resistive drain. Most modern processes minimize drain resistance by using a composite structure consisting of a lightly doped fringe surrounding a heavily doped core. The fringe depletes at relatively low voltages, forming a *drift region* bounded by the metallurgical junction on one side and by the heavily doped core, or *extrinsic drain,* on the other. The width of the drift region sets the width of the drain depletion region x_d. The drift region should be made just wide enough to support the desired operating voltage, and no wider. Any additional width would increase drain resistance without providing any corresponding benefit. The optimal width of the drift region usually represents no more than a small fraction of the channel length.

Several device structures have been developed that control the drift region width through various forms of self-alignment. The drift region must self-align to both the extrinsic drain and to the poly gate in order to minimize overlap capacitance. Figure 12.2

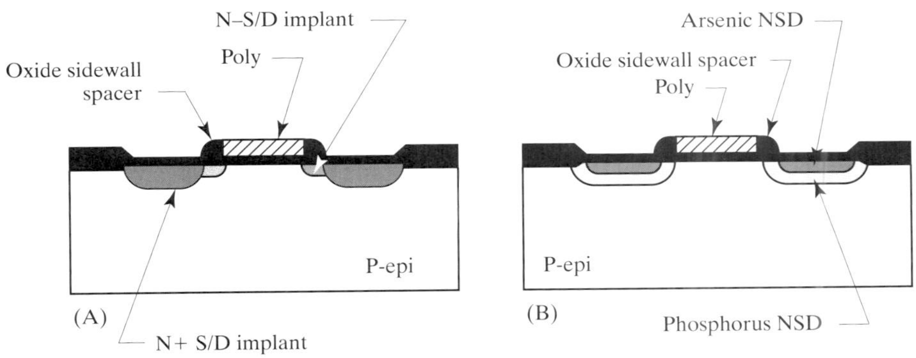

FIGURE 12.2 Extended-voltage transistors: (A) lightly doped drain and (B) double-diffused drain.

shows two structures that fulfill these requirements. Both employ a special feature called an *oxide sidewall spacer* formed by isotropically depositing and anisotropically etching an oxide layer (Section 3.2.4). By its very nature, the oxide sidewall spacer self-aligns to the poly gate. The sidewall spacer is approximately as wide as the poly is thick. A limited range of spacer widths can be fabricated by adjusting poly thickness and etch conditions.

A *lightly doped drain*, or LDD, uses oxide sidewall spacers to define the width of the drift region (Figure 12.2A). The LDD structure requires two separate source/drain implants: one performed before spacer formation and one afterward. The first implant forms a lightly doped drift region aligned to the polysilicon gate, while the second forms a heavily doped extrinsic drain aligned to the oxide sidewall spacer.[2] This technique requires no additional masking steps, but it does require an oxide deposition, an oxide removal, and a second drain implant. The performance advantages gained from higher operating voltages offset the cost of the additional processing.

The process that forms the LDD structure does not discriminate between source and drain. The source resistance can be reduced if the LDD structure appears only on the drain end of the device. The resulting transistor is said to be *asymmetric* because its source and drain terminals are not interchangeable. Asymmetric LDD transistors require additional masking and processing steps to remove the sidewall spacer from the source end of the transistor. The benefits produced by this additional processing are rarely sufficient to justify the additional cost.

The *double-diffused drain*, or DDD, uses two implants driven through the same oxide opening to form a composite drain structure (Figure 12.2B).[3] These two implants require dopants of widely differing diffusivities, most commonly arsenic and phosphorus. A brief drive causes the phosphorus to diffuse outside the boundaries of the arsenic implant to form a lightly doped drift region. An oxide sidewall spacer minimizes overlap capacitances by preventing the drift region from diffusing underneath the poly gate.

The double-diffused drain is not readily applicable to PMOS transistors because of the absence of a slow-diffusing acceptor for silicon. NMOS transistors sometimes use the DDD structure in preference to the LDD structure because it can fabricate very narrow drift regions with great precision. The DDD structure also increases the avalanche voltage of the source/drain implants by grading their junctions. It is difficult to construct wide DDD drift regions because the drive required to force the phosphorus under a wide sidewall spacer also interferes with threshold voltage control. The LDD structure is therefore favored for wider drift regions and the DDD structure for narrower ones. The cost and complexity of the two techniques are comparable, as both require oxide sidewall spacers and both employ two drain implants.

The electric field intensity required to generate hot holes is two or three times larger than that required to generate hot electrons, so many applications that require LDD or DDD NMOS transistors can still use ordinary PMOS transistors. These PMOS devices only require a single source/drain implant, so they are called *single-diffused drain* (SDD) transistors. If the process includes an LDD or DDD NMOS, then the SDD PMOS also receives oxide sidewall spacers. The spacers can actually transform a buried-channel PMOS transistor into an LDD device. The contact po-

---

[2] S. Ogura, P. J. Tsang, W. W. Walker, D. L. Critchlow, and J. F. Shepard, "Design and Characteristics of the Lightly Doped Drain-Source (LDD) Insulated Gate Field-Effect Transistor," *IEEE Trans. on Electron Devices*, Vol. ED-27, #8, 1980, pp. 1359–1367.

[3] E. Takeda, H. Kume, T. Toyabe, and S. Asai, "Submicrometer MOSFET Structure for Minimizing Hot-carrier Generation," *IEEE Trans. on Electron Devices*, Vol. ED-29, #4, 1982, pp. 611–618.

tential of the doped poly gate inverts the portions of the buried channel underneath it. The portions of the buried channel protruding under the oxide sidewall spacers do not invert because they do not experience the full electric field generated by the gate electrode. The stubs of the buried channel therefore form two lightly doped P-type regions that act as lightly doped drains (Figure 12.3). This type of structure is sometimes called a *buried-channel lightly doped drain* (BCLDD).[4]

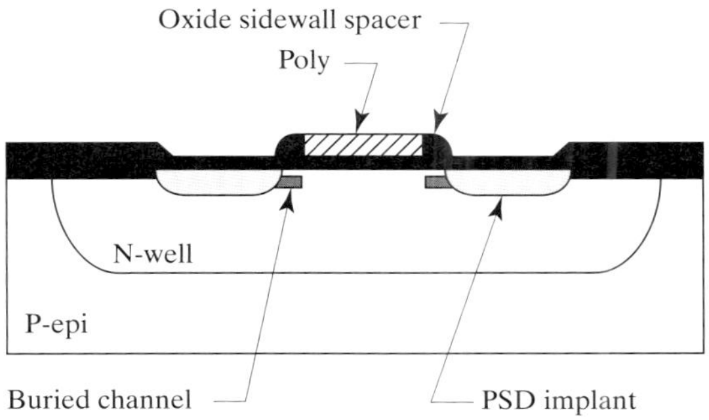

FIGURE 12.3 PMOS buried-channel lightly doped drain (BCLDD).

12.1.2 Extended-drain Transistors

Higher-voltage LDD and DDD transistors require thicker oxide sidewall spacers. The difficulties inherent in constructing wide spacers limit these structures to voltages of about 15 to 20V. Much higher voltages can be achieved using non-self-aligned composite drains that do not employ sidewall spacers to delimit the drift regions. The simplest structure of this sort is called an *extended drain.* It consists of a shallow, heavily doped diffusion entirely contained within a deeper, more lightly doped one. The inner diffusion forms the extrinsic drain, while the fringes of the outer diffusion act as a drift region. For example, an NMOS extended drain might consist of NSD within N-well; the NSD implant then acts as the extrinsic drain, and the N-well as the drift region.

One can often create extended-drain transistors from existing diffusions. The resulting devices are usually larger than purpose-built devices such as lateral DMOS transistors (Section 12.2.2). On the other hand, the extended-drain devices do not require additional processing steps or masks. If an integrated circuit only requires a few small high-voltage transistors, then the most economical solution probably consists of extended-drain transistors constructed from existing masks. On the other hand, large, low-resistance devices are better constructed using purpose-built devices that can achieve lower specific on resistances and overlap capacitances.

Extended-drain NMOS Transistors

Figure 12.4A shows the cross section of a typical extended-drain NMOS transistor constructed in an N-well CMOS process. The extended drain consists of an NSD plug contained inside a larger N-well geometry. The N-well diffuses outward to produce a very lightly doped drain capable of withstanding high voltages. These voltages will rupture the thin gate oxide unless the drain incorporates a special *field-relief structure.* This structure consists of a section of thick-field oxide placed just inside the metallurgical drain-backgate junction. As the depletion region intrudes further into the

---

[4] R. H. Eklund, R. A. Haken, R. H. Havemann, and L. N. Hutter, "BiCMOS Process Technology," in *BiCMOS Technology and Applications,* 2nd ed., A. R. Alvarez, ed. (Boston: Kluwer Academic Publishers, 1993, pp. 93–95.

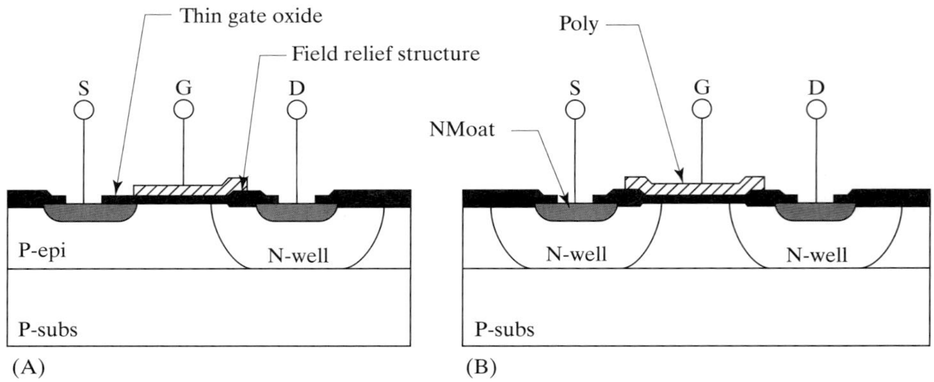

FIGURE 12.4 Cross sections of (A) asymmetric and (B) symmetric extended-drain NMOS transistors.

drain, it passes underneath the bird's beak. The highest drain-to-source voltages appear across the thick-field oxide over the drift region. As long as the field-relief structure is properly positioned relative to the metallurgical junction, the thin gate oxide experiences the full drain-to-source voltage. The field-relief structure has no effect on the transconductance or the threshold voltage of the transistor because these depend solely on the channel, all of which remains underneath the thin gate oxide.

Figure 12.4A illustrates an *asymmetric extended-drain NMOS* transistor. Only one end of this transistor receives an extended-drain structure. This produces a relatively compact layout, but one that is not suitable for applications where either end of the transistor may see high voltages. The *symmetric extended-drain NMOS* in Figure 12.4B equips both ends of the transistor with extended drains. The symmetric transistor can withstand large drain-to-source voltages regardless of which end of the transistor acts as its drain. It cannot simultaneously withstand large voltage differentials between the source and both source/drain regions, because one of these must serve as its source. Transistors that must withstand large gate-to-source voltages require thicker-gate oxides (Section 12.1.3).

Figure 12.5 shows the layout of asymmetric and symmetric extended-drain NMOS transistors. In both cases, the drawn gate length L_d, equals the distance across the moat beneath the gate. The minimum drawn gate lengths of these transistors are relatively large (typically 4 to 6μm), but the effective gate lengths are much smaller because of outdiffusion of the wells. Asymmetric transistors with multiple gate fingers lend themselves to a compact layout in which source and drain fingers alternate; this allows efficient use of the N-well strips, which form the extended drains.

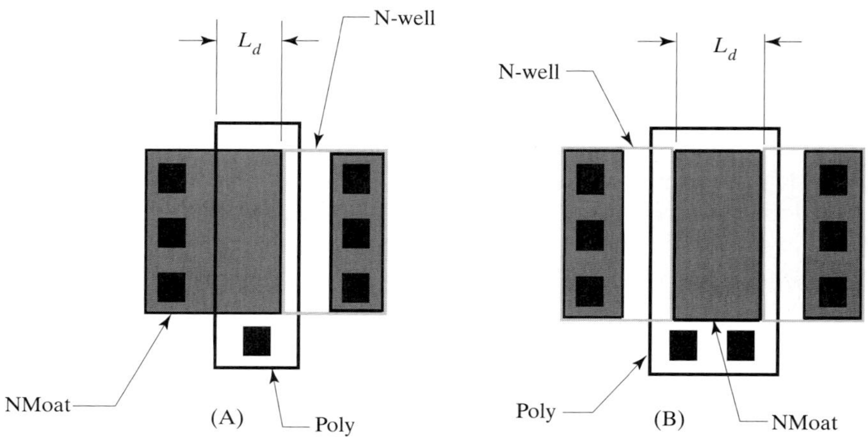

FIGURE 12.5 Layouts of (A) asymmetric and (B) symmetric extended-drain NMOS transistors.

The extremely light doping of the N-well suppresses hot electron generation, so the operating voltage ratings of properly designed extended-drain NMOS transistors are limited only by the avalanche voltage of the well-substrate junction and by the effectiveness of the field-relief structures. It is not particularly difficult to design extended-drain transistors that can withstand operating voltages two or three times larger than those of regular NMOS and PMOS transistors. Such voltages generally exceed the thick-field threshold of the process, requiring the addition of field plates or channel stops (Section 4.3.2).

Extended-drain PMOS Transistors

Figure 12.6A shows the cross section of an *asymmetric extended-drain PMOS* transistor constructed in an N-well CMOS process. The drift region of the extended drain consists of channel-stop implant. The CMOS process flow presented in Section 3.2 uses a patterned phosphorus channel-stop implant to counterdope a blanket boron channel-stop implant. If the channel-stop mask is modified to block the phosphorus implant from the vicinity of the extended drain, then this region receives only the boron implant. The boron outdiffuses during the field oxidation to form a deep, lightly doped P-type diffusion suitable for use as a drift region. Figure 12.6B shows a *symmetric extended-drain PMOS* that uses the channel-stop implant to form drift regions for both source/drain terminations.

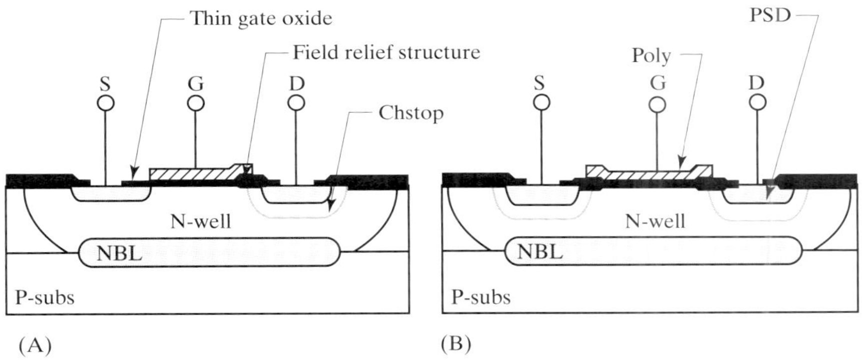

FIGURE 12.6 Cross sections of (A) asymmetric and (B) symmetric extended-drain PMOS transistors.

Extended-drain PMOS transistors use a special coding layer called *Chstop* to block the patterned phosphorus channel-stop implant. The geometries on the Chstop layer are added to the channel-stop mask during mask generation. Different processes may employ other coding techniques, but the principles remain broadly the same.

Figure 12.7 shows the layout of asymmetric and symmetric extended-drain PMOS transistors. In both cases, the drawn gate length equals the distance across the moat region beneath the gate. The transistors must contain NBL to stop vertical depletion from punching through the lightly doped bottom of the N-well and shorting the drain to the substrate.

12.1.3. Multiple Gate Oxides

The previous two sections have described transistors that can operate at large drain-to-source voltages. These transistors can also operate at large drain-to-gate voltages as long as they incorporate suitable field-relief structures. If they must operate simultaneously at large gate-to-source voltages, then no remedy exists but to increase the thickness of the gate oxide. Any increase in gate oxide thickness also reduces device transconductance. Circuit designers take an understandably dim

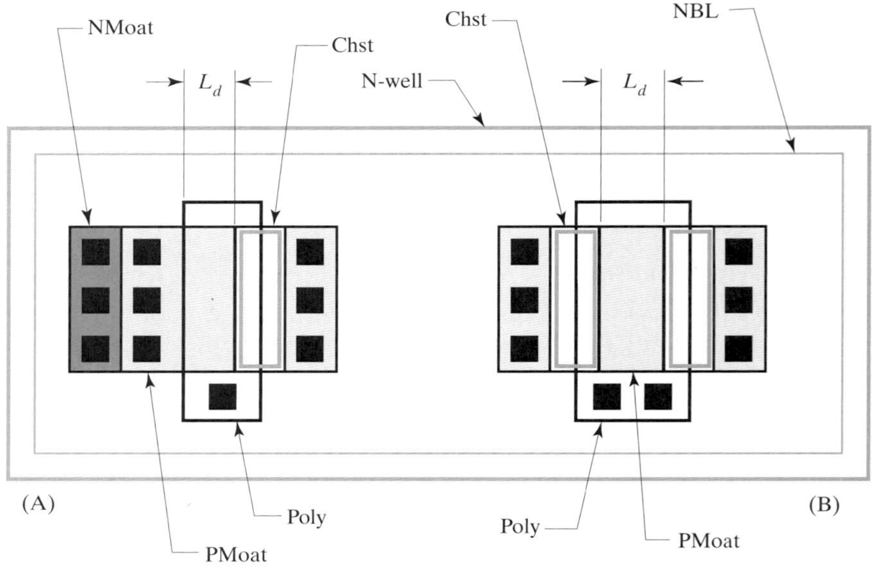

FIGURE 12.7 Layouts of (A) asymmetric and (B) symmetric extended-drain PMOS transistors.

view of increasing the gate oxide thickness of all transistors merely to accommodate increased voltages on a few. This conflict can be resolved by producing two separate thicknesses of gate oxide. The thinner gate oxide provides high transconductance for low-voltage applications, while the thicker gate oxide can withstand higher voltages. The circuit designer chooses which gate oxide each transistor receives based on expected operating conditions.

Multiple gate oxides are fabricated using either staged oxidation or etch-and-regrowth techniques. *Staged oxidation* requires a separate polysilicon deposition for each gate oxide. The thinnest gate oxide is grown first, followed by the deposition of the first polysilicon layer (Figure 12.8A). Once patterned, the poly acts as an oxidation mask for a continuation of the gate oxidation (Figure 12.8B). After the gate oxidation is complete, the second poly layer is deposited and patterned (Figure 12.8C). Any transistor with a poly-1 gate receives the thin gate oxide, and any transistor with a poly-2 gate receives the thick gate oxide.

Processes with only one layer of poly can use the *etch-and-regrowth* technique instead of staged oxidation. The etch-and-regrowth process requires an additional masking step instead of an additional poly deposition. The extra mask patterns a layer of photoresist spun on top of a thin-gate oxide. The exposed oxide regions are etched away (Figure 12.8D), after which the gate oxidation is resumed. A thin-gate oxide forms over the areas that were etched back, while a thicker layer of oxide forms over the areas where the initial oxide was left undisturbed (Figure 12.8E). A single layer of polysilicon can now form the gates of both thin-oxide and thick-oxide transistors (Figure 12.8F).

Some processes use staged oxidation, while others use etch-and-regrowth. Insofar as the layout designer is concerned, the main difference between the two lies in the number of polysilicon layers required. Figure 12.9 shows a comparison of layouts required by staged oxidation and etch-and-regrowth. Both processes use a *Moat-2* geometry coded around the gate region, but this geometry performs different functions for each process. In the staged oxidation process, Moat-2 defines the region receiving the threshold adjusts for the thick-oxide devices. In the etch-and-regrowth process, the Moat-2 geometry defines both the regions protected from etchback and the regions receiving the thick-oxide threshold adjusts.

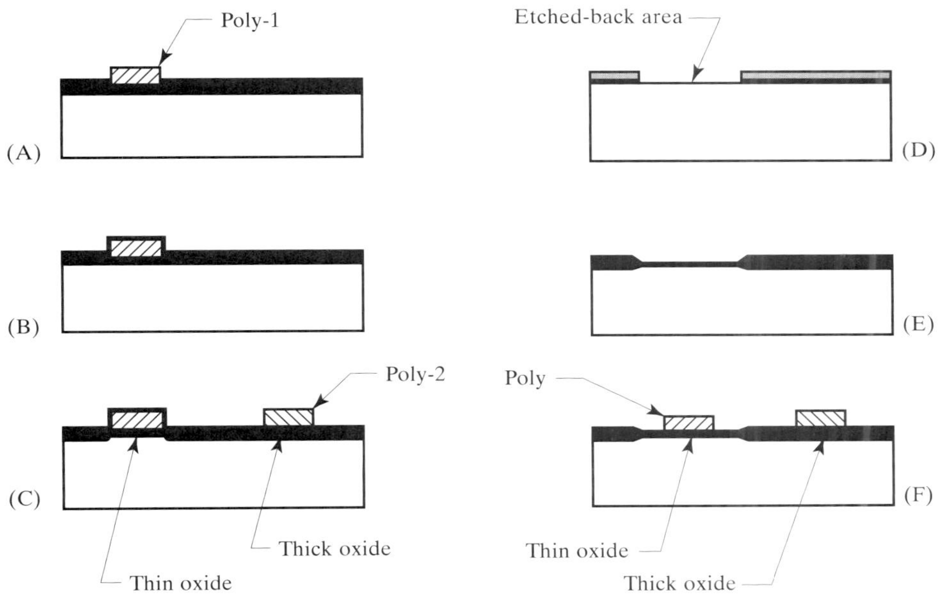

FIGURE 12.8 Process steps for growing multiple thicknesses of oxide using staged oxidation (A-B-C) and etch-and-regrowth (D-E-F) techniques.

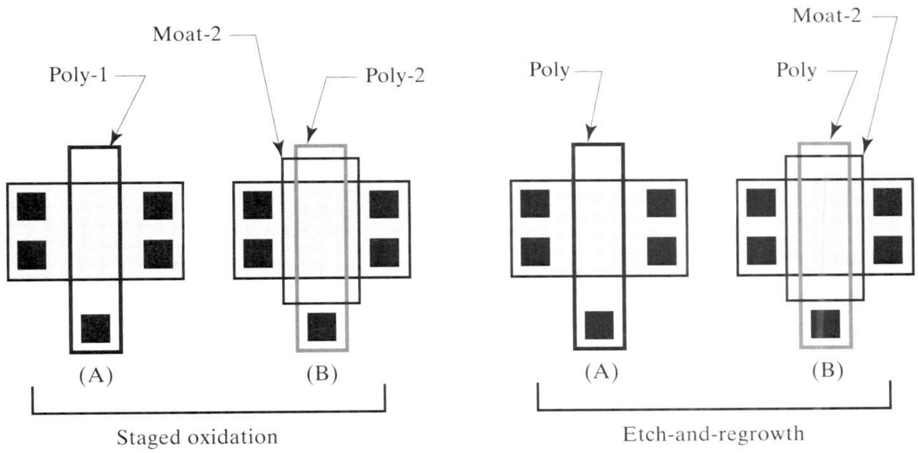

FIGURE 12.9 Comparison of layouts of (A) thin-oxide and (B) thick-oxide transistors constructed for a staged-oxidation process and an etch-and-regrowth process.

12.2 POWER MOS TRANSISTORS

MOS transistors can switch or regulate large amounts of power. Devices specifically designed for such applications are called *power transistors* to distinguish them from low-power, or *small-signal,* devices. MOS power transistors have several advantages over their bipolar counterparts. The forward-bias safe operating area of an MOS transistor is not constrained by thermal runaway or secondary breakdown, as is the safe operating area of a bipolar transistor. MOS transistors also make superior switches because they are not subject to the saturation effects that plague bipolar transistors. MOS transistors also have several significant limitations, most notably low transconductance and poor transient power handling capability.

Thermal Runaway

The collector current of a bipolar transistor increases with temperature, making these devices vulnerable to a positive feedback mechanism called *thermal runaway* (Section 8.1.3). MOS transistors are almost immune to thermal runaway because their drain currents actually decrease with temperature. The positive temperature coefficient of the bipolar collector current stems from the temperature coefficient of V_{BE} (–2mV/°C). Threshold voltages also decrease with temperature, but the low transconductance of the MOS transistor prevents this temperature coefficient from having much effect on drain current. The negative temperature coefficient of the device transconductance causes the drain current to decrease with temperature, as stated above.

Secondary Breakdown

Secondary breakdown (Section 8.1.3) clips off a portion of the forward-biased safe operating area (FBSOA) curve of a bipolar transistor (Figure 12.10A). Since MOS transistors do not experience secondary breakdown, the MOS FBSOA curve is limited only by breakdown voltage, power dissipation, and current-handling capability (Figure 12.10B). The safe operating area of the MOS transistor not only exceeds that of an equivalent bipolar transistor, but it is also more predictable because it is limited only by packaging, metallization, and breakdown voltages. Bipolar transistors may have more robust characteristics than equivalent MOS transistors under reverse breakdown or when handling transient power pulses.

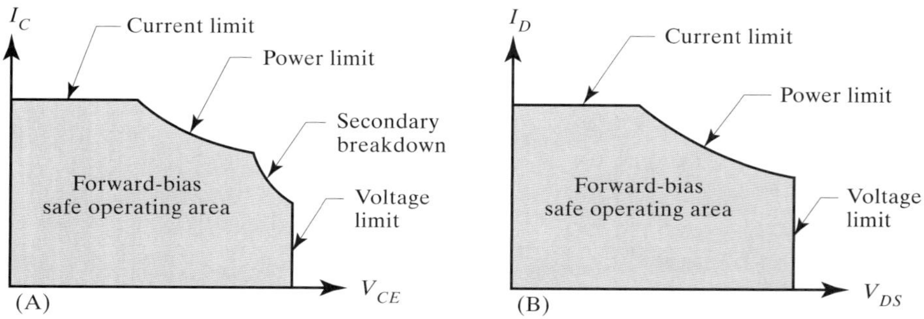

FIGURE 12.10 Forward-biased safe operating area (FBSOA) curves for (A) a power bipolar transistor and (B) a power MOS transistor.

Rapid Transient Overload

Although MOS transistors generally do not experience thermal runaway or secondary breakdown, they do sometimes experience current focusing problems caused by debiasing in long, narrow polysilicon gate fingers. The resistance of these fingers can become quite large, especially if the poly is not silicided. This resistance forms a distributed RC network with the gate capacitance. When the gate voltage slews rapidly, the ends of the transistor nearest the gate connection turn on and off before the rest of the device. This progressive turn-on characteristic sometimes leads to localized overheating and device failure.

Figure 12.11 shows a simplified model of a single gate finger of an MOS power device. M_1 represents the portion of the gate finger nearest the gate connection, while M_2 and M_3 represent more distant portions. Resistors R_1 and R_2 model the resistance of the polysilicon gate. If the gate drive voltage V_G rises rapidly, then transistor M_1 turns on before transistors M_2 and M_3 due to the RC time delays caused by gate resistance and capacitance. The portion of the finger nearest its termination (M_1) begins to discharge load capacitance C_L before the other portions of the transistor can take up their share of the load. If the voltage across C_L is large, then the current density through M_1 may become large enough to damage the device. This type of failure only occurs if the rise time of the gate drive voltage V_G is smaller than the gate time delay, which is typically about a few nanoseconds. Rapid transient overloads of this sort often occur in ESD protection circuits and MOS gate drivers.

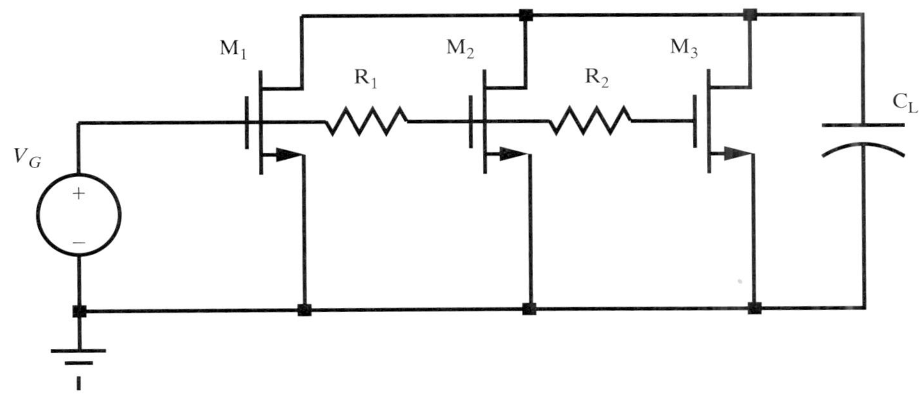

FIGURE 12.11 Model of a long MOS gate finger driving a capacitive load.

Bipolar transistors can usually withstand rapid transient overloading because they dissipate energy into the relatively large volume of their collector-base depletion regions. MOS transistors are more vulnerable because they dissipate energy within the small volume of their pinched-off channel. The bipolar transistor also has the advantage of dissipating energy deep within the silicon, far away from contacts that are prone to damage from overheating. Bipolar transistors are therefore favored for pulsed-power applications such as gate drivers and ESD protection circuitry.

MOS Switches versus Bipolar Switches

MOS transistors can conduct large currents at remarkably low drain-to-source voltages. The behavior of the transistor under these conditions can be derived from the Shichman-Hodges equation for the linear region. The equation is first rearranged in the form

$$I_D = k\,(V_{GS} - V_t)\,V_{DS} + \frac{kV_{DS}^2}{2} \qquad [12.2]$$

The quadratic term becomes negligible at low drain-to-source voltages ($V_{DS} << V_{GS} - V_t$). The above equation then simplifies to

$$I_D \cong k\,(V_{GS} - V_t)\,V_{DS} \qquad [12.3]$$

This equation reveals a linear relationship between the drain-to-source voltage V_{DS} and the drain current I_D. The transistor therefore behaves as if it were a resistor whose value $R_{DS(on)}$ equals

$$R_{DS(on)} \cong \frac{1}{k(V_{GS} - V_t)} \qquad [12.4]$$

The *on resistance* $R_{DS(on)}$ varies inversely with device transconductance and inversely with effective gate voltage V_{gst}. In theory, the on resistance can be reduced to arbitrarily small values by increasing *W/L*. In practice, the resistance of the metallization system and the bondwires places a lower limit on the achievable on resistance. Typical values of $R_{DS(on)}$ for large integrated power transistors are 50 to 500mΩ.

MOS transistors can achieve much lower forward voltage drops than bipolar transistors. The difference between the built-in potentials of the collector-base and the emitter-base junctions sets a lower limit on the achievable saturation voltage of a bipolar transistor. This quantity, called the *intrinsic saturation voltage,* usually equals at least 50mV for vertical bipolar transistors. Lateral bipolar transistors have zero intrinsic saturation voltages, but no junction-isolated transistor can approach zero saturation voltage without experiencing substrate injection (Section 8.1.4). Depending on junction temperature and current density, practical saturation voltages range from 0.1V to as much as 1V. This performance is much poorer than that of a 250mΩ MOS transistor with a forward voltage drop of only 25mV at 100mA. Clearly, MOS transistors make much better switches than do bipolar transistors.

MOS power switches also require much less sophisticated predrive circuitry than bipolar power switches. A bipolar switch must have adequate base drive to ensure that it remains in saturation at the maximum load current. The excess base drive is wasted when the transistor operates at lower currents. The most efficient types of base drive circuits monitor the collector current and provide just enough base drive to ensure that the transistor always remains saturated. These *proportional base drive* circuits are generally rather complicated. Most base drive circuits incorporate additional components to extract stored base charge from the saturated transistor, but even with these additional components the switching speed of power bipolar transistors remains less than 500kHz.[5] MOS power transistors require neither proportional drive nor special turnoff circuitry, and they can easily achieve switching speeds of several megahertz. MOS transistors are therefore overwhelmingly superior to bipolar transistors for high-power, high-speed switching applications such as switched-mode power supplies.

12.2.1. Conventional MOS Power Transistors

MOS transistors generally do not require ballasting because they do not experience thermal runaway or secondary breakdown. The same finger layouts used to construct small-signal transistors therefore serve equally well for power applications.

MOS power transistors are usually specified in terms of their on-resistance $R_{DS(on)}$ measured at a specified gate voltage V_{GS} and junction temperature. The $R_{DS(on)}$ of a power MOS transistor typically increases by 50% when the junction temperature rises from 25°C to 125°C, and it varies about ±30% over process. The metallization resistance becomes significant for on resistances of less than an Ohm, and the equation for $R_{DS(on)}$ then becomes

$$R_{DS(on)} \cong \frac{1}{k(V_{GS} - V_t)} + R_M \qquad [12.5]$$

where R_M is the sum of the resistance of the source and drain metallization. This metallization resistance is difficult to compute because it depends on transistor geometry. Many designers avoid the need for determining R_M by relying on measured $R_{DS(on)}$ data. This method requires that one measure the $R_{DS(on)}$ of a sample device whose layout resembles that of the proposed power device. The measured $R_{DS(on)}$ is then used to compute a figure of merit called the *specific on-resistance* R_{SP}

$$R_{SP} = A_d R_{DS(on)} \qquad [12.6]$$

where A_d represents the drawn area of the sample layout. The specific on-resistance is usually given units of $\Omega \cdot mm^2$. Smaller values of R_{SP} indicate increasingly area-efficient layouts.

Once the specific on-resistance has been determined, one can use equation 12.6 to compute the area required to obtain any desired on resistance. The biggest prob-

[5] This switching speed applies only to large power devices operated in deep saturation. Small-signal devices can operate at much faster speeds, especially if they do not enter saturation.

lem with this technique lies in obtaining an accurate estimate of the specific on-resistance. This is much more difficult than it might seem. The measured value of R_{SP} should not include the resistances of bondwires and leadframe because these do not scale with device area. This is best done by providing Kelvin connections (Section 14.3.2) to the sample device. Alternatively, one can measure the resistance of the leads and bondwires of a dummy unit that contains no die. $R_{DS(on)}$ computations based on R_{SP} do not include the bondwire and leadframe resistance, so these must be added to obtain the total $R_{DS(on)}$.

The specific on-resistance also varies with device area and aspect ratio. Any one value of R_{SP} only applies to a limited range of device sizes and aspect ratios. In practice, one cannot rely on an empirical value of R_{SP} to scale the $R_{DS(on)}$ of a device by more than a factor of two or three. If R_{SP} values are available for a range of device sizes, then one can interpolate between these measurements to find the R_{SP} value for a device of intermediate size. A similar process can be used to account for variations in aspect ratio.

Alternatively, one can attempt to compute the metallization resistance based on an analysis of the geometry of the transistor. Hand calculations yield only approximate results because of the large number of assumptions and simplifications required to render the problem tractable. A computerized finite-element analysis provides more accurate results because it takes into account a larger number of geometric factors. In either case, the computations require a detailed knowledge of the metallization pattern. The following sections describe two of the most popular metallization patterns for MOS power transistors. Many other patterns have been proposed for specific applications, but most of these are not sufficiently general to merit further discussion.

The Rectangular Device

Figure 12.12 shows a diagram of a simple double-level-metal pattern that produces a compact rectangular device. The top portion of the diagram shows only the interdigitated metal-2 patterns for the source and the drain fingers. The lower portion of the diagram shows these metal-2 patterns superimposed over the metal-1 fingers. Each of these fingers consists of a narrow strip of metal-1 spanning the entire width of the transistor and containing one row of contacts and vias. The metal-2 buses running up the left and right sides of the transistor collect the currents from all of the fingers and feed the source and drain terminations, which may either lie at the top or the bottom of the transistor.

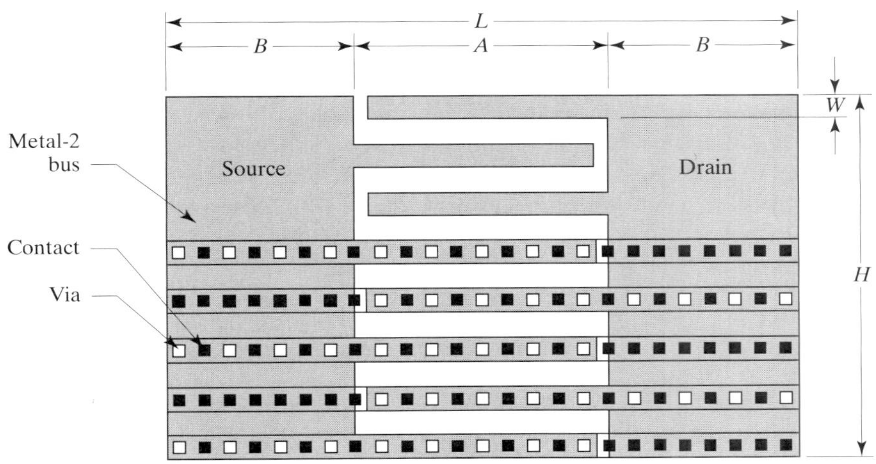

FIGURE 12.12 Metallization pattern for a rectangular MOS power device.

The metallization pattern of Figure 12.12 provides the maximum possible amount of metallization for both the source and the drain. To understand how this has been achieved, consider a single finger in isolation from the rest of the device—for example, the bottom finger. Current flows along this finger from right to left, finally exiting into the metal-2 bus connecting to the source. No vias exist on the rightmost third of the finger. The current flowing through this portion of the finger must pass entirely through metal-1. Each contact feeds a small amount of current into the metal, so the magnitude of the current increases as it flows leftward. Metal debiasing becomes an increasingly serious concern as the magnitude of the current increases. Once the current reaches the middle third of the finger, a portion of it flows upward through vias to reach a strip of metal-2. The current now flows through a sandwich of metal-1 and metal-2. The magnitude of the current continues to increase as it flows leftward. Once the current reaches the leftmost third of the finger, it flows up into the metal-2 bus and out to the termination of the transistor.

The magnitude of the voltage drop along a source finger, V_{SM}, increases from right to left, while the magnitude of the voltage drop along a drain finger, V_{DM}, increases from left to right (Figure 12.13). The sum of these voltage drops $V_{DM} + V_{SM}$, varies less than either of the terms that comprise it. This not only ensures approximately equal conduction through all parts of the transistor, but it also reduces the overall $R_{DS(on)}$. This principle can be applied to the metallization patterns for almost any type of power device, but it is particularly applicable to MOS power transistors where metallization resistance plays such an important role.[6]

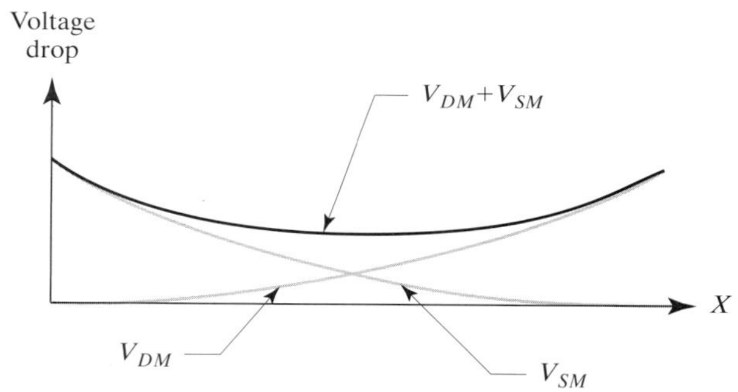

FIGURE 12.13 Graph of voltage drops across a lateral section of the power transistor of Figure 12.12.

The metal-2 buses along the left and right sides of the transistor collect the currents flowing from the individual source and drain fingers. These currents flow vertically to the terminations of the device. A variety of different termination arrangements are possible, some of which perform better than others. Figure 12.14A shows a common arrangement in which both terminations lie on the same end of the transistor. The paired terminations allow connections to adjacent bondpads, but they produce excessive voltage drops and an uneven distribution of current in the device. Figure 12.14B shows a better arrangement where the source and drain terminations lie at opposite ends of the transistor. This arrangement delivers a more even distribution of current and exhibiting a lower total resistance than the arrangement in Figure 12.14A.

---

[6] Krieger analyzes the distribution of currents for both parallel and antiparallel current flow in G. Krieger, "Nonuniform ESD Current Distribution Due to Improper Metal Routing," *EOS/ESD Symposium Proc.*, EOS-13, 1991, pp. 104–108.

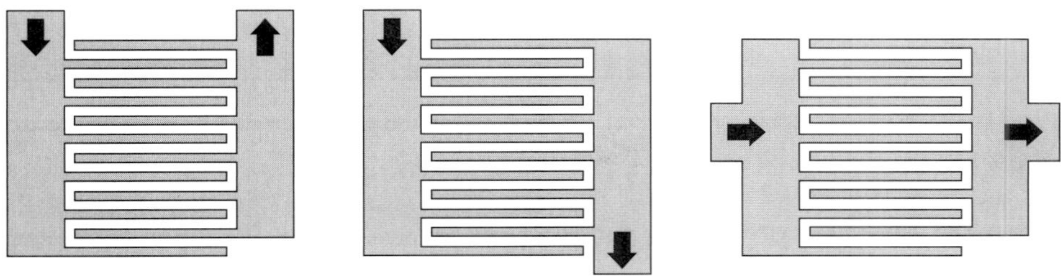

FIGURE 12.14 Three different metallization patterns for rectangular power transistors. The arrows indicate the flow of current to the termination points.

Figure 12.14C shows another method sometimes used to minimize metallization resistance. The termination points lie midway along either bus, so the current does not have to flow through the full length of the buses. This arrangement does minimize resistance, but it also produces an uneven current distribution. For most purposes, the layout in Figure 12.14B is superior to those in both Figure 12.14A and 12.14C.

The Diagonal Device

The rectangular layout in Figure 12.14 has one glaring flaw. The width of the metal-2 buses remains constant, yet the current flowing through them varies. Tapered buses can minimize debiasing and can provide a more uniform distribution of current among the fingers of the transistor. Figure 12.15 shows a layout employing tapered buses. The fingers of the transistor are arranged in a diagonal pattern that naturally produces trapezoidal metal-2 buses on either side of the device. The source and drain terminations must lie on opposite ends of the transistor. This device is more difficult to construct than the rectangular layout shown in Figure 12.12, and computer simulations are required to determine its optimum dimensions. The triangular areas beneath the metal-2 buses must be filled with circuitry in order to obtain the full packing density promised by this layout. Many designers prefer to use rectangular layouts, such as the one in Figure 12.14B, which are easier to construct and to optimize.

Computation of R_M

Accurate calculations of the metallization resistance become very complex and generally require computer modeling. The metallization pattern in Figure 12.14B

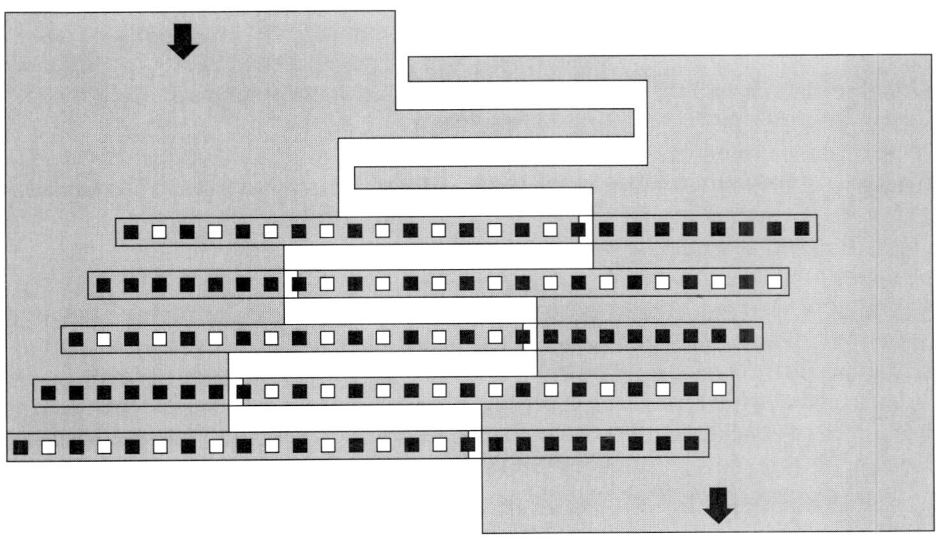

FIGURE 12.15 Metallization pattern for a diagonal MOS power device.

represents an exception, as it is possible to estimate its metallization resistance using the formula

$$R_M = \frac{B^2 R_{S1}}{2W\ N_D L} + \frac{A R_{S12}}{2W\ N_D} + \frac{H R_{S2}}{2B} \qquad [12.7]$$

where N_D equals the number of drain fingers (or half the total number of source/drain fingers), R_{S1} equals the sheet resistance of metal-1, R_{S2} equals the sheet resistance of metal-2, and R_{S12} equals the sheet resistance of a parallel combination of metal-1 and metal-2. Figure 12.12 shows the relationship between dimensions *A*, *B*, *W*, *L*, and *H*. This derivation assumes that each source/drain finger conducts an equal amount of current and that the current flowing through a finger increases linearly along its length. The formula also neglects the variation in voltage across the width of the metal-2 buses.

The above equation can be analyzed (Appendix D) to determine the optimum width, *B*, of the metal-2 buses, which equals

$$B = \left(\frac{R_{S12}}{R_{S1}}\right) L \qquad [12.8]$$

Assuming that both metal-1 and metal-2 have the same composition, equation 12.8 becomes

$$B = \left(\frac{t_1}{t_1 + t_2}\right) L \qquad [12.9]$$

where t_1 and t_2 are the thicknesses of metal-1 and metal-2, respectively. This equation provides a means of sizing the metal-2 buses. If the thickness of metal-1 equals or exceeds the thickness of metal-2, then the buses should each extend across half of the transistor. In this case, the interdigitated region described by dimension *A* vanishes entirely. If the thickness of metal-2 exceeds the thickness of metal-1, then the buses should not entirely cover the transistor, and an interdigitated region should exist between them. Although equations 12.8 and 12.9 were derived specifically for the structure in Figure 12.14B, they concern only a single finger and so apply also to the structures in Figures 12.14A and 12.14C.

Other Considerations

The metal leads and bondwires that connect a power transistor to its load can substantially increase $R_{DS(on)}$. These resistances depend on the general size and shape of the transistor and its placement relative to its bondpads. The calculations form part of the floorplanning process discussed in Section 14.2.

The connection of the gate lead also merits consideration. The resistance of long stretches of polysilicon substantially slows the switching of large power transistors. This resistance can be minimized by connecting the individual gate fingers with metal jumpers. Connecting both ends of the gate further reduces the gate resistance by a factor of about four (Figure 12.16).

Other factors worth considering when laying out power transistors include the placement of backgate contacts and guard rings. Power transistors may use either interdigitated or distributed backgate contacts, depending on which technique provides the smallest overall area (Section 11.2.7). Devices requiring independent connection to the backgate must use interdigitated backgate contacts spaced away from the neighboring source fingers. PMOS transistors constructed in analog BiCMOS processes may use NBL to obtain a solid backgate connection without the need for interdigitated or distributed backgate contacts.

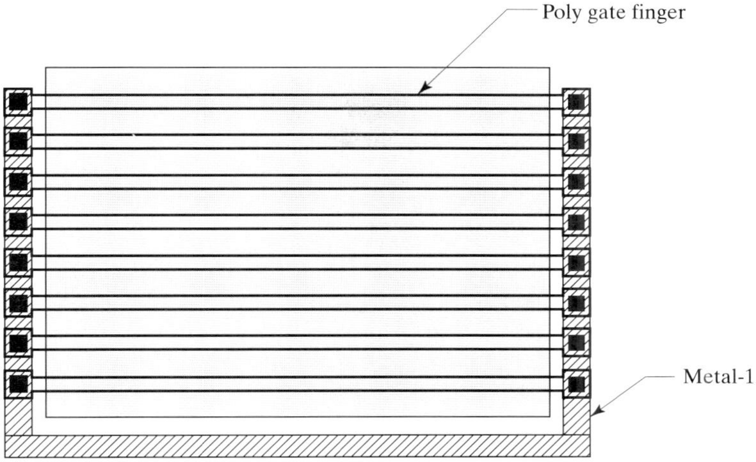

FIGURE 12.16 Gate metallization connecting both ends of the gate fingers reduces gate resistance.

Some applications momentarily forward-bias the backgate diode of a power MOS transistor. If this occurs, then the transistor not only requires an extensive network of backgate contacts to provide recombination current, but it also requires an efficient system of minority carrier guard rings.

Transistors constructed in the epi pose a particular problem because no way exists to block the flow of minority carriers to the substrate. This situation occurs when NMOS transistors are constructed in N-well CMOS processes. In BiCMOS processes, one can sometimes isolate these NMOS transistors using a combination of deep-N+ sinker and NBL (Section 11.2.2). An isolated NMOS cannot inject electrons into the substrate because the isolation structure acts as an electron-collecting guard ring. If an isolated structure is not feasible, then the designer must fall back on guard rings placed around the periphery of the transistor. These guard rings are often quite effective when combined with a heavily doped substrate. If the process uses a lightly doped substrate, then the guard rings should be made as wide as possible to prevent minority carriers from passing underneath them. Substrate contacts should be placed on the far side of the guard ring from the point of injection. Majority carrier current flowing underneath the guard ring through the lightly doped substrate generates an electric field that opposes the flow of minority carriers underneath the guard ring. One can sometimes arrange the wells of adjacent power transistors so they also act as guard rings.

Some circuits use a small transistor to sense the current passing through a much larger one. Ideally, the sense transistor should consist of a number of segments scattered throughout the transistor, so the average of these segments represents the average operating conditions of the power transistor. It is difficult to embed sense transistors within the interior of the power transistor. Instead, these devices usually lie at the ends of gate fingers along one or two sides of the device. If only one sense transistor segment can be used, then this should occupy the center of one side of the device. Two sense segments should occupy the center of opposite sides of the device. Four sense segments should occupy pairs of sites located symmetrically around an axis passing through the centroid of the power transistor. If the sense segments can lie within the confines of the power transistor, they should occupy locations approximately half way from the center of the transistor to its periphery, and should be located in symmetric locations as discussed above.

Figure 12.17 shows a typical example of a single embedded sense transistor located on the end of a gate finger. In practice, the power transistor would have a much larger number of fingers, and the one chosen for the sense device would lie as near to an axis of symmetry of the device as possible. The sense device shares common gate, source, and backgate connections, but it has an independent drain connection.

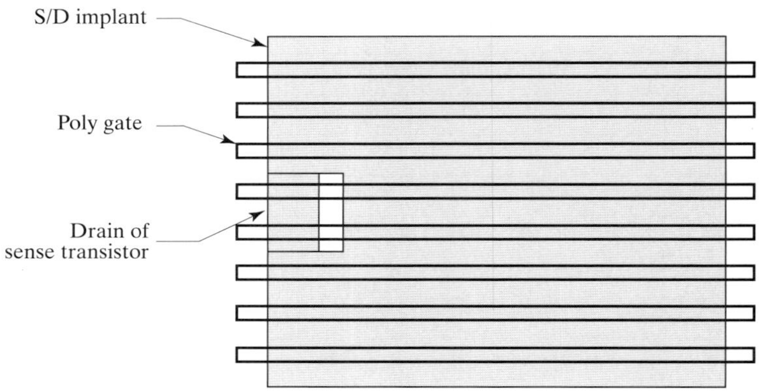

FIGURE 12.17 Construction of an embedded sense transistor (contacts and metallization not shown).

Nonconventional Structures

The conventional self-aligned poly-gate transistor consists of a series of interdigitated source and drain fingers. Although this arrangement possesses the virtue of simplicity, it does not produce the densest possible layout. Other designs can achieve lower specific on-resistances by tightly packing arrays of cleverly shaped source and drain elements. The *waffle transistor* in Figure 12.18A exemplifies this concept. It uses a mesh of horizontal and vertical poly strips to divide the source/drain implant into an array of squares. Each square contains a single contact. By alternately connecting these contacts to the source and drain metallization, one can arrange four drains around each source and four sources around each drain. The drain and source metallization consists of a series of diagonal strips of metal-1, which are usually combined with an interdigitated metal-2 pattern similar to those used for conventional transistors.

An analysis of the W/L ratios achieved for a given device area will show that the waffle transistor provides an increase in packing density equal to

$$\frac{(W/L)_w}{(W/L)_c} = \frac{2S_d}{L_d + S_d} \qquad [12.10]$$

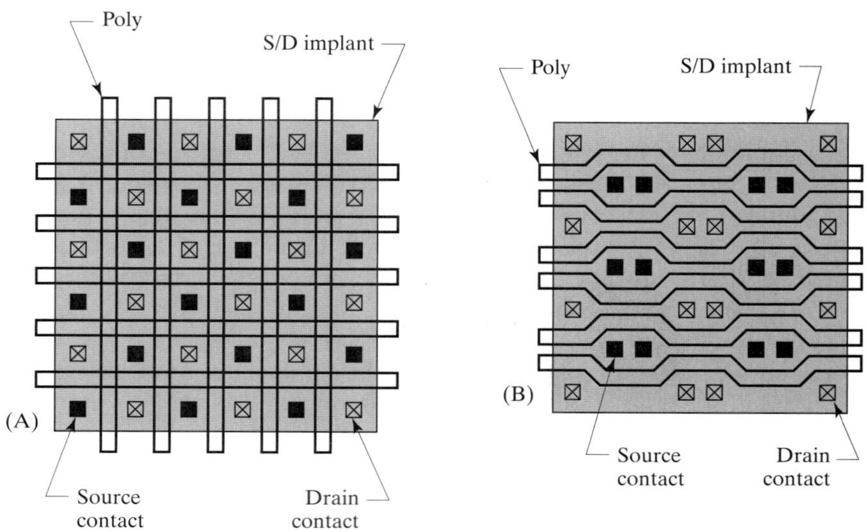

FIGURE 12.18 Nonconventional MOS transistor layouts: (A) waffle and (B) bent-gate. The drain contacts have been drawn differently than the source contacts to aid in their identification.

where $(W/L)_w$ of the waffle transistor and $(W/L)_c$ of the conventional interdigitated transistor are measured from two devices consuming equal die areas. The waffle transistor offers better packing than the conventional interdigitated transistor as long as the spacing between the gates S_d exceeds the gate length L_d. Almost all power transistors meet this requirement. Suppose the layout rules specify a minimum drawn gate length of 2μm, a minimum contact width of 1μm, and a minimum spacing poly-to-contact of 1.5μm. Using these rules, equation 12.10 indicates that the waffle transistor provides approximately 33% more transconductance than the interdigitated-finger transistor. A more precise estimate of the benefits of the waffle transistor would account for the differences between drawn gate length and effective gate length and would include corner conduction terms for the waffle transistor, neither of which are included in Equation 12.10.

The waffle transistor has three crucial defects. First, the above analysis does not consider the effects of metallization resistance. The metallization invariably contributes a significant portion of the $R_{DS(on)}$ of the transistor, and in thin CMOS metal systems it often becomes the dominant factor. If one assumes that the metallization contributes about half the total $R_{DS(on)}$, then the improvement gained by using the waffle layout drops by half, or from 33% to 16% for this example. The situation is actually even worse because the waffle layout is difficult to properly metallize. The metal-1 fingers must repeatedly cross the gate poly, which almost certainly introduces significant step-induced metal thinning. Second, the waffle transistor contains a large number of bends in its channels. These bends produce sharp corners in the source/drain regions that avalanche at lower voltages than the remainder of the transistor. Localized avalanche limits the amount of energy the waffle transistor can dissipate. This limitation becomes apparent in ESD testing, where the performance of waffle transistor may fall short of that of the conventional layout. Fillets or chamfers applied to the corners of the source/drain squares will largely eliminate this problem, and a transistor that includes them may provide even better ESD performance than a conventional layout.[7] Third, the waffle transistor makes no provision for backgate contacts. Unless the transistor is used in combination with a heavily doped substrate or a buried layer to provide backgate contact, it is quite susceptible to backgate debiasing and latchup. No simple way exists to add interdigitated or distributed backgate contacts to a waffle layout.

Figure 12.18B shows a *bent-gate transistor* that avoids most of the difficulties of the waffle transistor, while offering some unique advantages. The bent gates increase the gate width while simultaneously allowing the gate strips to pack more closely together. This layout readily accommodates distributed backgate contacts without sacrificing undue die area. It also avoids the use of 90° bends in favor of gentler 135° bends, which are less prone to localized avalanche. The diagonal arrangement of the source and drain contacts also provides additional source/drain ballasting that can improve robustness under extreme conditions, such as those encountered during ESD testing. These benefits, combined with the ease of inserting extensive networks of distributed backgate contacts, make this device ideal for applications that routinely experience transient overloads.

12.2.2. DMOS Transistors

High-voltage transistors require short, heavily doped backgates and wide, lightly doped drift regions. These are more-conveniently produced by diffusing the backgate into the drift region than *vice versa*. The *double-diffused MOS*, or DMOS, uses

---

[7] L. Baker, R. Currence, S. Law, M. Le, C. Lee, S. T. Lin, and M. Teene, "A 'Waffle' Layout Technique Strengthens the ESD Hardness of the NMOS Output Transistor," *EOS/ESD Symposium Proc.*, EOS-11, 1989, pp. 175–181.

this approach to produce short-channel, high-voltage transistors optimized for use as power devices.

Like the DDD transistor, the DMOS relies on the self-alignment of two diffusions driven through a common oxide opening. An N-channel DMOS is fabricated by diffusing boron and arsenic into lightly doped N-type silicon (Figure 12.19). Boron outdiffuses more rapidly than arsenic, producing a moderately doped P-type region enclosing a shallower and more-heavily doped N-type region. The heavily doped arsenic core forms the source of the DMOS transistor, while the surrounding, moderately doped boron diffusion forms the backgate. The channel length of the transistor equals the difference between the surface outdiffusion distances of the boron and arsenic implants, which depends solely on doping concentrations and drive times.

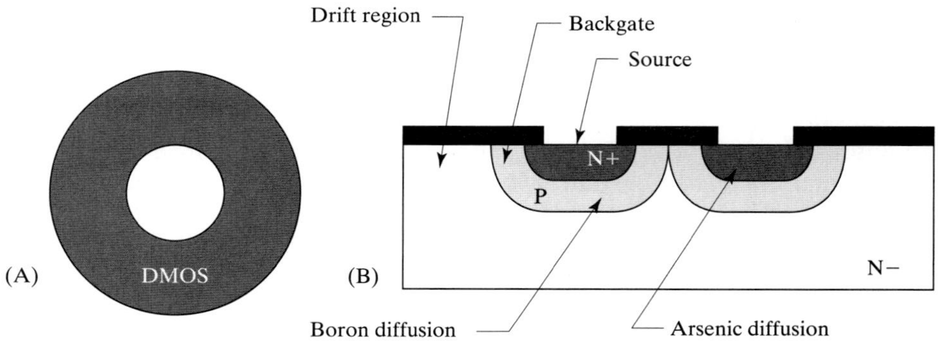

FIGURE 12.19 Layout of (A) DMOS mask geometry and (B) the resulting pattern of diffusions.

The lightly doped N-type region surrounding the backgate serves as the drift region of the DMOS transistor. This drift region must be contacted by means of a heavily doped N-type region. Discrete DMOS transistors often occupy a lightly doped N-type epi deposited on a heavily doped N-type substrate. The drain current flows down through the epi to the substrate and exits through the backside of the die. The epi thickness determines the width of the drift region and hence the maximum operating voltage of the transistor. In order to integrate this *vertical DMOS,* the drain region must be isolated from the substrate. This can be achieved by placing the transistor in an N-well furnished with NBL and deep-N+. This resolves the isolation issue, but the transistor still exhibits excessive drain resistance at low forward voltages because of incomplete depletion of the drift region. Drain resistance represents a major challenge for constructing low-voltage DMOS transistors. The epi thickness cannot be reduced too far or the tail of the NBL will intersect the DMOS diffusions, so another approach must be tried.

The Lateral DMOS Transistor

Most integrated DMOS transistors use a shallow, heavily doped N-type diffusion placed next to the DMOS backgate to extract the drain current. This type of device is called a *lateral* DMOS, or LDMOS.[8] The separation of the backgate and drain contact diffusions determines the width of the lateral drift region. This drift region is designed to fully deplete through at a relatively low voltage. This type of transistor does not require NBL or deep-N+, although these are often added to minimize substrate injection in the event that the backgate forward-biases into the drain.

---

[8] J. D. Plummer and J. D. Meindl, "A Monolithic 200-V CMOS Analog Switch," *IEEE J. Solid-State Circuits,* Vol. SC-11, #6, 1976, pp. 809–817.

The DMOS backgate contact represents something of a problem. Not only is the backgate relatively lightly doped, but it is also extremely narrow. A heavily doped P-type diffusion can contact the backgate, but only if it also contacts the N+ source. The resulting P+/N+ junction may leak so badly that the backgate cannot be isolated from the source. Most DMOS transistors use an annular geometry containing a central P+ plug that serves as a backgate contact. The P+ plug shorts to the source of the transistor through a single contact opening that covers both (Figure 12.20A). DMOS transistors are considered asymmetric devices because their backgate usually connects to their source and because the diffused backgate is more lightly doped near the drain than near the source.

Figure 12.20 shows a simple LDMOS transistor constructed in an N-well analog BiCMOS process. An annular DMOS implant consisting of arsenic and boron defines the source and backgate regions of the device, and an enclosing N-well serves as its drift region. The P+ plug contacting the backgate consists of a PSD implant whose outer edge coincides with the inner edge of the DMOS implant. The extrinsic drain consists of a ring of NSD implanted around the transistor. The spacing between the NMoat and the DMOS geometries determines the width of the drift region. The poly gate overlaps the thick-field oxide over the drift region to form a field-relief

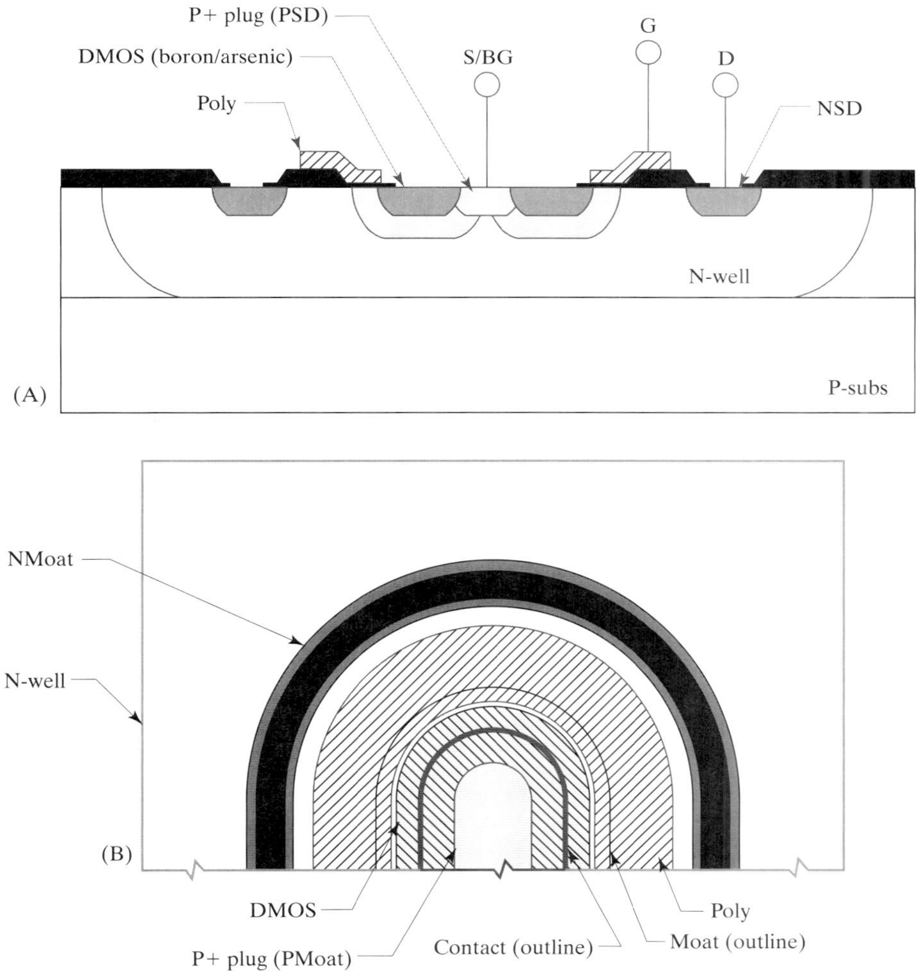

FIGURE 12.20 Layout and cross section of a lateral DMOS transistor constructed in analog BiCMOS.

structure, allowing the transistor to withstand large drain-to-gate voltage differentials without requiring a thick gate oxide. Many elaborations upon this structure exist.[9]

The smallest DMOS transistor uses a circular ring of DMOS implant, like that shown in Figure 12.19. Larger widths can be obtained by connecting many small annular devices in parallel, but such arrays of annular transistors pack rather loosely. The transistor in Figure 12.20 uses an elongated annular geometry to improve packing density. A large power device consists of an array of annular DMOS transistors interdigitated with drain contacts. This type of structure resembles a conventional interdigitated transistor and can use any of the metallization patterns discussed in Section 12.2.1.

NBL often forms part of a DMOS transistor, but its presence does not always prove beneficial. Consider the case of a DMOS transistor without NBL. The drain-backgate and drain-substrate depletion regions both widen as the drain voltage increases. If the well is sufficiently shallow, then it will punch through from backgate to substrate long before the well-backgate junction avalanches. This punchthrough only causes problems if the source of the transistor connects to a voltage other than substrate potential. NBL can prevent punchthrough by constraining the depletion region, but in so doing it intensifies the electric field and reduces the well-backgate avalanche voltage. The DMOS transistor therefore has two separate drain-to-source voltage ratings. A device whose source connects to substrate potential can omit NBL and can obtain a higher operating voltage. Devices whose sources connect to voltages other than substrate potential require NBL and are therefore constrained to lower operating voltages. Various improved versions of the lateral DMOS structure have been proposed.[10]

The DMOS NPN

The DMOS structure in Figure 12.20 contains a parasitic NPN transistor. The source of the DMOS acts as the emitter of this transistor, the backgate as its base, and the drain as its collector. This parasitic NPN has a heavily doped emitter that enhances its emitter injection efficiency, a thin, moderately doped base that reduces its Gummel number, and a wide, lightly doped collector that minimizes the Early effect. The performance of the DMOS NPN can approach that of a conventional CDI NPN, making it a useful alternative to the latter device.

Figure 12.21 shows a layout and cross section of a typical DMOS NPN. This structure uses a circular DMOS implant to form the base and emitter of the transistor. The drawn emitter area equals the drawn area of the DMOS implant. The emitter is contacted by a central plug of NSD, and the base is contacted by a ring of PSD surrounding the DMOS implant. The boron DMOS implant must overlap the PSD implant sufficiently to allow for misalignment. This usually requires that the two implants practically abut one another. The extrinsic collector consists of NBL and deep-N+, just as in a conventional NPN transistor.

---

9 The RESURF (reduced surface field) structure eliminates avalanche breakdown due to electric field intensification at the corners of the drain: J. A. Appels and H. M. J. Vaes, "High Voltage Thin Layer Devices (RESURF Devices)," *IEEE 25th Int. Electron Devices Meeting,* 1979, pp. 238–241. For a typical integrated bipolar/CMOS/DMOS process, see A. Andreini, C. Contiero, and P. Galbiati, "A New Integrated Silicon Gate Technology Combining Bipolar Linear, CMOS Logic, and DMOS Power Parts," *IEEE. Trans. Electron Devices,* Vol. ED-33, #12, 1986, pp. 2025–2030.

10 For example, the RESURF structure has been touted for reducing R_{sp}. See C.-Y. Tsai, J. Arch, T. Efland, J. Erdeljac, L. Hutter, J. Mitros, J.-Y. Yang, and H.-T. Yuan, "Optimized 25V, 0.34mΩ · cm ^{2} Very-Thin-RESURF (VTR), Drain-Extended IGFETs in a Compressed BiCMOS Process," *IEDM,* 1996, pp. 469–472.

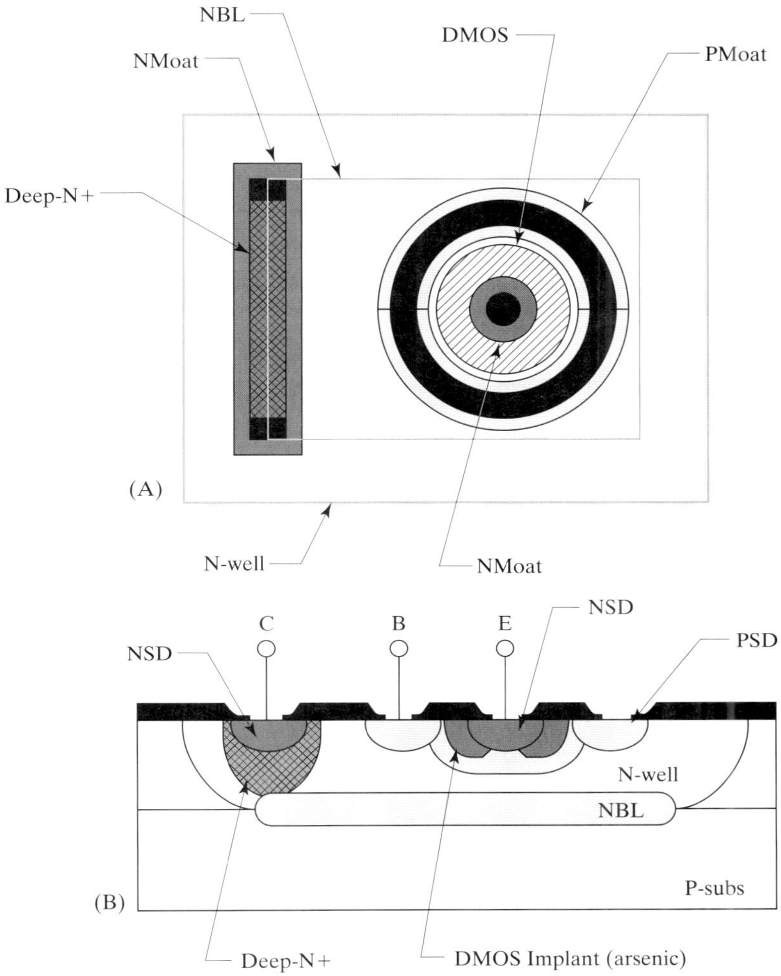

FIGURE 12.21 Layout and cross section of a DMOS NPN (omitting poly field plate).

A P+/N+ junction appears in the conventional DMOS structure between the P+ backgate contact and the N+ DMOS implant. The potential for leakage across this junction is of no concern in a DMOS transistor because the source always shorts to the backgate. The same is not true in the DMOS NPN because these diffusions form its base and emitter. Leakage can be avoided by reducing the dosage of the arsenic DMOS implant. NSD must now be added to the source regions to allow Ohmic contact to the lightly doped arsenic implant.

The structure in Figure 12.21 omits the moat geometry normally covering the DMOS implant, and allows thick-field oxide to grow over it. Dopant segregation and oxidation-enhanced diffusion drive the arsenic emitter deeper into the boron base, reducing the base width of the transistor. Conducting a field oxidation over the DMOS implant thus increases the beta of the DMOS NPN.

All DMOS NPN transistors contain a parasitic DMOS transistor connected between collector and emitter. The structure in Figure 12.21 does not show the poly gate electrode required to suppress this parasitic device. The poly electrode, or *field plate,* must cover the exposed boron DMOS implant with sufficient overlap to allow for misalignment. This field plate is usually connected to the emitter, since this connection shorts the gate and source of the parasitic DMOS.

12.3 THE JFET TRANSISTOR

Junction field-effect transistors (JFETs) were used throughout the 1970s and the early 1980s as substitutes for the less-reliable MOS devices of that era. JFETs were often used in the input stages of operational amplifiers to obtain input leakage currents several orders of magnitude smaller than those generated by the best bipolar circuits.[11] JFETs were also used as analog switches and as current sources.

Standard bipolar easily accommodated the steps required to construct simple JFET structures. The resulting *BiFET* processes merged bipolar and JFET transistors in much the same way that modern BiCMOS processes merge bipolar and CMOS. These BiFET processes were primarily used to construct low-input-current and low-noise operational amplifiers. The older BiFET processes have become largely obsolete because modern BiCMOS processes generally offer better performance (although low-noise BiMOS amplifiers can still outperform their BiCMOS counterparts).

JFET transistors remain of interest because they can be constructed on many existing processes without requiring any additional masking steps, and the resulting devices can replace high-value resistors in startup circuits. The following sections provide a brief overview of the operation and construction of JFETs, with an emphasis on structures compatible with standard bipolar and analog BiCMOS processes.

12.3.1. Modeling the JFET

Although the I-V characteristics of the JFET broadly resemble those of the depletion-mode MOS transistor, the underlying physics of the two devices are quite different. Most textbooks derive the JFET equations from fundamental principles, but so many assumptions are made along the way that the results have little practical value. This section discusses only those aspects of the theoretical model required to understand the sizing of JFET transistors and leaves the remaining details to other texts.[12]

The *pinchoff voltage* V_P of a JFET equals the minimum drain-to-source voltage V_{DS} required to pinch off the drain end of the channel when the gate-to-source voltage V_{GS} equals zero. In theory, the pinchoff voltage of an ideal JFET equals

$$V_P \cong 1.9 \cdot 10^{-16} N_C t^2 \qquad [12.11]$$

where N_C equals the doping concentration of the channel in atoms/cm^3 and t equals the channel thickness in microns.[13] In practice, the channel doping usually varies with depth and the pinchoff voltage must be determined empirically. This is done by examining the I-V characteristics of the JFET for $V_{GS} = 0$. The drain current I_D remains approximately constant at high drain-to-source voltages. As V_{DS} decreases, a point is eventually reached at which the drain current begins to diminish (Figure 12.22B). The pinchoff voltage equals the drain-to-source voltage at this inflection point.

The *saturation current* I_{DSS} of a JFET equals the drain current at $V_{GS} = 0$ and $V_{DS} = V_P$. If one assumes a uniformly doped channel with resistivity ρ, width W, length L, and thickness t, then the saturation current equals

$$I_{DSS} = \frac{V_P t}{3\rho}\left(\frac{W}{L}\right) \qquad [12.12]$$

---

[11] The advantages of JFET input stages vanish at higher temperatures because the input current of a JFET increases exponentially with temperature, while the input current of a base-current-compensated bipolar circuit increases somewhat more slowly.

[12] R. S. Muller and T. I. Kamins, *Device Electronics for Integrated Circuits,* 2nd ed. (New York: John Wiley & Sons, 1986), p. 202ff.

[13] The full equation is $V_P = qN_C t^2/2\varepsilon$, where q is the charge on the electron and ε is the permittivity of silicon.

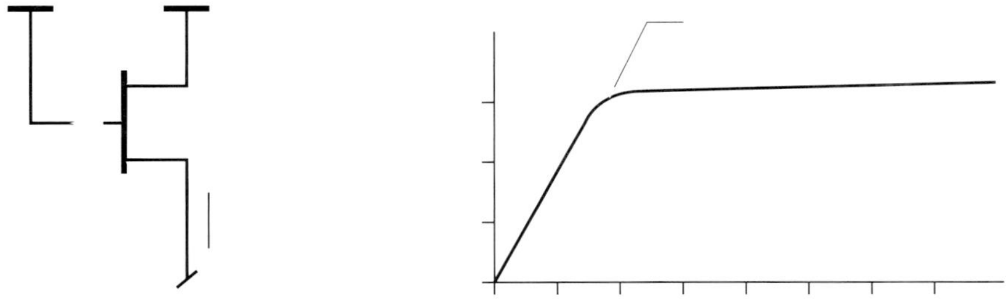

FIGURE 12.22 (A) A schematic of a P-JFET connected as a current source and (B) an I-V plot of the same device showing $V_P = 8\text{V}$ and $I_{DSS} = 16\mu\text{A}$.

This equation can also be used for nonuniformly doped channels providing that one empirically determines the effective channel resistivity ρ by measuring devices of different widths and lengths and fitting these measurements to the equation. Several factors complicate the extraction of I_{DSS}. The width and length used in the equation do not exactly correspond to the drawn dimensions of the device, any more than the effective width and length of MOS transistors exactly correspond to their drawn dimensions (Section 11.2.1). Correction factors δW and δL relate the effective width W_{eff} and effective length L_{eff} to the drawn width W_d and drawn length L_d

$$W_{eff} = W_d + \delta W \quad [12.13A]$$

$$L_{eff} = L_d + \delta L \quad [12.13B]$$

For devices with channel widths of less than 10μm, the value of I_{DSS} can be accurately determined only by measuring a device having the desired channel width. Devices that channel lengths of less than 10μm are better avoided because a variety of short-channel effects complicate the task of sizing the transistors.

12.3.2. JFET Layout

Practical JFET devices can be created using existing layers of a standard bipolar or an analog BiCMOS process. Figure 12.23 shows one type of N-channel JFET compatible with standard bipolar processing. This device is sometimes called an *epi-FET* because its channel consists of a portion of the N-type epitaxial layer. The epi-FET is also called an *epi pinch resistor,* particularly when it operates in its linear region (Section 5.5.5). The thickness of the channel has been greatly reduced by placing a base diffusion over the epi. The updiffusion of the underlying substrate causes substantial grading of the backgate-body junction and renders the constant-doping approximations that underlie equations 12.11 and 12.12 of questionable validity. These devices are usually sized by interpolating between the I_{DSS} currents measured on an array of test devices.

The tank geometry determines the drawn width W_d while the base pinch plate determines the drawn length L_d. The effective width of an epi-FET is substantially smaller than its drawn width because of isolation outdiffusion. The relationship between effective width and drawn width becomes nonlinear for small widths because of diffusion interactions between the opposing sidewalls of the channel. The width correction factor δW may therefore vary with width, especially for small widths. The length correction factor δL also varies with length because of the presence of the extrinsic source/drain regions on either end of the channel, but δL has little effect upon devices having channel lengths of at least 50μm.

The base pinch plate extends into the surrounding isolation and shorts the gate of the epi-FET to substrate. Most epi-FETs are used as startup devices in which the grounded-gate configuration is quite acceptable. If the drain-to-source voltage across the epi-FET is large enough, then its drain will draw a current equal to the

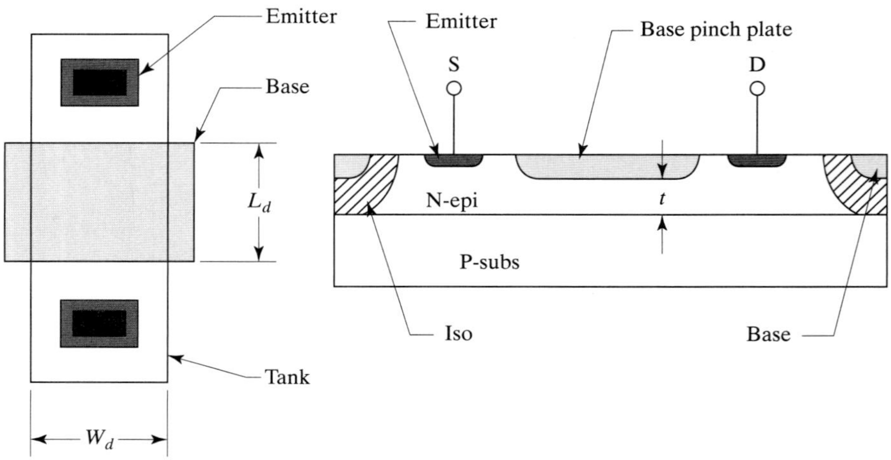

FIGURE 12.23 Layout and cross section of an N-channel JFET constructed in standard bipolar. The gate connects to the substrate and is accessed through an adjacent substrate contact.[14]

saturation current I_{DSS}. In practice, most epi-FETs have such large pinchoff voltages that they do not fully saturate under normal operating conditions, and they therefore resemble pinch resistors. The main advantages of the epi-FET include high breakdown voltage and low transconductance, which together allow it to replace a much larger pinch resistor. The operating voltage of an epi-FET is limited only by the breakdown of the epi-base junction. JFETs are immune to hot-carrier-induced threshold shifts because they do not contain a gate dielectric. They are also immune to the parasitic channel formation and conductivity modulation because the base pinch plate serves as a field plate covering the active region of the device.

Epi-FETs are designed for compactness rather than for precision. These transistors normally use the minimum channel width, even though wider devices exhibit less variability. The channel is frequently serpentined to fit into unused areas in the layout. Contacts are usually placed over the base pinch plate and connected to substrate potential. Although not strictly necessary, these contacts help minimize variations in epi-FET current caused by substrate debiasing. Any contact to the base pinch plate also serves as a substrate contact in its own right and helps extract stray substrate currents flowing near the epi-FET. Some designers use rounded bends in serpentine epi-FETs believing that these increase the breakdown voltage by preventing electric field intensification. Although this practice causes no harm, it provides little or no benefit because the exposed edge of the base pinch plate usually breaks down before the isolation sidewalls.

Analog BiCMOS processes can construct an *N-well JFET* analogous to the epi-FET in Figure 12.23 by substituting N-well for the tank and NMoat for emitter (Figure 12.24). The resulting device usually has a lower pinchoff voltage than its epi-FET counterpart due to the graded nature of the well. The pinchoff voltage can be reduced still further by growing field oxide over the base pinch plate, as the resulting oxidation-enhanced diffusion drives the base deeper into the N-well.

N-well JFETs and epi-FETs vary in several ways. One critical difference concerns the overlap of the base pinch plate over the channel. In the epi-FET, the isolation diffuses inward and the base pinch plate need overlap the tank only slightly, if at all. The base pinch plate of the N-well JFET must overlap the well by a much greater distance because the N-well diffuses outward rather than inward. The high resistance of the P-epi also makes it desirable to add contacts directly to the base pinch plate

[14] A similar device is discussed in D. J. Hamilton and W. G. Howard, *Basic Integrated Circuit Engineering* (New York: McGraw-Hill, 1975), p. 170.

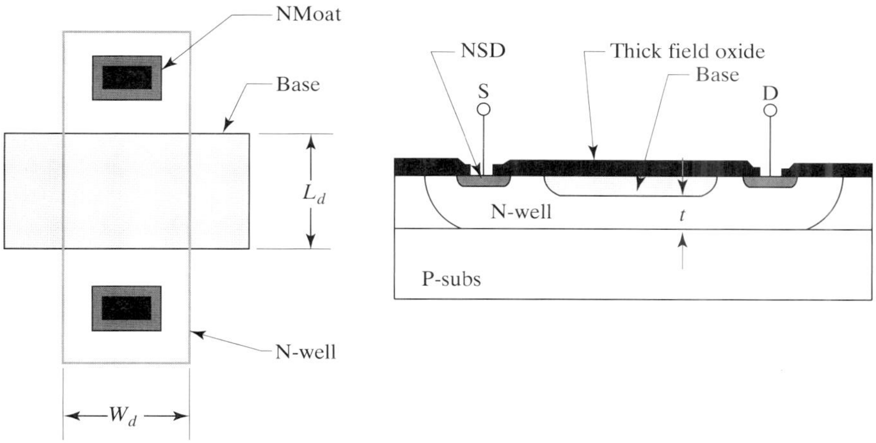

FIGURE 12.23 Layout and cross section of an N-well JFET constructed in an analog BiCMOS process (the contacts to the base plate have been omitted for clarity).

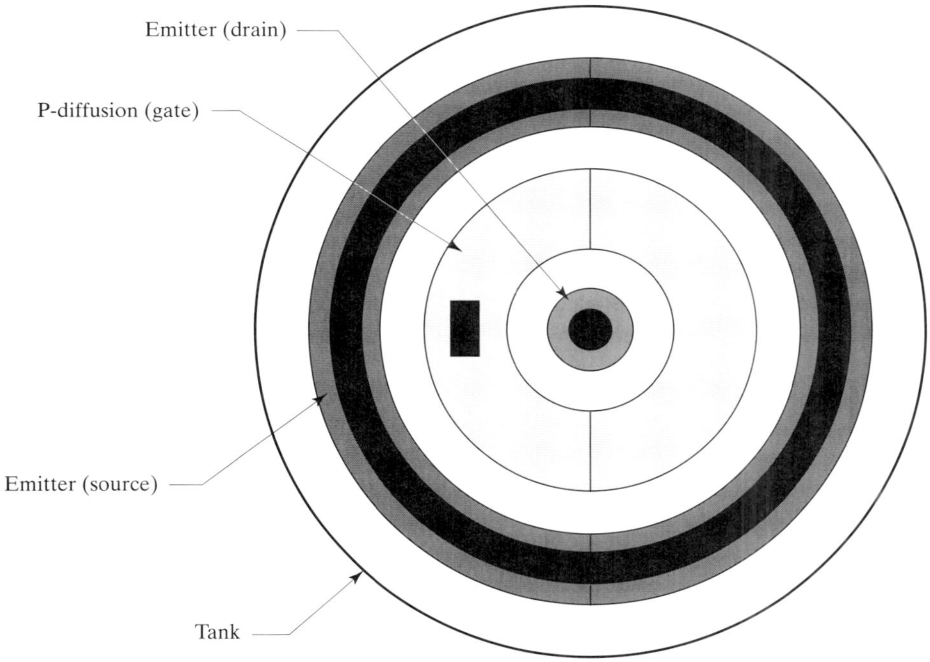

FIGURE 12.24 Annular circularly symmetric N-channel JFET.

rather than to rely on the presence of substrate contacts elsewhere on the die. These contacts should not reside over the channel of the N-well JFET because the moat region required for the contact alters the thickness of the channel. Instead, the contacts should be located next to the device and should connect to it by a strip of base or PSD diffusion.

An N-well JFET with a minimum-width channel has a much lower pinchoff voltage than one with a wider channel. If the channel is covered by thick-field oxide, the narrowest devices may even pinch off entirely and become unusable. These effects occur because the dopant in a narrow N-well diffuses laterally as well as vertically, leaving a lower overall doping concentration within the channel. The dopant in a wider well diffuses laterally near the edges, but the center of the well still retains a high dopant concentration. The wider device thus has a higher pinchoff voltage than the narrow one. If necessary, the pinchoff voltage of an N-well JFET can be increased by placing a moat region above the base or by substituting a PSD implant for the base implant. The PSD implant usually gives so high a pinchoff voltage that the device cannot saturate under normal operating conditions. Thus it behaves as a nonlinear pinch resistor rather than as a true FET (Section 5.5.9).

P-channel JFETs can be constructed in both standard bipolar and analog BiCMOS processes, but those constructed from existing diffusions leave much to be desired. The standard bipolar device has the same structure as a base pinch resistor (Figure 3.15). The analog BiCMOS device has a similar structure, consisting of base pinched by NSD rather than by emitter. The operating voltages of these devices are limited by the avalanche of the base-emitter and base-NSD junctions, respectively. The pinchoff voltages of both devices greatly exceed their respective breakdown voltages, so neither device ever saturates. Both of these devices are really nothing more than nonlinear pinch resistors (Section 5.5.3). A special N-type implant must be added to the process to construct a true P-JFET capable of operating in saturation. This implant must have a slightly shallower junction depth than the base diffusion, and a doping concentration just sufficient to invert the base diffusion. A shallower diffusion yields too large a pinchoff voltage, and a more heavily doped one produces too low a breakdown voltage. No suitable diffusion exists in either standard bipolar or analog BiCMOS, although one can be added as a process extension. Previous-generation BiCMOS processes were generally derived from standard bipolar by the addition of just such an extension. The P-channel transistors constructed in this way are called *double-diffused JFETs* because their gates are produced by the diffusion of the N-implant into the base. The layout and cross section of the double-diffused P-JFET are essentially the same as that of the base pinch resistor in Figure 3.15, with the substitution of the new N-implant for the emitter. New processes rarely support the P-JFET extension because CMOS transistors have largely supplanted JFETs.

All of the layouts previously discussed short the gate to the backgate, which in the N-channel device consists of the substrate. In order to use the N-channel JFET in any application other than as a grounded current source, one must first separate the gate and the backgate electrodes by using an annular structure similar to that in Figure 12.24. The gate of the annular N-JFET consists of a ring-shaped P-type diffusion placed inside an N-epi tank. Tank contacts placed inside and outside this ring serve as the drain and source, respectively. This arrangement minimizes the drain capacitance at the cost of increased source capacitance.

The schematic symbol used for the annular N-JFET is exactly the same as that used for the conventional N-JFET (Figure 1.30A). The two can be differentiated by examining the connection of the gate electrode. The conventional layout in Figure 12.23 should be used if the gate connects to substrate potential. If the gate connects to any other potential, the transistor must use the annular layout shown in Figure 12.24. The substrate forms the backgate of the annular device. The width and length of annular JFET devices are computed using the rules presented for annular MOS transistors (Section 11.2.6).

12.4 MOS TRANSISTOR MATCHING

A wide variety of analog circuits use matched MOS transistors. Some circuits, such as differential pairs, rely on matching of gate-to-source voltages, while others, such as current mirrors, rely on matching of drain currents. The biasing conditions required to optimize voltage matching differ from those required to optimize current matching. One can optimize MOS transistors either for voltage matching or for current matching, but not simultaneously for both.

The relationship between biasing and voltage matching is easily derived from the Shichman-Hodges equations (Section 11.1.1). Suppose two matched MOS transistors operate at the same drain current I_D. If the transistors were ideal devices, then they would develop exactly the same gate-to-source voltage V_{GS}. In practice, mismatches cause the gate-to-source voltages of the two transistors to differ by an amount $\Delta V_{GS} = V_{GS1} - V_{GS2}$. Assuming that the transistors operate in saturation, as is usually the case, then the offset voltage ΔV_{GS} equals

$$\Delta V_{GS} \cong \Delta V_t - V_{gst1}\left(\frac{\Delta k}{2k_2}\right) \qquad [12.14]$$

where ΔV_t equals the difference between the threshold voltages of the two transistors, Δk equals the difference between their device transconductances, V_{gst1} equals the effective gate voltage of the first transistor, and k_2 equals the device transconductance of the second (Appendix D). The offset voltage ΔV_{GS} depends on device dimensions due to the presence of the device transconductance k_2 in the denominator. Similarly, the offset voltage depends on biasing conditions because of the presence of the effective gate voltage V_{gst1} in the equation. These dependencies are unique to MOS transistors and are not shared by bipolar transistors (Section 9.2).

The MOS designer can minimize the offset voltage ΔV_{GS} by reducing the effective gate voltage V_{gst} of the matched transistors. MOS circuits that depend on voltage matching therefore benefit from the use of large W/L ratios and low operating currents. The improvements obtainable in this manner are limited by the onset of subthreshold conduction and by the presence of threshold mismatches. As a practical matter, reducing V_{gst} below about 0.1V produces little improvement in voltage matching.

MOS circuits relying on current matching behave quite differently. The mismatch between two drain currents, I_{D1} and I_{D2}, can be specified in terms of a ratio I_{D2}/I_{D1} equal to

$$\frac{I_{D2}}{I_{D1}} \cong \frac{k_2}{k_1}\left(1 + \frac{2\Delta V_t}{V_{gst1}}\right) \qquad [12.15]$$

The mismatch in drain currents actually increases at low effective gate voltages due to a larger contribution from the threshold mismatch ΔV_t (Appendix D). MOS circuits relying on current matching should operate at reasonably large effective gate voltages to avoid exacerbating threshold voltage variations. The optimal value of V_{gst} depends on many factors and is difficult to quantify. As a practical matter, one should endeavor to maintain a nominal V_{gst} of at least 0.3V (and preferably 0.5V) in MOS transistors generating matched currents. Larger effective gate voltages may provide some additional benefit, but most applications cannot spare the headroom to support a higher V_{gst}.

In summary, MOS circuits that generate matched voltages should operate at low effective gate voltages, while MOS circuits that generate matched currents should operate at high effective gate voltages. For most purposes, a nominal V_{gst} of 0.1V or less will provide optimal voltage matching, and a nominal V_{gst} of 0.3V or more will provide optimal current matching. Assuming that the circuit designer adjusts the biasing of the transistors to these values, the matching now depends almost entirely on the care taken in transistor layout. The next three sections discuss layout considerations that affect MOS matching.

12.4.1. Geometric Effects

The size, shape, and orientation of MOS transistors all affect their matching. Large transistors match more precisely than small ones because increased gate area helps minimize the impact of localized fluctuations. Long-channel transistors match more precisely than short-channel ones because longer channels reduce linewidth variations and channel-length modulation. Transistors oriented in the same direction match better than those oriented in different directions because of the anisotropic nature of monocrystalline silicon. This section discusses the impact of these and other geometric factors on MOS transistor matching.

Gate Area

MOS mismatches have been experimentally measured for a number of processes. These measurements reveal that the magnitude of the threshold voltage mismatch varies inversely with the square root of the active gate area. This relationship can be expressed in terms of the effective channel dimensions W_{eff} and L_{eff} as follows

$$s_{Vt} = \frac{C_{Vt}}{\sqrt{W_{eff}\,L_{eff}}} \qquad [12.16]$$

where s_{Vt} is the standard deviation of the threshold voltage mismatch and C_{Vt} is a constant.[15] The value of C_{Vt} is empirically determined by measuring the random mismatch between pairs of transistors of different sizes. The results only apply to transistors closely resembling the test devices used to derive C_{Vt}. The relationships between drawn dimensions and effective dimensions are not always known, and sometimes the drawn dimensions W_d and L_d must be substituted for the effective dimensions W_{eff} and L_{eff}. This substitution will have little effect on the accuracy of the predictions as long as both dimensions of the transistor are several times greater than minimum.[16]

Strictly speaking, equation 12.16 only applies to MOS transistors that have been carefully laid out to ensure optimal matching. Poorly matched transistors often exhibit gross defects that do not scale as predicted. Once these gross defects have been eliminated, the residual threshold mismatches usually follow equation 12.16 quite precisely. Theoretical studies suggest that residual threshold mismatches stem mostly from statistical fluctuations in the distribution of backgate dopants.[17] Statistical fluctuations in the distribution of fixed oxide charge may also play a minor role.

Random short-range variations also appear to determine the residual transconductance mismatches observed between well-matched devices. If the transconductance mismatch is described as a normalized ratio s_k/k, then it varies with effective dimensions, W_{eff} and L_{eff}, as follows

$$\frac{s_k}{k} = \frac{C_k}{\sqrt{W_{eff}\,L_{eff}}} \qquad [12.17]$$

where C_k is a constant. Possible causes for short-range variations in transconductance include linewidth variation, gate oxide roughness, and statistical variations in mobility. The relative importance of these causes is not known, although several authors have suggested that mobility variations predominate.

Gate Oxide Thickness

Many designers believe that MOS transistors with thin gate oxides match better than those with thick gate oxides. At first glance, the evidence seems to support this hy-

---

[15] K. R. Lakshmikumar, R. A. Hadaway, and M. A. Copeland, "Characterization and Modeling of Mismatch in MOS Transistors for Precision Analog Design," *IEEE J. Solid-State Circuits,* SC-21, #6, 1986, pp. 1057–1066.

[16] Substituting drawn for effective dimensions will have very grave effects if either the width or the length of the matched devices is small; see S. J. Lovett, M. Welten, A. Mathewson, and B. Mason, "Optimizing MOS Transistor Mismatch," *IEEE J. Solid-State Circuits,* Vol. 33, #1, 1998, pp. 147–150.

[17] M. J. M. Pelgrom, A. C. J. Duinmaijer, and A. P. G. Welbers, "Matching Properties of MOS Transistors," *IEEE J. Solid-State Circuits,* Vol. SC-24, #5, 1989, pp. 1433–1439. Also see Lakshmikumar, *et al.,* p. 1059.

pothesis, but factors other than oxide thickness are probably at work. The low-voltage, thin-oxide transistors are usually produced by some form of constant-field scaling (Section 11.2.5) that affects not only gate oxide thickness but also backgate doping. If, as research seems to indicate, backgate doping is the dominant cause of threshold voltage mismatch, then constant-field scaling should decrease mismatch by a factor of S, where S is the scaling factor. Constant-field scaling also decreases oxide thickness by a factor of S. This coincidence may account for the empirically observed relationship between oxide thickness and threshold voltage mismatch. Regardless of the exact cause, scaling MOS transistors to smaller dimensions does seem to improve their threshold voltage matching. This effect does not extend to transconductance matching, which appears to remain largely independent of scaling.

Channel Length Modulation

Channel length modulation can cause severe mismatches between short-channel transistors operating at different drain-to-source voltages. The systematic mismatch between the transistors is proportional to the difference between their drain-to-source voltages, and inversely proportional to their channel length. Drawn lengths of 15 to 25μm are generally adequate for noncritical applications such as current distribution networks. Greater precision can be obtained by operating the matched transistors at similar drain-to-source voltages, for example, through the addition of cascodes. MOS designers rarely use source degeneration to combat channel length modulation because the low transconductance of MOS transistors makes it difficult to obtain adequate degeneration without using extremely large resistors.

Orientation

The transconductances of MOS transistors depend on carrier mobilities, and these in turn exhibit orientation-dependent stress sensitivities. MOS transistors oriented along different crystal axes will therefore exhibit different transconductances under stress. Since all packaged devices experience some stress, these mismatches can only be avoided by orienting matched transistors in the same direction. The devices in Figure 12.25A, which are oriented along the same crystal axis, match better than the devices in Figures 12.25B and 12.25C, which are not. Stress-induced mobility variations can induce current matching errors of several percent between rotated devices.[18] The use of tilted wafers may induce current matching errors of as much as 5%.[19]

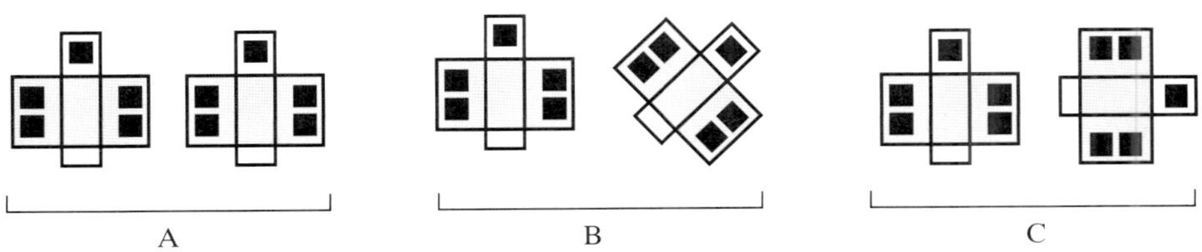

FIGURE 12.25 (A) Devices oriented in the same direction match more precisely than (B,C) those oriented in different directions.

Editing can easily introduce orientation errors if the design has not been properly partitioned. Consider a circuit that contains two matched transistors: M_1, located in cell X_1; and M_2, located in cell X_2. During top-level layout, the designer decides to rotate cell X_1 by 90°. Although this operation seems innocuous, it actually introduces a 90° difference between the orientations of M_1 and M_2. Errors of this sort can be prevented by grouping matched devices together in the same cells. This can some-

---

[18] Pelgrom, *et al.*, p. 1436.

[19] J. E. Chung, J. Chen, P.-K. Ko, C. Hu, and M. Levi, "The Effects of Low-Angle Off-Axis Substrate Orientation on MOSFET Performance and Reliability," *IEEE Trans. on Electron Devices*, Vol. 38, #3, 1991, pp. 627–633.

times make the schematic more difficult to comprehend, but it greatly reduces the risk of inadvertently introducing matching errors during editing.

MOS transistors that do not self-align must follow very strict orientation rules. Consider the asymmetric extended-drain NMOS transistors, M_1 and M_2, in Figure 12.26A. Each of these transistors is a mirror image of the other. The channel lengths of M_1 and M_2 are both defined by the overhang of their poly gates beyond their respective N-well regions. Suppose that photolithographic misalignment causes the poly gates to shift to the right. This misalignment increases the channel length of M_1 and decreases the channel length of M_2. These mismatches are easily eliminated by ensuring that the matched devices are superimposable, as are M_3 and M_4 in Figure 12.26B. Even fully self-aligned transistors may experience slight orientation-dependent mismatches due to diagonal shifting of the source/drain implants (Section 12.4.4).

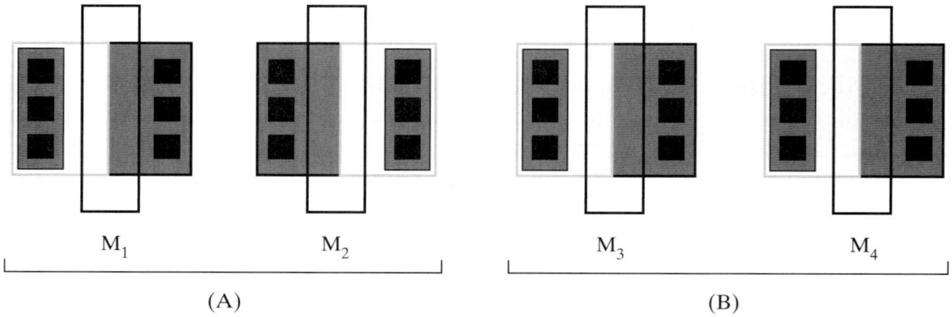

FIGURE 12.26 Extended-drain transistors that are (A) mirror images of one other experience mismatches that do not affect (B) superimposable transistors.

12.4.2. Diffusion and Etch Effects

The previous section examined sources of mismatch that depended solely upon geometry. Certain other types of mismatch are caused by the presence or absence of other structures near the matched transistors. For example, the presence of poly regions near the gate electrodes can cause slight variations in polysilicon etch rates. These variations produce mismatches in the effective widths and lengths of the matched transistors. Similarly, the placement of other diffusions near the channel may influence the backgate dopant concentration and may therefore cause variations in both threshold voltage and transconductance. This section discusses these and other sources of interaction-dependent mismatch.

Polysilicon Etch Rate Variations

Polysilicon does not always etch uniformly. Large poly openings clear more quickly than small ones because etchant ions have freer access to the sides and bottom of the large opening. The edges of the large opening therefore exhibit some degree of overetching by the time the smaller openings clear. This effect can cause variations in the gate lengths of poly-gate MOS transistors. Consider the layout in Figure 12.27A. The gate of transistor M_2 faces adjacent gates on both sides, but the gates of transistors M_1 and M_3 face an adjacent gate on only one side. The outside edges of the gates of M_1 and M_3 experience more erosion than the corresponding edges of the gate of M_2, so the gate lengths of M_1 and M_3 are slightly shorter than the gate length of M_2.

The etch rate variations experienced by MOS transistors are usually smaller than those experienced by poly resistors (Section 7.2.4), because poly gates do not lie as close together as poly resistor segments do. Many MOS transistors also use relatively long channel lengths. Even so, transistors that must achieve moderate or precise current matching should use dummy gates to ensure uniform etching. Failure to do so may produce current mismatches of 1% or more. Figure 12.27B shows an example

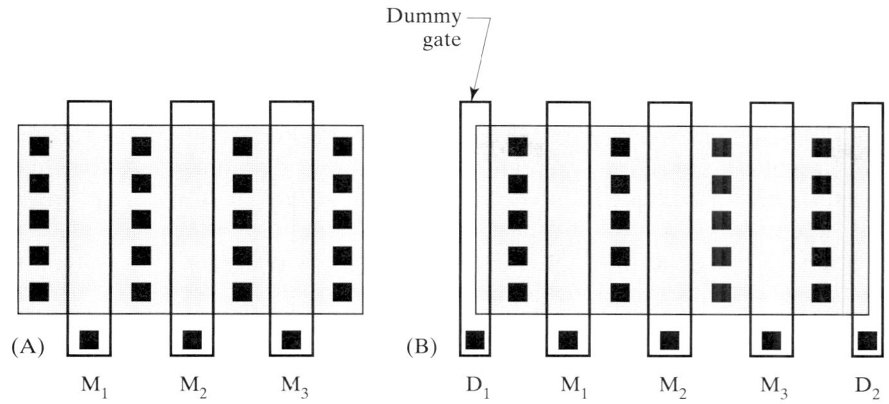

FIGURE 12.27 MOS transistor arrays (A) without dummy gates and (B) with dummy gates.

of an array of MOS transistors incorporating dummies. Most designers make the dummy gates the same width as the active ones, but this precaution is not strictly necessary because the width of the poly strips is far less significant than their spacing. Dummies D_1 and D_2 are therefore made as narrow as possible while still allowing space for a contact. The spacing between the dummies and the actual gates must exactly equal the spacing between the actual gates themselves.

Since the dummies are not actual transistors, they do not require the presence of source/drain regions along their outside edges. The source/drain implant can therefore terminate on top of the dummies, as it does in Figure 12.27B. This should not introduce significant mismatches as long as the moat geometry extends beyond the inner edge of the dummy gate electrodes by a few microns to ensure that the edge of the dummy rests on thin gate oxide.

The dummy gate electrodes should be electrically connected to prevent them from floating at unknown potentials. Although this precaution is not strictly necessary, it helps ensure that the electrical characteristics of the transistors are not affected by the formation of spurious channels or depletion regions beneath the dummies. Some designers connect the dummies to the adjacent gate electrodes, but this practice is not recommended because it increases terminal capacitances and leakage currents. A better practice consists of connecting the dummies to the backgate potential.

Many designers interconnect multiple gate electrodes with a strip of polysilicon. While this is undeniably convenient, it may introduce etch rate variations due to the presence of an adjacent polysilicon geometry. For the best possible matching, one should use simple rectangular strips of polysilicon connected by metal.

Contacts Over Active Gate

For reasons not well understood, the placement of contacts over the active gate regions of MOS transistors sometimes induces significant threshold voltage mismatches. One possible explanation for this effect is the presence of metal above the active gate (see Section 12.4.3). Another potential mechanism for contact-induced mismatches involves the localized silicidation of contacts. In processes where the gate poly is sufficiently thin, some silicide may actually penetrate entirely through the gate poly. The presence of silicide at the oxide interface drastically alters the work function of the gate electrode in the vicinity of the contact and can cause gross threshold voltage mismatches. Changes in grain size, dopant distributions, and stress patterns may also play a role in generating contact-induced mismatches. Figure 12.27 illustrates the proper placement of gate contacts in extensions of the poly gate electrodes. This precaution ensures that the contacts reside over thick-field oxide, where they cannot significantly alter transistor properties.

Annular transistors such as those in Figure 11.19 present a special problem because they require contacts to be placed over active gate regions. Matched annular transistors should only be used if absolutely necessary. If they are used, then they should incorporate identical arrangements of minimal numbers of small gate contacts. In annular extended-drain transistors, the gate contacts should reside over the field-relief regions so they rest on field oxide rather than on gate oxide. This precaution effectively locates the contacts outside the active gate region. In cases in which the field relief region is not wide enough to accommodate the contacts, they should still be located as far inside it as possible to take advantage of the zone of intermediate oxide thickness just inside the edges of the moat region (the bird's beak).

Diffusions Near the Channel

Deep diffusions can affect the matching of nearby MOS transistors. The tails of these diffusions extend a considerable distance beyond their junctions, and the excess dopants they introduce can shift the threshold voltages and alter the transconductances of nearby transistors. The deep-N+ sinker of the analog BiCMOS process represents one example of a deep diffusion. All sinkers and similar diffusions should be spaced away from matched channels by at least twice their junction depth.

Wells also qualify as deep diffusions. N-well geometries should not be placed near matched NMOS transistors to prevent the tail of the N-well dopant distribution from intersecting the channels of the matched transistors. PMOS transistors should be placed far inside the edges of their enclosing N-well regions to prevent outdiffusion from causing variations in backgate doping. In all cases, a spacing from the active gate regions equal to or greater than twice the junction depth of the deep diffusion should limit interactions to negligible levels (Figure 12.28).

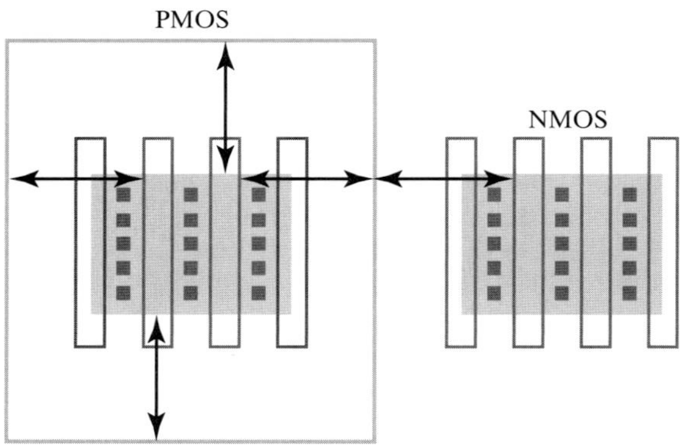

FIGURE 12.28 Spacings between drawn well boundaries and active gate regions.

Because MOS transistors are surface devices, they are vulnerable to the surface discontinuities produced by the NBL shadow. The channels of matched MOS transistors should be placed far enough away from NBL boundaries to allow for both misalignment and pattern shift. If the pattern shift has not been characterized, assume that it can displace the NBL shadow by up to 150% of the epi thickness. Thus the spacing from the active gate regions to the edge of the nearest NBL region should equal at least 150% of the epi thickness. Although this substantially increases the overlap of NBL over the matched transistor, much of this space is already required to satisfy the increased well spacings discussed previously.

PMOS versus NMOS Transistors

NMOS transistors usually match more precisely than PMOS transistors. This phenomenon has been observed on a number of different processes, including both

P-well and N-well variants. Several authors have reported that PMOS transistors exhibit 30 to 50% more transconductance mismatch than comparable NMOS transistors.[20] Some studies have also found increased threshold mismatches in PMOS transistors, although these do not appear to be as significant as the differences in transconductance matching.

The mechanisms responsible for the differences between PMOS and NMOS transistors are not well understood. Possible culprits include increased backgate doping variability, the presence of buried channels, and orientation-dependent stress effects. Several authors have suggested that the increased variability stems (at least in part) from differences in the threshold adjust implants, but this seems an unlikely explanation since so many different processes behave similarly.

12.4.3. Thermal and Stress Effects

Another important category of mismatches stems from long-range variations called *gradients.* The magnitude of gradient-induced mismatches depends on the separation between the effective centers, or *centroids,* of the matched devices. Providing that the devices are placed relatively close to one another, the variation ΔP in parameter P between two matched devices equals the product of the distance d between the centroids and the gradient ∇P along a line connecting the two centroids:

$$\Delta P \cong d \nabla P \qquad [12.18]$$

The impact of the gradient on matching depends on both the magnitude of the gradient and the distance between the centroids of the matched devices. Gradients that affect MOS matching include those of oxide thickness, stress, and temperature.

Oxide Thickness Gradients

The thickness of a grown oxide film depends on the temperature and composition of the oxidizing atmosphere used to create it. Although modern oxidation furnaces are very precisely controlled, slight variations of temperature and atmospheric composition still occur within the furnace tube. Thick oxide layers often exhibit a pattern of concentric rainbow-colored rings that betray the presence of a radial oxide thickness gradient. Gate oxides are too thin to exhibit interference colors, but they also tend to exhibit radial oxide thickness gradients. Devices placed close to one another have very similar oxide thicknesses, while devices placed further apart exhibit greater differences in oxide thickness. These differences directly affect threshold voltage matching.

Stress Gradients

Stress affects the device transconductance of MOS transistors by causing variations in carrier mobilities. As discussed in Section 7.2.6, the effects of stress on mobility depend on orientation. In bulk silicon, holes experience maximum stress dependence along the <110> axis and minimum stress dependence along the <100> axis. Similarly, electrons in bulk silicon experience maximum stress dependence along the <100> axis and minimum stress dependence along the <110> axis. Dice are oriented to the major wafer flat, which lies perpendicular to a <110> axis. Therefore electrons experience minimum stress-induced bulk mobility variation in directions

---

[20] Lakshmikumar, *et al.*, pp. 1060, 1062; Pelgrom, *et al.*, p. 1437.

aligned with the X- and Y-axes of a (100)-oriented die, while holes experience minimum stress-induced bulk mobility variations in directions oriented 45° to these axes.

The stress dependence of bulk mobilities drops to nearly zero along the preferred orientations, but unfortunately the same is not true of the *effective mobilities* of carriers confined to a channel. The stress dependence of the effective mobilities does decrease along the directions predicted by theory, but these minima are nowhere near as pronounced as in the case of bulk mobilities. A diagonal placement of a PMOS transistor may reduce the stress dependence of its device transconductance by only 50%, rather than by the 90% or more one would expect based on bulk mobility data.[21] The randomizing effects of carrier collisions with the oxide/silicon interface probably account for the reduced orientation dependence of effective mobilities, but not all researchers agree on the details of this mechanism. Given these uncertainties, there seems little reason to diagonally orient PMOS transistors. One should instead rely on proper design of common-centroid layouts to minimize stress sensitivity.

Stress has relatively little effect on voltage matching because the threshold voltages of MOS transistors are largely independent of stress. What small stress dependencies do exist are probably caused by stress-induced changes in the bandgap voltage of silicon. The threshold voltage generally does not exhibit more than a few millivolts of stress-induced variation, which can be reduced still further by using common-centroid layout techniques.

Metallization-induced Stresses

The routing of metal leads over the active gate region of MOS transistors has been shown to produce stress-induced mismatches of several percent.[22] Metallization may cause even larger mismatches if the wafers are not sufficiently annealed in a reducing atmosphere, as the deposition of metal above the gate oxide appears to introduce a surface state charge into the gate oxide (see Section 11.1.1).

Ideally, leads should never be routed across the active gate regions of matched MOS transistors. If leads must cross the MOS transistors, then consider adding dummy leads so that each MOS transistor is crossed by an identical segment of metallization at the same position along its channel. This precaution will minimize the impact of the metallization on matching but will not totally eliminate it, so for the highest precision one should entirely avoid routing leads across the active gate regions.

Thermal Gradients

The voltage matching of MOS transistors depends primarily on the matching of threshold voltages. Threshold voltages decrease with temperature at roughly the same rate as base-emitter voltages of bipolar transistors—about –2mV/°C. Most of the temperature coefficient stems from variations in the work functions of the gate and backgate materials with temperature, and it is therefore virtually independent of drain current.[23] Voltage-matched MOS transistors therefore exhibit about the same sensitivity to thermal gradients as bipolar transistors.

MOS and bipolar input differential pairs respond quite differently to offset trimming. In bipolar circuits, trimming the offset voltage to zero also trims its temperature dependence to zero. This happens because the equation for the offset voltage between two matched bipolar transistors contains only one significant source of tem-

---

[21] H. Mikoshiba, "Stress-sensitive Properties of Silicon-gate MOS Devices," *Solid-State Elect.,* Vol. 24, #3, pp. 221–232.

[22] H. Tuinhout, M. Pelgrom, R. P. de Vries, and M. Vertregt, "Effects of Metal Coverage on MOSFET Matching," *IEDM,* 1996, pp. 735–738.

[23] F. M. Klaasen and W. Hes, "On the Temperature Coefficient of the MOSFET Threshold Voltage," *Solid-State Elect.,* Vol. 29, #8, 1986, pp. 787–789.

perature variation: the thermal voltage V_T. The temperature dependence therefore scales directly with ΔV_{BE}, and when ΔV_{BE} has been trimmed to zero, it vanishes. The input offset voltages of MOS transistors are trimmed by adjusting drain current densities. This operation attempts to cancel the mismatch in threshold voltages by introducing a compensating offset in device transconductance. The temperature coefficient of threshold voltage is caused by different mechanisms than the temperature coefficient of transconductance is, so the two are not equal, and the trimming operation does not reduce the temperature coefficient to zero. Thus, while trimmed bipolar input differential pairs retain their very low offset voltages over temperature, trimmed MOS input differential pairs do not. Trimmed bipolar amplifiers and comparators therefore provide much better performance over temperature than do their MOS counterparts.

The current matching of MOS transistors depends primarily on the matching of device transconductances. These transconductances are directly proportional to effective carrier mobilities, which exhibit rather large temperature coefficients. At temperatures near 25°C, MOS device transconductances typically exhibit temperature coefficients of about +7000ppm/°C. Temperature variations in threshold voltage have little effect on current matching as long as the transistors operate at a relatively large effective gate voltage V_{gst}. The low transconductance of MOS transistors makes them much less sensitive to thermal gradients than bipolar transistors, but it also makes it difficult to improve matching by source degeneration. Instead of relying on degeneration resistors, one should use common-centroid layout techniques.

12.4.4. Common-centroid Layout of MOS Transistors

Gradient-induced mismatches can be minimized by reducing the distance between the centroids of the matched devices. Some types of layout can actually reduce the distance between the centroids to zero. These *common-centroid* layouts can entirely cancel the effects of long-range variations as long as these are linear functions of distance. Even if the variations contain a nonlinear component, they still remain approximately linear over short distances. The more compact the common-centroid layout can be made, the less susceptible it becomes to nonlinear gradients. The best layouts for MOS transistors combine exact alignment of the centroids with compactness.

The active gate region of an MOS transistor usually takes the form of a long, narrow rectangle. As in the case of resistors, MOS transistors are usually divided into segments, or *fingers,* to allow the construction of a compact array. The simplest types of arrays involve the placement of multiple device fingers in parallel. If these fingers are properly interdigitated, then the centroids of the matched devices will align at a point midway along the axis of symmetry bisecting the array. Figure 12.29 shows an example of a pair of matched MOS transistors laid out as an interdigitated array.

This layout uses the interdigitation pattern ABBA to ensure exact alignment of the centroids (Section 7.2.6). If source and drain fingers are denoted by subscripts, then the pattern becomes ${}_{D}A_{S}B_{D}B_{S}A_{D}$. Notice that the A-segment on the right has its drain on the right, while the A-segment on the left has its drain on the left. Similarly, the B-segment on the right has its source on the right, while the B-segment on the left has its source on the left. Each transistor thus contains one segment oriented in either direction. The reason for this precaution is rather subtle. Suppose one transistor consists entirely of segments with drains on the left, while a second transistor consists entirely of segments with drains on the right. If left-oriented and right-oriented segments differ in any way, then the two transistors will not match. If both transistors consist entirely of segments oriented in the same direction, then

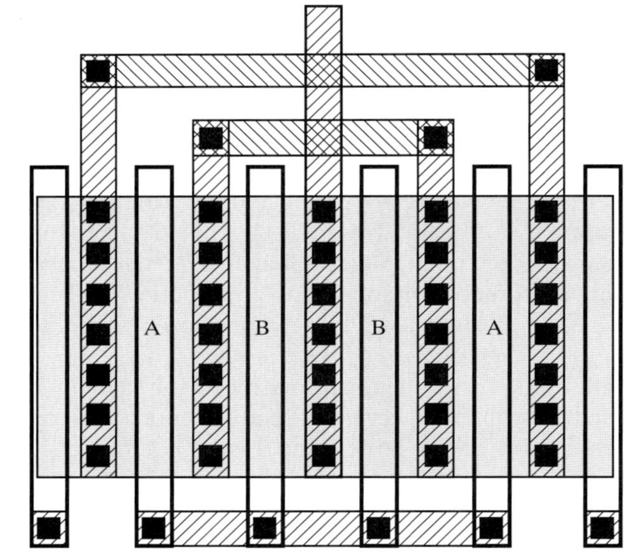

FIGURE 12.29 Interdigitated MOS transistors.

the effect of orientation on each transistor will be the same (Section 12.4.1). If each transistor consists of an equal number of left-oriented and right-oriented segments, then the effects of orientation will cancel and the transistors will again match.

More generally, if we define the *chirality* of a transistor as the fraction of right-oriented segments it contains minus the fraction of left-oriented segments it contains, then transistors having equal chirality will not experience orientation-dependent mismatches.[24] For example, a transistor having three right-oriented segments and one left-oriented segment has a chirality of $3/4 - 1/4 = 1/2$. Similarly, a transistor having nine right-oriented and three left-oriented segments has a chirality of $9/12 - 3/12 = 1/2$. Since these transistors have equal chirality, they do not exhibit any orientation-dependent mismatch. Most designers prefer to use transistors having chiralities of zero; in other words, transistors that consist of equal numbers of left- and right-oriented segments.

Orientation-dependent mismatches can develop in MOS transistors due to diagonal shifts in the source/drain implants. Such diagonal shifts occur when ion implantation is performed at an angle to prevent channeling.[25] Such *tilted implants* cause the source/drain regions on the left side of the gates to differ from the source/drain regions on the right side (Figure 12.30). If the matched devices are arranged in a pattern such as $_DA_SB_D$, then the drain of the left-hand device differs from the drain of the right-hand device. Similarly, the source of the left-hand device differs from the source of the right-hand device. Tilted implants have little effect upon the matching of transistors operated in the linear region, but saturated devices sometimes experience small transconductance differences. These mismatches become worse as the voltage drop across the device approaches maximum because tilted implants have an especially strong impact on hot-carrier generation.[26] These orientation dependencies cancel as long as the matched transistors have equal chirality.

---

[24] The term *chirality* refers to the asymmetry, or handedness, of an object. The term is most commonly encountered in stereochemistry.

[25] J. F. Gibbons, "Ion Implantation in Semiconductors—Part I: Range Distribution Theory and Experiment," *Proc. IEEE,* Vol. 56, #3, 1968, pp. 296–319.

[26] F. K. Baker and J. R. Pfiester, "The Influence of Tilted Source-Drain Implants on High-Field Effects in Submicrometer MOSFETs," *IEEE Trans. on Electron Devices,* Vol. 35, #12, 1988, pp. 2119–2124.

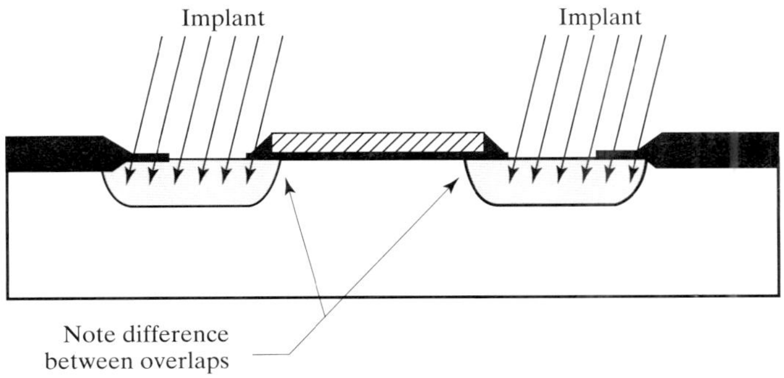

FIGURE 12.30 Diagonal shift in the source/drain regions of an implanted transistor due to the use of a tilted implant. The angle of implantation has been exaggerated for clarity.

Some newer ion implantation systems support on-axis implantation to help minimize the problems caused by tilted implants. The addition of an oxide layer helps minimize channeling, but some still occurs. BiCMOS processes sometimes use *tilted wafers* that have been cut off-axis to minimize pattern distortion. The tilted silicon lattice channels a portion of the ion beam, and this diagonal channeling may cause slight device asymmetries despite the use of on-axis implants.

The interdigitation patterns for common-centroid MOS transistor arrays are often difficult to construct, as it is not easy to satisfy all of the rules of common-centroid layout. MOS transistors must obey not only all four rules given in Section 7.2.6 but also a fifth rule—that of *orientation*. This additional rule ensures that tilted implants (and other device asymmetries) do not affect matching. The full set of rules for MOS devices are as follows:

1. **Coincidence:** The centroids of the matched devices should at least approximately coincide. Ideally, the centroids should exactly coincide.
2. **Symmetry:** The array should be symmetric around both the X- and Y-axes. Ideally, this symmetry should arise from the placement of segments in the array and not from the symmetry of the individual segments themselves.
3. **Dispersion:** The array should exhibit the highest possible degree of dispersion; in other words, the segments of each device should be distributed throughout the array as uniformly as possible.
4. **Compactness:** The array should be as compact as possible. Ideally, it should be nearly square.
5. **Orientation:** Each matched device should consist of an equal number of segments oriented in either direction; more generally, the matched devices should possess equal chirality.

Table 12.1 shows a few of the simpler interdigitation patterns used for MOS transistors. Source and drain fingers are denoted by subscripts, and sequences of segments that may be repeated are enclosed in parentheses: $({}_{S}A_{D}A)$. When a pattern includes more than one repeated sequence, each portion of the sequence in

1. $({}_{S}A_{D}A)({}_{S}B_{D}B_{S}B_{D}B)({}_{S}A_{D}A)_{S}$
2. $({}_{D}A_{S}B_{D}\text{–}{}_{D}B_{S}A_{D})\text{–}({}_{D}A_{S}B_{D}\text{–}{}_{D}B_{S}A_{D})$
3. $({}_{D}A_{S}B_{D}B_{S}A)_{D}$
4. $({}_{S}A_{D}A_{S}B_{D}B)_{S}(B_{D}B_{S}A_{D}A_{S})$
5. $({}_{S}A_{D}A_{S}B_{D}B_{S}A_{D}A)_{S}$
6. $({}_{S}A_{D}A_{S}B_{D}\text{–}{}_{S}A_{D}A_{S}\text{–}{}_{D}B_{S}A_{D}A)_{S}$
7. $({}_{S}A_{D}A_{S}B_{D}B_{S}C_{D}C)_{S}(C_{D}C_{S}B_{D}B_{S}A_{D}A_{S})$

TABLE 12.1 Sample interdigitation patterns for MOS transistor arrays.

parentheses must be replicated the same number of times. Certain patterns contain locations where the source/drain fingers cannot merge with one another; these are denoted by dashes. All of the entries in this table obey the rules of coincidence, symmetry, and orientation, but many of them are not as disperse nor as compact as possible. For example, consider patterns 1 to 4, all of which provide a 1:1 ratio between two matched devices. Pattern 1 lacks dispersion because it contains long runs of segments belonging to the same device. Pattern 2 contains gaps that make it less compact than the others. Patterns 3 and 4 both exhibit considerable dispersion because the segments appear in pairs throughout most parts of the array. However, the middle of pattern 4 contains a run of four segments belonging to the same device. The middle of pattern 3 contains a run of only two segments, so it provides better dispersion than pattern 4. In summary, pattern 3 should exhibit more precise matching than patterns 1, 2, and 4. The device of Figure 12.29 uses pattern 3.

Interdigitated MOS transistors do not provide the best possible cancellation of gradients because they rely on the symmetry of individual device segments to provide one of their two axes of symmetry. A two-dimensional common-centroid array provides a higher degree of symmetry because both axes of symmetry arise from the layout of the array, rather than from the segments comprising it. Two-dimensional common-centroid arrays are particularly useful for matching pairs of transistors of equal size, such as differential pairs. Layouts of this sort are called *cross-coupled pairs.* As with other common-centroid MOS layouts, care must be taken to ensure that orientation dependencies cancel.

Figure 12.31 shows the simplest possible cross-coupled pair. This layout follows the interdigitation pattern ${}_{D}A_{S}B_{D}/{}_{D}B_{S}A_{D}$, where the slash (/) separates the segments that occupy the upper two quadrants from those that occupy the lower two.[27] Not only does this produce a very compact layout, but it also satisfies the rule of orientation, because the two segments belonging to each matched device are oriented in opposite directions. This layout is especially suited for pairs of relatively small MOS transistors.

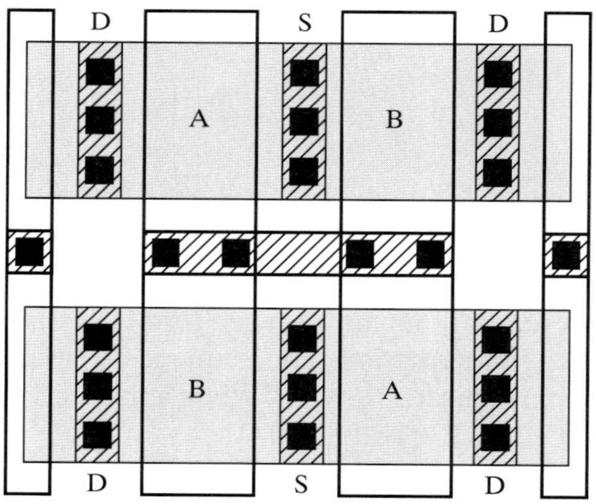

FIGURE 12.31 Cross-coupled MOS transistors.

Larger cross-coupled pairs are more difficult to construct. Most designers simply divide each transistor into two equal halves and place these halves in diametrically opposite corners of the array. A layout of this sort can be represented by the pattern XY/YX, where X and Y are subarrays composed entirely of segments of transistors A and B, respectively. A typical implementation of such an array is $({}_{S}A_{D}A)_{S}$

[27] Tsividis discusses a ${}_{D}A_{S}B_{D}/{}_{D}B_{S}A_{D}$ array, but without the dummies: Y. Tsividis, *Mixed Analog-Digital VLSI Devices and Technology* (New York: McGraw-Hill, 1997), p. 233.

$(B_DB_S)/({}_SB_DB)_S(A_DA_S)$. While this pattern satisfies most of the rules of interdigitation, it does not provide optimal dispersion. As the array grows larger, its lack of dispersion renders it increasingly susceptible to mismatches caused by nonlinear components of the variation. A much better pattern for large cross-coupled pairs is $({}_DA_SB_DB_SA)_D/{}_D(B_SA_DA_SB_D)$. If the array becomes very large, then additional dispersion can be introduced by elaborating this array in the vertical dimension, as shown in the following examples:

${}_DA_SB_DB_SA_D$	${}_DA_SB_DB_SA_D$	${}_DA_SB_DB_SA_DA_SB_DB_SA_D$
${}_DB_SA_DA_SB_D$	${}_DB_SA_DA_SB_D$	${}_DB_SA_DA_SB_DB_SA_DA_SB_D$
	${}_DA_SB_DB_SA_D$	${}_DA_SB_DB_SA_DA_SB_DB_SA_D$
	${}_DB_SA_DA_SB_D$	${}_DB_SA_DA_SB_DB_SA_DA_SB_D$
		${}_DA_SB_DB_SA_DA_SB_DB_SA_D$
		${}_DB_SA_DA_SB_DB_SA_DA_SB_D$

The main drawback to the more elaborate patterns of this type lies in the difficulty of connecting the various segments together to form the full device. This becomes particularly difficult in cases where the gates of the two matched devices do not connect together. The simpler—and hence easier to connect—patterns generally serve for all except the most demanding applications.

12.5 RULES FOR MOS TRANSISTOR MATCHING

This section summarizes the previously given information in the form of a set of qualitative rules. These rules allow designers to construct matched MOS transistors, even if no quantitative matching data exists for the process in question. The rules use the terms *minimal, moderate,* and *precise* to denote increasingly precise degrees of matching, which may be interpreted as follows:

- **Minimal matching:** Typical three-sigma drain current mismatches of several percent. Minimal matching is often used for constructing bias current networks that do not require any particular degree of precision. This level of matching corresponds to typical offsets in excess of ±10mV and is therefore inadequate for voltage matching applications.
- **Moderate matching:** Typical three-sigma offset voltages of ±5mV or drain current mismatches of less than ±1%. Useful for constructing input stages of noncritical op-amps and comparators, where untrimmed offsets of ±10mV can be maintained.
- **Precise matching:** Typical three-sigma offset voltages of less than ±1mV or drain current mismatches of less than ±0.1%. This level of matching usually involves trimming, and the resulting circuit will probably meet specification within only a limited range of temperatures due to the presence of uncompensated temperature variations.

The following rules summarize the most important principles of MOS transistor matching:

1. Use identical finger geometries.
   Transistors of different widths and lengths match very poorly. Even minimally matched devices must have identical channel lengths. Most matched transistors require relatively large widths and are usually divided into sections, or *fingers.* Each of these fingers should have the same width and length as all others. Do not

attempt to match transistors of different widths and lengths, because the width and length correction factors, δW and δL, vary substantially from lot-to-lot.

2. Use large active areas.
   The active area of an MOS transistor equals the product of its channel width and length. Assuming that all other matching considerations have been addressed, the residual offset due to random fluctuations scales inversely with the square root of device area. Moderate matching usually requires active areas of several hundred square microns, while precise matching requires thousands of square microns.
3. For voltage matching, keep V_{gst} small.
   The offset voltage of a pair of matched MOS transistors contains a term dependent on device transconductance. This term scales with V_{gst}, so smaller values of V_{gst} provide better voltage matching. Reducing the V_{gst} below 0.1V provides little additional benefit because threshold voltage variations begin to dominate the offset equation. Most designers decrease V_{gst} by using larger W/L ratios because these simultaneously increase the active area of the transistors.
4. For current matching, keep V_{gst} large.
   The current mismatch equation contains a term dependent upon threshold voltage. This term scales inversely with V_{gst}, so large values of V_{gst} minimize its impact upon current matching. Circuits relying upon current matching should maintain a nominal V_{gst} of at least 0.3V. Moderately matched transistors should maintain a nominal V_{gst} of at least 0.5V whenever headroom allows. Precisely matched transistors should use the largest value of V_{gst} allowed by the configuration of the circuit, but in any event should equal at least 0.5V.
5. Orient transistors in the same direction.
   Transistors that do not lie parallel to one another become vulnerable to stress- and tilt-induced mobility variations that can cause several percent variation in their transconductance. This effect is so severe that even minimally matched transistors should lie parallel to one another. Matched transistors, especially those that are not fully self-aligned, should have equal chirality. This condition can be met by ensuring that each transistor contains an equal number of segments oriented in each direction.
6. Place transistors in close proximity.
   MOS transistors are vulnerable to gradients in temperature, stress, and oxide thickness. Even minimally matched devices should reside as close as possible to one another. Moderately or precisely matched transistors should be kept next to one another to facilitate common-centroid layout.
7. Keep the layout of the matched transistors as compact as possible.
   MOS transistors naturally lend themselves to long, spindly layouts that are extremely vulnerable to gradients. Common-centroid layout cannot entirely eliminate this vulnerability, so the designer should strive to create as compact an arrangement of matched devices as possible. This usually requires that each device be divided into a number of fingers.
8. Where practical, use common-centroid layouts.
   Moderately and precisely matched MOS transistors require some form of common-centroid layout. This can often be achieved by dividing each transistor into an even number of fingers and by then arranging these fingers in an

interdigitated array. Pairs of matched transistors should be laid out as cross-coupled pairs to take advantage of the superior symmetry of this arrangement.

9. Place dummy segments on the ends of arrayed transistors.
Arrayed transistors should include dummy gates at either end. These dummies need not have the same length as the actual gates, but the spacing between the dummy gates and the actual gates must equal the spacing between actual gates. The moat diffusion should extend at least several microns into the dummies to prevent the edge of the dummies from resting on the bird's beak. The dummy gates should be connected, preferably to a potential that prevents channel formation beneath them. This is most easily achieved by connecting the dummies to the backgate potential.

10. Place transistors in areas of low stress gradients.
The stress gradients reach a broad minimum in the center of the die. Any location ranging from the center of the die out halfway to the edges will fall within this broad minimum. Whenever possible, precisely matched transistors should reside within this low-stress area. Moderately and precisely matched transistors should reside at least 10mils (250μm) away from any side of the die. The stress distribution reaches a maximum in the die corners, so avoid placing any matched transistors near corners. PMOS transistors may experience slightly less stress dependence when oriented along [100] directions. This effect is not sufficiently pronounced to justify placing minimally or moderately matched transistors diagonally, but precisely matched transistors might benefit from this unconventional orientation. NMOS transistors should always be oriented horizontally and vertically.

11. Place transistors well away from power devices.
For purposes of discussion, any device dissipating more than 50mW should be considered a power device, and any device dissipating more than 250mW should be considered a major power device. Precisely matched transistors should reside on an axis of symmetry of the major power devices using an optimal symmetry arrangement (see Section 7.2.7). Moderately and precisely matched transistors should reside no less than 10 to 20mils (250 to 500μm) away from the closest power device. Minimally matched devices may be placed next to power devices, but only if they use some form of common-centroid layout.

12. Do not place contacts on top of active gate area.
Whenever possible, extend the gate poly beyond the moat and place the gate contacts over thick-field oxide. When this is not possible, minimize the number and size of the gate contacts and place them in the same location on each transistor. Consider placing the gate contacts of high-voltage annular transistors over the field-relief structure because this is not part of the active gate.

13. Do not route metal across the active gate region.
Whenever possible, avoid routing metal across the active gate region of precisely matched MOS transistors. Leads may route across moderately matched MOS transistors, but additional dummy leads should be added so that every section of the array of matched devices is crossed at the same location along its channel by an identical length of lead.

14. Keep all junctions of deep diffusions far away from active gate area.
The minimum spacing between a drawn well boundary and a precisely matched MOS transistor should equal at least twice the well junction depth.

Moderately and minimally matched transistors need only obey the applicable layout rules. Similar considerations apply to deep-N+ sinkers and other deep diffusions.

15. Place precisely matched transistors on axes of symmetry of the die.
Arrays of precisely matched transistors should be placed so that the axis of symmetry of the array aligns with one of the two axes of symmetry of the die. If the design contains large numbers of matched transistors, then reserve the optimal locations for the most critical devices.
16. Do not allow the NBL shadow to intersect the active gate area.
The NBL shadow should not fall across the active gate region of any precisely matched transistor. If the direction of the NBL shift is unknown, allow adequate overlap of NBL over the transistor on all sides. If the magnitude of the NBL shift is also unknown, then overlap NBL over the active gate region by at least 150% of the maximum epi thickness.
17. Connect gate fingers using metal straps.
Connect the gate fingers of moderately and precisely matched transistors using metal rather than poly. Minimally matched transistors can use a poly comb structure to simplify the connection of the gate electrodes.
18. Use thin-oxide devices in preference to thick-oxide devices.
Some processes offer multiple thicknesses of gate oxides. The transistors with thinner gate oxides generally exhibit better matching characteristics than those with thick gate oxides. Whenever circuit considerations allow, consider using thin-oxide transistors in preference to thick-oxide transistors.
19. Consider using NMOS transistors rather than PMOS transistors.
NMOS transistors generally match better than PMOS transistors. Whenever circuit considerations allow, consider using NMOS transistors rather than PMOS transistors.

12.6 SUMMARY

Many circuit designers think of MOS transistors primarily as building blocks for digital logic. While they are indispensable in this role, they also have many other important applications. Modern mixed-mode integrated circuits rely heavily upon MOS transistors for power switching and low-current analog functions.

MOS power transistors have revolutionized power switching. The new generation of high-efficiency switch-mode power supplies relies almost exclusively upon MOS power transistors. Similarly, virtually all low-voltage power distribution circuits use MOS transistors. Now that BiCMOS processes offer integrated power MOS transistors with specific on-resistances approaching those of discrete devices, it has become economically feasible to integrate many power switches into a single integrated circuit. The low on-resistances of the power switches also minimize power dissipation, allowing the use of compact surface-mount packages.

MOS transistors have also found many applications in analog signal processing circuitry. Although bipolar transistors remain entrenched in a few applications, the vast majority of analog circuitry can be implemented using MOS transistors. MOS circuits are usually smaller and consume less power than their bipolar counterparts. Even though most modern analog circuits are implemented in BiCMOS processes, the vast majority of the circuitry consists of MOS transistors. Circuit designers continue to develop new applications for MOS transistors, so future integrated circuits will probably incorporate an even larger percentage of MOS circuitry.

12.7 EXERCISES

Refer to Appendix C for layout rules and process specifications.

12.1. Suppose an extended-voltage transistor with a drain depletion width x_d equal to 10% of the pinched-off region width x_p can withstand a drain-to-source voltage of 10V. What drain-to-source voltage could a similar device withstand if x_d were increased to 50% of x_p?

12.2. Suggest a structure for a self-aligned extended-voltage PMOS transistor. Draw a cross section of a representative transistor using this structure.

12.3. If the thin gate oxide of an extended-drain NMOS having no field-relief structure can withstand 10V, then which of the following biasing conditions are allowable, and why?

a. Asymmetric NMOS, $V_{GS} = 6\text{V}, V_{DS} = 10\text{V}$.
b. Asymmetric NMOS, $V_{GS} = 7\text{V}, V_{DS} = 16\text{V}$.
c. Asymmetric NMOS, $V_{GS} = 3\text{V}, V_{DS} = 16\text{V}$.
d. Symmetric NMOS, $V_{GS} = -13\text{V}, V_{DS} = -16\text{V}$.
e. Symmetric NMOS, $V_{GS} = 20\text{V}, V_{DS} = 0\text{V}$.

12.4. Lay out asymmetric and symmetric extended-drain NMOS transistors, each having drawn dimensions of 2(15/10). The N-well drain geometry should abut the N-moat source geometry beneath the gate as shown in Figure 12.5. The overlap of the poly gate over the N-well should equal exactly 3μm. Include abutting backgate contacts for the asymmetric transistor. Why can't abutting backgate contacts be used for the symmetric transistor?

12.5. Compute the maximum theoretical $R_{DS(on)}$ for a 50000/2 NMOS power transistor operating at $5 < V_{GS} < 15\text{V}$ and $-40 < T_j < 150°\text{C}$. Assume that the device's threshold voltage equals 0.7 ± 0.2V with a temperature coefficient of −2mV/°C, and that its process transconductance equals 35uA/V^2 ± 20% at 150°C.

12.6. Determine the specific on-resistance (in $\Omega \cdot \text{mm}^2$) for a power device having an $R_{DS(on)}$ of 165mΩ and an area of 2.26mm^2. Use this information to determine the area required for a 100mΩ power transistor. Assume both $R_{DS(on)}$ values do not include bondwire or leadframe resistance.

12.7. Compute the ideal ratio B/L for a metal system consisting of a first layer of metal that is 7500Å thick and a second layer of metal that is 14000Å thick. Lay out a 20,000/2 PMOS transistor using this ratio and the analog BiCMOS layout rules in Appendix C. Divide the transistor into sufficient fingers to produce a roughly square aspect ratio. Fill the well with NBL and ring the outer edge of the well with deep-N+ to provide a backgate contact for the transistor. Include all necessary metallization.

12.8. Compute the approximate metallization resistance of the transistor in Exercise 12.7. Do not include the resistance of the metal-2 buses extending beyond the interdigitated region of the transistor.

12.9. Assume the source and drain leads of the transistor in Exercise 12.7 run 25μm from the drawn edge of the well to the edge of their respective bondpads, and that each bondpad connects to a 600μm-long 1mil-diameter gold bondwire. Calculate the total metallization resistance of the transistor, including bondwires. Assume that the resistance between the edge of the bondpad and the bondwire is negligible.

12.10. Lay out a minimum-size, circular-annular, lateral DMOS transistor using the analog BiCMOS rules in Appendix C, supplemented by the following rules for the DMOS layer:

1. **DMOS** width — 5μm
2. **DMOS** spacing to **DMOS** — 4μm
3. **DMOS** spacing to **PMOAT** — 0μm
4. **POLY** extends into **DMOS** — 2μm
5. **POLY** overhang of **DMOS** — 4μm
6. **MOAT** overlap of **DMOS** — 2μm
7. **CONT** extends into **DMOS** — 2μm

A DMOS to PMOAT spacing of 0μm implies that the outer edge of the PSD plug should coincide with the inner edge of the annular DMOS ring. Include all necessary metallization.

12.11. If the length of the DMOS channel equals 1μm and the inner edge of the channel coincides with the outer edge of the drawn DMOS geometry, then what is the drawn width of the transistor constructed in Exercise 12.9?

12.12. Lay out a standard bipolar epi-FET having drawn dimensions of 30/8. Assume that the base pinch plate must extend at least 2μm into the isolation.

12.13. Construct a minimum-size circularly symmetric epi-FET. Include all necessary metallization. What are the drawn width and length of this device?

12.14. A cross-coupled NMOS differential pair of transistors, each having dimensions 100/10, has a three-sigma random mismatch of ±2.85mV. Estimate the three-sigma random mismatch of a similar differential pair where the transistors each have dimensions of 1000/5.

12.15. Lay out a pair of differential NMOS transistors, each having dimensions of 1000/5, to obtain the best possible matching. The transistors may be divided into as many or as few segments as desired. Assume that backgate contacts are required only along the edges of the array. Include all necessary metallization, including the links connecting individual source/drain fingers and the links connecting the gate fingers.

12.16. Lay out the MOS operational amplifier shown in Figure 12.32 following the recommendations for optimal matching. Use the poly-gate CMOS rules in Appendix C, and include all necessary backgate and substrate contacts. Assume that all PMOS transistors have backgates connecting to VDD.

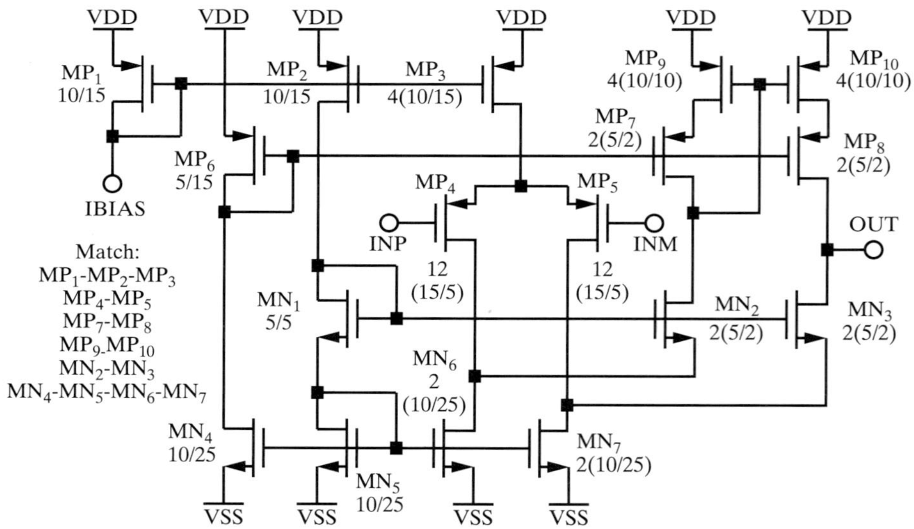

FIGURE 12.32 Folded-cascode MOS operational amplifier for Exercise 12.16.

12.17. Compare the matching of the following interdigitation patterns. Which pattern provides the best matching, and why?

a. ${}_{S}A_{D}A_{S}B_{D}-{}_{D}A_{S}A_{D}-{}_{D}B_{S}A_{D}A_{S}$

b. ${}_{D}A_{S}A_{D}A_{S}B_{D}B_{S}A_{D}A_{S}A_{D}$

c. ${}_{D}B_{S}A_{D}A_{S}A_{D}A_{S}A_{D}A_{S}B_{D}$

12.18. What are the chiralities of each of the following interdigitation patterns? Which patterns exhibit orientation-dependent mismatches?

a. ${}_{D}A_{S}B_{D}B_{S}A_{D}$

b. ${}_{S}A_{D}-{}_{D}B_{S}B_{D}B_{S}-{}_{D}A_{S}$

c. ${}_{S}A_{D}A_{S}-{}_{S}B_{D}-{}_{D}A_{S}A_{D}$

13 Special Topics

The previous chapters have presented the details of constructing and matching resistors, capacitors, diodes, and transistors. Integrated circuits also contain a number of more specialized components, including merged devices, guard rings, tunnels, bondpads, and ESD protection devices.

Merged devices appear separate from one another in schematics, but they are combined in the layout. Mergers not only save space but in some cases also improve performance. The designer must weigh the benefits of mergers against the possibility of introducing unexpected interactions between merged devices.

Guard rings prevent minority carriers injected by one device from interfering with the operation of another device. Not only do guard rings prevent latchup, but they also block noise coupling that might otherwise interfere with the operation of low-power circuitry.

Tunnels are low-value resistors used as signal crossing points. Single-level-metal layouts usually include at least a few tunnels. Multiple-level-metal layouts do not require them, but they occasionally offer a convenient way to route leads to otherwise-inaccessible areas of the die. Tunnels introduce parasitics that can degrade circuit performance, so each proposed tunnel must be carefully analyzed for potential problems.

Bondpads allow the connection of the integrated circuit to the external world. Most bondpads require electrostatic discharge (ESD) protection circuitry. *Trimpads* and *testpads* are accessible only during wafer-level probing, so they do not require ESD protection.

13.1 MERGED DEVICES

The largest spacings in standard bipolar are those associated with the isolation diffusion. Most circuits contain components whose tanks are connected to the same potential. Considerable space can be saved by placing these devices in common tanks. Figure 13.1A shows three minimum-geometry NPN transistors laid out side-by-side, while Figure 13.1B shows the same three transistors merged into a common tank.

FIGURE 13.1 (A) Three separate NPN transistors and (B) the same transistors merged into one tank.

The merged devices require only about 70% of the area of the separate devices. Additional area can be saved by reducing the size of the collector contact.

The largest spacings in CMOS and BiCMOS processes are those associated with the isolated well (or wells). The merger of components into common wells can again save considerable die area. This is particularly true for designs containing large numbers of small components, such as minimum-size MOS transistors.

The injudicious merger of devices that should have remained separate has caused many circuit malfunctions, some of which have proved remarkably difficult to diagnose and fix. Consequently, designers have become somewhat reluctant to merge devices, even when mergers would obviously result in significant area savings.

Virtually every failure caused by merged devices can be traced to one of three sources: minority carrier injection, Ohmic debiasing, or capacitive coupling. If a designer understands these three mechanisms, then he or she can discern which mergers are safe and which should be avoided. Minority carrier injection is by far the most serious problem encountered in designing merged devices. Trouble can occur whenever a device injects minority carriers into a shared region such as a tank or a well. Some of the injected minority carriers transit to other devices, where they are collected by reverse-biased junctions. The flow of minority carriers between devices that should remain isolated causes leakage currents to appear at unexpected points in the circuit. These currents can cause circuit malfunctions ranging from subtle parametric shifts to catastrophic latchup. The following section discusses several device mergers that are known to have minority carrier cross-injection problems.

13.1.1. Flawed Device Mergers

Figure 13.2A shows a split-collector lateral PNP transistor. This structure essentially consists of a pair of merged lateral PNP transistors sharing common emitter and base regions. Under normal operating conditions, collectors C_1 and C_2 both remain

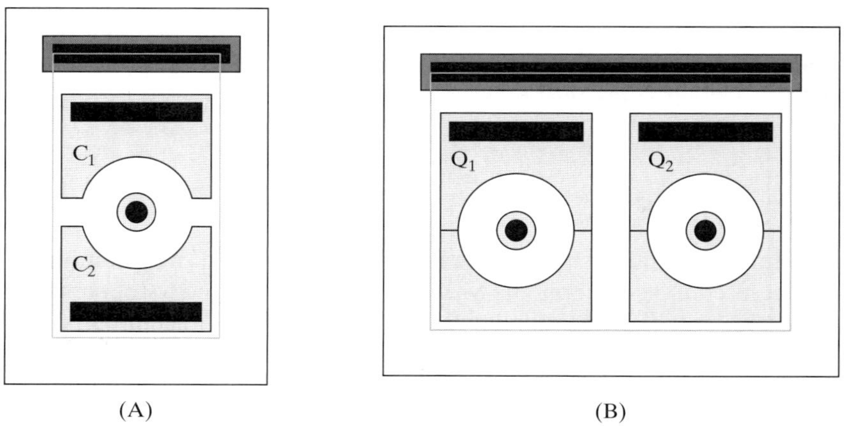

FIGURE 13.2 Merged lateral PNP structures susceptible to cross-injection: (A) split-collector lateral PNP and (B) two lateral PNP transistors merged in a common tank.

reverse-biased. Holes flow radially from the shared emitter to the two collectors, and each collector intercepts about half of the total emitter current. Now suppose that collector C_1 saturates. The minority carriers that should have been absorbed by C_1 are now re-injected from its surfaces. Most of these re-injected carriers flow to the isolation sidewalls and thence to the substrate, but some also flow from collector C_1 to collector C_2. This *cross-injection* increases the current flowing out of C_2 and imbalances the apparent ratio between the two collectors.

Most designers know that the saturation of a split-collector PNP causes cross-injection, but many overlook the possibility of cross-injection between separate lateral PNP transistors occupying a common tank. Figure 13.2B shows a tank containing two lateral PNP transistors, Q_1 and Q_2. Normally the collectors intercept virtually all of the minority carriers injected by their respective emitters, so the two transistors remain effectively isolated from each other, even though both occupy the same tank. Now suppose that Q_1 saturates while Q_2 continues to operate in the normal active region. The holes injected by the emitter of Q_1 transit to its collector, but when this collector forward-biases, they are re-injected back into the tank. Since the collectors of Q_1 and Q_2 face each other across a relatively narrow gap, most of the carriers launched from Q_1 toward Q_2 reach its collector. Hence when Q_1 saturates, about a quarter of its emitter current flows to Q_2.

The simplest way to prevent cross-injection between lateral PNP transistors is to place each transistor in its own tank. However, this wastes so much space that designers have devised other methods of preventing (or at least minimizing) cross-injection between merged devices. For example, two lateral PNP transistors can be separated from one another by a P-bar or an N-bar (Section 4.4.2). These bars block most, but not all, of the cross-injected minority carriers. The designer should consider what would happen to the circuit if, for example, 5% of the current injected by the saturating device were to reach adjacent devices. If this amount of current could cause a malfunction, then the devices would require separate tanks.

Another example of minority carrier injection involves the merger of an NPN transistor, Q_2, driving a lateral PNP, Q_1 (Figure 13.3).[1] The collector of Q_2 is electrically common to the base of Q_1. This merged device will probably operate satisfactorily as long as Q_1 does not saturate. If it does, then holes re-injected by its collector will flow to the base of NPN Q_2. This additional base current drives Q_2 harder than before and provides additional collector current. The increased collector current feeds Q_1 and increases its emitter current. This situation is a classic example of

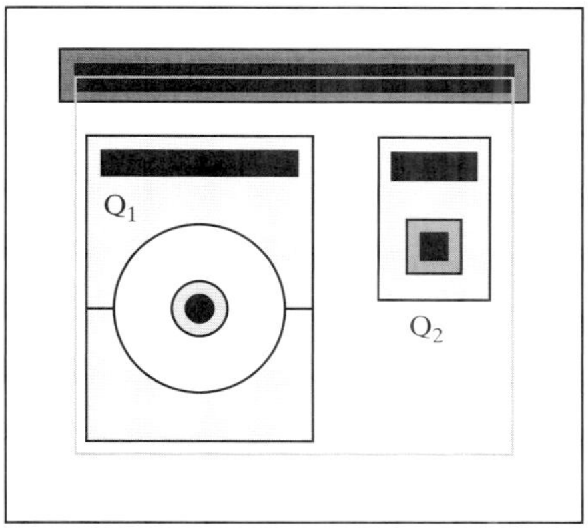

FIGURE 13.3 Example of a device merger prone to latchup due to minority-carrier injection.

[1] C. Jones, "Bipolar Parasitics," unpublished manuscript, 1987, pp. 27–29.

SCR latchup. Once latchup has been triggered, it will continue until the power is interrupted. The die may overheat and self-destruct at high supply voltages; otherwise it simply malfunctions and consumes excessive supply current. The potential for latchup makes this structure risky, even if the lateral PNP never saturates during normal operation. If some transient condition saturates the lateral PNP, then the structure will latch up.

The use of P-bars or N-bars to suppress device latchup is potentially risky. In order for latchup to occur, the product of the beta of the PNP, β_P, and the beta of the NPN, β_N, must exceed unity. The addition of a bar between the two transistors adds a third term, η_c, representing the fraction of minority carriers intercepted by the bar. The condition for latchup (Section 11.2.7) then becomes

$$\beta_N \beta_P (1 - \eta_c) > 1 \qquad [13.1]$$

The beta of a standard bipolar NPN transistor may equal 300 or more. The beta of the lateral PNP is lower than that of a conventional transistor because the actual basewidth equals the separation between the transistors. Even so, the beta of the lateral PNP could easily equal ten. In order to prevent latchup with these betas, the efficiency of the bar must exceed 0.997, or 99.7%. N-bars will almost certainly fall short of this efficiency, and P-bars may not always achieve it. If there is the slightest chance that a lateral PNP may saturate, then it should not occupy the same tank as an NPN transistor.

Figure 13.4 illustrates another pair of merged devices prone to latchup.[2] This example is particularly noteworthy because the latchup stems from the interaction of two separate mechanisms, namely Ohmic debiasing and minority carrier injection. This structure places an NPN transistor Q_1 in the same tank as a base resistor R_1. The NPN is configured as an emitter follower and one end of the base resistor connects to the supply. The tank contact serves as the collector contact of Q_1 and the tank contact of R_1. Both the base-collector junction of Q_1 and the base-tank junction of R_1 remain reverse-biased under normal conditions. This situation may change if Q_1 draws enough collector current through the shared tank contact. If the voltage drop between the tank contact and the intrinsic collector of Q_1 becomes large enough, then the positive end of R_1 will forward-bias into the tank. This is an example of *Ohmic debiasing*. Some of the minority carriers injected by R_1 reach the base

FIGURE 13.4 Another example of a device merger prone to latchup due to minority-carrier injection.

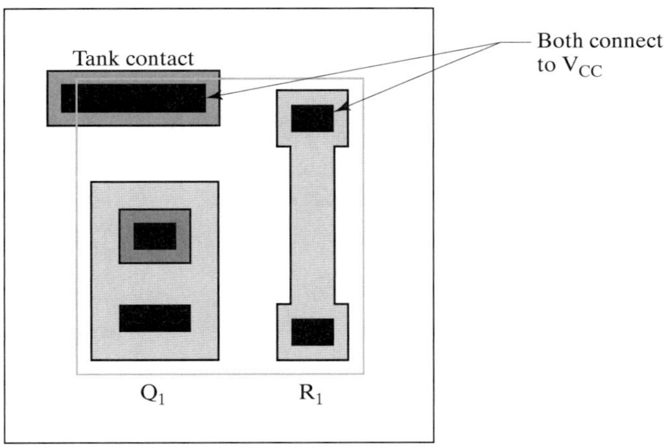

[2] W. F. Davis, *Layout Considerations,* unpublished manuscript, 1981, pp. 32–33.

of Q_1, where they provide additional base drive. Transistor Q_1 now pulls even more collector current. The additional collector current increases the Ohmic debiasing experienced by R_1, causing R_1 to inject additional holes into the tank. The resulting positive feedback causes the circuit to latch up. As with the previous example, latchup occurs because of the presence of an SCR, which in this case consists of base resistor R_1, the shared tank, and the base and emitter of Q_1.

Although the structure in Figure 13.4 contains an SCR, it will not latch up unless triggered by a voltage drop produced by Ohmic debiasing within the tank. The voltage drop required to trigger the SCR equals about 0.3V at 150°C (see Section 4.4.1). If the NPN transistor conducts an average of 100μA, then the tank resistance must equal 3kΩ to produce 0.3V of debiasing. The vertical resistance between the tank contact and the NBL can equal hundreds, if not thousands, of Ohms if the transistor lacks a deep-N+ sinker. The inclusion of even a minimum plug of deep-N+ reduces the tank resistance to no more than a few hundred Ohms. The structure in Figure 13.4 is therefore likely to latch up without deep-N+, but it is very unlikely to do so if a sinker is present.

Ohmic debiasing can also cause capacitive coupling of noise into sensitive nodes. Using the merger of Figure 13.4 as an example, suppose that the debiasing is insufficient to actually trigger latchup. Even so, the current flowing through Q_1 still causes some voltage drop within the tank. If the operation of Q_1 causes a rapid fluctuation of the collector current, then this will in turn generate a high-frequency ripple on the tank voltage. This signal can couple through the capacitance of the reverse-biased junction surrounding R_1. If R_1 is part of a sensitive circuit, then the noise injected by tank voltage fluctuations can cause problems. Designers should avoid merging noisy circuitry and sensitive circuitry in the same tank. Although some such mergers function satisfactorily, many do not.

Figure 13.5 shows another pair of problematic merged devices. This example merges an NPN transistor, Q_1, with a Schottky diode, D_1. The collector of Q_1 connects to the cathode of D_1 through the tank. Since Schottky diodes are majority-carrier devices, this might appear a safe merger. Unfortunately, this is not the case. Most Schottky diodes incorporate a field-relief guard ring consisting of a P-type diffusion. This guard ring begins to inject minority carriers into the tank as soon as the voltage across the Schottky rises above the forward voltage of the guard-ring junction. The series resistance of small Schottky diodes can equal hundreds or thousands of Ohms, so their guard rings are easily debiased into conduction.

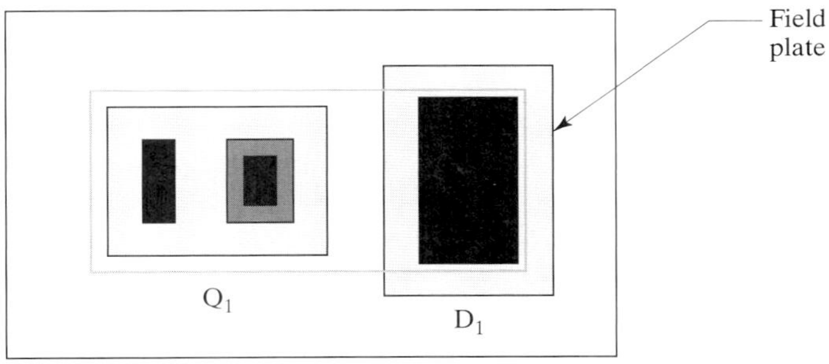

FIGURE 13.5 Another structure prone to minority-carrier cross-injection, consisting of an NPN transistor, Q_1, and a Schottky diode, D_1.

The structure in Figure 13.5 uses a field-plated Schottky diode instead of a guard-ringed Schottky. This eliminates the possibility of the guard ring forward-biasing into the tank, but low levels of cross-injection can still occur due to the presence of the Schottky barrier itself. Although a rectifying Schottky junction conducts primarily by means of majority carriers, small numbers of minority carriers are also injected into the semiconductor side of the junction. The minority carriers injected by Schottky diode D_1 can travel to the base of Q_1, where they appear as additional base drive. This mechanism causes parametric shifts and could potentially trigger latchup.

The type of malfunction that occurs in the circuit in Figure 13.5 can potentially occur in an ordinary NPN transistor. If a heavily doped diffusion does not entirely enclose the collector contact, the portion of the contact touching the lightly doped epi forms a Schottky barrier. This barrier can inject minority carriers in much the same way as the Schottky diode in Figure 13.5. This problem normally occurs in structures that have been incorrectly laid out, but misalignments caused by poor photolithography has been known to produce the same result in structures that pass all applicable design rules.

13.1.2. Successful Device Mergers

This section presents two device mergers of the sort often encountered in standard bipolar layouts. There are countless possible mergers, and the examples given here only provide a general impression of what a skilled designer can achieve. Additional mergers can be discovered in the layout of many standard bipolar integrated circuits.

The *Darlington pair* in Figure 13.6A consists of a power NPN transistor, Q_1, and a smaller predrive transistor, Q_2, both sharing a common collector connection. Each transistor also has an associated base turnoff resistor. The layout of Figure 13.6B shows how all four of these components can occupy the same tank. The tank contact consists of a bar of deep-N+ placed along the left edge of the tank. The deep-N+ sinker does not encircle the power device, Q_2, because such an arrangement would consume additional die area. In general, only saturating NPN transistors and large power devices require full deep-N+ rings. Q_2 is unlikely to saturate because the extrinsic collector-to-emitter voltage V_{CE} cannot drop below the sum of the extrinsic V_{BE} of Q_2 and the extrinsic V_{SAT} of Q_1. The extrinsic V_{BE} of Q_2 probably approaches 1V at high current levels, so the tank contact can debias by the better part of a volt before Q_2 begins to saturate. The deep-N+ bar in Figure 13.6B probably exhibits no more than 5 to 10Ω of vertical resistance, so it can handle several hun-

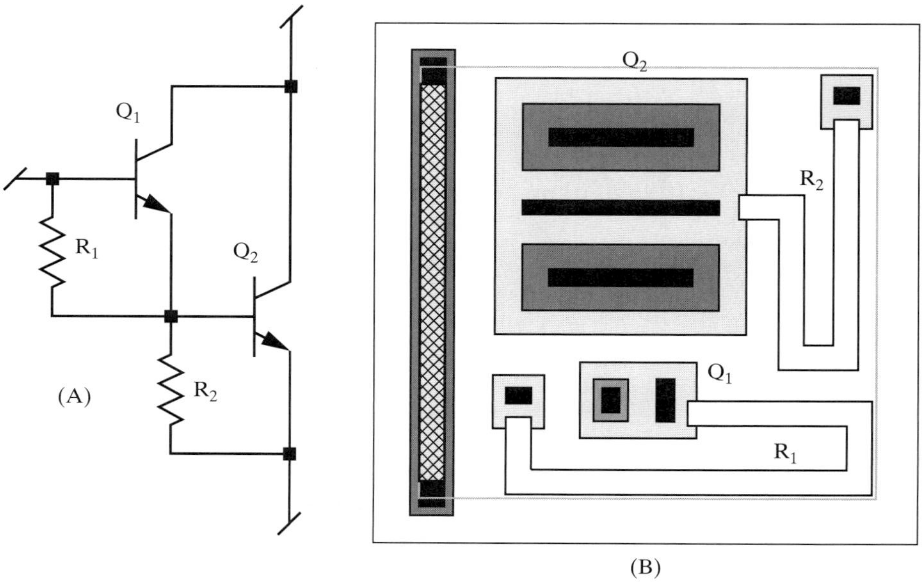

FIGURE 13.6 (A) Schematic and (B) layout of a merged Darlington pair (metal omitted for clarity).

dred milliamps of current without debiasing enough to allow Q_2 to saturate. Q_1 can saturate, but the current flow through this device is small enough to limit substrate injection to manageable levels.

Even if Q_1 saturates, it will not interfere with the operation of the Darlington. When Q_1 saturates, it will be delivering as much current as possible into Q_2. Holes injected by Q_1 into the common tank will be collected by R_1, R_2, or Q_2. The holes collected by R_1 will either return to the base of Q_1 or will flow to the base of Q_2. The influx of additional base drive into either transistor causes no harm because both transistors are already conducting all the current they can. Holes collected by Q_2's base simply represent additional base drive for Q_2. Holes collected by R_2 will probably flow to the emitter of Q_2, there to combine with the much larger current flowing through this transistor. Again, the influx of extra current causes no harm. In conclusion, it makes little difference whether or not Q_1 saturates.

The base contact of each NPN transistor also serves as a contact head for its respective base turnoff resistor. This merger can save considerable area, but the HSR implant must be spaced far enough away from the emitter to prevent this implant from raising the doping concentration in the intrinsic base region of the NPN. This layout achieves the necessary separation without enlarging the transistor by running the HSR implant into the transistor base behind the base contact.

The devices in Figure 13.6 have been arranged to enable interconnection with one level of metal. The reader may wish to mentally trace the connections between the contacts. The collector lead enters the tank from the left and the emitter lead exits from the right. These leads can be as wide as desired. The lead connecting the emitter of Q_1 to the base of Q_2 passes between the tank contact and the body of Q_2.

Figure 13.7A shows another example of a circuit that can benefit from device mergers. Transistors Q_1 and Q_2 act as a differential pair. Three-fourths of the emitter

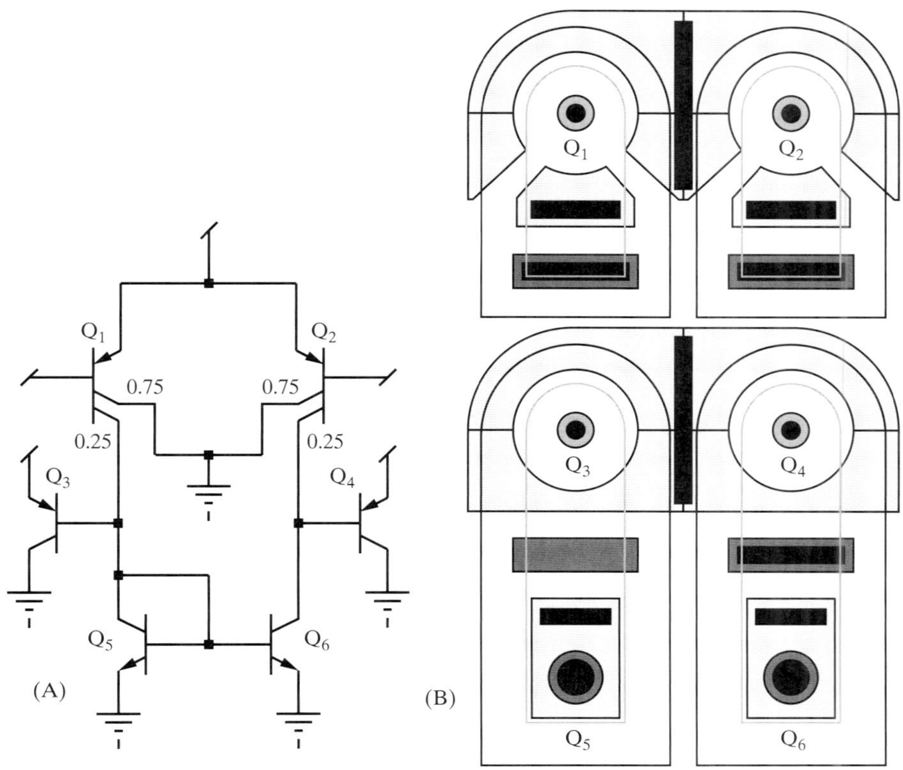

FIGURE 13.7 (A) Schematic and (B) layout of a merged operational amplifier input stage. The metallization has been omitted for clarity.

current of Q_1 and Q_2 is shunted to ground, while one-fourth feeds a current mirror consisting of Q_5 and Q_6. The output of this circuit is taken through a substrate PNP emitter follower, Q_4. An identical substrate PNP, Q_3, balances the load on the circuit and eliminates a systematic offset otherwise created by the base current of Q_4.

The layout of this circuit requires no fewer than four tanks. Q_1 and Q_2 have separate base connections and therefore require individual tanks. Q_3 and Q_5 can reside in a common tank, as can Q_4 and Q_6. The larger collectors of Q_1 and Q_2 connect to ground through an extension of the collector into the surrounding isolation. This saves considerable area by eliminating the spacing between the collector and the isolation. The size of these transistors can be further reduced by aligning the smaller collector segments to allow them to fit within a narrower tank (Figure 13.7B). The narrower tank cannot contain enough NBL to fully floor the active region of the transistor, but NBL outdiffusion prevents any significant leakage of minority carriers to the substrate.

The mergers of Q_3–Q_5 and Q_4–Q_6 present more significant problems. Each of these structures places a substrate PNP in the same tank as an NPN. The substrate PNP injects holes that could trigger latchup if they reached the merged NPN transistors. The illustrated layout circumvents this problem by substituting lateral PNP's for the substrate PNP's normally used in this role. The lateral PNP collector functions as a P-bar. This collector also extends out into the isolation to save space. The P-bar may not entirely block minority carrier flow, but several standard bipolar designs have successfully used this layout. Latchup has not been observed, and any low level of hole collection experienced by Q_5 is also experienced by Q_6. Since Q_5 and Q_6 balance one another, the circuit inherently tolerates low levels of cross-injection between Q_3–Q_5 and Q_4–Q_6.

The mergers of Q_3–Q_5 and Q_4–Q_6 are not quite identical because Q_4–Q_6 requires a tank contact while Q_3–Q_5 does not. Both mergers incorporate identical strips of emitter to ensure matching, but only Q_4–Q_6 includes a tank contact. The contact and its associated metallization have little, if any, impact on matching, so Q_3–Q_5 omits these.

13.1.3. Low-risk Merged Devices

Some device mergers are easy to perform and pose few risks. These mergers should be used whenever possible because the benefits they provide greatly outweigh their disadvantages. Examples of low-risk mergers include:

1. *Multiple segments of a single matched device.*
   Matched devices often consist of multiple segments connected in series or parallel. These segments can frequently share a tank or well, and sometimes they can share other diffusions as well. For example, a matched NPN transistor can consist of several separate emitters occupying a common base region inside a common tank. The compactness of merged layouts makes them less sensitive to gradients than conventional layouts. On the other hand, resistor segments merged in a common tank may become vulnerable to tank modulation.
2. *Multiple segments of power devices.*
   Power devices often consist of multiple segments, or *fingers.* These fingers normally share a tank, and may share other diffusions as well. For example, a power NPN transistor frequently consists of multiple emitter fingers occupying a common base region inside a common tank. Merged power devices are more compact, although this is not always a desirable feature. If the device dissipates considerable power, then a more dispersed structure may actually reduce peak temperatures within the device. Some power bipolar transistors

interdigitate base regions with strips of deep-N+ to spread power dissipation over a larger area while simultaneously reducing collector resistance.

3. *Sense transistors and their associated power transistors.*
   A power transistor sometimes has an associated sense transistor. The current passing through the sense transistor is smaller than, but proportional to, the current passing through the power transistor. The sense and power transistor must match fairly precisely despite the presence of large thermal gradients. Ideally, the sense device should consist of two equal segments located on an axis of symmetry passing through the power device. Each segment should reside approximately halfway between the center of the power device and its periphery so that its temperature roughly matches the average temperature of the whole power device. If the sense device cannot be divided into segments, then it should reside on the periphery of the power device on one of its axes of symmetry.
4. *Schottky clamps and their associated NPN transistors.*
   Schottky-clamped NPN transistors are usually constructed as merged devices. The merged Schottky clamp can use the same deep-N+ sinker as the transistor, and if necessary it can also use an extension of the NPN base region as a guard ring.
5. *NPN Darlington transistors.*
   Darlington transistors can use a layout similar to that in Figure 13.6. Most Darlington transistors are power devices that require custom layouts. Only a little additional time and effort are required to incorporate the predrive transistor and turnoff resistors in the same tank.
6. *Base turnoff resistors for NPN transistors.*
   NPN transistors often require base turnoff resistors. If these resistors are diffused devices, then they can occupy the same tanks as their associated NPN transistors.

13.1.4. Medium-risk Merged Devices

Another class of device mergers is easy to perform and offers substantial area savings, but these mergers are not without an element of risk. The risks involved are understood, and they can usually be avoided without much difficulty. Examples of moderate-risk device mergers include the following:

1. *MOS transistors in common wells.*
   Designers routinely merge MOS transistors into common wells to save space. This practice has become so widespread that digital designers frequently take it for granted. Actually, such mergers pose a certain amount of risk because cross-injection can occur between the merged source/drain regions, whether these reside in a common well or in the epi. Any problems that occur usually stem from output transistors whose source/drain regions connect to pins other than supply or ground. Externally induced transients on such pins can forward-bias source/drain regions into backgates. All output transistors require their own minority carrier guard rings (Section 13.2). If possible, output transistors should not be merged with other transistors that do not connect to the same pin. For example, an output PMOS transistor in an N-well CMOS process should occupy its own well. On the other hand, the two PMOS transistors of a NAND gate both connect to the same output and can therefore occupy the same well.

   When constructing merged transistors, the designer should always provide as much backgate contact as possible to minimize debiasing should a merged transistor forward-bias. If the process includes a suitable buried layer, then the well should contain as large an area of the buried layer as possible. Well contacts become more important as the size of the well and the number of transistors it contains increase. Very large MOS transistors often require integrated backgate contacts (Section 11.2.7).

Some circuits contain MOS transistors that regularly forward-bias into their backgate regions during normal operation. For example, some charge pumps contain devices that forward-bias during startup. These devices do not necessarily connect to pins and are not always easily identified. The circuit designer should clearly identify all such devices so that the layout designer can protect them using guard rings and separate wells.

2. *Diffused resistors in common tanks.*
   Diffused resistors are frequently merged in common tanks to save space. This practice allows the construction of compact resistor arrays that are less vulnerable to stress and thermal mismatch. Cross-injection cannot occur between the merged resistors as long as none of them forward-bias into the tank. Problems can occur if any of the resistors connect to a pin that is neither power nor ground. Such resistors should occupy their own tanks, and they may also require guard rings if the process is prone to latchup or if the transients occur during normal operation. Some circuits operate one or more resistors under biasing conditions that might cause minority carrier generation. The circuit designer should clearly mark each such resistor so that it can receive its own tank, along with any necessary guard rings. The layout designer should also be wary of merging noise-sensitive resistors with components carrying high-frequency signals, because of the potential for capacitive coupling. If in doubt, use separate tanks.
3. *Lateral PNP transistors.*
   Lateral PNP transistors sharing the same base connection can occupy the same tank. Many bipolar designs make extensive use of lateral PNP mergers. As long as none of the merged transistors saturate, their collectors act as P-bars isolating them from one another. Minority carrier cross-injection can occur if any of the transistors saturates. P-bars and N-bars (Section 4.4.2) can at least partially block cross-injection between adjacent lateral PNP transistors, but the safest solution consists of placing the saturating transistors in their own tanks.
4. *Split-collector lateral PNP transistors.*
   A split-collector lateral PNP is really a type of merged lateral PNP. A single split-collector transistor can perform the role of several ordinary lateral transistors while consuming much less die area. As long as none of the collectors saturates, few holes escape between the segments of a split-collector device. The saturation of any of the split collectors causes the currents flowing through the remaining collectors to increase. No way exists to block cross-injection in split-collector transistors, short of replacing the offending split-collector devices with separate transistors.
5. *Zener diodes.*
   Emitter-base Zener diodes can be merged with other components in a common tank as long as the tank voltage always equals or exceeds the voltage on the Zener's cathode. This condition ensures that the parasitic NPN transistor does not conduct. Series-connected Zener diodes can also occupy a common tank biased to a potential equal to or greater than that of the cathode end of the Zener string.

13.1.5. Devising New Merged Devices

Any imaginative designer will find many additional opportunities to merge devices. Before implementing a proposed device merger, determine whether it can pass the following three tests:

1. *Can any of the merged devices inject minority carriers into the shared tank or well?*
   If not, then the merged devices are safe from cross-injection and latchup. Possible sources of minority carriers include saturating bipolar transistors, forward-biased Schottky diodes, and any diffusion connecting to an external pin. Designers should be particularly wary of merging NPN transistors with other devices that can potentially inject minority carriers into the tank, because the resulting PNPN structure may latch. Potential sources of minority carriers should either reside in their own tanks or should be guarded by P-bars or N-bars unless the designer can show that cross-injection will not upset the operation of the circuit.
2. *Can any of the merged devices pull substantial current through the tank or well contact?*
   If so, then these devices may cause debiasing. Those that pull relatively small amounts of current (up to a few milliamps) rarely cause objectionable debiasing as long as the tank or well contact contains a plug of deep-N+. Higher-current devices require more extensive deep-N+ regions to prevent debiasing.
3. *Can noise coupling upset the circuit?*
   If the merged tank or well contains both noisy devices and noise-sensitive devices, then capacitive coupling between these can degrade circuit performance. The potential for noise coupling is particularly great if the noisy device pulls significant current through the tank contact.

13.2 GUARD RINGS

Of all the many types of failures that plague integrated circuits, none is so frustrating and so elusive as latchup. Devices that operate properly in one circuit latch up the moment they are inserted into another. Sometimes a device operates properly for hundreds or thousands of hours before it latches. Simulation rarely uncovers latchup problems, and neither do most forms of testing.

The most frequent causes of device latchup are external transients that pull device pins above supply or below ground. Common sources of such transients include low-level ESD events; momentary power interruptions; inductive kick-back from relays, motors, and solenoids; and inductive spiking of rapidly switched signals. Proper board-level design minimizes, but does not eliminate, these transients. Circuit designers must ensure that their designs can withstand at least moderate levels of transient injection without latching up or otherwise malfunctioning.

Power supply pins and substrate connections rarely trigger latchup, but any other pin (including grounds not connected to substrate) can cause problems. The designer should trace every lead from such a pin back through the circuit to determine whether or not it connects to any diffusions. Each diffusion connecting directly to the pin can inject minority carriers when the pin flies above supply or below ground. Diffusions connecting to pins through deposited resistors still pose a concern if the series resistance is less than about 50kΩ. Larger-value deposited resistors reduce injected currents so much that they no longer pose any significant threat.

Latchup can be suppressed by enclosing each vulnerable diffusion (or device) in a suitable minority carrier guard ring. Multiple diffusions connecting to a common pin can share a common guard ring. ESD devices residing around the periphery of the die can often share a common guard ring separating the core of the die from the ESD devices and bondpads. Many older standard bipolar designs omitted guard rings from some or all pins, but new designs should not follow this practice because it sometimes results in costly redesigns.

13.2.1. Standard Bipolar Electron Guard Rings

Any tank connecting to a device pin can inject electrons into the substrate. No way exists to block these electrons from reaching the substrate. Guard rings do exist that can collect electrons that touch them, but nothing stops electrons from burrowing underneath the guard rings and escaping. The structure in Figure 13.8A is arguably the best electron guard ring that can be constructed in standard bipolar. It consists of a strip of deep-N+ residing in an N-tank augmented by both NBL and emitter.[3] This combination of diffusions forms the deepest possible guard ring and therefore collects the largest possible fraction of electrons. The presence of deep-N+ also helps prevent Ohmic debiasing. This guard ring would ideally connect to the highest available supply voltage to drive the depletion region as deeply as possible into the substrate. This style of guard ring also functions if it is connected to ground, but grounded guard rings are more susceptible to debiasing. Grounded guard rings are sometimes used to minimize power dissipation caused by minority carrier injection, which sometimes becomes a concern in high-current designs. If a grounded guard ring is used to minimize power dissipation, it can be supplemented by a second guard ring connected to the supply and placed outside of the grounded guard ring. This secondary guard ring will provide protection if the grounded guard ring saturates.

One can sometimes take advantage of adjacent tanks that connect to the supply. If these tanks are strategically situated between the point of minority carrier injection and adjacent sensitive circuitry, then they become very effective guard rings. All tanks used for this purpose should contain as much NBL as possible and should use deep-N+ sinkers to minimize debiasing. The efficiency of an electron-collecting guard ring also increases if it is placed next to the source of injected carriers to take advantage of the proximity effect.

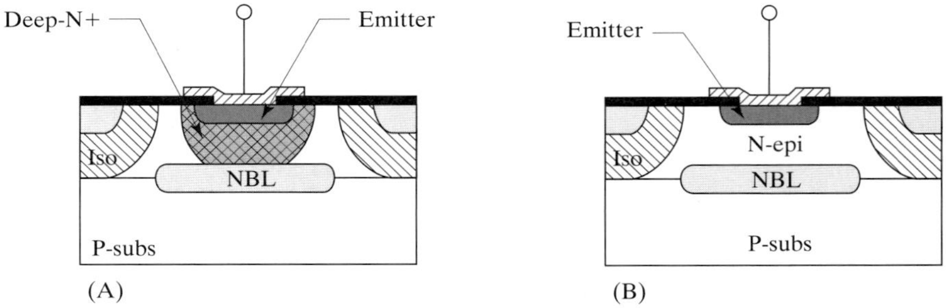

FIGURE 13.8 Electron-collecting guard rings for standard bipolar: (A) preferred structure and (B) alternate structure.

The guard ring in Figure 13.8B should only be used if deep-N+ is not available. The vertical resistance of the epi layer separating the NBL from the emitter diffusion makes this guard ring extremely vulnerable to Ohmic debiasing. This structure is still marginally effective as long as it connects to a power supply, but it is virtually useless when connected to ground.

---

[3] A similar guard ring for a BiCMOS process is presented in E. Bayer, W. Bucksch, K. Scoones, K. Wagensohner, J. Erdeljac, and L. Hutter, "A 1.0μm Linear BiCMOS Technology with Power DMOS Capability," *BCTM Proceedings,* 1995, pp. 137–141.

Electron guard rings constructed in standard bipolar are only marginally effective, yet they consume vast amounts of die area. Most designers omit these guard rings to save space and rely instead on large spacings and strategically placed hole guard rings to prevent latchup. These measures usually suffice for linear circuits such as operational amplifiers and voltage regulators. Devices that switch inductive loads are another matter entirely, as these loads can generate extremely energetic transients during normal operation. Even if these transients do not cause latchup, they can still inject noise into sensitive circuitry. High-frequency MOSFET gate drivers can also experience severe transients caused by resonance in the gate lead. The output circuitry of MOSFET gate drivers and inductive load drivers must be carefully shielded by electron guard rings to minimize noise coupling and latchup sensitivity.

Electron guard rings become much more effective if the process incorporates a P+ substrate. The P+/P– interface generates an electric field that traps most of the injected electrons in the P– epi. Those few that do penetrate into the P+ substrate quickly recombine. The P+ substrate makes deep guard rings such as the one in Figure 13.8A extremely effective, particularly if they are biased to a high-enough potential to drive a depletion region down to meet the P+ substrate.

13.2.2. Standard Bipolar Hole Guard Rings

Any P-type region can inject holes into a tank. Hole guard rings can prevent these carriers from flowing to adjacent P-type regions or to the sidewalls of the tank. Two types of hole guard rings exist: the *hole-collecting guard ring* and the *hole-blocking guard ring*. Figure 13.9A shows a typical hole-collecting guard ring deployed to prevent holes from reaching the sidewalls of a tank. The presence of NBL prevents the holes from flowing down to the substrate and instead forces them to flow laterally. The guard ring consists of a reverse-biased base diffusion surrounding the point of injection. This diffusion acts as the collector of a lateral PNP transistor. Any holes reaching the depletion region surrounding the guard ring are drawn into it. Hole-collecting guard rings are normally grounded to maximize the reverse bias between the tank and the guard ring. This not only drives the depletion region deeper into the tank but also minimizes the effects of Ohmic debiasing within the guard ring itself. Grounded hole guard rings tie to the same potential as the isolation system, so they can be merged to save space. Examples of such merged guard rings include the P-bar in Figure 4.24 and the grounded collectors of transistors Q_3 and Q_4 in Figure 13.7. A hole-collecting guard ring can also be tied to the tank potential, but this reduces its effectiveness and does not save any appreciable space.

Figure 13.9B shows a typical example of a hole-blocking guard ring.[4] This type of guard ring surrounds the point of injection with heavily doped N-type regions. The

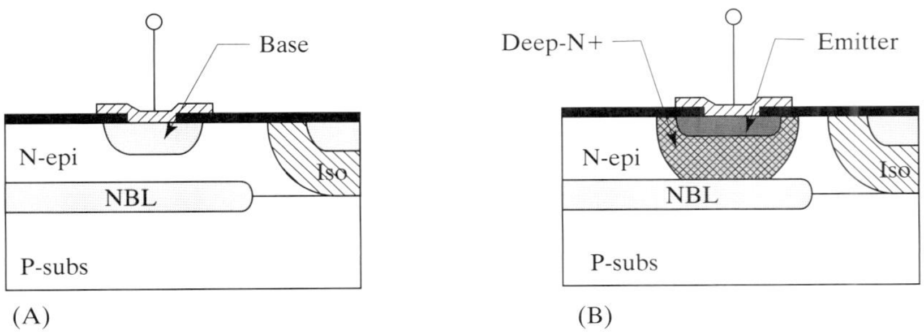

FIGURE 13.9 Hole guard rings for standard bipolar: (A) hole-collecting guard ring and (B) hole-blocking guard ring.

[4] *Ibid.*

N+/N– interface generates an electric field that acts as a barrier to the passage of holes. Any holes overcoming this barrier typically recombine before they can traverse the N+ region. The electron current required to sustain recombination flows through contacts to the N+ region. The structure in Figure 13.9B relies on NBL to block the downward travel of holes, and deep-N+ to block their lateral movement. In order to be fully effective, a hole-blocking guard ring must contain no gaps or holes. The only practical way to achieve this goal consists of entirely encircling the injector with a ring of deep-N+. Partial hole-blocking rings such as the N-bar in Figure 4.25 may allow a substantial fraction of the holes to escape around the gaps at either end. In some processes, these gaps can be eliminated by extending the deep-N+ bar into the isolation on either side of the tank. Most processes do not allow this configuration because of leakage across the isolation/deep-N+ junction. Both hole-collecting and hole-blocking guard rings can provide efficiencies of 95% or better in typical standard bipolar processes. A combination of both types of guard rings (Figure 4.24) can achieve an efficiency in excess of 99%.

Standard bipolar designs use relatively few hole guard rings because this process rarely requires them. Standard bipolar designs seldom experience latchup due to external transients because the deep-P+ isolation and the large spacing between components both help reduce the beta product of the parasitic SCR (Section 11.2.7). Hole injection into the substrate only becomes a problem if it overwhelms the capability of the substrate contact system. This is unlikely to occur because the grid of P+ isolation diffusion helps to magnify the effective area of substrate contacts, while the P– isolation helps to limit the maximum injected current and to contain substrate debiasing within relatively limited regions of the die. Hole guard rings are usually employed only to prevent cross-injection between merged components (Section 13.2.1). P-type regions connecting to external pins that are neither power supplies nor substrate ground are usually isolated by placing them in their own tanks. This practice requires about the same amount of space as the construction of hole guard rings and requires less effort.

13.2.3. Guard Rings in CMOS and BiCMOS Designs

CMOS designs are more prone to latchup than standard bipolar. This vulnerability results in part from the smaller dimensions of modern CMOS and BiCMOS processes and in part from differences between isolation systems. CMOS-derived processes usually substitute a lightly doped epitaxial layer for the vertical P+ isolation of standard bipolar. The light doping increases the gain of the lateral bipolar transistor formed across the isolation and makes it more probable that minority carrier injection will trigger SCR action. The light doping of the P-epi also makes it more difficult to extract substrate current. Most of these processes rely on the presence of a P+ substrate to reduce their vulnerability to latchup through the substrate, but scrupulous care must be taken to block lateral conduction by using guard rings.

Figure 13.10A shows an electron-collecting guard ring implemented in a CMOS process. This structure consists of an NMoat ring placed in the P-epi surrounding the source of injected electrons. The NSD implant is relatively shallow, so it can only intercept a fraction of the carriers. This type of guard ring relies on the P+ substrate beneath the wells to prevent minority carriers from bypassing the guard ring by burrowing through the substrate. Unfortunately, the presence of an electric field across the P+/P– interface tends to repel electrons from the substrate and to channel them laterally toward adjacent wells. This phenomenon makes it difficult to construct truly effective barriers against minority carrier injection into the substrate. Connecting the NMoat ring to a power supply rather than to ground helps only marginally, since the

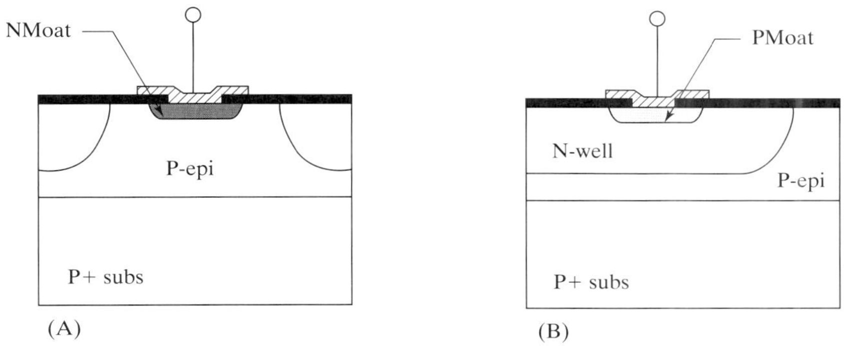

FIGURE 13.10 Minority-carrier guard rings for N-well CMOS: (A) electron-collecting guard ring and (B) hole-collecting guard ring.

added depth of the depletion region does not penetrate more than a small fraction of the epitaxial layer. In low-voltage CMOS processes where the NMoat resides in a P-well, the guard ring should connect to substrate potential rather than to a power supply. The increased surface doping of the low-voltage P-well reduces the width of the NSD/P-well depletion region and causes the electric field across it to increase. High electric-field intensities can trigger avalanche multiplication within the depletion region. The resulting debiasing actually improves the collection efficiency of the guard ring, but it may cause problems elsewhere on the die.

Figure 13.10B shows a hole-collecting guard ring implemented in a CMOS process. This guard ring consists of a ring of PMoat placed in the N-well around the source of injected holes. This type of guard ring is not very effective because most of the holes flow down to the substrate rather than laterally to the guard ring. Increasing the width of the guard ring does little to improve its effectiveness.

The minority carrier guard rings that can be constructed in a CMOS process usually have very limited effectiveness. The two types of guard rings tend to reinforce one another, so the best design practice consists of using both electron and hole collecting guard rings around every device that might inject minority carriers. Since CMOS devices do not inject any substantial level of minority carriers during normal operation, this requirement is satisfied if every device connecting to an output pin receives a guard ring. The designer should examine each pin that neither connects to a power supply nor to substrate potential. Each source/drain region connecting to such a pin requires a guard ring. PMOS transistors require hole-collecting guard rings even when placed in their own wells. NMOS transistors require electron-collecting guard rings. A combination of guard rings and backgate contacts should suppress most forms of latchup, but they may prove inadequate for handling the severe minority carrier injection problems associated with inductive kickback and resonance.

Analog BiCMOS processes normally include NBL and deep-N+. The presence of these layers allows the construction of deep electron-collecting guard rings similar to the one in Figure 13.8A. These guard rings are especially effective on designs using a P+ substrate because the built-in potential of the P-epi/substrate interface helps confine the electrons within the epi. A deep-N+ guard ring on a thin-epi P+ process may collect 90% or more of the electrons injected into the epi.[5]

Although analog BiCMOS supports the construction of hole-blocking and hole-collecting guard rings, these may not be as effective as their standard bipolar coun-

[5] R. R. Troutman, "Epitaxial Layer Enhancement of n-Well Guard Rings for CMOS Circuits," *IEEE Electron Device Letters*, Vol. EDL-4, #12, 1983, pp. 438–440.

terparts. Analog BiCMOS processes often use a lower NBL doping concentration to minimize lateral autodoping. The lower doping decreases the built-in potential at the NBL/N-well interface and allows more carriers to enter the NBL. The lighter NBL also decreases the Gummel number of the parasitic substrate PNP, so a substantial fraction of the carriers may actually penetrate through the NBL/N-well interface to the substrate underneath. This problem becomes even more acute on low-voltage processes using heavily doped wells, since the increased well doping further degrades the built-in potential at the NBL/N-well interface.

The efficiency of hole guard rings suffers if the NBL cannot efficiently block hole flow to the substrate. The addition of a hole blocking guard ring may actually increase substrate injection through a "leaky" NBL.[6] This seemingly paradoxical behavior probably results from a reduction in the effective volume of the N-well. The hole-blocking guard ring repels holes from the portion of the well it occupies, concentrating them within the remaining volume of the well. The higher concentration of holes near the NBL/N-well interface increases the injection rate of carriers into the substrate. This *reduction in volume effect* should not affect hole-collecting guard rings, but the presence of a "leaky" NBL still reduces their collection efficiency.

Analog BiCMOS designs also exhibit excessive substrate resistance. Even if the design uses a P+ substrate, the presence of a lightly doped P-epi makes it difficult to establish a low-resistance substrate contact. Even relatively low levels of substrate injection can produce substantial substrate debiasing. Substrate debiasing can be prevented by blocking minority carriers before they can reach the substrate through the use of hole guard rings. All high-current saturating NPN transistors should incorporate such guard rings to prevent substrate debiasing and noise coupling.

Analog BiCMOS designs sometimes use a P– substrate to avoid the necessity of growing two epitaxial layers. Designs constructed on a P– substrate are even more susceptible to latchup because electron guard rings no longer benefit from the presence of an electron barrier at the P–/P+ interface. Many designs can still achieve satisfactory levels of immunity to transient-induced latchup providing that every potential source of minority carrier injection is surrounded by a suitable guard ring. Even the most conservatively designed guard rings may prove unable to handle the severe minority carrier injection problems associated with inductive kickback and resonance. Such designs may require the use of a P+ substrate despite the additional cost associated with the second epitaxial deposition.

13.3 SINGLE-LEVEL INTERCONNECTION

Most modern processes offer at least two levels of metallization. Since leads can freely cross one another, the placement of components becomes constrained only by matching and packing. Routing almost never presents a problem as long as the designer leaves a little space between components. Given time, almost any designer can compress the wasted space out of such a layout to produce a reasonably densely packed design.

Interconnection becomes much more difficult if the process offers only one level of metallization. The lack of second metal makes it difficult to cross leads. Although low-value resistors can be inserted to create crossing points, these *tunnels* consume die area and add resistance and capacitance that degrades the performance of the circuit. A properly arranged layout contains surprisingly few tunnels. The compo-

---

6 N. Gibson, unpublished report, 1998.

nents in such a layout are arranged to minimize the number of crossing points. Leads often route across resistors or between the terminals of transistors, and sometimes they even tunnel through tanks or base diffusions.

Single-level interconnection requires far greater skill and ingenuity than multi-level interconnection. The designer must be able to anticipate possible blockages between components and be able to mentally shuffle the components to clear the blockages. A move that clears one blockage often creates others. Skilled designers have a sort of "geometric intuition" that aids them in placing components and routing leads. This intuition seems largely an innate talent and not a learned skill. There are, however, a number of specific skills and techniques that can help any designer better cope with single-level-metal designs. Although these skills may not seem to have any application in modern multi-level-metal processes, many designs dedicate the upper metal layers for power routing, electrostatic shielding, or optical shielding. A skilled designer must therefore understand how to route designs with a minimum number of layers of interconnection.

13.3.1. Mock Layouts and Stick Diagrams

The greatest challenge of single-level routing lies in properly arranging the components to minimize the number of tunnels required. The presence of matched components often complicates this task so that even skilled designers have to try several arrangements before finding a suitable one. These trial arrangements usually take the form of rough sketches, or *mock layouts*, similar to that in Figure 13.11. The transistors appear in this sketch as rectangles with emitter, base, and collector marked. The resistors appear as strips with connections to either end. Dummies and resistor tanks do not appear in the sketch. The tank contacts are marked "TC." Merged devices occupying a common tank are shown abutting one another, as in the case of Q_3 and Q_4. Although crude, this sketch illustrates all of the important features of the proposed layout.

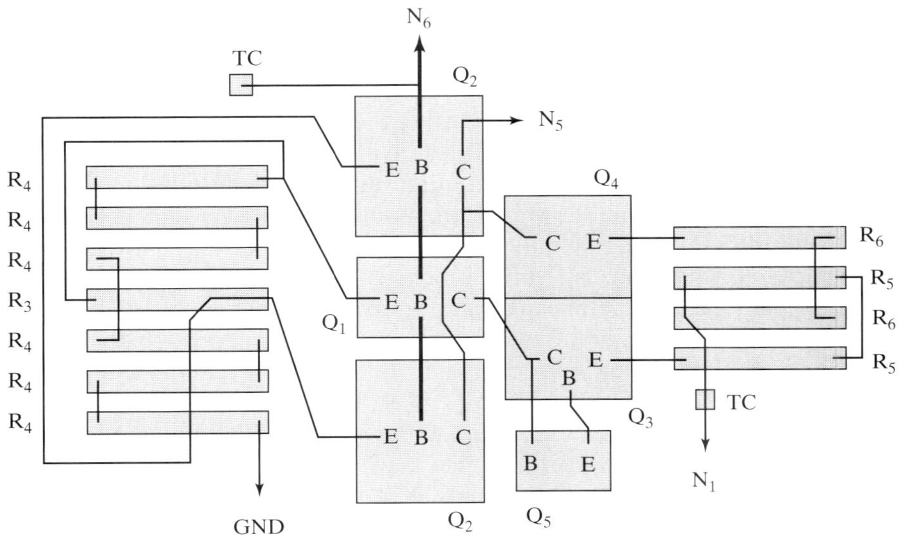

FIGURE 13.11 Mock layout for a portion of the circuit shown in Figure 14.2. The following components must match one another as accurately as possible: R_3–R_4, Q_1–Q_2, Q_3–Q_4, and R_5–R_6.

This particular layout contains a large number of matched devices. In order to obtain the best possible matching, each set of these components has been arranged symmetrically around one axis of the layout as advocated in Section 7.2.6. The axis of symmetry passes horizontally through the middle of the sketch.

Resistors R_3 and R_4 consist of 160Ω/□ base material and have values of 621Ω and 4kΩ, respectively. These resistors are not in simple integer ratio ($R_4/R_3 = 6.441$). This complicates, but does not prohibit, the construction of a sectioned and interdigitated array. Suppose that R_3 is taken as the unit resistor for the array. R_4 then requires a minimum of seven segments. Unfortunately, the centroid of a one-segment resistor cannot perfectly align to the centroid of a seven-segment resistor. In order to achieve a true common-centroid layout, R_4 must consist of eight partial segments. This arrangement is not particularly compact because all of the segments are relatively short (R_3 contains 3.88 squares). A better arrangement consists of six segments of 666.7Ω each for R_4 and a partial segment for R_3. The sliding contact in R_3 also allows the resistor ratio to be tweaked. R_3 occupies the center of the array to ensure that the centroids align (Section 7.2.6). The six segments of R_4 have been interconnected to cancel thermoelectrics (Section 7.2.7). Resistors R_5 and R_6 each consist of 18.75 squares of 160Ω/□ base diffusion. The mock layout shows that each resistor consists of two segments of 9.375 squares that are interdigitated to form a compact array.

Transistors Q_1 and Q_2 form a 6:1 ratioed pair. The layout of such transistors normally involves splitting the larger transistor into two halves placed on either side of the smaller transistor (Figure 9.19A). Transistors Q_3 and Q_4 are matched, minimum-size lateral PNP transistors. These transistors can reside in a common tank because they share the same base connection. They are placed side-by-side to improve matching and to simplify interconnection. P-bars and N-bars are unnecessary because neither transistor saturates in normal operation.

Although this circuit contains several crossing points, none requires a tunnel. The lead interconnecting the emitters of Q_2 routes through the resistor array R_3–R_4. The collector lead could follow the same route, but this requires separating the resistor segments. Instead, transistors Q_1 and Q_2 have been stretched to allow Q_2's collector lead to route between the base and collector of Q_1. This configuration actually requires little or no elongation of the transistor tanks because the presence of deep-N+ in Q_1 and Q_2 already necessitates a large base-to-collector spacing.

The mock layout in Figure 13.11 has not been drawn to scale, but the rectangles representing the components have roughly the same proportions as the components themselves. Sometimes designers carry this type of sketch one step further by using paper plots of the actual components. All of the components are plotted to the same scale, typically either 100:1 or 250:1. The individual components are cut out and shuffled about on a large sheet of paper until a suitable arrangement appears, and then the components are glued down and the interconnections are marked in pencil or ink. Cut-and-paste mock layouts (*paper dolls*) are especially useful for designing tight-packed layouts, since all of the dimensions of the components are to scale.

CMOS designers sometimes use another type of mock layout called a *stick diagram.* Although stick diagrams were originally intended to portray digital logic cells, they can represent analog circuitry. Figure 13.12 shows a schematic and a stick diagram of a CMOS NAND gate. The thick, black horizontal lines represent PMoat and NMoat regions. The NMoat usually lies at the bottom of the diagram and the PMoat at the top. The thick, gray vertical lines represent poly traces. A transistor forms wherever poly crosses either PMoat or NMoat. Contacts are represented by X-marks and metal leads by thin, black lines. Stick diagrams of analog circuits usually show NMoat and PMoat regions in different colors to aid in distinguishing them from one another. The names of nodes and devices may both appear on the stick diagram, and additional notations may be added to identify resistors and capacitors.

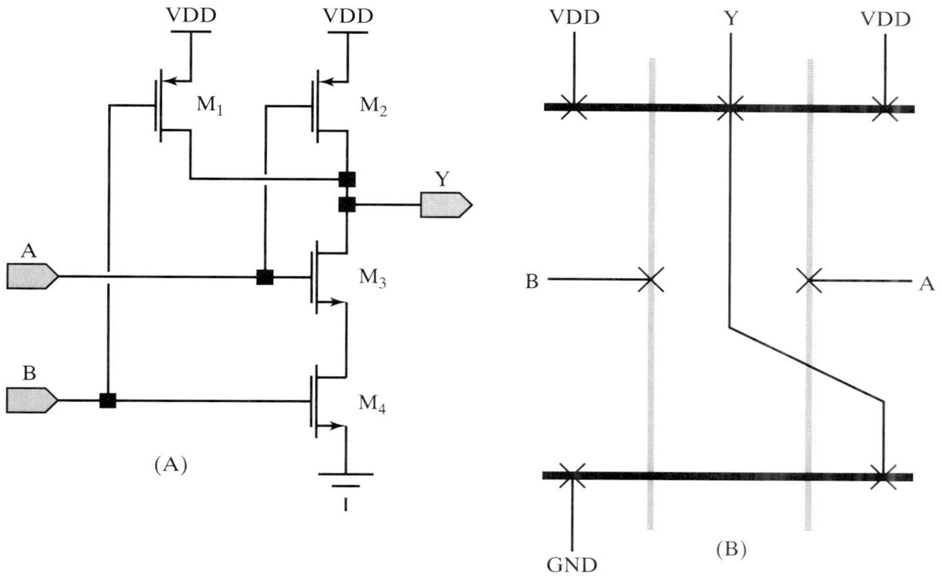

FIGURE 13.12 (A) Schematic and (B) stick diagram of a CMOS NAND gate.

13.3.2. Techniques for Crossing Leads

No matter how hard one tries to avoid them, most single-level metal layouts still require crossing points. The following rules summarize the techniques available for crossing leads using only one level of metal. These techniques were originally developed for standard bipolar designs, but they are also applicable to multiple-level metallization designs where the upper metal layers are required for power routing, electrostatic shielding, or some other purpose.

1. *Cross leads over resistors.*
   A lead routed across a resistor provides a crossing point without consuming additional area, but not every lead can safely cross every resistor. Some resistors require field plating that restricts or even prevents lead crossings. Other resistors are susceptible to noise coupling from overlying leads. Lightly doped materials such as 2kΩ/□ HSR may also experience voltage modulation effects.
2. *Rearrange the terminals of a device.*
   Crossing points can often be eliminated by altering the arrangement of device terminals. For example, the CEB layout for an NPN transistor places the emitter terminal between the collector and base, while the alternate CBE layout places the base terminal between the collector and emitter (Figure 8.14). The CBE layout exhibits slightly more collector resistance than the CEB layout, but the difference rarely affects circuit operation.
3. *Stretch devices to allow leads to pass through them.*
   Most types of devices can stretch to accommodate one or more leads between their terminals. Figure 8.15 shows three examples of stretched NPN transistors, and Figure 5.14 shows an example of a stretched HSR resistor. Stretched devices usually possess more parasitic resistance and capacitance than unstretched devices, so their use sometimes affects circuit operation. If one of a group of matched devices employs a stretched layout, then so should all the others.

4. *Connect signals through merged devices.*
Certain types of devices lend themselves to use as tunnels. For example, the NPN transistor in Figure 8.15C contains a stretched base region with two contacts. This component actually merges an NPN transistor and a base tunnel into the same tank. A similar type of merger uses multiple tank contacts rather than multiple base contacts. These types of merged tunnels insert parasitic resistances and capacitances that can affect device operation. If large currents flow through the tunnel, the resulting debiasing may also affect circuit operation.
5. *Insert tunnels.*
Tunnels, or *cross-unders,* are low-value resistors incorporated into the layout to allow leads to cross one another. Various types of tunnels exist, but all share similar disadvantages. They not only consume die area, but they also insert parasitic resistance and capacitance into the tunneled lead. The insertion of a tunnel into a high-current lead can cause excessive voltage drops and power dissipation. Tunnels can also upset matching by introducing voltage drops where they cannot be tolerated. Since the placement of tunnels affects circuit operation, the circuit designer must ultimately approve or reject each potential tunnel. When all of the tunnels have been placed, the circuit designer should add their resistances and capacitances to the circuit and resimulate to see if any critical parameters have shifted.
6. *Rearrange the bondpads.*
If high-current leads must cross one another to reach their respective bondpads, consider rearranging the bondpads to eliminate the crossing point. Sometimes one bondpad arrangement may lend itself to interconnection much more readily than others.

The layout designer usually requires guidance from the circuit designer to determine what types of stretches and tunnels are allowable. One simple and effective means of communicating this information consists of an annotated schematic prepared by the circuit designer. This diagram only requires a few minutes to prepare, yet it can save many hours of layout effort. Table 13.1 describes a simple annotation scheme using different colors to highlight components and signals. The annotated schematic should also include lists of matched components, guard rings, and other special requirements that might influence the routing of the leads.

TABLE 13.1 A simple schematic annotation scheme.

Category	Precautions	Marking
Power leads	No tunnels allowed; leads must equal or exceed a certain width.	Highlight in **red;** mark width over lead.
Noisy leads	Do not cross sensitive devices.	Highlight in **yellow.**
Sensitive leads	Do not tunnel. Do not place substrate contacts in sensitive grounds leads.	Highlight in **green.**
Sensitive devices	Do not cross noisy leads over sensitive devices.	Highlight in **green.**

13.3.3. Types of Tunnels

Tunnels can be constructed from any diffusion having a relatively low sheet resistance. In standard bipolar, the candidates include base, emitter, deep-N+, and NBL. CMOS and BiCMOS processes often use gate poly jumpers instead of tunnels. The base diffusion usually has a sheet resistance of 100 to 200Ω/□, while the other three materials usually have sheet resistances of about 10Ω/□. Of these, only the base diffusion can occupy a common tank with other components. Most standard bipolar designs contain a number of merged base tunnels and a few stand-alone tunnels of other types.

All tunnels add series resistance. In the case of base tunnels, this resistance usually equals a few hundred Ohms. Although this may not seem like much resistance, it is sufficient to cause some circuits to malfunction. A typical tunnel also exhibits a few hundred femtofarads of parasitic junction capacitance. Some high-speed circuits contain nodes that would be seriously slowed by even this small amount of capacitance. Any attempt to reduce the series resistance of the tunnel by widening also increases its shunt capacitance. Larger tunnels also become more vulnerable to junction leakage and to minority carrier collection.

The sheet resistance of the emitter diffusion is an order of magnitude lower than that of base. The simplest sort of emitter tunnel simply consists of a strip of emitter diffusion placed in a tank (Figure 13.13). The tank-substrate junction provides the necessary isolation between the signal and the underlying substrate. The tank requires no contact other than that provided by the tunnel itself. The addition of NBL does not significantly reduce the tunnel resistance or improve latchup immunity. NBL actually increases the parasitic shunt capacitance, so emitter tunnels generally omit it.

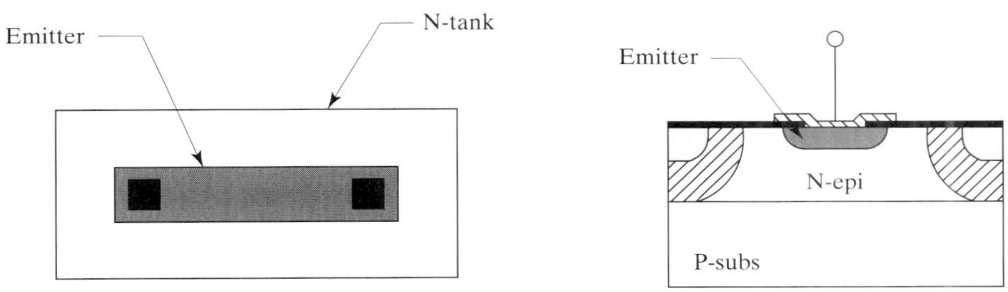

FIGURE 13.13 Layout and cross section of a conventional emitter tunnel.

The emitter-in-iso tunnel in Figure 13.14 saves considerable area by eliminating a tank geometry. The emitter diffusion can counterdope the isolation, but the breakdown of the resulting N+/P+ junction may equal only a few volts. Junctions with breakdown voltages this low tend to leak, but emitter-in-iso tunnels can still be used to route the substrate return line around the die.[7] Processes with emitter-iso breakdown voltages of 6V or higher can often employ emitter-in-iso tunnels to route other signals, but the capacitance of the emitter-iso junction is relatively large (typically

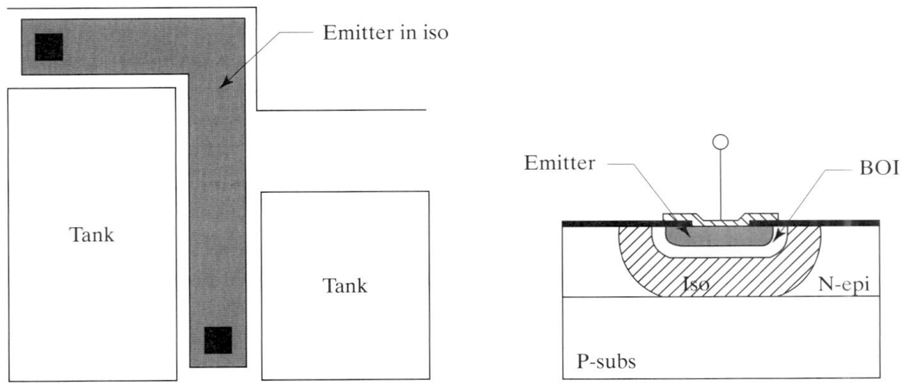

FIGURE 13.14 Layout and cross section of an emitter-in-iso tunnel.

[7] Davis, pp. 16–17.

1pF/mil^2, or 1.6fF/μm^2). Some circuits have taken advantage of this high capacitance to fabricate emitter-iso junction capacitors. These capacitors can occupy unused areas of the isolation, allowing the construction of large junction capacitors with little or no increase in die area.

Some applications require a resistance that is lower than emitter alone can provide. A combination of all available N-type regions (N-epi, NBL, deep-N+, and emitter) will produce a slightly lower sheet resistance (Figure 13.15). A stack of this sort usually has a sheet resistance of about 5Ω/□. NBL provides little benefit without deep-N+, so designs that do not use deep-N+ cannot make effective use of stacked tunnels.

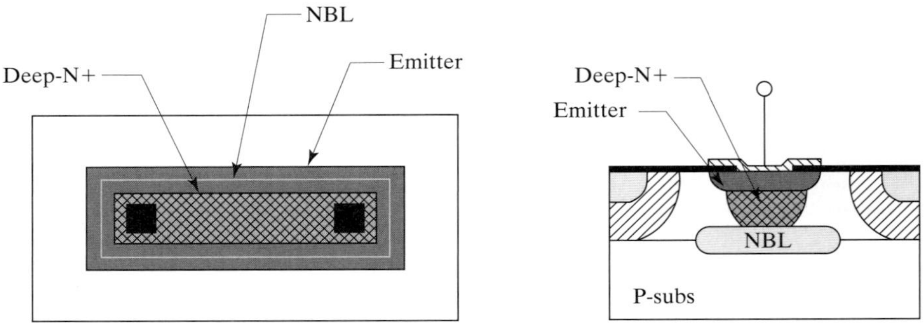

FIGURE 13.15 Layout and cross section of a low-resistance, stacked tunnel containing emitter, deep-N+, and NBL.

Several other types of tunnels are occasionally used. Of these, the deep-N+ tunnel merits at least a brief mention. These tunnels are sometimes used in place of emitter tunnels on processes with thin emitter oxides. An ESD strike could rupture the thin emitter oxide if an emitter tunnel were placed underneath a lead connecting to a bondpad. Deep-N+ tunnels are covered by a thick oxide that is virtually immune to ESD damage. There are few applications for deep-N+ tunnels in modern designs because the newer processes generally use thick emitter oxides, and because modern design practices dictate the use of ESD structures on virtually every pin.

13.4 CONSTRUCTING THE PADRING

The *padring* of an integrated circuit consists of scribe streets, pads, ESD structures, and guard rings. Each of these plays a vital role in determining the success or failure of the design. Many circuits have failed because of inadequate ESD protection, misplaced bondpads, or missing guard rings. The following sections provide guidance that should help the layout designer avoid most of these mistakes.

13.4.1. Scribe Streets and Alignment Markers

Scribe streets must surround the die to provide room for the passage of the sawblade used to separate the dice. The saw consumes a strip of silicon about 25μm (1mil) wide, but the scribe streets must be three or four times wider to provide room for misalignment. Oxide and nitride tend to crack during sawing, and metal clogs the sawblade; therefore most scribe streets consist of bare silicon. The edges of the die abutting the scribe street are often fitted with special structures called *scribe seals* to prevent contaminants from seeping underneath the exposed edges of the protective overcoat (Section 4.2.2). Additional structures may reside within the scribe street itself. Some fabs place *alignment markers* within the streets. These markers are used to align photomasks to previous steps of the process and are destroyed during sawing. Sometimes arrays of test devices are also placed in the scribe streets. These devices can be used to evaluate the performance of the wafer before it is sawn and assembled. The test devices also provide a means of characterizing large numbers of devices for statistical device modeling. The test devices can be destroyed during sawing because they already will have been tested and so will have served their purpose.

Most wafer fabs specify the scribe streets required for their process. Sometimes the scribe street occupies a separate database prepared by the fab and sent to the mask shop independently of the main design. Alternatively, the scribe street structures may be provided to the layout designer to place around the edges of the main layout. Regardless of whether they appear in the layout, the scribe seals abut the die on all four sides. Since these seals usually incorporate substrate contacts, they can form a useful addition to the substrate contact system. A thin strip of metal placed around the edges of the pad ring will make contact to the scribe seal metallization. In addition to providing substrate contacts, the scribe seal metallization also provides a convenient method of routing the substrate potential around the periphery of the die. The width of the metallization in the scribe seal adds to the width of metal in the padring, producing a relatively wide lead that can conduct significant current without debiasing or electromigration failure (Figure 13.16). Because of its convenient placement around the periphery of the die, the substrate metallization often forms the return path for the ESD structures (Section 13.4.3).

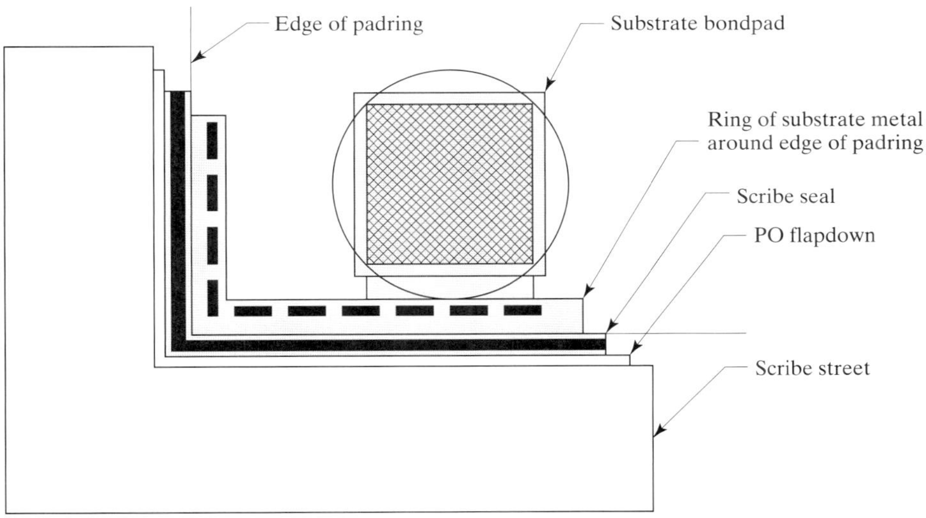

FIGURE 13.16 Diagram illustrating the relationship between the scribe street and the padring.

It is not possible to provide generalized guidance for constructing scribe seals, because every wafer fab has different requirements. The layout designer must seek the fab's guidance to determine the proper scribeline selection and placement. Some fabs may impose additional requirements, including limitations on the size and aspect ratio of the die. For example, some older steppers only accept dice whose dimensions are an integer number of mils. The efficient use of photolithographic equipment may also dictate that the die dimensions fall within a certain range of values. These issues should be resolved during die floorplanning so the designer can maximize the profitability of the design. These issues become much more difficult to resolve as the layout nears completion, so do not wait until the last moment to obtain the necessary information!

13.4.2. Bondpads, Trimpads, and Testpads

Most integrated circuits connect to the external world through bondwires. These wires consist of either gold or aluminum, and they range in diameter from 0.8 to 10.0 mils (20 to 250μm). The most common form of bonding uses gold wires approximately 1 mil (25μm) in diameter attached to the die by means of ball bonds. The ball bonding process uses a hydrogen flame to create a tiny gold ball on the end of the bondwire. A capillary tube presses this ball down against the exposed aluminum metal with enough force to cause the two metals to alloy together (Section 2.7.1). The bonding process deforms the soft gold ball into a pancake-like structure (Figure 13.17). The actual area of gold-aluminum alloying is usually about the same diameter as the original bondwire, but the metal pad on which the ballbond rests must be two to three times larger to account for the misalignments that inevitably occur during automated bonding. Each ballbond thus requires an exposed metal pad several mils across. These specialized structures are called *bondpads.*

FIGURE 13.17 Cross section and layout of a double-level-metal bondpad intended for ballbonding.

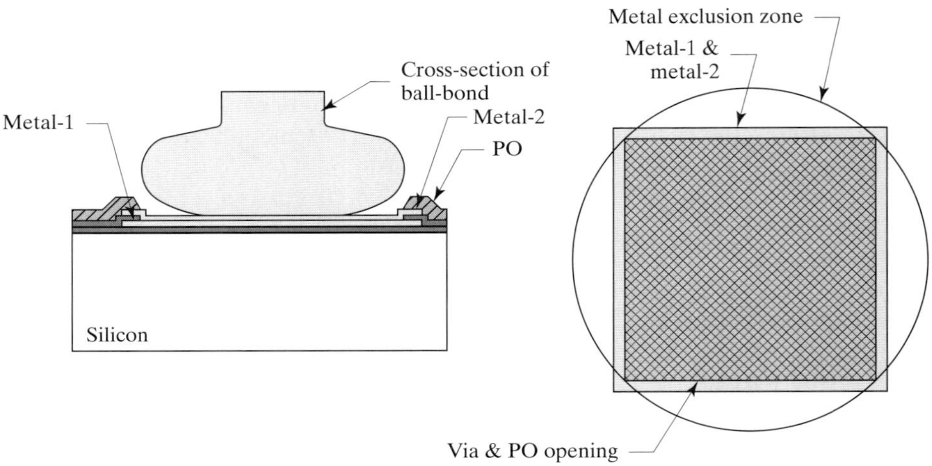

The simplest conceivable bondpad consists of a square of metal placed underneath a matching opening in the protective overcoat (PO). The metal must overlap the opening sufficiently to seal the die against the ingress of mobile ions (Section 4.2.2), even if overetching and misalignment occur. If the process offers more than one level of metal, then the bondpad typically includes plates of each metal placed coincident to one another. The interlevel oxides should be removed from the area of the bondpad opening so the ballbond lands on the stacked metal layers. Figure 13.17 shows a typical double-level-metal bondpad constructed according to these principles. The bondpad consists of a square of metal-2 placed over an identical square of metal-1. These metal plates overlap a PO opening and a via, both of which also coincide to one another.

Dice are usually bonded using the smallest possible diameter of wire, since this enables the use of the smallest bondpads. Most assembly sites offer either a 1.0mil (25μm) or an 0.8mil (20μm) gold wire as a minimum wire diameter. A gold bondwire packaged in plastic can conduct approximately one amp of continuous current per mil of diameter (Section 14.3.3). Thus an 0.8mil gold wire can conduct a continuous current of about 800mA. Higher currents require either larger-diameter wires or multiple bondwires connected in parallel. Every pin of a typical package can accommodate two (or possibly three) bondwires. Some surface-mount packages are so small that they can only accommodate one bondwire per pin, and these same packages are usually so thin that they can only use the finest wire diameters. The designer should verify that the package can handle the required number and diameter of bondwires during the earliest stages of floorplanning.

Small-diameter bondwires have relatively large resistances. The resistance of a bondwire can be estimated using the equation

$$R_w \cong \frac{\omega L}{D^2} \qquad [13.2]$$

where R_w is the resistance of the bondwire in Ohms, L is its length in mils, and D is its diameter in mils. The constant of proportionality ω equals approximately 1.1mΩ·mil for gold and 1.4mΩ·mil for aluminum.[8] Bondwires are typically about 30mils long, so a typical 1mil gold bondwire exhibits a resistance of 30mΩ. Equation 13.1 does not consider the resistance of the leadframe, nor the resistance of the bond contact, each of which may add a few additional milliohms of resistance. Gold bondwires are usually limited to a maximum of 2mils in diameter, but much larger aluminum bondwires are available. Although it is technically possible to bond a die using different types or diameters of wire, this requires a corresponding number of passes through the bonding equipment. The additional time and expense are rarely justifiable unless the design requires the use of extremely large-diameter wire.

Historically, gold ballbonds have required square bondpad openings about three times the diameter of the wire. The increasing precision of modern bonding equipment has enabled many assembly sites to accept smaller bondpads. The layout designer should obtain the current guidelines from the assembly site for the specific type and diameter of wire that will be used to bond the die. Aluminum wire must be wedge bonded rather than ballbonded, and this usually requires an elongated bondpad placed at a specific angle relative to the fingers of the leadframe. The rules for aluminum wirebonding can become quite complex, and the designer should seek guidance from the assembly site before attempting to lay out a design employing it.

The locations of the bondpads must simultaneously satisfy several conflicting requirements. The bondpads cannot lie too close to one another, or the capillary tube will damage one bond while placing the next. The bondwires must not pass too close to adjacent pads lest the capillary damage the wires while placing subsequent bonds. Long bondwires can short to one another or to adjacent bondpads due to a phenomenon called *wiresweep*. The injection-molding process forces molten plastic over the die, and the viscous drag of the plastic on the bondwires causes them to move. Larger-diameter wires are more rigid and can better resist wiresweep, allowing them to span larger distances. Wiresweep also makes it inadvisable to cross one bondwire over another. The best bonding arrangements consist of a ring of pads placed around the periphery of the die in locations that allow the shortest and most direct wirebonds. Parts packaged in ceramic or metal can ignore the limitations imposed on plastic packages due to wiresweep, but they must still follow the spacing rules required to prevent capillary damage.

It is often difficult to find a suitable bonding arrangement for a die having a large number of bonds. Some assembly sites provide software tools that can evaluate a proposed bonding arrangement to ensure manufacturability. Others require that any potential bonding arrangement pass through a review process in order to obtain production approval. The layout designer should always check to make sure that the bondpad arrangement meets the approval of the assembly site before beginning the top-level layout. If this is not done and the finished layout does not meet

---

[8] These values are only approximations based on the bulk resistances of gold and aluminum. The actual resistance of a bondwire is affected by impurities and work hardening.

the assembly site's requirements, substantial time and effort will be required to correct the problems.

The placement of bondpads also restricts the routing of adjacent metal leads. Misalignment or excessive bonding force may cause the bond to press against the protective overcoat adjacent to the bondpad opening. The resulting stress can crack the protective overcoat and can even damage underlying leads. Many assembly sites state that metal leads that do not connect to the bondpad must not pass within a certain distance of it. For ballbonds, this requirement usually takes the form of a circular exclusion zone centered on the bondpad. Many designers mark this exclusion zone with a circle placed on a special drawing layer. The layout in Figure 13.17 shows an example of one of these so-called *bondpad circles.* The assembly site will usually specify the dimensions of the exclusion zones. If no guidelines are available, assume that the bondpad circle passes through the four vertices of a square bondpad opening of minimum dimensions. Wedge bonding also requires exclusion zones, but the dimensions of these zones depend on the placement of the pads in relation to the leadframe.

There was a time when many assembly sites recommended placing tanks (or wells) underneath all bondpads and probepads. These tanks were generally left unconnected. They were intended to protect the die against shorts caused by the probe needles scratching through the bondpad metallization and field oxide during wafer-level testing. If such shorts occurred, they would connect the bondpad to a tank rather than to the substrate. This would—theoretically—prevent the device from failing. The placement of unconnected tanks under bondpads is actually a very questionable practice. If the bondpad shorts to the tank, then the tank may inject electrons into the substrate. This means that the tank requires an electron-collecting guard ring, which wastes considerable space. Most modern designs do not place tanks or wells under pads unless they form part of some adjacent device whose tank or well connects to the pad.

Some assembly sites also require that the bondpad for pin #1 be visually distinct from all of the others. This requirement originally arose from the limitations of early machine vision systems used in automated bonding equipment. Even though most modern machines no longer require a distinct pad #1, it still provides a convenient visual reference point for operators who must inspect the mounted dice. A variety of different techniques have been used to mark pad #1, the simplest of which notches the four corners of the protective overcoat (PO) opening (Figure 13.18A). These notches are usually about 0.5mil (~10μm) deep. If possible, the metal pattern should also contain notches corresponding to those of the PO opening on at least two corners of the bondpad. Another technique marks pad #1 with an octagonal PO opening (Figure 13.18B), while a third employs a circular opening (Figure 13.18C). The requirements of the assembly site may dictate a choice among these options. Otherwise, the designer should probably follow the conventions established by previous designs.

Since trimpads and testpads only allow access for probe needles, they need not follow all of the requirements that apply to bondpads. The size of trimpads and test-

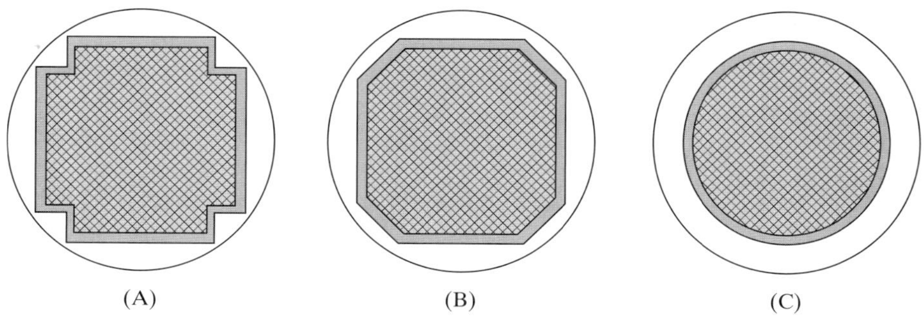

FIGURE 13.18 Three unique styles of bondpads sometimes used to identify pin #1.

pads depends on the diameter of the probe needles and on the alignment tolerances of the probing equipment. These requirements usually prove somewhat less restrictive than those associated with bonding, so trimpads and testpads are usually smaller and more closely spaced than bondpads. The relatively high currents associated with trimming sometimes necessitate the use of larger-diameter needles, so trimpads may require larger PO openings than do testpads. Both types of pads are customarily placed around the periphery of the die to simplify the design of the probe card. Additional testpads are sometimes placed in the interior of the die, but these are usually removed before the design reaches production to minimize the ingress of contaminants.

Probing does not generate the same level of mechanical stresses as bonding, so many assembly sites allow the placement of trimpads and testpads over active circuitry. This concession virtually eliminates the area penalty associated with the use of small numbers of fuses or Zener zaps. Trimpads placed over active area consist of a square of top-level metal and an associated PO opening. They require neither vias nor lower-level metal, so they can reside over any portion of the circuit free of top-level metal. Some processes use a very thick layer of top-level metallization that provides enough mechanical compliance to dissipate the mechanical stresses that are induced during bonding. These processes may also allow the placement of bondpads over active area. This practice can result in substantially more compact designs, but only if the metallization system can support the resulting stresses.

13.4.3. ESD Structures

In addition to bondpads, the padring also contains structures intended to safely dissipate the energy of ESD events. These *ESD structures* must reside near their respective bondpads to minimize lead resistances and inductances that might otherwise interfere with their operation. The ESD structures usually require a low-impedance return path to the substrate terminal. The scribe seal metallization can provide this return path without consuming much additional die area. In order to take full advantage of the scribe seal, the ESD structures must either reside between the bondpads and the scribe seal, or between adjacent bondpads.

ESD structures are intended to limit the peak voltages seen at the bondpads in order to protect the remainder of the integrated circuit. The stresses imposed on ESD structures differ depending on the testing conditions chosen. The two most common tests are the *human-body model* (HBM) and the *machine model* (MM). The human body model charges a 150pF capacitor, C_1, to a known voltage, V_{esd}. The capacitor then discharges into ESD structure, D_1, through a 1.5kΩ series resistor, R_1 (Figure 13.19A).[9]

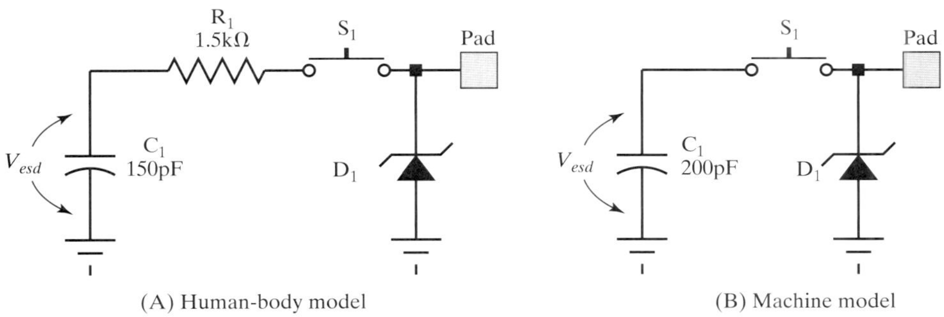

FIGURE 13.19 Equivalent circuits for (A) the human-body model and (B) the machine model.

[9] T. M. Madzy and L. A. Price II, "Module Electrostatic Discharge Simulator," *EOS/ESD Symposium Proc. EOS-1*, 1979, pp. 36–40. MIL-STD-883 is identical to Madzy's model except that C_1 = 100pF. The 150pF value is now widely accepted.

The resistor absorbs most of the energy, but the ESD device must still conduct relatively large peak currents without failure. For example, a 2kV HBM strike produces a peak current of about 1.3A. Higher ESD voltages generate proportionately larger peak currents. Modern integrated circuits must generally pass 2kV HBM, while certain pins on selected devices must stand as much as 20kV HBM.

The machine model employs a 200pF capacitor C_1 charged to a specified voltage V_{esd}. This capacitor discharges directly through ESD device D_1 (Figure 13.19B). The peak currents flowing through the ESD device are primarily limited by the parasitic inductance of the external circuit, which is usually less than 500nH. The ESD structure must not only withstand the extremely high peak currents that result, but it must also dissipate the energy of the ESD strike without permanent damage. An ESD device can typically withstand only about one-tenth as much voltage in the machine model test as it can withstand in the human-body model test. Thus modern integrated circuits are often specified to withstand 2kV HBM and 200V MM.

The newest ESD test is called the *charged-device model* (CDM). This test involves charging the integrated circuit's package to a specified voltage relative to a ground plate, and then discharging one pin through a low-impedance probe to the plate.[10] The charged-device model generates even higher peak currents than the machine model, but the transients contain less energy because the capacitance of an integrated circuit package is much smaller than the 200pF capacitance specified by the machine model. The charged-device model is gradually replacing the machine model throughout the industry. A typical testing requirement is 2kV HBM and 1kV CDM.

The effects of ESD vary depending on the type of component involved.[11] PN junctions are usually destroyed by overheating. Since considerable energy is required to melt even a few cubic microns of silicon, diffused junctions are generally quite robust. An avalanching junction dissipates most of its heat within its depletion region. Since lightly doped junctions have wider depletion regions, they can dissipate more energy than can heavily doped junctions. A lightly doped junction also has additional series resistance that dissipates some of the ESD energy. Large junctions are more robust than small ones because they contain a correspondingly greater volume of silicon within their depletion regions. The collector-base and collector-substrate junctions of standard bipolar processes are so large and so lightly doped that they can withstand most ESD transients without damage. Base-emitter junctions are more vulnerable because of their smaller dimensions and heavier doping. The base-emitter junctions of NPN transistors are also susceptible to avalanche-induced beta degradation (Section 8.1.2). The shallow, heavily doped junctions used in CMOS processes are more easily damaged than the deep, lightly doped junctions of standard bipolar. CMOS transistors with silicided source/drain regions (*clad moats*) are particularly fragile because of the lack of ballasting and the presence of silicide immediately adjacent to the depletion region.

Thin insulating films, such as those used in MOS transistors and deposited capacitors, are extremely fragile. High voltages will rupture these films within nanoseconds. Even if the insulating film does not rupture, it may suffer degradation that causes it to fail during normal operation due to time-dependent dielectric breakdown (TDDB). These delayed failures are very difficult to detect during high-speed automated testing. The only proven way to prevent such failures is to limit the voltage across the di-

---

[10] Y. Fukuda, S. Ishiguro, and M. Takahara, "ESD Protection Network Evaluation by HBM and CDM (Charged Package Method)," *EOS/ESD Symposium Proc. EOS-8*, 1986, pp. 193–199.

[11] A. Amerasekera, W. van den Abeelen, L. van Roozendaal, M. Hannemann, and P. Schofield, "ESD Failure Modes: Characteristics, Mechanisms and Process Influences," *IEEE Trans. on Electron Devices*, Vol. 39, #2, 1992, pp. 430–436.

electric to safe values. Deposited resistors can also suffer dielectric breakdown through the thick-field oxide or the interlevel oxide (ILO), but hundreds of volts are required to rupture these layers. Any diffusion connected to a bondpad will avalanche long before the field oxide or the ILO ruptures, and will therefore protect them.

Early bipolar integrated circuits rarely incorporated any intentional ESD protection, but they generally withstood the rigors of ordinary handling because their junctions were sufficiently robust to absorb and dissipate low-level ESD strikes without damage. A few parts undoubtedly suffered damage during handling, but most of these failures were attributed to processing defects or infant mortality. CMOS integrated circuits proved much more fragile. Large numbers of early devices were destroyed through gate oxide rupture during normal handling. Once the mechanisms responsible for these failures were identified, designers began to recognize that even bipolar circuits were potentially vulnerable. A variety of protective structures were proposed, some of which worked and some of which did not. The reasons for success and failure were poorly understood, so ESD protection gained a reputation of being a "black art." This reputation is largely undeserved because ESD devices obey the same principles that govern other components. The following sections examine several types of ESD devices often used to protect analog integrated circuits. The strengths and weaknesses of each device are explained in terms of their structures and electrical properties. Using this information, the layout designer can construct ESD structures for a variety of applications.

Zener Clamp

The simplest ESD device consists of a Zener diode connected between the bondpad and the substrate return line (Figure 13.20A). Possible choices include the emitter-base Zener of standard bipolar and the NSD/P-epi and PSD/N-well Zeners of analog CMOS. An ideal Zener diode would impose a positive clamp voltage equal to its reverse breakdown voltage and a negative clamp voltage equal to its forward drop. Most Zener diodes contain enough internal series resistance to make the clamp voltages much larger than these ideal values. A minimum-size emitter-base Zener has from 100 to 300Ω of internal series resistance, and NSD/P-epi and PSD/N-well diodes have even more. These resistances actually increase the robustness of the Zener by spreading the ESD energy over a larger volume of silicon, but in so doing they cause the bondpad voltage to rise above the theoretical clamp voltage by some tens of volts. This consideration severely limits the usefulness of Zener diodes as ESD structures.

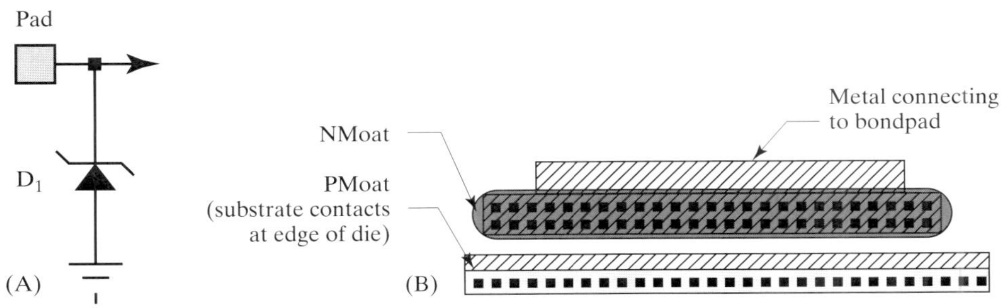

FIGURE 13.20 (A) Schematic diagram and (B) layout of the simple Zener clamp ESD circuit.

The source/drain regions of NMOS and PMOS transistors can sometimes protect themselves against ESD damage. Consider the case of a large NMOS transistor whose drain connects to a pin. Negative ESD transients forward-bias the NSD/P-epi junction. Most of the resulting voltage drop occurs within the P-epi. Positive ESD transients avalanche the NSD/P-epi junction, dumping energy into its depletion region. Since the depletion region contains less silicon than the P-epi, NMOS transistors are more vulnerable to positive transients than to negative ones. A similar analysis shows that PMOS transistors are most susceptible to negative transients. If a pin connects to both PMOS and NMOS transistors, then both conduct some portion of the ESD pulse and the more fragile device will determine the circuit's survival or failure. Larger transistors are more robust than small ones because they can dissipate energy within a larger volume of silicon. A large number of small transistors usually provides the same degree of protection as a single large transistor. A 10V nonsilicided, single-diffused drain MOS device with a drawn drain area of about $1000\mu m^2$ will usually withstand 2kV HBM and 200V MM. Since PMOS and NMOS transistors avalanche under different biasing conditions, a large NMOS will not necessarily protect a small PMOS, or *vice versa.* If both PMOS and NMOS transistors connect to a pin, then the total PMOS drain area and the total NMOS drain area must both suffice to independently withstand the ESD strike.

Low-voltage CMOS processes are much more vulnerable to ESD damage than previous-generation, higher-voltage processes, due in part to the extreme shallowness of the low-voltage source/drain diffusions and in part to the higher backgate doping required to prevent punchthough. Both of these factors reduce the volume of silicon in the depletion regions. Clad-moat devices sometimes exhibit localized breakdown due to their lack of source/drain ballasting. Once localized breakdown occurs, ESD performance no longer improves with increasing device area. The ESD performance of clad-moat devices can be improved by removing silicide from the perimeter of the source/drain implants in order to provide a small amount of ballasting. Robustness can be further improved by increasing the overlap of the (unsilicided) source/drain diffusions over their respective contacts on any diffusions that might avalanche due to ESD.[12] Most researchers attribute the increased robustness to ballasting, but some studies suggest that minority carrier injection from the source/drain contacts may also play a role.[13]

If the area of the source/drain regions connected to a pad is insufficient to provide adequate ESD protection, then a dedicated ESD protection device can be connected to the pad. In an N-well CMOS process, an NSD/P-epi diode (often called a *thick-oxide device* in the literature) usually offers the best protection for a given die area. Figure 13.20B shows a typical layout for such a diode.[14] The elongated NMoat region is placed alongside a strip of substrate contacts forming part of the scribe seal. The close proximity of the substrate contacts minimizes the series resistance of the device. The aspect ratio of this device allows it to be placed between a bondpad and the scribe seal metallization, or between adjacent bondpads. The corners of the NMoat diffusion are filleted to prevent premature avalanche breakdown. The radius of these fillets should equal or exceed the junction depth of the diffusion. The overlap of the NMoat region over its contacts should exceed the minimum layout dimension by 1 to $2\mu m$ to provide additional ballasting. In clad-moat processes, the silicide block mask should block silicide for a distance of at least 1 to $2\mu m$ from the drawn junction. The diode should contain at least $500\mu m^2$ of NMoat, and it should be surrounded by an electron-collecting guard ring and by as many substrate con-

[12] T. L. Polgreen and A. Chatterjee, "Improving the ESD Failure Threshold of Silicided n-MOS Output Transistors by Ensuring Uniform Current Flow," *IEEE Trans. on Electron Devices,* Vol. 39, #2, 1992, pp. 379–388.

[13] T. J. Maloney, "Contact Injection: A Major Cause of ESD Failure in Integrated Circuits," *EOS/ESD Symposium Proc. EOS-8,* 1986, pp. 166–172.

[14] A somewhat similar diode appears in R. J. Antinone, P. A. Young, D. D. Wilson, W. E. Echols, M. G. Rossi, W. J. Orvis, G. H. Khanaka, and J. H. Yee, *Electrical Overstress Protection for Electronic Devices* (Park Ridge, NJ: Noyes Publications, 1986), p. 19.

tacts as area permits. This device provides a reasonable degree of protection for NMOS and PMOS source/drain regions that are not large enough to protect themselves. In some cases, a low-value series resistor may be required to ensure that the ESD current flows through the Zener clamp rather than through the protected device.

Two-stage Zener Clamps

Even a large protection Zener has an internal series resistance well in excess of 10Ω. A 2kV HBM strike produces peak currents of about 1.3A, which in turn produce voltage drops of tens of volts across the series resistance of the Zener. These voltages are sufficient to rupture a thin gate oxide. Although the Zener cannot protect the gate dielectric by itself, it can reduce the peak voltage of the ESD transient from thousands of volts to tens of volts. A second protection structure connected in series with the first can provide enough additional clamping to protect the thin gate oxide. The schematic diagram in Figure 13.21A shows the conceptual arrangement of the resulting *two-stage ESD clamp.* Zener diode D_1 clamps the pad voltage to a maximum of perhaps 100V. A second Zener, D_2, connects to the pad through a series limiting resistor, R_1. The presence of R_1 limits the current flow through D_2, enabling this second Zener to limit the voltage across the gate oxide to safe levels. In order for the circuit to function properly, the resistance of R_1 should equal at least several times the series resistance of D_2. A relatively small Zener diode may exhibit several hundred Ohms of internal series resistance, so R_1 typically equals several kilohms. The inclusion of this resistance limits the slew rate of the gate voltage, but this is often desirable since excessively large transient currents can damage gate dielectrics. R_1 also adds time delays of a few nanoseconds that may interfere with certain high-speed applications.

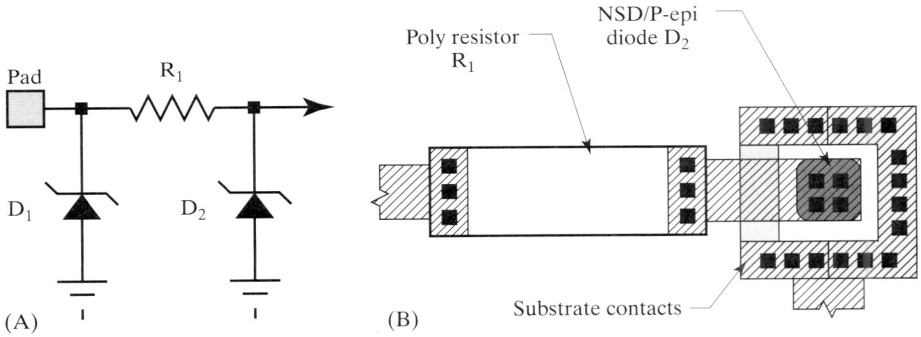

FIGURE 13.21 (A) Schematic diagram and (B) partial layout of a two-stage Zener clamp. The layout of primary protection diode D_1 is identical to that in Figure 13.20B.

Figure 13.21B shows one possible layout for the series limiting resistor R_1 and the secondary protection Zener D_2. Resistor R_1 consists of a wide strip of lightly doped polysilicon with multiple small contacts on either end. The relatively large physical dimensions of this resistor help to ensure that it can successfully dissipate the energy dumped into it during an ESD transient. Diffused resistors are more robust than poly resistors because they dissipate a portion of their energy through a distributed avalanche mechanism, so many authors suggest using them instead of poly resistors.[15] Poly resistors will survive 2kV HBM and 200V MM providing their resistance equals at least several hundred Ohms, they are at least 5 to 8μm wide, and each end of the resistor includes at least 6 to 8 minimum contacts (fewer contacts are required for higher-value resistors). Regardless of the type of resistor chosen, it should not include

[15] A. R. Pelella and H. Domingos, "A Design Methodology for ESD Protection Networks," *EOS/ESD Symposium Proc. EOS-7,* 1985, pp. 24–40.

any bends because current focuses near the inner corner of the bend, producing a hot spot that will fail before the remainder of the resistor. Zener diode D_2 consists of a relatively small plug of NMoat placed inside a concentric ring of substrate contacts. These substrate contacts not only minimize the series resistance of the Zener but also help minimize substrate debiasing near the secondary protection device. If these substrate contacts were omitted, or if they were located further away from the secondary Zener, then the large transient currents flowing through the primary Zener D_1 could induce tens of volts of substrate debiasing near D_2. This debiasing would add to D_2's clamp voltage, potentially resulting in the destruction of the gate oxide that the structure is intended to protect. If possible, the secondary Zener should be placed 50 to 100μm away from the primary one. One common arrangement places D_1 between the bondpad and the scribe R_1 alongside the bondpad, and D_2 on the inner side of the bondpad. Both D_1 and D_2 should be enclosed within an electron-collecting guard ring. The charged device model generates such extreme currents that the secondary protection device is often placed next to the source of the transistor that is to be protected to prevent substrate debiasing from generating destructive voltage drops.[16]

Two-stage ESD structures similar to that in Figure 13.21 have successfully protected MOS gates on many moderate-voltage CMOS processes. Since this type of ESD structure contains too much series resistance to allow its use on anything other than a high-impedance input terminal, it is often called an *input ESD device.* A similar two-stage ESD circuit can be constructed for some types of low-impedance applications, such as for the protection of the outputs of relatively small CMOS logic gates. This alternative type of structure uses a primary Zener diode D_1 identical to that employed in the input ESD circuit. The series resistance R_1 is reduced to 50 to 500Ω, and the source/drain diffusions of the output MOS transistors serve as the secondary Zener diode, D_2. Although the source/drain junctions avalanche, the presence of the series resistance limits the current to safe levels. Larger output transistors can employ proportionally smaller series resistors. This type of ESD structure is sometimes called an *output ESD device.* These devices have successfully protected even minimum-area source/drain implants.

Sometimes a circuit includes both source/drain diffusions and gate electrodes connected to the same bondpad. A combination of input and output ESD circuitry will successfully protect this circuit. A single primary protection device connects from the bondpad to the substrate return line. Two separate limiting resistors are required: a low-value one for the source/drain implants and a much higher-value one for the gate electrodes. The source/drain implants serve as their own secondary protection, but the gate electrodes require the addition of a secondary Zener.

Buffered Zener Clamp

The availability of bipolar transistors allows the construction of very robust ESD circuits. Figure 13.22A shows a *buffered Zener clamp* that uses an NPN transistor to reduce the effective series resistance of a Zener diode. Emitter-base Zener D_1 provides base drive to a much larger NPN transistor Q_1. This transistor multiplies the current through the Zener by its own effective beta. The positive clamp voltage of this structure equals the sum of an emitter-base breakdown and a diode drop. Assuming a standard bipolar V_{EBO} of 6.8V, the positive clamp voltage lies near 8V. The collector-substrate junction of Q_1 clamps negative ESD transients to one diode drop (plus substrate debiasing).

[16] L. R. Avery, "ESD Protection Structures to Survive the Charged Device Model (CDM)," *EOS/ESD Symposium Proc. EOS-9,* 1987, pp. 186–191.

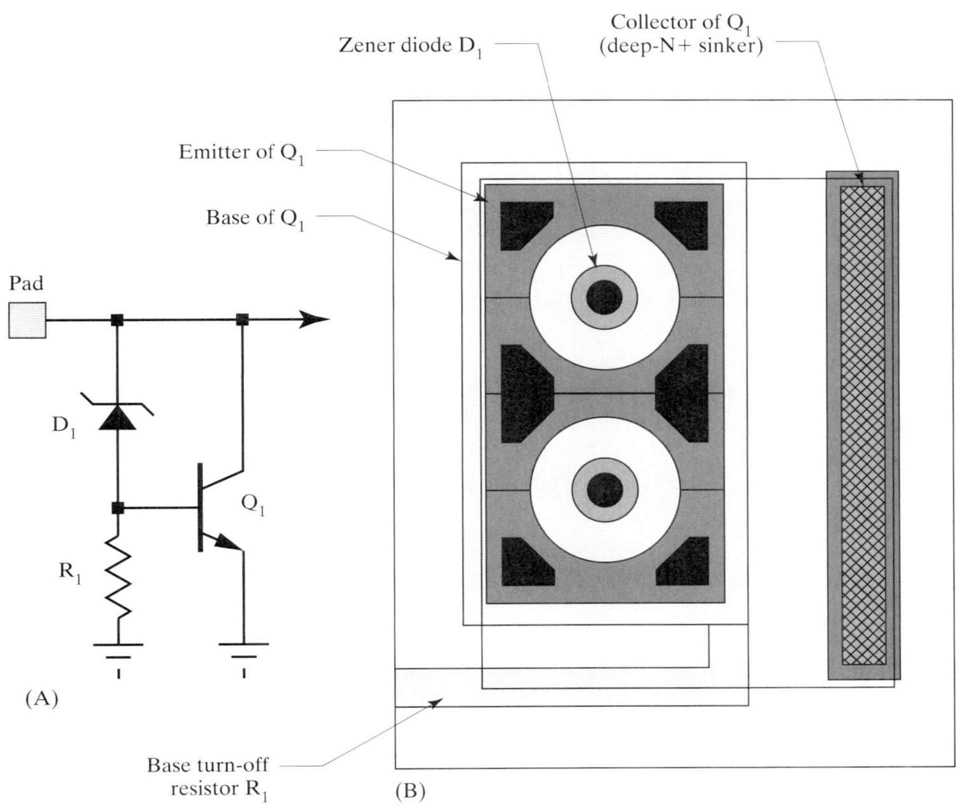

FIGURE 13.22 Schematic and layout of a buffered Zener clamp (metallization omitted for clarity).[17]

The positive clamp voltage of the buffered Zener remains roughly constant as long as the voltage across the collector resistance of Q_1 does not exceed about 7V. Even relatively small NPN transistors have collector resistances of less than 10Ω, so it is quite possible for a buffered Zener clamp to protect a gate dielectric without requiring a secondary protection device. If necessary, the clamp voltage can be increased by the inclusion of a second Zener diode, or by the addition of one or more diode-connected transistors in series with the Zener. The maximum clamp voltage this structure can safely support equals the $V_{CEO(sus)}$ of the NPN transistor. If one attempts to obtain a higher clamp voltage, the NPN transistor will avalanche and snap back to $V_{CEO(sus)}$. This type of snapback characteristic forms the basis of the V_{CES} clamp discussed in the next section.

The buffered Zener clamp dissipates most of its energy in its large base-collector depletion region. An NPN transistor with an emitter area of 300 to 600μm^2 will usually provide 2kV HBM and 200V MM protection. Larger NPN transistors can provide protection against proportionately higher ESD voltages. The ultimate limits of this structure are probably determined more by the metallization and the bondwires than by the ability of the silicon to absorb ESD energy.

All of the components of the buffered Zener clamp can occupy a common tank. In the structure in Figure 13.22B, the emitter of the power transistor Q_1 consists of a series of annular emitter geometries. Inside each hole are smaller plugs of emitter diffusion that form the cathodes of Zener diode D_1. Both the emitters of Q_1 and the

---

[17] M. Corsi, R. Nimmo, and F. Fattori, "ESD protection of BiCMOS Integrated Circuits which need to operate in the Harsh Environment of Automotive or Industrial" *(sic), EOS/ESD Symposium Proc. EOS-15,* 1993, pp. 209–213.

cathodes of D_1 are enclosed in a common base region, a portion of which extends out into the isolation to form base resistor R_1. All of these merged devices reside within a common tank having a single shared deep-N+ sinker. The buffered Zener clamp operates as follows: when the cathodes of Zener diode D_1 avalanche into the shared base region, they inject holes into it. These holes forward-bias the base-emitter junction of Q_1 and cause the large NPN transistor to conduct. The ESD transient only lasts for a few hundred nanoseconds, which is not enough time for a hot spot to form and collapse. Since thermal runaway cannot occur, the shape of the emitter diffusion can be tailored to improve performance in other ways. The annular shape of the emitters of Q_1 ensures rapid and even turn-on of all portions of Q_1's emitter. Base resistor R_1 holds Q_1 off during normal operation; its value is not particularly critical to the operation of the whole device.

Although Figure 13.22B illustrates a buffered Zener clamp for a standard bipolar process, these structures are actually better suited to the protection of analog BiCMOS circuitry, because the smaller spacings of the latter process reduce the size of the structure to the point where it can reside in the pad ring. This structure has successfully protected the gate oxide of a 20V analog BiCMOS process against 2kV HBM and 200V MM ESD strikes. The extremely low series resistance of the buffered Zener often eliminates the need for a secondary breakdown ESD device for gate oxides with rupture voltages of 20V or more. Lower-voltage gate oxides usually require a secondary protection structure similar to that in Figure 13.21A. The buffered Zener clamp can be adapted to higher-voltage applications by inserting additional Zener diodes or diode-connected transistors in series with the anode of Zener D_1. These additional devices can also be merged into a common tank with the other portions of the ESD circuit.

V_{CES} *Clamp*

Figure 13.23A shows an ESD circuit that uses the collector-base breakdown of an NPN transistor to clamp positive ESD transients. The initial breakdown voltage of this circuit equals the V_{CES} rating of transistor Q_1. Once conduction has begun, it does not cease until the voltage across the transistor drops below $V_{CEO(sus)}$. These two thresholds are sometimes called the *trigger voltage* (or *strike voltage*) and the *sustain voltage*. A typical 40V standard bipolar transistor has a nominal trigger voltage of about 65V and a nominal sustain voltage of about 45V. The snapback from the higher trigger voltage to the lower sustain voltage decreases the voltage drop across the NPN and helps reduce the energy dissipated in the transistor. Despite the rela-

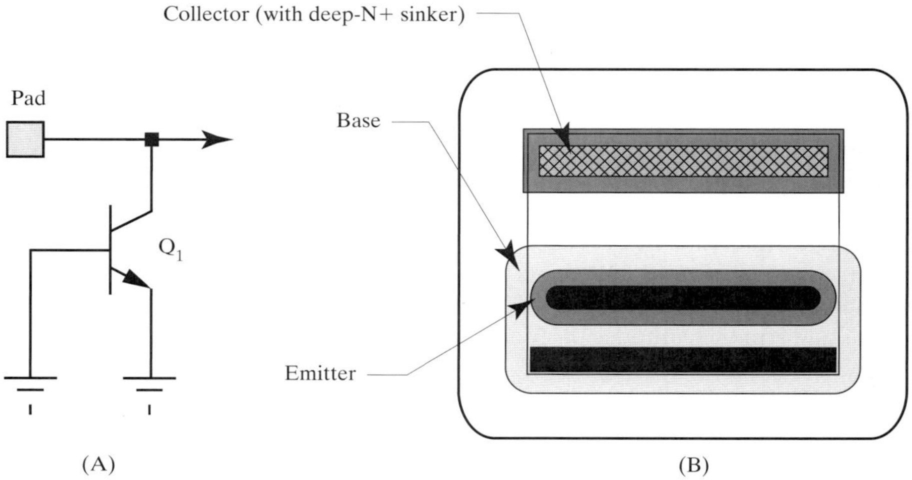

FIGURE 13.23 (A) Schematic diagram of the V_{CES} clamp and (B) a layout of a suitable NPN transistor.

tively high breakdown voltage of this structure, it is easily capable of withstanding 2kV HBM and 200V MM ESD strikes. An emitter area of about 300 to 500μm^2 provides this level of protection in standard bipolar. Larger emitter areas provide proportionately higher levels of ESD protection. This structure has successfully served as the primary protection device for a 20V analog BiCMOS gate oxide against 2kV HBM and 200V MM ESD events.[18]

ESD devices with snapback characteristics cannot safely protect low-impedance pins operating at or beyond their sustain voltage. If a transient triggers snapback, and the external circuit can supply enough current to sustain conduction, then the ESD device will continue to conduct indefinitely. The resulting power dissipation quickly overheats and destroys the integrated circuit. If the external circuitry cannot provide enough current to sustain conduction, then the ESD device can protect a pin even if it operates at a voltage in excess of the device's sustain rating. This type of application should not be contemplated unless the designer has full characterization data for the ESD device and can confidently state that the application will never deliver enough current to sustain conduction.

Figure 13.23B shows a typical layout of an NPN transistor used as a V_{CES} clamp. The filleted corners on the base diffusion raise the strike voltage of the device slightly, and also help make conduction slightly more uniform. Many designers also fillet other diffusions as shown in the illustration, but these fillets make no substantial difference in the operation of the transistor as a V_{CES} clamp. The transistor can be made more rugged by increasing the overlap of the base and emitter diffusions over their respective contacts by 1 to 2μm. This precaution is particularly useful if the process does not include silicide or refractory barrier metal, since pure aluminum contacts are more vulnerable to alloying failures than other types of contacts.

V_{CES} *Clamp*

If the emitter and collector terminals of a bipolar transistor are swapped, the device will continue to operate as a bipolar transistor. When such a transistor is biased into conduction, it is said to operate in the *reverse active,* or *inverse active,* mode. The collector-base junction forward-biases and injects minority carriers into the base, which are collected by the emitter-base junction. An NPN transistor operated in reverse active mode has a very low beta because the substitution of the lightly doped collector for the heavily doped emitter drastically reduces emitter-injection efficiency. The heavily doped base-emitter junction also avalanches at a much lower voltage than the lightly doped collector-base junction. Because of this reduction in breakdown voltage, a transistor operated in reverse active mode makes an excellent low-voltage ESD device. Suppose the transistor in Figure 13.23B is used as an ESD clamp with its emitter connected to a bondpad and its base and collector connected to ground (Figure 13.24A). The trigger voltage of the V_{ECS} *clamp* equals the V_{EBO} of the NPN transistor, and its sustain voltage equals approximately 60 to 80% of this voltage. Since analog BiCMOS NPN transistors typically have a V_{EBO} of 8 to 10V, these devices can serve as primary protection devices for pins that do not operate at more than 5V. This type of ESD device has a very low series resistance due to the presence of the NBL and the deep-N+ sinker. Devices with emitter areas of less than 600μm^2 have successfully withstood 2kV HBM and 200V MM ESD strikes, and somewhat larger emitter areas have protected circuits to 10kV HBM. These devices present a low-impedance path to negative ESD transients as well as positive

[18] J. Z. Chen, X. Y. Zhang, A. Amerasekera, and T. Vrotsos, "Design and Layout of a High ESD Performance NPN Structure for Submicron BiCMOS/Bipolar Circuits," *International Electron Devices Meeting Proc.*, 1995, pp. 337–342.

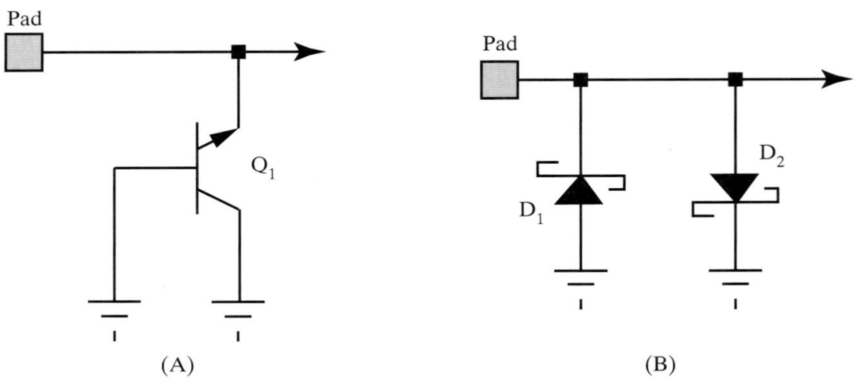

FIGURE 13.24 Schematic diagrams of (A) the V_{ECS} clamp and (B) the antiparallel-diode clamp.

ESD transients, a benefit that few other ESD devices offer. The V_{ECS} clamp does not require electron-collecting guard rings because its tank does not connect to a pin. This structure does contain a parasitic substrate PNP transistor that can inject large majority-carrier currents into the substrate, so the clamp should be surrounded by as many substrate contacts as possible to minimize debiasing in the surrounding substrate.

The V_{ECS} structure is an extremely useful device for protecting low-voltage pins. Higher-voltage ESD circuits can be created by stacking V_{ECS} clamps in series, but this arrangement increases both the area required for the device and its series resistance. A buffered Zener or a V_{CES} clamp will probably provide better performance than stacked V_{ECS} clamps.

Antiparallel Diode Clamps

Many integrated circuits require multiple ground pins. Sometimes all of the ground pins connect to the substrate, but some are frequently separated to minimize noise coupling and substrate injection. Those ground pins that do not connect to substrate require some form of ESD protection. The most common configuration consists of a pair of back-to-back (or *antiparallel*) diodes. One can employ diode-connected NPN transistors, diode-connected substrate PNP transistors, or even Schottky diodes (Figure 13.24B). Since the voltage drop across these devices is relatively small, they experience much less internal heating than do other types of ESD devices, and therefore they can be made somewhat smaller. Schottky diodes with areas of several thousand square microns will generally provide protection against 2kV HBM and 200V MM ESD strikes, and diode-connected transistors can be made even smaller than this. All of these structures require enclosure within electron-collecting guard rings.

Additional ESD Structures for CMOS Processes

The first ESD structures for CMOS processes relied on Zener clamps, but these were found to have objectionably large series resistance. A variety of other structures have been proposed. A lateral NPN transistor can be constructed by placing two NMoat regions next to one another. The NSD diffusions act as the collector and the emitter of the transistor, while the P-epi separating them acts as its base. If one of the two NMoat regions connects to the bondpad and the other to the substrate return line, then the bipolar transistor will operate as a V_{CES} clamp. Historically, most of the ESD devices constructed along these lines also included a gate electrode on top of the thick-field oxide separating the two NSD diffusions. Some designers connected this gate electrode to the bondpad, while others connected it to the substrate return. Since the thick-field threshold usually exceeds the breakdown of the NSD/P-epi junction, the gate electrode has little effect regardless of its connection. Most designers included it anyway and treated the resulting device as an NMOS transistor (Figure 13.25A). Since the gate dielectric consisted of thick-field oxide, these devices were popularly called *thick-field transistors.*

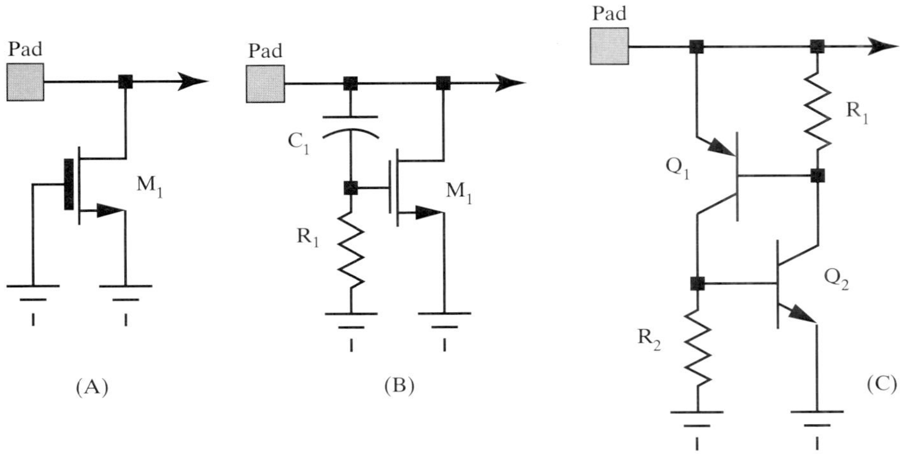

FIGURE 13.25 Three types of CMOS ESD: (A) the thick-field device, (B) the rate-triggered NMOS clamp, and (C) an SCR clamp.

The thick-field transistor has proved something of a disappointment. Its shallow, heavily doped junctions are prone to overheating during ESD transients, and incremental junction damage often causes excessive leakage, even if the device does not catastrophically fail. Ironically, the large series resistance of the NSD/P-epi Zener improves its robustness by spreading energy dissipation throughout a larger volume of silicon. The lower series resistance of the thick-field device actually renders it more fragile than an NSD/P-epi Zener of similar size. The snapback characteristic of the thick-field device also causes problems. The sustain voltage of this structure is usually about 60% of the NSD/P-epi breakdown voltage, a value that is often less than the maximum operating voltage of the NMOS transistor. When thick-field devices are used to protect low-impedance pins, the operating voltage of the pin must never exceed the sustain voltage of the thick-field device. Before using thick-field transistors, sample devices must be constructed and characterized to ensure that they can withstand ESD strikes, and to determine the actual sustain voltage of the structure. These devices should not be used unless such data is available, because they are often relatively marginal devices. A number of variants on the thick-field transistor have been developed, including thin-oxide transistors that break down by punchthrough instead of avalanche.[19] These devices do not exhibit snapback, but their breakdown voltages have proved rather difficult to control because of the extremely short channel lengths required to achieve punchthrough.

The two salient features of ESD transients are their high peak voltages and their rapid slew rates. Although most ESD structures are voltage-triggered, one can also construct circuits that respond to the voltage slew rate of a signal. These types of ESD structures are called *rate-triggered clamps* (or *dV/dt-triggered clamps*). Figure 13.25B shows a simple rate-triggered clamp in which a capacitor C_1 couples the input signal to the gate of an NMOS transistor M_1. Resistor R_1 provides a bleed path from the gate of M_1 to the substrate return line to ensure that the transistor remains off during normal operation. This structure is often called a *gate-coupled NMOS* in

[19] J. K. Keller, "Protection of MOS Integrated Circuits From Destruction by Electrostatic Discharge," *EOS/ESD Symposium Proc. EOS-2,* 1980, pp. 73–80.

the literature.[20] An ESD strike will couple sufficient energy through C_1 to turn M_1 on, allowing it to absorb the remainder of the ESD energy. In order for M_1 to properly clamp the ESD transient, it must have an $R_{ds(on)}$ of no more than a few Ohms. This usually results in a rather large device, and a variety of other rate-triggered devices have been proposed as smaller alternatives.

Rate-triggered devices have become more popular as the operating voltages of CMOS processes have decreased below 5V. The avalanche voltage of the source/drain junctions cannot drop below about 5V without causing excessive junction leakage due to tunneling. The gate oxides of low-voltage CMOS processes are therefore difficult, if not impossible, to protect using avalanche-triggered structures. Rate-triggered devices are susceptible to false triggering caused by rapidly slewing signals. Ironically, most digital inputs and outputs qualify as rapidly slewing signals. Rate-triggered devices should not be used unless sufficient characterization data are available to determine whether or not they will be triggered by the slew rates encountered in normal operation.

Many low-voltage processes use some variant of the *silicon-controlled rectifier* (SCR) for ESD protection.[21,22] An SCR is best described as a pair of back-to-back bipolar transistors, connected as shown in Figure 13.25C. In CMOS processes, these two bipolar transistors are both parasitic lateral devices. Substrate PNP Q_1 consists of a PSD diffusion inside an N-well placed in the P-epi. Lateral NPN Q_2 consists of the N-well, the P-epi, and an adjacent NSD diffusion. R_1 represents the resistance of the N-well, and R_2 represents the resistance of the P-epi and the substrate. This circuit is triggered into conduction by the collector-base avalanche of either Q_1 or Q_2. Suppose Q_2 avalanches first. Carriers injected into the base of Q_2 cause it to conduct. Q_2 now pulls current from the base of Q_1, causing it to turn on and provide additional base drive for Q_2. Each of the two transistors now provides base drive to the other, and conduction continues until the input voltage drops so low that R_1 and R_2 can extract more current than the transistors can supply. If R_1 and R_2 are both large, then the SCR may have a sustain voltage of less than 2V. Smaller values of resistance will produce higher sustain thresholds, but the relationship between resistance and sustain voltage is difficult to predict. In practice, a multitude of SCR structures are constructed and measured to determine which geometry gives the desired strike and sustain voltages.

The SCR clamp is surprisingly robust. Under HBM conditions, its low sustain voltage forces the external 1.5kΩ resistor to dissipate almost all of the ESD energy. SCRs can also dissipate remarkable amounts of energy during machine-model testing, perhaps due in part to the wide depletion region at the N-well/P-epi junction. A properly constructed SCR clamp can often withstand several times as much ESD energy as other CMOS ESD structures.

The trigger voltage of a simple SCR clamp is often too high to protect the gate dielectric of a low-voltage CMOS process. Rate-triggered SCR clamps have been developed that include capacitors connected from the pad to the base of Q_2, or from the base of Q_1 to ground. Rapidly slewing transients cause these capacitors to trigger the SCR. Rate-triggered SCR circuits can provide excellent protection, but the

[20] C. Duvvury and C. Diaz, "Dynamic Gate Coupling of NMOS for Efficient Output ESD Protection," *International Reliability Physics Symposium Proc.,* 1992, pp. 141–150.

[21] L. R. Avery, "Using SCR's as Transient Protection Structures in Integrated Circuits," *EOS/ESD Symposium Proc. EOS-5,* 1983, pp. 177–180.

[22] J. Z. Chen, A. Amerasekera, and T. Vrotsos, "Bipolar SCR ESD Protection Circuit for High-Speed Submicron Bipolar/BiCMOS Circuits," *IEDM,* 1995, pp. 337–340.

only way to ensure that they will operate properly is to construct and characterize them before using them in a design.

13.4.4. Selecting ESD Structures

Pins connected directly to the substrate, or connected only to relatively robust diffusions, can usually survive without the addition of dedicated ESD structures. Most other pins require some form of ESD protection. The following guidelines offer some specific advice for several commonly encountered situations:

1. *Pins connecting to base or emitter diffusions.*
   The relatively low sheet resistances of base and emitter diffusions render them vulnerable to ESD damage. Larger diffusions may spread the energy over sufficient area to protect themselves, but localized heating often damages smaller diffusions. The minimum diffusion area capable of self-protection depends on process parameters and testing conditions, but it is probably safe to say that a $500\mu m^2$ $160\Omega/\square$ base diffusion will survive 2kV HBM and 200V MM. Smaller diffusions should include some form of ESD clamp that avalanches before the base diffusion, such as a V_{CES} clamp or a V_{ECS} clamp. Series limiting resistors are rarely necessary because either the diffusion or the region enclosing it is usually quite resistive.
2. *Pins connecting to the emitters of NPN transistors.*
   The emitters of vertical NPN transistors are vulnerable to avalanche-induced beta degradation. If possible, the circuit should be designed to eliminate any direct connection between an emitter and a bondpad other than substrate ground. Otherwise, an ESD clamp device must be connected to the bondpad and a series resistance of several hundred Ohms placed between the bondpad and the emitter. The circuit designer must consider the impact of this resistance on circuit operation. The emitters of power NPN transistors sometimes operate at substrate potential, but return through a separate pin. In this case, an antiparallel diode clamp will provide adequate protection without requiring the insertion of any series resistance.
3. *Pins connecting to CMOS gates.*
   CMOS gate dielectrics are so fragile that they usually require some form of two-stage ESD protection. The primary protection device need only limit the voltage at the pad to a few hundred volts. The secondary ESD protection device should clamp the gate voltage to no more than 75% of the oxide rupture voltage. If the secondary ESD device returns through the substrate, then its clamp voltage must include any substrate debiasing generated either by itself or by the primary device. The series limiting resistor between the primary and secondary devices should have a resistance several times larger than that of the secondary protection structure. The resistor may consist either of a diffusion or of polysilicon, but poly resistors should be at least 5 to 8μm wide and should contain at least six or eight contacts at either end to help prevent excessive localized heating. Resistors used in ESD devices should not contain any bends, as these generate localized hot spots that may fail before the remainder of the resistor. Zeners used as secondary protection devices may require series limiting resistors of several kilohms. The secondary protection device and limiting resistor can be omitted if the primary protection device can clamp the voltage at the pad to approximately 75% of the gate oxide rupture voltage. The high currents generated by machine-model testing make this very difficult to achieve, but V_{ECS} clamps have successfully protected a 20V gate oxide against 2kV HBM and 200V MM. CDM testing will almost certainly

require secondary protection, and these devices frequently have to reside near the device to be protected in order to prevent substrate debiasing from developing excessive voltage drops. Some low-voltage CMOS processes have oxide rupture voltages below the trigger voltages of conventional avalanche-triggered ESD structures, in which case rate-triggered devices or SCRs must be used.

4. *Pins connecting to moat regions.*
   Some types of moat regions will protect themselves against ESD, while others will not. Silicided moats almost always require some form of additional ESD protection, as do moats with breakdown voltages of less than 5 to 8V. Nonsilicided moats of transistors with breakdown voltages of 10V or more will probably protect themselves against 2kV HBM and 200V MM transients, providing the total drawn area of each type of moat diffusion exceeds 500μm^2. A large NSD diffusion will not necessarily protect a small PSD diffusion, or *vice versa.* The exact moat areas required to provide self-protection vary depending on processing parameters and testing conditions. Small moat regions, particularly clad ones, generally require some form of additional ESD protection. A single-stage ESD circuit will suffice if this structure can clamp the voltage at the bondpad to less than the avalanche voltage of the moat diffusions. V_{ECS} clamps and buffered Zener clamps can sometimes provide this level of protection, but Zener clamps usually have too much internal series resistance. A series limiting resistance of a few hundred Ohms enables the use of a Zener clamp as a protection device for small moat regions. Large silicided moats often exhibit localized breakdown due to lack of ballasting. Consider using a silicide block mask to remove the silicide from the periphery of moat regions connected to bondpads. The unsilicided moat periphery slightly increases the resistance of the transistor, but one can compensate by increasing the size of the device.
5. *Pins connecting both to moat regions and to CMOS gates.*
   The moats may serve as a primary protection device if they are sufficiently large; otherwise a primary protection device must be connected to the bondpad. Small moats, or ones made especially vulnerable by silicidation, may require a series limiting resistor of 50 to 200Ω. Unless the primary protection device has a very low series resistance, it cannot protect the gates without the addition of a secondary protection device. A resistor of several hundred Ohms to several kilohms should be connected between the pad and the gates, and a suitable secondary protection structure should be placed after this resistor. This structure now has separate conduction paths for gates (which require large series resistances) and moats (which do not).
6. *Pins connecting only to polysilicon.*
   The voltages generated during human-body model testing are sufficient to rupture the thick-field oxide and interlevel oxide surrounding polysilicon resistors and leads. If a bondpad does not directly connect to any diffusion, then the voltages across the oxide surrounding the polysilicon may rise to destructive levels. An N-well geometry placed beneath the bondpad and connected to it by means of a ring of NMoat contacts encircling the bondpad will provide adequate protection while consuming very little die area.
7. *Pins connecting to capacitors.*
   Thin oxide or nitride dielectrics require the same type of protection as gate dielectrics. Junction capacitors usually contain a thin, heavily doped diffusion that requires protection similar to an emitter region.

8. *Pins connecting to Schottky diodes.*
   Field-plated Schottky diodes should not operate in avalanche breakdown because their depletion regions are very thin and are located immediately adjacent to a silicide layer. Large Schottky diodes can be protected by adding a field-relief guard ring that avalanches before the Schottky contact. Smaller Schottky diodes may be protected by a large area of moat diffusion, forming part of another device connected to the same pin. If no suitable moat region exists, then a field-relief guard ring and a series resistance of a few hundred Ohms should provide adequate protection.
9. *Bondpads operating at substrate potential but not connected to substrate.*
   These bondpads are usually ground returns isolated from substrate to minimize noise coupling. ESD protection is not required if these pads are bonded to the same pin as the substrate pad. Otherwise, an antiparallel diode clamp connected between the pad and the substrate return will provide sufficient protection for most applications.
10. *Multiple bondpads connecting to the same pin through multiple bondwires.*
    Many dice use multiple bondwires attached to a common pin. If two or more bondpads connect to the same pin through separate bondwires, then only one of these pads requires a primary ESD device. Series limiting resistors and secondary protection devices must be placed on every bondpad requiring them, as secondary protection placed on one bondpad cannot protect circuitry connected to another bondpad.
11. *Test pads and probe pads.*
    Test pads and probe pads normally do not require ESD protection because they are encapsulated within the package and therefore do not experience ESD transients.

When placing ESD structures, always remember to include any necessary guard rings and substrate contacts. Of the types of ESD structures just discussed, only the V_{ECS} clamp does not require guard rings. The guard rings should be placed in the padring during its construction. They require so much room that they are often very difficult to add later.

13.5 EXERCISES

Refer to Appendix C for layout rules and process specifications.

13.1. Lay out three minimum-size standard bipolar NPN transistors without deep-N+ sinkers. Place the transistors side-by-side as shown in Figure 13.1A, and measure the area of a rectangle enclosing all three devices. Now lay out a merged device similar to that in Figure 13.1B. Assume the area consumed by the merged device equals the area of its tank. What is the ratio of the area of the merged device to the area of the three separate devices?

13.2. Describe the risks posed by each of the following mergers:

a. Two HSR resistors placed in the same tank, one of which connects to a bondpad.
b. A base resistor merged in the same tank as a Schottky diode.
c. A lateral PNP transistor merged with an NPN transistor.
d. A junction capacitor merged with a Darlington NPN transistor pair.
e. Two substrate PNP transistors merged in the same tank.

13.3. Propose measures that will minimize the risks associated with each of the mergers in Exercise 13.2.

13.4. Lay out a merged Darlington NPN transistor capable of conducting 100mA. Use standard bipolar layout rules and assume a maximum emitter current density of $8\mu A/\mu m^2$. Use a wide-emitter, narrow-contact structure with an emitter overlap of contact of $6\mu m$ for the power device, and connect a $5k\Omega$ base turnoff resistor between its emitter and collector. Size the predrive transistor, assuming that the power transistor has a minimum beta of 20 at 100mA. Assume that the predrive transistor does not require a minimum-width base turnoff resistor. Include all necessary metallization.

13.5. Assume that the Darlington transistor in Exercise 13.4 must operate at voltages exceeding the thick-field threshold of the process. Modify the layout to include all necessary field plates and channel stops.

13.6. Lay out the totem pole driver circuit in Figure 13.26 using standard bipolar layout rules. Use wide-emitter, narrow-contact transistors with an emitter overlap of contact of 6μm for Q_1, Q_2, and D_1. The power supply voltage VCC can exceed the thick-field threshold of the process, and both OUT and PGND may go below substrate potential during switching transients. The leads connecting to VCC, OUT, and PGND must have widths of at least 15μm. Include all necessary metallization and label all leads and devices.

FIGURE 13.26 Schematic of totem pole output stage. All dimensions are emitter areas.

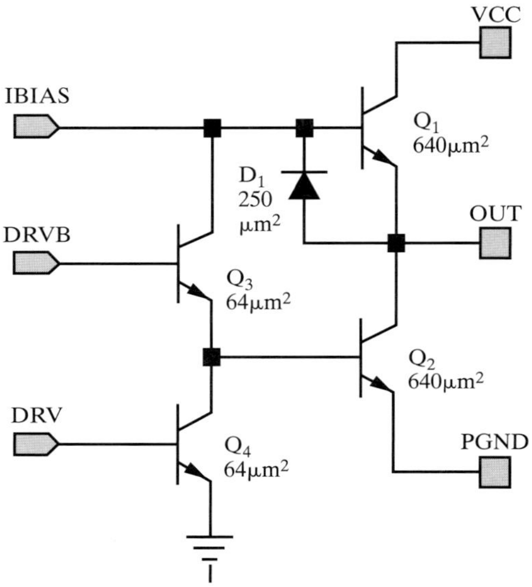

13.7. Modify the MOS operational amplifier in Exercise 12.16 to provide latchup protection, assuming that only the output pin (OUT) connects directly to a bondpad. Make the circuit as robust as possible without using either deep-N+ or NBL.

FIGURE 13.27 Schematic of D-type flip-flop.

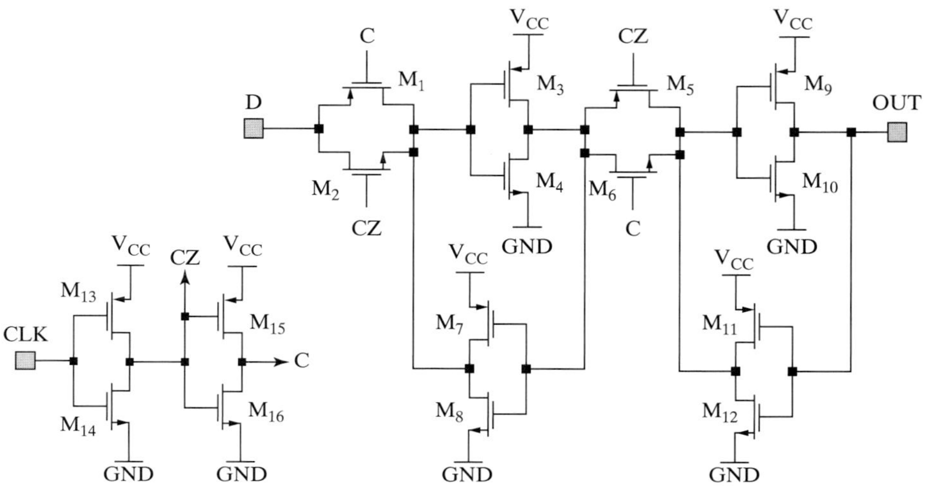

13.8. (A) Draw a stick diagram of the flip-flop of Figure 13.27. Use a single VDD bus across the top of the cell and a single GND bus across the bottom of the cell. All NMOS transistors are 4/3 and all PMOS transistors are 7/3. Do not use any metal-2 within the cell. Poly can be used to route gate leads, and short poly jumpers can be placed in source and drain leads if necessary. (B) Following the stick diagram as closely as possible, lay out the flip-flop using the CMOS rules in Appendix C. Label all leads and devices.

13.9. Construct a padring for an analog CMOS layout. The die must be exactly 110mils wide by 86mils high, including scribe. The lower left-hand corner of the die must reside at the origin (0, 0). Denote the extents of the die using a rectangle on a layer called BOUNDARY. The scribe streets lie along the left side and bottom of the die, and each is exactly 110μm wide. Denote the locations for the scribe by drawing two rectangles on the BOUNDARY layer. Draw substrate metal coincident with the edge of the padring, as shown in Figure 13.16. This metal should include a 12μm strip of metal-1 and a 12μm strip of metal-2 that exactly coincide with one another. Place PMOAT underneath the substrate metallization. Provide contacts between the substrate metallization and the PMOAT, and vias between the two layers of substrate metallization.

13.10. Construct bondpads for the padring in Exercise 13.9. Assume that the pads require a square nitride opening 75μm across, and that they must contain both metal-1 and metal-2. The two metal geometries should exactly coincide. Place a single, large via 75μm across coincident to the nitride opening. Denote the metal exclusion zone using a circle with a diameter of 90μm centered on the bondpad opening. Place eight of these bondpads in the padring of Exercise 13.9 in the following locations: top left, top center (2), top right, bottom left, bottom center (2), and bottom right. The pad at the bottom right connects to pin #1. Choose a suitable way of denoting this pad, and modify the layout accordingly. Number the pads in counterclockwise order and label each.

13.11. Construct a Zener clamp similar to that in Figure 13.20 using the CMOS layout rules in Appendix C. The NMoat region should have a total area of at least 650μm^2, and should contain two rows of contacts. Overlap NMoat over contact by 3μm and fillet both ends of the diode. Place two of these Zener clamps in the layout in Exercise 13.10 to protect pins 3 and 5. These clamps should reside between the respective pads and the substrate metallization. Include all necessary guard rings.

13.12. Construct a two-stage Zener clamp using the diode of Exercise 13.11 as a primary protection device. The series resistor should equal 1kΩ and have a width of 8μm. Place at least eight contacts on either end of the resistor. The secondary protection structure requires an NMoat area of 100μm^2. Overlap the NMoat over the contacts by at least 3μm and encircle the device with a ring of substrate contacts. All metallization within the ESD structure should have a width of at least 6μm, and if vias are used in any of these leads, use a minimum of eight vias. Place two of these clamps in the layout of Exercise 13.11 so that they protect pins 1 and 7. Include all necessary guard rings.

13.13. Construct a V_{ECS} clamp using the analog BiCMOS layout rules in Appendix C. Assume that the clamp requires an emitter area of 350μm^2. Instead of a single contact for the emitter, use two rows of minimum contacts. Overlap the emitter over the contact by 4μm and fillet both ends of the emitter. The deep-N+ sinker should completely encircle the clamp to form a hole-blocking guard ring. Compare the area of this structure to the area consumed by all of the components of the two-stage Zener clamp in Exercise 13.12 (excluding guard ring).

13.14. Lay out a thick-field transistor for use as an ESD structure in a CMOS design. The source and drain consist of strips of NMoat, each 20μm long and just wide enough to contain two rows of minimum contacts. Overlap NMoat over the contacts by 2μm and include fillets on both source and drain. Separate the two moat regions by 6μm and place a metal-1 plate over the region between the source and drain to act as a "gate." Connect this gate to the source. Compare the area consumed by this clamp to the area consumed by the Zener clamp in Exercise 13.11.

14 Assembling the Die

The first step in laying out an integrated circuit is estimating the die area. Layout designers should not rely on preliminary area estimates, as these are seldom accurate. The area of each circuit block, or *cell,* should be computed separately. The total die area equals the sum of the areas of all the cells plus the area required for wiring, bondpads, scribe seals, and scribe streets. As area estimates tend toward optimism, the prudent designer always includes a generous safety margin.

Once the dimensions of the die and the areas of all of the cells have been determined, a floorplan should be constructed. A good floorplan includes an outline of the die, placements of all pads, and the sizes and locations of all the major cells. The initial floorplan often requires revision as the layout progresses.

The completed floorplan serves as a template for constructing the individual cells. Each cell requires a multitude of resistors, capacitors, transistors, and diodes. When these components have been laid out, the designer must arrange them to optimize matching, packing, and ease of interconnection. The components are then connected to form a cell, and the cells are connected to form the completed die. Previous chapters have examined the design of individual components and their placement relative to one another. This chapter examines the process of planning a die layout, constructing a die floorplan, and interconnecting the finished cells to form the complete die.

14.1 DIE PLANNING

The layout of an integrated circuit requires considerable planning and forethought. An experienced designer knows what tasks must be accomplished and in what order; the layout progresses smoothly and all of the components fit into their assigned places. A novice attempting the same feat soon discovers that it is not as easy as it seems. Days or weeks of effort often come to naught because of unforeseen complications. Most of these difficulties are usually due to incorrect die area estimates, misplaced components, and inadequate wiring channels. The cautious designer can avoid most of these problems by spending a few hours planning the layout.

Using the information gathered during the planning phase, one can estimate the total die area and the cost of manufacture. Assuming that the design appears profitable, a floorplan can be developed showing the size, shape, and location of each cell. This floorplan forms the basis for the top-level layout of the die. Floorplans become particularly valuable on larger designs where many people must simultaneously create portions of the layout.

14.1.1. Cell Area Estimation

The first phase of the planning effort consists of compiling a list of all the cells used in the design. If detailed schematics are available, then this task amounts to little more than listing the cells found in the top-level schematic. If no schematics exist, then the circuit designer must prepare a list based on a careful examination of the specifications. The list should only contain cells appearing in the top-level schematic, and should exclude cells occupying the lower levels of the schematic hierarchy. The top-level cells may number as few as three or four, or as many as thirty or forty, depending on the scale of the design. The list should also include any power devices that require specific locations due to routing or matching considerations.

The designer must now estimate the area required by each cell. Some cells may have already been laid out for previous designs, in which case accurate area estimates are easily obtained by measurement. If a previous design contains a similar cell, then its layout may provide a close approximation of the area required by the new cell. If no previous layout exists, then the area of the cell must be computed from the areas of the individual components. The following sections explain how to rapidly estimate the size of the areas required by various types of components. These estimates are, of necessity, somewhat imprecise, but planners are generally allowed a margin of error of at least $\pm 20\%$. Area estimates are usually given in either square millimeters (mm^2) or in thousands of square mils ($kmil^2$), where $1kmil^2 = 0.645mm^2$ and $1mm^2 = 1.55kmil^2$.

Resistors

The area, A, required to construct one or more resistors can be estimated using the formula

$$A \cong \frac{1.2RW_r(W_r + S_r)}{R_s} \qquad [14.1]$$

where R equals the desired resistance, R_s is the applicable sheet resistance, W_r is the width of the resistor, and S_r is the spacing between adjacent resistor stripes. The factor of 1.2 helps account for the space consumed by dummy resistors, contact heads, and less-than-ideal layouts. For example, 122kΩ of 2kΩ/□ HSR with a width of 6μm and a spacing of 12μm will consume an estimated 7900μm^2 of die area. Resistors of different widths or materials must be computed separately.

Capacitors

The area required by capacitors depends on the capacitance per unit area of the dielectric. For finger-style junction capacitors, an average capacitance per unit area can be computed using an existing capacitor as a guideline. For example, suppose that a 50pF finger-style junction capacitor has a measured area of 27,500μm^2. This capacitor has an average capacitance per unit area of 1.8fF/μm^2.

Vertical Bipolar Transistors

The area of vertical NPN and substrate PNP transistors must be computed separately, but the same principles apply to both types of devices. The area required for a minimum-emitter device equals the area consumed by its tank; this is best measured from the layout of an existing device. The device area will not scale linearly with emitter area because the emitter forms only a small part of the transistor. It is usually not worth the effort to obtain exact area values for small transistors. One can safely assume that transistors with emitter areas of two to five times the minimum require 150% of the minimum device area. Larger transistors should be roughly sketched out, and their area estimated on this basis. Figure 14.1 shows a sketch of a wide-emitter, narrow-contact transistor. On the basis of the dimensions indicated, this device would consume an area of 38,800μm^2 and would contain 4,000μm^2 of emitter.

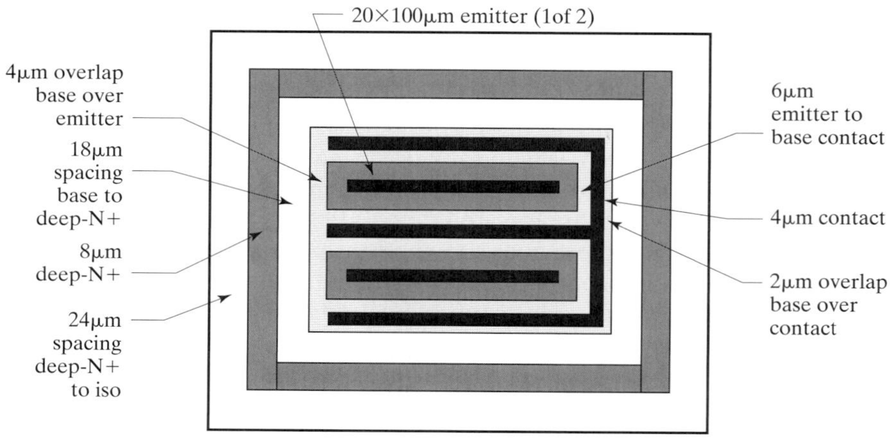

FIGURE 14.1 Sketch of a wide-emitter, narrow-contact transistor used for estimating its area.

Lateral PNP Transistors

The area required by a minimum lateral PNP transistor can be obtained by measuring the tank area of an existing device. Larger transistors are normally constructed either by placing multiple copies of a single device in a common tank or by stretching the transistor along one axis. In either case, the device area scales approximately linearly with collector size. Split-collector transistors require additional area, so count each such device as 150% of the minimum device area.

MOS Transistors

The area, A, required for a finger-style MOS transistor can be approximated by

$$A \cong 1.3W_g(L_g + S_{gg}) \tag{14.2}$$

where W_g equals the width of the gate, L_g equals the length of the gate, and S_{gg} equals the spacing between adjacent gate stripes of a multiple-finger transistor. The factor of 1.3 helps account for the space consumed by the terminations on either end of the transistor array, well spacings, and less-than-ideal packing. This formula usually underestimates the area required by small transistors, particularly if they require guard rings or separate wells.

MOS Power Transistors

MOS power transistors are usually specified in terms of their on-resistance, $R_{ds(on)}$. Area estimations based on device models or SPICE simulations do not properly account for metallization resistance. Estimates based on measured specific on-resistance R_{sp} give much better results. The area required to obtain a desired $R_{ds(on)}$ equals

$$A \cong \frac{R_{sp}}{R_{ds(on)} - R_p} \qquad [14.3]$$

The variable R_p accounts for the resistance of the package, including bondwires and leadframe. The bondwires contribute the largest component of the package resistance. A typical 1mil gold bondwire contributes about 25 to 50mΩ (Section 13.4.2). Larger-diameter bondwires or multiple bondwires placed in parallel can greatly reduce this resistance.

The accuracy of equation 14.3 depends on how closely the proposed transistors resemble the test devices. The proposed transistor should have the same gate length, and the R_{sp} and $R_{ds(on)}$ figures should be measured at the same gate-to-source voltage. Because R_{sp} varies with device area, the area of the proposed transistors should not differ from the area of the test devices by more than a factor of five. Further, the dimensions of the finger structure and the metallization pattern of the proposed transistor and the test device should closely resemble one another.

Computing Cell Area

The area of a cell A_{cell} can be estimated using the following formula

$$A_{cell} = P_f \Sigma A \qquad [14.4]$$

where ΣA equals the sum of the areas of all of the individual components. The packing factor, P_f, accounts for the area consumed by isolation and device interconnection as well as the area wasted by imperfect packing. Standard bipolar designs employing single-level metal typically have packing factors of 1.5 to 3.0. Values toward the lower end of this range represent well-packed designs using custom-crafted devices and extensive device mergers. The values toward the upper end of this range represent designs using standardized components and few or no device mergers. Standard bipolar designs using double-level metal require less area and will typically have packing factors of 1.5 to 2.0. Double-level-metal analog CMOS or BiCMOS designs using standardized components usually achieve packing factors of 1.5 to 2.0. Triple-level metal provides little improvement unless the cell contains an extraordinary amount of interconnection.

Suppose that the bandgap circuit in Figure 14.2 will be laid out in a standard bipolar process using single-level metallization. Three of the transistors require individual sketches: Q_1, Q_2, and Q_{10}. The first two form the ratioed pair of the Brokaw bandgap cell, while Q_{10} is a small power device. The area of the other devices can be estimated by the procedures discussed. Table 14.1 shows the results of this calculation.

14.1.2. Die Area Estimation

Three factors contribute to the overall die area: the circuitry it contains, the ring of pads around its periphery, and the scribe streets separating it from adjacent dice. The circuitry resides in the middle of the die, forming the *core*. The pads lie around the periphery of the die, forming the *padring*. Ideally, both the core and the padring should contain no wasted space. Practical designs almost never meet this goal. In a *core-limited* design, the core packs tightly into the space inside the padring, but there

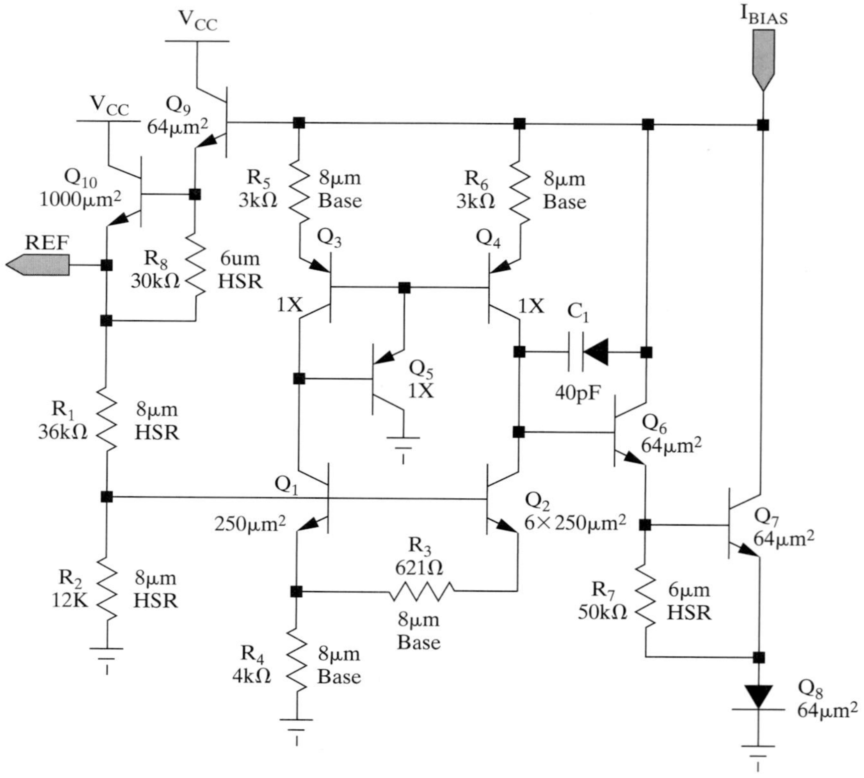

FIGURE 14.2 Schematic of the sample circuit block.

TABLE 14.1 Estimated area for the simple Brokaw bandgap.

Device	Amount	Area
8μm 160Ω/□ base resistance	10.621kΩ	14,200μm^2
8μm 2kΩ/□ HSR resistance	48.0kΩ	4,600μm^2
6μm 2kΩ/□ HSR resistance	80.0kΩ	5,200μm^2
Junction capacitance, 1.8fF/μm^2	40pF	22,200μm^2
Minimum NPN transistors @ 2,200μm^2	4	8,800μm^2
Minimum PNP transistors @ 4,100μm^2	3	12,300μm^2
Bandgap NPN transistors @ 3,100μm^2	7	21,700μm^2
Output NPN transistors @ 6,600μm^2	1	6,600μm^2
Total area of components		95,400μm^2
Estimated cell area ($P_f = 2$)		0.19mm^2

are not enough pads to fill the ring (Figure 14.3A). The gaps between the pads are often used for ESD structures and trim circuitry. A *pad-limited* design has so many pads that the space remaining inside the padring exceeds the area required by the core (Figure 14.3B). The estimation procedure must determine whether the design is core-limited or pad-limited before a final area estimate is possible.

The first step in computing the estimated die area consists of computing the core area A_c

$$A_c \cong R_f P_f \Sigma A_{cell} + P_f \Sigma A_{pwr} \qquad [14.5]$$

where ΣA_{cell} represents the sum of the areas of all the individual cells, and ΣA_{pwr} represents the sum of the areas of all the power devices not contained in any cell. The core does not include bondpads, trimpads, ESD devices, scribe seals, and scribe

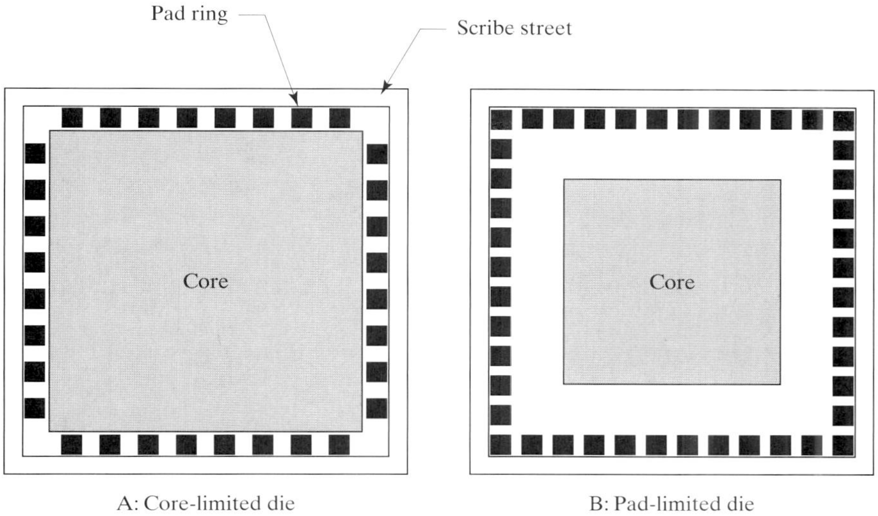

FIGURE 14.3 Comparison of core-limited and pad-limited dice.

streets. The *routing factor* R_f accounts for the area consumed by top-level wiring. Typical routing factors for a design with several hundred top-level signals are 1.3 to 1.5 for single-level metal, 1.2 to 1.3 for double-level metal, and 1.1 to 1.2 for triple-level metal. Designs making extensive use of poly routing fall somewhere between these values. For example, a design routed on metal-1 and poly-1 might achieve a routing factor of 1.3, as compared with a routing factor of 1.4 for metal-1 alone. The *packing factor* P_f represents the areas of wasted space between cells. A die containing twenty or thirty moderate-sized cells should achieve a packing factor of 1.1 to 1.2. Layouts incorporating very large or very odd-shaped cells often have much higher packing factors. Conversely, hand-optimized designs can achieve a packing factor close to one. However, hand optimization requires additional time and effort, particularly for larger designs.

The following formula computes the estimated area of the entire die A_{die} based upon a square aspect ratio:

$$A_{die} = (\sqrt{A_c} + 2W_{pr} + W_s)^2 \qquad [14.6]$$

The design rules should specify the scribe width W_s, typically 75 to 125µm. The width of the padring W_{pr} usually equals about 130% of the width of a bondpad. The design rules generally specify the minimum allowed dimensions of the bondpads. For ballbonded gold wires, the minimum bondpad width equals about three times the diameter of the wire. Following these guidelines, a 1mil (25µm) gold wire requires a 75µm bondpad and a 100µm padring. These approximations are adequate for preliminary area estimates, but the final area estimate should only be made after the padring has been constructed (Section 13.4).

The above computations assume that the padring contains enough space to contain all of the required pads. The minimum die perimeter P_{min} required to place the pads approximately equals

$$P_{min} \cong (N_p + 4)(W_p + S_p) + 4W_s \qquad [14.7]$$

where N_p equals the total number of pads, W_p equals the width of a pad, W_s equals the width of the scribe, and S_p equals the minimum spacing allowed between adjacent pads (measured between the facing edges of adjacent pads). This formula assumes that the pads can be placed relatively close to the corners of the die and that circuitry cannot reside underneath the bondpads. If the layout rules specify a minimum allowed distance from the corner of the die to the edges of the nearest bondpads, then add eight times this distance to the estimate of P_{min} computed above.

The *perimeter utilization factor* P represents the fraction of available pad perimeter used by the pads. Assuming that the die has a square aspect ratio, P equals

$$P = \frac{P_{min}}{4\sqrt{A_{die}}} \qquad [14.8]$$

If P is less than one, then the design is core-limited and the estimated die area is probably correct. The following formula will determine the approximate number of ESD devices N_e that will fit between the pads

$$N_e \cong \frac{N_p(W_p + S_e)}{P(W_p + S_p)} \qquad [14.9]$$

where S_e equals the minimum spacing between pads required to fit one ESD structure. If N_e exceeds the number of ESD devices used by the design, then they should all fit into the padring. If there are too many ESD devices, then add the area of the devices that will not fit into the padring to the area of the core and repeat the calculations using equations 14.5 to 14.8.

If the perimeter utilization factor P exceeds one, then the design is pad-limited and the die area must increase. The total die area A_{die} for a pad-limited die with a square aspect ratio equals

$$A_{die} = \frac{P^2_{min}}{16} \qquad [14.10]$$

The pad-limited die will have wasted space equal to the difference between the area estimates of equations 14.6 and 14.10. It is possible to slightly increase the amount of usable die periphery by elongating the die, but reasonable aspect ratios rarely provide enough additional perimeter to transform a pad-limited die into a core-limited one.

14.1.3. Gross Profit Margin

Managers and marketers use die-area estimates to determine the profitability of a design. The figure of merit most often used for this purpose is the *gross profit margin* (GPM), defined as the percentage of the sales price remaining after manufacturing costs have been subtracted. The procedure used to determine GPM is worth examining, as it provides some insight into the economics of integrated circuit manufacture.

The first step consists of computing the number of dice obtainable from one wafer N_d

$$N_d = \frac{\eta \pi d^2}{4A_d} \qquad [14.11]$$

where d represents the diameter of the wafer in millimeters (or mils) and A_d represents the area of the die in square millimeters (or square mils). The *wafer utilization factor* η represents the fraction of the wafer's surface covered by potentially usable dice. Some of the dice around the edges will be incomplete, either because they extend beyond the edge of the wafer or because they fall outside the field of exposure. Wafer utilization factors range from 0.7 to 0.9, depending on die size and photolithography techniques. As an example of the use of this equation, consider a 10mm^2 die constructed on a 6″ (150mm) wafer with a wafer utilization factor of 0.8. This wafer will yield 1414 dice.

The cost of a functional die C_d can be determined using the following formula

$$C_d = \frac{C_w}{N_d Y_p} \qquad [14.12]$$

where C_w represents the cost of one wafer, including probing and sawing, and Y_p represents the *probe yield,* defined as the fraction of the potentially usable dice that pass wafer probe. Probe yields depend on the area of the die, the complexity of the process, and the robustness of the design. Probe yields of analog integrated circuits usually range from 0.7 to 0.9, although both lower and higher figures are possible. Continuing the previous example, suppose that each 6″ wafer costs $600 and the probe yield equals 85%. Assuming 1414 potential dice per wafer, each good die costs 50¢.

The total cost C_t of an integrated circuit equals

$$C_t = \frac{C_d + C_a}{Y_a} \qquad [14.13]$$

where C_d represents the die cost and C_a the assembly cost (including packaging, symbolization, final testing, storage, and shipping). The *assembly yield* Y_a usually exceeds 0.95 because the vast majority of the defective dice have already been rejected during wafer probe. Continuing the previous example, if each good die costs 50¢, assembly costs 20¢, and the assembly yield equals 0.95, the cost of the finished integrated circuit equals 74¢.

The *gross profit margin* (GPM) is computed from the total cost C_t and the sales price S:

$$GPM = \frac{S - C_t}{S} \cdot 100\% \qquad [14.14]$$

If an integrated circuit costs 74¢ to produce and sells for $1.50, then it will have a GPM of 50.7%. The gross profit margin must cover all the costs associated with running a large company, including sales and distribution, engineering, research and development, fixed overhead, and administrative costs. A GPM of at least 50% is desirable.

14.2 FLOORPLANNING

The final phase of the planning process consists of creating a sketch of the layout, called a *floorplan*, showing the placement of the bondpads and the locations and shapes of all the cells. During the layout, the floorplan serves as a guide for constructing the padring. If any cell requires significantly more or less space than allocated, the floorplan should be revised accordingly. Once most or all of the cells have been completed, the floorplan can be used as a template for assembling the top level layout.

The information required to construct a floorplan includes area estimates for each cell as well as an area estimate for the whole die. The designer must also obtain a complete listing of all pads and the order of their placement. Table 14.2 shows a sample worksheet containing the information needed to produce a floorplan of a small analog integrated circuit.

Having obtained the necessary information, the next step consists of sketching out the padring. Assuming a square die aspect ratio, a 1.33mm^2 die will have dimensions of 1153 × 1153μm. This distance must be rounded off to the nearest increment allowed by the stepper, as determined by the photomask vendor. Many older steppers require the die size to equal an integer number of mils. A die that is 1153μm on a side would become 46mils (1168.4μm) on a side.

TABLE 14.2 Sample floorplanning worksheet.

Device:	Dual operational amplifier
Process:	Standard bipolar, double-level metal
Dimensions in:	Mils
Package type:	8-pin DIP
Die area estimate:	1.33 mm^2 ($P_f = 1, R_f = 1.2$)
Bondpad width:	3mil (75μm)
Padring width:	100μm estimated
Scribe width:	3mil (75μm)

Circuit Block	Area	Dedicated Pins	Shared Pins
AMP1	0.32mm^2	IN1+, IN1–, OUT1	V+, V–
AMP2	0.32mm^2	IN2+, IN2–, OUT2	V+, V–
BIAS	0.13mm^2	None	V+, V–

Note: AMP1 and AMP2 are identical to each other.

Pin #	Pin Name	Function of Pin
1	OUT1	Output of first amplifier
2	IN1–	Inverting input of first amplifier
3	IN1+	Noninverting input of first amplifier
4	VEE	Negative supply (connect to substrate)
5	IN2+	Noninverting input of second amplifier
6	IN2–	Inverting input of second amplifier
7	OUT2	Output of second amplifier
8	VCC	Power supply

Once the dimensions of the die have been determined, a leadframe must be chosen. Figure 14.4 shows a drawing of a typical 8-pin DIP leadframe. The large square tab in the middle of the drawing is called the *mount pad*. The die must be slightly smaller than the mount pad to allow for misalignment. Given an allowance of 5mils (125μm) per side, an 80 × 100mil mount pad can accommodate a die with maximum dimensions of 70 × 90mil. This leadframe can easily accommodate a 46mil-square die. Given a choice of several leadframes, choose the smallest one that will accommodate the die. Excessively large leadframes may reduce the assembly yield because of wiresweep and sag.

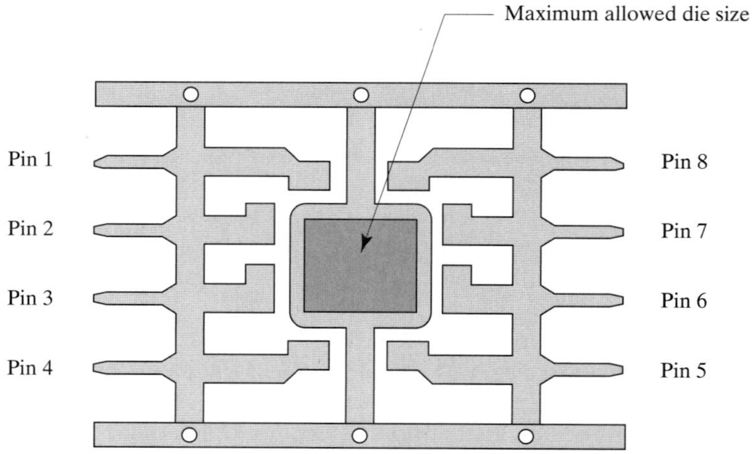

FIGURE 14.4 Sample leadframe drawing for an 8-pin DIP package.

The leadframe in Figure 14.4 illustrates a common arrangement in which the lead fingers actually encircle the mount pad. Thus, while pin one emerges from the top left of the package, the bondwire attaches to its lead finger directly above the mount pad. The bondpads should be placed to obtain the shortest and most direct bondwire routings. This precaution not only minimizes the chance of wires shorting to one another but also reduces the amount of gold wire required to bond the die.

The floorplan should also show the location of the scribe streets. Some processes require the scribe street on the bottom and left sides, others on the top and right sides, and still others place half-width scribe streets on all four sides of the die. All of these arrangements result in the same pattern of scribe streets on the wafer, but each requires a slightly different layout. For the purposes of this example, we will assume that the scribe occupies the top and right sides of the die and that the origin of the layout resides at the lower left corner.

The floorplan sketch includes a rectangle marking the extents of the die and a second smaller rectangle delimiting the area reserved for the core (Figure 14.5). A strip along the top and right-hand sides of the die shows where the scribe fits. The individual cells are represented by rectangles having the appropriate areas. Since this design contains two identical amplifiers, it makes sense to use mirror-image placements of a single amplifier cell. This not only saves layout effort but also ensures that the two amplifiers will have similar electrical characteristics. The placement of the amplifiers must allow wires to route to the appropriate pins. AMP1 connects to pins 2, 3, and 4, while AMP2 connects to pins 6, 7, and 8. AMP1 should therefore reside on the left side of the die and AMP2 on the right. The BIAS block fits into the middle between the two amplifiers. This arrangement produces rather elongated circuit blocks, but there is no reason why this should cause problems. The die's aspect ratio remains square, and the elongation of the amplifier blocks actually helps improve matching by allowing the sensitive input circuitry to be placed

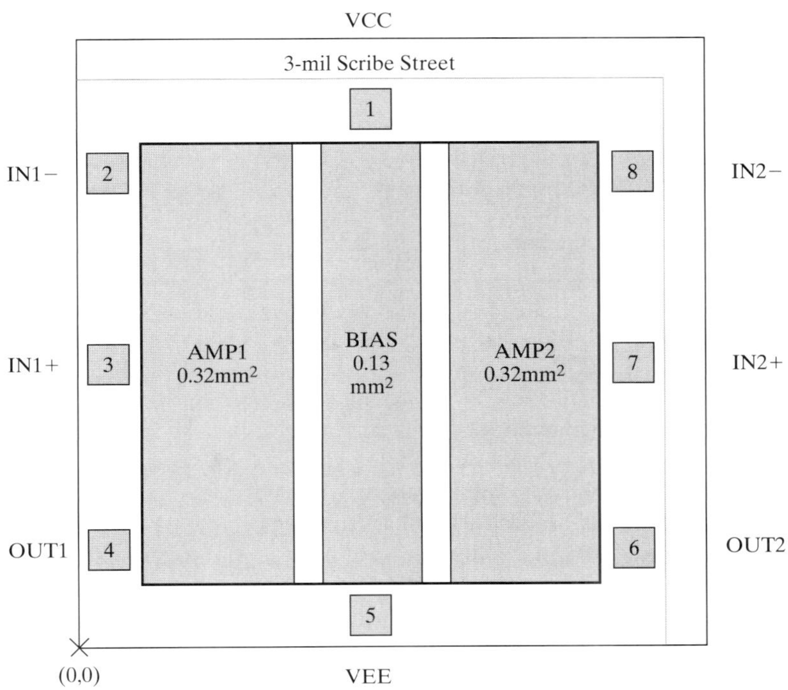

FIGURE 14.5 Floorplan of the eight-pin dual op-amp die.

far away from the power devices. The shape of the bias circuit is somewhat awkward, but not unworkable.

The die area estimate reserved 20% of the core area for routing. This space has been incorporated into the floorplan in the form of two narrow strips running vertically across the entire die. The actual placement of the leads will be determined later; these strips merely reserve room for them.

The next step consists of placing the bondpads. This requires that the floorplan be superimposed on a copy of the leadframe drawing. The bondpads should initially sit adjacent to their respective leadframe fingers (Figure 14.6A). This arrangement results in the shortest bondwires and the largest separation between the wires, but it does not necessarily provide the best interconnection between pads and circuit blocks. The pads can move slightly to accommodate layout, but if they are moved too far the design becomes unmanufacturable. In order to allow automated bonding, the bondwires must not intersect, nor should they even closely approach one another. Figure 14.6B shows an acceptable bonding arrangement, while Figure 14.6C shows an unacceptable one. This arrangement causes the wire for pin #2 to pass too close to the ball bond for pin #1. Similarly, the wire for pin #3 comes too close to the ball bond for pin #2. These faults make it almost impossible to bond the die without damaging some of the wires. Once the bondpad arrangement has been tentatively decided, the designer should forward a bonding diagram to the assembly site to allow them to check for potential bonding problems.

The floorplan in Figure 14.5 uses the pad arrangement shown in Figure 14.6B, and places the three input/output pads of AMP2 directly opposite the three corresponding input/output pads of AMP1. This symmetric placement helps simplify the connection of these pads to their respective amplifiers. The two power pads move to the top and bottom of the die. These locations not only help minimize interference between adjacent bondwires but also help to ensure that the power leads can run to all three blocks.

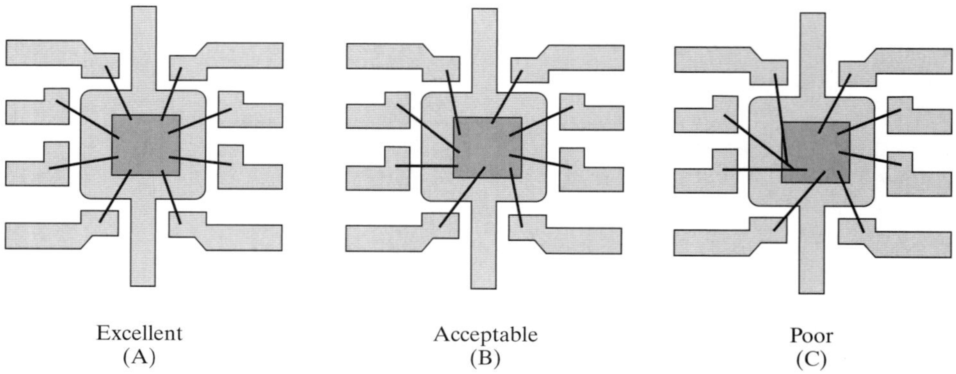

FIGURE 14.6 Three possible placements of bondpads for an 8-pin leadframe.

If the design incorporates high-current circuitry, then the designer should check the routing of high-current leads. Electromigration sets a lower limit on the width of a high-current lead, but metal resistance often forces the use of much wider leads. All high-current leads should be kept as short as possible to minimize unnecessary metal resistance. The anticipated locations of each high-current lead should be marked on the floorplan, along with the equivalent DC current that they must conduct. The resulting diagram will show whether any awkward or unnecessarily long leads exist.

Figure 14.7 shows a diagram of the dual op-amp highlighting the locations of the high-current leads. The VCC lead routes directly across the top of the bias cell. This is an acceptable arrangement for a double-level-metal layout because the lead can be routed in metal-2 and the circuitry in metal-1. The lead routing does place a limi-

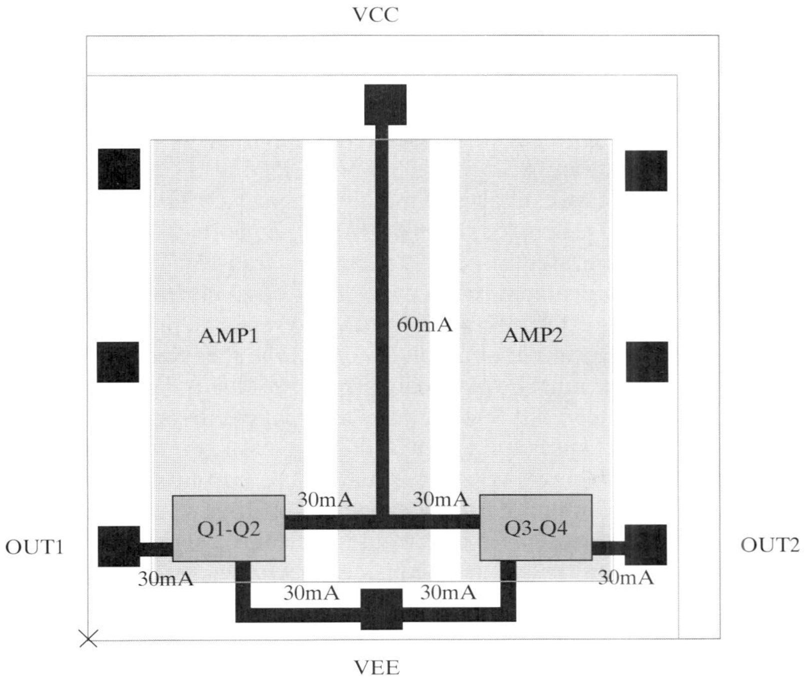

FIGURE 14.7 Floorplan of dual op-amp showing the routing of leads to transistors Q1 to Q4.

tation on the layout of the BIAS block, which might not have otherwise been apparent. If the BIAS block must use metal-2, then the power lead can slide laterally into either of the two routing channels.

Although the floorplan in Figure 14.7 explicitly shows the VEE leads connecting to each amplifier, in practice these leads form part of the scribe seal metallization. Most dice arrange their power and ground return signals to take advantage of the scribe seal metallization as much as possible, and this design is no exception. An additional width of metal placed around the periphery of the die abutting the scribe seal allows the metal to carry additional current. Substrate contacts placed under this metallization augment those in the scribe seal and form a significant portion of the substrate contact system of the die.

As previously mentioned, electromigration determines the minimum width of high-current leads. The minimum allowed lead width W_{min} (in microns) equals

$$W_{min} = \frac{10^{12} I_{DC}}{J_{max} t} \qquad [14.15]$$

where I_{DC} is the equivalent DC current in amps, J_{max} equals the maximum allowed current density in A/cm^2, and t is the thickness of the metal in Angstroms (Å). The values for J_{max} and t depend on the process and operating conditions (Section 14.3.3). If the lead passes over oxide steps, then the minimum thickness of the lead at the crossing point should be used for electromigration calculations.

For example, suppose $J_{max} = 5 \cdot 10^5$ A/cm^2 and $t = 8$kÅ. Given these values, Equation 12.14 indicates that a lead carrying 60mA requires a width of 15µm. The sheet resistance R_s for a metal lead can be computed using the following formula

$$R_s = \frac{10^8 \rho}{t} \qquad [14.16]$$

where ρ is the resistivity of the metal in Ω·cm and t is the thickness of the metal in Angstroms (Å). The resistivity of aluminum containing 0.5% copper and 2% silicon equals about 2.8μΩ·cm. Refractory barrier metals are much more resistive than aluminum alloys, so the sheet resistance of the metallization should be computed on the basis of the thickness of the aluminum and not on the thickness of the full metallization. The thinning of the metal system over oxide steps can be ignored when computing lead resistance. An 8kÅ aluminum-copper-silicon metal system has a sheet resistance of about 35mΩ/□. The 15μm VCC lead in Figure 14.7 runs about 1000μm and contains about 67 squares of metal that have a total resistance of 2.3Ω and generate a total voltage drop of about 140mV. Since both amplifiers connect to this lead, voltage drops generated by one amplifier can interfere with the performance of the other amplifier, a problem called *crosstalk*. Crosstalk can be reduced by making the lead wider, but a better solution is to run two separate power leads, one for each amplifier. This is an example of a *Kelvin connection* (see Section 14.3.2).

Sometimes vias must be placed in large power leads. The vias not only increase the resistance of the lead but also limit the amount of current that it can conduct before electromigration causes the vias to fail. A typical 1μm^2 via exhibits 100mΩ of series resistance and can safely conduct 25mA of DC current. A 1-amp lead would therefore require a minimum of forty vias, which together would add about 4mΩ of resistance. An array of vias that large would consume a significant amount of area, and their presence would have to be considered while constructing the floorplan.

The floorplan for the dual op-amp has now been completed. Although this example was fairly simple, the same principles apply to larger and more complex dice. The arrangement of the circuit blocks becomes more difficult, and often begins to resemble the construction of a jigsaw puzzle (Figure 14.8). The placement and width of routing channels also become much more critical. The careless placement of a block can easily constrict a routing channel. The resulting *choke points* can needlessly complicate routing, especially if they are not found until the routing has begun. The layout of Figure 14.8 contains two obvious choke points.

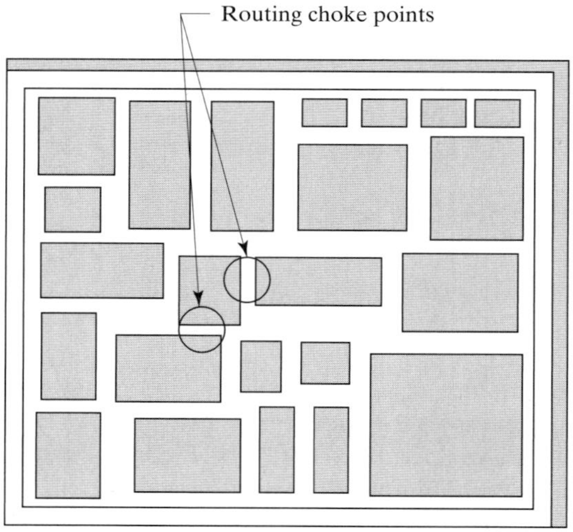

FIGURE 14.8 Floorplan of a complex die showing two potential choke points in the routing.

14.3 TOP-LEVEL INTERCONNECTION

Most analog and mixed-signal layouts actually benefit from manual interconnection. Autorouting software can interconnect blocks very quickly, but a human designer better understands such factors as wiring resistance, electromigration, noise coupling, and heat distribution. A skilled layout designer also takes advantage of every opportunity to insert substrate contacts, bypass capacitors, probe points, and test pads. Even if an autorouter is used, the layout designer must still verify that each analog signal has been properly routed. This process of verification (and the inevitable corrections) may require as much time as manual interconnection.

Some designers advocate running most of the wires across or through the circuit blocks themselves, a technique called *maze routing*. Others advocate running the wires alongside the individual blocks, a technique known as *channel routing*. Maze routing can save 5 to 10% of the initial die area, but it often requires two or three times as long to complete. Most modern designs use channel routing because it is quicker to implement and easier to modify.

14.3.1. Principles of Channel Routing

Channel routing requires at least two levels of interconnection. These consist, at a minimum, of one level of metal and one level of polysilicon. Unsilicided gate poly usually has a sheet resistance of 20 to 50Ω/□, while silicided poly can routinely achieve sheet resistances of less than 5Ω/□. Many signals can tolerate the insertion of short poly jumpers, especially if these are silicided. On the other hand, most signals cannot tolerate the resistance of long poly runs. This consideration makes it difficult to produce a compact wiring arrangement using only one level of metal, so most modern designs use at least two. Multiple metal layers reduce the need for poly routing, but the use of limited amounts of poly can still help clear congested wiring channels. As long as the designer carefully chooses which signals to route in poly, the presence of long polysilicon traces has little or no impact on circuit performance.

In a double-level-metal design, channel routing allocates one layer of metallization for vertical routing and a second for horizontal routing. It matters little which layer runs horizontally and which vertically. Figure 14.9 shows portions of two routing channels and an intersection between them. As this example shows, leads can exit from either routing channel at any point by jumping from one metal layer to the other. This orderly arrangement can only be maintained as long as all of the leads reside on the designated layers. If each signal is routed on whichever metal layer seems convenient at the time, then the design soon becomes a tangle of unwieldy (and unnecessary) metal jumpers.

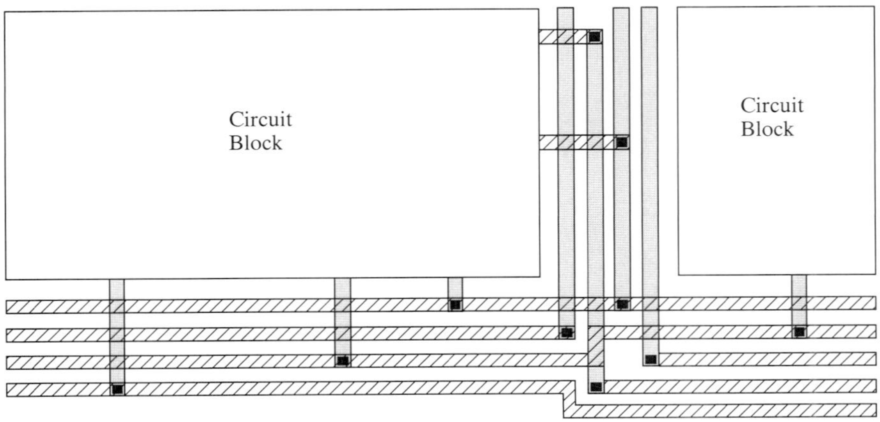

FIGURE 14.9 A portion of a channel-routed layout.

The leads on successive layers should run at right angles to one another. For example, metal-3 leads should run at right angles to metal-2 leads. Similarly, poly should run at right angles to metal-1. Thus, if poly runs horizontally, metal-1 should run vertically, metal-2 horizontally, and metal-3 vertically. This arrangement minimizes the need for jumpers and results in the best utilization of channel space.

Metal systems are often specified in terms of their *metal pitch.* The metal pitch P_m equals the sum of the minimum drawn metal width W_m and the minimum drawn metal spacing S_m:

$$P_m = W_m + S_m \qquad [14.17]$$

For example, a process that can fabricate 2μm leads spaced 1.5μm apart has a metal pitch of 3.5μm. The width of a wiring channel can be determined using the formula

$$W_c = NP_m + S_m \qquad [14.18]$$

where W_c equals the width of a wiring channel that can just accommodate N minimum-width leads. Continuing the previous example, a six-lead wiring channel would require a width of 22.5μm. The width W_c includes the space on either side of the channel between the leads and adjacent metallization.

In order to ensure optimal packing, the vias should not require enlarged metal heads. This requirement will be satisfied if the following inequality holds

$$W_m \geq W_v + 2O_{mv} \qquad [14.19]$$

where W_v equals the minimum width of a via and O_{mv} equals the minimum overlap of metal over via. If this inequality does not hold, then W_m should be increased. For example, suppose a process is capable of fabricating a 2μm metal lead, but it requires 1.5μm vias and a metal overlap of via of 0.5μm. The width of the metal head required to contain the via is 2.5μm, which is 0.5μm larger than the specified metal width. In order to maintain proper packing in the wiring channels, the designer should increase the width of the metal leads in the channel to 2.5μm. The increased lead width wastes a little area, but it greatly simplifies wiring. Figure 14.10A shows a layout employing enlarged via heads, while Figure 14.10B shows the same layout reworked using wider leads. Despite the presence of a little wasted space, the layout of Figure 14.10B is obviously preferable to that of Figure 14.10A.

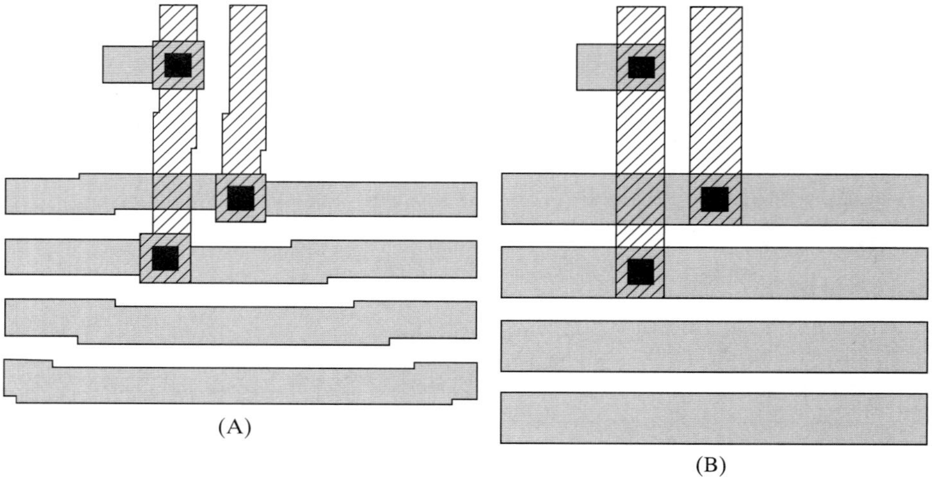

FIGURE 14.10 (A) A layout that requires enlarged via heads (B) becomes much simpler when the lead width is slightly increased.

In most designs, two or three routing channels carry a large percentage of the signals. These *primary routing channels* usually intersect somewhere near the center of the die. These intersections may easily become choke points unless the channels are made quite wide. As a conservative rule, each primary routing channel should contain space for about 20% of the top-level signals. A design containing 100 top-level signals requires room for about 20 leads in each of its primary routing channels. The width of the primary wiring channels can decrease as they approach the edges of the die. Similarly, the feeder channels branching off the primary channels can be made proportionately narrower. The resulting pattern of routing channels resembles the path of watercourses across a plain. Even the narrowest channels should always allow capacity for 3 to 5 signals, as it is very difficult to guess exactly where leads need to route, and it is very difficult to increase channel widths once the top-level interconnection has commenced.

Poly routing can reduce the width of the primary routing channels by perhaps 30%. This reduction in width assumes that about a third of the leads are routable in poly. The use of poly routing under both horizontal and vertical routing channels can lead to gridlock in the vicinity of major intersections. This type of gridlock is best avoided by running poly leads in only one direction, preferably in the same direction as metal-2. If possible, one should align the largest (or longest) primary routing channel so that it contains poly.

14.3.2. Special Routing Techniques

Certain leads require special consideration during routing. Some leads are particularly sensitive to voltage drops or noise coupling, while others require wider metal to minimize voltage drops and to prevent electromigration. This section discusses some of the techniques used to handle these concerns.

Kelvin Connections

The metallization resistance, although small, is not always negligible. Consider an amplifier circuit containing a pair of matched bipolar transistors whose emitters connect to a ground return line carrying 100μA (Figure 14.11A). The two emitter leads do not tap into the ground return at the same point, so the ground current generates a small voltage differential between them. Suppose the ground return lead between points A and B contains ten squares of 30mΩ/□ metal. A 100μA current flowing through 0.3Ω develops a voltage of 30μV. Any variation in ground current produces a corresponding variation in this voltage drop, which the circuit amplifies. If the ground current fluctuates by ±10% and the amplifier has 80dB of voltage gain, then the output will fluctuate by 0.3V! This is clearly an unacceptable situation.

The magnitude of the voltage drop in the ground lead depends on the distance between points A and B. Figure 14.11B shows the same circuit with one modification: both emitter leads now return to a common point C. Since the leads return to the same point, the ground current cannot generate any differential between them. Ground drops elsewhere along the lead cause the voltage on both emitters to vary in unison, but most circuits exhibit a high degree of immunity to such *common-mode variations.*

The common point C is called a *star node* or a *Kelvin connection.* These points are often represented on schematics by leads entering a solder dot at 45° angles, as

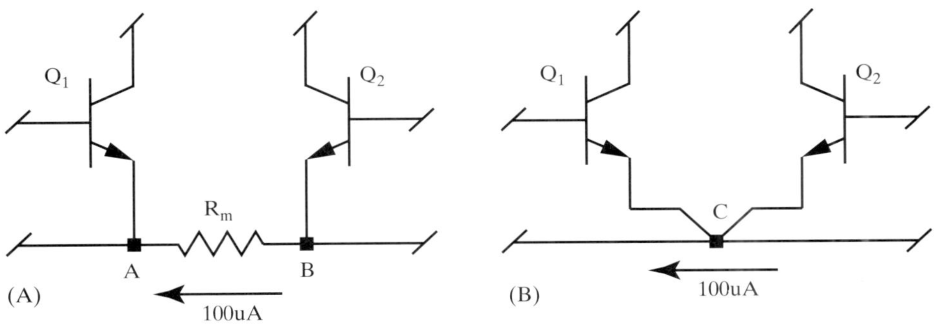

FIGURE 14.11 (A) The effects of voltage drops in a ground return line (B) can be eliminated through the use of a Kelvin connection.

in Figures 4.11 and 4.12. Kelvin connections have many other applications. Figure 14.12A shows a pair of Kelvin connections that allow accurate sensing of the voltage developed across a metal resistor. Leads F_1 and F_2 carry a large current through resistor R_1, while leads S_1 and S_2 connect the resistor to a sensing circuit. F_1 and F_2 are called *force leads,* while S_1 and S_2 are called *sense leads.* Figure 5.16 shows sample layouts of metal resistors with force and sense leads. The sense leads carry very little current to minimize the voltage drops occurring in them. In high-precision applications, the currents flowing through the sense leads are carefully balanced, and the layouts of the leads are adjusted so that each contains the same number of squares. This precaution ensures that any voltage drops that do occur in the sense leads are common-mode variations.

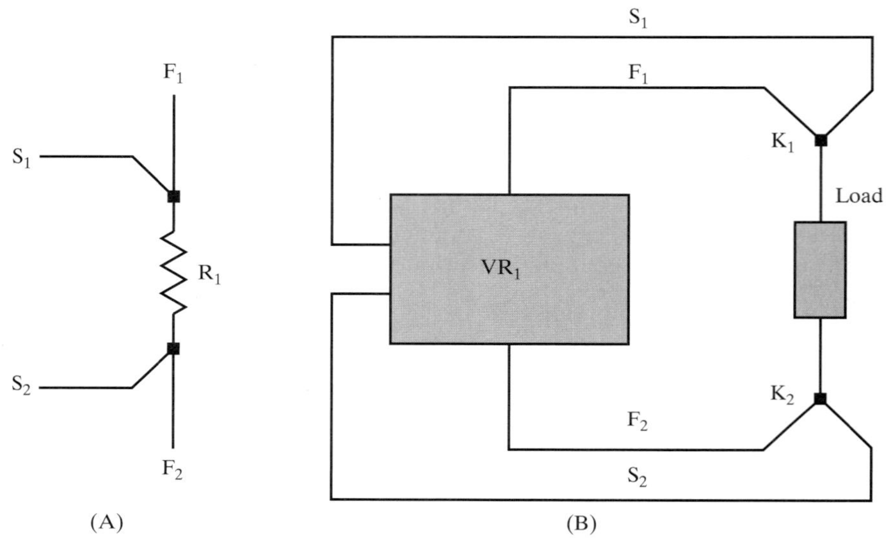

FIGURE 14.12 Additional applications of Kelvin connections: (A) metal sense resistor and (B) remote sensing.

Figure 14.12B shows another application for Kelvin connections. A voltage regulator, VR_1, must provide a large amount of current to an external circuit. Voltage drops inevitably occur in power lead F_1 and in ground lead F_2. A separate set of sense leads, S_1 and S_2, is employed to sense the voltage across the load at Kelvin connections K_1 and K_2. This arrangement ensures that voltage drops in the force leads do not degrade the regulation of the circuit. The manufacturers of commercial power supplies call this arrangement *remote sensing.* Integrated voltage regulators sometimes support remote sensing by providing separate force and sense pins. If the package does not provide enough pins to accommodate separate force and sense connections, two bondwires can be connected to the same pin to eliminate voltage drops in the IC metallization and the bondwires. Even if the package cannot accommodate double bonding, the designer can still run separate force and sense leads to the output and ground bondpads.

Noisy Signals and Sensitive Signals

Broadly speaking, electrical noise consists of any unintended or undesired signal. All electrical devices generate some noise, but the level of this *device noise* is so low that it only affects very sensitive circuits. Most of the noise problems encountered in integrated circuits are caused by capacitive coupling of signals from one circuit node to another. A capacitor appears whenever leads cross or run alongside one another. Although these capacitors are very small, the amount of energy coupled through them increases with frequency. At sufficiently high frequencies, even intersections between minimum-width leads can couple objectionable amounts of noise energy from one circuit to another.

In practice, capacitively coupled noise becomes a concern only if one or more signals contain appreciable energy at frequencies in excess of 1MHz. Most analog signals have relatively little high-frequency content, but digital signals are quite another matter. The crisp transitions so prized by digital designers generate high-frequency harmonics extending well beyond 1GHz. Each digital signal generates a burst of noise every time it switches states. All digital signals that experience state transitions during normal circuit operation should be considered potential noise sources. All power switching transistors should be considered noise sources, as should signals connecting to pins seeing rapid transients during normal operation.

The mere existence of any number of noisy signals is harmless. Problems only arise if the integrated circuit also contains one or more signals that are unusually sensitive to capacitively coupled noise. Analog signals are far more noise-sensitive than digital signals, but not all analog signals are equally sensitive. The most sensitive nodes are those that carry very low-level signals at high impedance levels. For example, the input of an amplifier is far more noise sensitive than its output because any signal present at the input is amplified by the gain of the amplifier. Furthermore, the output of an amplifier is usually low-impedance, while its inputs are often very high-impedance. The following types of signals are among the most noise sensitive:

- Inputs to high-gain amplifiers and precision comparators
- Inputs to analog-to-digital converters (DACs)
- Outputs of precision voltage references
- Analog ground lines to high-precision circuitry
- Precision high-value resistor networks
- Very low-level signals, regardless of impedance
- Very low-current circuitry of any sort

Most layout designers do not possess the knowledge and experience required to correctly identify all of the noisy and sensitive nodes in a complicated analog circuit. Instead, the circuit designer must identify these signals, preferably by marking them on the schematic in some distinctive manner. Once the layout designer understands which signals are noisy and which are sensitive, the signals can be routed appropriately.

Noisy signals should not run on top of sensitive signals, or *vice versa*. If a crossing must occur, the area of intersection should be minimized. The circuit designer should examine each crossing to determine if it requires electrostatic shielding (Section 7.1.7). The usual method of constructing such shielding in double-level-metal designs is to run one signal in poly and the other in metal-2. A plate of metal-1 interposed between the two signals acts as an electrostatic shield. The shield should connect to a quiet low-impedance node such as an analog ground line. The shielding plate should extend beyond the area of intersection by 2 to 3μm to block fringing fields.

Noisy signals should not run adjacent to sensitive signals. If such an arrangement seems unavoidable, then the layout designer should run another signal between the two. The shield lead may consist of some other relatively low-noise, low-impedance signal, such as the output of a digital logic gate that seldom changes states. Alternatively, it may consist of an extra ground line or supply line added to the layout specifically to shield the noisy signal from the sensitive one. In order for the shield lead to perform its function, it must connect to a fairly low-impedance node in the circuit. Supply and ground lines satisfy this requirement, as will the output of digital logic gates.

Whenever possible, the designer should place noisy circuitry as far away from sensitive circuitry as possible. This usually requires that noisy circuitry occupy one portion of the die and sensitive circuitry another. This type of arrangement often proves beneficial from other standpoints because it separates sensitive circuitry from large power devices.

Sensitive signals should never run farther than necessary. The shorter these signals, the fewer opportunities they provide for noise coupling. A layout designer should always try to place analog circuits so that the sensitive signals running between them do not pass through other circuit blocks. Although this is not always possible, the fewer long sensitive leads a design contains, the easier it is to prevent capacitive coupling problems.

14.3.3. Electromigration

As discussed in Section 4.1.2, electromigration limits the current density that can safely flow through metallization. If the current density becomes too large, the pressure of carrier collisions on the aluminum metal atoms causes a slow displacement of the metal. This eventually results in voiding or lateral extrusions (*whiskers*) that can short adjacent signals.[1]

Electromigration obeys a modified Arrhenius relationship that was first discovered by J. R. Black, and is therefore called *Black's Law:*[2]

$$MTF = \frac{1}{AJ^2} e^{\frac{E_a}{kT_j}} \qquad [14.20]$$

where MTF is the mean time to metallization failure, in hours, A is a rate constant having units of $cm^4/A^2/hr$, J is the current density passing through the lead in A/cm^2, E_a is the activation energy in electron volts (eV), k is Boltzmann's constant ($8.6 \cdot 10^{-5}$eV/K), and T_j is the absolute junction temperature in Kelvin.

Since the rate of electromigration varies exponentially with temperature, high-temperature testing can accelerate the failure process to allow experimental determination of constants A and E_a. On the basis of such testing, one can derive a maximum current density, J_{max}, corresponding to any desired operating life. As one might expect, J_{max} is a strong function of temperature. A typical value for copper-doped aluminum is $5 \cdot 10^5 A/cm^2$ at 85°C. This value is widely accepted throughout the industry for computing the required width of metallization in devices operating at junction temperatures of 85°C or less. The current that an arbitrary lead can carry equals

$$I_{max} = 10^{-9} J_{max} W t \qquad [14.21]$$

where I_{max} is the maximum current (in mA) that the lead can safely carry, J_{max} is the maximum allowed current density in A/cm^2, W is the width of the lead in microns, and t is the thickness of the metallization in Angstroms (Å). A factor of 10^{-9} is inserted in equation 14.21 to balance the units. As an example of the use of this equation, suppose a 10kÅ lead is constructed of copper-doped aluminum having $J_{max} = 5 \cdot 10^5 A/cm^2$. If this lead is 10μm wide, then it can conduct no more than 50mA.

---

1 J. R. Black, "Electromigration—A Brief Survey and Some Recent Results," *IEEE Trans. Electron Devices,* Vol. ED-16, #4, 1969, pp. 338–348.

2 J. R. Black, "Mass Transport of Aluminum by Momentum Exchange with Conducting Electrons," *Proc. 1967 Ann. Symp. on Reliability Physics,* IEEE Cat 7-15C58, 1967.

Metal leads passing over oxide steps often experience less-than-perfect step coverage. Most processes state a minimum guaranteed step coverage in terms of a percentage of normal metal thickness. The carrying capacity of any lead crossing an oxide step must be derated by the percentage step coverage. Standard bipolar processes typically quote a 50% step coverage figure. Thus, if the 10μm lead discussed above crosses an oxide step, then its current carrying capacity would drop to 25mA.

Most parts are assumed to experience junction temperatures of 85°C or less for the majority of their operating life, even if the parts are rated for higher junction temperatures. Certain types of devices, particularly those that experience substantial differences between junction and ambient temperatures, may actually operate at high temperatures for long periods of time. If such conditions are anticipated, then the current carrying capacity of the leads must be multiplied by a derating factor, D, equal to

$$D = e^{\frac{E_a}{2k}\left(\frac{1}{T_j} - \frac{1}{T_o}\right)} \qquad [14.22]$$

where T_o is the junction temperature at which I_{max} was computed in degrees Kelvin, T_j is the anticipated maximum operating temperature in Kelvin, E_a is the activation energy in electron volts, and k is Boltzmann's constant. A typical activation energy for electromigration-induced voiding in pure aluminum is 0.5eV, while values for copper-doped aluminum[3] are closer to 0.7eV. Suppose we wish to compute the derating factor for 125°C (398K), given an original current density calculation performed at 85°C. Assuming $E_a = 0.7$eV, the derating factor D equals 0.32. Therefore a lead that can safely carry 25mA at 85°C can safely carry only 32% of this current, or 8mA, at 125°C.

If a lead does not carry a constant current, but rather a time-varying current, then its current-handling capability is increased. One type of time-varying current, consisting of short pulses repeated at frequent intervals, often occurs in digital logic and MOS gate drivers. The derating factor D for pulsed-current operation equals[4]

$$D \cong \frac{1}{d^2} \qquad [14.23]$$

where d is the duty cycle of the signal and D is the derating factor. For example, a lead conducting current only 50% of the time has a duty cycle of 0.5 and a derating factor of 4. If the lead can safely carry 25mA DC, then it can handle 50mA pulses at a duty cycle of 0.5. The derating factor computed in equation 14.23 assumes that the pulsed current flows only in one direction. The rate of AC electromigration is somewhat slower than the rate of DC electromigration because some of the displacement occurring during one phase reverses during the other phase. The magnitude of the derating factor for AC operation is difficult to determine because it varies with the exact waveforms involved, but a conservative estimate is $D = 1.5$. Thus, if a lead could handle 50mA unipolar pulses, then it could handle 75mA bipolar pulses.

Another common type of time-varying signal consists of a sinusoid, or sine wave. A lead can handle a sinusoidal current with a peak value equal to about three times its DC rating. Thus, if a lead can handle 25mA DC, then it can handle a sinusoidal current with a peak value of 75mA (which equals 150mA peak-to-peak or 106mA root-mean-square).

---

[3] H. V. Schreiber, "Activation Energies for the Different Electromigration Mechanisms in Aluminum," *Solid-State Elect.*, Vol. 24, 1981, pp. 583–589.

[4] J. S. Suehle and H. A. Schafft, "Current Density Dependence of Electromigration t_{50} Enhancement Due to Pulsed Operation," *Proc. International Reliability Physics Symposium*, 1990, pp. 106–110.

Bondwires also have limited current-carrying capability. If too much current flows through a wire, its internal temperature may rise to the point where it fails, either gradually through electromigration, or suddenly through fusing or burning. Military specification Mil-M-38510 sets a maximum allowed current, I_{max}, (in mA) equal to[5]

$$I_{max} = kd^{3/2} \quad [14.24]$$

where d is the bondwire diameter in mils and k is a constant equal to approximately 480mA·mil$^{2/3}$ for aluminum and 650mA·mil$^{2/3}$ for gold. This law is derived from the classical three-halves power law of radiative processes.[6] While it applies to long bondwires suspended in air, it does not apply to bondwires embedded in plastic or other encapsulating materials. These substances act as thermal insulators and prevent radiative heat transfer. The temperature of the wire is therefore determined solely by conduction of heat through the wire to its endpoints and through the encapsulation. Assuming that the wire is relatively long and that the encapsulation is a good thermal insulator, then the temperature of the wire depends only on the amount of power dissipated and not on its surface area or volume. Under these conditions, the current-handling capability of a bondwire scales linearly with its diameter. Large-diameter bondwires are placed under a disadvantage when they are encapsulated because they do not gain the benefit of their larger surface area.

In practice, bondwires that are less than about 100mil long conduct enough heat to their ends to produce a noticeable increase in their current-carrying capacity.[7] For this reason, most designers use slightly larger current values than Mil-M-38510 suggests. Gold wires encapsulated in plastic are often scaled using the rule of "one amp per mil of diameter," and aluminum wires are usually rated to carry about half the current of an equivalent gold wire. Aluminum wires are not allowed to carry as much current as gold wires because they are more vulnerable to electromigration and high-temperature corrosion. Conservative designers may wish to apply the Mil-M-38510 values of 650mA/mil for gold and 480mA/mil for aluminum, and to allow these to scale linearly with diameter if the wires are encapsulated, and to scale with the three-halves power of diameter if they are not.

14.3.4. Minimizing Stress Effects

The coefficients of thermal expansion of packaging materials rarely match that of silicon, and encapsulation usually takes place at an elevated temperature (Section 7.1.5). The stresses that accumulate as the die cools become permanently frozen into the finished part. On larger dice, or on those mounted using solder or gold eutectic, the stress levels at the corners of the die may become severe enough to damage the die.[8] Common forms of damage include sheared bondwires, broken metal traces, and delamination of the protective overcoat from the underlying metallization.

Some assembly sites prohibit the placement of circuitry in the stress-prone regions around the corners of the die. The prohibited regions usually take the form of *stress triangles* extending some 5 to 10 mils (125 to 250μm) from each corner of the die (Figure 14.13). The designer should not place circuit components, leads, or bond-

---

[5] "Maximum Current in Wires," *Semiconductor Reliability News,* Vol. I, #10, 1989, p. 9.

[6] W. H. Preece, "On the Heating Effects of Electric Currents," *Proc. Roy. Soc.(London),* April 1884, December 1887, April 1888.

[7] From bondwire fusing data given in "Fusing Currents of Bond Wires," *Semiconductor Reliability News,* Vol. VIII, #12, 1996, p. 8. See also B. Krabbenborg, "High Current Bond Design Rules Based on Bond Pad Degradation and Fusing of the Wire," *Microelectronics Reliability,* Vol. 39, 1999, pp. 77–88.

[8] J. R. Dale and R. C. Oldfield, "Mechanical Stresses Likely to be Encountered in the Manufacture and Use of Plastically Encapsulated Devices," *Microelectronics and Reliability,* Vol. 16, 1977, pp. 255–258.

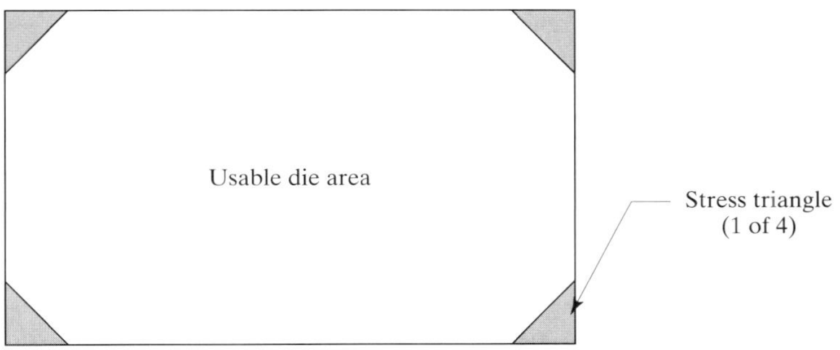

FIGURE 14.13 Layout of a die showing the placement of stress triangles.

pads in these triangles. Test structures can still reside in the stress triangles, as can identification markings and alignment structures.

Stress triangles are usually added to large dice, but not to small ones. The benefits of adding stress triangles to a large die outweigh the relatively small percentage of die area they consume. The stress levels on a small die are lower, yet the stress triangles consume a larger fraction of the available area. At some point, the triangles become more burdensome than they are worth. Most dice under 10kmil^2 (6.5mm^2) do not incorporate stress triangles, and those under 20kmil^2 (13mm^2) often leave them off unless the die experiences unusually stressful conditions, such as solder mounting or gold eutectic bonding.

Certain precautions should always be taken when placing components in the corners of a die. Matched components should never occupy the corners of a die (Section 7.2.6). If possible, one should also avoid placing bondpads in the corners of a die to minimize the possibility of sheared bonds. This prohibition does not apply to trimpads and testpads since neither of these receives bondwires.

Leads should not make right angles near a corner of the die. Stresses concentrate on the outside vertex of such a lead, and this can, in turn, lead to delamination and cracking of the protective overcoat. The designer should insert a short 45° segment into each such lead to help distribute the stresses more evenly (Figure 14.14A). This precaution helps prevent delamination and subsequent damage to the metal system.

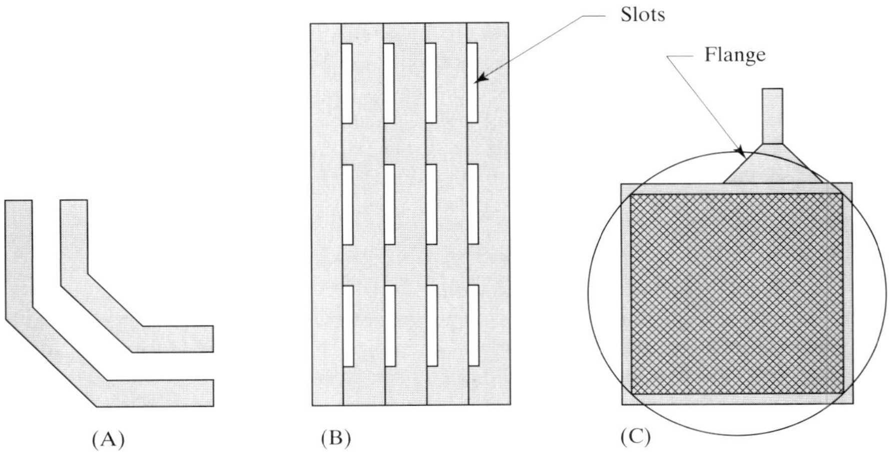

FIGURE 14.14 Various stress-reduction techniques: (A) 45° bends in leads, (B) slots in a wide metal-2 lead, and (C) flanged bondpad.

The presence of large areas of top-level metal near the corners of a die can also cause delamination. The planarization techniques that improve the step coverage of upper-level metal also eliminate the irregularities that help lock the protective overcoat to the die. The addition of an array of narrow slots in the metal helps restore these irregularities and consequently discourages delamination. The slots should be oriented in the same direction that the current flows through the lead so that they do not greatly reduce its effective cross-sectional area (Figure 14.14B). Leads less than 1mil (25μm) wide generally do not require slots. An array of vias placed between two levels of metal also produces surface irregularities that will lock the protective overcoat in much the same way that the addition of slots will. Thus arrays of vias do not require the addition of slots.

The bonding process also produces high levels of stress. Although this stress only lasts for an instant, it can still damage adjacent metal traces or components. Most processes therefore define a circular exclusion zone around each ball bond and a rectangular or trapezoidal exclusion zone around each wedge bond (Section 13.4.2). Leads that do not connect to the bondpad should not enter this zone, and active circuitry should also reside outside it. Many designers also add a triangular or rectangular *flange* of metal to each lead entering a bondpad (Figure 14.14C). This helps ensure that the stresses associated with bonding do not sever the lead from the bondpad. Modern bonding equipment has minimized the need for these flanges by more accurately controlling the forces applied during bonding, but the flanges consume so little area that it seems unreasonable to eliminate them.

14.4 CONCLUSION

A layout designer must understand not only how to design individual components but also how to interconnect these components to form a complete die. This chapter has touched upon the skills required to predict the size and structure of the die, and those required to actually assemble it. These are perhaps the most difficult of all of the skills the layout designer must learn. They partake of the essential character of all analog layout, in that they are largely arts and not sciences. As such, they require a certain degree of intuition and sound judgment that can only be learned through experience. The information presented here represents only a foundation, and the designer must continue to learn through actually practicing the art of layout.

14.5 EXERCISES

Refer to Appendix C for layout rules and process specifications.

14.1. Estimate the area of the following components in a standard bipolar process. Show all computations and state all assumptions used in the estimations.
  a. 250kΩ of 8μm-wide HSR resistance.
  b. A minimum-area lateral PNP transistor.
  c. 100pF of junction capacitance (assume 1.8fF/μm^2).
  d. A power NPN transistor with an emitter area of 1200μm^2.
  e. A bondpad for 1.2mil ballbonded gold wire.

14.2. Estimate the total area of a circuit laid out in standard bipolar using single-level metallization containing 410kΩ of 6μm HSR resistance, 55kΩ of 8μm base resistance, 50pF of junction capacitance (at 1.8fF/μm^2), eleven minimum NPN transistors, seven minimum lateral PNP transistors, two split-collector lateral PNP transistors, one 4X lateral PNP, and one power NPN transistor requiring an emitter area of 660μm^2. Justify your choice of packing factor.

14.3. Compute the core area of a design containing six cells having the following areas: 0.35, 0.27, 0.21, 0.18, 0.10, and 0.08mm^2. The design also includes two power transistors that each consume 0.77mm^2. The design will be laid out in double-level metal and contains about fifty top-level signals. Justify your choices of routing and packing factors.

14.4. Compute the estimated die area for each of the following designs. Assume that all designs require a bondpad width of 75μm, a spacing between adjacent bondpads of 65μm, and a scribe width of 75μm. State whether each design is core-limited or pad-limited.

a. Core area of $10.1mm^2$ using 23 pads.
b. Core area of $1.49mm^2$ using 42 pads.
c. Core area of $8.8mm^2$ using 11 pads.

14.5. Using the parameters specified in Exercise 14.4, estimate the number of bondpads that can be packed around the periphery of a square die having a total area of $3.9mm^2$. How many additional pads could be obtained if the die had an aspect ratio of 1.5:1?

14.6. A device with a die area of $7.6mm^2$ is fabricated on $150mm^2$ wafers. Assume a wafer utilization factor of 0.85. (A) Approximately how many dice will each wafer yield? (B) If each wafer costs $650 and yields 77% functional dice, what is the cost of a single functional die? (C) If packaging and final testing costs 11.5¢ per device and 97% of devices pass the final test, what is the cost of a completed integrated circuit? (D) If the device is sold for 83¢, what is the GPM? (E) What would be a more reasonable sale price, and why?

14.7. Suppose an improved testing technique increases the probe yield of the device in Exercise 14.6 to 92%. This technique adds 4¢ to the cost of each die probed. Is it worthwhile to implement the new technique?

14.8. A certain integrated circuit contains four operational amplifiers, each with a cell area of $0.41mm^2$. The circuit also contains a bias circuit that consumes $0.15mm^2$. This circuit requires fourteen bondpads: VCC, VEE, IN1+, IN1–, OUT1, IN2+, IN2–, OUT2, IN3+, IN3–, OUT3, IN4+, IN4–, and OUT4. VCC and VEE are the power and ground pins, respectively. The remaining pins are the inputs and outputs of each of the four amplifiers. The package has fourteen pins, with pin #1 bonded to the top center of the die and the remaining pins arranged in counterclockwise order. (A) Suggest a reasonable pinout for this device. (B) Draw a die floorplan showing the placement of each of the five cells of this design. (C) Each amplifier requires high-current leads to VCC, VEE and its output pin. Assuming that the design uses only a single level of metallization, draw a diagram illustrating a suitable routing pattern.

14.9. A voltage reference circuit contains a reference cell occupying $0.13mm^2$ and a small power transistor requiring $0.16mm^2$. (A) Assuming the die uses single-level metallization and a hand-packed layout, estimate the core area of the die. (B) Estimate the complete die area assuming a 110μm-wide scribe and a total of three bondpads, each requiring a total area of 9000μm (pad plus ESD protection). (C) Draw a die floorplan for this device. The power transistor must connect to the VIN and VOUT pads, which are at the top right and bottom right of the die, respectively. The reference cell must connect to the GND pin, which must lie somewhere along the left side of the die. (D) Indicate the preferred location for critical matched devices.

14.10. An octal buffer circuit contains eight buffers, each of which require an area of $0.09mm^2$. Each buffer has an input and an output, and all require connections to VCC and GND. (A) Assuming that the die uses single-level metallization and must be assembled quickly, estimate the core area of the die. (B) Estimate the total die area assuming a 110μm-wide scribe and a total of eighteen pads, each requiring 9000μm of area. (C) Draw a die floorplan for this device. Pin #9 must be GND and pin #18 must be PWR; suggest a reasonable arrangement for the eight inputs and eight outputs. Assume pad #1 must occupy the top left corner of the die and the remaining pads are arranged in counterclockwise order. (D) Each buffer requires high-current connections to VCC, GND, and its output pin. Draw a diagram of a routing pattern that will connect all eight cells without requiring any tunnels in power leads.

14.11. Calculate the width required to construct a wiring channel capable of holding twelve leads using the CMOS layout rules of Appendix C.

14.12. Suppose the output lead of a buffer carries a digital signal with a duty cycle of 50%. When the output is on, it sources 360mA of current, and when it is off it sinks negligible current. (A) If the metallization consists of 10kÅ of copper-doped aluminum capable of carrying a constant current of $5 \cdot 10^5 A/cm^2$ at 85°C, how wide must the output lead be made in order for it to withstand electromigration? (B) How wide must the lead be made to safely cross an oxide step that may induce 30% metallization thinning? (C) How wide must the lead be made if the device will operate continuously at a junction temperature of 125°C? Assume an activation energy of 0.7eV.

14.13. The power output of an amplifier circuit conducts an average current of 3.1A under worst-case conditions. (A) Assuming the part is packaged in plastic, how many 1.2mil gold bondwires are required to safely conduct this current? (B) How many 2mil aluminum wires would be required to conduct the same current?

Appendix A
Table of Acronyms Used in the Text

AC	Alternating Current
A/D	Analog-to-Digital
BCLDD	Buried-Channel Lightly Doped Drain
BiCMOS	Bipolar and Complementary Metal-Oxide-Semiconductor
BiFET	Bipolar and junction Field-Effect Transistor
BJT	Bipolar Junction Transistor
BOI	Base Over Isolation
BPSG	BoroPhosphoSilicate Glass
CDI	Collector Diffused Isolation
CDM	Charged Device Model
CMOS	Complementary Metal-Oxide-Semiconductor
CTE	Coefficient of Thermal Expansion
CVD	Chemical Vapor Deposited
D/A	Digital-to-Analog
DC	Direct Current
DCML	Differential Current-Mode Logic
DDD	Double-Diffused Drain
DIP	Dual In-line Package
DLM	Double-Level Metal
DMOS	Double-diffused Metal-Oxide-Semiconductor
DRAM	Dynamic Random-Access Memory
DSW	Direct Step on Wafer
ECL	Emitter-Coupled Logic
EEPROM	Electrically Erasable Programmable Read-Only Memory

EOS	Electrical OverStress
ESD	ElectroStatic Discharge
FBSOA	Forward-Biased Safe Operating Area
FET	Field-Effect Transistor
GPM	Gross Profit Margin
HBM	Human-Body Model
HF	Hydrofluoric acid (chemical formula)
HSR	High-Sheet Resistor
IC	Integrated Circuit
ILO	InterLevel Oxide
IPTAT	current Proportional To Absolute Temperature
JFET	Junction Field-Effect Transistor
JI	Junction Isolation
LDD	Lightly Doped Drain
LDMOS	Lateral Double-diffused Metal-Oxide-Semiconductor
LED	Light-Emitting Diode
LOCOS	LOCal Oxidation of Silicon
LPCVD	Low-Pressure Chemical Vapor Deposition
LSTTL	Low-power Schottky-clamped Transistor-Transistor Logic
MLO	MultiLevel Oxide
MM	Machine Model
MOS	Metal-Oxide-Semiconductor
MOSFET	Metal-Oxide-Semiconductor Field-Effect Transistor
NBL	N-type Buried Layer
NMOS	N-channel Metal-Oxide-Semiconductor
NSD	N-type Source/Drain
ONO	Oxide-Nitride-Oxide
OR	Oxide Removal
PBL	P-type Buried Layer
PG	Pattern Generation
PMOS	P-channel Metal-Oxide-Semiconductor
PO	Protective Overcoat
PPM	Parts Per Million
PSD	P-type Source/Drain
PSG	PhosphoSilicate Glass
RIE	Reactive Ion Etch(ing)
SCL	Space Charge Layer
SCR	Silicon Controlled Rectifier
SDD	Single-Diffused Drain (or Single-Doped Drain)
SI	*Systéme Internationale* (the metric system)
SLM	Single-Level Metal
SOA	Safe Operating Area

SOIC	Small-Outline Integrated Circuit
SPICE	Simulation Program with Integrated Circuit Emphasis
SSA	Super Self-Aligned
TCR	Temperature Coefficient of Resistivity
TDDB	Time-Dependent Dielectric Breakdown
TEOS	TetraEthOxySilane
TTL	Transistor-Transistor Logic
UV	UltraViolet
VLSI	Very Large-Scale Integration
VPTAT	Voltage Proportional To Absolute Temperature

Appendix B

The Miller Indices of a Cubic Crystal

A *crystal* consists of an orderly arrangement of atoms or molecules extending indefinitely in all directions. This arrangement is periodic in the sense that the same patterns of atoms or molecules reappear at regular intervals along certain axes. One can imagine the crystal consisting of a large number of submicroscopic building blocks called *unit cells*. In the case of crystals belonging to the cubic system, the unit cells are tiny cubes that stack in rows and columns to create a rectilinear crystal lattice (Figure B.1). As the illustration suggests, the resulting crystal is not always cubic in shape. In every case, the underlying periodic structure of the crystal remains invariant regardless of its external form or dimensions.

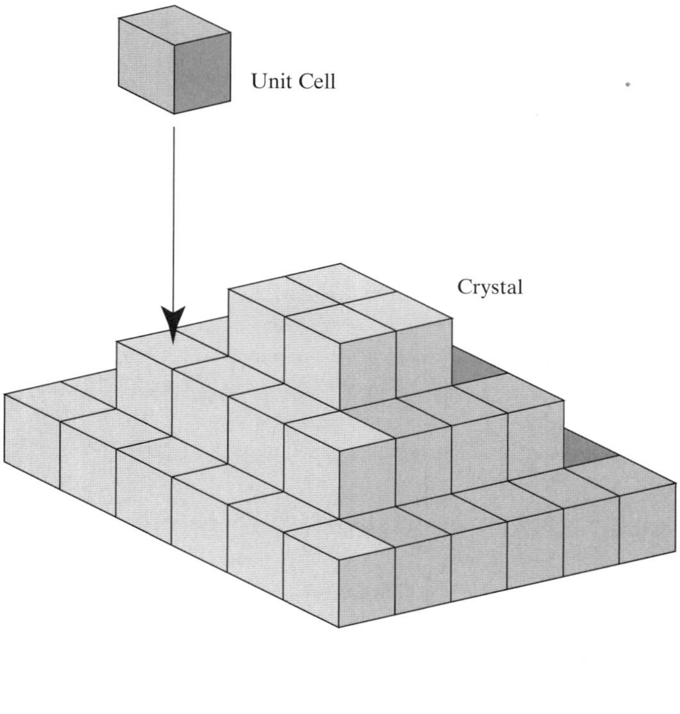

FIGURE B.1 Relationship between a cubic crystal and its unit cell.

A silicon ingot is a single cubic crystal, and each wafer cut from the ingot constitutes a slice cut from this crystal. The properties of the wafer depend on the angle of the cut relative to the axes of the crystal lattice. A set of numbers called *Miller indices* are useful for labeling various cuts of silicon and for labeling directions on the surface of a wafer. This appendix discusses the system of Miller indices used for cubic crystals, including those of silicon and germanium.

A plane intersecting a cubic crystal can be assigned a set of three Miller indices that together specify its orientation relative to the crystal axes. In order to compute these indices, one must first identify the three crystal axes. These lie orthogonal to one another and correspond to the X-, Y-, and Z-axes of the Cartesian coordinate system. The cubic unit cells stack in neat rows and columns aligning to these three axes. The location of any point can be given in terms of multiples of the width of the unit cell along each of the three crystal axes. A plane intersecting the lattice can now be described in terms of its X, Y, and Z intercepts. For example, the intercepts of the plane of Figure B.2A are X = 1, Y = 3, Z = 3. Similarly, the intercepts of the plane of Figure B.2B are X = 3, Y = 2, Z = 2.

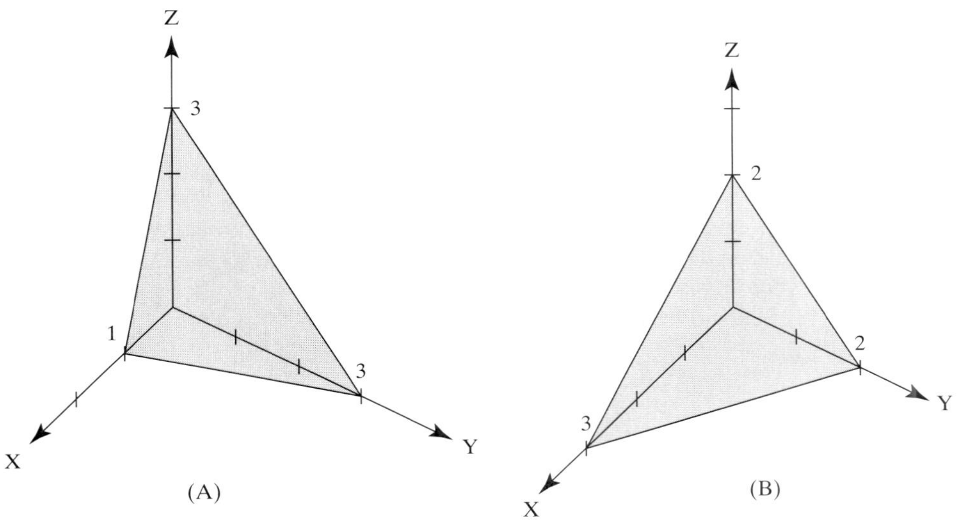

FIGURE B.2 Examples of two planes that intersect a cubic lattice. The tick marks represent multiples of the unit cell dimension.

A crystal plane may lie parallel to one or more of the crystal axes, in which case its intercepts with those planes lie at infinity. For example, the intercepts of a horizontal plane are X = ∞, Y = ∞, Z = 1. Miller indices avoid infinite entries by using the reciprocals of the intercepts rather than the intercepts themselves. For example, the reciprocal intercepts of the horizontal plane are X = 0, Y = 0, Z = 1. A set of Miller indices consists of three integer numbers corresponding to the reciprocal intercepts for the X, Y, and Z axes, respectively. These three reciprocal intercepts are always expressed in terms of the smallest possible integer values and are enclosed in parentheses. For example, the Miller indices of the horizontal plane are (001). The Miller indices of any arbitrary plane can be computed using the following rules:

1. Determine the X, Y, and Z intercepts of the plane. The intercepts of the plane of Figure B.2A are X = 1, Y = 3, Z = 3.
2. Take the reciprocals of the three intercepts. If one or more of the intercepts are infinite, assume that their reciprocals equal zero. The reciprocal intercepts for the plane of Figure B.2A are X = 1, Y = 1/3, Z = 1/3.
3. Multiply the reciprocal intercepts by their lowest common denominator to obtain three integers. The lowest common denominator for the reciprocal

intercepts of the plane of Figure B.2A is three, so the new reciprocal intercepts are X = 3, Y = 1, Z = 1.

4. Enclose the resulting reciprocal intercepts in parentheses to form the Miller indices. The Miller indices for the plane of Figure B.2A are (311). Those for the plane of Figure B.2B are (233). If one or more of the numbers happens to be negative, then a bar is placed over this number in the Miller indices.

Planes whose Miller indices are permutations of one another are said to be *equivalent.* For example, the (001), (010), and (100) planes are all equivalent. This does not mean that these are all one and the same plane; in fact these three planes lie at right angles to one another. Rather, it means that all three of these planes have identical crystallographic properties, which also implies that they have identical chemical, mechanical, and electrical properties. Equivalent planes are denoted by a trio of Miller indices enclosed in braces. Thus, the set of equivalent planes {100} includes the (001), (010), and (100) planes. Remember that Miller indices enclosed in braces refer to a set of planes, and not to any one specific plane. There is no such thing as a {100} wafer—the surface of the wafer might be a (001) plane or a (010) plane or a (100) plane, but it cannot be all three at the same time!

Miller indices can also be used to describe directions relative to the crystal lattice. The line passing perpendicularly through the plane (ABC) has the Miller indices [ABC]. The Miller indices for the X, Y, and Z axes are [100], [010], and [001], respectively. Directions whose Miller indices are permutations of one another are also said to be equivalent. For example, the directions [100], [010], and [001] are equivalent to one another. Equivalent directions are denoted by a trio of Miller indices placed in angle brackets. For example, the set of equivalent directions <100> includes the [100], [010], and [001] directions. Since an axis actually aligns to two direction vectors—that point in opposite directions—the Miller indices for an axis should be enclosed in angle brackets.

Appendix C

Sample Layout Rules

The rules given in this appendix were developed for use with the problems presented in the text. The numerical values given here do not correspond to any single process, but they are broadly representative of industry practices during the late 1980s. No attempt has been made to bring the rules in line with the current state of the art because this would greatly increase their complexity without providing any additional insight. For much the same reasons, many seldom-used rules have been omitted entirely or have been relegated to individual problem descriptions.

SECTION C.1 STANDARD BIPOLAR RULES

The standard bipolar process presented here is a 30V process using a single P+ isolation. The isolation spacings for this process are significantly wider than the illustrations in the text might suggest. Modern processes generally use up/down isolation to reduce these spacings, bringing them more in line with the illustrations. Table C.1 lists key electrical parameters of this process.

The baseline process uses eight coding layers: NBL, TANK, DEEPN, BASE, EMIT, CONT, METAL1, and POR. The TANK and POR (protective overcoat removal) layers code for openings in the isolation diffusion and protective overcoat, respectively. Base-over-isolation (BOI) is automatically generated from the geometries coded on the TANK layer. Two process extensions are described, one producing 2k$\Omega/\square$ HSR resistors and one generating Schottky diodes. Table C.2 lists the baseline layout rules, while Tables C.3 and C.4 list the rules for the HSR and Schottky contact process extensions, respectively. All of the rules are given in microns, assuming a coding grid of 2μm. Section C.3 explains the syntax used to describe the layout rules.

TABLE C.1 Standard bipolar parametric specifications.

Parameter	Min	Mean	Max	Units
NPN beta	100	200	300	
NPN V_{EBO}	6.4	6.8	7.2	V
NPN V_{CBO}	40			V
NPN V_{CEO}	30			V
Base sheet	130	160	190	Ω/□
Emitter sheet	5	7	10	Ω/□
Pinched base sheet	1.5	3	4.5	kΩ/□
HSR sheet	1.6	2	2.4	kΩ/□
Thick-field threshold	35			V

TABLE C.2 Standard bipolar baseline rules.

1. NBL width	8μm
2. TANK width	8μm
3. TANK spacing to TANK	6μm
4. TANK overlap NBL	22μm
5. DEEPN width	8μm
6. TANK overlap DEEPN	24μm
7. BASE width	6μm
8. BASE spacing to DEEPN	18μm
9. BASE spacing to BASE	14μm
10. TANK overlap BASE	22μm
11. EMIT width	6μm
12. BASE spacing to EMIT	12μm
13. EMIT spacing to EMIT	6μm
14. TANK overlap EMIT	18μm
15. BASE overlap EMIT	4μm
16. CONT width	4μm
17. EMIT spacing to CONT	6μm
18. CONT spacing to CONT	4μm
19. BASE overlap CONT	2μm
20. EMIT overlap CONT	2μm
21. METAL1 width	6μm
22. METAL1 spacing to METAL1	4μm
23. METAL1 overlap CONT	2μm
24. POR width	10μm
25. POR spacing to POR	10μm
26. METAL1 overlap POR	4μm

TABLE C.3 Standard bipolar HSR extension rules.

27. HSR width	6μm
28. HSR spacing to DEEPN	16μm
29. HSR spacing to BASE	14μm
30. HSR spacing to EMIT	10μm
31. HSR spacing to CONT	4μm
32. HSR spacing to HSR	12μm
33. BASE overhang HSR	2μm

34. HSR extends into BASE	2μm
35. TANK overlap HSR	20μm
36. SCONT spacing DEEPN	12μm
37. SCONT spacing to BASE	6μm
38. SCONT spacing to EMIT	4μm
39. SCONT spacing to HSR	4μm
40. SCONT spacing to CONT	4μm
41. SCONT spacing to SCONT	4μm
42. SCONT overlap CONT	2μm
43. METAL1 overlap SCONT	2μm

TABLE C.4 Standard bipolar Schottky extension rules.

SECTION C.2 POLYSILICON-GATE CMOS RULES

This section describes a 10V, N-well, poly-gate CMOS process. The LDD NMOS has a minimum allowed channel length of 4μm, while the SDD PMOS allows channel lengths as short as 3μm. Both transistors use the same N-type gate poly, and a single boron threshold adjust sets both threshold voltages simultaneously. A combination of boron and phosphorus channel stops ensure that both the NMOS and the PMOS thick-field thresholds lie safely above the operating voltage of the process. CMOS latchup is minimized by the use of a P+ substrate and by optional availability of NBL as part of the analog BiCMOS process extension. This process cannot fabricate Schottky diodes because it uses titanium silicide to minimize contact resistance. Table C.5 lists key electrical parameters of this process.

Parameter	Min	Mean	Max	Units
NMOS V_t	0.5	0.7	0.9	V
NMOS k	50	70	90	$\mu A/V^2$
NMOS V_{DS}	10			V
NMOS V_{GS}	12			V
PMOS V_t	–0.9	–0.7	–0.5	V
PMOS k	17	25	33	$\mu A/V^2$
PMOS V_{DS}	12			V
PMOS V_{GS}	12			V
Thick-field thresholds	15			V
Poly-1 sheet	20	30	40	Ω/□
Poly-2 sheet (w/PSD)	160	200	240	Ω/□
Base sheet	400	500	600	Ω/□
Gate oxide capacitance	0.85	0.95	1.05	$fF/\mu m^2$
Poly-poly capacitance	1.3	1.5	1.7	$fF/\mu m^2$
NPN beta	40	80	120	
NPN V_{EBO}	7	8	9	V
NPN V_{CBO}	15			V
NPN V_{CEO}	12			V

TABLE C.5 Polysilicon-gate CMOS parametric specifications.

The baseline process as described in Table C.6 uses eleven masks: NWELL, MOAT, NSD, PSD, CHST, POLY1, CONT, METAL1, VIA, METAL2, and POR. These eleven masks are normally coded using nine drawing layers: NWELL, NMOAT, PMOAT, POLY1, CONT, METAL1, VIA, METAL2, and POR. The NMOAT drawing layer simultaneously produces geometries on the MOAT and NSD masks. The PMOAT drawing layer simultaneously produces geometries on the MOAT and PSD masks. The information for the CHST mask is obtained from the

TABLE C.6 Poly-gate CMOS baseline rules.

1. NWELL width	5.0μm
2. NWELL spacing to NWELL	15.0μm
3. NMOAT width	3.0μm
4. NMOAT spacing to NWELL	9.5μm
5. NMOAT spacing to NMOAT	5.5μm
6. NWELL overlap NMOAT	1.0μm
7. PMOAT width	3.0μm
8. PMOAT spacing to NWELL	7.0μm
9. PMOAT spacing to NMOAT *(note 1)*	4.0μm
10. PMOAT spacing to PMOAT	5.5μm
11. NWELL overlap PMOAT	2.0μm
12. POLY1 width	2.0μm
13. POLY1 spacing to NMOAT	2.0μm
14. POLY1 spacing to PMOAT	2.0μm
15. POLY1 spacing to POLY1	2.0μm
16. POLY1 overhang NMOAT	1.0μm
17. POLY1 overhang PMOAT	1.0μm
18. NMOAT overhang POLY1	4.0μm
19. PMOAT overhang POLY1	4.0μm
20. CONT width	1.0μm exactly
21. CONT spacing to POLY1	2.0μm
22. CONT spacing to CONT	2.0μm
23. NMOAT overlap CONT	1.0μm
24. PMOAT overlap CONT	1.0μm
25. POLY1 overlap CONT	1.0μm
26. METAL1 width	2.0μm
27. METAL1 spacing to METAL1	2.0μm
28. METAL1 overlap CONT	1.0μm
29. VIA width *(note 2)*	1.0μm exactly
30. VIA spacing to CONT *(note 3)*	2.0μm
31. VIA spacing to VIA	2.0μm
32. METAL1 overlap VIA	1.0μm
33. METAL2 width	2.0μm
34. METAL2 spacing to METAL2	2.0μm
35. METAL2 overlap VIA	1.0μm
36. POR width	4.0μm
37. POR spacing to POR	4.0μm
38. METAL2 overlap POR	2.0μm

Notes: [1] PMOAT allowed to abut NMOAT if the two are connected by METAL1.
[2] Except for bondpads.
[3] VIA must not touch CONT.

NWELL and MOAT coding layers. The geometries on the POR mask represent openings in the protective overcoat. All of these layout rules are given in microns and presume a coding grid of 0.5μm. Section C.3 explains the syntax used to describe the layout rules.

This process supports two extensions, the first of which adds a second layer of polysilicon to the process using the POLY2 mask. This poly is deposited in a near-intrinsic state and is doped with PSD using the PSD drawing layer to produce poly resistors. A thin oxide-nitride-oxide dielectric deposited between the two polysilicon layers provides poly-poly capacitors. Table C.7 lists all of the rules associated with this extension.

TABLE C.7 Poly-2 extension rules.

39. POLY2 width	2.0μm
40. POLY2 spacing to POLY2	2.0μm
41. NSD spacing to POLY2	2.0μm
42. PSD spacing to POLY2	2.0μm
43. POLY2 overlap CONT	1.0μm
44. POLY1 overlap POLY2	1.5μm
45. NSD overlap POLY2	1.5μm
46. PSD overlap POLY2	1.5μm
47. PSD spacing to NMOAT	2.0μm
48. NSD spacing to PMOAT	2.0μm

The second process extension adds analog BiCMOS functionality using three additional masks: NBL, DEEPN, and BASE. Data coded on the BASE drawing layer automatically generates geometries on both the MOAT and the BASE masks. The addition of an N-buried layer forces the process to incorporate a second epitaxial layer for compatibility with a P+ substrate. This process extension allows the creation of vertical NPN transistors, lateral PNP transistors, and substrate PNP transistors. Table C.8 lists the layout rules required to construct analog BiCMOS structures.

TABLE C.8 BiCMOS extension rules.

47. NBL width	5.0μm
48. NBL spacing to NBL	19.0μm
49. NWELL spacing to NBL	16.5μm
50. NWELL overlap NBL	5.0μm
51. DEEPN width	5.0μm
52. DEEPN spacing to NMOAT	6.0μm
53. DEEPN spacing to PMOAT	7.5μm
54. DEEPN spacing to DEEPN	9.0μm
55. NWELL overlap DEEPN	2.0μm
56. BASE width	3.0μm
57. BASE spacing to NMOAT	4.5μm
58. BASE spacing to PMOAT	6.0μm
59. BASE spacing to DEEPN	8.0μm
60. BASE spacing to BASE	6.5μm
61. BASE overlap NSD	1.5μm
62. PSD overlap BASE	1.0μm
63. NWELL overlap BASE	3.0μm
64. NSD spacing to CONT	2.5μm
65. NSD spacing to PSD	3.0μm
66. NSD spacing to PSD	3.5μm
67. NSD overlap CONT	1.0μm
68. PSD overlap CONT	1.0μm
69. BASE overlap CONT	1.0μm

Note: Used for constructing NPN transistors, but not for lateral PNPs.

SECTION C.3 LAYOUT RULE SYNTAX

The layout rules listed in this appendix follow a format that is similar (but not identical) to that of Chameleon, a layout verification program developed at Texas Instruments beginning in 1976 and currently supported and marketed by K2 Technologies. This notation resembles plain English and is therefore particularly easy for novices to understand. Those employing other verification programs should have little difficulty translating the rules into the appropriate format, particularly since all of the rules fall into one of five elementary categories: width, spacing, overlap, overhang, and extent into. The following sections explain each of these types of rules.

Width

A width check verifies that all dimensions of every geometry on a given layer equal or exceed a minimum feature size. The syntax for a width check is

LAYER1 width N µm

where N is the minimum allowed dimension of all geometries on LAYER1. In order for the geometry of Figure C.1 to pass the above width check, dimensions A to D must all equal or exceed the minimum width, N. Occasionally the dimension will be followed by the notation "exactly." In such cases, the width of every figure on the indicated layer must equal exactly N microns, neither more nor less.

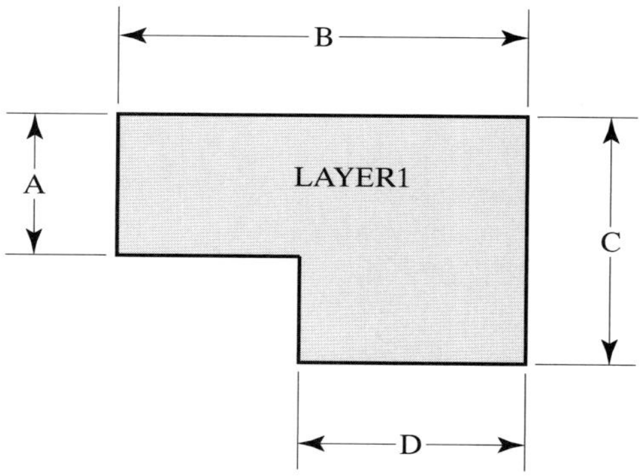

FIGURE C.1 Dimensions checked by "LAYER1 width."

Spacing

A spacing check verifies that all geometries on the specified layers maintain a minimum separation from one another. The syntax for a spacing check is

LAYER1 spacing to LAYER2 N µm

This check determines the minimum distance from each geometry on LAYER1 to each geometry on LAYER2. A violation occurs if any of these distances is less than N. Elements that touch or overlap do not violate the spacing. A spacing check may also be applied to a single layer using the syntax

LAYER1 spacing to LAYER1 N µm

In this case, the check applies not only to the separation between any two geometries on LAYER1 but also to separate portions of a single geometry, such as the turns of a serpentine resistor (for an example, see dimension C in Figure C.2).

Overlap

An overlap check only applies to geometries on a first layer that are partially or wholly enclosed by geometries on a second layer (Figure C.3). The syntax of an overlap check is

LAYER1 overlap LAYER2 N µm

FIGURE C.2 Dimensions checked by "LAYER1 spacing to LAYER1."

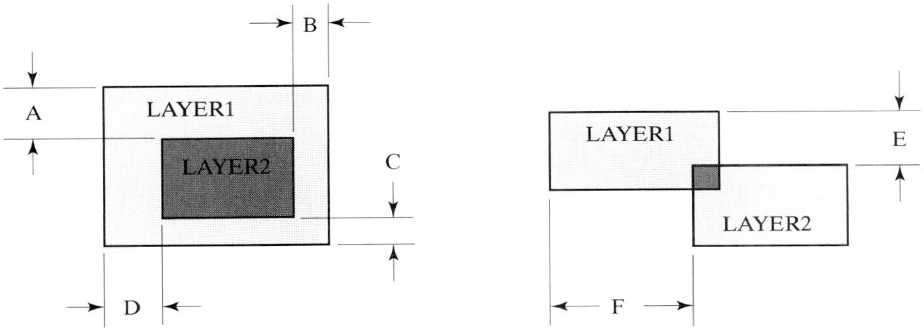

FIGURE C.3 Dimensions checked by "LAYER1 overlap LAYER2."

Wherever a geometry on LAYER1 encloses a geometry on LAYER2, the amount by which the former geometry overlaps the latter must equal or exceed *N*. If the geometries do not completely overlap, the sides that do not overlap automatically pass the check. Figure C.3 shows examples of fully and partially overlapped geometries and the dimensions checked in each case.

Overhang

An overhang check only applies to cases where a geometry on one layer partially overlaps a geometry on another (Figure C.4). The syntax of an overhang check is

LAYER1 overhang LAYER2 $\quad N$ μm

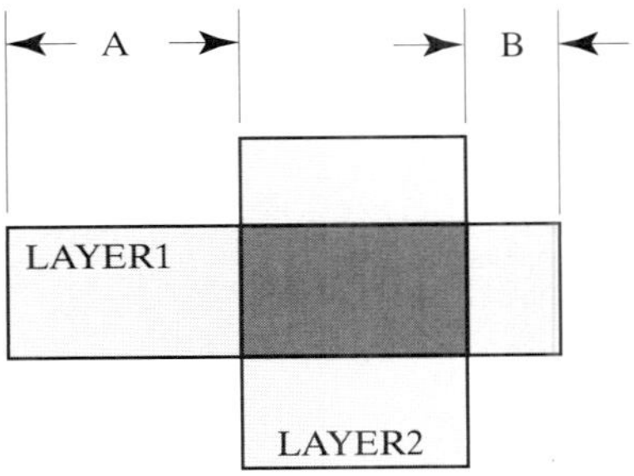

FIGURE C.4 Dimensions checked by "LAYER1 overhang LAYER2."

Wherever a geometry on LAYER1 partially overlaps a geometry on LAYER2, the amount by which the former geometry extends beyond the latter must equal or exceed *N*. Cases where the former geometry is fully enclosed by the latter automatically pass the check.

Extent Into

An extent into check only applies to cases where a geometry on one layer partially overlaps a geometry on another (Figure C.5). The syntax of an extent into check is

LAYER1 extends into LAYER2 $\quad N$ μm

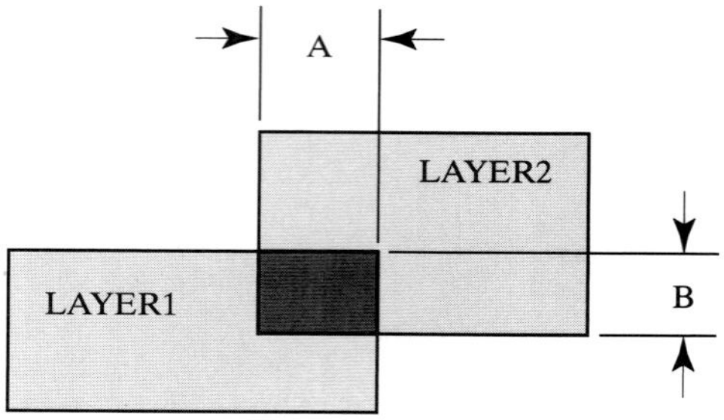

FIGURE C.5 Dimensions checked by "LAYER1 extends into LAYER2."

Wherever a geometry on LAYER1 partially overlaps a geometry on LAYER2, the amount by which the former geometry extends into the latter must equal or exceed *N*. Cases where the former geometry is fully enclosed by the latter automatically pass the check.

Appendix D Mathematical Derivations

Equations 7.5, 11.9A–B, 12.7–12.9, 12.14, and 12.15 have been derived specifically for this text. This appendix contains the details of these derivations, which were deemed too complicated for inclusion within the body of the text.

EQUATION 7.5

The random variance of the resistor is assumed to consist of an areal component and a peripheral component. If we assume the random fluctuations occur on an infinitesimal scale of distance and the variance of resistivity of a unit area element equals σ_a^2, then the areal variance σ_A^2 of a resistor having width W and length L equals

$$\sigma_A^2 = \frac{\sigma_a^2}{WL} \quad \text{[D.1]}$$

since $R \equiv R_s W / L$, $L = R_s W / R$. Substituting this into equation D.1 gives

$$\sigma_A^2 = \frac{\sigma_a^2 R}{W^2 R_s} \quad \text{[D.2]}$$

If the variance of width of a section of the resistor having unit length equals σ_w^2, then this section's resistance will have a variance of σ_w^2 / W^2 and a resistor of length L will have a peripheral variance σ_P^2 of

$$\sigma_P^2 = \frac{\sigma_w^2}{W^2 L} = \frac{\sigma_w^2 R}{W^3 R_s} \quad \text{[D.3]}$$

Summing the areal and peripheral variances and taking the square root of the resulting sum, one obtains the standard deviation of resistance σ_R

$$\sigma_R = \sqrt{\frac{\sigma_w^2 R}{W^3 R_s} + \frac{\sigma_a^2 R}{W^2 R_s}} \quad \text{[D.4]}$$

Let $k_p \equiv \sigma_w^2 / R_s$ and $k_a \equiv \sigma_a^2 / R_s$. Normalizing σ_R by dividing by R results in

$$\frac{\sigma_R}{R} = \frac{1}{W\sqrt{R}}\sqrt{k_a + \frac{k_p}{W}} \qquad \text{[D.5]}$$

EQUATION 11.9A–B

In a circularly symmetric device, current flows radially from source to drain. The length of the channel therefore equals the difference between the radii of the outer and inner edges of the channel, or $L = 1/2\ (B–A)$. The width of the channel W can be determined using the Shichman-Hodges equation for the linear region. Rearranging this equation in terms of L gives

$$L = k'\frac{W}{I_D}\left(V_{gst} - \frac{V_{DS}}{2}\right)V_{DS} \qquad \text{[D.6]}$$

Differentiating to find dL/dV_{DS} gives

$$\frac{dL}{dV_{DS}} = k'\frac{W}{I_D}(V_{gst} - V_{DS}) \qquad \text{[D.7]}$$

For any infinitesimal length of channel dL the corresponding width W equals $2\pi L$, where $L = 0$ denotes the center of the annular structure. Separating terms and integrating gives

$$\int_0^{V_{DS}} (V_{gst} - V_{DS})dV_{DS} = \frac{I_D}{k'}\int_{A/2}^{B/2} \frac{dL}{2\pi L} \qquad \text{[D.8]}$$

The limits of these integrals assume that the source is at the inner edge of the channel ($L = A/2$) and the drain is at the outer edge of the channel ($L = B/2$). Integrating gives

$$V_{gst}V_{DS} - \frac{V_{DS}}{2} = \frac{I_D}{2\pi k'}\ln(B/A) \qquad \text{[D.9]}$$

Gathering terms,

$$I_D = \frac{2\pi k'}{\ln(B/A)}\left(V_{gst} - \frac{V_{DS}}{2}\right)V_{DS} \qquad \text{[D.10]}$$

This equation is analogous to the Shichman-Hodges equation for the linear region with a W/L ratio equal to

$$\frac{W}{L} = \frac{2\pi}{\ln(B/A)} \qquad \text{[D.11]}$$

Substituting the previously determined value of L gives

$$W = \frac{\pi(B - A)}{\ln(B/A)} \qquad \text{[D.12]}$$

EQUATION 12.7

Let a source/drain finger consist of a uniform rectangular strip having width W, length L, and sheet resistance, R_s. Let variable x define position along the length of the finger. Assume that an equal amount of current flows into the finger at every point along its length, so the current $I(x)$ increases linearly from $x = 0$ to $x = L$. Let I_{max} equal the current $I(x)$ at $x = L$. The voltage drop V from $x = 0$ to $x = L$ equals

$$V = \int_0^L \frac{R_s I(x)}{W} dx \quad [D.13]$$

Since $I(x) = I_{max} x / L$, this reduces to

$$V = \frac{R_s I_{max}}{WL} \int_0^L x dx = \frac{R_s I_{max} L}{2W} \quad [D.14]$$

$$V = \frac{R_s L}{2W} I_{max} \quad [D.15]$$

The resistance of any source/drain finger may be divided into three components: (1) the portion of the finger consisting only of metal-1, (2) the portion of the finger consisting of a sandwich of metal-1 and metal-2, and (3) the portion of the finger under a plate of metal-2. If we assume that the metal-2 buses are equipotential across their width, then the resistance of the finger consists of only components (1) and (2). The voltage drop V_1 across component (1) equals

$$V_1 = \frac{R_{s1} B}{2W} I_1 \quad [D.16]$$

where I_1 equals the current flowing through component (1) of a single source/drain finger. Since component (1) has a length of B, while the full length of one finger equals L, and since there are N_D pairs of source/drain fingers, the current I_1 equals

$$I_1 = \frac{B}{LN_D} I_D \quad [D.17]$$

where I_D equals the total drain current of all fingers. Substituting equation D.17 into equation D.16 yields

$$V_1 = \frac{R_{s1} B^2}{2WLN_D} I_D \quad [D.18]$$

The voltage drop V_2 across component (2) can likewise be computed

$$R_2 = \frac{R_{s12} A}{2W} I_2 + \frac{R_{s12} A}{W} I_1 \quad [D.19]$$

where I_2 equals the total current entering the finger within component (2) and I_1 equals the current flowing into component (2) from component (1). I_2 can be computed in a similar manner to that used to determine I_1. Substituting I_1 and I_2 into equation D.19,

$$V_2 = \frac{R_{s12} A}{2W} \left(\frac{A}{LN_D} \right) I_D + \frac{R_{s12} A}{W} \left(\frac{B}{LN_D} \right) I_D \quad [D.20]$$

$$V_2 = \frac{R_{s12} A}{2W} \left(\frac{A + 2B}{LN_D} \right) I_D \quad [D.21]$$

But since $L \equiv A + 2B$,

$$R_2 = \frac{R_{s12} A}{2WN_D} I_D \quad [D.22]$$

The total voltage drop across the fingers of the transistor equals twice the sum of V_1 and V_2 because there are both source and drain fingers involved. To this must be added the voltage drop across the metal-2 buses V_B

$$V_B = \frac{HR_{s2}}{2B} I_D \quad [D.23]$$

The total voltage drop V_M therefore equals

$$V_M = \left(\frac{R_{s1}B^2}{WLN_D} + \frac{R_{s12}A}{WN_D} + \frac{HR_{s2}}{2B} \right) I_D \qquad \text{[D.24]}$$

Since the metallization resistance $R_M \equiv V_M / I_D$,

$$R_M \frac{R_{s1}B^2}{WLN_D} + \frac{R_{s12}A}{WN_D} + \frac{HR_{s2}}{2B} \qquad \text{[D.25]}$$

EQUATION 12.9 AND 12.10

To determine the optimum value of B, take $\partial R_M / \partial B$ and set this equal to zero. This determines an inflection point in the function $R_M(B)$. Begin by replacing A by $(L - 2B)$ and differentiating

$$\frac{\partial R_M}{\partial B} = \frac{2BR_{s1}}{WN_D L} - \frac{2R_{s12}}{WN_D} \qquad \text{[D.26]}$$

Setting this equal to zero, we obtain

$$\frac{BR_{s1}}{L} = R_{s12} \qquad \text{[D.27]}$$

$$\frac{B}{L} = \frac{R_{s12}}{R_{s1}} \qquad \text{[D.28]}$$

Assuming both metal layers have resistivity ρ, $R_{s1} = \rho / t_1$ where t_1 is the thickness of metal-1, and $R_{s12} = \rho / (t_1 + t_2)$, where t_2 is the thickness of metal-2. Substituting these relationships into equation D.28 gives

$$\frac{B}{L} = \frac{t_1}{t_1 + t_2} \qquad \text{[D.29]}$$

EQUATION 12.14

This equation is derived from the Shichman-Hodges equation for saturation. Let transistor M_1 have drain current I_{D1}, transconductance k_1, and effective gate voltage V_{gst1}. Let transistor M_2 have drain current I_{D2}, transconductance k_2, and effective gate voltage V_{gst2}. Since $I_{D1} \equiv I_{D2}$,

$$k_1 V_{gst1}^2 = k_2 V_{gst2}^2 \qquad \text{[D.30]}$$

Rearranging

$$\frac{k_1}{k_2} = \left(\frac{V_{gst2}}{V_{gst1}} \right)^2 \qquad \text{[D.31]}$$

Let $\Delta V_{gs} \equiv V_{gs1} - V_{gs2}$; then $V_{gs2} = V_{gs1} - \Delta V_{gs}$. Let $\Delta V_t \equiv V_{t1} - V_{t2}$; then $V_{t2} = V_{t1} - \Delta V_t$. Substituting these relationships into equation D.31 yields

$$\frac{k_1}{k_2} = \left(\frac{V_{gs2} - V_{t2}}{V_{gst1}} \right)^2 = \left(\frac{V_{gs1} - \Delta V_{gs} - V_{t1} + \Delta V_t}{V_{gst1}} \right)^2 \qquad \text{[D.32]}$$

$$\frac{k_1}{k_2} = \left(1 + \frac{\Delta V_t - \Delta V_{gs}}{V_{gst1}} \right)^2 \qquad \text{[D.33]}$$

$$\sqrt{\frac{k_1}{k_2}} = 1 + \frac{\Delta V_t - \Delta V_{gs}}{V_{gst1}} \quad \text{[D.34]}$$

$$\left(\sqrt{\frac{k_1}{k_2}} - 1\right)V_{gst1} = \Delta V_t - \Delta V_{gs} \quad \text{[D.35]}$$

Collecting ΔV_{gs}

$$\Delta V_{gs} = \Delta V_t - V_{gst1}\left(\sqrt{\frac{k_1}{k_2}} - 1\right) \quad \text{[D.36]}$$

Let $\Delta k \equiv k_1 - k_2$; then $k_1 = k_2 + \Delta k$. Substituting this into equation D.36 gives

$$\Delta V_{gs} = \Delta V_t - V_{gst1}\left(\sqrt{1 + \frac{\Delta k}{k_2}} - 1\right) \quad \text{[D.37]}$$

By the binominal expansion

$$\sqrt{1 + x} = \sum_{n=0}^{\infty}\binom{1/2}{n}x^n = 1 + \frac{x}{2} - \frac{x^2}{8} + \frac{3x^3}{48} + \ldots \quad \text{[D.38]}$$

If x is sufficiently small, then only the first two terms are significant. Applying this expansion to equation D.37 gives

$$\Delta V_{gs} \cong \Delta V_t - V_{gst1}\frac{\Delta k}{2k_2} \quad \text{[D.39]}$$

EQUATION 12.15

This equation is derived from the Shichman-Hodges equation for saturation. Let transistor M_1 have drain current I_{D1}, transconductance k_1, and effective gate voltage V_{gst1}. Let transistor M_2 have drain current I_{D2}, transconductance k_2, and effective gate voltage V_{gst2}. The ratio of the two drain currents I_{D2} / I_{D1} equals

$$\frac{I_{D2}}{I_{D1}} = \frac{k_2}{k_1}\left(\frac{V_{gst2}}{V_{gst1}}\right)^2 \quad \text{[D.40]}$$

Let $\Delta V_t \equiv V_{t1} - V_{t2}$; then $V_{t2} = V_{t1} - \Delta V_t$. Substituting this into equation D.40

$$\frac{I_{D2}}{I_{D1}} = \frac{k_2}{k_1}\left(\frac{V_{gs2} - V_{t2}}{V_{gst1}}\right)^2 = \frac{k_2}{k_1}\left(\frac{V_{gs2} - V_{t1} + \Delta Vt}{V_{gst1}}\right)^2 \quad \text{[D.41]}$$

But since $V_{gs1} \equiv V_{gs2}$,

$$\frac{I_{D2}}{I_{D1}} = \frac{k_2}{k_1}\left(\frac{V_{gs1} - V_{t1} + \Delta V_t}{V_{gst1}}\right)^2 = \frac{k_2}{k_1}\left(\frac{V_{gst1} + \Delta V_t}{V_{gst1}}\right)^2 \quad \text{[D.42]}$$

Expanding

$$\frac{I_{D2}}{I_{D1}} = \frac{k_2}{k_1}\left(\frac{V_{gst1}^2 + 2\Delta V_t V_{gst1} + \Delta V_t^2}{V_{gst1}^2}\right) \quad \text{[D.43]}$$

So long as $\Delta V_t << V_{gst1}$,

$$\frac{I_{D2}}{I_{D1}} \cong \frac{k_2}{k_1}\left(1 + \frac{2\Delta V_t}{V_{gst1}}\right) \quad \text{[D.44]}$$

Appendix

E Sources of Layout Editor Software

This textbook does not discuss layout software, because these programs constantly evolve and any information given here would soon become obsolete. The following are several sources of layout editors and documentation:

- J. P. Uyemura, *Physical Design of CMOS Integrated Circuits using L-EDIT™* (Boston: PWS Publishing Company, 1995). This text provides an excellent introduction to layout editing software, using the L-EDIT™ package developed and sold by Tanner Research. The text also includes a student version of the program on a floppy disk.
- Tanner Research, located at **http://www.tanner.com.** Tanner Research sells a professional version of L-EDIT™ that provides a much more capable editing environment than the student version included with *Physical Design of CMOS Integrated Circuits using L-EDIT™*.
- Cadence Design Systems, located at **http://www.cadence.com.** Cadence sells Virtuoso, which is one of the most widely used layout editors in the industry.
- Mentor Graphics, located at **http://www.mentorg.com.** Mentor sells IC Graph, another widely used layout editor.

Index